Alliant International University
Los Angeles Campus Library
1000 South Fremont Ave., Unit 5
Alhambra, CA 91803

Self Control in Society, Mind, and Brain

OXFORD SERIES IN SOCIAL COGNITION AND SOCIAL NEUROSCIENCE

Series Editor
Ran R. Hassin

Editorial Board
Mahzarin Banaji
John A. Bargh
John Gabrieli
David Hamilton
Elizabeth A. Phelps
Yaacov Trope

The New Unconscious
Edited by Ran R. Hassin, James S. Uleman, and John A. Bargh

Oxford Handbook of Human Action
Edited by Ezequiel Morsella, John A. Bargh, and Peter M. Gollwitzer

Forthcoming **Social Neuroscience: Toward Understanding the Underpinnings of the Social Mind**
Edited by Alexander Todorov, Susan T. Fiske, and Deborah Prentice

Forthcoming **Oxford Handbook of Social Neuroscience**
Edited by Jean Decety and John T. Cacioppo

Self Control in Society, Mind, and Brain

Edited by

Ran R. Hassin
Kevin N. Ochsner
Yaacov Trope

2010

OXFORD
UNIVERSITY PRESS

Oxford University Press, Inc., publishes works that further
Oxford University's objective of excellence
in research, scholarship, and education.

Oxford   New York
Auckland   Cape Town   Dar es Salaam   Hong Kong   Karachi
Kuala Lumpur   Madrid   Melbourne   Mexico City   Nairobi
New Delhi   Shanghai   Taipei   Toronto

With offices in
Argentina   Austria   Brazil   Chile   Czech Republic   France   Greece
Guatemala   Hungary   Italy   Japan   Poland   Portugal   Singapore
South Korea   Switzerland   Thailand   Turkey   Ukraine   Vietnam

Copyright © 2010 by Oxford University Press

Published by Oxford University Press, Inc.
198 Madison Avenue, New York, New York 10016
www.oup.com

Oxford is a registered trademark of Oxford University Press

All rights reserved. No part of this publication may be reproduced,
stored in a retrieval system, or transmitted, in any form or by any means,
electronic, mechanical, photocopying, recording, or otherwise,
without the prior permission of Oxford University Press.

Library of Congress Cataloging-in-Publication Data
CIP data on file
ISBN 978-0-19-539138-1

9 8 7 6 5 4 3 2 1

Printed in the United States of America
on acid-free paper

PREFACE

How do we—societies and individuals alike—(sometimes) manage to act in line with our high priority goals when we are faced with tempting yet conflicting alternatives? Put differently, how do we (sometimes) resolve a conflict between super ordinate, global goal, and subordinate, local one in favor of the former? Self-control dilemmas of this kind can be found in many contexts and may be described at many levels of analysis. Examples include the dilemmas faced by groups fostering cooperation among group members, when defection is more beneficial for each individual; dieters resisting a tempting cake in order to stay on their diet, and the cognitive system trading off speed for accuracy in conflict tasks like the Stroop.

All of the aforementioned dilemmas have a common structure: A goal that is higher in one's hierarchy conflicts with a lower level goal. Given this goal hierarchy, one of the hallmarks of self control conflicts is that they have a correct, or desired answer. Although one is tempted to eat the cake, the "correct" behavior of a dieter is to refrain from eating it; although the individual is tempted to defect, one knows that the "correct" social behavior is to cooperate; and, lastly, while one is tempted to respond quickly in an incongruent Stroop trial, one knows that the "correct" answer is to name the ink. Self-control success is responding in line with one's higher order goals, and self-control failure is responding in line with one's lower order goals. This common structure is the essence of self-control and the theme underlying the present volume on the process of self-control, its antecedents and consequences for individuals and societies.

This book represents social, cognitive, and neuroscientific approaches to the question of self-control. Our goal was to bring together multiple perspectives on these kinds of dilemmas, connecting recent work in cognitive, and social psychology with recent advances in cognitive and social neuroscience. The book emphasizes integrative, multi-disciplinary approaches to self-control. It offers a single reference volume comprising contributions from leading researchers within various allied disciplines.

The book consists of three sections: the Neural, the Mental, and the Social. The Neural section looks at brain processes that underlie self control attempts and that speak directly to the level of the mental. The Mental is the book's anchor: it comprises cognitive and social-cognitive contributions, which examine within-individual self-control processes at all levels; from low level attention, to motivation, and motivational systems. The Social section looks at group processes, broadly defined, and examines how groups and societies (attempt to) resolve conflicts between their global goals and the individual's self interest.

We hope that by crossing these disciplinary boundaries within a single volume, the book will promote cross-talk between theoretical and data-driven approaches at various levels of analyses. We believe that this will help improve our understanding of one of the most crucial aspects of human behavior; our capacity to overcome the influence of local, short-term concerns, and act in line with our long-term, overarching goals.

Before you turn the page to begin reading this book, we wish to thank the authors for contributing a set of deep, thorough, and thought provoking chapters. We would also like to thank Catharine Carlin, senior editor at OUP, for her assistance and support throughout this project. Ran Hassin would like to dedicate this book to the memory of Eliahu Hassin, a father, a guide, an inspiration, and a dear friend.

Ran R. Hassin
Kevin N. Ochsner
Yaacov Trope

CONTENTS

Contributors   xi

NEURAL

1. **Anterior Cingulate Cortex Contributions to Cognitive and Emotional Processing: A General Purpose Mechanism for Cognitive Control and Self-Control**   3
   Marie K. Krug and Cameron S. Carter

2. **Damaged Self, Damaged Control: A Component Process Analysis of the Effects of Frontal Lobe Damage on Human Decision Making**   27
   Lesley K. Fellows

3. **Working Hard or Hardly Working for those Rose-colored Glasses: Behavioral and Neural Evidence for the Automatic Nature of Unrealistically Positive Self-Perceptions**   38
   Jennifer S. Beer

4. **Control in the Regulation of Intergroup Bias**   49
   David M. Amodio and Patricia G. Devine

5. **Integrating Research on Self-Control across Multiple Levels of Analysis: Insights from Social Cognitive and Affective Neuroscience**   76
   Ethan Kross and Kevin N. Ochsner

6. **Using the Stroop Task to Study Emotion Regulation**   93
   Jason Buhle, Tor Wager, and Ed Smith

7. **Motivational Influences on Cognitive Control: A Cognitive Neuroscience Perspective**   114
   Hannah S. Locke and Todd S. Braver

8. **The Common Neural Basis of Exerting Self-Control in Multiple Domains**   141
   Jessica R. Cohen and Matthew D. Lieberman

MENTAL

9. Working Memory Capacity: Self-Control Is (in) the Goal — 163
   James M. Broadway, Thomas S. Redick, and Randall W. Engle

10. The Dynamic Control of Human Actions — 174
    Florian Waszak, Anne Springer, and Wolfgang Prinz

11. Task Switching: Mechanisms Underlying Rigid vs. Flexible Self-Control — 202
    Nachshon Meiran

12. Unconscious Influences of Attitudes and Challenges to Self-Control — 221
    Deborah L. Hall and B. Keith Payne

13. Self-Control Over Automatic Associations — 243
    Karen Gonsalkorale, Jeffrey W. Sherman, and Thomas J. Allen

14. Perish the Forethought: Premeditation Engenders Misperceptions of Personal Control — 260
    Carey K. Morewedge, Kurt Gray, and Daniel M. Wegner

15. The Power of Planning: Self-Control by Effective Goal-striving — 279
    Peter M. Gollwitzer, Caterina Gawrilow, and Gabriele Oettingen

16. Unpacking the Self-Control Dilemma and Its Modes of Resolution — 297
    Arie W. Kruglanski and Catalina Kŏpetz

17. Conflict and Control at Different Levels of Self-Regulation — 312
    Abigail A. Scholer and E. Tory Higgins

18. Getting Our Act Together: Toward a General Model of Self-Control — 335
    Eran Magen and James J. Gross

19. Implicit Control of Stereotype Activation — 354
    Gordon B. Moskowitz and Peizhong Li

20. Ego Depletion and the Limited Resource Model of Self-Control — 375
    Nicole L. Mead, Jessica L. Alquist, and Roy F. Baumeister

21. Walking the Line between Goals and Temptations: Asymmetric Effects of Counteractive Control — 389
    Ayelet Fishbach and Benjamin A. Converse

22. Seeing the Big Picture: A Construal Level Analysis of Self-Control — 408
    Kentaro Fujita, Yaacov Trope, and Nira Liberman

23. From Stimulus Control to Self-Control: Toward an Integrative Understanding of the Processes Underlying Willpower — 428
    Ethan Kross and Walter Mischel

SOCIAL

24. **Self-Control in Groups** — 449
    John M. Levine, Kira M. Alexander, and Thomas Hansen

25. **Justice as Social Self Control** — 473
    Tom R. Tyler

26. **System Justification and the Disruption of Environmental Goal-Setting: A Self-Regulatory Perspective** — 490
    Irina Feygina, Rachel E. Goldsmith, and John T. Jost

27. **Teleological Behaviorism and the Problem of Self-Control** — 506
    Howard Rachlin

    Author Index — 523

    Subject Index — 543

CONTRIBUTORS

Kira M. Alexander
University of Pittsburgh
Pittsburgh, PA

Thomas J. Allen
University of California
Davis, CA

Jessica L. Alquist
Florida State University
Tallahassee, FL

David M. Amodio
New York University
New York, NY

Roy F. Baumeister
Florida State University
Tallahassee, FL

Jennifer S. Beer
University of Texas at Austin
Austin, TX

Todd S. Braver
Washington University
St. Louis, MO

James M. Broadway
Georgia Institute of Technology
Atlanta, GA

Jason Buhle
Columbia University
New York, NY

Cameron S. Carter
Pittsburgh University
Pittsburgh, PA

Jessica R. Cohen
UCLA
Los Angeles, CA

Benjamin Converse
University of Chicago
Chicago, IL

Patricia G. Devine
University of Wisconsin-Madison
Madison, WI

Randall W. Engle
Georgia Institute of Technology
Atlanta, GA

Lesley K. Fellows
McGill University
Montréal, Canada

Irina Feygina
New York University
New York, NY

Ayelet Fishbach
University of Chicago
Chicago, IL

Kentaro Fujita
Ohio State University
Columbus, OH

Caterina Gawrilow
University of Hamburg
Hamburg, Germany

Rachel E. Goldsmith
Mt. Sinai School of Medicine
New York, NY

Peter M. Gollwitzer
New York University and Universität Konstanz
New York, NY

Karen Gonsalkorale
University of Sydney
Sydney, Australia

Kurt Gray
Harvard University
Cambridge, MA

James J. Gross
Stanford University
Palo Alto, CA

Deborah L. Hall
Duke University
Durham, NC

Thomas Hansen
University of Pittsburgh
Pittsburgh, PA

Ran R. Hassin
The Hebrew University
Jerusalem, Israel

E. Tory Higgins
Columbia University
New York, NY

John T. Jost
New York University
New York, NY

Catalina Kőpetz
University of Maryland
College Park, MD

Ethan Kross
University of Michigan
Ann Arbor, MI

Marie K. Krug
Pittsburgh University
Pittsburgh, PA

Arie W. Kruglanski
University of Maryland
College Park, MD

John M. Levine
University of Pittsburgh
Pittsburgh, PA

Peizhong Li
University of Wisconsin-Stout
Menomonie, WI

Nira Liberman
Tel Aviv University
Tel Aviv, Israel

Matthew D. Lieberman
UCLA
Los Angeles, CA

Hannah S. Locke
Washington University
St. Louis, MO

Eran Magen
Stanford University
Palo Alto, CA

Nicole L. Mead
Florida State University
Tallahassee, FL

Nachshon Meiran
Ben-Gurion University of the Negev
Beer-Sheva, Israel

Walter Mischel
Columbia University
New York City, NY

Carey K. Morewedge
Carnegie Mellon University
Pittsburgh, PA

Gordon B. Moskowitz
Lehigh University
Bethlehem, PA

Kevin N. Ochsner
Columbia University
New York, NY

Gabriele Oettingen
New York University
New York, NY
and
University of Hamburg
Hamburg, Germany

CONTRIBUTORS

B. Keith Payne
University of North Carolina
Chapel Hill, NC

Wolfgang Prinz
Max Planck Institute for Human Cognitive
and Brain Sciences
Leipzig, Germany

Howard Rachlin
SUNY Stony Brook
Stony Brook, NY

Thomas S. Redick
Georgia Institute of Technology
Atlanta, GA

Abigail A. Scholer
Columbia University
New York, NY

Jeffrey W. Sherman
University of California
Davis, CA

Ed Smith
Columbia University
New York, NY

Anne Springer
Max Planck Institute for Human Cognitive
and Brain Sciences
Leipzig, Germany

Yaacov Trope
New York University
New York, NY

Tom R. Tyler
New York University
New York, NY

Tor Wager
Columbia University
New York, NY

Florian Waszak
Université Paris Descartes
Paris, France

Daniel M. Wegner
Harvard University
Cambridge, MA

PART I

Neural

CHAPTER 1

Anterior Cingulate Cortex Contributions to Cognitive and Emotional Processing: A General Purpose Mechanism for Cognitive Control and Self-Control

Marie K. Krug and Cameron S. Carter

ABSTRACT

This chapter addresses the topic of self-control from the perspective of conflict theory, a well-studied framework for understanding the behavioral and neural adaptation effects seen during the performance of a selective attention task. We begin with an in-depth explanation of conflict theory and a review of recent literature in support of this theory. We explain how the anterior cingulate cortex (ACC) monitors for processing or response conflict and recruits dorsolateral prefrontal cortex (DLPFC) to resolve these conflicts, increasing attention to goal-related stimuli and adaptively improving behavioral performance. Next, we review alternative theories and explanations of cognitive control and compare them to conflict theory. Finally, we focus on the recent application of conflict theory to the understanding of a wide range of mental processes including emotion regulation and appraisal as well as social cognitive phenomena such as moral reasoning and attitudes, social exclusion, and cognitive dissonance. We conclude that conflict theory, a mechanistic framework originally designed to account for cognitive control functions related to attention, also shows promise in its ability to elucidate higher-level emotional and social behaviors and their associated neural activity. We propose that this model should be considered in future studies of processes related to self-control.

Keywords: Anterior cingulate cortex, attention, cognitive control, cognitive dissonance, conflict adaptation, conflict theory, dorsolateral prefrontal cortex, emotional conflict, emotional processing, fMRI, response conflict, social cognition

INTRODUCTION

Self-control can be defined as the effortful attempts by a person to change thoughts, feelings, and behaviors in order to reach long-term interests or goals (Muraven & Baumeister, 2000). This chapter focuses on some of the specific elementary mechanisms of cognitive control, the implementation of which is necessary for successful self-control in many situations.

Cognitive control allows us to coordinate or direct lower-level or more automatic processes and ensure that our resulting actions will be in line with our goals (Miller & Cohen, 2001).

Cognitive control supports a wide range of cognitive functions from attention and memory retrieval to language production and comprehension. Through cognitive control mechanisms, these processes are managed so that self-control can be attained and, subsequently, goals can be reached.

This chapter particularly focuses on the role of the anterior cingulate cortex (ACC) in the cognitive control processes that are needed to resolve conflict between automatic and goal-directed actions or responses. Recent computational and neuro-imaging findings have suggested that rather than contributing directly to control, the ACC monitors for processing or response conflicts, recruiting control systems, such as the dorsolateral prefrontal cortex (DLPFC), when conflict levels are high and more control is needed to maintain performance. We discuss this theory in detail and review the recent cognitive neuroscience literature related to this topic. We also review alternative theories of the contributions of the ACC to cognitive control and contrast them with conflict theory. Finally, we address the range and scope of conflict monitoring functions that may be supported by the ACC, including the contribution of this region to emotional processing and social cognition. Although the cognitive control literature provides evidence that the ACC and DLPFC detect and resolve conflict that occurs at the response level (or earlier), the application of this theory in studies of emotion and social cognition suggest that these lower-level cognitive control mechanisms form the basis of the higher-level behaviors and decision-making processes that are essential for successful self-control and underlie complex human behaviors.

Conflict Theory

Cognitive control is an essential and widely investigated function of the human brain. It forms the basis of higher cognitive functions and can help us to accomplish goals and perform difficult tasks despite interference from distracting or irrelevant stimuli (Miller & Cohen, 2001). Since the beginnings of cognitive neuroscience the neural mechanisms underlying cognitive control have been the subject of intense investigation. The ACC has featured prominently as a region of the brain whose activity seems strongly correlated with the level of cognitive control demanded during performance. For example, the first positron emission tomography (PET) studies conducted during the 1980s showed that the ACC was active during verb generation, stem completion, free recall, and performance of willed actions such as word generation and finger movements (Petersen et al., 1988; Frith et al., 1991; Grasby et al., 1993; Buckner & Tulving, 1995; Drevets & Raichle, 1998).

Pardo et al. (1990) conducted the first study to look at the brain regions activated during performance of the color-word Stroop task, which has often been used to study the behavioral and neural underpinnings of cognitive control. In this task, subjects are required to read a word and name the color ink in which it is printed (Stroop, 1935). Responses are quick and accurate when the ink color and word meaning are the same, such as when the word "red" is written in red ink. However, on incongruent trials the color of the ink and the meaning of the word are not the same, such as when the word "red" is written in blue ink. Performance is slower and less accurate on incongruent trials because of the presence of response conflict. During these types of trials cognitive control is needed to respond with the correct color-naming response as opposed to the incorrect and conflicting automatic word-reading response. Pardo et al. (1990) used PET scanning to contrast brain regions active during correct incongruent trials in comparison to correct congruent trials. The most robust region of activation was the ACC and the authors hypothesized that the ACC is involved in the selection of task-relevant processes (color-naming) despite interference from competing processes (word-reading).

Although early neuro-imaging studies (Posner et al., 1988; Pardo et al., 1990) of the ACC suggested a role for this region in cognitive control, with increasing demands for control consistently correlating with increasing activity in this region of the brain (Posner & Petersen, 1990), human electrophysiology was suggesting a more specific, perhaps less direct role for the ACC in cognitive control. Gehring et al. (1990) and Hohnsbein et al. (1989) described the "error negativity", a negative deflection of the average event-related potential (ERP) seen in response-locked incorrect trials, that appeared to have a medial frontal, possibly anterior cingulate source (Dehaene et al., 1994). In studies this component appeared related to the behavioral adjustments that have been associated with error commission (Gehring et al., 1993). This led to the hypothesis that the ACC was involved in monitoring and compensating for errors. Elegant in its conceptualization and specification, this theory was difficult to reconcile with activation of the ACC during correct responding which was so widely observed during neuro-imaging studies.

An early event-related fMRI study suggested that the ACC may have a different role in cognitive control that could account both for ACC activity found during correct incongruent trials and for ACC activity following errors (Carter et al., 1998). In this study, subjects performed the Continuous Performance Test (AX-CPT) under conditions that elicited high error rates or induced high levels of response conflict. Results showed that an identical region of the ACC responded to incorrect trials and to correct trials with response conflict. In other words, the ACC was activated during error trials *and* during trials where an incorrect automatic response was overcome in favor of responding correctly.

Carter et al. (1998) suggested a novel hypothesis, namely that errors were a unique instance of response conflict. The occurrence of conflict during correct incongruent trials on a task such as the Stroop is quite intuitive. For example, when the word "red" is presented in blue ink, the incorrect response "red" becomes active because of the automaticity of word reading, along with the correct color-naming response "blue" through effortful attention to the task-relevant dimension of the stimulus. When a correct response is made, the incorrect "red" response does not reach threshold. However, prior to making the correct "blue" response, both responses are activated simultaneously, creating response conflict (Fig. 1–1a). During errors, the

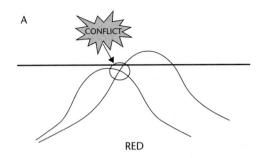

Figure 1–1a. During correct incongruent trials, conflict occurs when both responses are simultaneously active prior to responding with the correct color-naming ("blue") response.
See also figure in color insert.

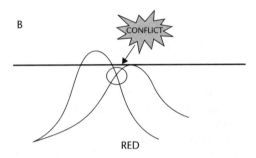

Figure 1–1b. During error trials, the incorrect word-reading response quickly surpasses the response threshold. Conflict occurs after the incorrect response has been made because of later processing of the correct color-naming response.
Red curve = "red" response
Blue curve = "blue" response
Black line = response threshold
See also figure in color insert.

automatic "red" response surpasses the response threshold and an error is committed. However, even as the incorrect response is executed the correct response is also activated as a result of ongoing stimulus evaluation. This co-activation of the incorrect and correct response during and also immediately after error commission is an instance of response conflict and leads to ACC activation on error trials (Fig. 1–1b). According to this theory, during correct high conflict trials and error trials, the ACC detects response conflict and contributes *indirectly* to cognitive control by signaling the need to be more strongly engaged to maintain performance during future incongruent trials and to avoid further errors.

Results from later ERP studies have also supported this theory. According to conflict theory, both incorrect responses and correct incongruent responses are associated with high levels of conflict and dorsal ACC activity (Carter et al., 1998; Botvinick et al., 2001). However, the timing of ACC activation for these two different types of events would differ. For errors, ACC activity should peak after the response has been made, although on correct high conflict trials the ACC conflict monitor should be activated prior to response (see Fig. 1-1) (Gratton et al., 1988; Botvinick et al., 2001). Van Veen and Carter (2002b) measured ERP during performance of the Eriksen Flanker Task. They found a negative peak after error commission that had previously been termed the error-related negativity (ERN) and is typically found following errors (Gehring et al., 1993; Falkenstein et al., 2000). They also found a negativity 340 to 380 milliseconds after trial onset (called the N2) that was enhanced for correct incongruent trials. Source localization showed that both the N2 and the ERN could be attributed to the same ACC source region (Fig. 1-2). This implies that conflict theory can accurately account for the timing of dorsal ACC activity associated with both high conflict correct trials and errors (van Veen & Carter 2002a; van Veen & Carter 2002b).

To further test and refine this hypothesis, Botvinick et al. (1999) designed an fMRI study to specifically determine whether the ACC was involved in selection-for-action, response conflict monitoring, or both functions. Subjects performed a Flanker Task in which they indicated which direction a center arrow surrounded by "flankers" was pointing. On compatible trials, all arrows were pointing in the same direction (Example: <<<<<). In an incompatible trial, the flankers pointed in the opposite direction and prompted an incorrect response (Example: <<><<). Their key analysis compared activity on incompatible trials preceded by another incompatible trial (iI trials) and incompatible trials preceded by compatible trials (cI trials). Previous behavioral studies have shown that although incompatible trials are slower and less accurate than compatible trials (much like

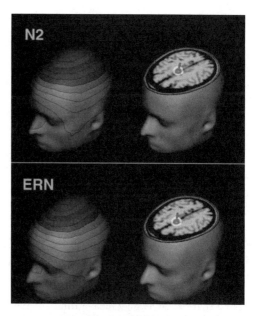

Figure 1–2. BESA 2000 source localization shows that the N2 and the ERN can be attributed to a common ACC source (van Veen & Carter, 2002a; van Veen & Carter, 2002b).
See also figure in color insert.

incongruent trials in the Stroop Task), iI trials are considerably faster and more accurate than cI trials (Gratton et al., 1992). Performance on iI trials is improved because of a strengthening of selection-for-action or control. In other words, on iI trials, subjects focus better on the central arrow and the flanker arrows are no longer distracting. On iI trials, control is high and response conflict is low. If the ACC is involved in control, it should show increased activity on high control iI trials in comparison to cI trials. However, if the ACC monitors response conflict, it should show greater activity during cI trials. During cI trials, flanker arrows are more likely to prompt an incorrect response. Response conflict is high and control is low. Botvinick et al. (1999) found increased ACC activity on high conflict cI trials in comparison to high control iI trials, indicating that the ACC is involved in conflict monitoring as opposed to implementation of control. They also found that the difference in reaction time between cI and iI trials was positively correlated with the amount of ACC activity on the cI trial.

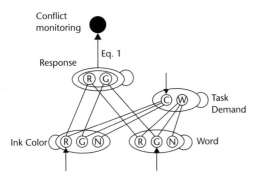

Figure 1–3. Model of conflict detection and cognitive control in the color word Stroop task (from Botvinick, 2001).

A subsequent fMRI study provided further evidence that the ACC is a conflict monitor (Carter et al., 2000). In this experiment, subjects performed several blocks of the color-word Stroop task. In some blocks, incongruent trials occurred frequently (80% of trials) whereas in other blocks, incongruent trials were infrequent (20% of trials). Functional MRI results indicated that ACC activity on incongruent trials was much higher when they occurred infrequently within a block than when incongruent trials occurred frequently. On high-frequency blocks, control is assumed to be high, as the first few incongruent trials in these blocks would recruit control mechanisms. Subsequent incongruent trials would cause less conflict. In infrequent blocks, however, control mechanisms are not established at the time an incongruent trial appears, resulting in high conflict and a strong ACC response.

In a theoretical paper, Botvinick et al. (2001) provided a computational model implemented in the PDP framework that provides a formal account of how the ACC measures conflict and engages cognitive control. Figure 1–3 shows their model of the color-word Stroop task.

In this model (Fig. 1–3), the bottom layer represents the various aspects of the stimulus, where "R" stands for "red," "G" stands for "green," and "N" stands for "neutral." (Neutral refers to a noncolor-related stimulus dimension. For example, the word "car" written in red ink would be associated with the "R" for ink color and "N" for word.) Outputs from the stimulus ink color and word reading feed forward to the response units. Energy at the response level activates the conflict monitor. If only one response is prompted by the stimulus (as occurs on congruent trials), then there will be zero energy at the response unit and the conflict monitor will not be activated. If two different responses are induced by the stimulus (as would occur on an incongruent trial), the energy detected by the conflict monitor would be greater than zero. Energy at the response level increases as the two responses are simultaneously activated. The conflict monitor links to a control unit, which enhances attention to a particular task dimension based on the task instructions or goals. In the Stroop task, if conflict, as detected by the conflict monitor, is high, attention to the color ("C") is enhanced. Simulations using this model predicted the effects of incongruent trial frequency on ACC activity that were found in Carter et al. (2000). Botvinick et al. (2001) also described a similar model for the Flanker task that predicts the enhanced ACC activity on cI trials and improved performance and decreased ACC activity on iI trials described in Botvinick et al. (1999).

Although the Botvinick model was able to provide a mechanistic account of a range of behavioral effects reported in response-conflict tasks, as well as ACC activity on high-conflict trials, direct evidence of ACC recruitment of control centers had not yet been shown in neuro-imaging studies. In an fMRI study of a task switching version of the Stroop task, MacDonald et al. (2000) showed evidence that the DLPFC and ACC contribute to distinct aspects of cognitive control. They used an event-related fMRI design to dissociate areas active during task preparation and areas active during responding. In their experiment, subjects performed the color-word Stroop task. Prior to each trial, subjects received an instruction that told them whether the upcoming trial required a word-reading response or a color-naming response. The color-word stimulus following the instruction could be either congruent or incongruent. Results showed that DLPFC was active during

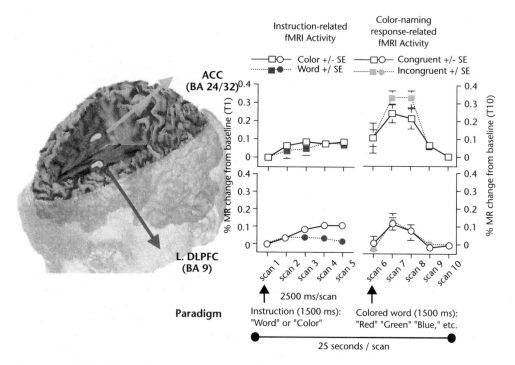

Figure 1–4. Results from MacDonald et al. (2000) showing increased DLPFC activity while preparing for a color-naming trial of the Stroop task and increased ACC activity during incongruent color-naming trials. See also figure in color insert.

the preparation phase when the task instruction was for color-naming. The ACC was active on incongruent color-naming trials. Additionally, greater DLPFC activity during the color-naming preparation phase was correlated with a reduced Stroop interference effect (Fig. 1–4). This means that subjects who activated DLPFC prior to color-naming trials showed less of a reaction time slowing effect in response to incongruent trials. On the other hand, subjects that showed larger reaction time interference effects had higher levels of ACC activity on incongruent trials. These results suggest that DLPFC recruitment can reduce ACC activity and reaction time slowing on subsequent difficult incongruent trials.

Whereas MacDonald et al. (2000) provided some evidence that the ACC conflict monitor recruits the DLPFC for control, Kerns et al. (2004a) used a fast event-related design to separate neural activity on high conflict (cI) and high control (iI) trials. They looked at trial-to-trial adjustments in both behavior and neural activity while subjects performed the color-word Stroop task. They showed that ACC activity was higher on high conflict cI trials in comparison to iI trials. Anterior cingulate cortex activity on one trial was positively correlated with both the amount of DLPFC activity on the following trial and advantageous behavioral adjustments in task performance (i.e., fast reaction time). In other words, a fast iI trial was associated with high DLPFC activity and high ACC activity on the preceding incongruent (cI) trial (Fig. 1–5). This study further established the role of the ACC as the conflict monitor and the role of a separate and distinct area, the DLPFC, in control. Activity in these areas was also tightly correlated with the characteristic trial-to-trial adjustments in behavioral performance on incongruent trials as predicted by the Botvinick model (2001).

A later study done by Kerns (2006) showed similar results in experiments using the Simon task, providing evidence that the role of the ACC and the DLPFC in conflict adaptation generalizes to other response conflict tasks.

ACC CONTRIBUTIONS TO COGNITIVE AND EMOTIONAL PROCESSING

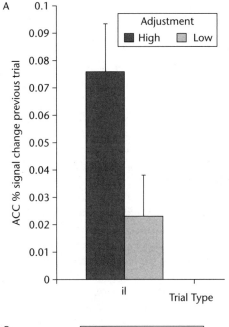

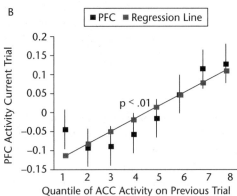

Figure 1–5. A high adjustment iI trial is defined by a fast reaction time whereas a low adjustment iI trial is defined by a slower reaction time. High adjustment iI trials were preceded by higher amounts of ACC activity than low adjustment iI trials (Kerns et al., 2004). ACC activity on the previous trial is positively correlated with DLPFC activity on the current trial (Kerns et al., 2004) See also figure in color insert.

Despite Kerns et al.'s (2004a) evidence that the ACC detects response conflict and recruits control from the DLPFC, an important unanswered question involved *how* the DLPFC resolves response conflict to improve task performance. Egner & Hirsch (2005) used a variation of the color-word Stroop task to determine whether control in response conflict tasks is mediated by increased processing of task-relevant information, inhibition of task-irrelevant or conflicting information, or both. They used a facial Stroop task where subjects viewed faces of various actors or politicians. The name of a particular actor or politician was superimposed over each face. In "face-target" blocks subjects were required to indicate with a button-press whether the face depicted a famous actor or politician. In congruent trials, the face and the name were from the same category, whereas in the incongruent trials the face and the name were from different categories to create response conflict. Incongruent trials preceded by an incongruent trial were associated with improved behavioral performance and increased DLPFC activity. Incongruent trials preceded by an incongruent trial were also characterized by increased activity in face-processing areas. In "face-distracter" blocks, subjects were required to respond according to the category of the word. Incongruent trials preceded by an incongruent trials were, once again, associated with DLPFC activity and improved performance. However, there was no change in activity in face processing areas when iI trials were compared to cI trials. In other words, DLPFC-mediated control can enhance processing of face stimuli when they are task-relevant, but does not decrease processing of face stimuli when they are distracters. The authors concluded that conflict resolution is mediated by increased processing of task-relevant information, and does not appear to be influenced by inhibition of task-irrelevant stimuli.

ALTERNATIVE CONCEPTUALIZATIONS OF ANTERIOR CINGULATE CORTEX FUNCTION AND CRITIQUES OF CONFLICT THEORY

Repetition Priming Effects

Despite the evidence in favor of conflict theory, it has not been without criticism. Mayr et al. (2003) suggested that the conflict adaptation effect is not the result of conflict detection and cognitive control, but rather the result of simple stimulus repetition priming. In the Flanker task, 50% of iI trials are exact stimulus repetitions. If the first incongruent stimulus is "<<><<," then there is a 50% chance that the second incongruent stimulus will also be "<<><<<." This could explain why iI trials are faster and more accurate than cI trials. Stimulus repetitions never occur for cI trials, which could explain why these trials are slower and less accurate. In their first experiment, Mayr et al. (2003) had subjects perform the standard Flanker task. They found no evidence of conflict adaptation effects when they removed stimulus repetitions from their subjects' data. In their second experiment, they used a modified version of the Flanker task where trials alternated between arrows pointing up and down and arrows pointing left and right. This design eliminated stimulus and response repetitions. Mayr et al. (2003) did not find reaction time (RT) conflict adaptation effects in this modified version of the Flanker task.

In response to Mayr et al. (2003), Ullsperger et al. (2005) reported evidence showing robust conflict adaptation in two versions of a Flanker task *after* removing stimulus repetitions. They suggest several reasons why Mayr et al. (2003) did *not* observe conflict adaptation. In Mayr et al.'s first experiment, negative priming may have masked conflict adaptation effects. In that experiment, when stimulus repetitions are removed, the second incongruent trial is always the reverse stimulus of the preceding trial. If the first incongruent (cI) trial is "<<><<," then the second incongruent (iI) trial will be ">><>>." Stimulus reversals could cause response slowing on the iI trial that counteracts the effects of conflict adaptation (Ullsperger et al., 2005). Mayr et al. (2003) also used an unspeeded version of the Flanker task that may have diminished the need for conflict adaptation. For their second experiment, Mayr et al. (2003) may not have found conflict adaptation because trial switching between up- and down-pointing arrows and right- and left-pointing arrows may be treated by subjects as two different tasks. At the time it was unknown if trial-to-trial effects would be observed across trials of different tasks. A later study showed that this is not the case (Egner et al., 2007); trials with high conflict in one stimulus dimension do not enhance performance on a following trial that has conflict in another stimulus dimension (this study will be discussed in greater detail below).

As a result of Mayr et al. (2003), behavioral data after removal of stimulus repetitions are now often reported. Conflict adaptation effects have consistently been observed even after the removal of exact stimulus repetitions (Kerns et al., 2004a; Egner & Hirsch 2005; Etkin et al., 2006; Kerns 2006). In facial Stroop tasks, exact stimulus repetitions can be avoided because of the availability of a larger stimulus set. However, reaction time data are often analyzed after the removal of category repetitions. In iI trials, the category of the face and word are often repeated from one trial to the next and could result in a category priming effect. For example, in Egner and Hirsch (2005), all iI trials that consist of two consecutive actor-face/politician-word trials or two consecutive politician-word/actor-face trials would need to be removed to control for category repetition effects. Conflict adaptation effects are still found in Facial Stroop tasks despite the absence of stimulus repetitions and the removal of category repetitions (Egner & Hirsch 2005; Etkin et al., 2006).

Results from Lesion Studies

If the ACC is essential to conflict monitoring, then disruption of this region of the brain by injury should result in disruption of the ability of subjects to show behavioral evidence of conflict-related adjustments in cognitive control. Results from human lesion studies regarding the role of the ACC have to date been mixed.

Fellows and Farah (2005) found that four patients with dorsal ACC lesions showed similar behavioral effects in comparison to controls in response to manipulations of incongruent trial frequency in a color-word Stroop task. One criticism of this paper is that only one of the four patients had a bilateral cingulate lesion, whereas the other three patients had damage restricted to only one side. Intact performance could be attributed to an intact cingulate (on one side) in three of the four subjects.

Swick and Turken's (2002) patient R.N., who had a left hemisphere dorsal ACC lesion, performed a Stroop task while undergoing ERP recording. Behaviorally, R.N. was slower than

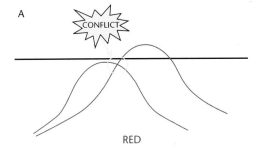 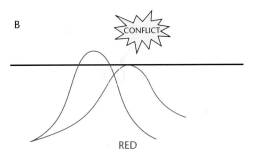

Figure 1-1A During correct incongruent trials, conflict occurs when both responses are simultaneously active prior to responding with the correct color-naming ("blue") response.

Figure 1-1B During error trials, the incorrect word-reading response quickly surpasses the response threshold. Conflict occurs after the incorrect response has been made because of later processing of the correct color-naming response. Red curve = "red" response; Blue curve = "blue" response; Black line = response threshold.

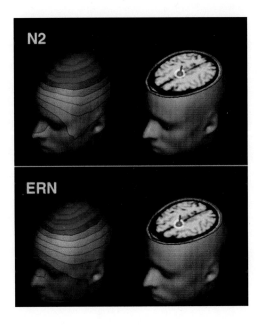

Figure 1-2 BESA 2000 source localization shows that the N2 and the ERN can be attributed to a common ACC source (Van Veen & Carter, 2002; van Veen & Carter, 2002).

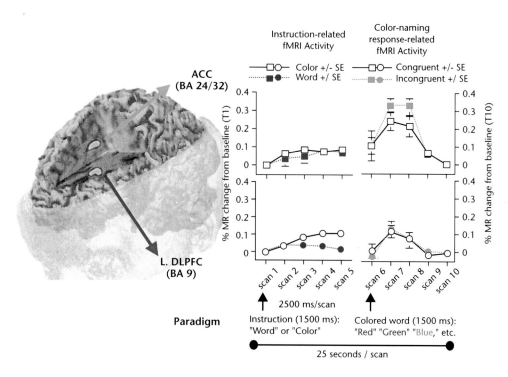

Figure 1-4 Results from MacDonald et al. (2000) showing increased DLPFC activity while preparing for a color-naming trial of the Stroop task and increased ACC activity during incongruent color-naming trials.

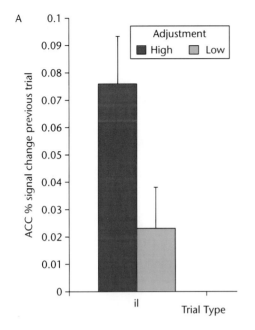

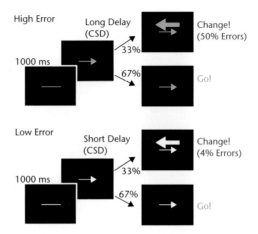

Figure 1-6 Change signal task used in Brown and Braver (2005).

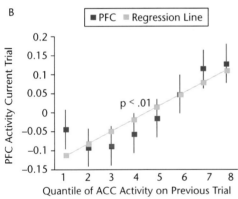

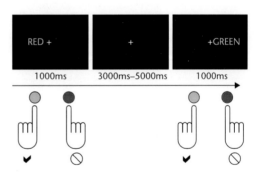

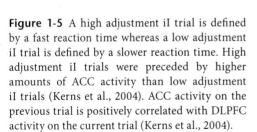

Figure 1-5 A high adjustment iI trial is defined by a fast reaction time whereas a low adjustment iI trial is defined by a slower reaction time. High adjustment iI trials were preceded by higher amounts of ACC activity than low adjustment iI trials (Kerns et al., 2004). ACC activity on the previous trial is positively correlated with DLPFC activity on the current trial (Kerns et al., 2004).

Figure 1-7 The first trial is incongruent in the Stroop stimulus dimension, whereas the second trial is incongruent with respect to the Simon stimulus dimension (Egner et al., 2007).

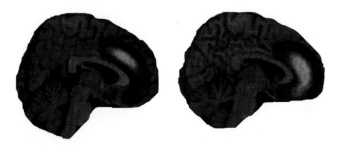

Figure 1-8 Figure from Steele and Lawrie (2004) shows probability maps for ACC activation in cognitive studies (left) and emotion studies (right).

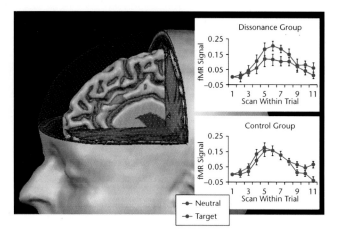

Figure 1-9 Figure adapted from van Veen (submitted). A whole-brain Group (dissonance, control) by Sentence type (target, neutral) interaction identified activation in bilateral ACC. Extracted average time-courses for this region are shown on the right for both the dissonance and control group. Time-course data show that the effect results from ACC activation in response to target sentences by subjects in the dissonance group.

controls, and showed a greater Stroop interference effect in reaction time and accuracy. ERP data indicated that for patient R.N., the N2 was intact while the ERN was decreased in comparison to normal controls. The authors concluded that dorsal ACC may be the neural generator of the ERN, but is not responsible for the N2 and hence that the ACC is not necessary for conflict monitoring to occur in the brain. In response to this finding, Yeung and Cohen (2006) used computational modeling to investigate whether in fact conflict theory could accommodate the apparently discrepant results found by Swick and Turken (2002). They showed that the N2 is sensitive to processing of incorrect or irrelevant stimulus information (as would occur on high conflict incongruent trials), whereas the ERN is sensitive to conflict resulting from processing the task-relevant stimulus dimension during errors and would only be produced if subjects continue to process the correct response as the incorrect response is made. They were able to model R.N's behavioral and ERP results by assuming that conflict detection itself was intact and mediated by the undamaged right ACC, but that the ACC conflict signal could not properly recruit cognitive control. They concluded that R.N.'s lesion disrupts the ACC's ability to recruit cognitive control, which results in the increased interference from incongruent stimuli that underlies his intact N2 signal while reducing his ERN as a result of impaired target processing.

Ochsner et al. (2001) administered several cognitive tasks to a bilateral cingulotomy patient 2 days before and again 3 days following the cingulotomy procedure. Her performance was compared to eight matched normal controls. Results showed that the patient was impaired on incongruent trials of a color-word Stroop task following cingulotomy. The cingulotomy patient also showed impairment on several tasks that required control mechanisms such as mental imagery, mental rotation, and selection of mental images. She was not impaired on simple perceptual tasks that involved a comparison of two presented images.

In the largest ACC lesion study to date, 11 normal controls, six patients with nonfrontal lobe lesions, and eight patients with ACC lesions performed the Simon response incompatibility task (di Pellegrino et al., 2007). Although healthy subjects and patients with nonfrontal lobe lesions showed typical conflict adaptation effects, ACC lesion patients did not, failing to show reduced interference effects following high conflict incongruent trials as predicted by conflict theory.

When considering the effects of cingulate lesions on performance of cognitive tasks, it is important to also consider the role of the ACC in autonomic functions. The role of the ACC in autonomic regulation is generally well-supported. Mean arterial blood pressure was associated with right ACC activity during a mental arithmetic task (Critchley et al., 2000), and ACC activity is associated with both changes in galvanic skin response (GSR) and the cognitive negative variation (CNV), which is an ERP often indicative of response anticipation (Nagai et al., 2004). Hoshkikawa and Yamamoto (1997) found changes in various autonomic measures while subjects performed the Stroop task, although fMRI was not used to attribute a source to these changes.

Critchley et al. (2003) had normal subjects perform a cognitive n-back task and a motor task during fMRI scanning. ECG data was also collected. They found that dorsal ACC and ventral ACC activity during both tasks coincided with heart rate variability (HRV), in particular the low frequency (LF), or sympathetic component of heart rate. Next, three patients with dorsal ACC damage performed an effortful cognitive task involving rapid serial subtraction of seven. Although their performance was not significantly different from normal controls, patients did not show typical autonomic cardiovascular changes during task performance. Additionally, two of the three patients showed greater HRV than controls during a mental arithmetic test. The authors concluded that studies attributing ACC activation to cognitive functions such as conflict monitoring may be wrong. Instead, the ACC may be involved in changing and regulating autonomic states to cope with task demands.

However, Matthews et al. (2004) tell a different story. They collected ECG and fMRI as subjects performed a counting Stroop task and found dorsal ACC activity for incongruent trials in comparison to congruent trials. A separate ventral area of ACC was active during incongruent trials under speeded instructions. This ventral ACC activity was correlated with peak high frequency HRV measures. These results suggest that ACC involvement in cognition and autonomic arousal may be dissociated anatomically within the ACC, with dorsal and ventral ACC involved in conflict detection and autonomic regulation, respectively.

In summary, results of lesion studies are mixed with some negative and some positive findings with regard to the relationship between integrity of the ACC and whether subjects showed evidence of trial to trial adjustments. Given the substantial limitations of each of these studies, one could argue that there is a need for larger, well-controlled studies of the impact of focal lesions in this area on the ability of subjects to show conflict-related adjustments in cognitive control. The idea that the ACC is involved in autonomic regulation is not necessarily in opposition to conflict theory, as autonomic functions may mediate recruitment of cognitive control. Additionally, a separate and distinct region of ACC may be responsible for autonomic functions. Future lesion studies will need to take into account cognitive impairment, autonomic regulation impairment, and extent (in particular, whether the lesion is bilateral or unilateral) and location of the cingulate lesion.

Reinforcement Learning Theory of ACC Error Detection

Competing theories have attempted to explain the role of the ACC in cognitive control. Holroyd and Coles (2002) suggest that the ACC response to errors is the result of a reinforcement learning signal. According to reinforcement learning theory, unexpected or unpredicted errors result in a decrease in dopamine release from the mesencephalic dopamine system. This in turn disinhibits dorsal ACC. The activated ACC is then able to select a proper motor response to optimize subsequent task performance. They propose that this reinforcement learning signal accounts for the ERN following errors (response ERN) and also the negativity following negative feedback (feedback ERN).

Holroyd and Coles (2002) had subjects perform a two-choice decision task. Subjects saw a picture of an object and were required to respond with one of two possible button presses. In the 100% mapping condition, a subject was always rewarded if they pressed a particular button in response to that particular object. In this condition, subjects could easily learn which response yielded a reward. Because this association would become highly learned, subjects would know immediately upon responding if they were correct or not, and the feedback provided after each trial would not be of much use. In the 50% mapping condition, a subject was rewarded on 50% of the trials associated with that particular object, regardless of the response given. In this condition, the subject could not "learn" which response would be rewarded and feedback would be needed for the subject to know the outcome of their response.

In both simulations and experimental data from human participants, Holroyd and Coles (2002) found that in the 100% mapping condition there was a large response ERN, whereas in the 50% mapping condition there was a large feedback ERN. Additional analyses showed that the response ERN in the 100% mapping condition increased throughout the block, showing that once the response mapping had been learned, the response ERN to unexpected errors increased. For the 50% mapping condition, the feedback ERN was larger when the feedback on the current trial was of different valence than the feedback on the previous trial, suggesting that the ERN is sensitive to feedback that is in opposition to a prediction from the preceding trial. These results nicely show that an ERN can occur after an incorrect response or after negative feedback, but that it always occurs in concert with a negative outcome.

According to reinforcement learning theory, both the feedback ERN and the response ERN are the result of the same underlying process: ACC activity as a result of dopaminergic reduction in response to unexpected negative outcomes. Miltner et al. (1997) suggested that the feedback

ERN (like the response ERN) also has an ACC generator. Although conflict theory predicts dorsal ACC activity for both response incongruent trials and errors (Carter et al., 1998; Botvinick et al., 2001; van Veen & Carter 2002a; van Veen & Carter 2002b), it does not predict ACC activation following negative feedback in tasks that do not otherwise have conflict (van Veen et al., 2004).

Van Veen et al. (2004) used fMRI to determine if response conflict, errors, and negative feedback activated the same area of the ACC. They found that incongruent trials on a Stroop task and errors trials on a Stroop task activated an overlapping area of dorsal ACC. However, the same subjects did not show any ACC activity in response to negative feedback in a time estimation task nearly identical to that used in Miltner et al. (1997) (despite comparable behavioral performance). Holroyd et al. (2004) later used fMRI to show that error trials and negative feedback responses activate the same region of dorsal ACC. However, their task was similar to Holroyd and Coles (2002).

Van Veen et al. (2004) argue that some negative feedback tasks may evoke conflict at an intermediate level by violating expectancy. In other words, the ACC could be activated in response to negative feedback as a result of conflict between the expected outcome and the outcome that actually occurs. In Holroyd and Coles (2002), for example, subjects respond by pressing one of two response buttons. Following an "incorrect" response they may respond in a corrective fashion, setting them up to expect certain feedback based on the feedback from the preceding trial.

In van Veen et al. (2004), the time estimation task from Miltner et al. (1997) was specifically chosen because it involves error feedback in the absence of response conflict or conflict from intermediate sources and is better for discerning feedback responses that are not confounded with conflict processing. In a time-estimation task, subjects are required to press a button 1000 milliseconds after a cue was shown. Subjects are given "correct" feedback if their reaction time falls within a certain window of time around 1000 milliseconds. This time window is made smaller following correct responses and widened following incorrect responses. Thus, each subject will be "correct" 50% of the time regardless of how good they are at performing the task. Van Veen et al. (2004) argue that in this task, subjects have no opportunity to develop feedback expectancy and are most likely unable to judge how they are performing.

Functional MRI results from van Veen et al. (2004) were later replicated using a similar task (Nieuwenhuis & Yeung 2005). Nieuwenhuis and Yeung (2005) used the same time-estimation task but reduced the delay between trials. Once again there was no region in dorsal ACC that responded preferentially to negative feedback. They also collected EEG results for the same task and found that a feedback ERN was produced following negative feedback. Source-modeling data constrained by the fMRI results instead suggest that the feedback ERN is generated by the rostral ACC, posterior cingulate, and right superior frontal gyrus. Despite the controversy regarding the source of the feedback ERN, Holroyd and Krigolson (2007) have recently shown that feedback ERN amplitude is positively correlated with behavioral RT changes. Regardless of whether the feedback ERN is generated by the ACC or other brain region(s), it may be involved in optimizing task performance.

Error Likelihood Theory

In a theory that has similarities to both the reinforcement learning theory of error detection and the conflict monitoring theory of indirect control, Brown and Braver (2005) proposed that the ACC responds to error likelihood. In other words, ACC activity is high during conditions that are likely to produce errors. Brown and Braver (2005) designed a change signal task to test their theory. In their task, subjects were first presented with a colored difficulty cue, which signaled whether the upcoming trial would be associated with a high likelihood of making an error or with a low likelihood of making an error. Next, an arrow pointing either right or left was presented. Subjects were required to make a response regarding the direction of the arrow. In one-third of the trials, a second "change" arrow appeared over the original arrow. This arrow

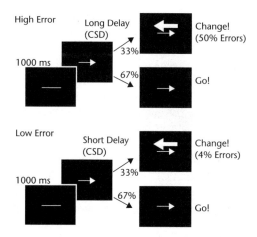

Figure 1-6. Change signal task used in Brown and Braver (2005).
See also figure in color insert.

pointed in the opposite direction and required subjects to change the response they had been preparing in regards to the first arrow. In the high error condition, the change arrow appeared after a long delay, resulting in high error rates. In the low error condition, the change arrow appeared shortly after the first arrow. On these trials, subjects made few errors as they were able to change their responses if the second arrow appeared soon after the first arrow (Fig. 1-6).

According to their model, the ACC receives signals from an error unit and the stimulus representation associated with that trial after an error has been committed. This event strengthens the weight between that particular stimulus representation and the ACC. The ACC learns that a particular stimulus has been associated with an error, and will activate strongly to that type of stimulus in the future. Although both the error-likelihood model and the conflict model predict similar behavioral results, they differ in their predictions regarding ACC activity. The error likelihood model predicts ACC activity following high error cues, even if a second arrow is not presented because the ACC will activate in response to the cue, which is associated with a high-error condition. Functional MRI results showed that the dorsal ACC was activated more during non-change trials following a high error cue in comparison to non-change trials following a low error cue. This effect gradually increased throughout the experiment, suggesting that the ACC learns to respond to trials that are associated with high error likelihood, as predicted by their model.

Although Brown and Braver (2005) offered fMRI results in support of the error-likelihood theory, their results have not, to date, been replicated. Nieuwenhuis et al. (2007) reevaluated error likelihood theory using three different experiments. In their first experiment they used fMRI to investigate ACC activity in response to error-likelihood cues during a visual search task. Visual search could be difficult (high error-likelihood) or easy (low error-likelihood), and these two types of trials were preceded by corresponding error-likelihood cues (as in Brown & Braver, 2005). They found no brain regions that were activated more to the high error-likelihood cue in comparison to the low error-likelihood cue.

In their second experiment they used the same stop-change task used by Brown and Braver (2005; Nieuwenhuis et al., 2007). They used a longer cue-target interval to better dissociate cue-related activity from target processing and response-related activity. Functional MRI results indicated that high error-likelihood cues did not show stronger ACC activity than low error-likelihood cues. Nieuwenhuis et al. (2007) suggest that Brown and Braver's ACC activation could be attributed to performance-related differences on high error-likelihood trials in comparison to low error-likelihood trials. In Brown and Braver's study, subjects were slower on the high error-likelihood non-change trials in comparison to the low error-likelihood non-change trials. These strategic differences may have produced the ACC activation, despite Brown and Braver's efforts to control for reaction time differences (Brown & Braver 2005; Nieuwenhuis et al., 2007).

In their final experiment, Nieuwenhuis et al. (2007) measured ERP while subjects performed the stop-change task exactly as described in Brown and Braver (no timing changes). They found no difference in ERP waveforms in response to high error-likelihood cues versus low error-likelihood cues. They also did not find any difference in ERP waveforms during non-change trials following high error-likelihood cues in comparison to non-change trials following low error-likelihood cues. However, the N2 was stronger in response to change trials in comparison to non-change trials, suggesting that the ACC responds to conflict and not error-likelihood predictions.

Conflict Theory: Different Types of Conflict and Control

Although the studies described earlier provide evidence in support of conflict theory, many questions regarding the specific mechanisms underlying conflict monitoring and cognitive control still remained. The behavioral and neural underpinnings of response conflict and subsequent cognitive control on a color-word Stroop or Flanker task are fairly well-understood. Next, the behavioral and neural underpinnings of other forms of conflict and control were investigated to determine how conflict theory can be extended as a more general model for cognitive and self-control.

In Botvinick's (2001) model, conflict was detected at the response level. Van Veen et al. (2001) investigated whether the ACC detects conflict present at early levels of processing. Van Veen et al. created a flanker task with two types of incongruent stimuli: response incongruent (RI) and stimulus incongruent (SI). In their version of the flanker task, the letters "S" and "M" were mapped to one response while the letters "H" and "P" were mapped to another response. Thus, the stimulus "HHMHH" would produce response conflict because the "H" and the "M" would prompt different responses. The stimulus "HHSHH", although incongruent at the stimulus-identification level, would not be considered incongruent at the response level since both the "H" and the "S" map to the same response. Results showed that although both RI and SI trials result in slowed reaction time, only RI trials activated the ACC.

However, in another study, van Veen et al. (2005) manipulated the type of response conflict in a color-word Stroop task. In this study, the colors red and yellow mapped to one response button while the colors blue and green mapped to the other response button. Thus, the word "red" written in green ink would be response incongruent, whereas the word "red" written in yellow ink would be congruent at an intermediate or semantic level. While both response conflict and semantic conflict activated the ACC and the DLPFC, there were no regions of overlap in activity. Response conflict also activated superior temporal cortex and thalamus, whereas semantic conflict trials activated inferior parietal cortex. The authors concluded that although both semantic and response conflict activate the ACC and DLPFC, these networks include distinct and nonoverlapping brain regions.

In a version of the color-word Stroop that included neutral trial types (e.g., the word LOT printed in red ink) a more posterior region of the ACC was more active on congruent and incongruent trials in comparison to neutral trials (Milham & Banich, 2005). They concluded that the posterior ACC may detect the presence of a possible task-relevant response in the irrelevant task dimension (regardless of whether it is conflicting with the correct response).

Egner et al. (2007) addressed the issue of whether control is context-specific. They created a Stroop/Simon combination task. In this task, the word "RED" or "GREEN" was presented in either red or green ink. The word could appear on either the right or left side of the screen. If the ink color was red, subjects were to respond with a right-hand button press. If the ink color was green, subjects were to respond with a left-hand button press. This task design created two types of incongruent trials. If the word and ink color were different, it was an incongruent Stroop trial. However, if a word written in green ink appeared on the right side of the screen (or a word written in red appeared on the left side of the screen) it would be considered an incongruent Simon trial caused by the conflict between the correct response and the spatial location of the stimulus (Fig. 1–7).

Egner et al. (2007) found typical conflict adaptation effects in the reaction time data when both previous and current trials were incongruent in the Stroop dimension and when both previous and current trials were incongruent in the Simon dimension. In other words, an iI trial would be faster than a cI trial if the preceding incongruent trial was incongruent in the same dimension. However, if an iI trial was

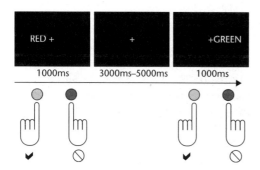

Figure 1–7. The first trial is incongruent in the Stroop stimulus dimension, whereas the second trial is incongruent with respect to the Simon stimulus dimension (Egner et al., 2007). See also figure in color insert.

incongruent in the Simon dimension and the preceding incongruent trial was incongruent in the Stroop dimension (or vice versa), conflict adaptation did not occur. Control in the Stroop dimension activated superior parietal cortex whereas control in the Simon dimension recruited ventral premotor cortex. There were no areas of activity associated with conflict detection or resolution that were common to both Stroop and Simon task dimensions, suggesting the brain has separate and nonoverlapping brain regions for different types of conflict and control. These results suggest that conflict recruits context-specific control mechanisms that do not generalize to the resolution of different types of conflict.

Conflict Theory and Self-Control: Broader Implications

Although not without controversy, after almost 10 years, the conflict theory has received a great deal of support in the cognitive neuroscience literature and has been often invoked as a general purpose mechanism that supports cognitive control processes across a broad range of higher cognitive functions (Kerns et al., 2004b; Kuhl et al., 2007). This theory and role for the ACC has also been increasingly invoked in affective and social cognitive processing domains. We will now review this more recent literature so as to address the broader implications of conflict theory for our understanding of the neural basis of human cognition, behavior, and self-control.

Emotion and the ACC

In addition to its role in cognitive processing, there is a wealth of literature that implicates the ACC in emotion processing. Macaque monkeys with bilateral cingulate lesions show disrupted social behavior and emotion, characterized by fewer social interactions, fewer vocalizations and facial expressions, and increased time with inanimate objects, whereas bilateral cingulate ablation in cats is associated with abnormal emotional behavior that includes excessive rage, snarling, and growling (Kennard 1955; Hadland et al., 2003). It is important to note that in the monkey study (Hadland et al., 2003) the lesion was done using aspiration, which may have caused damage not only to the cingulate cortex but also to the neighboring cingulum bundle, which contains several different fiber components with projections to various prefrontal, premotor, and limbic regions (Mufson & Pandya 1984). An aim of future studies will be to determine more specifically which regions of the ACC and/or cingulum bundle contribute to emotional behavior.

Abnormalities in ACC structure and function have been associated with mood disorders, such as unipolar and bipolar depression and posttraumatic stress disorder (PTSD). For example, bipolar and unipolar depression patients show decreased blood flow to and reduced cortical volume of subgenual ACC (Drevets et al., 1997; Drevets et al., 1998) and hypometabolism in this region is associated with negative treatment response for depression (Mayberg et al., 1997). However, depressed patients show increased activation in subgenual ACC in response to positively and negatively valenced stimuli in comparison to controls (Gotlib et al., 2005). Posttraumatic stress disorder is characterized by reduced ACC recruitment in response to negatively valenced stimuli (Shin et al., 2001).

Cingulotomy, a surgical procedure where the anterior cingulate cortex is lesioned bilaterally, is an option for patients who have not responded to medications or behavioral treatments.

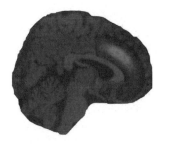 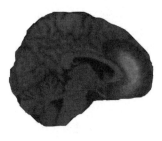

Figure 1-8. Figure from Steele and Lawrie (2004) shows probability maps for ACC activation in cognitive studies (left) and emotion studies (right).
See also figure in color insert.

Cingulotomy is successful in reducing symptoms in many patients with obsessive compulsive disorder, unipolar depression, bipolar depression, and generalized anxiety disorder that previously had not responded to conventional treatments and has relatively few negative side effects (Spangler et al., 1996; Dougherty et al., 2002; Dougherty et al., 2003) However, some patients with ACC lesions show impairments on recognizing emotion in voices and faces. They also report post-lesion changes in subjective experience of emotion and many described increased intensity of positive and negative emotions (Hornak et al., 2003). More recent work has investigated the effects of the extent and location of cingulate/cingulum bundle lesions on clinical outcome (Steele et al., 2008).

Based upon analysis of anatomical connectivity, a number of investigators have proposed that separate regions of the ACC serve cognitive (such as conflict monitoring) and emotional functions. The proposed cognitive subdivision is dorsal and caudal to the affective subdivision, and includes Brodmann areas 24a', 24b', 24c', and 32'. It has strong connections with the prefrontal, premotor, and supplementary motor areas. The affective subdivision is comprised of Brodmann areas 24a, 24b, 24c, 32, 33 and 25 and projects to the amygdala, orbitofrontal cortex (OFC), and autonomic brainstem nuclei (Devinsky et al., 1995; Vogt et al., 1995; Bush et al., 2000). Results from neuro-imaging studies do provide further support for these hypothesized cognitive and affective subdivisions of the ACC, with the cognitive division showing activation during the Stroop and other response conflict tasks, while the affective division is activated during emotional tasks (Bush et al., 1998; Whalen et al., 1998; Bush et al., 2000; Steele & Lawrie 2004).

The exact role of the rostral ACC during emotional tasks is uncertain. One study reported enhanced rostral ACC activation to task-irrelevant emotional distracters. However, it could not be determined whether this activation resulted from processing of the emotional distracters, detection of processing conflict, or to implementation of control to block or inhibit processing of the emotional distracters (Vuilleumier et al., 2001). Whalen et al. (1998) found that an emotional version of the counting Stroop task recruited rostral ACC, whereas the same subjects recruited dorsal ACC during incongruent blocks of a non-emotional counting Stroop task (Bush et al., 1998; Bush et al., 2000). During counting Stroop tasks, subjects must indicate how many word items are shown. In the non-emotional counting Stroop, trials are incongruent when the words prompt a numeric response that is different from the number of items displayed (e.g., the word "three" written four times). In the emotional counting Stroop, words of negative valence must be counted (e.g., the word "murder" written four times). The authors suggested that the rostral ACC and the dorsal ACC are recruited to perform a similar function, with the rostral ACC recruited only when the information is emotional, and the dorsal ACC recruited when information is nonemotional.

Bishop et al. (2004) specifically tested the role of the rostral ACC in "emotional conflict" detection. In their task, subjects were shown two houses and two faces on each trial. They were instructed to respond whether the two houses were the same or different while ignoring the faces. The two faces were either neutral or fearful. In the "infrequent blocks" the fearful faces appeared on 30% of trials, whereas in the "frequent blocks" the distracters were fearful faces on 70% of trials. They showed that the rostral ACC was activated more strongly in response to infrequent threat-related distracters in comparison to frequent threat-related distracters, suggesting that the rostral ACC comes online to detect processing conflict from emotional distracters.

In all of these studies (Bush et al., 1998; Whalen et al., 1998; Bush et al., 2000; Vuilleumier et al., 2001; Bishop et al., 2004), a lack of behavioral effects on trials with emotional content prohibited analysis of trial-to-trial adjustments and corresponding ACC and prefrontal activations that could be associated with conflict monitoring and subsequent control. Without these types of analyses, it is difficult if not impossible to determine if the rostral ACC is involved in detecting conflict that arises from emotional stimuli.

Etkin et al. (2006) designed an emotional Facial Stroop that would allow for these trial-to-trial analyses. In their task subjects were required to respond whether a face was happy or fearful, while ignoring the word "happy" or "fear" superimposed over the face. High conflict (cI) trials were associated with heightened activity in the amygdala, dorsomedial prefrontal cortex, and dorsolateral prefrontal cortex. Activity in these three regions was predictive of rostral ACC activity on the following trial. Rostral ACC recruitment was associated with a reduction in amygdala activity and enhanced task performance on that trial. They concluded that amygdala activity is heightened when two conflicting emotional states are represented in the same trial, and that the rostral ACC is involved in resolving this conflict via control over the amygdala. Likewise, in a review of 55 PET and fMRI studies Phan et al. (2002) reported that emotional tasks with cognitive demand (such as tasks requiring the subject to rate an emotional stimulus) activate the rostral ACC more often than emotional tasks that require passive viewing only. These tasks were associated with less amygdala activation in comparison to passive viewing tasks. They suggest that during cognitive tasks, the rostral ACC modulates or controls the amygdala response to emotional stimuli.

In a more recent follow-up study, Egner et al. (2008) directly compared conflict and control-related activity in the emotional Facial Stroop task to a similar but non-emotional version of the task. Their emotional task was the same as that used in Etkin et al. (2006). In their non-emotional task subjects judged whether happy and fearful faces were male or female in gender while ignoring the words "male" and "female" superimposed over the face.

Both tasks elicited comparable behavioral conflict adaptation effects. As in their previous study, high control trials in the emotional task were associated with activity in rostral ACC, and connectivity analyses showed that rostral ACC activity was associated with decreased amygdala activity. In the non-emotional task, high-control trials were associated with activity in right lateral prefrontal cortex (LPFC). Connectivity analyses indicated that increased LPFC activity was associated with enhanced activity in fusiform face processing areas. However, in this study, they found that the amygdala was not significantly more active during high conflict trials in the emotional task in comparison to the non-emotional task. Thus, they could not conclude that the amygdala responds specifically to emotional response conflict. However, they did find overlapping regions of dorsal ACC in response to high conflict trials in both the emotional and the non-emotional task. These results suggest that although different prefrontal areas are needed to resolve response conflict in emotional and non-emotional tasks, the dorsal ACC detects response conflict *regardless* of whether the conflict arises from emotional or non-emotional stimuli. Future studies will be needed to further investigate how the dorsal ACC can differentially recruit rostral ACC or LPFC to resolve emotional or non-emotional response conflict, respectively.

Dorsal ACC may also be involved in reappraisal. Reappraisal is a type of emotion regulation where conscious cognitive reasoning is used to either enhance or dampen emotional experience. Ochsner et al. (2002) trained subjects to reappraise negative photographs by generating a story about each photograph to interpret the negative scene in a less negative manner. In the scanner, subjects were presented with negative pictures. On each trial, they were required to either reappraise the negative picture or to attend to the picture without engaging in reappraisal. They were then required to rate their negative affect in response to the preceding picture. Behavioral results showed that reappraisal works; subjects reported weaker negative affect following reappraised pictures in comparison to attended pictures. Functional MRI results showed increased activity in dorsolateral, ventrolateral, and medial PFC and decreased activation in emotion processing areas such as the right amygdala and OFC during reappraisal in comparison to attending. A "reappraisal success" score was calculated for each participant by comparing the amount of negative affect following a reappraised picture and the amount of negative affect following an attended picture. Reappraisal success correlated with increased activity in dorsal ACC during the reappraisal trials. In other words, subjects who had strong dorsal ACC activity during the reappraisal trials were better at decreasing their negative affect. Ochsner et al. (2002) suggest that during reappraisal, dorsal ACC may be monitoring for conflicts between emotional processing of the picture and the reappraisal strategy. Future studies will need to show exactly *how* dorsal ACC conflict monitoring enhances reappraisal success.

It is important to know which prefrontal areas dorsal ACC is recruiting and the targets of these prefrontal regions. Ochsner et al. (2002) show that left VLPFC activity during reappraisal is inversely correlated with activity in the right amygdala and left medial OFC, suggesting that this region may be directly modulating activity in areas involved in emotion processing. However, in a review, Ochsner and Gross (2005) show that reappraisal studies and other types of emotional control tasks also show activity in DLPFC. Because this area has few direct connections to emotion-processing areas, it may aid emotion control indirectly by maintaining representations of emotional responses to stimuli, which in turn could recruit other ventral control areas or posterior brain regions that could directly modify emotion processing (Ochsner & Gross 2005).

Although dorsal ACC and rostral ACC have been described as being involved in conflict detection and emotional control, respectively, there is evidence that both areas of the ACC may play a role in emotional awareness. In one study (Lane et al., 1998), subjects viewed emotional and non-emotional films and recalled emotional and nonemotional experiences. They were also administered the Levels of Emotional Awareness Scale (LEAS). During the LEAS, subjects read passages and then describe the feelings and emotional experiences of the characters in the passage. Scores are calculated based on the complexity in the description of the emotional content of the passages, with high scores indicative of higher levels of emotional awareness. Results showed that dorsal ACC was commonly activated in both the emotional film and emotional recall conditions in comparison to the non-emotional film and non-emotional recall conditions. Activity in this region was correlated with LEAS score. Lane et al. (1998) suggest that dorsal ACC activity during emotional experience may help redirect attentional resources in order to optimize behavior in response to emotional stimuli. However, during these tasks subjects were instructed to focus on feeling the target emotion of each film or recalled experience. Lane et al. (1998) also suggest that dorsal ACC activity could reflect the attentional demand of feeling the target emotion despite other distracting, non-target emotions. Subjects with higher LEAS scores recruit more dorsal ACC activity because they feel a greater complexity of emotions, and thus need more attentional resources to inhibit competing emotions. We would argue that this dorsal ACC activity could also be indicative of conflict. The ACC may be responding to conflict between various emotional responses and could be recruiting other prefrontal areas (DLPFC or perhaps rostral ACC) to enhance attention to target emotions and/or inhibit attention to competing emotions.

Rostral ACC may also play a role in emotional awareness. Lane et al. (1997) had subjects view blocks of pleasant and unpleasant pictures. For each block subjects were instructed to either rate their emotional response to the picture or indicate whether the picture depicted a scene that was occurring indoors or outdoors. When activation to these tasks was compared, the blocks in which subjects were required to rate their emotional experience activated a region of rostral ACC (BA 32) that extended anteriorly into medial PFC (BA 9). They concluded that this rostral ACC/medial PFC region represents subjective emotional experiences. Lane maintains that activity extending from rostral ACC into BA 9 during emotional experiences is both anatomically and theoretically plausible, and rostral ACC and BA 9 are not necessarily functionally different (Lane, 2000). However, in light of the more recent findings that rostral ACC is involved in control over emotions, future studies will need to determine if representation of emotional response and emotional control are mediated by common areas, or if representation of emotional response is mediated by a more anterior and superior region.

Social Cognition and the Anterior Cingulate Cortex

Conflict theory has recently been used to explain a variety of social cognitive phenomena, suggesting that interactions between the ACC and prefrontal areas may underlie a variety of more complex behaviors.

Greene et al. (2004) found that conflict monitoring and cognitive control may be involved in moral reasoning. In their study, subjects were presented with moral dilemmas. After reading each scenario they were then asked to respond whether a particular utilitarian action in response to the dilemma was "appropriate" or "inappropriate." An action is defined as utilitarian if a committed moral violation contributes to the overall welfare of those involved (e.g., smothering a crying baby to save the lives of five people who are hiding from an approaching enemy). Greene et al. (2004) measured how long it took participants to respond regarding the action taken. They found that difficult moral dilemmas (those associated with a long response time) activated the anterior DLPFC and the ACC more than easy moral dilemmas (those that the subjects responded to very quickly). Next, they divided the difficult moral dilemmas into two types: those in which the subject deemed a utilitarian judgment "appropriate" and those in which the subject deemed a utilitarian judgment "inappropriate." They found that utilitarian judgments deemed appropriate were characterized by increased DLPFC activity in comparison to utilitarian judgments considered inappropriate. Greene et al. (2004) suggest that difficult moral dilemmas are associated with ACC and DLPFC activity because there is a conflict between moral principles and associated emotions (it is wrong to kill a baby) and cognitive processes that can overcome these moral violations if they are appropriate in a particular context (when killing a baby is necessary to save the lives of many people). The ACC may be detecting the conflict that arises from these two competing behavioral responses. The DLPFC activity in response to approval of utilitarian decisions also makes sense in light of conflict theory. To overcome the automatic emotional response to a moral violation and respond in a utilitarian manner, cognitive control is needed. Future studies will need to investigate trial-to-trial adjustments in moral decision-making, as conflict monitoring and subsequent control could not be separated on a trial-to-trial basis as in Kerns et al. (2004a).

Luo et al. (2006) concluded that responding in opposition to moral attitudes evokes activity from conflict and cognitive control areas. They used a modified version of the IAT (Implicit Association Test) to look at the neural correlates of implicit moral attitude. In their task, subjects responded to pictures of legal and illegal activities and good and bad animals. In congruent blocks, subjects responded to good animals and legal activities with one response button, and bad animals and illegal activities with the other response button. In the incongruent blocks, illegal activities and good animals mapped to one response, whereas legal activities and bad animals required the other response. Implicit moral attitude was measured as the amount of reaction time slowing in the incongruent blocks in comparison to the congruent blocks. In other words, a person with strong moral attitudes would have greater difficulty in responding to illegal scenes when this response is paired with "good" attributes. Incongruent trials were associated with activity in the anterior cingulate gyrus (BA 24), the subgenual/rostral ACC (BA 25), the ventrolateral prefrontal cortex, and the caudate. The authors speculate that the ACC may be detecting the conflict between differently valenced items that are associated with the same response. Lateral PFC may be resolving this conflict by controlling motor-response areas such as the caudate.

In a fascinating fMRI study, Eisenberger et al. (2003) investigated the neural correlates of social rejection. Subjects had high ACC activity and self-reported distress when they were suddenly excluded from a ball tossing game they believed they were playing with two other players. The amount of self-reported distress in response to social exclusion was positively correlated with ACC activity. Ventrolateral prefrontal cortex (VLPFC) was also active during social exclusion. VLPFC activity was negatively correlated with self-reported distress and ACC activity. The authors attribute the ACC activity to engagement of pain processing regions, and suggest that VLPFC is involved in regulation of pain from distress. In a later paper, the authors suggest that dorsal ACC activity during social exclusion is similar to the conflict-related activity shown in Stroop tasks. They suggest that in cognitive tasks and in their social exclusion task, the dorsal ACC may act as a "neural alarm system" that detects problems and alerts control areas of the brain that can then work to resolve the problem. The distress or pain of social exclusion may activate the dorsal ACC neural alarm system so that prefrontal regions can be recruited to relieve the distress caused by social exclusion (Eisenberger & Lieberman 2004). Future studies will need to investigate how LPFC areas resolve the distress caused by social exclusion.

However, dorsal ACC activity during exclusion trials could be a result of response conflict in this task. When the ball is tossed to the subject, the automatic response is to throw the ball to the other player. Dorsal ACC activity on an exclusion trial could be a result of conflict between an automatic response (throwing the ball to the other player) and the correct response (not throwing the ball because you do not have it). Prefrontal activity could be recruited to enhance attention to the game so that response to a ball throw (or lack thereof) will be appropriate. In Eisenberger et al. (2003) exclusion/inclusion was manipulated by block. In future studies, exclusion and inclusion could be manipulated on a trial-to-trial basis. Conflict theory would predict that exclusion trials preceded by inclusion trials should be high in dorsal ACC activity, whereas exclusion trials preceded by exclusion trials should be higher in PFC activity. Regardless of whether dorsal ACC activity during social exclusion is the result of response conflict or distress/social pain, it is evident that dorsal ACC and PFC control networks are implicated in many different tasks and behaviors in addition to the more commonly studied Stroop, Flanker, and Simon tasks.

Finally, van Veen et al. (2009) have recently shown that dorsal ACC activity during cognitive dissonance is predictive of subsequent attitude change. According to Festinger's cognitive dissonance theory, people experience the psychologically uncomfortable state of cognitive dissonance when beliefs and attitudes are not consistent with behavior. When people cannot attribute their counter-attitudinal behavior to an outside force (such as payment or coercion), then they often adjust their beliefs and attitudes to align with their behavior (Festinger, 1957; Festinger & Carlsmith, 1959).

In their study, van Veen et al. (2009) adapted a standard cognitive dissonance paradigm for use in the MRI scanner. For the first part of the experiment, all subjects performed a long (about 1 hour) and tedious cognitive task in the MRI scanner. They were then presented with a second task in the MRI scanner, where they were instructed to rate on a six-point scale how much they agreed or disagreed with sentences that would appear on the screen. One-third of the sentences (target sentences) would be about their enjoyment of the scanning process with questions such as "The scanner feels very comfortable" and "The loud noise of the scanner is too obnoxious." The remaining sentences (neutral control sentences) would ask about various preferences and attitudes and current bodily states that were not related to scanner enjoyment such as "I like to eat macaroni and cheese"

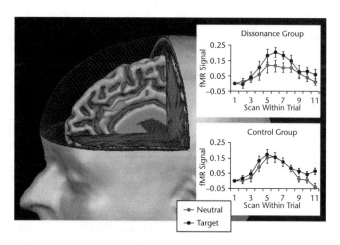

Figure 1–9. Figure adapted from van Veen (2009). A whole-brain Group (dissonance, control) by Sentence type (target, neutral) interaction identified activation in bilateral ACC. Extracted average timecourses for this region are shown on the right for both the dissonance and control group. Time-course data show that the effect results from ACC activation in response to target sentences by subjects in the dissonance group. See also figure in color insert.

and "My feet are touching each other right now." For this second task, half of the subjects were assigned to the "control group." They were told that they would be paid additional money if they responded to the scanner-related questions as if they were having a positive and enjoyable experience. The other subjects were assigned to the "dissonance group." They were told that there was a nervous patient waiting in the scanner suite who would be able to see their answers to the scanner questions that would soon appear on the screen. These subjects were asked to respond as if they are enjoying the scanner experience to help put the nervous patient at ease. Following scanning, all subjects were taken to a separate room and asked to provide honest ratings in response to the target questions. A final attitude score was calculated for each participant based on their post-scan final ratings.

As predicted by cognitive dissonance theory, subjects in the control group rated the scanner more negatively than subjects in the dissonance group. Because subjects in the dissonance group could not justify their counterattitudinal argument by an outside source (such as payment) their opinions of the scanner were adjusted to match their behavior and reduce dissonance.

Functional MRI results showed bilateral dorsal ACC activity in response to the target sentences for subjects in the dissonance group. In other words, when subjects were rating their scanner experience, the dorsal ACC (and also the anterior insula) was active, but only if they were in a situation that would induce cognitive dissonance (rating that the scanner experience was positive in the presence of a patient as opposed to rating the scanner experience as positive to earn money) (Fig. 1–9).

Final attitude score was correlated with peak ACC amplitude during target sentences for subjects in the dissonance group. In other words, subjects who had the greatest amount of ACC activity when responding to target sentences while inside the scanner later reported a more positive scanner experience.

Dorsal ACC activity during a dissonance-inducing counter-attitudinal argument suggests that dorsal ACC responds to the conflict between attitudes and behavior or action. ACC activity during conditions of cognitive dissonance has been predicted by Harmon-Jones' action-based model of cognitive dissonance (Harmon-Jones 2004; Harmon-Jones, 2008). Their group also showed that the ACC-generated ERN was enhanced following race-biased responses during a priming task, suggesting that the ACC responds to accidental race-prejudiced actions that are not in line with a self-concept of being a non-prejudiced person (Amodio et al., 2004; Harmon-Jones, 2008).

Van Veen et al. (2009) found that the amount of ACC activity is predictive of the extent of attitude change. However, the neural mechanism of this attitude change was not shown in this study. Harmon-Jones & Harmon-Jones (2008) argue that left frontal cortical areas involved in resolution of response conflict also mediate reductions in cognitive dissonance, and provide extensive experimental evidence in support of this hypothesis (Harmon-Jones, 2008). This dissonance reduction ensures that the proper (formerly counterattitudinal) behavior is effectively carried out without regret or negative affect (Harmon-Jones, 2008).

Conclusion

Many functional neuro-imaging and EEG studies implicate the ACC in cognitive control processes. Conflict theory provides a mechanistic framework that explains and predicts many behavioral effects and a wide range of neural activity observed in this region across many different cognitive and emotional tasks. Although the theory and its variants will continue to be challenged and refined, we would argue that it provides a valid and useful heuristic for understanding the neural implementation of dynamic control mechanisms across the cognitive, emotional, and social systems involved in self-control.

The term "self-control" encompasses a broad range of the diverse and dauntingly complex behaviors that are an essential part of human experience. Conflict theory, at a first glance, appears only to explain the behavioral and neural effects during the performance of a very specific task in a non-naturalistic setting. However, we have shown that a well-defined and tested theory of a simple cognitive control mechanism has the potential to account for the neural underpinnings of increasingly more complex human behaviors.

References

Amodio, D. M., Harmon-Jones, E., Devine, P. G., Curtin, J. J., Hartley, S. L., & Covert, A. E. Neural signals for the detection of unintentional race bias. Psychol Sci 2004; 15(2): 88–93.

Bishop, S., Duncan, J., Brett, M., & Lawrence, A. D. Prefrontal cortical function and anxiety: Controlling attention to threat-related stimuli. Nat Neurosci 2004; 7(2): 184–188.

Botvinick, M., Nystrom, L. E., Fissell, K., Carter, C. S., & Cohen, J. D. Conflict monitoring versus selection-for-action in anterior cingulate cortex. Nature 1999; 402(6758): 179–181.

Botvinick, M. M., Braver, T. S., Barch, D. M., Carter, C.S., & Cohen, J.D. Conflict monitoring and cognitive control. Psychol Rev 2001; 108(3): 624–652.

Brown, J. W., & Braver, T. S. Learned predictions of error likelihood in the anterior cingulate cortex. Science 2005; 307(5712): 1118–1121.

Buckner, R. L., & Tulving, E. Neuroimaging studies of memory: Theory and recent PET results. In: Boller, F., & Grafman, J. (Eds.), Handbook of neuropsychology. Amersterdam, Elsevier, 1995: pp. 439–466.

Bush, G., Luu, P., & Posner, M. I. Cognitive and emotional influences in anterior cingulate cortex. Trends Cogn Sci 2000; 4(6): 215–222.

Bush, G., Whalen, P. J., Rosen, B. R., Jenike, M. A., McInerney, S. C., & Rauch, S. L. The counting Stroop: An interference task specialized for functional neuroimaging—validation study with functional MRI. Hum Brain Mapp 1998; 6(4): 270–282.

Carter, C. S., Braver, T. S., Barch, D. M., Botvinick, M. M., Noll, D., & Cohen, J. D. Anterior cingulate cortex, error detection, and the online monitoring of performance. Science 1998; 280(5364): 747–749.

Carter, C. S., Macdonald, A. M., Botvinick, M., et al. Parsing executive processes: strategic vs. evaluative functions of the anterior cingulate cortex. Proc Natl Acad Sci U S A 2000; 97(4): 1944–1948.

Critchley, H. D., Corfield, D. R., Chandler, M. P., Mathias, C. J., & Dolan, R. J. Cerebral correlates of autonomic cardiovascular arousal: A functional neuroimaging investigation in humans. J Physiol 2000; 523 Pt 1: 259–270.

Critchley, H. D., Mathias, C. J., Josephs, O., et al. Human cingulate cortex and autonomic control: converging neuroimaging and clinical evidence. Brain 2003; 126(Pt 10): 2139–2152.

Dehaene, S., Posner, M. I., & Tucker, D. M. Localization of a neural system for error detection and compensation. Psychol Sci 1994; 5(5): 303–305.

Devinsky, O., Morrell, M. J., & Vogt, B. A. Contributions of anterior cingulate cortex to behaviour. Brain 1995; 118(Pt 1): 279–306.

di Pellegrino, G., Ciaramelli, E., & Ladavas, E. The regulation of cognitive control following rostral anterior cingulate cortex lesion in humans. J Cogn Neurosci 2007; 19(2): 275–286.

Dougherty, D. D., Baer, L., Cosgrove, G. R., et al. Prospective long-term follow-up of 44 patients who received cingulotomy for treatment-refractory obsessive-compulsive disorder. Am J Psychiatr 2002; 159(2): 269–275.

Dougherty, D. D., Weiss, A. P., Cosgrove, G. R., et al. Cerebral metabolic correlates as potential predictors of response to anterior cingulotomy for treatment of major depression. J Neurosurg 2003; 99(6): 1010–1017.

Drevets, W. C., Ongur, D., & Price, J. L. Neuroimaging abnormalities in the subgenual prefrontal cortex: Implications for the pathophysiology of familial mood disorders. Mol Psychiatry 1998; 3(3): 220–226, 190–191.

Drevets, W. C., Price, J. L., Simpson, J. R., Jr., et al. Subgenual prefrontal cortex abnormalities in mood disorders. Nature 1997; 386(6627): 824–827.

Drevets, W. C., & Raichle, M. E. Reciprocal suppression of regional cerebral blood flow during emotional versus higher cognitive processes: Implications for interactions between emotion and cognition. Cogn Emot 1998; 12(3): 353–385.

Egner, T., Delano, M., & Hirsch, J. Separate conflict-specific cognitive control mechanisms in the human brain. Neuroimage 2007; 35(2): 940–948.

Egner, T., Etkin, A., Gale, S., & Hirsch, J. Dissociable neural systems resolve conflict from emotional versus nonemotional distracters. Cereb Cortex 2008; 18(6): 1475–1484.

Egner, T., & Hirsch, J. Cognitive control mechanisms resolve conflict through cortical amplification of task-relevant information. Nat Neurosci 2005; 8(12): 1784–1790.

Eisenberger, N. I., & Lieberman, M. D. Why rejection hurts: A common neural alarm system for physical and social pain. Trends Cogn Sci 2004; 8(7): 294–300.

Eisenberger, N. I., Lieberman, M. D., Williams, K. D. Does rejection hurt? An FMRI study of social exclusion. Science 2003; 302(5643): 290–292.

Etkin, A., Egner, T., Peraza, D. M., Kandel, E. R., & Hirsch, J. Resolving emotional conflict: A role for the rostral anterior cingulate cortex in modulating activity in the amygdala. Neuron 2006; 51(6): 871–882.

Falkenstein, M., Hoormann, J., Christ, S., & Hohnsbein, J. ERP components on reaction errors and their functional significance: A tutorial. Biol Psychol 2000; 51(2–3): 87–107.

Fellows, L. K., & Farah, M. J. Is anterior cingulate cortex necessary for cognitive control? Brain 2005; 128(Pt 4): 788–796.

Festinger, L. *A theory of cognitive dissonance.* Stanford, CA: Stanford University Press, 1957.

Festinger, L., & Carlsmith, J. M. Cognitive consequences of forced compliance. J Abnorm Psychol 1959; 58(2): 203–210.

Frith, C. D., Friston, K., Liddle, P. F., & Frackowiak, R. S. Willed action and the prefrontal cortex in man: A study with PET. Proc Biol Sci 1991; 244(1311): 241–246.

Gehring, W. J., Coles, M. G., Meyer, D. E., & Donchin, E. The error-related negativity: An event-related potential accompanying errors. Psychophysiology 1990; 27(4a): S34.

Gehring, W. J., Goss, B., Coles, M. G., Meyer, D. E., & Donchin, E. A neural system for error detection and compensation. Psychol Sci 1993; 4: 385–390.

Gotlib, I. H., Sivers, H., Gabrieli, J.D., et al. Subgenual anterior cingulate activation to valenced emotional stimuli in major depression. Neuroreport 2005; 16(16): 1731–1734.

Grasby, P. M., Frith, C. D., Friston, K. J., Bench, C., Frackowiak, R. S., & Dolan, R. J. Functional mapping of brain areas implicated in auditory—verbal memory function. Brain 1993; 116(Pt 1): 1–20.

Gratton, G., Coles, M. G., & Donchin, E. Optimizing the use of information: Strategic control of activation of responses. J Exp Psychol Gen 1992; 121(4): 480–506.

Gratton, G., Coles, M. G., Sirevaag, E. J., Eriksen, C. W., & Donchin, E. Pre- and poststimulus activation of response channels: A psychophysiological analysis. J Exp Psychol Hum Percept Perform 1988; 14(3): 331–344.

Greene, J. D., Nystrom, L. E., Engell, A. D., Darley, J. M., & Cohen, J. D. The neural bases of cognitive conflict and control in moral judgment. Neuron 2004; 44(2): 389–400.

Hadland, K. A., Rushworth, M. F., Gaffan, D., & Passingham, R. E. The effect of cingulate lesions on social behaviour and emotion. Neuropsychologia 2003; 41(8): 919–931.

Harmon-Jones, E. Contributions from research on anger and cognitive dissonance to understanding the motivational functions of asymmetrical frontal brain activity. Biol Psychol 2004; 67(1–2): 51–76.

Harmon-Jones, E., & Harmon-Jones, C. Action-based model of dissonance: A review of behavioral, anterior cingulate, and prefrontal cortical mechanisms. Soc Personal Psychol Compass 2008; 2/3: 1518–1538.

Hohnsbein, J., Falkenstein, M., & Hoorman, J. Error processing in visual and auditory choice reaction tasks. J Psychophysiol 1989; 3: 320.

Holroyd, C. B., & Coles, M. G. The neural basis of human error processing: Reinforcement learning, dopamine, and the error-related negativity. Psychol Rev 2002; 109(4): 679–709.

Holroyd, C. B., & Krigolson, O. E. Reward prediction error signals associated with a modified time estimation task. Psychophysiology 2007; 44(6): 913–917.

Holroyd, C. B., Nieuwenhuis, S., Yeung, N., et al. Dorsal anterior cingulate cortex shows fMRI response to internal and external error signals. Nat Neurosci 2004; 7(5). 497–498.

Hornak, J., Bramham, J., Rolls, E. T., et al. Changes in emotion after circumscribed surgical lesions of the orbitofrontal and cingulate cortices. Brain 2003; 126(Pt 7): 1691–1712.

Hoshikawa, Y., & Yamamoto, Y. Effects of Stroop color-word conflict test on the autonomic nervous system responses. Am J Physiol 1997; 272(3 Pt 2): H1113–H1121.

Kennard, M. A. Effect of bilateral ablation of cingulate area on behaviour of cats. J Neurophysiol 1955; 18(2): 159–169.

Kerns, J. G. Anterior cingulate and prefrontal cortex activity in an FMRI study of trial-to-trial adjustments on the Simon task. Neuroimage 2006; 33(1): 399–405.

Kerns, J. G., Cohen, J. D., Mac Donald, A. W., 3rd, Cho, R. Y., Stenger, V. A., & Carter, C. S. Anterior cingulate conflict monitoring and adjustments in control. Science 2004a; 303(5660): 1023–1026.

Kerns, J. G., Cohen, J. D., Stenger, V. A., & Carter, C. S. Prefrontal cortex guides context-appropriate responding during language production. Neuron 2004b; 43(2): 283–291.

Kuhl, B. A., Dudukovic, N. M., Kahn, I., & Wagner, A. D. Decreased demands on cognitive control reveal the neural processing benefits of forgetting. Nat Neurosci 2007; 10(7): 908–914.

Lane, R. Neural correlates of conscious emotional experience. In: Lane, R., & Nadel, L. (Eds.), *Cognitive neuroscience of emotion*. New York: Oxford University Press; 2000.

Lane, R. D., Fink, G. R., Chau, P. M., & Dolan, R. J. Neural activation during selective attention to subjective emotional responses. Neuroreport 1997; 8(18): 3969–3972.

Lane, R. D., Reiman, E. M., Axelrod, B., Yun, L. S., Holmes, A., & Schwartz, G. E. Neural correlates of levels of emotional awareness. Evidence of an interaction between emotion and attention in the anterior cingulate cortex. J Cogn Neurosci 1998; 10(4): 525–535.

Luo, Q., Nakic, M., Wheatley, T., Richell, R., Martin, A., & Blair, R. J. The neural basis of implicit moral attitude—an IAT study using event-related fMRI. Neuroimage 2006; 30(4): 1449–1457.

MacDonald, A. W., 3rd, Cohen, J. D., Stenger, V. A., & Carter, C. S. Dissociating the role of the dorsolateral prefrontal and anterior cingulate cortex in cognitive control. Science 2000; 288(5472): 1835–1838.

Matthews, S. C., Paulus, M. P., Simmons, A. N, Nelesen, R. A., & Dimsdale, J. E. Functional subdivisions within anterior cingulate cortex and their relationship to autonomic nervous system function. Neuroimage 2004; 22(3): 1151–1156.

Mayberg, H. S., Brannan, S. K., Mahurin, R. K. et al. Cingulate function in depression: A potential predictor of treatment response. Neuroreport 1997; 8(4): 1057–1061.

Mayr, U., Awh, E., & Laurey, P. Conflict adaptation effects in the absence of executive control. Nat Neurosci 2003; 6(5): 450–452.

Milham, M. P., & Banich, M. T. Anterior cingulate cortex: An fMRI analysis of conflict specificity and functional differentiation. Hum Brain Mapp 2005; 25(3): 328–335.

Miller, E. K., & Cohen, J. D. An integrative theory of prefrontal cortex function. Annu Rev Neurosci 2001; 24: 167–202.

Miltner, W. H. R., Braun, C. H., & Coles, M. G. Event-related brain potentials following incorrect feedback in a time-estimation task: Evidence for a generic neural system for error detection. J Cogn Neurosci 1997; 9(6): 788–798.

Mufson, E. J., & Pandya, D. N. Some observations on the course and composition of the cingulum bundle in the rhesus monkey. J Comp Neurol 1984; 225(1): 31–43.

Muraven, M., & Baumeister, R. F. Self-regulation and depletion of limited resources: Does self-control resemble a muscle? Psychol Bull 2000; 126(2): 247–259.

Nagai, Y., Critchley, H. D., Featherstone, E., Fenwick, P. B., Trimble, M. R., & Dolan, R. J. Brain activity relating to the contingent negative variation: An fMRI investigation. Neuroimage 2004; 21(4): 1232–1241.

Nieuwenhuis, S., Schweizer, T. S., Mars, R. B., Botrinick, M. M., & Hajcak, G. Error-likelihood prediction in the medial frontal cortex: A critical evaluation. Cereb Cortex 2007; 17(7): 1570–1581.

Nieuwenhuis, S., & Yeung, N. Neural mechanisms of attention and control: Losing our inhibitions? Nat Neurosci 2005; 8(12): 1631–1633.

Ochsner, K. N., Bunge, S. A., Gross, J. J., & Gabrieli, J. D. Rethinking feelings: An FMRI study of the cognitive regulation of emotion. J Cogn Neurosci 2002; 14(8): 1215–1229.

Ochsner, K. N., & Gross, J. J. The cognitive control of emotion. Trends Cogn Sci 2005; 9(5): 242–249.

Ochsner, K. N., Kosslyn, S. M., Cosgrove, G. R., et al. Deficits in visual cognition and attention following bilateral anterior cingulotomy. Neuropsychologia 2001; 39(3): 219–230.

Pardo, J. V., Pardo, P. J., Janer, K. W., & Raichle, M. E. The anterior cingulate cortex mediates processing selection in the Stroop attentional conflict paradigm. Proc Natl Acad Sci U S A 1990; 87(1): 256–259.

Petersen, S. E., Fox, P. T., Posner, M. I., Mintun, M., & Raichle, M. E. Positron emission tomographic studies of the cortical anatomy of single-word processing. Nature 1988; 331(6157): 585–589.

Phan, K. L., Wager, T., Taylor, S. F., & Liberzon, I. Functional neuroanatomy of emotion: A meta-analysis of emotion activation studies in PET and fMRI. Neuroimage 2002; 16(2): 331–348.

Posner, M. I., & Petersen, S. E. The attention system of the human brain. Annu Rev Neurosci 1990; 13: 25–42.

Posner, M. I., Petersen, S. E., Fox, P. T., & Raichle, M. E. Localization of cognitive operations in the human brain. Science 1988; 240(4859): 1627–1631.

Shin, L. M., Whalen, P. J., Pitman, R. K., et al. An fMRI study of anterior cingulate function in posttraumatic stress disorder. Biol Psychiatr 2001; 50(12): 932–942.

Spangler, W. J., Cosgrove, G. R., Ballantine, H. T., Jr., et al. Magnetic resonance image-guided stereotactic cingulotomy for intractable psychiatric disease. Neurosurgery 1996; 38(6): 1071–1076; discussion 1076–1078.

Steele, J. D., Christmas, D., Eljamel, M. S., & Matthews, K. Anterior cingulotomy for major depression: Clinical outcome and relationship to lesion characteristics. Biol Psychiatr 2008; 63(7): 670–677.

Steele, J. D., & Lawrie, S. M. Segregation of cognitive and emotional function in the prefrontal cortex: A stereotactic meta-analysis. Neuroimage 2004; 21(3): 868–875.

Stroop, J. R. Studies of interference in serial verbal reactions. J Exp Psychol 1935; 18: 643–662.

Swick, D., & Turken, A. U. Dissociation between conflict detection and error monitoring in the human anterior cingulate cortex. Proc Natl Acad Sci U S A 2002; 99(25): 16354–16359.

Ullsperger, M., Bylsma, L. M., & Botvinick, M. M. The conflict adaptation effect: It's not just priming. Cogn Affect Behav Neurosci 2005; 5(4): 467–472.

van Veen, V., & Carter, C. S. The anterior cingulate as a conflict monitor: fMRI and ERP studies. Physiol Behav 2002a; 77(4–5): 477–482.

van Veen, V., & Carter, C. S. The timing of action-monitoring processes in the anterior cingulate cortex. J Cogn Neurosci 2002b; 14(4): 593–602.

van Veen, V., & Carter, C. S. Separating semantic conflict and response conflict in the Stroop task: A functional MRI study. Neuroimage 2005; 27(3): 497–504.

van Veen, V., Cohen, J. D., Botvinick, M. M., Stenger, V. A., & Carter, C. S. Anterior cingulate cortex, conflict monitoring, and levels of processing. Neuroimage 2001; 14(6): 1302–1308.

van Veen, V., Holroyd, C. B., Cohen, J. D., Stenger, V. A., & Carter, C. S. Errors without conflict: Implications for performance monitoring theories of anterior cingulate cortex. Brain Cogn 2004; 56(2): 267–276.

van Veen, V., Krug, M. K., Schooler, J. S., & Carter, C. S. Neural activity predicts attitude change in cognitive dissonance. Nat Neurosci 2009; 12(11): 1469–1474.

Vogt, B. A., Nimchinsky, E. A., Vogt, L. J., & Hof, P. R. Human cingulate cortex: Surface features, flat maps, and cytoarchitecture. J Comp Neurol 1995; 359(3): 490–506.

Vuilleumier, P., Armony, J. L., Driver, J., & Dolan, R. J. Effects of attention and emotion on face processing in the human brain: An event-related fMRI study. Neuron 2001; 30(3): 829–841.

Whalen, P. J., Bush, G., McNally, R. J., et al. The emotional counting Stroop paradigm: A functional magnetic resonance imaging probe of the anterior cingulate affective division. Biol Psychiatr 1998; 44(12): 1219–1228.

Yeung, N., & Cohen, J. D. The impact of cognitive deficits on conflict monitoring. Predictable dissociations between the error-related negativity and N2. Psychol Sci 2006; 17(2): 164–171.

CHAPTER 2

Damaged Self, Damaged Control: A Component Process Analysis of the Effects of Frontal Lobe Damage on Human Decision Making

Lesley K. Fellows

ABSTRACT

Frontal lobe damage can disrupt judgment, decision making, and self-control, often with devastating impact on the everyday life of the affected person. Studies of these phenomena can identify the specific brain regions important for self-control and can specify the component processes for which these regions are necessary. This chapter provides an overview of recent neuropsychological work on regional frontal lobe contributions to reinforcement learning and decision making in humans. These findings argue that self-control can be understood in terms of simpler component processes, including the ability to flexibly learn from reward and punishment, to track the value of potential choices, or to predict future events. Further, these processes have been shown to rely on particular brain regions, an important step in delineating the neural mechanisms underlying self-control.

Keywords: Lesion, reversal learning, future thinking, prefrontal cortex, preference judgments, neuroeconomics

VARIETIES OF SELF-CONTROL

One of the central challenges in understanding a complex process like self-control is knowing how best to dissect this complexity. This becomes particularly important if the aim is to relate these aspects of behavior to their underlying brain substrates. Although at first glance self-control might seem like a tidy and easily operationalized construct, the existing cognitive neuroscience literature indicates the contrary. There are clearly multiple levels and aspects of behavioral control (Stuss & Alexander, 2007; Yin & Knowlton, 2006), as well as multiple processes combining to generate the sense of a subjective "I" somehow in charge of that behavior (Gillihan & Farah, 2005). This complexity is not specific to self-control; similar challenges of definition and operationalization have been identified for other higher-order behaviors, such as impulsivity (Evenden, 1999). From this perspective, self-control might be optimistically considered an emergent property of the combined actions of these various component processes, or, more cynically, a *post hoc* gloss applied to give a sense of coherence to the mix of habit and stochastic choice tendencies that together constitute our everyday behaviors.

In part because of this complexity, cognitive neuroscience is not currently in a position to provide a "grand unifying theory" of self-control. However, it has the potential to guide our thinking about the nature of self-control, at the least providing insights into the potentially relevant component processes of this complex construct. This chapter describes potential component processes of self-control suggested by recent cognitive neuroscience studies of decision making. This vantage point means that self-control is here considered in relation to choice. I would argue that self-control is invoked only when there is more than one possible behavior, even if the choice in question is only the couch potato's dilemma of whether to act at all. It follows that adaptive, authentic *choices* are the expression of self-control. Within that framework, I will explore concepts that bear on self-control and on the relation of self-control to the brain.

Whereas I remain uncertain about the mappings between self-control and the brain, others have been less circumspect. Perhaps the dominant view of self-control and the brain is a hierarchical model of "top-down" (rational) control over "instinctive" or "impulsive" behaviors. I will begin by explaining why this model is unsatisfactory, and argue instead for a more integrative view of the neural substrates of decision making. I will then discuss recent findings from the cognitive neuroscience of decision making, primarily from studies of patients with frontal lobe damage, and suggest how these findings may be relevant to a brain-based understanding of self-control.

Hierarchies of Control

Human behavior is often framed as a struggle between "bottom-up" and "top-down" processes. When this terminology is applied in a neuroanatomical sense, the "bottom" refers to structures such as the brainstem, basal ganglia, and limbic system, including the amygdala and hypothalamus. These areas are, in a literal sense, under the cortex, and are also phylogenetically older (and so more primitive) parts of the brain. The "top" refers to cerebral cortex in general, and often refers to the highest order association cortices—notably prefrontal cortex (PFC)—in particular (Mesulam, 2003). In such accounts, PFC grabs the top spot, lording it over not only subcortical structures but also over lower-order cortical regions. Even within PFC a hierarchy is often implied, with lateral PFC literally and figuratively above orbitofrontal cortex (OFC), the "seamy underside" of PFC that is most closely linked to the limbic system (Barbas, 2000), and hierarchies within lateral PFC have also been proposed (Badre, 2008).

The top-down/bottom-up dichotomy can also be expressed at the level of behavior. Here, these terms refer to a struggle between reflexive, environmentally triggered, or emotionally driven acts and abstract, rational, goal-directed pursuits that may require multiple steps, distant in both time and place, from the final intended outcome. The processes that support the latter are often subsumed under the umbrella term "executive functions," a term that summons up reassuring images of business organograms and orderly chains of command.

Executive functions are, of course, linked to PFC, whereas reflexive, emotional, or habitual responses are related to subcortical and limbic systems. These relationships would seem to boil down nicely to a model of brain function in which simpler, more primitive circuits clamor to satisfy basic desires (or passing whims), with the prefrontal-executive system more-or-less successfully riding herd over this turmoil to allow the achievement of long-term, rational goals. Such a model would seem to provide a convenient framework in which to situate self-control. Indeed, the model appears to owe its very terms of reference to the concept of self-control. Sadly, this dichotomous representation of the brain basis of behavioral control is at best overly simplistic and arguably so misleading that it would be better to abandon it entirely.

On neural grounds, executive function is increasingly being viewed as a fractionated set of interacting processes, rather than a monolithic entity (Stuss & Alexander, 2007). Conceptually, the choices that reflect self-control need not involve top-down "struggles" over limbically mediated temptations. Limbic systems alone may have quite enough to struggle with, as potentially incommensurate homeostatic and environmental cues—never mind abstract goal states—prompt a variety of choice behaviors. Importantly, such competing behavioral options can, in principle, be resolved without recourse to an "executive," and nevertheless still be instantiations of self-control.

Untangling More Distributed Models of Self-Control

If more nuanced models are needed, then cognitive neuroscience approaches are likely to be useful for generating them. Such work does double duty, providing new perspectives on self-control as a construct, and insights into the instantiation of these processes in the brain. A brain-based, component approach to self-control requires only that putative component processes be neurobiologically plausible; they need not fall within some pre-ordained and mutually exclusive "rational" or "emotional" category. The extent to which putative component processes are in fact distinct can then be tested with neuroscience methods. These methods can also specify the brain mechanisms that underlie these processes. For example, if injury to a particular region of PFC disrupts one component of executive function but leaves another intact, this at the least argues that two distinct processes are being measured and is a beginning to understanding the circuitry that underlies them (Bates et al., 2003; Chatterjee, 2005).

Among the several cognitive neuroscience methods that may be helpful in tackling the general question of the nature of (and neural substrates of) self-control, loss-of-function techniques can be particularly useful. These experimental approaches involve measuring the effects of a disruption of brain function on behavior. With sound design, these methods can support inferentially powerful "necessity" claims—that is, they can show that a particular brain region is necessary for producing a particular behavior (Fellows et al., 2005; Rorden & Karnath, 2004). Relatedly, as discussed above, they can speak to whether particular behavioral measures are capturing dissociable processes, and so provide strong tests of component process hypotheses.

There are three such methods commonly used in humans: lesion studies, transcranial magnetic stimulation, and pharmacological manipulations. The latter help to specify the neurochemical, rather than the neuroanatomical systems supporting behavior, and are not the focus of this chapter. I have employed the lesion method to identify component processes of decision making and to determine the role of particular regions of human PFC in these processes. In the pages that follow, I will review this and related work, both to illustrate the general approach and to describe some specific, prefrontally mediated component processes of decision making relevant to developing a neurobiologically based understanding of self-control.

Frontal Lobe Damage Affects Aspects of Self-Control

Decision making is one lens through which to view self-control. Although decision making need not involve self-control, self-control—at least as I have defined it here—involves decision making. Non-arbitrary choices require generating or identifying options, evaluating one or more of these options, and deciding a course of action (Fellows, 2004). Of these stages, cognitive neuroscience work to date has focused primarily on "evaluation." Subjective value is not a fixed property of a stimulus. It depends on factors external to the organism (What else is available?), and internal information (Am I hungry?) (Izquierdo et al., 2004; O'Doherty et al., 2000; Padoa-Schioppa & Assad, 2006). It can change rapidly and may continue to change even after a decision is made. Value (or utility) is, of course, a central concept in economics. Economics and decision science have provided formal models of how value ought to be adjusted to take into account factors such as risk, ambiguity, and delay, and empirical work has described how these factors actually influence decision making (Baron, 1994).

These concepts and frameworks are being applied to inform neural models of decision making. This nascent field has been heavily influenced by clinical reports of the effects of frontal lobe damage (Eslinger & Damasio, 1985; Loewenstein et al., 2001) and by experimental

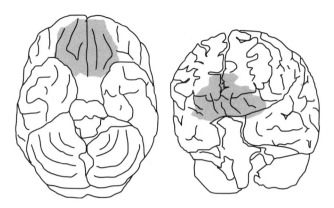

Figure 2–1. Schematic diagram of the brain, showing the region referred to in the text as the ventromedial frontal lobe (VM) in grey. The left panel shows a ventral view of the brain surface; VMF encompasses the medial aspect of orbitofrontal cortex. The right panel shows an oblique view, with the frontal poles cut away to show how VMF includes medial orbitofrontal cortex and the adjacent ventral aspect of the medial wall of prefrontal cortex.

findings from human lesion studies (Bechara et al., 2000; Fellows, 2007), both of which have provided clues that the ventromedial region of PFC (VMF; *see* Fig. 2–1) may play a critical role in decision making. Efforts to understand the role VMF plays in decision making have provided interesting perspectives on the basic building blocks of self-control.

In the mid-1990s, Bechara and colleagues developed an experimental task in an effort to capture the clinically evident decision-making deficits of patients who had suffered damage to the ventral aspect of the frontal lobes (Bechara et al., 1997). Now known as the Iowa gambling task (IGT), the task succeeded in that patients with VMF damage had difficulty performing it, but the nature of that deficit has been much debated in the subsequent literature (Bechara et al., 2005; Dunn et al., 2006; Maia & McClelland, 2004). The task is a relatively complex card game, involving learning, ambiguity, risk, and integration of reward and punishment both within and across trials. This complexity has fuelled the debate about the interpretation of these findings. The task was initially framed as a test of the ability to anticipate future consequences, and the impairment of patients with VMF damage was taken as a reflection of "myopia for the future," a deficit in turn related to an impaired ability to "feel" risk—that is, to generate somatic signals in anticipation of potential losses (Bechara et al., 1997).

FLEXIBLE REINFORCEMENT LEARNING AS A MODEL OF ADAPTIVE SELF-CONTROL

Although "myopia for the future" has obvious resonance with ideas about self-control, the IGT is not a good test of such an ability. It is probably more straightforward to think of the task not in terms of anticipating the future, but rather of learning from past experience. The subject chooses from four decks of cards, learning by trial and error which decks are advantageous, which disadvantageous. Two decks are disadvantageous overall, in that choosing from those decks will lead to frequent wins but occasionally more substantial losses. The advantageous decks hold small wins, but even smaller losses, and so are the best bets over the long term. Importantly, card order in each deck is fixed, meaning the contingencies associated with each deck change partway through the task. The large losses associated with the disadvantageous decks are only experienced after several choices from those decks, and after both healthy and VMF-damaged subjects have established a preference for those decks. As these losses accrue, healthy subjects tend to shift their preference to the lower win but overall advantageous decks, but VMF damage impedes this shift.

This pattern of performance is reminiscent of observations that had been made in non-human primates performing simple instrumental learning tasks (Butter, 1969; Jones & Mishkin, 1972). As far back as the 1960s a particular learning deficit had been observed after lesions to orbitofrontal cortex. Animals with such damage learn stimulus–reward associations as well as control animals but have difficulty adjusting their choices when the contingencies change. The paradigm that best illustrates this deficit is reversal learning. The usual design has two initially neutral stimuli, one paired with reward and the other with non-reward (or outright punishment). Once learning of these contingencies is demonstrated, the reinforcement contingencies are switched. Orbitofrontal cortex damage selectively impairs performance in this reversal phase of learning in rats, macaques, and humans (Murray et al., 2007). The process of learning new reinforcement contingencies associated with previously reinforced stimulus features is also termed "affective shifting," in that it involves reassigning the motivational (or "emotional") value of stimuli (Dias et al., 1996).

We asked whether this phenomenon explained the performance of patients with VMF damage on the IGT, and tested this possibility by administering a shuffled variant of the IGT to nine patients with VMF damage as well as a group with PFC damage-sparing VMF and a demographically matched healthy control group. The shuffled task was identical to the original with the exception that the card order did not lead to an initial tendency to favor the ultimately disadvantageous decks. Patients with VMF damage were impaired in the original task but performed as well as healthy subjects when the underlying requirement for reversal learning was eliminated from the task (Fellows & Farah, 2005a). This finding illustrates how components of a process as complex as the decision making required in the IGT can be identified and shows that such components can be relevant to understanding both the behavior and its relation to the brain. Reversal learning has been extensively studied in animal models, so connecting human decision making to this more basic construct allows this literature to be brought to bear.

A second important point arising from this and related work is that individuals may be impaired on the IGT because of fundamental deficits in various more basic processes. For example, patients with lateral PFC damage also have trouble with this task (Manes et al., 2002), but *not* because of its reversal learning requirement (Fellows & Farah, 2005a). It is likely that the deficit in these patients is related, at least in part, to the working memory requirement of the task, although this remains to be definitively shown (Bechara et al., 1998). I emphasize this point because it highlights an inferential problem that dogs efforts to translate the lesion study results using this task to understand the neural substrates of "disordered self-control" in neuropsychiatric conditions such as addiction (Bechara, 2003; Bechara et al., 2002). Simply put, abnormal performance on the IGT does not necessarily mean that an individual has VMF dysfunction.

This work also serves as something of a cautionary tale, highlighting the potential pitfalls of moving from clinical observations to underlying mechanisms. Whereas some patients with VMF damage do have clinically evident difficulties in judgment and decision making that might be interpreted as a failure of self-control, and many with VMF damage show characteristic deficits on simple reversal learning tasks and, in turn, on the IGT, it remains unclear whether these clinical and experimental observations reflect a causal pathway or neuroanatomical coincidence (Fellows & Farah, 2003; Rolls et al., 1994).

Can these reversal learning findings inform thinking about self-control? These data highlight the existence of multiple neural systems for learning to make adaptive choices, acting over different timescales. When reinforcement contingencies change, well-learned rewarding "habits" can be quickly overridden, through a VMF-dependent mechanism. This flexibility to adjust to a rapidly changing reward and punishment "landscape" could be considered an aspect of self-control, and simple paradigms like reversal learning are a useful experimental tool for operationalizing self-control to understand the brain processes involved.

Surprise! Breaches of Expectation and Dynamic Self-Control

The ventromedial frontal lobe—particularly its OFC component—is heavily interconnected with two brain regions that play important roles in motivated behavior. One is the amygdala, a nucleus in the medial part of the temporal lobe that is involved in rapidly signaling the salience of emotionally laden stimuli. The other is the hypothalamus, a region important for signaling basic drives (e.g., hunger, thirst) and for coordinating basic behavioral and autonomic responses to stimuli relevant to satisfying those drives (Ghashghaei & Barbas, 2002). Although some work has been done to investigate how the amygdala and OFC might function together to support learning and decision making in humans (Bechara et al., 1999; Hampton et al., 2007), this question has been addressed more extensively in studies in rats and non-human primates (Baxter et al., 2000; Murray et al., 2007). A recent, provocative finding in this regard comes from a dual lesion study in rats. Rats with lesions to OFC were impaired on a reversal learning task. Remarkably, when the basolateral amygdala was additionally lesioned, this reversal learning impairment was abolished (Stalnaker et al., 2007). The authors interpreted this finding as evidence that OFC serves a "gating" function, allowing rapid updating of amygdala-mediated stimulus–reinforcement associations. They propose that OFC, in this context, serves to detect breaches of stimulus–outcome expectation, in turn allowing other neural circuits to "learn something new."

Expectations of this kind might, in principle, be breached in one of two ways: either by delivery of an unexpected punishment (or omitted reward) or by the delivery of an unexpected reward (or omitted punishment). In practice, most of the work on this issue has examined responses to unexpected punishment (or omitted rewards), a type of feedback put in high relief in reversal learning tasks. A reversal trial provides a maximal "breach of expectation" signal, and it is the ability to rapidly respond to this signal that seems to be affected after OFC damage. This raises the possibility that OFC is critically involved primarily in learning from unexpected negative feedback.

We administered a reinforcement-learning task that allows separate measures of how much is learned from negative and positive feedback to patients with VMF damage to test this idea. The task requires learning to choose between pairs of arbitrary stimuli (Japanese Hiragana characters) through trial-and-error, with probabilistic feedback provided after each choice. Overall, one character in each pair was more often associated with positive feedback and the other with negative feedback. Once learning was demonstrated, subjects moved to the test phase of the task. This phase allowed the measurement of whether they had learned from positive feedback (i.e., to choose the most "correct" stimulus), from negative feedback (i.e., to avoid the most "incorrect" stimulus), or both (Frank et al., 2004). Healthy control subjects, and patients with frontal lobe damage outside VMF, learned almost equally from positive and negative feedback. Those with VMF damage learned normally from positive feedback but were very impaired in how much they learned from negative feedback (Wheeler & Fellows, 2008). This finding suggests that the contribution of VMF to reversal learning may rest on an even simpler component process: that of detecting unexpected negative outcomes, in turn permitting rapid adjustments in behavior (i.e., learning) from these outcomes.

How can these results be related to ideas about self-control? At the least, they identify one potential mechanism underlying aberrant self-control in a particular clinical population: the ability to detect, and so rapidly adjust to, unexpected (negative) outcomes. This is a more nuanced perspective on the role of a particular region within PFC than the "top-down control over basic drives" caricature outlined earlier in this chapter. It is also interesting that the experimental paradigms that have been particularly useful in understanding VMF contributions to behavioral control have been reinforcement learning tasks. This provides a different perspective on self-control, underlining the importance of feedback (perhaps particularly negative feedback) and the fact that decisions take place

FUTURE THINKING AND THE FRONTAL LOBES

Having argued that one major line of research on the role of the frontal lobes in human decision making concerns how past experience is applied to influence present choice, I now turn to the question of whether the frontal lobes play a critical role in future thinking. The proverbial "ant-and-grasshopper" choice between immediate gratification and delayed reward is a classic self-control dilemma. This type of problem is an attractive experimental target for a cognitive neuroscience understanding of decision making. It has a long history of study at the behavioral level, and lends itself to a component analysis. One likely component, the effect of delay on subjective value (so-called "temporal discounting") comes complete with defined experimental measures and well-specified theoretical models (Ainslie, 2001). Other components suggest themselves. For example, future rewards will carry less sway if they are more heavily discounted by delay. They will also carry less sway if the time at which they are to be delivered is outside the window of time considered as "the future." As an extreme example, a decision maker is likely to consider a reward delivered beyond his/her expected lifespan as worthless. Interestingly, the period of time spontaneously contemplated when individuals think about their own futures varies quite widely (Kastenbaum, 1961; Lessing, 1968).

We undertook an exploratory study of these two aspects of future thinking in patients with frontal lobe injury, as well as both healthy controls and a control group with nonfrontal focal brain injury. Temporal discounting was measured with a widely used task that requires subjects to choose (hypothetically) from various sums of money delivered "now" or larger sums to be received after a delay of between 7 and 186 days. The task has previously been shown to detect tendencies to steeper discounting in drug addicts compared to non-addicted controls (Bickel & Marsch, 2001; Petry et al., 1998).

Although we found large individual differences in our study population, focal brain injury, whether affecting VMF, dorsolateral frontal, or nonfrontal regions, had no systematic effect on temporal discounting rate (Fellows & Farah, 2005b). Thus, although fMRI data indicate that PFC is involved in intertemporal choice (Kable & Glimcher, 2007; McClure et al., 2004), no PFC region appears to be critical in setting temporal discounting rates.

In contrast, VMF damage affected the subjective window of time viewed as "the future." A task known as the "future time perspective" task was used to measure this construct. This task requires subjects to list a fixed number of events likely to occur in their own futures. Once listed, subjects are asked when these are likely to occur. The time to the most distant future event is the measure of interest. All subjects with brain injury had shorter future time windows than did healthy controls, presumably reflecting the effects of having experienced a serious illness. However, those with VMF damage ($n = 12$) had significantly shorter time windows compared to those with damage elsewhere ($n = 26$), suggesting a specific role for that part of the brain in the ability to envisage the more distant future (Fellows & Farah, 2005b).

This finding argues that when patients with VMF damage make choices that suboptimally emphasize present or near-term considerations, that may reflect a different view of time, rather than a steeper discounting of the value of future rewards. More generally, these findings make a case for the dissociability of these two aspects of future thinking, both of which are relevant to self-control as it is played out over time, but only one of which seems to rely critically on VMF.

SUBJECTIVE VALUE AS AN ANCHOR FOR SELF-CONTROL

An early view of the role of OFC was that it encoded the subjective (and relative) value of potential choices. For example, single-unit recordings from OFC in macaque monkeys indicated that a particular population of neurons might respond to the sight of a banana, with that activity suppressed and replaced by a different population encoding a second, more preferred option, if it was made available (Rolls, 2000). Subsequent work, using different paradigms, have confirmed this basic idea. At least some neurons within OFC appear to be "tuned" to particular rewards, these rewards seem to be evaluated within the context of what else is available, and their value is modified by internal factors such as selective satiety (i.e., a banana's value is reduced after the monkey has had the opportunity to eat several bananas) (Padoa-Schioppa & Assad, 2006; Roesch & Olson, 2004; Tremblay & Schultz, 1999). Functional MRI studies in humans have at least partially supported this view, although preference judgment has been related to activity within the ventral aspect of the medial wall of PFC, rather than OFC, in most such studies (Cunningham et al., 2003; Paulus & Frank, 2003).

Whereas these single unit and fMRI data indicate a correlation between VMF activity and value, they do not specify whether this activity is necessary for decision making, and if so, in what way. Given that relative evaluation would seem to be the very crux of most non-arbitrary choices, if VMF does play a crucial role in representing relative value, then damage affecting this region ought to affect even very simple forms of decision making. Perhaps the simplest form of decision making is preference judgment (e.g., chocolate or vanilla?), a choice that need not involve risk, ambiguity, or intertemporal considerations but that nevertheless can be difficult. There is some, albeit conflicting, evidence that OFC damage may affect the consistency of preference judgments in macaques (Baylis & Gaffan, 1991; Izquierdo et al., 2004). We tested this idea in humans with VMF damage using a simple preference task adapted from this animal work.

Subjects chose between all possible pairs within three categories of stimuli: foods, colors, famous people, answering the question "Which of these do you prefer?" There can be no objectively "wrong" answers to questions of subjective preference, but there can be inconsistencies. The overall rank order of preferences that emerged from these pair-wise choices was determined, and the number of individual choices that deviated from this order was taken as a measure of inconsistent choice. As predicted by the hypothesis that OFC represents relative value in the service of decision making, subjects with damage to VMF made more of these inconsistent choices than did either healthy controls, or those with damage to the frontal lobes that spared VMF (Fellows & Farah, 2007). These experimental findings of inconsistent preference after VMF damage have yet to be systematically linked to real-life behavior in these patients. However, independent clinical accounts of either a dramatic incapacity to choose at all (Eslinger & Damasio, 1985), or a tendency to whimsical or capricious choice after such damage, stretch back decades (Ackerly, 2000).

These results frame self-control in yet another way. If the ability to determine or compare value is disrupted, then resulting choices could be considered as somehow inauthentic. If you can't reliably represent value, is the resulting choice really "yours?" I suggest that one of the features that leads to the subjective sense of self is an at least broadly coherent set of preferences and choice tendencies—that is, "values," with value defined in the broad sense, rather than the narrower, moral sense. Inconsistent preferences after VMF damage can thus be conceived of as a form of aberrant self-control that is primarily an impairment of "self" rather than an impairment of "control."

SELF-CONTROL BEYOND DECISION MAKING

A complete understanding of self-control requires that these findings concerning basic elements of decision making be placed in a broader context. Although self-control involves choice, clearly it also involves other processes. Other component processes, also linked to PFC, including working memory, allocation of cognitive resources, maintenance and shifting of selective attention, and the inhibition of prepotent response tendencies are likely to be involved in at least some forms of self-control (see Chapters 1 and 6). Whereas it is beyond the scope of this chapter to review these in detail, there is now substantial evidence that these are dissociable and rely on distinct sectors within PFC (Miyake et al., 2000; Stuss & Alexander, 2007; Stuss et al., 2001; Tsuchida & Fellows, 2009).

That these processes can be dissociated should not be taken to mean that they normally occur in isolation. It is important to emphasize that the different sectors within PFC are directly interconnected, and interact as well through cortico-subcortical circuits (Barbas, 2000; Haber, 2003; Price et al., 1996). They are also extensively modulated by neurochemical inputs (notably dopamine, norepinephrine, and serotonin) that have their own influences on processes relevant to self-control (Chamberlain et al., 2006; Cools et al., 2006; Cools & Robbins, 2004; Frank et al., 2004; Schweighofer et al., 2008; Tanaka et al., 2007). The complex interactions within PFC, and between PFC and other cortical and subcortical regions that occur under normal circumstances put the lie to dichotomous "rational vs. emotional" or "cortical vs. limbic" accounts of self-control. Emotional and motivational processes are intrinsically linked to more conventionally "cognitive" processes and are in turn dynamically modulated by them.

The work reviewed here reflects an effort to reduce the complexity of self-control to tractable components. Relating these relatively new neuroeconomic and decision-making findings to the better elaborated understanding of other aspects of frontal-executive function will be an important enterprise for research in this area in the medium term. This approach should lead to an integrated view of how humans more-or-less successfully navigate a real world offering a sometimes dizzying array of opportunities, temptations, and long-term prospects. Such a neurobiologically grounded understanding will put us in a position to analyze, and hopefully address, what goes wrong to produce self-defeating choices in illnesses like drug abuse or in more prosaic, but nonetheless important, contexts, such as overeating.

Acknowledgments

Martha Farah made substantial contributions to several of the studies reviewed here. I acknowledge support from the NIH (R21 NS045074, R21 DA22630) and CIHR (MOP 77583 and a Clinician Scientist award).

References

Ackerly, S. Prefrontal lobes and social development. 1950. Yale J Biol Med 2000; 73: 211–219.

Ainslie, G. *Breakdown of will*. Cambridge, UK: Cambridge University Press, 2001.

Badre, D. Cognitive control, hierarchy, and the rostro-caudal organization of the frontal lobes. Trends Cogn Sci 2008; 12: 193–200.

Barbas, H. Complementary roles of prefrontal cortical regions in cognition, memory, and emotion in primates. Adv Neurol 2000; 84: 87–110.

Baron, J. *Thinking and deciding*. Cambridge, U.K.: Cambridge University Press, 1994.

Bates, E., Appelbaum, M., Salcedo, J., Saygin, A. P., & Pizzamiglio, L. Quantifying dissociations in neuropsychological research. J Clin Exp Neuropsychol 2003; 25: 1128–1153.

Baxter, M. G., Parker, A., Lindner, C. C., Izquierdo, A. D., & Murray, E. A. Control of response selection by reinforcer value requires interaction of amygdala and orbital prefrontal cortex. J Neurosci 2000; 20: 4311–4319.

Baylis, L. L., & Gaffan, D. Amygdalectomy and ventromedial prefrontal ablation produce similar deficits in food choice and in simple object discrimination learning for an unseen reward. Exp Brain Res 1991; 86: 617–622.

Bechara, A. Risky business: Emotion, decision-making, and addiction. J Gambl Stud 2003; 19: 23–51.

Bechara, A., Damasio, H., & Damasio, A. R. Emotion, decision making and the orbitofrontal cortex. Cereb Cortex 2000; 10: 295–307.

Bechara, A., Damasio, H., Damasio, A. R., & Lee, G. P. Different contributions of the human amygdala and ventromedial prefrontal cortex to decision-making. J Neurosci 1999; 19: 5473–5481.

Bechara, A., Damasio, H., Tranel, D., & Anderson, S. W. Dissociation of working memory from decision making within the human prefrontal cortex. J Neurosci 1998; 18: 428–437.

Bechara, A., Damasio, H., Tranel, D., & Damasio, A. R. Deciding advantageously before knowing the advantageous strategy. Science 1997; 275: 1293–1295.

Bechara, A., Damasio, H., Tranel, D., & Damasio, A. R. The Iowa Gambling Task and the somatic marker hypothesis: Some questions and answers. Trends Cogn Sci 2005; 9: 159–162; discussion 162–164.

Bechara, A., Dolan, S., & Hindes, A. Decision-making and addiction (part II): Myopia for the future or hypersensitivity to reward? Neuropsychologia 2002; 40: 1690–1705.

Bickel, W. K., & Marsch, L. A. Toward a behavioral economic understanding of drug dependence: Delay discounting processes. Addiction 2001; 96: 73–86.

Butter, C. Perseveration in extinction and in discrimination reversal tasks following selective frontal ablations in macaca mulatta. Physiol Behav 1969; 4: 163–171.

Chamberlain, S. R., Muller, U., Blackwell, A. D., Clark, L., Robbins, T. W., & Sahakian, B. J. Neurochemical modulation of response inhibition and probabilistic learning in humans. Science 2006; 311: 861–863.

Chatterjee, A. A madness to the methods in cognitive neuroscience? J Cogn Neurosci 2005; 17: 847–849.

Cools, R., Altamirano, L., & D'Esposito, M. Reversal learning in Parkinson's disease depends on medication status and outcome valence. Neuropsychologia 2006; 44: 1663–1673.

Cools, R., & Robbins, T. W. Chemistry of the adaptive mind. Philos Transact A Math Phys Eng Sci 2004; 362: 2871–2888.

Cunningham, W. A., Johnson, M. K., Gatenby, J. C., Gore, J. C., & Banaji, M. R. Neural components of social evaluation. J Pers Soc Psychol 2003; 85: 639–649.

Dias, R., Robbins, T. W., & Roberts, A. C. Dissociation in prefrontal cortex of affective and attentional shifts. Nature 1996; 380: 69–72.

Dunn, B. D., Dalgleish, T., & Lawrence, A. D. The somatic marker hypothesis: A critical evaluation. Neurosci Biobehav Rev 2006; 30: 239–271.

Eslinger, P. J., & Damasio, A. R. Severe disturbance of higher cognition after bilateral frontal lobe ablation: Patient EVR. Neurology 1985; 35: 1731–1741.

Evenden, J. L. Varieties of impulsivity. Psychopharmacology (Berl) 1999; 146: 348–361.

Fellows, L. K. The cognitive neuroscience of decision making: A review and conceptual framework. Behav Cogn Neurosci Rev 2004; 3: 159–172.

Fellows, L. K. Advances in understanding ventromedial prefrontal function: The accountant joins the executive. Neurology 2007; 68: 991–995.

Fellows, L. K., & Farah, M. J. Ventromedial frontal cortex mediates affective shifting in humans: Evidence from a reversal learning paradigm. Brain 2003; 126: 1830–1837.

Fellows, L. K., & Farah, M. J. Different underlying impairments in decision-making following ventromedial and dorsolateral frontal lobe damage in humans. Cereb Cortex 2005a; 15: 58–63.

Fellows, L. K., & Farah, M. J. Dissociable elements of human foresight: A role for the ventromedial frontal lobes in framing the future, but not in discounting future rewards. Neuropsychologia 2005b; 43: 1214–1221.

Fellows, L. K., & Farah, M. J. The role of ventromedial prefrontal cortex in decision making: Judgment under uncertainty, or judgment per se? Cereb Cortex 2007; 17: 2669–2674.

Fellows, L. K., Heberlein, A. S., Morales, D. A., Shivde, G., Waller, S., & Wu, D. H. Method matters: An empirical study of impact in cognitive neuroscience. J Cogn Neurosci 2005; 17: 850–858.

Frank, M. J., Seeberger, L. C., & O'Reilly, R. C. By carrot or by stick: Cognitive reinforcement learning in parkinsonism. Science 2004; 306: 1940–1943.

Ghashghaei, H. T., & Barbas, H. Pathways for emotion: Interactions of prefrontal and anterior temporal pathways in the amygdala of the rhesus monkey. Neuroscience 2002; 115: 1261–1279.

Gillihan, S. J., & Farah, M. J. Is self special? A critical review of evidence from experimental psychology and cognitive neuroscience. Psychol Bull 2005; 131: 76–97.

Haber, S. N. The primate basal ganglia: Parallel and integrative networks. J Chem Neuroanat 2003; 26: 317–330.

Hampton, A. N., Adolphs, R., Tyszka, M. J., & O'Doherty, J. P. Contributions of the amygdala to reward expectancy and choice signals in human prefrontal cortex. Neuron 2007; 55: 545–555.

Izquierdo, A., Suda, R. K., & Murray, E. A. Bilateral orbital prefrontal cortex lesions in rhesus monkeys disrupt choices guided by both reward value and reward contingency. J Neurosci 2004; 24: 7540–7548.

Jones, B., & Mishkin, M. Limbic lesions and the problem of stimulus-reinforcement associations. Exp Neurol 1972; 36: 362–377.

Kable, J. W., & Glimcher, P. W. The neural correlates of subjective value during intertemporal choice. Nat Neurosci 2007; 10: 1625–1633.

Kastenbaum, R. J. The dimensions of future time perspective, an experimental analysis. J Gen Psychol 1961; 65: 203–218.

Lessing, E. E. Demographic, developmental, and personality correlates of future time perspective (FTP). J Pers 1968; 36: 183–201.

Loewenstein, G. F., Weber, E. U., Hsee, C. K., & Welch, N. Risk as feelings. Psychol Bull 2001; 127: 267–286.

Maia, T. V., & McClelland, J. L. A reexamination of the evidence for the somatic marker hypothesis: What participants really know in the Iowa gambling task. Proc Natl Acad Sci USA 2004; 101: 16075–16080.

Manes, F., Sahakian, B., Clark, L., et al. Decision-making processes following damage to the prefrontal cortex. Brain 2002; 125: 624–639.

McClure, S. M., Laibson, D. I., Loewenstein, G., & Cohen, J. D. Separate neural systems value immediate and delayed monetary rewards. Science 2004; 306: 503–507.

Mesulam, M. M. Some anatomic principles related to behavioral neurology and neuropsychology. In: Feinberg, T. E., & Farah, M. J. (Eds.), *Behavioral neurology and neuropsychology*. New York: McGraw-Hill; 2003: pp. 45–56.

Miyake, A., Friedman, N. P., Emerson, M. J., Witzki, A. H., Howerter, A., & Wager, T. D. The unity and diversity of executive functions and their contributions to complex "Frontal Lobe" tasks: A latent variable analysis. Cognit Psychol 2000; 41: 49–100.

Murray, E. A., O'Doherty, J. P., & Schoenbaum, G. What we know and do not know about the functions of the orbitofrontal cortex after 20 years of cross-species studies. J Neurosci 2007; 27: 8166–8169.

O'Doherty, J., Rolls, E. T., Francis, S., et al. Sensory-specific satiety-related olfactory activation of the human orbitofrontal cortex. Neuroreport. 2000; 11: 399–403.

Padoa-Schioppa, C., & Assad, J. A. Neurons in the orbitofrontal cortex encode economic value. Nature 2006; 441: 223–226.

Paulus, M. P., & Frank, L. R. Ventromedial prefrontal cortex activation is critical for preference judgments. Neuroreport 2003; 14: 1311–1315.

Petry, N. M., Bickel, W. K., & Arnett, M. Shortened time horizons and insensitivity to future consequences in heroin addicts. Addiction 1998; 93: 729–738.

Price, J. L., Carmichael, S. T., & Drevets, W. C. Networks related to the orbital and medial prefrontal cortex: A substrate for emotional behavior? Prog Brain Res 1996; 107: 523–536.

Roesch, M. R., & Olson, C. R. Neuronal activity related to reward value and motivation in primate frontal cortex. Science 2004; 304: 307–310.

Rolls, E. T. The orbitofrontal cortex and reward. Cereb Cortex 2000; 10: 284–294.

Rolls, E. T., Hornak, J., Wade, D., & McGrath, J. Emotion-related learning in patients with social and emotional changes associated with frontal lobe damage. J Neurol Neurosurg Psychiatry 1994; 57: 1518–1524.

Rorden, C., & Karnath, H. O. Using human brain lesions to infer function: A relic from a past era in the fMRI age? Nat Rev Neurosci 2004; 5: 813–819.

Schweighofer, N., Bertin, M., Shishida, K., et al. Low serotonin levels increase delayed reward discounting in humans. J Neurosci 2008; 28: 4528–4532.

Stalnaker, T. A., Franz, T. M., Singh, T., & Schoenbaum, G. Basolateral amygdala lesions abolish orbitofrontal-dependent reversal impairments. Neuron 2007; 54: 51–58.

Stuss, D. T., & Alexander, M. P. Is there a dysexecutive syndrome? Philos Trans R Soc Lond B Biol Sci 2007; 362: 901–915.

Stuss, D. T., Floden, D., Alexander, M. P., Levine, B., & Katz, D. Stroop performance in focal lesion patients: Dissociation of processes and frontal lobe lesion location. Neuropsychologia 2001; 39: 771–786.

Tanaka, S. C., Schweighofer, N., Asahi, S., et al. Serotonin differentially regulates short- and long-term prediction of rewards in the ventral and dorsal striatum. PLoS ONE 2007; 2: e1333.

Tremblay, L., & Schultz, W. Relative reward preference in primate orbitofrontal cortex. Nature 1999; 398: 704–708.

Tsuchida, A., & Fellows, L. K. Lesion evidence that two distinct regions with prefrontal cortex are critical for n-back performance in humans. J Cogn Neurosci 2009; 21: 2263–2275.

Wheeler, E. Z., & Fellows, L. K. The ventromedial frontal lobe is critical for learning from punishment, but not reward. Brain 2008; 131: 1323–1331.

Yin, H. H., & Knowlton, B. J. The role of the basal ganglia in habit formation. Nat Rev Neurosci 2006; 7: 464–476.

CHAPTER 3

Working Hard or Hardly Working for those Rose-colored Glasses: Behavioral and Neural Evidence for the Automatic Nature of Unrealistically Positive Self-Perceptions

Jennifer S. Beer

ABSTRACT

Even rose-colored glasses cannot hide the apparent discrepancy between models of self-control and the adaptive view of positive illusions. Most models of self-control suggest that accurate perceptions of the relation between behavior and goals are fundamental for goal attainment. However, the adaptive view of positive illusions suggests that individuals with unrealistically positive self-perceptions are more successful at achieving goals such as satisfying personal relationships, well-being, and professional accomplishment. If people fool themselves into thinking that their behavior is consistent with their goals (e.g., "Sleeping through class will help me get a good grade because I will be well-rested on the day of the exam") or fail to acknowledge conflict between goals (e.g., "Eating peanut butter cups is delicious and healthy because peanut butter has protein"), then how can they execute the self-control needed to adjust behavior or resolve goal conflicts? This chapter integrates these perspectives by examining the evidence for the adaptive view of positive illusions and mechanisms that underlie unrealistically positive self-perceptions. The extant research suggests that positive illusions may be advantageous for goal attainment in the short-term, particularly mood regulation, but do not promote successful self-control across time. The failure of positive illusions to promote successful self-control in a sustained manner may be explained by the shallow information processing that supports many unrealistically positive self-views. In other words, positive illusions may often reflect cognitive shortcuts that need to be corrected to serve the monitoring function described in models of self-control. The adaptive benefit of positive illusions for mood regulation suggests that this relation occurs in situations in which mood regulation is a priority or it is not too costly to sacrifice other goals at its expense.

Keywords: Anterior cingulate, bias, brain, frontal lobe, fMRI, orbitofrontal cortex, overconfidence, positive illusions, self, self-control, self-enhancement, social cognition

In *Willy Wonka and the Chocolate Factory*, Charlie Bucket found himself in a number of sticky situations that had less to do with chocolate rivers and everything to do with challenges to his self-control abilities. Self-control is the process by which individuals select, inhibit,

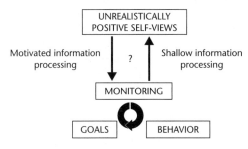

Figure 3–1. Unrealistically positive self-views may arise through motivated monitoring of the relation between behavior and goals in a self-serving manner. Alternatively, unrealistically positive self-views may be the result of shallow monitoring processes.

or otherwise modify their behavior to achieve their goals. Self-control helps individuals suppress maladaptive behavioral tendencies, select appropriate strategies for a particular goal, and resolve situations in which goals may conflict. For example, Charlie Bucket had to compromise between his desire to win a golden ticket and his family's need for shelter and nutritious food. Although there were only a few golden tickets in thousands of chocolate bars, Charlie and his grandfather still decided to spend some of their scarce financial resources on chocolate. Did Charlie's positive illusions about his chances of winning help him resolve that self-control dilemma in an optimal manner? If they did, then why do most models of self-control emphasize the need for accurate monitoring of discrepancies between behavior and goals? Some researchers have suggested that any effort put into maintaining unrealistically positive self-evaluations is worthwhile because they are beneficial for self-control and goal attainment (Taylor & Brown, 1988; Taylor, Lichtman, & Wood, 1984). This perspective suggests that maintaining positive illusions relies on motivational effects on the monitoring process (see Fig. 3–1); people distort information processing because they are motivated to feel good about themselves in the face of threatening information (Sedikides & Strube, 1995; Taylor & Brown, 1988; Taylor, Neter, & Wayment, 1995). Conversely, research examining the mechanisms underlying unrealistically positive self-views suggests that they

often arise from shallow information processing (e.g., Caputo & Dunning, 1995; Chambers & Windschitl, 2004; Klayman et al., 1999; Metcalfe, 1998; Swann, Pelham & Krull, 1989). Therefore, positive illusions might result when goals and behavior are not deeply monitored for consistency (see Fig. 3–1). This chapter integrates these lines of research by examining behavioral and neural evidence and suggests that unrealistically positive self-evaluations have limited, mostly short-term, benefits for self-control and goal attainment. Furthermore, unrealistically positive self-perceptions often arise from shallow information processing, that is, baseline cognitive shortcuts. Therefore, the benefits of positive illusions may be short-lived because over time these cognitive shortcuts begin to incur costs for self-control and goal attainment.

When Are Rose-Colored Glasses a Good Idea?

The adaptive and normative nature of positive illusions, particularly unrealistically positive self-evaluations, has been a longstanding debate (e.g., Colvin & Block, 1994; Colvin, Block, & Funder, 1995; Gramzow & Willard, 2006; Paulhus, 1998; Robins & Beer, 2001; Taylor & Brown, 1998; Taylor, Lichtman, & Wood, 1984). Positive illusions are theorized to be adaptive because they are associated with the achievement of goals such as maintaining personal well-being, social relationships, and professional achievement (Taylor & Brown, 1988); achieving these goals requires self-control. Much of the research supporting this view has been critiqued on the basis that achievement of these goals is often related to positive self-views and not specifically to positive self-views that are unrealistic (e.g., Colvin & Block, 1994). However, some of this research has directly measured unrealistically positive self-views (e.g., they are discrepant from objective measures or social consensus) and has found support for the adaptive role of positive illusions. For example, individuals with unrealistically positive self-perceptions are initially well-liked when meeting strangers (Paulhus, 1998) and do experience boosts in positive mood in the short-term (Robins & Beer, 2001). People who had unrealistically positive self-views of their contribution to a group project, when compared to group members' perceptions of those individuals' performances, experience a short-term boost in positive mood.

In contrast, research has shown that positive self-evaluations that are unrealistic are associated with failure to exert the self-control needed for goal achievement over time. For example, people with unrealistically positive self-perceptions of their status are often disliked by others (e.g., Anderson et al., 2006; Paulhus, 1998). People with unrealistically positive self-views may be disliked over time because they do not control condescending or hostile behavior (e.g., Colvin, Block, & Funder, 1995) and fail to maintain agreeableness towards others over time (Paulhus, 1998). A series of studies found that self-enhancement (i.e., an individual's self-view is more favorable than judges' ratings or peer-ratings of that individual) predicted poor interpersonal functioning such as increased hostility and condescension towards others (Colvin, Block, & Funder, 1995). Another longitudinal study found that self-enhancement (whether operationalized by trait levels of narcissism and self-deceptive enhancement or the discrepancy between self-view and peer-ratings of personality) was associated with a declining pattern of agreeableness over two months (Paulhus, 1998). Additionally, unrealistically positive self-perceptions are associated with poorer well-being over time (Robins & Beer, 2001). Together this research suggests that an unrealistically positive self-view is probably a successful goal attainment strategy in the short-term (e.g., immediate boosts in mood, social acceptance by strangers), but promotes self-control failures in the long-term (e.g., disinhibited and inappropriate social behavior, poor well-being).

Although originally proposed to be a normative aspect of human cognition, unrealistically positive self-evaluations have been shown to be normative in some domains but not in others. People generally see themselves as above average in domains such as personality characteristics, driving, morality, and life outcomes (e.g., Alicke, 1985; Baumhart, 1968; Dunning, Meyerowitz & Holzberg, 1989; Larwood, 1978; Svenson, 1981; Taylor & Brown, 1988). However, the above average effect is attenuated by ambiguity in the domain of interest. For example, people are much more likely to see themselves as above average on positive traits that lend themselves to a number of behavioral descriptors, such as being disciplined, but give more average ratings for positive traits that are behaviorally concrete, such as being neat (Dunning et al., 1989). Additionally, there is evidence that participants do not describe their personality more favorably when compared to descriptions by judges or peers (Colvin, Block, & Funder, 1995), nor do people inflate their self-perceptions of status (Anderson et al., 2006). An individual's self-perception of status within a group correlates significantly with the group members' perception of that individual's status within the group. The correlation between self-view and group-member ratings of status holds for minimally acquainted groups and across time. Together these studies suggest that unrealistically positive self-evaluations are only normative in certain domains.

Summary

Just as rose-colored glasses do not go with every outfit, positive illusions are not beneficial or normative across the board. The research suggests that unrealistically positive self-evaluations may help people feel good or attract the attention of strangers, but these benefits are short-lived. Over time, unrealistically positive self-perceptions are associated with self-control failures such as inappropriate and disinhibited behavior, making it difficult to achieve well-being and social acceptance in the long-term. Given the negative consequences of positive illusions, it is perhaps not surprising that they are not present in a number of domains. Most studies found that unrealistically positive self-views of social status were not normative and those individuals who did inflate their perceptions of status were most likely to exhibit the self-control failures in their social behavior.

Although not as pervasive as originally theorized, positive illusions do have some benefits for short-term goal attainment and, therefore, raises the question of how this research relates to models of self-control. Most models of self-control include a monitoring component through which individuals evaluate whether their actual or future behavior is consistent with achieving their goals (Baumeister & Heatherton, 1996; Carver & Scheier, 1990). Monitoring may occur implicitly or explicitly with the purpose of momentarily processing information about behavior in relation to goals in a deeper manner than occurs at the baseline. In the case of discrepancy, individuals can then exert self-control to adjust behavior toward goal achievement. If the purpose of self-monitoring is to motivate self-control when necessary, then accurate self-monitoring should arguably be critical for successful self-control. If people fool themselves into believing that their behavior is an appropriate strategy to achieve a goal when it is not, then they will not execute behavioral changes needed to achieve that goal. However, unrealistically positive perceptions can promote goal attainment (e.g., feeling good) in certain circumstances. This research suggests that optimal self-control is associated with judicious reliance on positive illusions, that is, self-serving monitoring.

Do We Work Hard or Hardly Work for our Unrealistically Positive Self-Evaluations?

To reap the short-term benefits of positive illusions, do we need to put forth cognitive effort to maintain unrealistically positive self-views or are they always there in the background and we need to put forth cognitive effort to appropriately depart from our inflated baseline perceptions? Research on positive illusions suggests that unrealistically positive self-evaluations arise when motivational influences are exerted on monitoring. Accuracy may result when this motivation is absent or when a motivation towards accuracy is invoked in its place (e.g., Sedikies & Strube, 1995; Swann, Pelham, & Krull, 1989; Taylor, Neter, & Wayment, 1995). "Illusion" was specifically chosen to distinguish these enduring, normative cognitive processes that are emotionally motivated from short-term or temporary errors that occur by chance (Taylor & Brown, 1988). From this perspective, unrealistically positive self-evaluations

arise from a motivation to feel good about one's self. This motivation or goal then influences the monitoring function so that the evaluations of behavior in relation to goals are viewed in a self-serving manner (see Fig. 3–1). Theoretically, any negative effects of failing to recognize real discrepancies between behavior and goals are counterbalanced by the positive effects of feeling good which may motivate individuals to persist in goal attainment. In contrast, researchers have also conceptualized unrealistically positive self-evaluations as cognitive shortcuts that represent shallow information processing (e.g., Dougherty, 2001; Gigerenzer, Hoffrage, & Kleinbolting, 1991; Klayman, 1995). Researchers taking this view often refer to unrealistically positive self-evaluations as overconfidence. Unrealistically positive self-evaluations arise when people focus on irrelevant information or fail to really consider a variety of information when making self-assessments. Therefore, unrealistically positive self-evaluations may arise when the monitoring component of self-control models is not engaged or is engaged in a faulty, weakened manner (Fig. 3–12). If the shallow information processing view is correct, then the positive illusions research and models of self-control do not necessarily contradict each other. In this case, it is possible that positive illusions are actually corrected by the periodic monitoring described in self-control models. The adaptive value of positive illusions may reflect situations in which mood regulation is the most important goal or when it is not too costly to prioritize mood regulation.

Behavioral Evidence

Much of the research on unrealistically positive self-evaluations has focused on the attenuating effect of depth of processing on self-serving monitoring. Unrealistically positive self-views are reduced when participants are made accountable for justifying why they endorsed particular attributes (e.g., Sedikides et al., 2002) or asked to introspect on their performance (Sedikides, Horton, & Gregg, 2007; Paulhus et al., 2003; Robins & John, 1997). For example, one study required participants to write an essay supporting or arguing against aspects of the United

States space program. After completing the essay, participants graded their own essays on a number of criteria. Half of the participants were randomly told that they would have to justify their grades to an expert on writing (a Master's level graduate student) whereas the other half were told that their grades would be kept confidential. Participants in the accountability condition gave their essays significantly lower grades than participants in the confidential condition. Another study found that requiring participants to introspect on aspects of their personality (e.g., generate reasons they may have the attribute and reasons why they may not have the attribute) significantly reduced their tendency to favor positive attributes over negative attributes (Sedikides, Horton, & Gregg, 2007). Together these studies suggest that unrealistically positive self-evaluations are reduced when people are asked to think through justifications for their evaluations. However, these studies do not provide strong evidence that self-serving perceptions are a baseline assumption of the monitoring system. The depth of processing manipulations may not represent a departure from a shallow information processing baseline, but may instead make the motivated search for self-serving information too difficult.

Although only a few studies have been conducted, stronger evidence for the baseline nature of self-serving monitoring comes from research examining the impact of mental load on self-evaluation (Alicke et al., 1995; Hughes & Beer, 2008; Paulhus, Graf, & Van Selst, 1989; Swann, Hixon, Stein-Seroussi, & Gilbert, 1990). For example, participants under mental load prefer interactional partners who provided positive feedback (Swann et al., 1990) and exhibit even more bias when estimating how their personality compares to that of other people (Alicke et al., 1995). Another series of studies found that in paradigms involving conditions that elicit biased or accurate judgments, mental load tended to shift accurate judgments towards bias (Hughes & Beer, 2008). For example, participants were asked to judge the self-descriptiveness of personality traits while monitoring digits that appeared on a computer screen. As the digit-monitoring task became more difficult, participants endorsed significantly more favorable traits and reduced their endorsement of negative traits. Another study found that mental load increased overconfidence in domains that are otherwise associated with relatively accurate self-evaluations (Hughes & Beer, 2008). Participants answered a series of general trivia questions and then rated their confidence that their answers were correct (modified from Klayman et al., 1999). In the absence of mental load, participants tended to be fairly accurate in their estimations of their performance on questions about state poverty levels. However, when mental load was added, participants began to significantly overestimate their ability to answer the same questions correctly. This effect held regardless of whether mental load was manipulated by shortened response windows (e.g., 1 second) or by counting backward by sevens from a nine-digit number. It is important to note that the mental load did not decrease performance suggesting that the load did not impair overall ability to perform the task. Together these studies suggest that load creates a preference for positive feedback and also exacerbates or introduces bias into judgments.

Neural Evidence

Another way to understand the automatic or controlled nature of unrealistically positive self-evaluations is to investigate the underlying neural systems. If unrealistically positive self-evaluations recruit neural regions associated with executive functioning, then it is likely that they arise from effortful information processing. In contrast, if executive function systems are recruited to attenuate unrealistically positive self-evaluations, then they may arise from automatic or shallow information processing. Very little neural research has been conducted on this topic (Beer, 2007), but the few extant studies are consistent with the behavioral research that many unrealistically positive self-evaluations are characterized by shallow information processing (Beer & Hughes, 2010; Beer et al., 2006a; Beer, Lombardo, & Bhanji, in press), which may be driven by attention to the valence of information that is processed about the self (Moran et al., 2006; Sharot et al., 2007).

Consistent with the view that positive illusions may arise from heuristic processing, two studies have suggested that bias is associated with ventral anterior cingulate activity (Moran et al., 2006; Sharot et al., 2007). In one study, participants judged the self-relevance of personality traits. Consistent with previous research, this study found that participants tended to favor positive personality traits as more self-descriptive than negative personality traits. This response pattern has been interpreted as reflecting unrealistically positive self-views because it suggests that individuals find their positive attributes more salient than their negative attributes. The ventral anterior cingulate was associated with judgments of positive personality traits compared to judgments of negative personality traits, especially for those that were self-descriptive (Moran et al., 2006). Another study found that a similar region of the anterior cingulate was associated with imagining future positive events compared to future negative events, especially for individuals who are dispositionally optimistic (Sharot et al., 2007). These studies equated bias with positive valence and accuracy with negative valence. Building on this research, a final study asked participants to judge positive and negative personality traits under conditions that either elicited accurate judgments or biased judgments because they either allowed participants to define the traits in idiosyncratic, self-serving ways or not (Beer & Hughes, 2010), making it possible to differentiate regions associated with valence versus bias. The results show that the ventral anterior cingulate cortex is important for distinguishing positive traits from negative traits, but does not support biased judgments (Beer & Hughes, 2010). These findings are consistent with previous research relating ventral anterior cingulate activity and the emotional valence of information (Bush, Luu & Posner, 2000; Rogers et al., 2004). It is important to note the ventral anterior cingulate cortex is distinct from the dorsal portion of the anterior cingulate associated with conflict monitoring and error detection in paradigms such as the Stroop task (e.g., Bishop et al., 2004; Botivinick, Cohen & Carter, 2004; Chapter 1). For example, the ventral anterior cingulate cortex is important for recognizing reward in gambling tasks. Just as participants recruit their anterior cingulate cortex for maximizing reward in gambling tasks, it may be that this region is important for quickly recognizing when information is likely to be rewarding and deemed self-descriptive.

Some neural research has found that accurate self-evaluations require the orbitofrontal cortex. Patients with orbitofrontal damage often fail to control their social behavior or are unable to identify the behaviors that are mostly likely to accomplish particular goals (e.g., Beer et al., 2003; Saver & Damasio, 1991). In one study, patients with orbitofrontal damage, patients with lateral prefrontal damage, and healthy comparison subjects were asked to evaluate their performance in a self-disclosure task with a stranger (Beer et al., 2006a). Pretesting found that all of the participants understood social norms regarding restrictions on self-disclosures in conversations. Patients with orbitofrontal lesions violated these social norms but evaluated their performance as socially appropriate. In another functional magnetic resonance (fMRI) study, participants had to answer questions from a general trivia test and then judge how confident they felt that their answers were correct. Participants who overestimated their performance on the general trivia test failed to recruit orbitofrontal cortex activity when making confidence judgments about their performance (Beer, Lombardo, & Bhanji, in press). In other words, orbitofrontal cortex activity was associated with accurate self-evaluations of test performance. In another study, participants judged how they compared to their average peer on personality traits that either permitted them to focus on self-serving, idiosyncratic definitions that yielded biased judgments, or more restricted definitions that yielded relatively accurate judgments (Beer & Hughes, 2010). Judgmental bias was associated with reduced engagement of the medial and lateral orbitofrontal cortex. The role of orbitofrontal cortex in accurate self-evaluation is consistent with research that has shown that orbitofrontal cortex is recruited for accurate predications about one's successes and failures on task

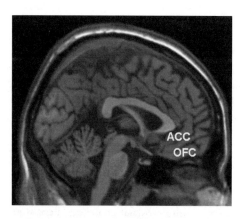

Figure 3–2. The orbitofrontal cortex is recruited to adjust positive illusions toward reality; the anterior cingulate may be recruited to support positive illusions. ACC = Anterior Cingulate cortex; OFC = Orbitofrontal cortex.

performance (Schnyer, Nicholls, & Verfaeille, 2005) and for discounting irrelevant, emotional information when making a gambling decision (Bechara, Bechara, & Damasio, 2000; Beer, Knight & D'Esposito, 2006b; DeMartino et al., 2006; Elliott, Dolan, & Frith, 2000; Elliott, Frith, & Dolan, 1997; Rolls, 2000). For example, DeMartino et al. (2006) found that participants recruited the orbitofrontal cortex when they successfully placed bets based on monetary principles rather than an emotionally salient, but irrelevant, framing of the bet. Participants in this study were endowed with monetary amounts and given the option to gamble the endowment or choose a guaranteed win or loss that was monetarily equivalent (e.g., guaranteed to keep $20 from a $50 endowment or to lose $30 from a $50 endowment, respectively). People tend to be loss averse and favor gambling in the loss framing although it is monetarily equivalent to the win framing. Participants who recognized the misleading aspect of the win and lose frames bet equivalently across the win and loss frames and recruited significantly greater orbitofrontal cortex activity. Just as these participants recruited orbitofrontal cortex to recognize the irrelevance of their automatic emotional reaction, it may be the orbitofrontal cortex is important for correcting automatic positive assumptions about the self.

Together, these studies suggest that bias recruits neural regions associated with detecting emotional valence (the ventral anterior cingulate cortex; ACC) and accuracy recruits neural regions associated with executive functioning (the orbitofrontal cortex; OFC) (*see* Fig. 3–2).

These findings are consistent with the perspective that *(a)* biases arise when individuals make judgments based on some kind of heuristic judgment and *(b)* accuracy requires that the heuristic judgment is evaluated for relevance to the task at hand. Specifically, participants may generate positively biased self-judgments because their self-judgments rely on neural circuitry (i.e., the ventral anterior cingulate cortex) that is helpful for quickly identifying whether a particular personality trait or event is positive. People may then use valence as a rule for claiming or dismissing the information as descriptive. Accuracy arises when participants engage neural circuitry (i.e., the orbitofrontal cortex) that is helpful for examining whether salient information is actually relevant for the task at hand. In this case, people are able to recognize that their judgments are skewed by a focus on only good information and are able to correct for this bias.

Summary

The behavioral and neural evidence suggest that many unrealistically positive self-evaluations arise from shallow or automatic information processing. Unrealistically positive self-evaluations are reduced by manipulations of deeper information processing (e.g., accountability, introspection) and the recruitment of the orbitofrontal cortex, a neural region associated with executive functioning (e.g., Beer, Shimamura, & Knight, 2004). Also consistent with this view are the findings that unrealistically positive self-evaluations are increased by manipulations of mental load (digit monitoring, speeded responses) which prevent elaborative processing because of time or cognitive resource restraints.

Conclusion

An integration of the research on positive illusions and models of self-control has several implications for understanding the self-monitoring processes that influence self-control and goal attainment. Models of self-control assume that the relation between goals and ongoing or future behavior is monitored on a periodic basis. The accurate nature of this monitoring is critical for identifying discrepancies which motivate self-control processes that modify the behavior to facilitate goal achievement. Consistent with this view, accurate self-perceptions tend to be normative in a number of domains and are associated with increased self-control and goal attainment across time. However, unrealistically positive self-perceptions have short-term benefits, particularly for mood regulation. The research on mechanisms underlying unrealistically positive self-evaluations suggests that they often reflect cognitive shortcuts; unrealistically positive self-evaluations are increased in circumstances of cognitive load and are decreased with the introduction of deeper information processing. Additionally, unrealistically positive self-evaluations are decreased in relation to the recruitment of the orbitofrontal cortex; the orbitofrontal cortex serves a monitoring function similar to those described in models of self-control. Therefore, the situations in which positive illusions are associated with short-term mood-regulation benefits may reflect circumstances in which mood regulation is the most important goal or its priority is not too costly for the achievement of other goals.

The integration of positive illusions and models of self-control also raises a number of questions that could be explored in future research. For example, the neural research suggests that a cognitive heuristic judgment may still be affected by a motivational influence. Dual process theories of judgment suggest that bias may arise either because the most easily accessible answer is incorrect or because of failures in the attempt to monitor the relevance of the easily accessible answer and correct any errors (e.g., Kahneman & Frederick, 2002). Although the neural research suggests that executive function regions such as the orbitofrontal cortex are responsible for the monitoring and control of self-evaluations that leads to accuracy, it is important to consider the psychological implications of the engagement of the ventral anterior cingulate cortex in self-evaluations. The ventral anterior cingulate cortex is typically associated with differentiating information on the basis of its valence. Although this region is not typically associated with control, it could be considered to support monitoring information for its valence. Why do people monitor information for its valence when making self-judgments? The neural findings suggest that the default state of the self-evaluation system is motivated to pay more attention to positive aspects of the self (Beer, 2007). Thus, the self-evaluation system is motivated towards flattering information, but this motivation may be accomplished in an automatic manner.

Further understanding about the default assumption of the self-evaluative process may be gained by paying more attention to examining positive illusions in children in relation to developing self-control abilities. Children's self-control suggests that it is not necessarily surprising to find that positive illusions may be a default mode that is corrected by a monitoring component supported by executive functioning. For example, think about children's overconfidence in their balance when climbing up on a precarious stool or standing at the edge of step. Is this self-control failure a result of a motivated, self-serving monitoring process or is it because the child does not stop and monitor the risks? Research assessing the prevalence of overconfidence in children will further test the baseline of shallow information processing that characterizes self-evaluation in much of the adult research. Additionally, research on children will be important for understanding how positive illusions develop over time. If it is the case that positive illusions represent the baseline self-assessment in children that is corrected by higher order monitoring functions, then future research may want to explore the implications of a baseline self-evaluation system that is inherently flawed and requires correction. Although this may seem counterintuitive, the visual system contains bottom-up assumptions that result in optical illusions, unless corrected by controlled processing (e.g., Eagleman, 2001). Not only is it reasonable to expect that something as ambiguous and difficult as self-evaluation may make flawed assumptions as well, it may be similar to the visual system in that its flaws exist for the sake of efficiency. Again, the term "positive illusions" was chosen to indicate that these beliefs are unrealistically positive, but are distinct from the costly and vastly discrepant delusions associated with psychopathology. It may be that permitting the motivation to favor positive information about the self to influence self-assessment is "good enough" much of the time. However, if corrective mechanisms are needed, they are available.

Finally, it is important to recognize that conceptualization of a default desire to feel good about one's self is accomplished through automatic processing does not preclude the execution of effortful processes for the purpose of boosting self-esteem. For example, behavioral research has shown that people will strive to put themselves in a good mood to buffer the effects of incoming negative information that may threaten their self-esteem (Raghunathan & Trope, 2002; Trope & Fishbach, 2000; Trope & Neter, 1994). Future research might examine what happens neurally when participants' self-esteem is threatened and the desire to boost self-esteem becomes conscious. Is it the case that neural systems supporting heuristic judgments based on the valence of information are engaged that much more and neural systems supporting monitoring are not engaged? Or are neural systems that support different kinds of executive function such as reappraisal (i.e., dorsolateral prefrontal cortex) engaged to actively seek out affirming information?

In conclusion, positive illusions can make you feel pretty good, but this positive outcome is short-lived because the cognitive shortcuts supporting positive illusions become too costly in the long run. Accurate self-monitoring is critical for successful self-control and goal attainment across time. Just as the Oompa Loompas cautioned the children in Willy Wonka's Chocolate Factory to consider the consequences of their actions, it is important to recognize when behavior is discrepant with goals or when goals conflict with one another.

References

Alicke, M. D. Global self-evaluation as determined by the desirability and controllability of trait adjectives. J Pers Soc Psychol 1985; 49: 1621–1630.

Alicke, M. D., Klotz, M. L., Breitenbecher, D. L., Yurak, T. J., & Vrendenburg, D. S. Personal contact, individuation, and the above-average effect. J Pers Soc Psychol 1995; 68: 804–825.

Anderson, C. P., Srivastava, S., Beer, J. S., Spataro, S. E., & Chatman, J. A. Knowing your place: Self-perceptions of status in social groups. J Personal Soc Psychol 2006; 91: 1094–1110.

Baumeister, R. F., & Heatherton, T. A. Self-regulation failure: An overview. Psycholog Inquiry, 1996; 7: 1–15.

Baumhart, R. An Honest Profit. New York: Prentice Hall, 1968.

Bechara, A., Damasio, H., & Damasio, A. R. Emotion, decision making, and the orbitofrontal cortex. Cerebral Cortex 2000; 10: 295–307.

Beer, J. S. The default self: Feeling good or being right? Trends Cogn Sci, 2007; 11: 187–189.

Beer, J. S., Heerey, E. H., Keltner, D., Scabini, D., & Knight, R. T. The regulatory function of self-conscious emotion: Insights from patients with orbitofrontal damage. J Pers Soc Psychol, 2003; 85: 594–604.

Beer, J. S., & Hughes, B. L. Neural systems of social comparison and the "above average" effect. NeuroImage, 2010; 49: 1810–1819.

Beer, J. S., John, O. P., Scabini, D., & Knight, R. T. Orbitofrontal Cortex and Social Behavior: Integrating Self-Monitoring and Emotion-Cognition Interactions. J Cogn Neurosci, 2006a; 18: 871–880.

Beer, J. S., Knight, R. T., & D'Esposito, M. Integrating Emotion and Cognition: The role of the frontal lobes in distinguishing between helpful and hurtful emotion. Psycholog Sci, 2006b; 17: 448–453.

Beer, J. S., Lombardo, M. V., & Bhanji, J. F. P. Roles of medial prefrontal cortex and orbitofrontal cortex in self-evaluation. J Cogn Neurosci, in press.

Beer, J. S., Shimamura, A. P., & Knight, R. T. Frontal lobe contributions to executive control of cognitive and social behavior. In M. S. Gazzaniga (ed.) The newest cognitive neurosciences (3rd Edition) (pp.1091–1104). Cambridge: MIT Press. 2004

Bishop, S., Duncan, J., Brett, M., Lawrence, A. D. Prefrontal cortical function and anxiety: controlling attention to threat-related stimuli. Nat Neurosci 2004; 7: 184–188.

Botvinick, M. M., Cohen, J. D., & Carter, C. S. Conflict monitoring and anterior cingulate cortex: An update. Trends Cogn Sci, 2004; 8: 539–546.

Bush, G., Luu, P., & Posner, M. I. Cognitive and emotional influences in anterior cingulate cortex. Trends Cogn Sci 2000; 4: 215–222.

Caputo, D., & Dunning, D. What you don't know: The role played by errors of omission in imperfect self-assessments. J Exp Soc Psychol 2005; 41: 488–505.

Carver, C. S., & Scheier, M. Principles of self-regulation: Action and emotion. In: Higgins, E. T., & Sorrentino, R. M. (Eds.). *Handbook of Motivation and Cognition: Foundations of Social behavior* New York, NY: The Guildford Press, 1990: pp.3–52.

Chambers, J. R., & Windschitl, P. D. Biases in social comparative judgments: The role of nonmotivated factors in above-average and comparative-optimism effects. Psycholog Bull, 2004; 130: 813–838.

Colvin, C. R., & Block, J. Do positive illusions foster mental health? An examination of the Taylor and Brown formulation. Psycholog Bull, 1994; 116:3–20.

Colvin, C. R., Block, J., & Funder D. Overly positive self evaluations and personality: Negative implications for mental health. J Personal Soc Psychol, 1995; 68: 1152–1162.

Critchley, H. D. Neural mechanisms of autonomic, affective, and cognitive integration. J Comp Neurol 2005; 493: 154–166.

DeMartino, B., Kumaran, D., Seymour, B., Dolan, R. J. Frames, biases, and rational decision-making in the human brain. Science 2006; 313: 684–687.

Dougherty, M. R. P. Integration of the ecological and error models of overconfidence using a multiple-trace memory model. J Exp Psychol: Gen 2001 130: 579–599.

Dunning, D. Trait importance and modifiability as factors influencing self-assessment and self-enhancement motives. Personal Soc Psychol Bull, 1995; 21: 1297–1306.

Dunning, D., Meyerowitz, J. A., & Holzberg, A. D. Ambiguity and self-evaluation: The role of idiosyncratic trait definitions in self-serving assessments of ability. J Personal Soc Psychol, 1989; 57: 1082–1090.

Eagleman, D. Visual illusions and neurobiology. Nat Rev Neurosci 2001; 2, 920–925.

Elliott, R., Dolan, R. J., & Frith, C. D. Dissociable functions in the medial and lateral orbitofrontal cortex: Evidence from human neuroimaging studies. Cerebral Cortex 2000; 10: 308–317.

Elliott, R., Frith, C. D., & Dolan, R. J. Differential neural response to positive and negative feedback in planning and guessing tasks. Neuropsychologia 1997; 35: 1395–1404.

Gigerenzer, G., Hoffrage, U., & Kleinbolting, H. Probabilistic mental models: A Brunswikian theory of confidence. Psycholog Rev 1991; 98: 506–528.

Gramzow, R. H., & Willard, G. Exaggerating current and past performance: Motivated self-enhancement versus reconstructive memory. Personal Soc Psychol Bull 2006; 32: 1114–1125.

Hughes, B. L., & Beer, J. S. Mental load influences on overconfidence and self-enhancement. Unpublished data: University of Texas at Austin, 2009

Kahneman, D., & Frederick, S. Representativeness revisited: Attribute substitution in intuitive judgment. In: Gilovich, T., Griffin, D., & Kahneman, D. (Eds.). Heuristics & Biases: The Psychology of Intuitive Judgment New York, NY: Cambridge University Press, 2002.

Klayman, J. Varieties of confirmation bias. In: Busemeyer, J., Hastie, R., & Medin, D. L. (Eds.). *Psychology of learning and motivation: Vol 32.* New York, NY: Academic Press, 1995: pp. 365–418.

Klayman, J., Soll, J. B., Gonzalez-Vallejo, C., & Barlas, S. Overconfidence: It depends on how, what, and whom you ask. Organizational Behav Human Decision Processes 1999; 79: 216–247.

Larwood, L. Swine flu: A field study of self-serving biases. J Applied Soc Psychol, 1978; 18: 283–289.

Metcalfe, J. Cognitive optimism: Self-deception or memory-based heuristic processing? Personal Soc Psychol Rev, 1998; 2: 100–110.

Moran, J. M., Macrae, C. N., Heatherton, T. F., Wyland, C. L., & Kelley, W. M. Neuroanatomical evidence for distinct cognitive and affective components of self. J Cogn Neurosci, 2006; 18: 1586–1594.

Ochsner, K. N., Beer, J. S., Robertson, E. A., et al. The neural correlates of direct and reflected self-knowledge. Neuroimage 2005; 28: 797–814.

Paulhus, D. L. Interpersonal and intrapsychic adaptiveness of trait self-enhancement: A mixed blessing? J Personal Soc Psychol 1998; 74: 1197–1208.

Paulhus, D. L., Harms, P. D., Bruce, M. N., & Lysy, D. The over-claiming technique: Measuring self-enhancement independent of ability. J Personal Soc Psychol 2003; 84: 890–904.

Paulhus, D. L., Graf, P., & Van Selst, M. Attentional load increases the positivity of self-presentation. Soc Cogn 1989; 7: 389–400.

Raghunathan, R., & Trope, Y. Walking the Tightrope Between Feeling Good and Being

Accurate: Mood as a Resource in Processing Persuasive Messages. J Personal Soc Psychol 2002; 83: 510–525.

Roberts, A. C., & Wallis, J. D. Inhibitory control and affective processing in the prefrontal cortex: Neuropsychological studies in the common marmoset. Cerebral Cortex 2000; 10: 252–262.

Roberts, N. A., Werner, K. H., Beer, J. S., et al. The impact of orbital prefrontal cortex damage on emotional reactivity during acoustic startle. Cogn Affective Behav Neurosci 2004; 4: 307–316.

Robins, R. W., & Beer, J. S. Positive illusions about the self: Short-term benefits and long-term costs. J Personal Soc Psychol 2001; 80: 340–352.

Robins, R. W., & John, O. P. Self-perception, visual perspective, and narcissism: Is seeing believing? Psycholog Sci 1996; 7.

Robins, R. W., & John, O. P. In: Hogan, R., Johnson, J., & Briggs, S. (Eds.) *Handbook of Personality Psychology*. New York, NY: Academic Press, 1997.

Rogers, R. D., Ramnani, N., Mackay, C., et al. Distinct portions of anterior cingulate cortex and medial prefrontal cortex are activated by reward processing in separable phases of decision-making cognition. Biolog Psychiatr 2004; 55: 594–602.

Rolls, E. T. The orbitofrontal cortex and reward. Cerebral Cortex 2000; 10: 284–294.

Rolls, E. T., Hornak, J., Wade, D., & McGrath, J. Emotion-related learning in patients with social and emotional changes associated with frontal lobe damage. J Neurol Neurosurg Psychiatr 1994; 57: 1518–1524.

Rushworth, M. F. S., Behrens, T. E. J., Rudebeck, P. H., & Walton, M. E. Contrasting roles for cingulate and orbitofrontal cortex in decisions and social behaviour. Trends Cogn Sci 2007; 11: 168–176.

Saver, J. L., & Damasio, A. R. Preserved access and processing of social knowledge in a patient with acquired sociopathy due to ventromedial frontal damage. Neuropsychologia 1991; 29: 1241–1249.

Schnyer, D. M., Nicholls, L., &Verfaellie, M. The role of VMPC in metamemorial judgments of content retrievability. J Cogn Neurosci 2005; 17: 832–846.

Sedikides, C., & Strube, M. J. The multiply motivated self. Personal Soc Psychol Bull 1995; 21: 1330–1335.

Sedikides, C., Herbst, K. C., Hardin, D. P, & Hardin, G. J. Accountability as a deterrent to self-enhancement: The search for mechanisms. J Personal Soc Psychol 2002; 83: 592–605.

Sedikides, C., Horton, R. S., & Gregg, A. P. The why's the limit: Curtailing self-enhancement with explanatory introspection. J Personal 2007 ; 75: 784–824.

Sharot, T., Riccardi, A. M., Raio, C. M., & Phelps, E. A. Neural mechanisms mediating optimism bias. Nature 2007; 450,102–105.

Swann, W. B., Hixon, J. G., Stein-Seroussi, A., & Gilbert, D. T. The fleeting gleam of praise: Cognitive processes underlying behavioral reactions to self-relevant feedback. J Personal Soc Psychol 1990; 59: 17–26.

Swann, W. B., Pelham, B. W., & Krull, D. S. Agreeable fancy or disagreeable truth? Reconciling self-enhancement and self-verification. J Personal Soc Psychol 1989; 57: 782–791.

Taylor, S. E., & Brown, J. D. Illusion and well-being: A social psychological perspective on mental health. Psycholog Bull 1988; 103: 193–210.

Taylor, S. E., Lichtman, R. R., & Wood, J. V. Attributions, beliefs about control, and adjustment to breast cancer. J Pers Soc Psychol, 1984; 46: 489–502.

Taylor, S. E., Neter, E., & Wayment, H. A. Self-Evaluation process. J Personal Soc Psychol Bull 1995; 21: 1278–1287.

Trope, Y., & Fishbach, A. Counteractive control processes in overcoming temptation. J Personal Soc Psychol 2000; 79: 493–506.

Trope, Y., & Neter, E. Reconciling competing motives in self-evaluation: The role of self-control in feedback seeking. J Personal Soc Psychol 1994; 66: 646–657.

CHAPTER 4

Control in the Regulation of Intergroup Bias

David M. Amodio and Patricia G. Devine

ABSTRACT

In the intergroup relations literature, theories of control concern the interplay of basic cognitive mechanisms of self-regulation with intrapersonal and societal-level goals and motivations. By integrating multiple levels of analysis, this literature has been uniquely positioned to advance our understanding of control as it operates in the complex social world that the human self-regulatory system has evolved to negotiate. In this chapter, we review research and theoretical models of control that have emerged from intergroup approaches. The first section of this chapter describes four theoretical models of control that have been central to research in prejudice, stereotyping, and intergroup relations. The next section highlights some of the recent exciting advances in the study of control in the intergroup domain and notes how they are refining and, in some cases, redefining basic conceptions of control as a self-regulatory process. The final section outlines what we see as some major challenges faced by current theories of control, both in the intergroup domain and in the broader psychological literature.

Keywords: control, self-regulation, prejudice, stereotyping, intergroup, implicit, brain, social neuroscience

Overt instances of racial prejudice are relatively rare in contemporary American society, having declined substantially over the past 60 years following government mandates for integration and equal rights (Schuman, Steeh, Bobo, & Krysan, 1997). However, according to several indicators in the social psychological literature, racial bias remains extremely prevalent in Americans' implicit, or automatic, tendencies (Jost et al., in press; Greenwald, Poehlman, Uhlmann, & Banaji, 2009). Although evidence that Americans show learned automatic race-biased associations is not new (e.g., Gaertner & McLaughlin, 1983), the immense prevalence of such biases in this modern age is surprising in light of declining racial prejudice in Americans' overt attitudes and behaviors. How is it that many Americans maintain egalitarian beliefs and manage to respond without bias in intergroup situations? According to theorists dating back to Allport (1954), nonbiased intergroup responses require control.

The notion of control is studied from many different perspectives and at multiple levels of analysis, ranging from studies of low-level response inhibition to the effects of broad and dynamic societal-level forces. The intergroup approach to studying control is notable for examining control at the nexus of these levels of analysis. By virtue of this integrative approach, the intergroup literature has been especially generative in contributing theoretical advances to the study of control in the field of psychology. Our goal in this chapter is to review the dominant theoretical ideas in the field and to discuss the current challenges and exciting new directions that this area of research currently faces.

Control in the Context of Prejudice and Stereotyping

Theoretical Roots

Contemporary theories of control in the intergroup relations literature reflect the culmination of multiple research traditions, and in thinking about modern theories of control, it is useful to consider the theoretical roots from which they developed. One major influence is the early research on person perception and *attribution*, the process through which we interpret the causes of others' behavior (Heider, 1958; Jones & Davis, 1965; Kelley, 1967). Research on these topics identified a consistent attributional bias, whereby behaviors are initially attributed to a target person's disposition, even when situational factors were known to influence the response (i.e., the Fundamental Attribution Error, Ross, 1977; *see also*, Uleman, 1999, for spontaneous trait inferences, and Nisbett & Wilson, 1977, for self-perception effects). Attributional biases appeared to operate without awareness or intention, and subsequent theorizing suggested that such biases may be corrected through the controlled (i.e., effortful and intentional) adjustment of one's perceptions (Gilbert, Pelham, & Krull, 1988; Trope, 1986; Tversky & Kahneman, 1974; Wilson & Brekke, 1994). Attribution-based models of control assumed that the correction of bias in one's perception occurred deliberatively and effortfully, and only after one acknowledged the potential influence of biasing factors, as when a school teacher might try to adjust for bias when grading the essay of a disliked student. This tradition of mental control is clearly evident in modern theories of corrective processes in prejudice and stereotyping (e.g., Brewer, 1988; Darley & Gross, 1983; Devine, 1989; Fiske & Neuberg, 1990).

A second major influence on contemporary models of control was the early research on parallel vs. serial visual search processes in studies of human factors and human cognition (e.g., Posner & Snyder, 1975; Shiffrin & Schneider, 1977). Parallel search processes were believed to operate automatically, whereas serial search processes operated in a controlled fashion. Unlike the corrective models of control proposed in attribution research, parallel and serial search processes were viewed as independent modes of information processing. Later research suggested that serial search processes corresponded to a set of controlled mechanisms that included deliberative attention (also working memory) and was coordinated by a "central executive" (Baddely, 1986; Norman & Shallice, 1986).

Although most early research on control in the cognitive literature focused on modes of information processing (McClelland & Rumelhardt, 1985), some work examined mechanisms through which habitual (i.e., automatic) behaviors are inhibited or changed (Logan & Cowan, 1984). It is notable that several recent dual-process models in social cognition follow in this tradition. For example, Smith and DeCoster (2000) proposed that independent slow- and fast-learning memory systems underlie automatic versus more controlled processes. These authors posited that the fast-learning system corresponds to a propositional mechanism that overrides the automatic processes associated with the slow-learning system (*see also* Sloman, 1996). Gawronski and Bodenhausen's (2006) Associational-Propositional Evaluation (APE) model suggests a similar explanation for attitudes and evaluations, whereby more controlled responses reflect a propositional process that operates independently of the automatic associational process. Finally, Strack and Deutsch's (2004) Reflective and Impulsive Determinants model extends this general line of theorizing to focus on the regulation of behavior, beyond the more common focus on intrapsychic effects involving attitudes and person perception. The influence of this general approach can also be seen in modern theories of behavioral inhibition and response override models of prejudice control (e.g., Devine, 1989; Montieth 1993).

Vestiges of Dualism?

When reviewing the theoretical roots of modern theories of control, it is also useful to acknowledge their more distal roots in the dualistic philosophies of the mind and body. For example, Descartes famously declared that reason is needed to overcome the unruly and impulsive passions of the body (cf. Damasio, 1994). More recently, Freud (1930, 1961) updated this view of human nature with his theory of the psyche as comprising the id, ego, and superego. With the superego as its guide, the ego functions to suppress the primitive impulses of the id. Although the views of Freud and Descartes have been largely discarded by modern theorists, the notion that behavior is regulated through a mechanism of top-down control remains pervasive in contemporary social psychology. Similarly, neuroscientific theories widely assume that the phylogenetically newer frontal cortex exerts top-down control over the "reptilian" limbic system, and although there are certainly interconnections between these general divisions of the brain, the extent to which frontal neocortical regions exert top-down "control" over limbic structures in the Freudian sense remains a matter of debate. Thus, the widely accepted assumption that self-regulation operates through a mechanism of top-down control may deserve re-evaluation, and the possibility of alternative forms of control should be explored. We shall return to this issue later in the chapter.

Definitions

In discussing issues of control, it useful to define one's terms at the outset. Of course, given the myriad perspectives on this topic, a useful definition of control is bound to be complex. In this chapter, we differentiate between *control* as an outcome and *controlled processing* as a theoretical process or mechanism. Hence, the term "control" refers to the implementation of an intended, goal-directed response (i.e., a "controlled response"). This definition of control is precise and amenable to clear interpretation because it is defined by a specific behavioral criterion. In studies of control, experimental tasks may be designed to elicit measurable responses that reflect an intentional (i.e., goal-directed) response, made in the context of other processes that either promote or interfere with the response (as in the color naming Stroop task). In the domain of intergroup relations, control pertains more specifically to responding without prejudice. That is, control refers to the implementation of a response that is not biased by stereotypes or negative evaluations, or by gut-level negative emotions (Devine, Monteith, Zuwerink, & Elliot, 1991). Thus, prejudice control refers specifically to the regulation of one's behavior in situations where a race-biased response is possible.

We use the phrase "controlled processing" to refer to the theoretical mechanisms that contribute to a controlled response. Unlike control (the outcome), controlled processing refers to one or more hypothetical processes that operate in the mind. The operation of controlled processes may be inferred from either behavior or neural activations using carefully designed experimental tasks. Because of its abstract nature and imperviousness to direct measurement, a controlled process is more difficult to define and interpret than a controlled response outcome. For these reasons, it is especially challenging to test hypotheses about the activation of controlled processes, the number of controlled processes involved in a psychological event, and their specific roles in producing an intended behavioral response. Thus, care is required when drawing inferences about the processes thought to underlie a controlled response. At the same time, the most interesting questions about control often concern underlying processes, and thus a major challenge in the study of control is to use careful experimentation and theorizing to constrain inferences about the mechanisms.

It is notable that the concepts of "implicit" and "explicit" are often invoked and sometimes associated with the concepts of automaticity and control. Although a detailed discussion of these concepts is beyond the scope of this chapter (see Amodio & Mendoza, in press), it is useful to distinguish between these sets of terms. Whereas control refers primarily to the implementation of an intended goal-driven response (with or without conscious deliberation), the terms implicit and explicit refer to one's level of awareness of a particular process or response (Squire & Zola, 1996). That is, a process or response may be automatic or controlled, independently of whether it is implicit or explicit. These two sets of terms may intersect in some instances, given that a person may be more likely to engage in deliberative controlled processing when a response is explicit than when it is implicit. These distinctions are often blurred in discussions of experimental tasks and measures of racial bias, such that implicit is assumed to be automatic, and explicit is assumed to be controlled. In keeping with our definitions, we examine issues of control and controlled processes, independently of whether they may be characterized as implicit or explicit.

Finally, although we have distinguished between control as an outcome and control as a process, our particular definitions may differ from those stated (or assumed) in the different programs of research that we will review. It is worth noting that differences in terminology and in conceptions of control often relate to important differences in the ways people design their research, and analyze and interpret their data. For the sake of consistency and clarity, however, we describe research in terms of our preferred definitions.

Models of Control in Stereotyping and Prejudice

Contemporary research on control in the context of intergroup processes incorporates a mixture of past approaches. In this section, we review the major theoretical conceptions of control in this literature. These include correction, inhibition, suppression, and override (Devine & Monteith, 1999; Gilbert, 1999). In most cases, these models have been applied to the control of racial stereotypes and evaluations of outgroup members. By contrast, relatively little research has explored the application of controlled processing to the regulation of *affective* responses to race (*but see* Amodio, Harmon-Jones, & Devine, 2003; *see also*, Amodio, 2008, for broader discussions of this issue). We supplement our review with critical analysis, weighing the complementary strengths and weaknesses of each model.

Correction

Following from attribution-based theories of person perception and attitudes, the *mental correction* model of control assumes that perceivers first form an initial impression of a person and then apply corrections to adjust for possible bias (e.g., Fazio, 1990; Wilson & Brekke, 1994; Wegener & Petty, 1997). A general model is illustrated in Figure 4–1. Successful correction depends on a set of preconditions: an individual must *(a)* be motivated and able to search for potential biases; *(b)* become subjectively aware of a source of bias; *(c)* have (or come up with) an implicit theory about how the bias may affect one's judgment (i.e., its direction and magnitude); and *(d)* be motivated and able to implement the degree of correction suggested by his or her theory. An important commonality across the majority of mental correction studies is a focus on deliberative and reflective judgments and the use of self-report measures. Indeed, two major tenets of the mental correction model assume one's subjective awareness of biasing factors and components of control (e.g., acknowledgement of a bias, having a theory of how the bias might affect behavior). Accordingly, studies conducted to test this model of control have used methods that capitalize on the central role of conscious deliberation. For example, correction processes (i.e., contrasting) have been shown to occur only following conscious awareness of a biasing cue, but not when the biasing cue is presented subliminally (Dijksterhuis & van Knippenberg, 1998; *see also* Stapel, Martin, & Schwarz, 1998). As such, mental correction pertains primarily to situations where one is consciously aware of a potential bias and when responses can be made deliberatively. For example, an employer reviewing the application of Black job candidate

Correction

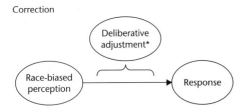

Goal: To adjust one's attitude or judgment

*requires motivation, awareness of bias, and implicit theory for how to correct

Figure 4–1. Diagram of the correction model of control, in which one's response or stored representation of an attitude object is adjusted through deliberative processes.

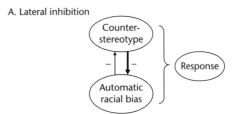

Goal: To dampen the activation of a thought or emotion

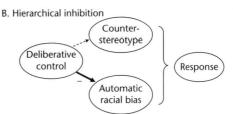

Goal: To dampen the activation of a thought or emotion

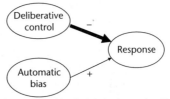

Goal: To inhibit the behavioral expression of bias

Figure 4–2. Diagram of inhibition models of control, illustrating lateral inhibition (A), hierarchical inhibition (B), and active behavior inhibition (C). Thicker arrows represent stronger connections.

may deliberatively adjust his evaluation to compensate for any biases associated with race.

The correction model of control has been applied to a range of phenomena, such as the evaluation of persuasive massages (Petty, Wegener, & White, 1998), social judgments (Petty & Wegener, 1993), evaluations of attitude objects (Wilson & Schooler, 1991), and self-assessment of one's mood (Schwartz & Clore, 1983; Strack & Hannover, 1996). Although the notion of mental correction is often invoked in intergroup research (e.g., Corneille, Vescio, & Judd, 2000; Harber, 1998; Kawakami, Dovidio, & van Kampe, 2005, 2007; Lambert, Kahn, Lickel, & Fricke, 1997; Olson & Fazio, 2004), few studies of intergroup bias have explicitly tested this model. In one such study, Gilbert and Hixon (1991) hypothesized and found that participants would form a less stereotypical impression of an Asian woman when they were given enough time to correct for an initially biased impression. This research suggests that mental correction is effective when making deliberative judgments when time is ample, but not particularly effective for other types of responses, such as spontaneous and quickly unfolding behaviors that occur in the midst of an interaction (*but see* Lepore & Brown, 2002).

Inhibition

Inhibition refers to a mental process through which the activity of a concept in the mind, such as a stereotype or evaluation, can be modulated by the activation of a related but countervailing concept(s) or by top-down influences. In their influential chapter, Bodenhausen and Macrae (1998) laid out a theoretical framework for stereotype inhibition that referred to three different mechanisms: lateral inhibition, hierarchical inhibition, and behavioral inhibition (Fig. 4–2).

Lateral inhibition refers to a process by which countervailing nodes in a semantic network are simultaneously activated and exert mutually inhibitory influences. This process is assumed to occur implicitly, and thus it is believed to constitute a "preconscious" or "spontaneous" mechanism of control. In a study by Macrae, Bodenhausen, and Milne (1995), participants were primed with the concept of either "Chinese" or "woman" (or a neutral control prime), through the parafoveal presentation of words for very short durations. Next, participants watched a video of a Chinese woman reading, ostensibly to check the editing quality of the video. Finally, participants completed a lexical decision task, in which target words included stereotypes of both women and Chinese people. The authors found that subjects who were initially primed with "Chinese" were faster to identify Chinese stereotype words and slower to identify female stereotype words, compared with subjects who were primed with "woman." The authors interpreted this effect as evidence for the lateral inhibition of stereotypes.

The notion of lateral inhibition has also been used to explain the effects of a situational context on the activation of racial stereotypes. For example, Wittenbrink, Judd, and Park (2001) primed subjects with a pictures of Black people superimposed over either a church scene or a dangerous-looking street corner. They found that participants' classification of stereotype-related words was facilitated by the street-corner background, but was slowed by the church scene (see also, Barden, Maddux, Petty, & Brewer, 2004). An explanation for these effects is that the simultaneous activation of "church" inhibited negative concepts that would otherwise become activated when primed by a Black face. Lateral inhibition may also provide a mechanism for recent theories of preconscious control, such as "chronic egalitarianism" (e.g., Moskowitz, Gollwitzer, Wasel, & Schaal, 1999; Moskowitz, Soloman, & Talyor, 2000). Moskowitz et al. (1999, 2000) suggested that chronic egalitarian goals may be activated automatically and then exert an inhibitory influence on a racial stereotype, resulting in a reduced net activation of the stereotype.

It is notable that although the notion of lateral inhibition is well-accepted in the social psychology literature, there is less agreement on the plausibility of this mechanism in other literatures. For example, research on the negative priming phenomenon, which provided the strongest test of lateral inhibition, suggests there may be alternative explanations other than inhibition (Milliken et al., 1998; Neill, 1997). In addition, arguments in favor of a lateral inhibition mechanism in social cognition often invoke research from neuroscience showing neuron-level inhibition within the visual system (e.g., in edge detection processes in the primary visual cortex). However, one cannot assume that inhibition at the neuronal level in the primary visual cortex operates in the same way as the processing of complex semantic concepts such as stereotypes and intergroup evaluations. The primary alternative to the idea of lateral inhibition is the possibility that attention shifts to the selected goal or cognition and, as a consequence, activity of the unselected goal or cognition wanes and exerts less influence on behavior. Although lateral inhibition continues to be an appealing and intuitive mechanism of control, more research will be needed to specify the processes that underlie inhibition within a semantic network.

Hierarchical inhibition refers to a top-down form of control in which the targets of control are lower-level thoughts and feelings (i.e., the control of psychological processes). Most contemporary theories of prejudice control assume a hierarchical mechanism of inhibition through which one's egalitarian beliefs "down-regulate" automatically activated stereotypes and race-biased emotions. For example, the higher-order goal to respond without prejudice is often thought to inhibit automatically activated stereotypes or prejudiced feelings (Devine, 1989). This inhibitory mechanism is consistent with recent sociocognitive theorizing on the role of inhibitory processing in hierarchical goal structures (Fishbach, Friedman, & Kruglanksi, 2003; Shah, Friedman, & Kruglanski, 2002).

Although there is a strong precedence for invoking inhibition as a mechanism of control in social cognitive theorizing, unequivocal evidence for lateral and hierarchical inhibition is difficult to obtain. For example, in a study by Macrae et al. (1995), a slowing in categorization of female-related words following the Chinese prime was interpreted as evidence for the inhibition of the "woman" stereotype. However, it is also possible that the response slowing was caused by conflict between co-activated woman and Chinese stereotypes that interfered with the implementation of the behavioral response (Bartholow, Riordan, Saults, & Lust, 2009).

Because the indicator of inhibition is typically behavioral, it is difficult to determine whether the target of inhibition is the psychological process (e.g., a cognitive representation), as assumed by models of hierarchical and lateral inhibition, or the implementation of a behavioral response. Consequently, it is difficult to know whether, at the psychological level, a particular concept was inhibited or the activity of an alternative concept was enhanced.

Behavioral inhibition whereas most research on the inhibition of racial bias has focused on the control of cognitive or affective processes, behavioral inhibition is distinguishable in that it refers to behavior as the target of regulation. There are two general types of behavioral inhibition. The first type refers to inhibitory control, which involves an intentional, top-down process of response control (e.g., Aron, Robbins, & Poldrack, 2004; Logan & Cowan, 1984). Research on inhibitory control has typically examined subjects' performance on tasks such as the Go/No-Go and stop-signal, in which a subject must withhold a prepotent response to a particular cue. Engagement in inhibitory control has been linked to activity in the right inferior prefrontal cortex (PFC) across several studies (Aron et al., 2004), a finding that is consistent with observations of right-frontal cortical asymmetry in situations calling for behavioral avoidance (Davidson, 1992; Harmon-Jones, 2003). Although mechanisms of inhibitory behavioral control have been studied extensively in the fields of cognitive psychology and cognitive neuroscience, little if any research has been conducted on the behavioral inhibition mechanism in the area of intergroup bias (note that some studies have focused on the right inferior PFC in associations with viewing pictures of racial outgroups, but these studies did not assess the regulation of intergroup behavioral responses; Cunningham et al., 2004; Lieberman et al., 2005; Richeson et al., 2003). The lack of research examining behavioral inhibition in the intergroup literature may reflect the field's primary focus on the regulation of cognitive and affective responses, with less attention given to the manifested behavior associated with an intergroup response. Furthermore, behavioral inhibition is typically assessed as withholding of a response, as in the go/no-go task, and therefore it is difficult to design a meaningful psychological task in which successful control of a racial bias can be clearly attributed to inhibition, as opposed to other potential mechanisms such as correction or override.

A second type of behavioral inhibition corresponds to the engagement of attentional processing of a potential threat or unexpected event (Gray & McNaughton, 2000). In this case, the inhibition of behavior does not result from one's intention to stop, but rather occurs as an indirect result of the attentional response. This type of inhibition characterizes Gray's (1987) Behavioral Inhibition System (BIS) and has been associated with conflict-related activity in the anterior cingulate cortex (ACC; Amodio et al., 2008). The ACC neural substrate of this form of control further distinguishes it as a passive response, as opposed to more active forms of control linked to the PFC. In the prejudice literature, the role of passive inhibition due to attentional processing has been examined in detail by Monteith and colleagues (Monteith, 1993; Monteith et al., 2002). According to Monteith's (1993) theory, when a person with low-prejudice attitudes realizes he or she has responded in a prejudiced manner, attention to situational cues associated with bias is increased, and this increase in attention is associated with an abatement of ongoing behavior (i.e., a passive inhibition). The inhibition of behavior is associated with feelings of guilt (Amodio, Devine, & Harmon-Jones, 2007; Monteith, 1993), which serves to promote more careful responding in future interracial situations so as to avoid expressions of bias. Interestingly, this passive form of behavioral inhibition is not a control mechanism in itself—that is, it stops behavior rather than controlling it. Yet it appears to play an important role by facilitating deliberation and goal setting, which then leads to more active forms of control (Amodio et al., 2004, 2007; Monteith, 1993; Monteith et al., 2002).

In summary, there are multiple forms of inhibition that differ on two main dimensions. The first dimension is the target of inhibition—for example, is the target a thought, emotion, or behavior? The second dimension is the degree to which inhibition involves top-down intentional control vs. a more passive form of inhibition. Although some of the specific assumptions of lateral and hierarchical inhibitory processes are difficult to observe experimentally, there is at least clear evidence for the inhibition of behavior.

Mental Suppression

Mental suppression refers to a deliberative effort to inhibit an unwanted thought or feeling (Fig. 4–3). The process of mental suppression may be distinguished from theories of inhibition by the critical role of conscious intention and deliberation. According to Wegner (1994), suppression involves the coordinated operation of two processes. First, a *monitoring* process checks for instances of the unwanted thought. This process is believed to operate continuously in the background of one's mental processing, and thus it is thought to be relatively implicit and automatic. When an unwanted thought is encountered, an *operating* process is engaged to replace the unwanted thought with one that is more acceptable. This second process is believed to be resource-demanding and associated with controlled processing.

Although mental suppression is effective for ridding the mind of an unwanted thought in the short term, it has been shown to fail in the long-term. As the monitoring system searches for the intrusive thought, it may ironically cause the thought to be more strongly activated (i.e., to monitor for thoughts about a "white bear," one must hold an active representation of a white bear in mind for the purpose of identifying it as being intrusive). When active efforts to monitor for unwanted thoughts are relaxed, the intrusive thought is able to express itself more strongly in the mind and behavior. This phenomenon is referred to as post-suppression *rebound*, and rebound effects have been demonstrated to occur on a range of outcome measures (Förster & Liberman, 2005; Wegner, 1994). In what follows, we describe how suppression may be applied to the control of stereotypic thoughts and affective responses to race.

Suppressing Stereotypic Thoughts

In research on intergroup bias, the effects of suppression have been examined primarily as a strategy for controlling the application of stereotypes (Macrae et al., 1994; Monteith, Sherman, & Devine, 1998). Studies of stereotype suppression typically involve two phases. The first phase involves the manipulation of suppression instructions, in which the subject is told to avoid thinking about or expressing racial stereotypes while writing a description of an outgroup member (e.g., an African American person). For example, in the influential series of studies reported by Macrae et al. (1994), participants viewed a picture of an outgroup member (in their case, a skinhead) and were asked to write about a typical day in the skinhead's life. Participants in the suppression condition were instructed to avoid any mention of stereotypes in their description. Participants rarely have trouble successfully responding without stereotypes on this task, and in this regard, the suppression strategy is effective.

The initial success of suppression comes with a cost in the form of stereotype rebound. Stereotype rebound effects are assessed in the second phase of this paradigm, in which participants move on to a new and ostensibly

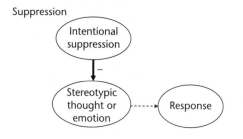

Goal: To rid one's mind of a thought or feeling

Figure 4–3. Diagram of the suppression model of control, in which the goal is to directly downregulate undesired thoughts and/or emotions. Thicker arrows represent stronger connections.

unrelated task. In some studies, participants were asked to read a short description of the daily activities of a race-unspecified person and rate his traits, a procedure adapted from Srull and Wyer's (1979, 1980) famous "Donald" paragraph (Wyer, Sherman, & Stroessner, 1998, 2000). Participants who had previously suppressed any skinhead stereotypes in the first phase were more likely to form stereotype-consistent impressions of the target in the second phase. Other evidence for postsuppression rebound has included faster categorization of stereotypic (vs. nonstereotypical) words in a lexical decision task, and farther seating distance from the purported member of the stereotyped group (Macrae et al., 1994). These studies typically show a pattern of increased bias (i.e., rebound) following suppression.

Research by N. Wyer and her colleagues have examined the roles of motivations to respond without prejudice on the use and effectiveness of stereotype suppression (Wyer, 2007; Wyer et al., 2000). This research has shown that people will spontaneously suppress the use of stereotypes in situations where the expression of stereotypes is inappropriate or might elicit social disapproval. For example, when subjects believed that their "day in the life" description would be evaluated by an African American political organization, they were more likely to suppress their use of stereotypes in the description and showed stronger rebound effects when later forming impressions of a race-unspecified individual, relative to a control condition (Wyer et al., 1998). These "spontaneous" suppressors exhibited a degree of rebound similar to participants who were explicitly instructed to suppress stereotypes. Additional research suggests that individuals who are highly sensitive to external social pressures to respond without prejudice are especially susceptible to rebound effects (Wyer, 2007).

As a strategy of control, suppression stands apart from other mechanisms as being particularly ineffective. Indeed, the point of most research on mental suppression has been to demonstrate that suppression is a faulty strategy for long-term control. Indeed, the postsuppression rebound effect is a well-replicated finding in the social psychology literature. In general, this work shows that people are generally ineffective in deliberately controlling their thoughts. It is important to note that although subjects are typically successful in suppressing stereotypes (i.e., during the suppression phase of the study), it is the behavioral expression that is being controlled, and not necessarily the stereotypic thought or emotion. In fact, research has shown that intrusive thoughts occur during one's active attempt at suppression, further suggesting that suppression operates on behavior but not thoughts (Wegner et al., 1987). Thus, it appears that suppression is not very effective as a means to control stereotypic thoughts in either the short or long term, although it has short-term benefits on controlling deliberative behaviors.

Suppressing Race-Biased Affect

Although very few studies have examined the suppression of prejudice-related affect (Burns, Isbell, & Tyler, 2008), work on emotion suppression outside this literature is relevant. The idea behind emotion suppression is similar to that of mental suppression in that it is a conscious strategy to reduce one's unwanted emotional feelings, such as fear, sadness, disgust, or anger (Gross, 1998; Gross & Levenson, 1993; sometimes contrasted with reappraisal). Like mental suppression, emotion suppression has been shown to be an ineffective means of control. For example, in a study Gross and Levenson (1993), subjects watched a disturbing video showing the surgical amputation of a limb. Subjects asked to suppress their emotional reactions were successful in controlling their outward facial expressions of emotion, yet their emotion-related physiological activity (heart rate, skin conductance) was significantly elevated compared with control subjects who were simply asked to view the video. Thus, as in studies of thought suppression, participants were successful in suppressing the behavioral expression of emotion but failed to regulate the emotion itself. However, it is notable that a preemptive form of regulation—cognitive reappraisal—may be used effectively to reframe the meaning of an emotion-inducing stimulus to modulate its impact (Gross & Levenson, 1993; Ochsner & Gross, 2005). Cognitive reappraisal achieves the broader goal of regulating an emotional response by changing one's perception of the emotion-eliciting stimulus, thereby preventing the unwanted emotional response from occurring in the first place.

Override

Early conceptions of automaticity and control in the cognitive psychology literature posited that automatic and controlled processes represented independent processes that competed for dominance over behavioral output (e.g., Logan, 1984). For example, Logan's (1984) influential model of response control focused on the mechanism through which an unwanted response is replaced by an intended response. That is, Logan's theorizing suggested a "race" model whereby two different action tendencies compete (or race) for the control of behavior. This type of "override," or "replacement," model (Fig. 4–4) suggests that controlled processing functions to strengthen the influence of one response tendency on behavior relative to a competing tendency (Cohen, Dunbar, & McClelland, 1990). In this way, control does not directly down-regulate the unwanted tendency, as suggested by the inhibition model, but rather obviates its effect on behavior. An important feature of the override model is that the behavioral outcome is the primary target of control, rather than thought or emotion.

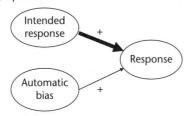

Figure 4–4. Diagram of the override/replace model of control, in which control functions to promote an intended response that overrides the influence of bias. Thicker arrows represent stronger connections.

Devine's (1989) early theorizing on the roles of automaticity and control in intergroup responses suggested an override model of control. Devine (1989) suggested that stereotypic associations of stigmatized group members (e.g., African Americans) were passively acquired in people's memories as a result of exposure to their ubiquitous presence in one's culture. As a result, stereotypes become activated automatically in response to a representation of an outgroup member. Once activated, stereotypes will influence behavior unless controlled processes intervene. Devine (1989) argued that, when in an intergroup situation, people who reject prejudice (i.e., hold low-prejudice attitudes) would engage controlled processes to override the influence of stereotypes on their behavior.

Subsequent research has examined the mechanisms of this override form of control in greater detail (e.g., Amodio et al., 2004, 2007; Monteith, 1993; Monteith et al., 2002). Monteith (1993) proposed a model of self-regulation in the intergroup context that describes how a person deliberately overrides unintended racial biases. In this work, control refers to one's behavioral responses, such as one's choice to laugh at an offensive joke. In our own research, we examined changes in participants' motivation as they transitioned from feeling guilty about having made a prejudiced response to engaging in prosocial responses toward the outgroup (Amodio et al., 2007). This research revealed an increase in a brain activity pattern suggesting that intentions to respond without prejudice are related to approach motivation and control.

Other models of stereotyping imply an override mechanism of control. For example, according to classic dual process models of person perception (e.g., Brewer, 1988; Brewer & Harasty, 1999; Fiske & Neurberg, 1990; Fiske, Lin, & Neuberg, 1999), perceivers' initial impressions of an outgroup member are influenced by category-based beliefs or stereotypes. The perceiver may then begin to form an individuated impression of the target, assuming he or she is sufficiently motivated and that individuating information is available. When individuating information or motivation is absent, stereotypes will continue to influence one's impressions and behavior. Although they differ in some details, these models generally contend that control occurs through the process of individuation, whereby social perceivers avoid stereotypic biases by replacing category-based impressions with more individuated, highly personalized information.

Summary of Traditional Models of Control

Models of correction, inhibition, suppression, and override each provide unique accounts of how intergroup biases may be controlled. Correction appears to operate in appraisals, which in turn are more likely to be expressed in deliberative responses and subjective reports. Inhibition comprises several different subprocesses, and although the mechanisms of lateral and hierarchical inhibition may be less clear, active and passive forms of behavioral inhibition have been shown to play an important role in managing responses online. Similarly, the deliberative process of suppression appears to operate effectively on behavior, but often leads to a rebound effect of the heretofore suppressed thoughts and emotions. Finally, the override model best describes the implementation of an intended (e.g., egalitarian) response in the face of unwanted biasing tendencies. In most real-life situations, however, it is likely that each of these processes operates concurrently and dynamically to regulate one's responses across different behavioral channels.

RECENT ADVANCES IN THEORIES OF CONTROL IN INTERGROUP RELATIONS

To date, the intergroup relations area has provided fertile ground for probing the functions of control in the context of social behavior. By examining behavior in the intergroup context, researchers can study the interplay of multiple psychological factors related to control, such as personal attitudes, normative concerns, social goals, and emotions (e.g., intergroup anxiety). The consideration of these various factors has led researchers in the field of intergroup relations to develop innovative approaches to assessing multiple and interacting components of control, such as new statistical techniques and social neuroscience methods. Here we describe a few of the recent advances in theories of control produced by intergroup researchers.

Mathematical Modeling of Control

Making inferences about underlying cognitive processes from behavior is an enduring challenge in the study of control. This challenge grows when one begins to consider that any given behavioral response may reflect multiple underlying processes. Several researchers have begun to address this challenge by taking a multinomial mathematical approach to analyzing behavioral data from reaction time assessments of race bias (Conrey et al., 2005; Payne, 2001). The mathematical modeling approach involves the transformation of behavioral data in ways that represent the different cognitive processes believed to be involved. An assumption of this approach is that behavioral responses, such as response latencies and error rates, always reflect some combination of various underlying processes. At the simplest level, one might assume that behavioral responses represent some combination of automatic and controlled processes. The process-dissociation procedure provides a relatively simple model designed to estimate separately the role of automatic and controlled processes in a pattern of observed behavior (Jacoby, 1991; Payne, 2001).

More recently, theorists have begun to acknowledge that control itself reflects multiple processes (Amodio et al., 2004; Botvinick et al., 2001; Sherman et al., 2008). For example, our research has dissociated the processes of detecting the need for control (i.e., conflict monitoring) from processes for implementing an intended response (Amodio et al., 2004, 2006, 2008), following Botvinick et al. (2001). Sherman and his colleagues proposed a quadruple-process model (the "Quad Model") that assumes independent processes for *(1)* activation of an association; *(2)* discriminability of the intended response; and *(3)* the ability to overcome bias. A fourth process represents a guessing bias that reflects a form of systematic error (Conrey et al., 2005).

Across a set of studies, Sherman and his colleagues have shown that these different parameters may be manipulated independently and that they have dissociable effects on a range of outcomes (Conrey et al., 2005).

It is notable that the process-dissociation and quadruple-process models generally assume an override form of control. That is, they assume that automatic and controlled processes operate independently, and that controlled processes influence behavior directly rather than through the down-regulation of automatic processes. Indeed, research that has integrated mathematical modeling and neuroscience approaches to examine intergroup responses further supports an override model of control (Amodio et al., 2004, 2008). As the mathematical modeling approach continues to develop in the social psychological study of control, it will need to adapt to advances in functional neuroanatomy and in sociocognitive theories of self-regulation.

Control of Racial Bias Without Awareness: Nondeliberative Mechanisms

Theories regarding the role of awareness in control have evolved substantially over the past ten years, in large part due to advances in neuroscience research examining the neurocognitive mechanisms underlying response control. Ten years ago, Devine and Monteith (1999) wrote that to engage control, "one must be aware of the potential influence of the stereotype" (p. 346). However, since that time, theorizing on this topic has developed, and there are now several studies suggesting that the mechanisms of control may be put into motion very rapidly, with little or no deliberation, and with minimal awareness.

Findings from ERP studies

In an effort to examine the role of deliberation in the engagement and implementation of control, we have examined the time course of controlled processing using ERPs across a set of studies (Amodio et al., 2004, 2006, 2008). In each of these studies, participants completed the weapons identification task (Payne, 2001), in which they quickly classified objects as either

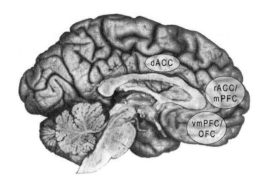

Figure 4–5. Medial view of the brain. Labeled regions include the dorsal anterior cingulate cortex (dACC), the rostral anterior cingulate cortex (rACC)/medial prefrontal cortex (mPFC), and the ventromedial prefrontal cortex (vmPFC)/orbital frontal cortex (OFC).

handguns or handtools after briefly viewing the face of a Black or White person. Consistent with stereotypes of Black people as violent and dangerous (Devine & Elliot, 1995), Black faces facilitated the correct classification of guns and, as a consequence, interfered with the classification of tools, relative to White faces. As a result, subjects also responded more accurately on Black-gun trials, on which the Black faces primed the correct "gun" response, but made more errors on Black-tool trials, on which the Black faces conflicted with the correct "tool" response. This pattern suggested that responding accurately on Black-tool trials requires greater controlled processing relative to Black-gun trials, due to the bias of African American stereotypes.

To assess the time course of controlled processes, we measured participants' brain activity using EEG as they completed the weapons identification task. Past research suggests that the dorsal region of the anterior cingulate cortex (dACC; Fig. 4–5) functions to monitor for conflicts between automatic tendencies and intentional responses (Botvinick et al., 2001). Activation of this region reflects the first step in the control process, which is thought to recruit activity of the PFC associated with higher-order control processes (Fig. 4–6). In the case of the weapons identification task, conflict between prepotent and intended responses occurs on Black-tool trials, where the Black face primes "gun" but the correct response is "tool." On Black-gun trials, by comparison, conflict is low and additional control is unnecessary. Therefore, we expected that ACC activity would be higher on Black-tool trials than on Black-gun trials. In addition, we expected to find evidence of conflict-related ACC activity very early in the response stream that would indicate that (a) mechanisms of control are recruited early and without conscious deliberation, and that (b) the process of controlling responses to race involves a conflict monitoring component that may be independent of more deliberative forms of control.

To test our hypotheses about the timing of the initiation of control, we examined the amplitudes of two ERP components known to arise from activity in the dorsal ACC, known as the error-related negativity (ERN) and the correct-related negativity (CRN; Fig. 4–7). The ERN has been examined in much past work. It is a large negative-polarity wave that is strongest over the fronto-central scalp region. It typically occurs in conjunction with response errors, which represent the strongest degree of conflict between an intended vs. unintended response. That is, by definition, task errors reflect a response conflict. Across studies, we found that ERNs for the high-conflict Black-tool trials were significantly larger than the low-conflict Black-gun trials. Although an ERN was observed on all error trials, as expected given that errors reflect an inherent response conflict, the fact that the ERN was larger on Black-tool trials suggested that the degree of response conflict was higher and that controlled processing was being recruited more intensely on these trials.

Analyses of the CRN component revealed a similar result. The CRN peaks about 100 milliseconds before a correct response is made on the weapons identification task. It may be interpreted as the strength of the signal indicating

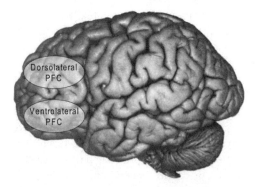

Figure 4–6. Lateral view of the brain indicating the prefrontal cortex (PFC) region. Note that approach-related action control is often associated with left-sided PFC activity, whereas response inhibition has been associated more specifically with right-sided ventrolateral regions of the PFC.

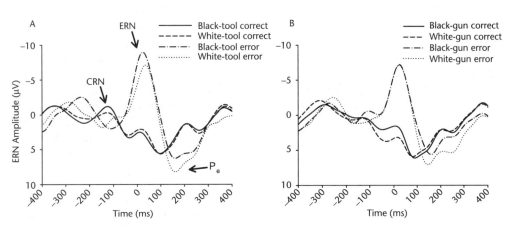

Figure 4–7. Response-locked event-related potential waveforms for correct and incorrect tool (A) and gun (B) trials as a function of accuracy and the race (Black vs. White) race of the face prime. The error-related negativity (ERN), correct-related negativity (CRN), and error positivity (Pe) components are labeled. On the x-axis, zero indicates the time of response.

that greater control is required prior to the implementation of the response (Bartholow et al., 2005). Here, we found that the CRN was largest on the high-conflict Black-tool trials and smallest on the low-conflict Black-gun trials. Furthermore, the timing of the CRN suggested it did not reflect deliberative control (see Fig. 4-7). Amplitudes of both the ERN and CRN were associated with more careful and accurate responding across trials on the task, indicating that subjects with neural systems more sensitive to stereotype-based conflict were better at responding without bias (i.e., in line with task goals). Taken together, these results suggest that the need for control is detected early as a response unfolds and with little, if any, conscious deliberation or awareness. Thus, the control of intergroup responses is not always dependent on deliberation, in contrast to many traditional social psychological models of control.

A further implication of our ERP research is that the mechanisms of control appear to be directed at behavior. That is, the ERN and CRN are response-locked ERP components, which means that they correspond to the implementation of a behavioral response. Additional analyses showed that the CRN and ERN waves were more tightly locked to behavioral responses rather than to the target stimulus, suggesting that these ERP signals reflect behavioral control rather than control over perceptual processing. Moreover, the dorsal ACC, from which these ERP components arise, is situated near the motor cortex and has strong reciprocal connections to regions of the brain involved in orchestrating motor responses. The findings of these ERN/CRN studies suggest that the target of control is behavior and that an override form of control is used.

Implementation Intentions to Respond Without Bias

Research in the goals literature has examined how particular goal strategies may facilitate preconscious forms of control, whereby response regulation is initiated and proceeds with little deliberation or effort. Mendoza, Gollwitzer, and Amodio (in press) explored the use of implementation intentions as a nondeliberative strategy for reducing expressions of implicit race bias. Implementation intentions are if-then plans that link a goal-directed response to a specific situational cue such as, "If I see a Black person, then I will respond more carefully" (Gollwitzer, 1993). The purpose of an implementation intention is to set one's goal and specific behavioral responses ahead of time, so that when a triggering cue is encountered, the intended response can be implemented without deliberation. In this way, implementation intentions engage vigilance for the opportunity to respond without bias, conceptually similar to Monteith's (1993) influential model of the self-regulation of intergroup responses.

In Mendoza et al. (in press), participants completed the Shooter Task, in which Black and White males appear in different background scenes holding either a gun or an innocuous object (Correll et al., 2002). Participants were instructed to "shoot" armed targets but not unarmed targets by pressing buttons labeled "shoot" or "don't shoot" on the computer keyboard. A pattern of race-biased responding is typically elicited by this task, such that unarmed targets are erroneously "shot" more often if they are Black than if they are White. Mendoza et al. provided some participants with an implementation intention ("If I see a gun, then I will shoot" and "If I see an object, then I will not shoot") prior to completing the task. Other participants received simple goal strategy ("I will always shoot a person I see with a gun" and "I will never shoot a person I see with an object") or only the basic task instructions. Compared with the other conditions, those who received the implementation intention showed significantly lower levels of implicit bias in their behavior on the task. Process-dissociation analyses determined that implementation intentions enhanced controlled patterns of responding toward Black targets, but did not affect automatic processes relative to the other strategy conditions. Furthermore, the increase in control afforded by the implementation intentions was not due to response slowing, and thus the execution of control did not appear to rely on deliberative processing. Although additional research will be needed to unpack the mechanisms underlying these effects, implementation intention research has been particularly useful in demonstrating that once an intention is conceived through deliberative processes, its implementation can be executed spontaneously (when the triggering cue is present) with little or no conscious deliberation.

Alternative Mechanisms for Internally vs. Externally Driven Control

So far, our review has focused on intrapersonal forms of control, through which a person attempts to overcome any automatic biases to respond consistently with his or her personal beliefs. Indeed, the literature on the control of intergroup responses has generally focused on this process. However, normative influences sometimes provide a powerful motivation to engage control (Cialdini & Trost, 1998). The literature on intergroup relations has provided an important testing ground for theories of control in response to the combination of personal and normative impetuses.

Cultural norms regarding the expression of prejudice in America have transformed over the last century, and where it was once acceptable—even encouraged—to discriminate against Black people and other social minorities, today's normative standards strongly proscribe such bias (Crosby et al., 1980; Plant & Devine, 1998; Plant et al., 2003). Thus, people may be motivated to control their intergroup responses for internal (personal) as well as external (normative) reasons. Plant and Devine (1998) developed a questionnaire to separately assess internal and external motivations to respond without prejudice, and their research has shown that these two motivations are conceptually dissociable. That is, a person may be motivated by primarily internal or primarily external motivations, by both motivations, or by neither (*see also* Dunton & Fazio, 1997). Furthermore, research suggests that internal vs. external motivations operate differently on behavior, with internal motivations having more stable and consistent effects than external motivations (Deci & Ryan, 2000; Ryan & Connell, 1989). In light of these findings, it is possible that internally and externally driven forms of control involve different underlying mechanisms that may relate to distinct neurocognitive processes.

Interestingly, the notion that different regulatory mechanisms may be involved in internally vs. externally driven forms of control has not been addressed by the neuroscience literature. Yet recent neuroscience studies on empathy and mentalizing are relevant to this issue because they concern the way an individual processes information about others (Frith & Frith, 1999). In neuroscience studies, empathy and mentalizing are typically associated with activity in regions of the mPFC and rostral (r)ACC (*see* Fig. 4–5; Harris & Fiske, 2006; Mitchell, Banaji, & Macrae, 2005; Singer et al., 2004; for a review, *see* Amodio & Frith, 2006). Amodio and Frith (2006) noted that this region of mPFC lies at the intersection of neural regions found to support the monitoring of internal and external self-regulatory signals, respectively. Past research has linked activity in the dorsal ACC to the process of monitoring of one's own actions for response conflicts (Botvinick et al., 2001). By contrast, more anterior regions of the medial frontal wall (i.e., the ventromedial and orbital frontal regions, *see* Fig. 4–5) have been associated with monitoring of reward and punishment signals from the external world (Elliott, Dolan, & Frith, 2000). Amodio and Frith (2006) proposed that activity in the region between these two ends of the medial frontal cortex—the mPFC/rostral ACC (also, anterior paracingulate)—that is typically associated with representations for self-other relationships (Mitchell et al., 2005), may function as a monitoring system for coordinating one's own actions with the anticipated actions of others.

In line with the theorizing of Amodio and Frith (2006), Amodio et al. (2006) hypothesized that activity in the mPFC/rACC may be especially important for externally driven forms of self-regulation, in comparison with dACC regions linked to the monitoring of internal regulatory cues. We tested this hypothesis by measuring ERPs while participants completed the weapons identification task either *(a)* in private or *(b)* while their responses were being monitored by the experimenter for signs of prejudice. As in past work, the ERN component was taken as an index of conflict monitoring processes. To assess activation of the rACC/mPFC, we examined the error-positivity (P_e) wave, a positive-polarity ERP component that immediately follows the ERN and is strongest at fronto-central scalp sites (see Fig. 4–7; Hermann et al., 2004; van Veen & Carter, 2002). It is notable that the P_e is associated with the conscious perception of an unintended response, whereas ERN responses have been shown to be independent of conscious awareness (Nieuwenhuis et al., 2001). Although awareness was not a central part of our theorizing, one might assume that the more complex task of monitoring one's own actions in relation to another's might be more reliant on conscious awareness than monitoring one's internally driven behavior.

In testing our hypothesis, we preselected low-prejudice participants who reported being either high or low in sensitivity to external (normative) pressures to respond without prejudice, using Plant and Devine's (1998) scale. We expected that greater ERN activity would be associated with better response control across all conditions, given that responses are always being monitored relative to internal goals. However, we expected that greater P_e amplitudes would be associated with better response control specifically among highly externally motivated participants who responded in public. Indeed, this pattern was observed (Amodio et al., 2006). These results provided the first evidence that internally vs. externally driven forms of control arise from different underlying neural mechanisms associated with the dACC and rACC/mPFC, respectively. These findings suggest that externally motivated self-regulation may be less stable (cf. Ryan & Connell, 1999) because it relies on a more complex set of monitoring processes for the engagement and guidance of control, compared with internally motivated self-regulation. It is further notable that in past fMRI research, mPFC activity was elicited by very deliberative types of tasks (e.g., self-reflection, introspecting about other people), but responses on our task were made quite rapidly and, most likely, with very little deliberation.

Control of Race Bias in Interracial Interactions: Intergroup Anxiety Effects on Control?

A long-standing question in the intergroup relations literature concerns how anxiety, which is often elicited in interracial situations (Stephan & Stephan, 1985), may affect mechanisms of control. In Black-White interactions, for example, a White person may worry about as appearing to be racist and may monitor him- or herself for unintentional expressions of bias (Gaertner & Dovidio, 1986; Ickes, 1984; Plant & Devine, 2003; Shelton, 2003). A Black person might worry about being the target of discrimination or social rejection, or about behaving in a way that might corroborate a stereotype (Schmader & Johns, 2003; Steele & Aronson, 1995). The experience of intergroup anxiety is also accompanied by changes in physiology associated with the stress response (Mendes, Blascovich, Lickel, & Hunter, 2002; Rankin & Campbell, 1955; Vrana & Rollock, 1996; for a review, see Guglielmi, 1999), which may further affect a person's behavior in intergroup situations.

Can intergroup anxiety impair control in interracial interactions? Nearly half a century's work suggests that anxiety can impact performance on a variety of tasks (Easterbrook, 1959; see also Baumeister & Showers, 1986). However, the idea that anxiety stemming from an interracial interaction might affect control had not been examined until recently (e.g., Richeson & Shelton, 2003; see also Amodio et al., 2006; Inzlicht, McKay, & Aronson, 2006; Lambert et al., 2003; Schmader & Johns, 2003; Vorauer, 2006). To date, findings from this literature have been somewhat mixed. Although manipulations intended to create intergroup anxiety have been associated with later impairments in response control (Lambert et al., 2003; Richeson & Shelton, 2003; Schmader & Johns, 2003; Trawalter & Richeson, 2005), self-report measures of subjective anxiety have not been found to predict these impairments (Lambert et al., 2003; in other studies, subjective anxiety was either not measured or not reported). For example, in a study by Lambert et al. (2003), participants completed the weapons identification task, which requires the inhibition of stereotype-driven responses on some trials. Participants who thought they would discuss their task responses with a group of peers immediately afterward reported heightened anxiety and showed impaired control in their task performance, compared with control participants. However, their degree of self-reported anxiety was not correlated with their degree of control on the task. Thus, the theoretical connection between anxiety and specific mechanisms of control remains unclear.

To address the question of how intergroup anxiety affects the control of intergroup bias, Amodio (2009) recently proposed and tested a physiological pathway through which intergroup anxiety may impair control. This research considered the pathways through which the neurochemical response to threat may modulate neural regions associated with the engagement of cognitive control (e.g., the ACC; Aston-Jones & Cohen, 2005; Morilak et al., 2005). To test this hypothesis, White participants interacted with either a Black or White partner to discuss issues concerning race and prejudice. As part of the interaction, participants completed the weapons identification task (described above), a reaction-time measure that provides indices of automatic and controlled processes associated with the inhibition of racial stereotypes. Measures of salivary cortisol were used as indicators of threat-related norepinephrine, given that cortisol release is triggered by norepinephrine signaling of hypothalamic-pituitary-adrenal axis (McEwen, 1998; Sapolsky, 1996). Amodio (2009) found that greater cortisol reactivity associated with the interracial interaction was associated with impaired control on the weapons identification task. This relationship was not found in the same-race interaction condition. Moreover, subjective reports of anxiety were not associated with response control, as in past work (e.g., Lambert et al., 2003). These results suggested a neuroendocrine model in which intergroup anxiety can affect control through physiological pathways, rather than through subjective experience.

Integration of Neuroscience with Research on Prejudice Control

One of the most exciting developments in research on control in the intergroup bias literature has been the integration of neuroscientific approaches (Amodio, 2008; Eberhardt, 2005; Ito & Bartholow, 2009); for general reviews of social neuroscience, see Blakemore, Winston, & Frith, 2004; Cacioppo, Visser, & Pickett, 2005; Lieberman, 2005; Ochsner, 2007). Within this new literature, researchers have begun to use neuroimaging methods in an attempt to assess aspects of intergroup bias and the engagement of controlled processes more directly. Much of the extant work on this topic has focused on mapping aspects of racial bias and control onto activity in specific brain structures. For example, initial social neuroscience research on intergroup bias examined the role of the amygdala in automatic or implicit evaluations of outgroup (vs. ingroup) members (Amodio, Harmon-Jones, & Devine, 2003; Phelps et al., 2000). These initial findings provided a foothold from which social psychologists could begin to use models of neural function to build theories of how intergroup responses may be controlled.

Building on previous findings from cognitive neuroscience suggesting that the ACC and lateral PFC were involved in control, many researchers sought to show that the control of prejudice was associated with activity in these frontal cortical areas. In some of these studies, participants passively viewed pictures of Black or White individuals, and activations within these brain regions were observed (e.g., Cunningham et al., 2004; Richeson et al., 2003). Although passive face viewing tasks are a commonly used method in fMRI studies, the limitation of this method is that it does not clearly require control, nor does it provide a behavioral measure that could be interpreted as involving control. Therefore, questions remain as to whether the frontal cortical activations observed in these studies were actually associated with control (as opposed to any of the many other processes associated with activity in the same region; Gilbert et al., 2006). Nevertheless, the activations observed in the ACC and PFC during such tasks have been relatively consistent.

As fMRI research in this area has advanced, it has begun to integrate "control" tasks that require participants to overcome any racial biases to respond in a desired way. Recent work by Lieberman et al. (2005) took an initial step toward examining control in the intergroup context. Participants in their study viewed a target face that was Black or White. On some trials, participants had to indicate whether the face matched one of two reference faces (one Black and one White). On other trials, the participant had to match the face with the appropriate ethnic label (African American or Caucasian American). Lieberman et al. (2005) surmised that matching a face to a label requires more control than matching a face to other faces and, in line with this view, observed stronger PFC activity during label vs. match trials. Although it is not clear whether the labeling task involves response control, or the control of racial bias more specifically, this study begins to approach a situation in which a person must override a bias to respond without bias. Future fMRI research will benefit from the inclusion of behavioral tasks that clearly involve the need to override race-biased tendencies that lead to unintended responses. Such tasks would allow researchers to infer that brain activations in "control" areas are in fact contributing to responses that can be interpreted as the regulation of intergroup bias. It is also notable that this broader class of studies is concerned with the engagement of controlled processes, even if the criterion of control (e.g., in behavior) is not always available, or if the type of control that is inferred from brain activity may not necessarily relate to the type of control with which the intergroup literature is concerned (e.g., the implementation of a nonprejudiced response).

To date, the most direct social neuroscience investigations of control have been conducted using neuroimaging methods combined with reaction-time tasks in which a participant must override automatic stereotypes to deliver an intended response. These tasks are appropriate for probing mechanisms of control because they require a clear response made in the context of response-incongruent racial stereotypes. As described above, we have conducted a program of ERP research examining the role of the ACC in detecting the need for control of intergroup bias (Amodio et al., 2004, 2006, 2008). We have applied this neuroscience model of control to begin to address some longstanding questions about prejudice in social psychology. For example, why do some people with non-prejudice attitudes show signs of bias in their spontaneous behaviors (e.g., when making rapid responses on a racial categorization task)? Our work has shown that these egalitarians vary in the sensitivity with which the ACC responds to the need for control. Thus, although all egalitarians are effective in responding without bias in deliberative responses (e.g., self-reports), only those who are highly sensitive to the activation of biased tendencies are effective in implementing control in more spontaneous responses. Other research by Bartholow, Dickter, and Sestir (2006) has begun to use ERPs to examine the role of the PFC in implementing a desired, non-prejudiced response, using experimental tasks that are appropriate for examining control in the context of racial bias.

Control in Intergroup Relations: Remaining Issues

Despite a large literature on the mechanisms of control in intergroup relations, several important theoretical issues concerning models of control remain to be addressed. Oftentimes, a model of control that was developed to explain a basic cognitive task, such as the Stroop Task, does not translate perfectly to the rich environment of social relations. As a result, the theoretical foundations of control in social cognition may not provide an accurate characterization of how behavior is regulated in actual social behavior. Here, we discuss three major issues concerning (a) the target of control, (b) the necessity of a top-down mechanism, and (c) the applicability of cognitive models of control to social behavior.

What's Being Controlled?

Traditional accounts of control do not always specify their target—that is, whether control operates on thoughts, emotions, or behavior (or on some other process), yet studies of control often focus on the control of either stereotypic thoughts, unwanted negative emotions, or discriminatory behavior. Interestingly, most theories of control in the ingroup literature, and in social psychology more broadly, do not take a strong stance on the specific target of controlled processing, with most models assuming that the same mechanisms are capable of regulating any type of psychological or behavioral process. However, we see the issue of what is being controlled as critical, for both the advancement of theory and the extension of research into practice in the field of intergroup relations.

As we discussed at the beginning of this chapter, modern social psychological theorizing often assumes that the chief function of the mind is to represent knowledge. As such, the chief function of control would be to adjust that knowledge to enhance its accuracy and utility. An alternative view is that the chief function of the mind is to coordinate action, in which case the primary function of control would be to orchestrate adaptive behavior. Research taking the former "cognitive" view often assumes that thoughts and emotions are the targets of control, and behavioral responses on experimental tasks are interpreted in terms of underlying psychological constructs. Research taking the latter "motivational" view often assumes that behavior is the target of control, and behavioral outcome variables are more often interpreted at face value. Both perspectives are important and there is clearly interaction between motivational and cognitive processes. But when it comes to understanding the target of control, is one view more correct?

Let us first consider research on suppression as a mechanism of control. Here, the target of control is an unwanted thought or emotion. Tests of mental suppression typically involve measures of behavioral as well as cognitive or emotional variables. Across studies, the evidence reveals that behavioral expressions are controlled quite effectively, whereas attempts to suppress thoughts and/or emotions fail, or worse, backfire in the form of "rebound" (Gross & Levenson, 1993; Wegner, 1994). Thus, despite that theoretical models of suppression focus on the control of thoughts and emotions, control in these paradigms appears to operate only on behavior.

As we described in the first section of this chapter, it is difficult to distinguish the target of control in inhibition paradigms because assessments of inhibition rely almost exclusively on behavioral indicators (e.g., reaction times). Although different patterns of behavior are often interpreted as providing indices of the underlying thought or emotion, it is possible that control operates on behavioral output rather than on the underlying intrapsychic process. Because of this ambiguity, the effects of inhibition on thoughts or emotions are unclear. Inhibition paradigms that assume that behavior is the target of control have the advantage of avoiding this interpretational ambiguity.

The mental correction model of control assumes that one's goal is to maintain an accurate cognitive representation of an object. Broadly speaking, corrective processes are propositional in nature and affect judgments through the application of a rule ("apples are healthy, and so I like apples more than chocolate"). Is this control, per se? It may be that the "controlled" aspect of mental correction is most evident when adjusting one's behavioral output to align with one's belief, such as when increasing a student's grade slightly to compensate for the fact that you find him annoying. Alternatively, "control" may reflect an added proposition to a representation of an attitude object that specifies the appropriate correction (i.e., a rule stored in memory that is to be applied when thoughts about the attitude object are activated).

Can neuroscience inform our question about the target of control? Several recent fMRI studies have sought to examine control in the context of racial bias. In most of these, regions of the PFC that are broadly associated with cognitive control (in addition to several other functions; Gilbert et al., 2006), have been linked to the simple viewing of Black (vs. White) faces (e.g., Cunningham et al., 2004; Richeson et al., 2003), or when attempting to assign an ethnic label to a face (Lieberman et al., 2005). Setting aside the issue of whether these tasks are appropriate for examining control, one can ask whether the "control"-related PFC activations target emotional reactions to the face or whether they target a behavioral response. Relevant to this question, multiple studies have observed a negative correlation between lateral regions of PFC and the amygdala, a neural structure involved in the perception of threat, on different task trials (Cunningham et al., 2004; Lieberman et al., 2005). Such correlations have been interpreted as evidence of control having a direct "down-regulatory" effect on an emotional response. Although the direction of this effect is widely assumed to run from the PFC to the amygdala, it is possible that the direction runs in the opposite way, or that a third variable affects both the PFC and amygdala. More importantly, however, neuroanatomical studies of the rhesus macaque, a monkey with a close homologue of the human brain, indicate that there are few if any direct connections between the lateral PFC and the amygdala, favoring the idea that a third variable is responsible for the correlation (Gabbot et al., 2005; Ghashghaei & Barbas, 2002). By contrast, the lateral PFC has very strong connections to regions involved in goal-directed behavior, such as the basal ganglia and the motor cortex (Lehéricy et al., 2004). This pattern of neuroanatomy corroborates the view from behavioral studies that control operates primarily on behavior rather than having direct effects on emotion or thought.

Does Control Always Involve a Top-Down Influence?

It is widely assumed that self-regulation in intergroup relations requires the top-down control of unwanted thoughts, emotions, and behaviors. However, in light of the previous section, one may wonder whether top-down control is the typical mechanism through which a person responds without prejudice. Research on humanization processes in intergroup relations suggests that an important aspect of prosocial behavior involves seeing another person as a fellow human being who is worthy of interaction (Harris & Fiske, 2006; Fiske et al., 2002). Other related research on Theory of Mind suggests that engaging in mentalizing may represent a mechanism of prosocial engagement that does not require the notion of a "top-down" form of control (Amodio & Frith, 2006).

Take for example a situation in which a White person is about to interact with a Black visitor. The White person may be worried about appearing biased in some way and may also feel intimidated by the Black visitor (vis-à-vis the negative Black stereotype). During their meeting, how does the White person manage to engage in a normal and respectful interaction? It seems unlikely that he would focus his efforts on suppressing any possible bias (e.g., telling himself not to think stereotypic thoughts about basketball and crime). Rather, prosocial intergroup interactions may simply involve engaging with the person. After one begins to build a personal rapport with the visitor, the visitor becomes individuated and less threatening, and any feelings of threat are likely to fade as a result. In this way, any unwarranted biases become irrelevant (Amodio, 2008). This "engagement" model of intergroup self-regulation builds on previous theories of individuation in person perception, but shifts the focus of this process from person perception to self-regulation. We expect that as more research is conducted on the interpersonal dynamics of intergroup relations, theories of control and self-regulation will continue to move beyond the confines of top-down models.

Studying Control in a Real Social Interaction

A tacit assumption in research on intergroup bias is that the same mechanisms of control that operate in "cognitive" tasks, such as the Stroop color-naming task, can be mapped onto the regulation of real-life social interactions. However, there are some critical differences. In typical cognitive tasks, a person has a clear response goal. Controlled processes allow individuals to make a goal-consistent response despite tendencies that may bias them toward unwanted responses. But in a social interaction, the specific goal is not always clear (Plant & Devine, 2008). Although one may have the goal of having a smooth interaction, how exactly is that accomplished? And how would successful control be evaluated? Oftentimes, one's goal in an interaction may be to avoid showing signs of prejudice. However, this type of goal—to avoid a mismatch to a desired end state (Higgins, 1997)—is notoriously difficult to achieve, in part because the only concrete index of progress is evidence of failure. Thus, applying traditional models of top-down control to the context of social behavior (e.g., an interracial interaction) is problematic. A major challenge for future work will be to find common ground between mechanisms of control operating in "cognitive" laboratory tasks and in actual social behavior. Researchers will also need to remain open-minded about the different ways that behavior may be regulated, as we may find that engagement forms of regulation are more relevant than top-down control in the context of intergroup relations.

Conclusion

Issues of control have been central to the study of intergroup relations for decades. Theories of control have advanced our understanding of how individuals judge and interact with members of stigmatized groups, despite the potential biasing effects of stereotypes and prejudices. At the same time, the domain of intergroup relations has provided a rich context for examining how mechanisms of control interact with a host of other psychological variables involved in social behavior, such as attitudes, social goals, and normative pressures. This line of research has refined and extended theories that were originally developed to understand basic cognitive processes and impression formation to understand mechanisms of self-regulation as they relate to real-life social interactions. Thus, the literature on intergroup relations has made (and continues to make) substantial contributions to psychologists' understanding of control.

In this chapter, we reviewed models of control that are most prevalent in the intergroup bias literature, noting their advantages and disadvantages as well as the situations in which they are most relevant. We also described what we view as cutting-edge advances in conceptions of control in this literature, including innovative methodological approaches and new mechanisms of externally driven control and pro-social engagement. As research begins to examine the role of control in actual social interactions, we expect a host of new advances in the quest to understand the mechanisms of control in social behavior. Most importantly, we see the intergroup domain as a valuable context for continued innovation in basic theories of control and the regulation of social behavior.

References

Allport, G. W. *The nature of prejudice*. Reading, MA: Addison-Wesley, 1954.

Amodio, D. M., & Frith, C. D. Meeting of minds: The medial frontal cortex and social cognition. Nat Rev Neurosci 2006; 7: 268–277.

Amodio, D. M. Intergroup anxiety effects on the control of racial stereotypes: A psychoneuroendocrine analysis. J Exp Soc Psychol 2009; 45: 60–67.

Amodio, D. M. The social neuroscience of intergroup relations. In: Stroebe, W., & Hewstone, M. (Eds.), *European review of social psychology*, Volume 19. Hove, UK: Psychology Press, 2008: pp. 1–54.

Amodio, D. M., Harmon-Jones, E., Devine, P. G., Curtin, J. J., Hartley, S. L., & Covert, A. E. Neural signals for the detection of unintentional race bias. Psychol Sci 2004; 15: 88–93.

Amodio, D. M., Kubota, J. T., Harmon-Jones, E., & Devine, P. G. Alternative mechanisms for regulating racial responses according to internal vs. external cues. Soc Cogn Affect Neurosci 2006; 1: 26–36.

Amodio, D. M., Devine, P. G., & Harmon-Jones, E. Individual differences in the regulation of intergroup bias: The role of conflict monitoring and neural signals for control. J Pers Soc Psychol 2008; 94: 60–74.

Amodio, D. M., Devine, P. G., & Harmon-Jones, E. A dynamic model of guilt: Implications for motivation and self-regulation in the context of prejudice. Psychol Sci 2007; 18: 524–530.

Amodio, D. M., Harmon-Jones, E., & Devine, P. G. Individual differences in the activation and control of affective race bias as assessed by startle eyeblink responses and self-report. J Pers Soc Psychol 2003; 84: 738–753.

Amodio, D. M., Master, S. L., Yee, C. M., & Taylor, S. E. Neurocognitive components of the behavioral inhibition and activation systems: Implications for theories of self-regulation. Psychophysiology 2008; 45: 11–19.

Amodio, D. M., & Mendoza, S. A. Implicit intergroup bias. In: Gawronski, B., & Payne, B. K. (Eds.), *Handbook of implicit social cognition*. New York: Guilford, in press.

Aron, A. R., Robbins, T. W., & Poldrack, R. A. Inhibition and the right inferior frontal cortex. Trends Cogn Sci 2004; 8: 170–177.

Aston-Jones, G., & Cohen, J. D. An integrative theory of locus coeruleus-norepinephrine function: Adaptive gain and optimal performance. Ann Rev Neurosci 2005; 28: 403–450.

Baddeley, A. D. *Working memory*. New York: Oxford University Press, 1986.

Bargh, J. A. The Four Horsemen of automaticity: Awareness, efficiency, intention, and control in social cognition. In: Wyer, R. S., Jr., & Srull, T. K. (Eds.), *Handbook of social cognition* (2nd ed.). Hillsdale, NJ: Erlbaum, 1994: pp. 1–40.

Bartholow, B. D., Pearson, M. A., Dickter, C. L., Sher, K. J., Fabiani, M., Gratton, G. Strategic control and medial frontal negativity: Beyond errors and response conflict, Psychophysiology 2005; 42: 33–42.

Bartholow, B. D., Riordan, M. A., Saults, J. S., & Lust, S. A. Psychophysiological evidence of response conflict and strategic control of responses in affective priming. J Exp Soc Psychol 2009; 45: 655–666.

Bartholow, B. D., Dickter, C. L., & Sestir, M. A. Stereotype activation and control of race bias: Cognitive control of inhibition and its impairment by alcohol. J Pers Soc Psychol 2006; 90: 272–287.

Baumeister, R. F., & Showers, C. J. A review of paradoxical performance effects: Choking under pressure in sports and mental tests. Eur J Soc Psychol 1986; 16: 361–383.

Bodenhausen, G. V., & Macrae, C. N. Stereotype activation and inhibition. In: Wyer, R. S. Jr. (Ed.), *Advances in social cognition*, Vol. 11. Mahwah, NJ: Erlbaum, 1998: pp. 1–52.

Botvinick, M. M., Braver, T. S., Barch, D. M., Carter, C. S., & Cohen, J. D. Conflict monitoring and cognitive control. Psychol Rev 2001; 108: 624–652.

Brewer, M. B., & Harasty, S. A. Dual processes in the cognitive representation of persons and social categories. In: Chaiken, S. & Trope, Y. (Eds.), *Dual process theories in social psychology*. New York, NY: Guilford Press, 1999: pp. 255–270.

Brewer, M. B. A dual process model of impression formation. In: Wyer, R. S., Jr., & Srull, T. K. (Eds.), *Advances in social cognition* (Vol. 1). Hillsdale, NJ: Erlbaum, 1988: p. 36.

Burns, K. C., Isbell, L. M., & Tyler, J. M. Suppressing emotions toward stereotyped targets: The impact on willingness to engage in contact. Soc Cogn 2009; 26: 276–287.

Cacioppo, J. T., Visser, P. S., & Pickett, C. L. *Social neuroscience: People thinking about thinking people*. Cambridge: MIT Press, 2005.

Cialdini, R. B., & Trost, M. R. Social influence: Social norms, conformity and compliance. In: Gilbert, D. T., Fiske, S. T., & Lindzey, G. (Eds.), *The handbook of social psychology*, Vol. 2. New York: McGraw-Hill, 1998: pp. 151–192.

Cohen, J. D., Dunbar, K., & McClelland, J. L. On the control of automatic processes: A parallel distributed processing model of the Stroop effect. Psychol Rev 1990; 97: 332–361.

Conrey, F. R., Sherman, J. W., Gawronski, B., Hugenberg, K., & Groom, C. Separating multiple processes in implicit social cognition: The Quad-Model of implicit task performance. J Pers Soc Psychol 2005; 89: 469–487.

Corneille, O., Vescio, T. K., & Judd, C. M. Incidentally activated knowledge and stereotype based judgments: A consideration of primed construct-target attribute match. Soc Cogn 2000; 18: 377–399.

Correll, J., Urland, G. R., & Ito, T. A. Shooting straight from the brain: Early attention to race promotes bias in the decision to shoot. J Exp Soc Psychol 2006; 42: 120–128.

Crosby, F., Bromley, S., & Saxe, L. Recent unobtrusive studies of Black and White discrimination and prejudice: A literature review. Psychol Bull 1980; 87: 546–563.

Cunningham, W. A., Johnson, M. K., Raye, C. L., Gatenby, J. C., Gore, J. C., & Banaji, M. R. Separable neural components in the processing of Black and White faces. Psychol Sci 2004; 15: 806–813.

Damasio, A. R. *Descartes' error*. New York, NY: Penguin Putnam, 1994.

Darley, J. M., & Gross, P. H. A hypothesis-confirming bias in labeling effects. J Pers Soc Psychol 1983; 44: 20–33.

Davidson, R. J. Emotion and affective style: Hemispheric substrates. Psychol Sci 1992; 3: 39–43.

Deci, E. L., & Ryan, R. M. The "what" and "why" of goal pursuits: Human needs and the self-determination of behavior. Psychol Inq 2000; 11: 227–268.

Devine. P. G., & Elliot, A. J. Are racial stereotypes really fading? The Princeton Trilogy revisited. Pers Soc Psychol Bull 1995; 21: 1139–1150.

Devine, P. G., & Monteith, M. M. Automaticity and control in stereotyping. In: Chaiken, S., & Trope, Y. (Eds.), *Dual process theories in social psychology*. New York: Guilford Press, 1999: pp. 339–360.

Devine, P. G. Prejudice and stereotypes: Their automatic and controlled components. J Pers Soc Psychol 1989; 56: 5–18.

Dijksterhuis, A., & van Knippenberg, A. The relation between perception and behavior, or how to win a game of Trivial Pursuit. J Pers Soc Psychol 1998; 74: 865–877.

Dunton, B. C., & Fazio, R. H. An individual difference measure of motivation to control prejudiced reactions. Pers Soc Psychol Bull 1997; 23: 316–326.

Easterbrook, J. A. The effect of emotion on cue utilization and the organization of behavior. Psychol Rev 1959; 66: 183–201.

Eberhardt, J. L. Imaging race. Am Psychol 2005; 60: 181–190.

Elliott, R., Dolan, R. J., & Frith, C. D. Dissociable functions in the medial and lateral orbitofrontal cortex: Evidence from human neuroimaging studies. Cereb Cortex 2000; 10: 308–317.

Fazio, R. H. Multiple processes by which attitudes guide behaviour: The MODE model as an integrative framework. In: Zanna, M. P. (Ed.), *Advances in experimental social psychology* (Vol. 23). New York, NY: Academic Press, 1990: pp. 75–109.

Fishbach, A., Friedman, R. S., & Kruglanski, A. W. Leading us not unto temptation: Momentary allurements elicit overriding goal activation. J Pers Soc Psychol 2003; 84: 296–309.

Fiske, S. T., & Neuberg, S. L. A continuum model of impression formation: Form category-based to individuating process as a function of information, motivation, and attention. In: Zanna, M. P. (Ed.), *Advances in experimental social psychology* (Vol. 23). San Diego, CA: Academic Press, 1990: pp. 1–108.

Fiske, S. T., Lin, M. H., & Neuberg, S. L. The Continuum Model: Ten years later. In: Chaiken, S., & Trope, Y. (Eds.), *Dual process theories in social psychology*. New York: Guilford, 1999: pp. 231–254.

Förster, J., & Liberman, N. A motivational model of post-suppressional rebound. *Eur Rev Soc Psychol* 2005; 15: 1–32.

Freud, S. *Civilization and its discontents*. Oxford: Hogarth, 1930.

Freud, S. *Civilization and its discontents* (J. Strachey, Ed. & Trans.). Oxford, England: Hogarth. (Original work published 1930), 1961.

Frith, C. D., & Frith, U. Interacting minds—a biological basis. Science 1999; 286: 1692–1695.

Gabbott, P. L., Warner, T. A., Jays, P. R., Salway, P., & Busby, S. J. Prefrontal cortex in the rat: Projections to subcortical autonomic, motor, and limbic centers. J Comp Neurol 2005; 492: 145–177.

Gaertner, S. L., & Dovidio, J. F. The aversive form of racism. In: Dovidio, J. F., & Gaertner, S. L. (Eds.), *Prejudice, discrimination, and racism*. San Diego: Academic Press, 1986: pp. 61–89.

Gaertner, S. L., & McLaughlin, J. P. Racial stereotypes: Associations and ascriptions of positive and negative characteristics. Soc Psychol Q 1983; 46: 23–30.

Gawronski, B., & Bodenhausen, G. V. Associative and propositional processes in evaluation: An integrative review of implicit and explicit attitude change. Psychol Bull 2006; 132: 692–731.

Ghashghaei, H. T., & Barbas, H. Pathways for emotion: Interactions of prefrontal and anterior temporal pathways in the amygdala of the rhesus monkey. Neuroscience 2002; 115: 1261–1279.

Gilbert, D. T. What the mind's not. In: Chaiken, S., & Trope, Y. (Eds.), *Dual-process theories in social psychology*. New York: Guilford Press, 1999: pp. 3–11.

Gilbert, D. T., & Hixon, J. G. The trouble of thinking: Activation and application of stereotypic beliefs. J Pers Soc Psychol 1991; 60: 509–517.

Gilbert, D. T., Pelham, B. W., & Krull, D. S. On cognitive busyness: When person perceivers meet persons perceived. J Pers Soc Psychol 1988; 54: 733–740.

Gilbert, S. J., Spengler, S., Simons, J. S., et al. W. Functional specialization within rostral prefrontal cortex (Area 10): A meta-analysis. J Cogn Neurosci 2006; 18: 932–948.

Gollwitzer, P. M. Goal achievement: The role of intentions. Eur Rev Soc Psychol 1993, 4: 141–185.

Gray, J. A., & McNaughton, N. *The neuropsychology of anxiety*. London: Oxford University Press, 2000.

Gray, J. A. *The psychology of fear and stress* (2nd ed.). London: Cambridge University Press, 1987.

Greenwald, A. G., Poehlman, T. A., Uhlmann, E., & Banaji, M. R. Understanding and using the Implicit Association Test: III. Meta-analysis of predictive validity. J Pers Soc Psychol 2009; 97: 17–41.

Gross, J. J., & Levenson, R. W. Emotional suppression: Physiology, self-report, and expressive behavior. J Pers Soc Psychol 1993; 64: 970–986.

Gross, J. J. Antecedent- and response-focused emotion regulation: Divergent consequences for experience, expression, and physiology. J Pers Soc Psychol 1998; 74: 224–237.

Guglielmi, R. S. Psychophysiological assessment of prejudice: Past research, current status, and future directions. Pers Soc Psychol Rev 1999; 3: 123–157.

Harber, K. Feedback to minorities: Evidence of a positive bias. J Pers Soc Psychol 1998; 74: 622–628.

Harmon-Jones, E. Clarifying the emotive functions of asymmetrical frontal cortical activity. Psychophysiology 2003; 40: 838–848.

Harris, L. T., & Fiske, S. T. Dehumanizing the lowest of the low: Neuroimaging responses to extreme out-groups. Psychol Sci 2006; 17: 847–853.

Heider, F. *The psychology of interpersonal relations*. New York, NY: Wiley, 1958.

Herrmann, M. J., Rommler, J., Ehlis, A. C., Heidrich, A., & Fallgatter, A. J. Source localization (LORETA) of the error-related-negativity (ERN/N_e) and positivity (P_e). Cogn Brain Res 2004; 20: 294–299.

Higgins, E. T. Beyond pleasure and pain. Am Psychol 1997; 52: 1280–1300.

Higgins, E. T., Rholes, W. S., & Jones, C. R. Category accessibility and impression formation. J Exp Soc Psychol 1977; 13: 141–154.

Ickes, W. Compositions in black and white: Determinants of interaction in interracial dyads. J Pers Soc Psychol 1984; 47: 330–341.

Ito, T. A., & Bartholow, B. D. The neural correlates of race. Trends Cogn Sci 2009; 13: 524–531.

Jacoby, L. L. A process dissociation framework: Separating automatic from intentional uses of memory. J Mem Lang 1991; 30: 513–541.

Jones, E. E., & Davis, K. E. From acts to dispositions: The attribution process in person perception. In: Berkowitz, L. (Ed.), *Advances in experimental social psychology*, Vol. 2. Orlando, FL: Academic Press, 1965.

Jost, J. T., Rudman, L. A., Blair, I. V., et al. The existence of implicit bias is beyond reasonable doubt: A refutation of ideological and methodological objections and executive summary of ten studies that no manager should ignore. Research in Organizational Behavior, in press.

Kawakami, K., Dovidio, J. F., & van Kamp, S. Kicking the habit: Effects of nonstereotypic association training and correction processes on hiring decisions. J Exp Soc Psychol 2005; 41: 68–75.

Kawakami, K., Dovidio, J. F., & van Kamp, S. The impact of counterstereotypic training and related correction processes on the application of stereotypes. Group Process Intergroup Relat 2007; 10: 139–156.

Kelley, H. H. Attribution in social psychology. Nebr Symp Motiv 1967; 15: 192–238.

Lambert, A. J., Payne, B. K., Jacoby, L. L., Shaffer, L. M., Chasteen, A. L., & Khan, S. R. Stereotypes as dominant responses: On the "social facilitation" of prejudice in anticipated public contexts. J Pers Soc Psychol 2003; 84: 277–295.

Lambert, A. J., Khan, S., Lickel, B., & Fricke, K. Mood and the correction of positive vs. negative stereotypes. J Pers Soc Psychol 1997; 72: 1002–1116.

Lepore, L., & Brown, R. The role of awareness: Divergent automatic stereotype activation and implicit judgment correction. Soc Cogn 2002; 20: 321–351.

Lieberman, M. D., Hariri, A., Jarcho, J. M., Eisenberger, N. I., & Bookheimer, S. Y. An fMRI investigation of race-related amygdala activity in African-American and Caucasian-American individuals. Nat Neurosci 2005; 8: 720–722.

Logan, G. D., & Cowan, W. B. On the ability to inhibit thought and action: A theory of an act of control. Psychol Rev 1984; 91: 295–327.

Macrae, C. N., Bodenhausen, G. V., Milne, A. B., & Jetten, J. Out of mind but back in sight:

Stereotypes on the rebound. J Pers Soc Psychol 1994; 67: 808–817.

Macrae, C. N., Bodenhausen, G. V., & Milne, A. B. The dissection of selection in person perception: Inhibitory mechanisms in social stereotyping. J Pers Soc Psychol 1995; 69: 397–407.

McClelland, J. L., & Rumelhart, D. E. Distributed memory and the representation of general and specific information. J Exp Psychol Gen 1985; 114: 159–188.

McEwen, B. S. Protective and damaging effects of stress mediators. N Engl J Med 1998; 338: 171–179.

Mendes, W. B., Blascovich, J., Lickel, B., & Hunter, S. Cardiovascular reactivity during social interactions with White and Black men. Pers Soc Psychol Bull 2002; 28: 939–952.

Mendoza, S. A., Amodio, D. M., & Gollwitzer, P. M. Reducing the expression of implicit stereotypes: reflexive control through implementation intentions. Pers Soc Psychol Bull, in press.

Milliken, B., Joordens, S., Merikle, P. M., & Seiffert, A. E. Selective attention: A re-evaluation of the implications of negative priming. Psychol Rev 1998; 105: 203–229.

Mitchell, J. P., Banaji, M. R., & Macrae, C. N. The link between social cognition and self-referential thought in the medial prefrontal cortex. J Cogn Neurosci 2005; 17: 1306–1315.

Monteith, M. J. Self-regulation of stereotypical responses: Implications for progress in prejudice reduction. J Pers Soc Psychol 1993; 65: 469–485.

Monteith, M. J., Ashburn-Nardo L., Voils, C. I., & Czopp, A. M. Putting the brakes on prejudice: On the development and operation of cues for control. J Pers Soc Psychol 2002; 83: 1029–1050.

Monteith, M. J., Sherman, J. W., & Devine, P. G. Suppression as a stereotype control strategy. Pers Soc Psychol Rev 1998; 2: 63–82.

Morilak, D. A., Barrera, G., Echevarria, D. J., et al. Role of brain norepinephrine in the behavioral response to stress. Prog Neuropsychopharmacol Biol Psychiatry 2005; 29: 1214–1224.

Moskowitz, G. B., Gollwitzer, P. M, Wasel, W., & Schaal, B. Preconscious control of stereotype activation through chronic egalitarian goals. J Pers Soc Psychol 1999; 77: 167–184.

Moskowitz, G. B., Salomon, A. R., & Taylor, C. M. Implicit control of stereotype activation through the preconscious operation of egalitarian goals. Soc Cogn 2000; 18: 151–177.

Neill, W. T. Episodic retrieval in negative priming and repetition priming. J Exp Psychol Learn Mem Cogn 1997; 23: 1291–1305.

Nieuwenhuis, S., Ridderinkhof, K. R., Blom, J., Band, G. P. H., & Kok, A. Error-related brain potentials are differently related to awareness of response errors: Evidence from an antisaccade task. Psychophysiology 2001; 38: 752–760.

Nisbett, R. E., & Wilson, T. D. Telling more than we can know: Verbal reports on mental processes. Psychol Rev 1977; 84: 231–259.

Norman, D., & Shallice, T., Attention to action: Willed and automatic control of behavior. In: Davidson, R., Schwartz, G., and Shapiro, D., (Eds.), *Consciousness and self regulation: Advances in research and theory*, Vol. 4. New York, NY: Plenum, 1986: pp. 1–18.

Ochsner, K., & Gross, J. J. The cognitive control of emotion. Trends Cogn Sci 2005; 9: 242–249.

Olson, M. A., & Fazio, R. H. Reducing automatically activated racial prejudice through implicit evaluative conditioning. Pers Soc Psychol Bull 2006; 32: 421–433.

Payne, B. K. Prejudice and perception: The role of automatic and controlled processes in misperceiving a weapon. J Pers Soc Psychol 2001; 81: 181–192.

Payne, B. K. Conceptualizing control in social cognition: How executive functioning modulates the expression of automatic stereotyping. J Pers Soc Psychol 2005; 89: 488–503.

Petty, R. E., & Wegener, D. T. Flexible correction processes in social judgment: Correcting for context induced contrast. J Exp Soc Psychol 1993; 29: 137–165.

Petty, R. E., Wegener, D. T., & White, P. Flexible correction processes in social judgment: Implications for persuasion. Soc Cogn 1998; 16: 93–113.

Phelps, E. A., O'Connor, K. J., Cunningham, W. A., et al. Performance on indirect measures of race evaluation predicts amygdala activation. J Cogn Neurosci 2000; 12: 729–738.

Plant, E. A., & Devine, P. G. Internal and external motivation to respond without prejudice. J Pers Soc Psychol 1998; 75: 811–832.

Plant, E. A., & Devine, P. G. The antecedents and implications of interracial anxiety. Pers Soc Psychol Bull 2003; 29: 790–801.

Plant, E. A., & Devine, P. G. Approach and avoidance motives in intergroup processes. In: Elliot, A. (Ed.), *Handbook of approach and avoidance motivation.* Hilldsdale, NJ: Erlbaum, 2008: pp. 563–576.

Posner, M. I., & Snyder, C. R. R. Facilitation and inhibition in the processing of signals. In: Rabbit, P. M. A., & Dornic, S. (Eds), *Attention and Performance V.* New York: Academic Press, 1975: pp. 669–681.

Rankin, R. E., & Campbell, D. T. Galvanic skin response to negro and white experimenters. J Abnorm Soc Psychol 1955; 51: 30–33.

Richeson, J. A., & Shelton, J. N. When prejudice does not pay: Effects of interracial contact on executive function. Psychol Sci 2003; 14: 287–290.

Richeson, J. A., Baird, A. A., Gordon, H. L., et al. An fMRI examination of the impact of interracial contact on executive function. Nat Neurosci 2003; 6: 1323–1328.

Ross, L. The intuitive psychologist and his shortcomings: Distortions in the attribution process. In: Berkowitz, L. (Ed.), *Advances in experimental social psychology* (vol. 10). New York, NY: Academic Press, 1977: pp. 173–220.

Ryan, R. M., & Connell, J. P. Perceived locus of causality and internalization: Examining reasons for acting in two domains. J Pers Soc Psychol 1989; 57: 749–761.

Sapolsky, R. M. Why stress is bad for your brain. Science 1996; 273: 749–750.

Schmader, T., & Johns, M. Converging evidence that stereotype threat reduces working memory capacity. J Pers Soc Psychol 2003; 85: 440–452.

Schuman, H., Steeh, C., Bobo, L., & Krysan, M. *Racial attitudes in America: Trends and interpretations.* Cambridge, MA: Harvard University Press, 1997.

Schwarz, N., & Clore, G. L. Mood, misattribution, and judgments of well-being: Informative and directive functions of affective states. J Pers Soc Psychol 1983; 45: 513–523.

Shah, J. Y., Friedman, R., & Kruglanski, A. W. Forgetting all else: On the antecedents and consequences of goal shielding. J Pers Soc Psychol 2002; 83: 1261–1280.

Shelton, J. N. Interpersonal concerns in social encounters between majority and minority group members. Group Process Intergroup Relat 2003; 6: 171–186.

Sherman, J. W., Gawronski, B., Gonsalkorale, K., Hugenberg, K., Allen, T. J., & Groom, C. J. The self-regulation of automatic associations and behavioral impulses. Psychol Rev 2008; 115: 314–335.

Sherman, J. W., & Amodio, D. M. Regulating implicit prejudice: Cognitive and motivational components of implicit egalitarianism. Invited symposium presented at the Annual Meeting of the Society for Personality and Social Psychology, Palm Springs, CA. 2006, January.

Shiffrin, R., & Schneider, W. Controlled and automatic human information processing: II. Perceptual learning, automatic attending, and a general theory. Psychol Rev 1977; 84: 127–190.

Singer, T., Seymour, B., O'Doherty, J., Kaube, H., Dolan, R. J., & Frith, C. D. Empathy for pain involves the affective but not sensory components of pain. Science 2004; 303: 1157–1162.

Sloman, S. A. The empirical case for two systems of reasoning. Psychol Bull 1996; 119: 3–22.

Smith, E. R., & DeCoster, J. Dual-process models in social and cognitive psychology: Conceptual integration and links to underlying memory systems. Pers Soc Psychol Rev 2000; 4: 108–131.

Srull, T. K., & Wyer, R. S. The role of category accessibility in the interpretation of information about persons: Some determinants and implications. J Pers Soc Psychol, 1979; 37: 1660–1672.

Srull, T. K., & Wyer, R. S. Category accessibility and social perception: Some implications for the study of person memory and interpersonal judgments. J Pers Soc Psychol 1980; 38: 841–856.

Stapel, D. A., Martin, L. L., & Schwarz, N. The smell of bias: What instigates correction processes in social judgments? Pers Soc Psychol Bull 1998; 24: 797–806.

Steele, C. M., & Aronson, J. Stereotype threat and the intellectual test performance of African Americans. J Pers Soc Psychol 1995; 69: 797–811.

Stephan, W. G., & Stephan, C. W. Intergroup anxiety. J Soc Issues 1985; 41: 157–175.

Strack, F., & Hannover, B. Awareness of influence as a precondition for implementing correctional goals. In: Gollwitzer, P., & Bargh, J. A. (Eds.), *The psychology of action: Linking cognition and motivation to behavior.* New York, NY: Guilford, 1996: pp. 579–595.

Strack, F., & Deutsch, R. Reflective and impulsive determinants of social behavior. Pers Soc Psychol Rev 2004; 8(3): 220–247.

Trawalter, S., & Richeson, J. A. Regulatory focus and executive function after interracial interactions. J Exp Soc Psychol 2006; 42: 406–412.

Trope, Y. Identification and inferential processes in dispositional attribution. Psychol Rev 1986; 93: 239–257.

Tversky, A., & Kahneman, D. Judgment under uncertainty: Heuristics and biases. Science 1974; 185: 1124–1131.

Uleman, J. S. Spontaneous versus intentional inferences in impression formation. In Chaiken, S., & Trope, Y. (Eds.), *Dual-process theories in social psychology*. New York, NY: Guilford, 1999: pp. 141–160.

van Veen, V., & Carter, C. S. The timing of action-monitoring processes in the anterior cingulate cortex. J Cogn Neurosci 2002; 14: 593–602.

Vorauer, J. D. An information search model of evaluative concerns in intergroup interaction. Psychol Rev 2006; 113: 862–886.

Vrana, S. R., & Rollock, D. Physiological response to a minimal social encounter: Effects of gender, ethnicity, and social context. Psychophysiology 1998; 35: 462–469.

Wegener, D. T., & Petty, R. E. The flexible correction model: The role of naïve theories of bias in bias correction. In: Zanna, M. P. (Ed.), *Advances in experimental social psychology* (Vol. 29). Mahwah, NJ: Erlbaum, 1997: pp. 141–208.

Wegner, D. M. Ironic processes of mental control. Psychol Rev 1994; 101: 34–52.

Wilson, T. D., & Brekke, N. Mental contamination and mental correction: Unwanted influences on judgements and evaluations. Psychol Bull 1994; 116: 117–142.

Wilson, T. D., & Schooler, J. W. Thinking too much: Introspection can reduce the quality of preferences and decisions. J Pers Soc Psychol 1991; 60: 181–192.

Wittenbrink, B., Judd, C. M., & Park, B. Spontaneous prejudice in context: Variability in automatically activated attitudes. J Pers Soc Psychol 2001; 81: 815–827.

Wyer, N. A. Motivational influences on compliance with and consequences of instructions to suppress stereotypes. J Exp Soc Psychol 2007; 43: 417–424.

Wyer, N. A., Sherman, J. W., & Stroessner, S. J. The spontaneous suppression of racial stereotypes. Soc Cogn 1998; 16: 340–352.

Wyer, N. A., Sherman, J. W., & Stroessner, S. J. The roles of motivation and ability in controlling the consequences of stereotype suppression. Pers Soc Psychol Bull 2000; 26: 13–25.

CHAPTER 5

Integrating Research on Self-Control across Multiple Levels of Analysis: Insights from Social Cognitive and Affective Neuroscience

Ethan Kross and Kevin N. Ochsner

ABSTRACT

Advances in neuroimaging methods and techniques and interest in understanding the neural bases of psychological phenomena are rapidly changing how the capacity for self-control is being addressed. An approach dubbed Social Cognitive and Affective Neuroscience (SCAN) integrates research across multiple levels of analysis, leading to important findings that link the basic social, cognitive, and affective processes underlying self-control to their neural substrates. This chapter illustrates how a SCAN approach can be useful for addressing questions including the problem of how to enable researchers from different areas with different types of expertise and interests in self-control to communicate with one another and most effectively use each other's (sometimes highly technical) theories and methods. Towards this end, we begin by describing the basic goals of SCAN and some of the key challenges facing researchers who adopt this approach. We then describe how this approach is currently being used to build an integrative understanding of the processes underlying a particular type of self-control process that involves actively reinterpreting the meaning of an emotionally evocative stimulus to meet and/or modulate ones' feelings. We conclude by discussing important future research directions in this area.

Keywords: Social cognitive and affective neuroscience; emotion; emotion regulation; fMRI; cognition-emotion interactions.

The famous American novelist Jack Kerouac once wrote, "My fault, my failure, is not in the passions I have, but in my lack of control of them." This quote poignantly illustrates a basic fact about human experience. Namely, that although people often cannot prevent themselves from experiencing certain "passions," they possess remarkable abilities to control them once they are aroused. This capacity for *self-control*– the ability to volitionally influence the content of one's thoughts, the nature of one's feelings, and expression of one's actions to align them with

---

Preparation of this chapter was supported by a grant from the National Institute of Mental Health (MH076137) and National Institute on Drug Abuse (DA022541).

one's long-term goals and standards—is fundamental to human survival and success in the modern world. Consequently, a critical challenge is to understand the processes that underlie it.

Although this basic issue has been the focus of psychological and philosophical inquiry for centuries, recent advances in neuroimaging methods and techniques, coupled with a surge of interest in understanding the neural bases of psychological phenomena, are rapidly changing how it is being addressed. Historical boundaries between different areas of psychology are rapidly being broken, and collaborative endeavors are being forged as part of new interdisciplinary Social Cognitive and Affective Neuroscience (SCAN) approaches that aim to integrate research across multiple levels of analysis (e.g., Blakemore & Frith, 2004; Blakemore, Winston & Frith, 2004; Cacioppo, 1994, 2002; Davidson, Jackson, & Kalin, 2000; Insel & Fernald, 2004; Lieberman, 2000, 2007; Ochsner, 2007; Ochsner & Lieberman, 2001; Ochsner & Schacter, 2000; Panksepp, 1998).

This movement towards integration has already led to a number of important findings that link the basic social, cognitive, and affective processes underlying self-control to their neural substrates (for reviews, see Cacioppo, 2007; Lieberman, 2007; Ochsner, 2007). However, it has also met a number of formidable challenges, including the problem of how to enable researchers from different areas with different types of expertise and interests in self-control to communicate with one another and most effectively use each other's (sometimes highly technical) theories and methods.

Defining the Social Cognitive & Affective Neuroscience Approach

Although social cognitive and affective neuroscience are often considered to be distinct disciplines (for discussion, see Ochsner, 2007), they share the goal of trying to understand human social and emotional behaviors at multiple levels of analysis and rely upon many of the same theories, methods, tools, and techniques for achieving that goal (Olsson & Ochsner, 2008). With this in mind, for present purposes, we explicitly insert the term "affective" into what we've previously described as the "social cognitive neuroscience approach" (e.g., Ochsner, 2007; Ochsner & Lieberman, 2001) to reflect the increasingly blurry lines that separate the sister disciplines of social cognitive and affective neuroscience.

As used here, the "social cognitive and affective neuroscience" approach describes attempts to explain psychological phenomena across three different levels of analysis: the social level of behavior and experience, the cognitive level of mental representations and process, and the neuroscience level of brain systems (see Fig. 5-1). The basic idea is that information from each of these levels of analysis is needed to build integrative models of the mechanisms underlying complex psychological phenomena, in part because models are more powerful when they can explain data across multiple levels of analysis, and in part because data from any one level can constrain inferences about variables cached at any other level of analysis. Thus, whereas work at the social level uses paradigms that simulate self-control dilemmas that people are likely to encounter in everyday life to reveal how different types of personal- or situation-level variables (e.g., situational forces, personality dispositions) impact self-control, work at the cognitive level uses more circumscribed tasks to shed light on the specific information processing mechanisms (i.e., psychological processes, such as attention, memory, language, emotion, attitudes) that underlie these social level processes, and findings at the neuroscience level use neuroimaging and other biological measures to indicate how these processes are instantiated in the brain. By putting these different levels of analysis together, an integrative understanding of the processes underlying self-control emerges.

It is important to recognize that SCAN's focus on integration, although novel in its current instantiation, is not new. Historically, two of the three levels of analysis described in Figure 5-1 have been linked. Specifically, beginning in the 1970s, social psychologists began to use the information processing concepts of cognitive psychology to describe the mechanisms

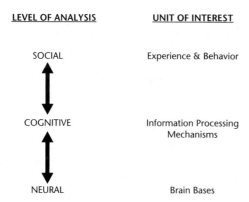

Figure 5–1. Units of Interest at the Social, Cognitive, and Neural Levels of Analysis.

underlying social phenomena (e.g., Higgins, Rholes, & Jones, 1977; Taylor, 1976). Adopting this information-processing framework provided social psychologists with a common language to compare and contrast different phenomena. Thus, rather than having to invent new terms to explain different effects, basic information-processing concepts (e.g., accessibility, schemas, and attention) began to be used, giving birth to the field of social cognition.

Roughly a decade and a half later, a similar merging took place between cognitive psychology and neuroscience. The impetus for this union was the advent and subsequent proliferation of functional neuroimaging methods such as positron emission tomography (PET) and functional magnetic resonance imaging (fMRI) that allowed researchers to examine how neural activity changed when healthy people engaged in different types of psychological tasks. The goals of this integration were twofold. First, to examine the information processing functions provided by different brain regions. Second, to use neuroscience methods as a tool to draw inferences about psychological processes (Ochsner & Kosslyn, 1999; Posner & DiGirolamo, 2000). To the extent that the brain is the organic material that underlies all information processing, psychologists working in cognitive psychology and neuroscience reasoned that important insights could be gleamed by investigating the patterns of brain activity engaged by different types of tasks (Kosslyn, 1999; Ochsner & Lieberman, 2001). Thus "cognitive neuroscience" was born as an approach that used a variety of neuroscience methods to study the neural bases of information processing.

The main purpose of SCAN is to move one step further along this path towards integration. It aims to use the methods and principles of cognitive neuroscience to shed light on the neural and cognitive underpinnings of social and emotional phenomena. In this vein, SCAN has two specific goals (Ochsner, 2007; Ochsner & Lieberman, 2001; *see also*, Sarter, Berntson, & Cacioppo, 1996). Goal one is to link specific patterns of brain activity to specific types of psychological processes. The focus is thus on drawing *functional inferences* about what different brain structures and systems do. Goal two is to use information about brain function to draw inferences about the psychological processes underlying social phenomena. The emphasis here is thus on drawing *psychological inferences* about the processes underlying a given behavior by using the activation of brain systems as markers for the occurrence of particular kinds of psychological processes. For example, consider the common case in which a researcher predicts that two behaviors are mediated by the same underlying mechanisms. One way to test this prediction is by using neuroimaging data. To the extent that this hypothesis is correct, one might expect to find similar patterns of neural activity underlying both types of behaviors. Alternatively, to the extent one predicts the opposite, that two types of ostensibly similar processes are mediated by different underlying mechanisms, than one would expect to observe the reverse—namely, different patterns of neural activity underlying the two behaviors (e.g., Kosslyn & Ochsner, 1994; Schacter, Alpert, Savage, Rauch, & Albert, 1996). Thus, to the extent that the psychological inferences one draws about psychological process from brain activity are valid, neuroimaging methods provide psychologists with a powerful window into how the mind operates when individuals engage in different kinds of carefully constructed psychological tasks that can be used to constrain psychological theorizing about how different processes interact.

Essential Challenges Associated With Adopting a SCAN Approach

Integrating across levels of analysis is not without its challenges. As one moves up and down the levels of analysis illustrated in Figure 5–1, different types of constraints appear that limit the inferences researchers can draw from their studies. In this section, we briefly describe what we perceive to be two of the key challenges facing researchers who adopt this approach.

The first challenge concerns a researcher's ability to draw inferences about the psychological processes that different patterns of brain activity reflect. Assuming that an experiment is well-designed so that it targets a particular psychological process, the degree to which a researcher can be confident that the pattern of neural activity observed in response to their task reflects a particular psychological process depends greatly on the reliability with which that particular function can be ascribed to a particular brain system (Ochsner, 2007; Poldrack, 2006). In the case of some psychological processes that have received a great deal of empirical attention at the neuroscience level of analysis, the reliability of the inferences one draws can be fairly strong. For example, a researcher can be reasonably confident that activation of the primary visual cortex (V1) reflects processes involved in early pattern recognition. As one moves up the ladder in Figure 5–1 to the social level of analysis, however, relatively less is known about the neural bases of many emotional and social cognitive processes. Thus, it is difficult for a researcher to draw strong psychological inferences about the processes reflected by brain activity observed in response to the tasks designed to focus on these processes. Consequently, researchers adopting this approach may be left in the precarious position of observing significant patterns of neural activity in their experiments, but not being able to confidently interpret their psychological meaning (for examples, see Ochsner, 2007).

A second challenge facing the integration-minded researcher concerns the issue of generalizability—how one can be confident that laboratory findings apply to people's real-world behaviors and experiences. There are multiple constraints associated with using neuroimaging methods, especially fMRI, to examine how brain activity changes in response to specific tasks. For example, individuals must lie motionless in an enclosed, cold, hollow tube for a prolonged period of time. In addition, fMRI studies typically require designs in which participants are repeatedly exposed to the same stimulus to acquire reliable estimations of neural activity (for overview of methods, see Wager et al., 2007). Thus, a study examining the neural processes underlying the generation of emotion might involve repeatedly exposing participants to gory pictures tens, if not hundreds, of times. Clearly, this is not the way most individuals typically experience negative affect under normal life circumstances. As such, as researchers design studies with the intent of drawing inferences not just about the social and cognitive level of analyses, but the neural level of analyses as well, the ecological validity of their experiments often declines, and along with it, the confidence with which they can generalize from the results of their experiments to more complicated, real-world situations involving self-control.

We describe these issues here not to discourage people from adopting a SCAN approach or to belittle researchers who actively employ it, but rather to make clear some of the trade-offs associated with it. In the next section, we discuss how a SCAN approach is currently being used to shed light on a particular form of self-control—the cognitive reappraisal of intense negative emotional responses–and how researchers engaged in this work are actively trying to overcome the challenges described above.

IMPLEMENTING THE SOCIAL COGNITIVE AND AFFECTIVE NEUROSCIENCE APPROACH: A CASE STUDY IN THE COGNITIVE REAPPRAISAL OF NEGATIVE EMOTION

Given the integrative goals of SCAN, the first step involved in adopting this approach involves identifying how the phenomenon one is interested in plays out across each of the three different levels of analysis that one ultimately hopes to integrate. Possessing this basic knowledge serves two vital functions. First, it provides the aspiring SCAN researcher with the raw ingredients needed to develop integrative models of self-control. Second, it gives researchers from different areas of psychology with different backgrounds and types of expertise a common foundation for discussing common interests, constructing experiments, and building collaborations. In this section, we simulate what this first step might look like for an integration-minded researcher with interests in understanding the processes underlying people's ability to exert self-control over their emotional responses by cognitively reinterpreting how they're feeling. In this vein, we begin by summarizing key findings from social cognitive and cognitive neuroscience research—two disciplines that together capture all three of the levels of analysis illustrated in Figure 5-1-that are relevant to understanding the processes underlying this specific type of self-control process.

Work Bridging the Social and Cognitive Levels of Analysis

Since the dawn of the cognitive revolution over 40 years ago, a major focus of psychological research has been to shed light on how a people's subjective construals of their experiences influence their feelings (Kelly, 1955). Lazarus was one of the first to empirically demonstrate this phenomenon. For example, in a now classic study he instructed participants to imagine that a stress-inducing video clip about penile circumcision depicted a fake event. Compared to participants who considered the event to be real, those adopting the "fake" perspective displayed significantly lower levels of physiological arousal (Lazarus & Alfert, 1964). This study established that simply providing participants with a different way of thinking about an aversive stimulus had dramatic influences on their subsequent emotional responses.

Subsequent work by Mischel and colleagues extended these initial findings by specifying the specific kind of reappraisal processes that facilitate and undermine adaptive self-control (for reviews, see Chapter 23; Mischel, Shoda, & Rodriguez, 1989). According to Mischel, whether reappraisal, or reconstrual, as this process is also commonly labeled, helped or hindered the individual trying to regulate their impulses and emotions depended critically on the type of reappraisal operation they engaged in—or in other words, how they mentally represented an affect-eliciting cue. For example, in the classic delay of gratification studies a child is presented with the option of receiving one treat immediately (e.g., a cookie or marshmallow) or two treats if they wait an undisclosed period of time. Critically, cueing a child to mentally represent the desired treat in terms of its concrete, "hot," consummatory features (i.e., think about how it would taste and smell), undermined their delay of gratification ability. In contrast, cueing a child to mentally represent the same stimulus in terms of its abstract, "cool", informative features (i.e., think about its shape and color) enhanced their delay of gratification ability.

The results from these early studies examining the effects of reappraisal on self-control have been extended in a variety of ways by contemporary research (for a review, see Gross, 1998). One particularly noteworthy development has been the development of a number of dual process theories of self-control, which provide a framework for conceptualizing how reappraisal processes impact people's ability to regulate their impulses and emotions. Although many dual process theories exist, most agree that two types of information processing systems govern self-control (for several examples of such theories, see Chaiken & Trope, 1999; Carver, 2005; Epstein, 1994; Lieberman, 2007; Metcalfe & Mischel, 1999; Strack & Deutsch, 2004). One system is automatic (or "bottom-up"), driven by the perception of perceptual inputs, operates outside of conscious awareness, and is specialized for quickly processing emotional stimuli. The other system is controlled (or "top-down"), requires cognitive resources to operate, and generates deliberate, reflective, and strategic behavior.

According to these models, managing negative emotions via reappraisal involves activating the controlled, deliberate system to modulate emotions generated by the automatic system. To illustrate, consider again the hungry child in the delay of gratification task. When presented with the image of a cookie, the child automatically experiences an impulse to consume the cookie. This impulse can be modulated, however, if the child thinks about the cookie as a dirty brown disc (i.e., a controlled system mediated reappraisal process), rather than a delicious cookie. In this example, a controlled system strategy is being generated to reduce the child's impulse to consume the cookie. It is important to note, however, that not all types of controlled system mediated strategies facilitate impulse and emotion control. People are capable of generating any number of reappraisal strategies via this system, and as Mischel and colleagues demonstrated over thirty years ago, whether reappraisal helps or hinders the individual in their attempts at self-control depends critically on the type of reappraisal process they engage in. Thus, the child in the example above who reappraises the chocolate chip cookie not as a dirty brown disc, but rather as the most delicious looking cookie ever baked, is likely to enhance rather than diminish their impulse to consume the item.

Work Linking the Cognitive and Neuroscience Levels of Analysis

Findings linking the social and cognitive levels of analysis reviewed above suggest that the effects of cognitive reappraisal on emotion are mediated by two sets of information processing systems, one that is automatic and specialized for generating automatic emotional responses and one that is controlled and capable of modulating automatically triggered emotional impulses. In this section, we describe research on the specific brain systems that may underlie these different information-processing systems.

Regions involved in generating emotion

One area that has consistently been found to play a role in emotional responding is the amygdala. The amygdalae are almond shaped subcortical structures located along the medial wall of each of the temporal lobes that have been implicated in a number of different types of processes that relate to making affective evaluations. For example, a number of studies indicate that this set of brain structures are critical for fear conditioning in animals (LeDoux, 2000) and humans (Buchel & Dolan, 2000; LaBar et al., 1995), and for consolidating explicit memory for emotionally arousing events (Cahill et al., 1995; Hamann et al., 1999). Although the amygdalae have been most extensively studied in the context of fear, they respond to the arousing properties of additional kinds of emotional stimuli, both positive and negative, as well (Adolphs & Tranel 1999; Hamann et al., 1999; Wager et al., 2008).

Beyond the amygdala, a number of other brain structures play an important role in affective evaluations, and are thus likely to be involved in the process of generating initial emotional responses. The insula, for example, is responsive to physical pain and signals the presence of disgust-related stimuli (Peyron, Laurent, & Garcia-Larrea, 2000). In addition, the orbitofrontal and ventromedial prefrontal cortex are sensitive to the same types of stimuli as the amygdala, but seem to play a special role in placing these associations under the control of situational goals. In most studies this is examined in the context of altering an existing affective association as the value of a stimulus changes over time. For example, orbitofrontal lesions in primates (Dias, Robbins, & Roberts, 1997) and in humans (Fellows & Farah, 2003) impair the ability to alter a stimulus-reinforcer association once it is learned, and single unit recording studies in rats and primates suggest that orbitofrontal cortex neurons change their firing properties to previously rewarded (but now not rewarded) stimuli more rapidly than do amygdala neurons (Rolls et al., 1996; *see also* Wager et al., 2008).

Regions involved in implementing reappraisal

In the same way that the experience of emotion is not dependent on a single brain region, it is likely that the process of reappraising an emotion to change how one feels should depend on a number of brain regions. In prior research, Ochsner and colleagues (2004) speculated that the ability to reappraise an emotionally evocative stimulus should depend on component processes that enable people to (a) actively generate strategies for reframing emotional stimuli and maintaining these strategies in working memory, (b) mediate interference between top-down interpretations of stimuli (e.g., reappraisals) and bottom-up appraisals that may continue to generate affective impulses, and (c) reinterpret the meaning of internal states with respect to the stimuli that elicited them. Prior research has connected these functions to a number of prefrontal and cingulate systems.

One structure that is believed to play a critical role in the first component process listed above—generating strategies and maintaining them in working memory—is lateral prefrontal cortex. Both neuropsychological and functional imaging studies have implicated this region as playing a critical role in executive functions that support working memory, reasoning, problem solving, and the ability to generate and organize plans of action (Barcelo & Knight, 2002; Cabeza & Nyberg, 2000; Miller & Cohen, 2001; Nielsen-Bohlman & Knight, 2000; Smith & Jonides, 1999).

The second region that is likely to play a role in reappraising emotional stimuli to alter one's feelings is the anterior cingulate cortex. A number of studies suggest that the dorsal portion of this region (dACC) may be essential for mediating interference between top-down regulation and bottom-up appraisals that generate competing emotional response tendencies (Botvinick et al., 2001; Ochsner & Feldman Barrett, 2001). In this vein, dACC activity has consistently been found in a variety of conditions that involve response conflict (Barch et al., 2001; Botvinick et al., 1999; for reviews, see Botvinick et al., 2001; Bush, Luu, & Posner, 2000), including tasks that require overriding prepotent response tendencies (Carter et al., 2000; Peterson et al., 1999). It has been suggested that dorsal ACC works together with PFC during cognitive control: whereas PFC implements control processes, ACC monitors the degree of response conflict or error and signals the need for control to continue (Botvinick et al., 2001; Gehring & Knight, 2000; MacDonald et al., 2000; Miller & Cohen, 2000).

Finally, dorsal regions of medial prefrontal cortex may also play an important role in reappraisal. Activations in this region have been observed when people are instructed to evaluate their own (Lane et al., 1997; Paradiso et al., 1999) or another person's (Gallagher et al., 2000; Happe et al., 1996) emotional state, and when judging the self-relevance of stimuli (Craik et al., 1999; Kelley et al., 2002). Thus to the extent that individuals think about the self when experiencing negative or positive emotions, as may often be the case given the relevance such emotions have for the individual, it is likely that this region may become increasingly active during reappraisal.

Work Linking the Social, Cognitive, and Neural Levels of Analysis

In the previous sections we outlined findings from social cognitive and cognitive neuroscience research that are relevant for building an integrative understanding of the mechanisms underlying people's ability to cognitively reappraise negative feelings. These findings suggest that the multiple networks of brain regions involved in the generation of emotion, detection of conflict, self-referential processing, and cognitive control are likely to be involved when an individual actively tries to reinterpret the meaning of an emotional stimulus to change the way they feel. In this section, we describe a program of research that has been directly exploring the role that these neural systems play in reappraising negative emotions.

Neural dynamics of cognitive reappraisal

In one of the first studies on the neural basis of cognitive reappraisal, fMRI was used to examine the regions of neural activity that underlie peoples' ability to cognitively reappraise images that arouse strong negative emotion (Ochsner et al., 2002). Two questions motivated this research. First, what processes are recruited when people reappraise a negative stimulus to reduce its aversive impact? Second, how do brain regions involved in reappraising emotional stimuli impact brain regions involved in generating emotional responses? Drawing from the cognitive neuroscience literature reviewed above, Ochsner and colleagues predicted that reappraisal would lead to increased levels of activity in prefrontal and cingulate regions implicated in cognitive control, and reduced levels of activity in regions of the brain involved in emotional processing such as the amygdala, insula, and orbitofrontal cortex.

To test these predictions, participants were scanned as they viewed a series of negative and neutral photos drawn from Peter Lang's International Affective Picture System (Lang, Bradley, & Cuthbart, 2005) for eight seconds each. Participants were instructed to simply view each image for the first four seconds. At the four-second mark, participants were cued to either reappraise the image in such a way that they would no longer feel a negative response, or on baseline trials, they were cued to attend to their feelings and let themselves respond naturally. Two contrasts were then performed to test the study hypotheses. A cognitive control contrast compared brain activity on trials when participants were instructed to reappraise the stimuli vs. trials in which they were instructed to simply attend to the stimulus (reappraise > attend). A second emotional processing contrast compared neural activity on trials in which participants simply attended to the photos to trials in which they reappraised the images (attend > reappraise).

Consistent with predictions, the cognitive control contrast revealed significantly more activity in left prefrontal regions implicated in cognitive control processes such as working memory and response selection. Moreover, subsequent analyses demonstrated that the more effective participants were at reappraising to reduce their negative feelings, the more a region of their right anterior cingulate cortex became activated. Given the cingulate's role in monitoring and evaluating the success of cognitive control, this finding suggested that individuals who more closely monitored the selection and application of reappraisal strategies were able to more successfully regulate their negative emotions. Also consistent with predictions, the emotional processing contrast showed increased levels of activity in the left medial orbitofrontal cortex, insula, and the amygdala—a network of brain structures identified in prior research as playing a key role in the processing of affective stimuli.

Taken together, findings from this first study provided preliminary evidence supporting the idea that cingulate and prefrontal systems play a key role in enabling individuals to cognitively reappraise emotional stimuli and regulate negative emotional responses. According to research from the social–cognitive level of analysis reviewed above, however, simply activating these brain systems should not necessarily lead to reductions in brain regions involved in generating emotional responses. Instead, whether reappraisal related activations lead to increases or decreases in such regions should depend critically on what the goal of the reappraisal strategy is–to increase *or* decease negative affect.

To examine whether engaging in reappraisal to achieve different types of regulatory goals influences brain activity, Ochsner and colleagues (2004) conducted a second fMRI study to examine the neural systems that become engaged when reappraisal is used to cognitively turn up (i.e., increase) or turn down (i.e., decrease) negative emotions (Ochsner et al., 2004). In this experiment, participants' use of reappraisal to either increase or decrease the negative affect they experienced as they were shown aversive IAPS images were directly compared. Participants completed three types of trials with aversive photos: baseline *Look* trials similar to those described previously, and *Increase* and *Decrease* trials in which participants appraised the context, affects, and outcomes depicted in photos in either increasingly negative or neutralizing ways. Contrast analyses examining activation on increase trials vs. look trials and decrease trials vs. look trials were then used to examine how engaging in these different types of reappraisal strategies differentially influenced brain activity.

The results from these contrasts indicated that regardless of whether individuals reappraised pictures to increase or decrease how they felt, increased levels of activity were observed in left lateral PFC, dACC, and dorsal MPFC. In addition, both the increase and decrease strategies modulated activity in the left amygdala, with amygdala activity becoming enhanced on increase trials and diminished on decrease trials. Thus consistent with predictions, a common set of structures became active when participants reappraised negative stimuli regardless of whether they did so to increase or decrease how upset they felt.

In spite of the similarities shared between these two strategies, however, direct comparisons of activity on increase and decrease trials (increase vs. decrease contrasts) revealed some notable differences between them as well. Specifically, increase trials differentially recruited a region of left dorsal MPFC associated with accessing the affective connotations of words and reasoning about one's own or other people's affective mental states (Cato et al., 2004; Ochsner et al., 2004). Decrease trials, on the other hand, differentially recruited right dorsolateral and orbitofrontal regions associated with response inhibition (Konishi et al., 1999) and with updating the motivational value of stimuli (O'Doherty et al., 2003).

The findings reviewed thus far provided evidence that is consistent with the general hypothesis that directing individuals to change the way they feel by directing them to think differently about a stimulus is mediated by prefrontal and cingulate control-systems-involved cognitive control processes. However, as the results just described make clear, different locations within these control systems were identified in the two studies described above, as well as between the up and down-regulate strategy. Ochsner and colleagues reasoned that one reason for this inconsistency could be variability in the specific kinds of reappraisal participants have been asked to employ. There are multiple ways that a person can reinterpret an evocative image to change the way they feel. To the extent that different types of reappraisal strategies involve different types of processing or different types of information, than they might recruit related but distinct neural systems.

To investigate this possibility, participants in the Increase vs. Decrease experiment described above were divided into two groups that achieved their emotion regulatory goals using one of two qualitatively distinct reappraisal strategies. Participants assigned to the *self-focus* group were asked to modulate their negative feelings by either increasing their sense of personal connection to the image (e.g., by imagining it could be a loved one or themselves depicted in the photo) or decreasing their sense personal connection to the image (i.e., by adopting a distant, detached, and clinical third-person perspective while viewing it). Participants assigned to the *situation-focus* group were asked to modulate their negative feelings by either reinterpreting the context, affects, and outcomes of pictured persons in increasingly or decreasingly negative ways. Ochsner and colleagues predicted that the self-focused strategy would differentially recruit MPFC systems involved in self-referential processing and monitoring (e.g. Kelley et al., 2002; Ochsner et al., 2004) because of the emphasis that this strategy placed on thinking about the self. In contrast, they predicted that the situation-focused strategy would more heavily recruit regions of lateral PFC involved in maintaining and manipulating perceptual information (Smith & Jonides, 1998) and retrieving information about emotion-eliciting contexts from semantic memory (Wagner et al., 2001) because of this strategy's emphasis on reinterpreting context.

The results from this study provided mixed support for these predictions. On one hand, results when participants were instructed to down-regulate negative feelings were generally consistent with these hypotheses. On down-regulate trials, participants in the self-focus group displayed significantly more activity MPFC, whereas participants in the situation-focus group displayed significantly more lateral PFC. However, no differences were observed between the situation-focused and self-focused groups when they were instructed to increase their negative emotions. This lack of a difference may have been a result of how participants in the self-focus group were told to increase their feelings. Specifically, they were told to reinterpret the outcomes and affects that they themselves or another person could experience, which is very similar to what participants in the situation-focus group were instructed to do. Future research is thus clearly needed to further unpack the specific mechanisms that underlie the regulatory effects of these difference strategies.

From Experimental Findings to Individual Differences

The findings from the experiments described thus far highlight a number of specific brain systems involved in reappraising negative experiences to make oneself feel better or worse.

Recently, a number of studies have begun to examine whether activity in these regions underlie individual differences associated with populations of people who experience difficulty regulating their emotions.

In one study designed to address this issue, Ray and colleagues (2005) examined whether individual differences in rumination were differentially associated with activity in brain regions involved in generating and regulating emotional responses via reappraisal. Rumination is a process that involves focusing repeatedly and passively on what one is feeling and why one is feeling a certain way (Nolen-Hoeksema, 1991). It has been shown to enhance anger and depression (e.g., Ayduk & Kross, 2008; Bushman, 2002; Kross, Ayduk, & Mischel, 2005; Kross & Ayduk, 2008, 2009; Rusting & Nolen-Hoeksema, 1998), lead to higher levels of depressive symptoms over time (e.g., Nolen-Hoeksema, & Morrow, 1991; Nolen-Hoeksema, Morrow, & Fredrickson, 1993), impair problem-solving ability (Lyubomirsky & Nolen-Hoeksema, 1995; Lyubomirsky et al., 1999), and both precipitate and maintain depressive disorders (Nolen-Hoeksema, 2000). Ray and colleagues predicted that individuals who ruminate may spontaneously reappraise their negative emotional experiences in ways that exacerbate how upset they feel. Consequently, they reasoned that people who score high on a trait measure of rumination might show significantly higher levels of activity in regions of the brain associated with generating emotional responses (e.g., amygdala) as well as brain regions involved in reappraisal (e.g., lateral prefrontal cortex) reflecting their use of reappraisal to enhance their negative emotions.

To test these predictions, Ray and colleagues examined how individual differences in rumination influenced brain activity when individuals reappraised aversive images to either increase or decrease their negative emotional response. This allowed the experimenters to observe whether ruminators were more or less effective at recruiting systems that support reappraisal, or at modulating the appraisal systems that generate the emotional response. Findings indicated that during reappraisal, rumination correlated positively with activity in the amygdala both when individuals were actively turning up or turning down their negative emotion. Interestingly, the tendency to ruminate did not correlate with the tendency to recruit dorsal lateral prefrontal or cingulate regions thought to implement reappraisal processes. Taken together, these results suggested that ruminators may get more "amygdalar bang" for their "prefrontal buck", or in other words, that they are able to efficiently modulate neural systems involved in generating emotional responses and don't need to recruit extra prefrontal resources to do so (for related results with depressed individuals, *see* Johnstone et al., 2007).

In another individual difference study, Kross and colleagues (2007) used fMRI to explore whether individual differences on another personality dimension, Rejection Sensitivity (RS), covaried with activity in brain regions involved in generating and reappraising emotional responses. RS is the tendency to anxiously expect, readily perceive, and intensely react to rejection (Downey & Feldman, 1996). Kross and colleagues predicted that individual differences between high and low RS individuals might result either from high RS individual's tendency to appraise emotional stimuli as more negative than low RS individuals or their failure to adaptively regulate their negative emotional responses when they become triggered. To examine the role that these different emotional appraisal and control processes play in distinguishing between these two groups of individuals, they scanned high and low RS participants as they viewed images designed to elicit feelings of rejection or acceptance.

Findings indicated that across all participants, the comparison of responses for rejection as opposed to acceptance images showed activation in regions of the brain involved in processing affective stimuli (posterior cingulate, insula) and cognitive control (dorsal anterior cingulate cortex; medial frontal cortex). Low and high RS individuals' responses to rejection vs. acceptance images were not, however, identical. Low RS individuals displayed significantly more activity in left inferior and right dorsal frontal regions, and activity in these areas correlated negatively with participants' self-report distress ratings. This suggested that the prefrontal regions may play a role in regulating responses to stimuli that convey themes of social rejection. In this vein, it is noteworthy that the activations observed in this study, when participants were free to appraise the meaning of stimuli in whatever way they chose, were very similar to those observed in studies described earlier, when participants were instructed to down-regulate negative responses to aversive images by reappraising their meaning in "cool" unemotional ways (Ochsner et al., 2002, 2004).

Taken together, these findings help to establish the generalizability of the findings on the neural bases of reappraisal. They demonstrate that brain regions involved in tracking reappraisal when people are instructed to engage in this process also become active during conditions that are diagnostic of theoretically relevant individual differences in which reappraisal processes are believed to play a prominent role.

From pictures to memories: Towards greater ecological validity

Although the studies reviewed above provide important insights into the neural bases of reappraisal, their reliance on standardized experimental stimuli (e.g., IAPS pictures) to generate emotional responses raises questions concerning the generalizability of the activations observed to "real world" situations. For example, it would be important to know whether the same network of brain regions that underlie people's ability to reappraise aversive images similarly underlie people's ability to reinterpret intense negative life experiences (e.g., being rejected in a romantic relationship; mourning the loss of a loved one) in ways that improve their feelings. Critical here is the question of to what extent do the processes engaged in these different types of emotion regulation paradigms are similar or different.

As a first step toward addressing this issue, we recently conducted an fMRI experiment using a novel memory-based emotion regulation paradigm. In this study, participants were instructed to recall a number of highly arousing negative emotional experiences from (e.g., rejection experiences, hostile arguments, etc.) that they had disclosed during a prior testing session. During the series of blocks completed in the scanner, participants first recalled each experience. They were then asked to use reappraisal strategies designed to either increase or decrease their negative emotional responses (Kross, Davidson, Weber, & Ochsner, 2009). Each block included a brief perceptual reasoning task in which they saw an arrow pointing left or right and were asked to indicate the direction the arrow was pointing. This task was meant to provide a working baseline condition against which reappraisal-related activations could be compared (*see* Raye et al., 2002).

The basic question motivating this study was whether instructing individuals to reappraise feelings associated with aversive memories results in patterns of neural activity similar to those observed when participants are instructed to reappraise how they feel when viewing aversive emotional images. To address this question, contrasts comparing brain activity on trials when participants were instructed to increase and decrease their negative feelings were compared against the perceptual baseline task and then subjected to a conjunction analysis. This type of analysis identifies overlaps in activation observed across different contrasts and thus provides a means of identifying the regions of neural activity that are common to both the increase and decrease reappraisal strategies used in this study.

Consistent with the fMRI studies of reappraisal reviewed above, this analysis revealed that both the increase and decrease reappraisal strategies were associated with increased levels of activity in regions of lateral prefrontal cortex, providing preliminary evidence supporting the idea that lateral prefrontal cortex may provide a common functional architecture for diverse types of reappraisal strategies. These similarities between the memory- and picture-based paradigms notwithstanding, a number of additional activations were also observed in this study. For example, contrasts comparing increase vs. decrease activity revealed significant activations in subgenual anterior cingulate cortex and ventromedial prefrontal cortex, whereas the reverse contrast revealed elevated levels in more dorsal regions of the medial prefrontal cortex. Thus, as one might expect, both differences and similarities characterize what people do in these different tasks. As we discuss in more detail below, one of the critical challenges for future research is to begin to specify what these different activations reflect.

Assessing Progress and Looking Toward the Future

In the previous section, we outlined the basic information needed to adopt a SCAN approach to examine the processes underlying the reappraisal of negative emotional stimuli, and described a program of research that is beginning to use this approach to build an integrative understanding of this phenomenon. In this penultimate section of the chapter, we simulate the final step involved in adopting a SCAN approach, which entails rigorously assessing how well one's studies are meeting their research goals and overcoming the challenges involved in implementing a SCAN approach.

What Have We Learned and What Do We Still Need to Know?

Perhaps the most robust finding to emerge from the research studies reviewed above concerns the role of lateral prefrontal cortex and dorsal anterior cingulate cortex in modulating emotional responses—either up or down—when people engage in reappraisal. These structures were consistently activated across the research studies described above, regardless of reappraisal strategy, reappraisal goals, or sample population. What's more, activity in these control regions often co-varies inversely with activity in brain regions involved in emotional appraisal. Over the past five years, converging evidence supporting these findings has come from a growing number of studies that also has begun to investigate related forms of cognitive reappraisal. In general, these studies have found that interactions between top-down prefrontal and cingulate systems and bottom-up emotion generation systems are involved when individuals maintain responses to aversive stimuli after they disappear (Schaefer et al., 2002), are instructed to 'suppress' sexual arousal (Beauregard, Levesque, & Bourgouin, 2001), sadness (Levesque et al., 2003; Levesque et al., 2004), or negative motion (Phan et al., 2005), or are instructed to distance themselves from painful inputs (Kalisch et al., 2005). Thus, taken together, the results of these studies suggest that these brain regions are critical to people's ability to cognitively reappraise negative emotional responses (Ochsner & Gross, 2007).

What remains unclear at the moment, however, is what specific psychological processes are reflected by activations in these different brain regions. For example, to what extent does activity in lateral prefrontal cortex activations reflect verbalization, working memory, response inhibition or some other process? Or, how does dorsal cingulate activity interact with lateral prefrontal activity to support reappraisal? Addressing questions such as these is critical to refining our understanding of the processes involved in this type of self-control process and building an integrative understanding of the mechanisms that underlie it. One research strategy that may be particularly helpful towards this end is the use of functional localizer tasks in future fMRI research on reappraisal. Functional localization is a research strategy in which brain regions that are involved in specific types of psychological processes (e.g., self-referential processing, conflict monitoring, cognitive control) are first identified functionally in subjects individually. Subsequent scans in the same subjects then examine whether activity in these functionally defined regions become active when participants engage in a process of interest (e.g., reappraisal). The use of such functional localizer tasks has a long and successful history of use in visual neurophysiology research (e.g., Epstein & Kanwisher, 1998; Saxe, Brett, & Kanwisher, 2005) and offers a route to systematically examining the role that different subprocesses play in reappraisal.

A second issue that remains unclear at present is whether the findings observed thus far on the neural basis of cognitive reappraisal apply to situations in which people are required to regulate their emotions in everyday life.

There is little question that repeatedly showing people aversive IAPS images and asking them to reinterpret how they feel in response to viewing them is not the way negative affect is typically triggered and regulated under typical life circumstances. Thus, a critical next step for building an integrative understanding of the processes underlying reappraisal is to examine how the basic findings revealed from the aforementioned studies extend to the range of situations and circumstances in which people are likely to use this strategy in everyday life. In this vein, it is encouraging that researchers are beginning to examine whether the brain systems involved in instructed reappraisal become active when participants may spontaneously engage in reappraisal processes on their own, in the absence of explicit experimental instructions to do so (Kross et al., 2006; Ray et al., 2005). Also encouraging are studies showing similarities between the region important for regulating responses to normatively negative imagistic stimuli and highly self-relevant autobiographical experiences (Kross, et al., 2009). Future research that continues along these lines and builds further is important for establishing the breadth of the findings reviewed in this chapter.

Conclusion

Charles Dickens famously opened his classic novel, *A Tale of Two Cities*, writing, "It was the best of times, it was the worst of times, it was the age of wisdom, it was the age of foolishness...." In many respects, we feel that this quote accurately describes the opportunities that currently face integration-minded researchers who are motivated to build cumulative, multi-leveled models of self-control processes. Current opportunities for conducting such integrative research are numerous, as reflected by the amount of funding that has been given to this area, as well as the amount of research that has been conducted recently (Cacioppo, 2007). Thus, times appear to be good in many regards, as measured by a number of different metrics. Times are bad, however, and run the risk of potentially becoming worse, to the extent that researchers who adopt this approach are not mindful of the trade-offs that one is forced to make when moving up and down to different levels of analysis in order to integrate, and do not take efforts to reduce the negative impact of these trade-offs. In this vein, we hoped to convey in this chapter the different kinds of trade-offs that an integrative researcher regularly faces when trying to construct experimental tasks to address the phenomena they care about and the steps that are needed to prevent "foolishness," and instead initiate the construction of an integrative science of self-control. Doing so requires multiple elements: a reasonable understanding of the processes underlying self-control as they have been studied in social-cognition and cognitive-neuroscience, an awareness of the trade-offs that one faces as they attempt to shift up or down different levels of analysis, and the recognition, above all else, that building a cumulative, integrative science of self-control requires the same basic elements as the construction of knowledge in any other research field–well-designed, theory-driven, programmatic studies that systematically examine and test predictions about phenomena of interest. Fortunately, at the moment there exist many examples of integrative researchers who subscribe to this approach. Thus the future seems bright and the best of times, we hope, are near.

References

Adolphs, R., & Tranel, D. Preferences for visual stimuli following amygdala damage. J Cogn Neurosci 1999; 11: 610–616.

Ayduk, O., & Kross, E. Enhancing the pace of recovery: Differential effects of analyzing negative experiences from a self-distanced vs. self-immersed perspective on blood pressure reactivity. Psycholog Sci. 2008; 19: 229–231.

Barcelo, F., & Knight, R. T. Both random and perseverative errors underlie WCST deficits in prefrontal patients. Neuropsychologia 2002; 40: 349–356.

Barch, D. M., Braver, T. S., Akbudak, E., Conturo, T., Ollinger, J., & Snyder, A. Anterior cingulate cortex and response conflict: effects of response modality and processing domain. Cerebral Cortex 2001; 11: 837–848.

Beauregard, M., Levesque, J., & Bourgouin, P. Neural correlates of conscious self-regulation of emotion. J Neurosci 2001; 21: RC165.

Bechara, A., Damasio, H., & Damasio, A. R. Emotion, decision making and the orbitofrontal cortex. Cerebral Cortex 2000; 10: 295–307.

Bench, C. J., Frith, C. D., Grasby, P. M., Friston, K. J., Paulesu, E., Frackowiak, R. S., & Dolan, R. J. Investigations of the functional anatomy of attention using the Stroop test. Neuropsychologia 1993; 31: 907–922.

Berns, G. S., McClure, S. M., Pagnoni, G., & Montague, P. R. Predictability modulates human brain response to reward. J Neurosci 2001; 21: 2793–2798.

Blakemore, S. J., & Frith, U. How does the brain deal with the social world? Neuroreport: For Rapid Communication of Neuroscience Research 2004; 15: 119–128.

Blakemore, S. J., Winston, J., & Frith, U. Social cognitive neuroscience: Where are we heading? Trends Cogn Sci 2004; 8: 216–222.

Botvinick, M. M., Braver, T. S., Barch, D. M., Carter, C. S., & Cohen, J. D. Conflict monitoring and cognitive control. Psychol Rev 2001; 108(3): 624–652.

Botvinick, M., Nystrom, L. E., Fissell, K., Carter, C. S., & Cohen, J. D. Conflict monitoring versus selection-for-action in anterior cingulate cortex. Nature 1999; 402: 179–181.

Buchel, C., & Dolan, R. J. Classical fear conditioning in functional neuroimaging. Curr Opin Neurobiol 2000; 10: 219–223.

Bush, G., Luu, P., & Posner, M. I. Cognitive and emotional influences in anterior cingulate cortex. Trends Cogn Sci 2000; 4: 215–222.

Cabeza, R., & Nyberg, L. Imaging cognition II: An empirical review of 275 PET and fMRI studies. J Cogn Neurosci 2000; 12: 1–47.

Cacioppo, J. T. Social neuroscience: autonomic, neuroendocrine, and immune responses to stress. Psychophysiology 1994; 31: 113–128.

Cacioppo, J. T. Social neuroscience: Understanding the pieces fosters understanding the whole and vice versa. Am Psychologist 2002; 57: 819–831.

Cacioppo, J. T. Social neuroscience: Progress and implications for mental health. Perspectives in Psycholog Sci 2007; 2: 99–123.

Canli, T., Desmond, J. E., Zhao, Z., Glover, G., & Gabrieli, J. D. Hemispheric asymmetry for emotional stimuli detected with fMRI. Neuroreport 1998; 9: 3233–3239.

Cahill, L., Babinsky, R., Markowitsch, H. J., & McGaugh, J. L. The amygdala and emotional memory. Nature 1995; 377: 295–296.

Carter, C. S., Macdonald, A. M., Botvinick, M., Ross, L. L., Stenger, V. A., Noll, D., & Cohen, J. D. Parsing executive processes: strategic vs. evaluative functions of the anterior cingulate cortex. Proc Natl Acad Sci 2000; 97: 1944–1948.

Carver, C. S. Impulse and constraint: Perspectives from personality psychology, convergence with theory in other areas, and potential for integration. Personal Soc Psychol Rev 2005; 9: 312–333.

Chaiken, S., & Trope, Y. *Dual-Process Theories in Social Psychology*. New York, NY: Guilford Press, 1999.

Craik, F. I. M., Moroz, T. M., Moscovitch, M., et al. In search of the self: A positron emission tomography study. Psycholog Sci 1999; 10: 26–34.

Critchley, H., Daly, E., Phillips, M., et al. Explicit and implicit neural mechanisms for processing of social information from facial expressions: A functional magnetic resonance imaging study. Hum Brain Mapping 2000; 9: 93–105.

Davidson, R. J., Jackson, D. C., & Kalin, N. H. Emotion, plasticity, context, and regulation: perspectives from affective neuroscience. Psycholog Bull 2000; 126: 890–909.

Dias, R., Robbins, T. W., & Roberts, A. C. Dissociable forms of inhibitory control within prefrontal cortex with an analog of the Wisconsin Card Sort Test: Restriction to novel situations and independence from "on-line" processing. J Neurosci 1997; 17: 9285–9297.

Epstein, S. Integration of the Cognitive and Psychodynamic Unconscious. Am Psycholog 1994; 49: 709–724.

Fellows, L. K., & Farah, M. J. Ventromedial frontal cortex mediates affective shifting in humans: evidence from a reversal learning paradigm. Brain 2003; 126: 1830–1837.

Gallagher, H. L., Happe, F., Brunswick, N., Fletcher, P. C., Frith, U., & Frith, C. D. Reading the mind in cartoons and stories: An fMRI study of 'theory of mind' in verbal and nonverbal tasks. Neuropsychologia 2000; 38: 11–21.

Gross, J. J. The emerging field of emotion regulation: An integrative review. Rev Gen Psycholog 1998; 2: 271–299.

Hamann, S. B., Ely, T. D., Grafton, S. T., & Kilts, C. D. Amygdala activity related to enhanced memory for pleasant and aversive stimuli. Nat Neurosci 1999; 2: 289–293.

Happe, F., Ehlers, S., Fletcher, P., et al. 'Theory of mind' in the brain. Evidence from a PET scan study of Asperger syndrome. Neuroreport 1996; 8: 197–201.

Higgins, E. T., Rholes, W.S., & Jones, C. R. Category accessibility and impression formation. J Ex Soc Psychol. 1977; 13: 141–154.

Hornak, J., Rolls, E. T., & Wade, D. Face and voice expression identification in patients with emotional and behavioural changes following ventral frontal lobe damage. Neuropsychologia 1996; 34: 247–261.

Insel, T. R., & Fernald, R. D. How the brain processes social information: Searching for the social brain. Ann Rev Neurosci 2004; 27: 697–722.

Johnstone, T., Reekum, C. M., Urry, H. L., Kalin, N. H., & Davidson, R. J. Failure to regulate: Counterproductive recruitment of top-down prefrontal-subcortical circuitry in major depression. J Neurosci 2007; 27: 8877–8884.

LaBar, K. S., LeDoux, J. E., Spencer, D. D., & Phelps, E. A. Impaired fear conditioning following unilateral temporal lobectomy in humans. J Neurosci 1995; 15: 6846–6855.

Lang, P. J., Bradley, M. M., & Cuthbert, B. N. International affective picture system (IAPS): Affective ratings of pictures and instruction manual. Technical Report A-6. University of Florida, Gainesville, FL, 2005.

LeDoux, J. E. Emotion circuits in the brain. Ann Rev Neurosci 2000; 23: 155–184.

Levesque, J., Eugene, F., Joanette, Y., Paquette, V., Mensour, B., Beaudoin, G., et al. Neural circuitry underlying voluntary suppression of sadness. Biol Psychiatr 2003; 53: 502–510.

Levesque, J., Joanette, Y., Mensour, B., et al. Neural basis of emotional self-regulation in childhood. Neuroscience 2004; 129: 361–369.

Kalisch, R., Wiech, K., Critchley, H. D., et al. Anxiety reduction through detachment: Subjective, physiological, and neural effects, 2005; 17: 874–883.

Kelly, G. A. *The psychology of personal constructs*. New York, NY: Norton. Reprinted by Routledge (London), 1991.

Kelley, W. M., Macrae, C. N., Wyland, C. L., Caglar, S., Inati, S., & Heatherton, T. F. Finding the self? An event-related fMRI study. J Cogn Neurosci, 2002; 14: 785–794.

Knutson, B., Adams, C. M., Fong, G. W., & Hommer, D. Anticipation of increasing monetary reward selectively recruits nucleus accumbens. J Neurosci 2001; 21: RC159.

Knutson, B., Westdorp, A., Kaiser, E., & Hommer, D. FMRI visualization of brain activity during a monetary incentive delay task. Neuroimage 2000; 12: 20–27.

Kosslyn, S. M. If neuroimaging is the answer, what is the question? Philos Trans R Soc Lond B, Biol Sci 1999; 354: 1283–1294.

Kross, E., & Ayduk, O. Boundary conditions and buffering effects. Does depressive symptomology moderate the effectiveness of self-distancing for facilitating adaptive emotional analysis? J Res Pers 2009; 43: 923–927.

Kross, E., & Ayduk, O. Facilitating adaptive emotional analysis: Short-term and long-term outcomes distinguishing *distanced*-analysis of depressive experiences from *immersed*-analysis and distraction. Personal Soc Psychol Bull, 2008; 34: 924–938.

Kross, E., Ayduk, O., & Mischel, W. When asking "why" does not hurt: Distinguishing rumination from reflective processing of negative emotions. Psycholog Sci 2005; 16: 709–715.

Kross, E., Davidson, M., Weber, J., & Ochsner, K. Coping with emotions past: The neural bases of regulating affect associated with negative autobiographical memories. Biol Psychiatry 2009; 65: 361–366.

Kross, E., Egner, T., Downey, G., Ochsner, K., & Hirsch, J. Neural dynamics of rejection sensitivity. J Cogn Neurosci 2007; 19: 945–956.

Lane, R. D., Fink, G. R., Chau, P. M., & Dolan, R. J. Neural activation during selective attention to subjective emotional responses. Neuroreport 1997a; 8: 3969–3972.

Lane, R. D., Reiman, E. M., Ahern, G. L., Schwartz, G. E., & Davidson, R. J. Neuroanatomical correlates of happiness, sadness, and disgust. Am J Psychiatr 1997b; 154: 926–933.

Lazarus, R. S., Alfert, E. Short-circuiting of threat by experimentally altering cognitive appraisal. J Abnormal Soc Psychol 1964; 69: 195–205.

Lieberman, M. D. Intuition: A social cognitive neuroscience approach. Psycholog Bull 2000; 126: 109–137.

Lieberman, M. D. Social cognitive neuroscience: A review of core processes. Ann Rev Psychol 2007; 58: 259–89.

Metcalfe, J., & Mischel, W. A hot/cool system analysis of delay of gratification: Dynamics of willpower. Psycholog Rev 1999; 106: 3–19.

Mischel, W., Shoda, Y., & Rodriguez, M. L. Delay of gratification in children. Science 1989; 244: 933–938.

Miller, E. K., & Cohen, J. D. An integrative theory of prefrontal cortex function. Ann Rev Neurosci 2001; 24: 167–202.

Morris, J. S., Frith, C. D., Perrett, D. I., et al. A differential neural response in the human amygdala to fearful and happy facial expressions. Nature 1996; 383: 812–815.

Nielsen-Bohlman, L., & Knight, R. T. Prefrontal cortical involvement in visual working memory. Cogn Brain Res 1999; 8: 299–310.

Nobre, A. C., Coull, J. T., Frith, C. D., & Mesulam, M. M. Orbitofrontal cortex is activated during breaches of expectation in tasks of visual attention. Nat Neurosci 1999; 2: 11–12.

Ochsner, K. N. Social cognitive neuroscience: Historical development, core principles, and future promise. In: Kruglanksi, A., & Higgins, E. T. (Eds.). *Social Psychology: A Handbook of Basic Principles* 2nd Ed. New York, NY: Guilford Press, 2007: pp. 39–66.

Ochsner, K. N., & Feldman Barrett, L. A multiprocess perspective on the neuroscience of emotion. In: Mayne, T. J. & Bonanno, G. A. (Eds.). *Emotions: Current issues and future directions.* New York, NY: Guilford Press, 2001: pp. 38–81.

Ochsner, K. N., Bunge, S. A., Gross, J. J., & Gabrieli, J. D. Rethinking feelings: An fMRI study of the cognitive regulation of emotion. J Cogn Neurosci 2002; 14: 1215–1229.

Ochsner, K. N., Ray, R. D., Robertson, E. R., et al. For better or for worse: Neural systems supporting the cognitive down-and up-regulation of negative emotion. Neuroimage 2004; 23(2): 483–499.

Ochsner, K. N., & Kosslyn, S. M. The cognitive neuroscience approach. In: Bly, B. M. & Rumelhart, D. E. (Eds.), *Cognitive science.* San Diego, CA: Academic Press, 1999: pp. 319–365.

Ochsner, K. N., & Lieberman, M. D. The emergence of social cognitive neuroscience. Ame Psychologist 2001; 56: 717–734.

Ochsner, K. N., Ray, R. D., Cooper, J. C., et al. For better or for worse: Neural systems supporting the cognitive down-and up-regulation of negative emotion. Neuroimage 2004; 23: 483–499.

Ochsner, K. N., & Schacter, D. L. A social cognitive neuroscience approach to emotion and memory. In: Borod, J. C. (Ed.), *The neuropsychology of emotion* London: Oxford University Press, 2000: pp. 163–193.

O'Doherty, J., Critchley, H., Deichmann, R., & Dolan, R. Dissociating outcome from response switching in human orbitofrontal cortex. Paper presented at the 10th Annual Meeting of the Cognitive Neuroscience Society, New York, NY, 2003.

O'Doherty, J., Rolls, E. T., Francis, S., Bowtell, R., & McGlone, F. Representation of pleasant and aversive taste in the human brain. J Neurophysiology 2001; 85: 1315–1321.

Olsson, A. & Ochsner, K. N. The relationship between emotion and social cognition. Trends in Cognitive Science, 2008, 12, 65–71.

Ongur, D., & Price, J. L. The organization of networks within the orbital and medial prefrontal cortex of rats, monkeys and humans. Cerebral Cortex 2000; 10: 206–219.

Panksepp, J. *Affective neuroscience: The foundations of human and animal emotions.* New York, NY: Oxford University Press, 1998.

Paradiso, S., Johnson, D. L., Andreasen, N. C., O'Leary, D. S., Watkins, G. L., Ponto, L.L., & Hichwa, R. D. Cerebral blood flow changes associated with attribution of emotional valence to pleasant, unpleasant, and neutral visual stimuli in a PET study of normal subjects. Am J Psychiatry 1999; 156: 1618–1629.

Peterson, B. S., Skudlarski, P., Gatenby, J. C., Zhang, H., Anderson, A. W., & Gore, J. C. An fMRI study of Stroop word-color interference: evidence for cingulate subregions subserving multiple distributed attentional systems. Biol Psychiatry 1999; 45: 1237–1258.

Peyron, R., Laurent, B., & Garcia-Larrea, L. Functional imaging of brain responses to pain. A review and meta-analysis. Neurophysiology Clin 2000; 30: 263–288.

Phan, K. L., Fitzgerald, D. A., Nathan, P. J., Moore, G. J., Uhde, T. W., & Tancer, M. E. Neural substrates for voluntary suppression of negative affect: A functional magnetic resonance imaging study. Biol Psychiatry 2005; 57: 210–219.

Phillips, M. L., Young, A. W., Senior, C., et al A specific neural substrate for perceiving facial expressions of disgust. Nature 1997; 389: 495–498.

Poldrack, R. A. Can cognitive processes be inferred from neuroimaging data? Trends Cogn Sci 2006; 10: 59–63.

Posner, M. I., & DiGirolamo, G. J. Cognitive neuroscience: origins and promise. Psychological Bull 2000; 126: 873–889.

Raye, C. L., Johnson, M. K., Mitchell, K. J., Reeder, J. A., & Greene, E. J. Neuroimaging a single thought: Dorsolateral PFC activity associated with refreshing just-activated information. NeuroImage 2002; 15: 447–453.

Roberts, A. C., & Wallis, J. D. Inhibitory control and affective processing in the prefrontal cortex: neuropsychological studies in the common marmoset. Cerebral Cortex 2000; 10: 252–262.

Rolls, E. T. The orbitofrontal cortex and reward. Cerebral Cortex 2000; 10: 284–294.

Rolls, E. T., Critchley, H. D., Mason, R., & Wakeman, E. A. Orbitofrontal cortex neurons: Role in olfactory and visual association learning. J Neurophysiology 1996; 75: 1970–1981.

Sarter, M., Berntson, G. G., & Cacioppo, J. T. Brain imaging and cognitive neuroscience. Toward strong inference in attributing function to structure. Am Psychologist 1996; 51: 13–21.

Schaefer, S. M., Jackson, D. C., Davidson, R. J., Aguirre, G. K., Kimberg, D. Y., & Thompson-Schill, S. L. Modulation of amygdalar activity by the conscious regulation of negative emotion. J Cogn Neurosci 2002; 14: 913–921.

Schneider, F., Grodd, W., Weiss, U., Klose, U., Mayer, K. R., Nagele, T., & Gur, R. C. Functional MRI reveals left amygdala activation during emotion. Psychiatry Research 1997; 76: 75–82.

Smith, E. E., & Jonides, J. Storage and executive processes in the frontal lobes. Science 1999; 283: 1657–1661.

Sprengelmeyer, R., Young, A. W., Calder, A. J., et al. Loss of disgust. Perception of faces and emotions in Huntington's disease. Brain 1996; 119: 1647–1665.

Strack, F., & Deutsch, R. Reflective and impulsive determinants of social behavior. Personal Soc Psychol Rev 2004; 8: 220–247.

Taylor, S. E. Developing a cognitive social psychology. Carroll, J. S. & Payne, J. W. (Ed). *Cognition and social behavior.* Oxford, England: Lawrence Erlbaum, 1976.

Wager, T. D., Hernandez, L., Jonides, J., & Lindquist, M. Elements of functional neuroimaging. In: Cacioppo, J. T., Tassinary, L. G., & Berntson, G. G. (Eds.). *Handbook of Psychophysiology* (4th ed.). Cambridge: Cambridge University Press, 2007: pp. 19–55.

Wager, T. D., Barrett, L. F., Bliss-Moreau, E., Lindquist, K., Duncan, S., Kober, H., Joseph, J., Davidson, M., & Mize, J. The neuroimaging of emotion. Chapter in M. Lewis, J. M. Haviland-Jones, & L.F. Barrett (Eds.), The handbook of emotion, 3rd Edition. New York: Guilford, 2008: pp. 249–271.

Zald, D. H., & Kim, S. W. Anatomy and function of the orbital frontal cortex, I: Anatomy, neurocircuitry; and obsessive-compulsive disorder. J Neuropsychiatry and Clin Neurosciences 1996; 8: 125–138.

CHAPTER 6

Using the Stroop Task to Study Emotion Regulation

Jason Buhle, Tor Wager, and Ed Smith

ABSTRACT

The Stroop task is among the most influential experimental paradigms for the study of cognitive control. Recent variants have sought to extend the Stroop task to the study of emotional regulation. To assess these emotional Stroop tasks, it is important to distinguish between those that seek to disrupt performance purely via distraction by emotional stimuli that engage attention, from those that do so by presenting emotional information that specifically conflicts with task-relevant judgments. The emotional stimuli in distraction-based Stroop tasks typically fail to disrupt the performance of healthy adults, and recent work suggests that when inference does occur, it lags behind goal-directed processing, primarily degrading performance on subsequent trials. Although early neuro-imaging research using the emotional distraction Stroop tasks gave rise to the influential hypothesis of distinct emotional and nonemotional processing regions in the anterior cingulate cortex, subsequent research has provided limited support. Other recent evidence suggests that interference in these distraction tasks might reflect a generic transient surprise rather than inherently emotional processes. In contrast to emotional distraction Stroop tasks, studies of emotional conflict have reported robust congruency effects, but it is unclear that the resolution of stimulus incompatibility is relevant to questions of how one controls one's emotions. Future research with emotional distraction Stroop tasks should seek to develop variants that evince more robust effects, whereas research on emotional stimulus incompatibility should leverage previous work with nonemotional conflict Stroop variants to explore topics such as the relationship between output modality and dimensional relevancy, and the distinction between categorization and identification task goals.

Keywords: Cognitive control, interference, emotional control, Stroop task, emotional Stroop, anterior cingulate cortex, conflict, distraction

Self-control broadly describes the goal-directed regulation of thought, feeling, and behavior. Although the history of experimental psychology offers countless paradigms developed to study self-control, perhaps none has achieved the renown or influence of the simple task Stroop described in 1935 (Stroop, 1935). Straightforward instructions guide the participant: identify the font color of a string of letters as quickly as possible, ignoring any meaning the letters may carry. This is easily done when the letters do not comprise words, or when they comprise words that lack a strong color association. But what happens if the to-be-ignored word names a color? The effect depends on the relationship between the irrelevant word and the relevant font color. If the word and font color are compatible, such as the word "blue" printed in blue, then participants often answer faster, suggesting facilitation. Conversely, when meaning and color are incompatible, such as the word "red" printed in blue, participants typically respond slower, indicating interference.

This phenomenon has inspired a great deal of work in the cognitive and neural sciences—the Web of Science database records nearly 4,000 citations of Stroop's original paper (www.isiwebofknowledge.com). To what can we attribute such enduring impact? Perhaps most obvious is the unusual reliability and magnitude of the interference effect (MacLeod, 1991). Stroop interference is evident to anyone who attempts the task. Furthermore, this interference is remarkably difficult to circumvent—only a handful of manipulations have been found to substantially reduce or eliminate the disruption (Alexander et al., 1989; Besner, 2001; Besner & Stolz, 1999; Raz et al., 2003, 2006, 2007; Raz, Shapiro, Fan, & Posner, 2002; Wenk-Sormaz, 2005).

Another major reason for the continued popularity of the Stroop task is likely its versatility. Interference has been demonstrated in Stroop variants using a wide range of stimuli (MacLeod, 1991; MacLeod & MacDonald, 2000), including pictures with embedded words, such as the word "cat" printed inside a drawing of a pig (Rosinski, Golinkoff, & Kukish, 1975); location words presented through headphones to a single ear, such as the word "left" played through the right ear channel (Pieters, 1981); and digit sets of varying size, such as the number "3" printed four times (Windes, 1968). Among these many variants exists a growing number in which emotional stimuli serve as the source of the interference. Although these emotional Stroop tasks are well-known in the emotional control literature, we are aware of only one published review that exclusively addresses this work (Williams, Mathews, & MacLeod, 1996). In the 12 years since this review, both the number and diversity of emotional Stroop variants have increased dramatically. Moreover, several recent experiments challenge the dominant theoretical accounts of interference and control offered earlier.

Given these developments, it seems appropriate to consider anew this body of work. We begin below by summarizing several current issues in the broader Stroop literature. Next, we differentiate emotional Stroop variants that seek to disrupt performance through dimensional conflict, from those that do so through distraction. Building on our earlier discussion, we evaluate the fit between prominent models of the traditional Stroop task and current findings from these emotional Stroop variants. This first section concludes with an examination of recent evidence that interference and control processes in both conflict and distraction tasks may persevere beyond the trial on which an emotional stimulus is presented, influencing performance and neural activity on subsequent trials.

We then turn our attention to a modern iteration of the classic question that asks, "How similar are emotional and nonemotional self-control?" Since the advent of neuro-imaging, researchers have increasingly looked to the brain for answers, asking, "How similar are the neural mechanisms that assert control in emotional and nonemotional contexts?" The second section first describes an early emotional Stroop brain imaging study (Whalen et al., 1998) and the influential hypothesis of distinct emotional and nonemotional processing regions in the anterior cingulate cortex (ACC) that it inspired (Bush, Luu, & Posner, 2000). Next, we assess the body of evidence that has amassed from both conflict and distraction emotional Stroop neuro-imaging studies in the years since the original proposal. We conclude that the emotional distinction hypothesis of the ACC requires significant modification.

The third section considers several features of emotional Stroop tasks that may limit the contribution they ultimately make to emotional control research. We begin by noting the fundamental constraints imposed by the quasi-experimental design of emotional distraction Stroop tasks, highlighting the problem of unequal lexical characteristics between word lists. Next, we consider evidence that suggests that interference in these distraction tasks might reflect a transient surprise caused by salience in general, rather than interference that is specific to the processing of emotional stimuli. The third section turns our attention to emotional conflict variants, examining the pertinence of the interference between representations these tasks model, to the question of how one controls one's emotions. Finally, the chapter concludes with suggestions to guide future research.

Adapting the Stroop Paradigm to the Study of Emotional Regulation

The Classic Stroop Task: An Experimental Model of Self-Control

A complete theoretical account of the Stroop task must address two core phenomena. First, it must explain why identifying font color takes longer on incongruent trials. Second, it must explain why congruency does not similarly influence speed when the task is reversed (i.e., the participant must identify the word and ignore font color). The models that have been put forth to explain these phenomena generally adopt a framework of hierarchical processing stages. Common in cognitive psychology, such models assume that across these stages, increasingly abstract stimulus representations mediate the transformation of incoming sensory information into a goal-directed motor response. Whereas lower processing stages perform perceptual analyses and object recognition, higher stages glean semantic and contextual meaning. Representations at these higher stages then interact with representations encoding task rules and goals that guide selection of the correct response. Finally, the response must be mapped to a motoric sequence dictated by the task, and this sequence must be programmed and executed.

Considering Stroop interference within such a framework demands specificity about how the processes interact to degrade performance, and when and where they do so. Most current approaches assume that representations of font color and word form are reflexively and simultaneously processed across early sensory and perceptual stages, but what happens to these two streams at higher levels is a matter of debate. One well-known view invokes the traditional distinction between automatic and controlled processes. An idea with over a century of history in psychological science (Cattell, 1886), automaticity describes highly practiced or hardwired processes that begin without intent, proceed by their own momentum, and persist in spite of volitional attempts to curtail them. In contrast, controlled processes must be initiated and sustained through the mobilization of attentional resources (Hasher & Zacks, 1979; Posner & Snyder, 1975; Shiffrin & Schneider, 1977). Automaticity-based accounts of Stroop interference consider word reading to be an automatic process in literate adults, and thus assume word analysis proceeds reflexively through the higher processing stages. Color identification is not considered to be an automatic process, but if the goal is to identify font color, then analysis of font color must also continue. Controlled attention is invoked to guide processing through these higher stages, yielding an abstract representation for font color that can compete with the automatically evoked word representation to determine the participant's response. As described further below, the presence of coactive representation at response levels is believed to thwart selection processes, ultimately delaying response execution.

For many years, the chief strength of automaticity accounts was the elegant explanation they provided for the performance asymmetry between the standard and reverse goal versions of the classic color-word task (MacLeod, 1991; MacLeod & MacDonald, 2000). From this perspective, setting the goal to word identification would not affect the reflexive processing of words, but higher-stage analysis of font color identification would lack the support of controlled attention and thus come to a halt. Regardless of congruency, only representations activated by the word meaning pathway would thus be made available for response selection, resulting in equivalent performance. More recently, translational models have challenged the conclusion that font color identification is inherently less automatic than word identification (Virzi & Egeth, 1985). This approach argues that asymmetrical interference emerges because word information is already represented in the required response modality, whereas color information must be translated into a verbal representation.[1] By analogy, it is not that word reading is automatic, but that the target this processing stream must reach is closer. But what if the task specifies a different target? According to the translational model, if changing the target differentially alters the translational demand required of representations from the two dimensions, then the symmetry of interference will also change. To test this, Durgin (2000, 2003) developed a Stroop variant in which the participant responds by pointing to color patches on the screen. Given that color identification no longer requires translation to the verbal domain, yet word identification now requires translation into the color domain, the translational model predicts the traditional asymmetry should be reversed (i.e., color identification performance should be unaffected by word meaning, but incongruent font color should disrupt word identification). Durgin found exactly that—a result apparently incompatible with automaticity accounts.

Although automaticity and translational accounts seek to explain the processes that lead to incompatible representations at higher stages of analysis, they do not explain why incompatible representations delay responding. Most accounts assume that response selection occurs when the activity level of a graded representation corresponding to that response is pushed above a preset threshold. For example, Logan's (1980) information accrual model envisions a response selection stage that sums output from parallel pathways processing each stimulus dimension. This output constitutes positive or negative support for each of the possible response options, so when the stimulus dimensions are congruent, both pathways contribute information in support of a particular response, allowing the preset threshold to be reached sooner. However, in the incongruent condition, information from the irrelevant dimension counts against the correct response, increasing the amount of information needed from the relevant dimension for the response threshold to be reached. The increase in reaction time reflects the additional time needed to accrue this information.

A similar accrual mechanism determines responding in the neural network models developed by Cohen and colleagues. In these models, the strength of a representation is governed not just by the strength of previous representations in that pathway, but also by within-level inhibitory and facilitatory connections with competing representations (Cohen, Dunbar, & McClelland, 1990). More recent versions of the model have incorporated mechanisms for the monitoring and control of interference (Botvinick, Braver, Barch, Carter, & Cohen, 2001; Botvinick, Cohen, & Carter, 2004). Information processing conflicts are first detected by a dedicated monitoring system, localized to the rostrodorsal ACC (rdACC), caudodorsal ACC (cdACC), and adjacent medial prefrontal cortex (MPFC). Often this monitoring system is described as multimodal, responding to conflicts across multiple levels of processing, although some have argued the ACC only monitors conflicts in response levels (Liu, Banich, Jacobson, & Tanabe, 2006; van Veen, Cohen, Botvinick, Stenger, & Carter, 2001). Regardless, when the monitoring system detects conflict, it cues the engagement of the task rule and goal representations in the lateral PFC that inhibit or facilitate processing at different processing stages, tipping the balance of activity in favor of the relevant processing stream (Botvinick et al., 2001, 2004; Kerns et al., 2004; Miller & Cohen, 2001). Crucially, the model does not allow these adjustments in control to take place quickly enough to impact the current trial. Thus, evidence for greater control in neural activity and performance can only be observed on the subsequent trial (Botvinick et al., 2001).

Conflict in Emotional Stroop Tasks

One approach to adapting the Stroop paradigm to questions of emotional control, has been to present stimuli with emotional dimensions that vary in congruency. Several studies have combined emotional words and photographs of facial expressions (Anes & Kruer, 2004; Egner et al., 2007; Etkin et al., 2006; Haas et al., 2006; Preston & Stansfield, 2008; Stenberg, Wiking, & Dahl, 1998) or scenes (Park et al., 2008). For example, Hirsch and colleagues printed the words "happy" and "fear" across happy and fearful facial expressions (Egner et al., 2007; Etkin et al., 2006). As in an earlier behavioral experiment (Anes & Kruer, 2004), participants were required to identify the expression, but incongruent words disrupted performance. Other picture-word studies have required participants to categorize a word by valence or emotion, while ignoring the valence or emotional category of the accompanying image (Haas et al., 2006; Park et al., 2008; Preston & Stansfield, 2008, experiment 2; Stenberg et al., 1998). For example, in one experiment participants categorized words such as "party" and "pride" as positive or negative, while ignoring happy and sad facial expressions (Haas et al., 2006). These studies have also consistently reported degraded incongruent performance.

Considering these findings in the context of the broader Stroop literature raises several questions. Taking into account that both words and images can interfere when serving as the goal-irrelevant dimension, might suggest the absence of a goal-dependent performance asymmetry akin to that of the classic color-word task. Unfortunately, to our knowledge no study has directly tested this possibility by alternating dimensional relevancy using a single set of stimuli and consistent experimental parameters. Given that reading and emotional face processing are widely believed to occur with a high degree of automaticity, an automaticity-based account might predict behavioral interference regardless of dimensional relevancy. A translational model might predict a different outcome. In the tasks of Hirsch and colleagues (Egner et al., 2007; Etkin et al., 2006), to indicate the emotion conveyed by the facial expression, a nonverbal representation of the expression would need to be translated to the corresponding verbal representation; in contrast, the word representation is inherently verbal, and thus requires no additional translation once it has been recognized. Thus, a translational model would likely predict little or no interference in a reversed-goal variant of the Hirsch task, in line with the lack of interference when the goal is reversed in the classic Stroop task.

Does the interference effect reported in word-relevant emotional Stroop variants then falsify the translational account? To provide an answer, we again turn to the broader Stroop literature. Stroop variants using nonemotional picture-word combinations first began to appear in the early 1970s (MacLeod, 1991). Typically the stimuli in these tasks featured words printed inside black and white line drawings. At first, the pattern of interference observed seemed to correspond to the color-word Stroop: when participants identified pictures, incongruent words disrupted performance, whereas word identification was unaffected by the congruency of the concurrent picture (Rosinski, Golinkoff, & Kukish, 1975). However, matters were complicated by the discovery that this performance asymmetry was reversed if participants were instructed to categorize the stimuli by object type—for example, responding "animal" when presented a picture of dog (Glaser & Dungelhoff, 1984; Smith & Magee, 1980). To account for these findings, Glaser and Glaser developed a model in which distinct processing units maintain semantic memory and the lexicon (Glaser & Glaser, 1989). They posited that word information has privileged access to the lexicon, whereas picture information has privileged access to semantic memory. Furthermore, they assumed that identification tasks rely on selection among lexical representations, but the critical selection for categorization tasks occurs at a semantic processing stage.

Combining this set of assumptions with a translation model of interference allows the Glaser and Glaser model to predict the observed outcomes of both emotional and nonemotional picture-word tasks. Incongruence disrupts performance when the irrelevant dimension has privileged access to the critical processing stage. Such is the case when the participant must identify the picture, as in emotional expression identification tasks (Anes & Kruer, 2004; Egner et al., 2007; Etkin et al., 2006), or categorize a word, as in word emotion and valence categorization tasks (Haas et al., 2006; Park et al., 2008; Preston & Stansfield, 2008, experiment 2; Stenberg et al., 1998). In contrast, incongruence has little or no effect on performance when the relevant dimension has privileged access to the critical processing stage. Such is the case in tasks that require determination of the semantic category of pictures or the identification of words. Although nonemotional studies provide ample confirmation of this prediction, we know of only one test in the context of an emotional Stroop task. In a preliminary experiment, Preston and Stansfield (2008, experiment 1) required participants to identify the words "sad," "angry," "happy," and "scared," while ignoring corresponding facial expressions. They found a small but significant 25-millisecond difference between the congruent and incongruent conditions. Although this is smaller than the 68-millisecond difference they saw in the second experiment, which required categorization of words, the design was changed in numerous other ways, so a direct comparison of effect size cannot be made. Nonetheless, that this effect is considerably smaller than both typical Stroop effects and other face-word effects is consistent with the Glaser and Glaser model.

Other emotional Stroop variants have manipulated the congruency of prosodic and lexico-semantic dimensions in speech stimuli. Most often stimuli consist of a set of valenced words, each of which carries a positive vocal tone in one recording and a negative tone in another (Grimshaw, 1998; Ishii, Reyes, & Kitayama, 2003; Kitayama & Ishii, 2002; Schirmer & Kotz, 2003; Schirmer, Zysset, Kotz, & Yves von Cramon, 2004). For example, one study recorded words such as "pretty" and "bitter" spoken with both "smooth and round" and "harsh and constricted" tones (Ishii, Reyes, & Kitayama, 2003). Less often, full sentences have been used (Mitchell, 2006a, 2006b; Rota et al., 2007; Vingerhoets, Berckmoes, & Stroobant, 2003). For example, one study recorded sentences such as, "The dog had to be put down" and "She won the lottery jackpot" spoken with happy and sad prosody (Mitchell, 2006a). All but one of these emotional speech Stroop studies required participants to categorize the relevant stimulus dimension. The singular exception might best be considered a hybrid identification-categorization task: participants either identified the words "mad," "sad," and "glad," or categorized the prosody these words carried (Grimshaw, 1998).

In contrast to the image-word variants discussed above, many of these studies examined interference in both dimensional relevance conditions (Grimshaw, 1998; Ishii et al., 2003; Kitayama & Ishii, 2002; Mitchell, 2006a; Schirmer & Kotz, 2003; Vingerhoets et al., 2003). Of the categorization studies reporting both conditions, all but one (Schirmer & Kotz, 2003) found bidirectional effects. These findings collectively suggest greater parity of access to the semantic processing stage between the prosodic and lexico–semantic stimulus dimensions than exists between the color and word dimensions of the classic Stroop task. A series of experiments by Ishii and colleagues suggest that even though prosodic–linguistic conflict might be bidirectional, interference effects show a culturally determined asymmetry (Ishii et al., 2003; Kitayama & Ishii, 2002). For example, American-born speakers of English demonstrated greater interference when identifying prosody valence, but native Japanese speakers and Tagalog-English bilinguals demonstrated greater interference when identifying word valence. These authors attributed the variability of the asymmetry in the populations to a broader difference in the emphasis eastern and western cultures place on contextual information, but the precise cognitive mechanisms that differentiate the processing streams remain unclear.

What results might the Glaser and Glaser (1989) model predict for the identification-categorization hybrid task described by Grimshaw (1998)? Given the parity of semantic access found in other studies, the model would likely anticipate interference during prosodic categorization. However, the privileged lexical access of printed words would presumably extend to spoken words, implying less or no interference in the identification portion of this experiment. In support of these predictions, robust conflict was found in the prosodic categorization condition in two experiments, whereas in the word identification condition, performance was disturbed only slightly (experiment 2), or not at all (experiment 1).

Distraction in Emotional Stroop Tasks

Although conflict Stroop tasks model the disruption of selection mechanisms at critical points in the processing hierarchy, distraction variants model nonspecific performance declines caused by processing a salient, but irrelevant, stimulus dimension. Emotional distraction Stroop tasks typically mimic the format of the traditional color-word Stroop task, requiring the participant to identify the font color of words while ignoring word meaning (Williams et al., 1996). However, in place of the semantically incongruent color words of the traditional task, these distraction-based variants substitute emotional words. Interference is assessed by comparing latency or accuracy on emotional word trials with performance on neutral word trials. Although such a contrast roughly parallels the comparison between incongruent and neutral trials in conflict Stroop tasks, distraction tasks lack putative facilitation trials, making it impossible to find a similar parallel of the more common comparison with congruent trials.

The impossibility of a congruency effect points to a pair of broader characteristics that distinguishes the distraction of these tasks from the conflict of the traditional Stroop and the emotional variants discussed above. First, the relationship between stimulus dimensions is arbitrary: the value of the goal-relevant dimension on a given trial has no bearing on the classification of that trial in the experimental design. For example, if "cancer" is chosen to serve as a valenced word, then a trial in which "cancer" is printed in red constitutes an interference trial, just as a trial in which it is printed in blue, green, or any other color used in the response set. As we will discuss in the following section, this characteristic profoundly influences the interpretation of results in these tasks.

Arbitrariness also characterizes the relationship between the irrelevant dimension and the goal the participant is instructed to achieve. In emotional distraction Stroop tasks, the task goal need only yield a performance measure that allows the experimenter to compare disruption in the emotional and nonemotional conditions. In contrast, assigning the goal of color identification is essential for the incompatible words of the traditional Stroop to disturb performance. This flexibility of task goal in emotional distraction tasks has been usefully exploited in one popular variant, the emotional counting Stroop (Whalen et al., 1998). Although the counting task differs from standard emotional distraction Stroop tasks, in that the participant is required to indicate the number of instances of a word on the screen, the same affective and nonaffective words may be used as in the original color identification version. In contrast, counting analogs of the traditional Stroop must use number words, not color words, in the irrelevant dimension (Bush et al., 1998). The arbitrariness of the assigned goal in emotional distraction Stroop tasks has been further explored with direct empirical investigation. Several studies have found that emotional word lists associated with reduced color identification performance were similarly associated with reduced performance when featured in a speeded reading or lexical decision task[2] (Algom, Chajut, & Lev, 2004; Larsen, Mercer, & Balota, 2006).

The dissimilarity of conflict and distraction has led several authors to question whether a distraction-related performance deficit should be described as a Stroop effect. For example, McKenna and Sharma suggested "emotional intrusion effect" would be more appropriate (McKenna & Sharma, 2004), whereas Algom and colleagues offered the alternative "generic slowdown" (Algom et al., 2004). Although we chose to consider both emotional Stroop variants in the present review, our point is not to suggest they engender a similar type of interference or call upon similar cognitive mechanisms. To the contrary, the empirical data we review below consistently follows the theoretical distinction we have drawn thus far.

Perhaps at no point is this difference as clear as when we compare the performance decrement observed in the two variants. Emotional conflict Stroop tasks consistently show robust effects, in line with the robust interference of the traditional Stroop task, but emotional distraction Stroop tasks typically report much smaller differences in performance, if performance between conditions differs at all. A recent meta-analysis of reaction time data from published accounts of color-word emotional distraction Stroop tasks, found that healthy, nonanxious adults responded with similar speed to both emotional and neutral words (Phaf & Kan, 2007). We do not take this to mean that distraction cannot be successfully modeled by these emotional Stroop variants. Several authors have observed small but consistent performance differences in carefully designed experiments (Algom et al., 2004; McKenna & Sharma, 2004). The meta-analytic null finding likely reflects a number of factors, including limited power at the level of the individual studies, and the inclusion of studies in which design factors inadvertently obscured an effect. However, the null finding also highlights the tenuous nature of distraction-based interference, further distinguishing distraction- and conflict-based Stroop tasks. As such, it is critical when considering emotional distraction Stroop tasks, to remain vigilant to the possibility that the emotional stimuli did not interfere with goal-directed processing, and to temper interpretation when no evidence of interference is provided by the performance data.

Subsequent Trial Interference in Emotional Distraction Stroop Tasks

Thus far, we have considered emotional interference and control only within the time frame of a single trial. Recent evidence from a number of sources suggests that a complete account will require that we expand our view, allowing for the possibility that these processes may exert influence not just within, but across trials also. An early indication of such extra-trial influence in emotional Stroop tasks came from the observation that blocked designs, in which trials are presented grouped by condition, showed a greater behavioral effect of emotional distraction than designs in which emotional and neutral trials were mixed (Algom et al., 2004; Cassiday, McNally, & Zeitlin, 1992; Jones-Chesters, Monsell, & Cooper, 1998; Kaspi, McNally, & Amir, 1995; Phaf & Kan, 2007; Richards et al., 1992; Waters & Feyerabend, 2000; Waters et al., 2005). Although the difference in performance between blocked and mixed designs could have resulted from nonspecific changes in motivation or some other aspect of mental set, these findings could also be accounted for by interference processes persisting beyond the trial on which the causal emotional stimulus appeared. If at least a portion of the interference effect from an emotional trial were spread across subsequent trials, then in mixed designs some of the subsequent latency might be attributed mistakenly to neutral trials, bringing the average reaction time down for the emotional condition and up for neutral condition.

Such temporally extended distraction could take one of at least three general forms. One possibility is that viewing the emotional stimuli induces a long-term change in emotional state in which performance is generally degraded compared to a neutral state. If a direct effect of the stimuli on the emotional state accounts for the greater performance disruption in blocked designs, we might expect this disruption to increase over time (Ehlers et al., 1988). A second possibility is that interference begins on a particular trial and persists into the next, but the disturbing influence wanes quickly. Evidence for this type of subsequent trial effect, which Waters and colleagues termed "carry-over effects" (Waters et al., 2005; Waters, Sayette, & Wertz, 2003), would be provided by data in which emotional stimuli independently accounted for decreased performance on both the presentation trial and subsequent trials. But what if one observed such a decrement subsequent to the presentation trial, but not on the presentation trial itself? Termed "slow effects" by McKenna and Sharma (2004), this might indicate that the presence of an emotional stimulus sets in motion a sequence of mental events that eventually lead to disruptive processes, but that goal-directed processing of the causal stimulus is completed before the disruptive portion of the sequence begins.[3] Finally, state, carryover, and slow effects might manifest on emotional trials exclusively, indicating a cumulative or potentiating influence, or such effects might manifest universally, increasing latency on emotional and neutral trials alike.

At present, only a handful of emotional distraction Stroop studies provide results capable of distinguishing among the above possibilities. We know of no studies that support the state effects hypothesis by showing an increase in interference across blocked emotional stimuli. In contrast, several studies found evidence of habituation (MacKay et al., 2004; McKenna & Sharma, 1995), or no change in interference (Sharma, Albery, & Cook, 2001), a direct contradiction of the state effects hypothesis. We know of four reports that directly examined performance as a function of previous trial type. In each instance, subsequent trial effects were identified, though the type of subsequent effects varied (Kunde & Mauer, 2008; McKenna & Sharma, 2004; Waters et al., 2003, 2005). Carry-over effects were found with smokers (Waters et al., 2003, experiment 1) and heroin addicts (Waters et al., 2005, experiment 1) using drug-related words, whereas slow-effects were found with smokers using drug-related words (Waters et al., 2003, experiment 2), and healthy controls using negative words (McKenna & Sharma, 2004), situation-specific stress words (Waters et al., 2005, experiment 2), and positive and negative pictures (Kunde & Mauer, 2008). Several studies reported an interaction between previous and current trial type, indicative of a cumulative effect or potentiation (Kunde & Mauer, 2008; Waters et al., 2005).

Although evidence for the existence of some kind of sequential effect at this point seems strong, the current set of mixed results limits further specification. Although it may be the case that the different experimental designs demonstrate different types of sequential effects because of an unknown variable, another possibility is that the use of unbalanced stimulus orders has in some cases yielded artifactual results. McKenna and Sharma pointed out that if trial types do not follow one another with equal frequency in an experiment, a lopsided distribution of sequential effects can distort observed effects (McKenna & Sharma, 2004).[4] In a series of careful experiments that maintained the frequency of sequential trial types, the authors observed only slow effects. Given the precision of the design, and that the participants consisted of healthy individuals in a standard experimental context, we believe the existence of slow effects seems to be the best supported finding at this time.

McKenna and Sharma further demonstrated that the slow effects they observed occurred on just the single trial following a negative word. Given the timing of stimulus presentation in their design, this implies the interference occurred within the second following response execution for the previous trial. This result is consistent with other work, showing that block design studies using very short gaps between trials, such as 32 milliseconds or 40 milliseconds, yielded greater interference effects than longer gaps, such as 1000 milliseconds or 500 milliseconds (Sharma & McKenna, 2001; van Hooff et al., 2008). However, other studies have reported subsequent effects with gaps between trials of 1000 milliseconds (Kunde & Mauer, 2008; Waters et al., 2005), 1500 milliseconds (Waters et al., 2005), and 2200 milliseconds (Waters et al., 2003). The extended interference in these studies may result in part from unique factors such as the use of picture stimuli (Kunde & Mauer, 2008), drug-addicted participants (Waters et al., 2005, Waters, 2003 #128), stress induction, context-specific words, or intoxication (Waters et al., 2005). Future studies will need to isolate and test the various possibilities to determine the precise temporal duration of the phenomenon and any factors that might modulate this duration.

Subsequent Trial Control in Emotional Conflict Stroop Tasks

To our knowledge, only Hirsch and colleagues have reported analyses of subsequent trial effects in an emotional conflict Stroop task (Egner et al., 2007; Etkin et al., 2006). In two neuroimaging studies, participants classified facial expressions as happy or fearful while ignoring the words "happy" or "fear" printed over each image. Following an approach developed in earlier work with nonemotional conflict tasks (Botvinick et al., 1999; Egner & Hirsch, 2005b; Gratton, Coles, & Donchin, 1992; Kerns et al., 2004), the authors compared performance on incongruent trials grouped according to the congruency of the previous trial. In line with findings in these earlier nonemotional studies, participants responded more quickly when the preceding trial was also incongruent (II trials) than when it was congruent (CI trials). Neural activity was also lower in the II trials in a portion of the midline rostrodorsal MPFC (Etkin et al., 2006) and in the right rdACC (Egner et al., 2007), again paralleling previous analyses with the classic color-word Stroop task (Kerns et al., 2004) and a nonemotional face-word variant (Egner & Hirsch, 2005a). The authors interpreted these findings in the conflict monitoring framework developed by Cohen and colleagues (Botvinick et al., 2001, 2004), positing that the detection of conflict on the preceding incongruent trial initiated a reactive increase in control. Increased control then reduced conflict on the subsequent trial, resulting in better performance and less detection-related activity in the ACC.

In contrast to the large literature examining the conflict monitoring hypothesis of Cohen and colleagues, the reactive control proposed by these models has been addressed only briefly. In previous work, Hirsch and colleagues reasoned that if greater activity in CI trials reflects conflict monitoring, then greater activity in II trials might correspond to the implementation of reactive control (Egner & Hirsch, 2005a, 2005b). In both emotional Stroop studies, this reverse contrast (II>CI) identified greater activity in left pgACC and adjacent MPFC. One possibility is that these clusters reflect representations of transient task rules or goals. In the models of Cohen and colleagues, reactive control is accomplished by strengthening PFC representations that code these rules and goals. As these representations are stronger, they alter connection weights across the processing hierarchy in favor of the goal-relevant dimension. In the emotional face-word task, rule or goal representations might serve to direct attention to the facial stimulus or otherwise facilitate processing in this pathway. However, this region did not show differential activity in a gender-identification task in which the same face images were instead combined with the words "male" and "female" (Egner et al., 2007). This suggests that in the emotion identification task, greater pgACC activity might specifically reflect the engagement of representations guiding assessment of the emotional aspect of the expressions. Consistent with such an interpretation, other work has associated this pregenual area with the perception of emotional stimuli in a wide variety of tasks (Wager et al., 2008).

However, interpreting activity differences in this pregenual region can be challenging. Another recent imaging study found decreases from baseline activity in this area during a word-scene valence categorization Stroop task, with greater decreases on incongruent than congruent trials (Park et al., 2008). Other work has reported activity reductions in the pgACC in both emotional (Gusnard et al., 2001) and nonemotional tasks (Gusnard et al., 2001; Mason et al., 2007), a pattern seen in a distributed set of brain regions. Paralleling this set of task-negative regions, goal-directed processing across many different tasks appears to engage a separate set of regions, including the rdACC and MPFC areas that Hirsch and colleagues identified in the CI>II contrast. Consistent with the findings of Hirsch and colleagues, activity in these goal-directed regions has been found to correlate negatively with activity in the task-negative (Fox & Raichle, 2007; Fox et al., 2005). Other work has found a greater activity reduction in the pregenual area during novel compared to practiced task blocks (Mason et al., 2007), and during blocks of speeded, incongruent trials compared to blocks of slower and congruent trials in a variant of the traditional Stroop task (Matthews et al., 2004). These findings suggest that the greater pregenual activity observed in the II trials might reflect the reduced challenge that increased control confers, rather than representations or processes specific to the instantiation of control.

These subsequent trial analyses support at least two important conclusions. First, the pattern of reaction time data, and to some degree the fMRI results as well, reinforce the similarity between emotional conflict Stroop tasks and other tasks that model conflict-based interference, including the classic color-word Stroop task. As we highlight throughout this chapter, this similarity confers a unique opportunity to utilize techniques developed in the long history of nonemotional conflict research, and the work of Hirsch and colleagues demonstrates this opportunity. Second, although no study yet has directly compared subsequent trial effects between conflict- and distraction-based emotional Stroop tasks, surveying the current literature suggests distinct phenomena. Although conflict may invoke control processes that result in improved performance on subsequent incongruent trials, emotional distraction seems to degrade subsequent performance, even on neutral trials. More broadly, this dissimilarity reinforces the contention that emotional distraction and conflict Stroop tasks involve dissimilar types of interference.

Are Emotional Control Mechanisms in MPFC Distinct from those That Implement Non-emotional Control?

A Neural Instantiation of a Traditional View

The cognitive sciences have long struggled to understand how the human brain balances two fundamental demands: *(1)* it must retain the flexibility to respond to an infinite diversity of possible situations and *(2)* it must do so efficiently as cranial space is limited and the metabolic needs of neural tissue are high. Although it is now clear that a substantial degree of localization characterizes neural mechanisms across domains ranging from perceptual to mnemonic and linguistic processing, the field continues to question the nature and degree of specificity among the control functions of the prefrontal cortex. For example, a good deal of work has debated whether the control functions of working memory are material-dependent (Smith & Jonides, 1999; Wager & Smith, 2003).

A similar question has asked whether emotional and nonemotional control rely on distinct neural mechanisms (Ochsner & Gross, 2005). One particularly influential answer to this question has claimed that such a distinction characterizes the ACC, with tissue extending posterior and superior from the rdACC dedicated to nonemotional processing, and tissue extending anterior and inferior from the rdACC dedicated to emotional processing (Bush et al., 2000). This emotional distinction hypothesis emerged from an early neuro-imaging experiment in which participants performed two separate counting Stroop tasks in a single scanning session. In a task modeling nonemotional conflict, the authors reported worse performance and greater cdACC activity in incongruent blocks, in line with expectations based on previous work (Bush et al., 1998). In contrast, a second task modeling emotional distraction revealed greater emotional compared to neutral word activity in the rdACC (Whalen et al., 1998).[5] The authors suggested this rdACC activity might reflect a regulatory or monitoring function for emotional material parallel to the putative role of the cdACC in nonemotional interference. In a well-known review that followed, a summary of results from a number of neuro-imaging studies was presented alongside the findings from these counting Stroop tasks to support the proposed emotional distinction hypothesis (Bush et al., 2000).

Although widely cited, the study by Whalen et al. poses several inferential challenges. First, as we have emphasized throughout, distraction-based interference differs theoretically and empirically from conflict-based interference. If we observe distinct regulatory activity in emotional distraction and nonemotional conflict, we cannot know if this divergence reflects differences in the type of interference or the emotional nature of the stimuli. Second, the nonemotional task was always the first of the two performed in the session. This lack of counterbalance further confounds emotional content with both time spent in the scanning environment and task exposure. Third, performance was equivalent in the emotional and neutral word blocks of the distraction task, but in the conflict Stroop task, performance varied with congruency. At best, this asymmetry might indicate that the neuro-imaging results are confounded by gross differences in the level of interference induced by the two tasks. At worst, the neuro-imaging results in the distraction task may not represent interference at all. As we argue throughout this chapter, reduced behavioral performance provides essential evidence for any claim of interference. A fourth inferential challenge is posed by the transient nature of the rdACC activity observed in the distraction task. The task was performed in two sets of alternating blocks, separated by a short break. The authors reported that they only observed a difference between the emotional and neutral conditions when they removed from analysis the second half of each of the two data sets collected. As we discuss further below, this transience might indicate that the neuro-imaging activity reflects a more general salience processing rather than emotional processing in particular. Finally, the report lacked formal tests of the emotional distinction hypothesis. The authors appealed to visual inspection of contrast maps of the two tasks, but informal assessment cannot replace statistical analysis of neuro-imaging data anymore than it can do so when comparing other types of data.

Emotional Interference in the Rostrodorsal Anterior Cingulate Cortex

A lack of consistency across studies further challenges the emotional distinction hypothesis. Of the six other studies we have identified that report imaging data of healthy adults performing the emotional distraction Stroop tasks (Compton et al., 2003; George et al., 1994; Herrington et al., 2005; Isenberg et al., 1999; Mohanty et al., 2007), only one found similar rdACC activity (Mohanty et al., 2007). This single replication is noteworthy, as the authors designed the study with the intention to overcome several of the limitations in the work of Bush and colleagues: the order of nonemotional conflict and emotional distraction Stroop blocks were counterbalanced across participants; analyses included the blocks in their entirety; and formal statistical tests confirmed the proposed region-by-task interaction. Despite these improvements, the design did not overcome other serious limitations of the earlier work of Bush and colleagues: most critically, the tasks compared distinct types of interference, distraction and conflict, and no performance decrement was observed in the emotional task.

A number of conflict-based neuro-imaging studies have also compared emotional and nonemotional Stroop tasks free from these serious limitations. As noted above, Hirsch and colleagues have found greater activity in left pgACC and MPFC contrasting II>CI trials in face-word conflict tasks (Egner et al., 2007; Etkin et al., 2006). The coordinates they report place these clusters some distance away from that found by Whalen et al. (Whalen et al., 1998), but within the expanse of tissue Bush and colleagues considered the emotional division of the ACC (Bush et al., 2000). Several other emotional conflict Stroop studies have also reported pregenual activity. Having asked participants to identify prosody as angry, sad, happy or neutral, Rota and colleagues compared incongruent sentences to sentences which were either congruent or contained at least one neutral dimension (Rota et al., 2007). The authors reported only two activation peaks, both very close to the midline in the pgACC and pregenual MPFC. However, the inclusion of both congruent and neutral element trials in this contrast muddies interpretation. In addition, the incongruent trials also contained a greater number of emotional elements, on average, and so the pregenual differences might simply reflect greater emotional input. As discussed above, a recent word-image conflict Stroop study also reported activity differences in the left pgACC and pregenual MPFC (Park et al., 2008), but in the opposite direction of Rota and colleagues; that is, greater activity was found in the congruent trials than in the incongruent trials.[6]

Emotional and Nonemotional Interference in the Dorsal Anterior Cingulate Cortex

Emotional conflict Stroop studies have also reported greater activity in the rdACC on incongruent compared to congruent blocks (Haas et al., 2006) and on CI>II trials (Egner et al., 2007). Although these conflict-related clusters appear to fall within an area of overlap among the emotional and nonemotional tasks classified by Bush and colleagues (Bush et al., 2000), the authors of both papers concluded the locations implied a domain-general conflict monitoring mechanism, in argument against the emotional distinction hypothesis. To bolster this claim, Egner and colleagues performed a similar CI>II analysis on data from a gender identification task in which conflict arose from incongruent gender labels. They identified a cluster of activated voxels with a peak in the left cdACC, consistent with previous research supporting a role for this area in nonemotional conflict monitoring. Furthermore, a portion of this cluster overlapped with the area identified in the emotional task. This shared area showed a main effect of CI>II across tasks, indicating consistent participation for the region regardless of the emotional nature of the conflicting dimensions.

Results from other emotional Stroop studies further challenge the emotional distinction hypothesis. In an earlier study, Hirsch and colleagues reported CI>II activity in their face-word task in a portion of dorsal MPFC, far from the proposed affective division (Etkin et al., 2006), and two emotional distraction Stroop studies also found emotion-related increases in dorsal MPFC (Compton et al., 2003; Isenberg et al., 1999). More direct evidence of emotion-related processing in the cdACC was provided by a rare human cellular recording study of OCD patients immediately prior to therapeutic cingulotomy targeting this region (Davis et al., 2005). The authors identified cdACC neurons that responded to both emotional distraction and nonemotional conflict counting Stroop tasks. Furthermore, the proportion of cells showing greater or exclusive response in negative word compared to the neutral word blocks (10/24), was much higher than the proportion of cells showing greater or exclusive response in the incongruent compared to neutral word blocks (2/20). A similar effect was seen when comparing a block of OCD-related words to the neutral condition (7/25). Although the relationship of such neuronal activity to the hemodynamic signal observed in fMRI remains unclear, this study provides unique evidence that neurons in the cdACC play a role in processing negative word stimuli in the context of a distraction Stroop task.

Taken as a whole, we believe the current set of emotional Stroop findings advocate revision of the emotional distinction hypothesis. The dorsal portion of the ACC that Bush and colleagues (2000) believed exclusive to nonemotional processing, has been repeatedly observed in both emotional and nonemotional Stroop neuro-imaging studies, in line with similar work comparing emotional and nonemotional interference (e.g., Fichtenholtz et al., 2004; Yamasaki, LaBar, & McCarthy, 2002). Although the precise computations supported by this tissue remain unclear, the overlapping activity suggests that they serve a common function regardless of the eliciting stimulus. The validity of the second half of the emotional distinction hypothesis, that the rostral and ventral portion of the ACC participate exclusively in emotional processing, also remains unclear. Although numerous emotional Stroop studies have identified interference-related activity in rdACC and pgACC, null findings are equally common. Furthermore, interpretation of the reported activations is difficult because they may reflect differential task-related activity reductions. Going forward, it will be important to determine if the relatively greater emotion-related activity that has been seen, reflects an active role in emotion-related processing, or if it is simply a passive byproduct of other processes. If this region does actively participate in emotional processing, further work should attempt to isolate the eliciting factors to account for the inconsistency of the current emotional Stroop literature. Finally, future work should consider the possibility that the rostrodorsal and pregenual regions serve distinct functions.

Is Interference a Result of Emotional Relevance or Other Stimulus Properties?

Quasi-Experimental Design in Emotional Distraction Stroop Tasks

A critical feature of experimental design is the control of extraneous variables that may obscure a true effect or confound an observed effect. Recently Larsen and colleagues argued the design of emotional distraction Stroop tasks is quasi-experimental because the words that comprise the valenced and neutral conditions cannot be assigned randomly (Larsen et al., 2006). Such nonrandom assignment incurs the risk that incidental differences between the lists might drive differences in performance. For example, given that word frequency has been shown to modulate the speed of font color identification (Burt, 1994, 1999, 2002; Monsell, Taylor, & Murphy, 2001), if the words in the emotional and nonemotional lists in a distraction Stroop task vary in frequency, then one might spuriously attribute a frequency-driven performance decrement to emotional distraction.

To test whether lexical differences between word lists might confound interpretation of emotional distraction Stroop tasks, Larsen and colleagues analyzed over 1000 unique word stimuli from 32 published reports (Larsen et al., 2006). Using a large collection of lexical decision and word identification data, the authors compared average performance on words culled from emotional and neutral Stroop lists. The emotional words were associated with slower and less accurate responding, consistent with previous findings that have linked performance in similar lexical and emotional distraction Stroop tasks (Algom et al., 2004). However, Larsen and colleagues also showed that these emotional words were longer, less frequently used, and had smaller orthographic neighborhoods[7] than the neutral words. Each of these factors is known to contribute to slower and less accurate word recognition and lexical assessment, independent of semantic content. When these factors were included in the statistical model, the performance difference between emotional and neutral words disappeared.

Quasi-experimental designs have played an important role in many research domains because they allow researchers to ask questions when experimental designs do not. Yet these results recommend caution when interpreting the existing emotional distraction Stroop literature. So that readers may exercise such caution judiciously, the authors provided tables with the lexical metrics that they calculated for each Stroop study in their analysis (Larsen et al., 2006). To help correct for such problems in future work, they further offered to provide word lists upon request that equate for the problematic lexical factors they identified. Although these efforts allow for needed improvements, the use of balanced lists does not change the underlying quasi-experimental nature of distraction Stroop tasks, and so leaves the manipulation open to the influence of unknown systematic bias. In contrast, the experimental design of conflict Stroop tasks avoids such ambiguity: using identical color words in both the incongruent and congruent conditions ensures equivalence.

Salience in Emotional Distraction Stroop Tasks

The construct validity of emotional distraction Stroop tasks is further challenged by other findings that suggest that the interference observed is not specific to emotional stimuli. Although the meta-analysis conducted by Phaf and Kan found no support for increased emotional word latency in healthy, non-anxious adults, an effect was found in several analyses with high-anxiety and clinical participants (Phaf & Kan, 2007). One possibility is that the poorer performance of these participants reflects pathological attentional control deficits. Alternatively, some have suggested that emotional words possess greater relevance to clinical and anxious participants, and that this greater relevance potentiates their emotional impact, increasing distraction and reducing performance. If so, the meta-analytic null finding in healthy adults might reflect insufficiently emotional stimuli, a problem that could be remedied by making the stimuli personally relevant. In support of this view, items individually tailored to affectively positive and negative concerns of healthy participants reduced performance compared to items that were not self-relevant (Giles & Cairns, 1989; Riemann & McNally, 1995), just as food words reduced performance in hungry, but not satiated, participants (Channon & Hayward, 1990).

However, the hypothesis that self-relevant emotional words interfere with performance because of heightened emotional impact, is challenged by findings in several nonemotional distraction Stroop Tasks. Interference has been found with marginally emotional but clearly self-relevant words, drawn from hobbies and interests (Dalgleish, 1995) and biographical information (Gronau, Ben-Shakhar, & Cohen, 2005; Gronau, Cohen, & Ben-Shakhar, 2003). In one attempt to resolve this ambiguity, participants performed a distraction Stroop task that included individually selected words culled from emotional and neutral memories, as well as standard emotional and neutral words (Gilboa-Schechtman et al., 2000). Reaction times were slower for standard emotional compared to neutral words, but only when congruent in valence to a film shown immediately before. Furthermore, reaction times were slower for individually selected compared to standard words regardless of whether they were emotional or neutral. Together, these results argue that relevance more profoundly determines distraction than emotional meaning.

Some evidence suggests this argument may need to be taken further. One set of distraction Stroop tasks has used salient words such as curses and out-group slurs (MacKay & Ahmetzanov, 2005; MacKay et al., 2004; Siegrist, 1995). Although some of these words may carry emotional or self-relevant meaning for certain participants, most seem unlikely to do so. Although fewer in number than emotional distraction Stroop studies, the existing literature indicates these salient words may more robustly disturb performance. The hypothesis that interference in distraction Stroop tasks primarily reflects salience or surprise is further supported by the transience of the effect. Rapid attenuation has been reported in both behavioral (MacKay et al., 2004; McKenna & Sharma, 1995) and brain (Whalen et al., 1998) difference measures.

Of course, describing distracting words as salient risks tautology. Predictive power rests on identifying the determinants of word salience. The distraction Stroop literature implicates a diverse set of factors, including situational and personal relevance, lexical qualities, and familiarity. Overall, the emotional content of a word seems to contribute only weakly to salience, if it independently contributes at all. Moving forward, a more fruitful approach might consider emotional distraction Stroop tasks as continuous with other distraction Stroop variants. Such an approach would call first for more comprehensive efforts to determine the relative contributions of the various factors that seem to slow color naming, followed by efforts to compare the mechanisms supporting performance as a function of both distraction degree and kind.

Is Emotional Stroop Conflict Truly Emotional?

Following closely the model of the classic Stroop task, emotional conflict variants avoid the ambiguities of quasi-experimental design. Yet the theoretical and empirical overlap between these tasks raises validity concerns of a different kind. As described above, variants of the Stroop have used a wide range of stimuli. The present chapter has followed previous work and used the term "emotional conflict" to describe interference between incompatible emotional dimensions. However, convention yields inconsistent terminology, as it prohibits the use of parallel terms such as "color conflict" or "number conflict" to describe interference between nonemotional stimuli. Uniquely terming emotional conflict according to stimulus type may betray an a priori bias of the emotional conflict Stroop literature to assume distinct emotional processes.

In our view, whether conflict should be considered distinct when arising from incompatible emotional dimensions, depends on where and how selection occurs. To warrant distinction, the critical point of selection must occur in a uniquely emotional processing stage or neural system, or the mechanisms supporting this selection must be unique. However, we do not believe that uniquely emotional conflict is a prerequisite for emotional conflict Stroop tasks to help advance our understanding of emotional processing. Just as word- and speech-based Stroop tasks have contributed extensively to knowledge about linguistic processing, so too could emotional Stroop tasks help define the hierarchy of processing stages that evaluate emotional stimuli.

Where to Go from Here?

As we have emphasized throughout this review, emotional conflict and distraction Stroop tasks model theoretically distinct types of interference. In conflict, the selection of a correct representation is thwarted by concurrent activation of an incorrect representation. This form of interference is specific: representations conflict because of the relationship between the relevant stimulus dimension, the irrelevant dimension, and the task goal. In contrast, distraction is a general form of interference, driven by inherent qualities of the irrelevant stimulus dimension. The relevant dimension and the goal are arbitrary: goal-related processes serve only to measure the disruptive influence of the irrelevant dimension.

This distinction is evident in the divergent findings of emotional conflict and distraction Stroop tasks. Although studies of emotional conflict report robust congruency effects, emotional stimuli in the distraction Stroop tasks typically fail to disrupt the performance of healthy adults (Phaf & Kan, 2007). When disruption does occur in these distraction tasks, recent work suggests the processes mediating interference lag behind goal-directed processing, perhaps only degrading performance when the gap between trials is shorter than this lag (Kunde & Mauer, 2008; McKenna & Sharma, 2004; Waters et al., 2003, 2005). Between-trial effects have also been found in emotional conflict tasks (Egner et al., 2007; Etkin et al., 2006), but rather than extended interference, conflict seems to invoke greater attentional control, resulting in better performance on subsequent incongruent trials.

It is especially important to recognize these differences when seeking to compare emotional and nonemotional interference resolution. Comparing matched emotional and nonemotional conflict variants, as in the recent work of Hirsch and colleagues (Egner et al., 2007), asks whether emotional stimuli conflict at distinct processing stages, and whether overlapping mechanisms are called upon to resolve this conflict. Other studies have compared emotional distraction and nonemotional conflict variants. This approach confounds changes in interference and stimulus type: if we observe unique mechanisms in emotional and nonemotional Stroop tasks, we cannot know if this divergence reflects a difference between nonemotional and emotional interference, or a difference between specific and general interference. The obvious solution is to contrast distracting emotional conditions with equally distracting nonemotional conditions. A few behavioral studies have taken this approach (e.g., Gilboa-Schechtman et al., 2000), however, we do not know of any imaging studies that have yet done so.

Both emotional conflict and distraction Stroop tasks offer exciting opportunities for future research. The path forward for work with emotional conflict Stroop tasks is clearer: researchers should follow in the footsteps left by many years of productive work with nonemotional conflict variants. The utility of such an approach already has been demonstrated in the subsequent trial analyses of Hirsch and colleagues (Egner et al., 2007; Etkin et al., 2006), which built on earlier work in traditional conflict tasks (Botvinick et al., 1999; Gratton et al., 1992; Kerns et al., 2004), and in the examination of goal-related performance asymmetries by Ishii and colleagues, which used the well-known technique of dimensional relevancy reversal from the classic color-word task to demonstrate cultural influences on speech processing (Ishii et al., 2003; Kitayama & Ishii, 2002). Especially promising opportunities for future research might build on nonemotional Stroop paradigms that relate output modality and dimensional relevancy (Durgin, 2000, 2003), and categorization and identification task goals (Glaser & Dungelhoff, 1984; Glaser & Glaser, 1989).

Research with emotional distraction Stroop tasks faces a more challenging path forward. Given evidence that interference in these paradigms may be driven less by the emotional nature of the stimuli than by personal relevance, familiarity, and lexical factors, future efforts should seek to determine more broadly the characteristics of stimuli, emotional or otherwise, which confer salience. Subsequent trial distraction effects offer another intriguing avenue for future work. However, the success of any effort to model emotional distraction rests on whether it provides clear evidence of inference through disrupted performance. Such evidence has been particularly lacking in neuro-imaging using emotional distraction tasks. Some have suggested that observed differences in neural activity might represent a liminal interference too subtle to impact cruder behavioral measures of performance, however, we caution against this approach. Such claims rely on a level of reverse inference that current neuro-imaging techniques do not provide, and perhaps never will. Moreover, the quasi-experimental nature of these tasks leaves a clear opening for alternative explanations. Yet, even if observed activity did reflect subtle distraction, we would question the value of studying interference of such minimal behavioral consequence. A greater contribution might be made by developing tasks that evince more robust effects.

Our primary goal in writing this chapter was to provide the first comprehensive review of emotional Stroop tasks. The valuable contributions these Stroop variants already have made to the study of emotional interference and control, testifies to the wisdom of adapting robust and versatile paradigms from nonemotional work. A second goal of this chapter was to leverage the detailed theoretical accounts of the broader Stroop literature in assessing current emotional Stroop research. In doing so, we have emphasized the distinction between conflict- and distraction-based interference, a prerequisite for meaningful review of this work. We have also drawn novel connections between recent emotional variants and the more extensive corpus of nonemotional Stroop research. These connections suggest important alternative explanations and offer compelling paradigms to test them. We hope that in highlighting just a handful of these possibilities we have revealed sufficient promise to inspire deeper integration between future emotional and nonemotional Stroop research.

Acknowledgments

We would like to thank Jennifer Silvers and the editors of this volume for helpful comments on earlier versions of the manuscript.

Notes

1 Note that the translational model assumes verbal representations mediate response selection, not just when a spoken answer is given: in line with other perspectives, if the task requires a motor act such as a button press be made to indicate the response, it is believed that the appropriate verbal representation is first selected, and then mapped to the motor response according to task-specific rules. Thus, whether button presses or speech are used to indicate the response, it does not change the prediction of the translational model.

2 First developed by the prominent cognitive psychologist David Meyer, lexical decision tasks compare performance when determining whether letter-strings constitute words (Meyer & Schvaneveldt, 1971).

3 Waters and colleagues do not appear to distinguish carry-over and slow effects. Thus, in experiment 2 of Waters et al. (2003), they describe a subsequent trial latency increase for smoking words as a carryover effect in the absence of a fast-effect. According to our nomenclature this effect would be a slow effect.

4 In a striking demonstration of this, the authors constrained trial order such that trials following a particular type were more likely to be of the opposite type, and the sequential effects of negative trials preferentially impacted neutral trials. Using words that had produced a 30ms emotional latency difference in an earlier blocked study (McKenna & Sharma, 1995), the constrained trial order now produced a reverse interference effect of 18 ms.

5 Similar to the pgACC activity elicited by the emotional scene-word conflict Stroop task we discussed earlier (Park et al., 2008), rdACC activity in this distraction counting Stroop task was lower in both the neutral and emotional conditions, than in the task baseline. Greater activity in the emotional blocks might therefore be more accurately described as less negative than that of the neutral condition.

6 As noted above, baseline activity was higher in both the congruent and incongruent conditions in this task.

7 The orthographic neighborhood of a target work is the number of other words that the target word can be transformed into by changing just one of its letters.

REFERENCES

Alexander, C. N., Langer, E. J., Newman, R. I., Chandler, H. M., & Davies, J. L. Transcendental meditation, mindfulness, and longevity: An experimental study with the elderly. J Pers Soc Psychol 1989; 57(6): 950–964.

Algom, D., Chajut, E., & Lev, S. A rational look at the emotional Stroop phenomenon: A generic slowdown, not a Stroop effect. J Exp Psychol Gen 2004; 133(3): 323–338.

Anes, M. D., & Kruer, J. L. Investigating hemispheric specialization in a novel face-word stroop task. Brain Lang 2004; 89(1): 136–141.

Besner, D. The myth of ballistic processing: evidence from Stroop's paradigm. Psychon Bull Rev 2001; 8(2): 324–330.

Besner, D., & Stolz, J. A. What kind of attention modulates the Stroop effect? Psychon Bull Rev 1999; 6(1): 99–104.

Botvinick, M. M., Braver, T. S., Barch, D. M., Carter, C. S., & Cohen, J. D. Conflict monitoring and cognitive control. Psychol Rev 2001; 108(3): 624–652.

Botvinick, M. M., Cohen, J. D., & Carter, C. S. Conflict monitoring and anterior cingulate cortex: An update. Trends Cogn Sci 2004; 8(12): 539–546.

Botvinick, M. M., Nystrom, L. E., Fissell, K., Carter, C. S., & Cohen, J. D. Conflict monitoring versus selection-for-action in anterior cingulate cortex. Nature 1999; 402(6758): 179–181.

Burt, J. S. Identity primes produce facilitation in a colour naming task. Q J Exp Psychol A 1994; 47A(4): 957–1000.

Burt, J. S. Associative priming in color naming: interference and facilitation. Mem Cogn 1999; 27(3): 454–464.

Burt, J. S. Why do non-color words interfere with color naming? J Exp Psychol Hum Percept Perform 2002; 28(5): 1019–1038.

Bush, G., Luu, P., & Posner, M. I. Cognitive and emotional influences in anterior cingulate cortex. Trends Cogn Sci 2000; 4(6): 215–222.

Bush, G., Whalen, P. J., Rosen, B. R., Jenike, M. A., McInerney, S. C., & Rauch, S. L. The counting stroop: An interference task specialized for functional neuroimaging—validation study with functional MRI. Hum Brain Mapp 1998; 6(4): 270–282.

Cassiday, K. L., McNally, R. J., & Zeitlin, S. B. Cognitive processing of trauma cues in rape victims with post-traumatic stress disorder. Cogn Ther Res 1992; 16: 283–295.

Cattell, J. M. The time it takes to see and name objects. Mind 1886; 11: 63–65.

Channon, S., & Hayward, A. The effect of short-term fasting on processing of food cues in normal subjects. Internat J Eat Disord 1990; 9(4): 447–452.

Cohen, J. D., Dunbar, K., & McClelland, J. L. On the control of automatic processes: A parallel distributed processing account of the Stroop effect. Psychol Rev 1990; 97(3): 332–361.

Compton, R. J., Banich, M. T., Mohanty, A., et al. Paying attention to emotion: An fMRI investigation of cognitive and emotional Stroop tasks. Cogn Affect Behav Neurosci 2003; 3(2): 81–96.

Dalgleish, T. Performance on the emotional Stroop task in groups of anxious, expert, and control subjects: A comparison of computer and card presentation formats. Cogn Emot 1995; 9(4): 341–362.

Davis, K. D., Taylor, K. S., Hutchison, W. D., et al. Human anterior cingulate cortex neurons encode cognitive and emotional demands. J Neurosci 2005; 25(37): 8402–8406.

Durgin, F. H. The reverse Stroop effect. Psychon Bull Rev 2000, 7(1): 121–125.

Durgin, F. H. Translation and competition among internal representations in a reverse Stroop effect. Percept Psychophys 2003; 65(3): 367–378.

Egner, T., Etkin, A., Gale, S., & Hirsch, J. Dissociable neural systems resolve conflict from emotional versus nonemotional distracters. Cereb Cortex. 2008; 18(6): 1475–1484.

Egner, T., & Hirsch, J. Cognitive control mechanisms resolve conflict through cortical amplification of task-relevant information. Nat Neurosci 2005a; 8(12): 1784–1790.

Egner, T., & Hirsch, J. The neural correlates and functional integration of cognitive control in a Stroop task. Neuroimage 2005b; 24(2): 539–547.

Ehlers, A., Margraf, J., Davies, S., & Roth, W. T. Selective processing of threat cues in subjects with panic attacks. Cogn Emot 1988; 2(3): 201–219.

Etkin, A., Egner, T., Peraza, D. M., Kandel, E. R., & Hirsch, J. Resolving emotional conflict: A role for the rostral anterior cingulate cortex in modulating activity in the amygdala. Neuron 2006; 51(6): 871–882.

Fichtenholtz, H. M., Dean, H. L., Dillon, D. G., Yamasaki, H., McCarthy, G., & LaBar, K. S. Emotion-attention network interactions during a visual oddball task. Brain Res Cogn Brain Res 2004; 20(1): 67–80.

Fox, M. D., & Raichle, M. E. Spontaneous fluctuations in brain activity observed with functional magnetic resonance imaging. Nat Rev Neurosci 2007; 8(9): 700–711.

Fox, M. D., Snyder, A. Z., Vincent, J. L., Corbetta, M., Van Essen, D. C., & Raichle, M. E. The human brain is intrinsically organized into dynamic, anticorrelated functional networks. Proc Natl Acad Sci USA 2005; 102(27): 9673–9678.

George, M. S., Ketter, T. A., Parekh, P. I., et al. Regional brain activity when selecting a response despite interference: An H2 15O PET study of the stroop and an emotional stroop. Hum Brain Mapp 1994; 1(3): 194–209.

Gilboa-Schechtman, E., Revelle, W., & Gotlib, I. H. Stroop interference following mood induction: Emotionality, mood congruence and concern relevance. Cogn Ther Res 2000; 24(5): 491–502.

Giles, M., & Cairns, E. Colour naming of violence-related words in Northern Ireland. Br J Clin Psychol 1989; 28(Pt 1): 87–88.

Glaser, W. R., & Dungelhoff, F. J. The time course of picture-word interference. J Exp Psychol Hum Percept Perform 1984; 10(5): 640–654.

Glaser, W. R., & Glaser, M. O. Context effects in Stroop-like word and picture processing. J Exp Psychol Gen 1989; 118(1): 13–42.

Gratton, G., Coles, M. G., & Donchin, E. Optimizing the use of information: Strategic control of activation of responses. J Exp Psychol Gen 1992; 121(4): 480–506.

Grimshaw, G. M. Integration and interference in the cerebral hemispheres: Relations with hemispheric specialization. Brain Cogn 1998; 36(2): 108–127.

Gronau, N., Ben-Shakhar, G., & Cohen, A. Behavioral and physiological measures in the detection of concealed information. J Appl Psychol 2005; 90(1): 147–158.

Gronau, N., Cohen, A., & Ben-Shakhar, G. Dissociations of personally significant and task-relevant distractors inside and outside the focus of attention: A combined behavioral and psychophysiological study. J Exp Psychol Gen 2003; 132(4): 512–529.

Gusnard, D. A., Akbudak, E., Shulman, G. L., & Raichle, M. E. Medial prefrontal cortex and self-referential mental activity: Relation to a default mode of brain function. Proc Natl Acad Sci USA 2001; 98(7): 4259–4264.

Haas, B. W., Omura, K., Constable, R. T., & Canli, T. Interference produced by emotional conflict associated with anterior cingulate activation. Cogn Affect Behav Neurosci 2006; 6(2): 152–156.

Hasher, L., & Zacks, R. T. Automatic and effortful processes in memory. J Exp Psychol Gen 1979; 108: 356–388.

Herrington, J. D., Mohanty, A., Koven, N. S., et al. Emotion-modulated performance and activity in left dorsolateral prefrontal cortex. Emotion 2005; 5(2): 200–207.

Isenberg, N., Silbersweig, D., Engelien, A., et al. Linguistic threat activates the human amygdala. Proc Nat Acad Sci USA 1999; 96(18): 10,456–10,459.

Ishii, K., Reyes, J. A., & Kitayama, S. Spontaneous attention to word content versus emotional tone: Differences among three cultures. Psychol Sci 2003; 14(1): 39–46.

Jones-Chesters, M. H., Monsell, S., & Cooper, P. J. The disorder-salient stroop effect as a measure of psychopathology in eating disorders. Int J Eat Disord 1998; 24(1): 65–82.

Kaspi, S. P., McNally, R. J., & Amir, N. Cognitive processing of emotional information in post-traumatic stress disorder. Cogn Ther Res 1995; 19: 433–444.

Kerns, J. G., Cohen, J. D., MacDonald, A. W., 3rd, Cho, R. Y., Stenger, V. A., & Carter, C. S. Anterior cingulate conflict monitoring and adjustments in control. Science 2004; 303(5660): 1023–1026.

Kitayama, S., & Ishii, K. Word and voice: Spontaneous attention to emotional utterances in two languages. Cogn Emot 2002; 16(1): 29–59.

Kunde, W., & Mauer, N. Sequential modulations of valence processing in the emotional Stroop task. Exp Psychol 2008; 55(3): 151–156.

Larsen, R. J., Mercer, K. A., & Balota, D. A. Lexical characteristics of words used in emotional Stroop experiments. Emotion 2006; 6(1): 62–72.

Liu, X., Banich, M. T., Jacobson, B. L., & Tanabe, J. L. Functional dissociation of attentional selection within PFC: Response and non-response related aspects of attentional selection as ascertained by fMRI. Cereb Cortex 2006; 16(6): 827–834.

Logan, G. D. Attention and automaticity in Stroop and priming tasks: Theory and data. Cogn Psychol 1980; 12(4): 523–553.

MacKay, D. G., & Ahmetzanov, M. V. Emotion, memory, and attention in the taboo Stroop paradigm. Psychol Sci 2005; 16(1): 25–32.

MacKay, D. G., Shafto, M., Taylor, J. K., Marian, D. E., Abrams, L., & Dyer, J. R. Relations between emotion, memory, and attention: Evidence from taboo stroop, lexical decision, and immediate memory tasks. Mem Cogn 2004; 32(3): 474–488.

MacLeod, C., & Rutherford, E. M. Anxiety and the selective processing of emotional information: Mediating roles of awareness, trait and state variables, and personal relevance of stimulus materials. Behav Res Ther 1992; 30(5): 479–491.

MacLeod, C. M. Half a century of research on the Stroop effect: An integrative review. Psychol Bull 1991; 109(2): 163–203.

MacLeod, C. M., & MacDonald, P. A. Interdimensional interference in the Stroop effect: Uncovering the cognitive and neural anatomy of attention. Trends Cogn Sci 2000; 4(10): 383–391.

Mason, M. F., Norton, M. I., Van Horn, J. D., Wegner, D. M., Grafton, S. T., & Macrae, C. N. Wandering minds: The default network and stimulus-independent thought. Science 2007; 315(5810): 393–395.

Matthews, S. C., Paulus, M. P., Simmons, A. N., Nelesen, R. A., & Dimsdale, J. E. Functional subdivisions within anterior cingulate cortex and their relationship to autonomic nervous system function. Neuroimage 2004; 22(3): 1151–1156.

McKenna, F. P., & Sharma, D. Intrusive cognitions: An investigation of the emotional Stroop task. J Exp Psychol Learn Mem Cogn 1995; 21(6): 1595–1607.

McKenna, F. P., & Sharma, D. Reversing the emotional Stroop effect reveals that it is not what it seems: The role of fast and slow components. J Exp Psychol Learn Mem Cogn 2004; 30(2): 382–392.

Meyer, D. E., & Schvaneveldt, R. W. Facilitation in recognizing pairs of words: Evidence of a dependence between retrieval operations. J Exp Psychol 1971; 90(2): 227–234.

Miller, E. K., & Cohen, J. D. An integrative theory of prefrontal cortex function. Ann Rev Neurosci 2001; 24: 167–202.

Mitchell, R. L. Does incongruence of lexicosemantic and prosodic information cause discernible cognitive conflict? Cogn Affect Behav Neurosci 2006a; 6(4): 298–305.

Mitchell, R. L. How does the brain mediate interpretation of incongruent auditory emotions? The neural response to prosody in the presence of conflicting lexico-semantic cues. Eur J Neurosci 2006b; 24(12): 3611–3618.

Mohanty, A., Engels, A. S., Herrington, J. D., et al. Differential engagement of anterior cingulate cortex subdivisions for cognitive and emotional function. Psychophysiology 2007; 44(3): 343–351.

Monsell, S., Taylor, T. J., & Murphy, K. Naming the color of a word: is it responses or task sets that compete? Mem Cognit 2001; 29(1): 137–151.

Ochsner, K. N., & Gross, J. J. The cognitive control of emotion. Trends Cogn Sci 2005; 9(5): 242–249.

Park, I. H., Park, H. J., Chun, J. W., Kim, E. Y., & Kim, J. J. Dysfunctional modulation of emotional interference in the medial prefrontal cortex in patients with schizophrenia. Neurosci Lett 2008; 440(2): 119–124.

Phaf, R. H., & Kan, K. J. The automaticity of emotional Stroop: A meta-analysis. J Behav Ther Exp Psychiatry 2007; 38(2): 184–199.

Pieters, J. M. Ear asymmetry in an auditory spatial Stroop task as a function of handedness. Cortex 1981; 17(3): 369–380.

Posner, M. I., & Snyder, C. R. R. Attention and cognitive control. In: Solso, R. L. (Ed.), *Information processing and cognition: The Loyola symposium*. Hillsdale, NJ: Erlbaum, 1975: pp. 55–85.

Preston, S. D., & Stansfield, R. B. I know how you feel: Task-irrelevant facial expressions are spontaneously processed at a semantic level. Cogn Affect Behav Neurosci 2008; 8(1): 54–64.

Raz, A., Kirsch, I., Pollard, J., & Nitkin-Kaner, Y. Suggestion reduces the stroop effect. Psychol Sci 2006; 17(2): 91–95.

Raz, A., Landzberg, K. S., Schweizer, H. R., et al. Posthypnotic suggestion and the modulation of Stroop interference under cycloplegia. Conscious Cogn 2003; 12(3): 332–346.

Raz, A., Moreno-Iniguez, M., Martin, L., & Zhu, H. Suggestion overrides the Stroop effect in highly hypnotizable individuals. Conscious Cogn 2007; 16(2): 331–338.

Raz, A., Shapiro, T., Fan, J., & Posner, M. I. Hypnotic suggestion and the modulation of Stroop interference. Arch Gen Psychiatry 2002; 59(12): 1155–1161.

Richards, A., French, C. C., Johnson, W., Naparstek, J., & Williams, J. Effects of mood manipulation and anxiety on performance of an emotional Stroop task. Br J Psychol 1992; 83(Pt 4): 479–491.

Riemann, B. C., & McNally, R. J. Cognitive processing of personally relevant information. Cogn Emot 1995; 9(4): 325–340.

Rosinski, R. R., Golinkoff, R. M., & Kukish, K. S. Automatic semantic processing in a picture-word interference task. Child Dev 1975; 46(1): 247–253.

Rota, G., Velt, R., Nardo, D., Weiskopf, N., Birbaumer, N., & Dogil, G. Processing of inconsistent emotional information: An fMRI study. Exp Brain Res 2008; 186(3): 401–407.

Schirmer, A., & Kotz, S. A. ERP evidence for a sex-specific Stroop effect in emotional speech. J Cogn Neurosci 2003; 15(8): 1135–1148.

Schirmer, A., Zysset, S., Kotz, S. A., & Yves von Cramon, D. Gender differences in the activation of inferior frontal cortex during emotional speech perception. Neuroimage 2004; 21(3): 1114–1123.

Sharma, D., Albery, I. P., & Cook, C. Selective attentional bias to alcohol related stimuli in problem drinkers and non-problem drinkers. Addiction 2001; 96(2): 285–295.

Sharma, D., & McKenna, F. P. The role of time pressure on the emotional Stroop task. Br J Psychol 2001; 92(3): 471–481.

Shiffrin, R. M., & Schneider, W. Controlled and automatic human information processing: II. Perceptual learning, automatic attending, and a general theory. Psychol Rev 1977; 84: 127–190.

Siegrist, M. Effects of taboo words on color-naming performance on a stroop test. Percept Mot Skills 1995; 81(3 Pt 2): 1119–1122.

Smith, E. E., & Jonides, J. Storage and executive processes in the frontal lobes. Science 1999; 283(5408): 1657–1661.

Smith, M. C., & Magee, L. E. Tracing the time course of picture-word processing. J Exp Psychol Gen 1980; 109(4): 373–392.

Stenberg, G., Wiking, S., & Dahl, M. Judging words at face value: Interference in a word processing task reveals automatic processing of affective facial expressions. Cogn Emot 1998; 12(6): 755–782.

Stroop, J. R. Studies of interference in serial verbal reactions. J Exp Psychol 1935; 18(6): 643–662.

van Hooff, J. C., Dietz, K. C., Sharma, D., & Bowman, H. Neural correlates of intrusion of emotion words in a modified Stroop task. Int J Psychophysiol 2008; 67(1): 23–34.

van Veen, V., Cohen, J. D., Botvinick, M. M., Stenger, V. A., & Carter, C. S. Anterior cingulate cortex, conflict monitoring, and levels of processing. Neuroimage 2001; 14(6): 1302–1308.

Vingerhoets, G., Berckmoes, C., & Stroobant, N. Cerebral hemodynamics during discrimination of prosodic and semantic emotion in speech studied by transcranial doppler ultrasonography. Neuropsychology 2003; 17(1): 93–99.

Virzi, R. A., & Egeth, H. E. Toward a translational model of Stroop interference. Mem Cogn 1985; 13(4): 304–319.

Wager, T. D., Barrett, L. F., Bliss-Moreau, E., et al. The neuroimaging of emotion. In: Lewis, M., Haviland-Jones, J. M., & Barrett, L. F. (Eds.), The handbook of emotions (3rd ed.). New York: Guilford, 2008; 249–271.

Wager, T. D., & Smith, E. E. Neuroimaging studies of working memory: A meta-analysis. Cogn Affect Behav Neurosci 2003; 3(4): 255–274.

Waters, A. J., & Feyerabend, C. Determinants and effects of attentional bias in smokers. Psychol Addict Behav 2000; 14(2): 111–120.

Waters, A. J., Sayette, M. A., Franken, I. A., & Schwartz, J. E. Generalizability of carry-over effects in the emotional Stroop task. Behav Res Ther 2005; 43(6): 715–732.

Waters, A. J., Sayette, M. A., & Wertz, J. M. Carry-over effects can modulate emotional Stroop effects. Cogn Emot 2003; 17(3): 501–509.

Wenk-Sormaz, H. Meditation can reduce habitual responding. Altern Ther Health Med 2005; 11(2): 42–58.

Whalen, P. J., Bush, G., McNally, R. J., et al. The emotional counting Stroop paradigm: A functional magnetic resonance imaging probe of the anterior cingulate affective division. Biol Psychiatry 1998; 44(12): 1219–1228.

Williams, J. M. G., Mathews, A., & MacLeod, C. The emotional Stroop task and psychopathology. Psychol Bull 1996; 120(1): 3–24.

Windes, J. D. Reaction time for numerical coding and naming of numerals. J Exp Psychol 1968; 78(2): 318–322.

Yamasaki, H., LaBar, K. S., & McCarthy, G. Dissociable prefrontal brain systems for attention and emotion. Proc Natl Acad Sci USA 2002; 99(17): 11447–11451.

CHAPTER 7

Motivational Influences on Cognitive Control: A Cognitive Neuroscience Perspective

Hannah S. Locke and Todd S. Braver

ABSTRACT

Motivation is an important component of self-regulation that helps set the effort level an organism is willing to expend to achieve a desired goal. However, motivation is an elusive concept in psychological research, with investigations typically targeting either very macro-level (e.g., effects of personality individual differences and experimental manipulations on global behavior) or very micro-level (e.g., physiological interventions targeting specific brain structures) processes. Thus, the current state of knowledge is very poor regarding the particular mechanisms by which motivation influences cognitive and neural systems to drive changes in specific components of behavior. This chapter reviews major perspectives on motivation arising from both the social-personality and neuroscience literatures, and then discuss how a cognitive neuroscience perspective might be fruitfully applied to fill the gaps between them. Specifically, the chapter reviews literature, including our own recent work, that suggests motivational manipulations impact brain regions associated with the exertion of specific cognitive control functions. The chapter concludes by outlining unresolved questions in motivation, and by suggesting directions for future progress in this domain.

Keywords: Dopamine, nucleus accumbens, basal ganglia, prefrontal cortex, orbitofrontal cortex, insula, amygdala, anterior cingulate cortex, reward, punishment, affect, incentive salience, individual differences, working memory, attention

MOTIVATIONAL INFLUENCES ON COGNITIVE CONTROL: A COGNITIVE NEUROSCIENCE PERSPECTIVE

Self-control involves bringing behavior in line with goals. Given that achieving one goal may preclude pursuing another, it is important to be able to prioritize competing goals, and allocate resources accordingly. We suggest that motivation is a central component of self-control, as it involves weighing benefits and costs of actions, deciding which course of action is most appealing, and setting effort levels in accordance with the value of the eventual goal. Importantly, motivational influences are not only of interest in the study of overt behavior, but have begun to be investigated in terms of their effects on cognition, such as learning and performance strategies and cognitive control processes (e.g., see Markman, Maddox, & Baldwin, 2005). The interaction between motivational and cognitive control systems is a key area of interest, as cognitive control resources, such as sustained attention, are costly but potentially useful self-regulatory mechanisms that can be brought to bear on task performance, given enough incentive to outweigh the costs. This chapter argues that the cognitive neuroscience approach, with its ability to bridge different levels of explanation and analysis–neurobiological, cognitive, behavioral, individual differences, and social-psychological–is well-suited to the study of motivation, and hence to gaining a more thorough understanding of self-control.

Although motivation has been the subject of psychological research for decades, there is not a consensus on what its core features are. Motivation is often described generally as the internal drive to act, as if it were a force of its own, akin to the concept of will. However, this vague definition ignores the complex chain of computations that take place when individuals perceive opportunities for either rewarding or aversive outcomes, and change their behavior to attain or avoid them. Part of the reason that motivation is such a complex subject to grasp is that it has been investigated from a number of different perspectives, and often, researchers have worked in parallel at different levels of analysis without endeavoring to link their work to other areas. Therefore, the aim of this chapter is not only to review the main findings within separate disciplines, but also to draw connections where possible.

The main sections of the literature review will look at the contributions of social and personality psychologists, neuroscientists, and finally the emerging work of cognitive neuroscientists, all of whom are interested in motivation, but are asking and answering different kinds of motivation-related questions. Social and personality psychologists ask how individuals are, or fail to be, motivated in different situations. Who are the people that come into the lab to participate in experiments, and how do their thoughts, goals, and personalities affect their motivation? Neuroscientists study the other end of the spectrum to ask how the brains of animals and humans process stimuli associated with rewards and punishment. Finally, the cognitive neuroscience approach to motivation focuses on the cognitive and neural mechanisms that underlie individuals' ability to enact behavior that matches their goals and motivations. Following the sections reviewing these three areas, a final section attempts to integrate these disparate views of motivation.

Social and Personality Perspectives

Social and personality psychologists argue that both individual variables, such as goals, affect, self-image, and personality, and task variables, such as the framing and social context, interact to influence the kind of rewards and punishments that will influence motivation at a high level. This body of research demonstrates that knowledge about individual differences can provide greater explanatory power to understand task behavior than can approaches that examine only average group responses. Various theorists from within this tradition have provided accounts of motivation that attempt to specify the nature of the interaction between goals, self-image, personality, and motivation.

Carver and Scheier view motivation as arising from the activity of feedback loops, using the individual's goal value as the set point, and working to reduce discrepancies between the current state and the goal value (Carver, Sutton, & Scheier, 2000). A second level in the feedback system monitors progress toward the goal value, and it is this rate of progress that influences affect. Positive affect occurs when the system detects greater progress toward the goal than expected. Conversely, negative affect occurs when less than expected progress is detected. No change in affect occurs when goal progress matches the expected value. Thus, when individuals have a goal in mind, there is a monitoring system linked to emotion that can exert influence to prompt increased effort if progress is not adequate. In this scenario, affect itself is part of the reinforcement process, and thus is a core component of motivational drive.

Goals are clearly an essential part of motivational processes, but goals are heterogeneous by nature. The intentions that people form in response to tasks can vary, and the type of intentions individuals have can affect the likelihood of success. Gollwitzer (1999) has postulated two distinct types of intention: goal intentions and implementation intentions. Goal intentions are focused on the end-point, the achievement of a particular outcome that is desired, such as getting a good grade on a test. Implementation intentions, by contrast, are focused on how the task is going to be accomplished, and have specific external cues associated with them—for example, after I get back from the cafeteria, I am going to study for the test by looking at the sample exam questions in the book. By breaking a larger goal intention down into the subset of activities that need to be done to reach it, and by setting environmental triggers, individuals who set implementation intentions achieve higher rates of goal attainment than individuals who do not (Gollwitzer, 1999).

Taking one step back from the structure of goals, it is necessary to ask what factors affect how goals are selected in the first place. One answer is that individuals' prediction of affect (affective forecasting) can be a motivator to engage in a task—that is, individuals may choose to take part in a task because they predict that the task's reward will make them feel happy (Gilbert & Wilson, 2000). Thus, the affective response that is generated when the person is thinking about the reward is taken as a predictor of what the actual reaction to the receipt of the reward will be. Gilbert and Wilson (2000) have suggested that this prediction can be inaccurate for a number of reasons, leading to the phenomenon of "miswanting"–wanting something that is not liked, or enjoyed, when actually received. For example, if people are in a particularly good mood when evaluating a reward, it may lead to an overestimation of how happy it will make them in the future, because they base this prediction on how they feel at the moment they are thinking about the reward.

Expectancy is an important variable in motivation, and not only where predictions of future happiness are concerned. According to Bandura's Self Efficacy Theory, individuals' confidence in their ability to accomplish a task at hand can affect their motivation to engage in it, and the effort they put forth (see Bandura, 1997). Bandura distinguishes between outcome expectations, the consequences of achieving versus not achieving a goal, and efficacy expectations, the competence with which individuals expect to pursue those goals. These factors are related, in that a high level of efficacy expectation usually results in a high outcome expectation. For example, a student with a high sense of academic efficacy may expect opportunities for good jobs in the future. However, high efficacy does not always lead to high outcome expectation, if the student in the aforementioned example is going into a field with relatively few available jobs.

Another important component of social and educational psychological theories is the distinction between intrinsic motivation, positive feelings engendered by performing a task per se, and extrinsic motivation, where performance on a task is reinforced in some way, such as by money, food, points, or verbal praise. For example, this distinction has played a key role in Deci & Ryan's Cognitive Evaluation Theory (CET; Deci & Ryan, 1985). According to CET, extrinsic incentives can be effective at improving performance, but ultimately harm intrinsic, or "authentic" motivation because they are perceived as controlling behavior and limiting autonomy. Participants are able to explain their engagement and effort on a task as caused by desire for the external reward rather than by their own intrinsic interest, even if the task was inherently interesting to them at the outset. Thus, adding incentives to a task has the potential to diminish, or even eliminate, intrinsic motivation.

Although this theory has been influential in motivation research and particularly in education, it is also controversial. Results of multiple meta-analyses by Cameron and colleagues (Cameron, 2001; Cameron & Pierce, 1994) showed that in general, rewards for good performance do not harm intrinsic motivation. Rather, a small subset of incentives was found to be harmful—namely, non-performance-contingent rewards, or rewards offered for mere participation rather than performance that met some criterion. This effect is only present when these non-performance-contingent rewards are expected, not when they are unexpected. Thus, harmful effects only occur for rewards offered in advance just for engaging in a task. In addition, verbal praise was found to increase intrinsic motivation.

In many of Deci & Ryan's experiments, intrinsic motivation is measured by time spent on the task in a free-choice situation. The choice of what activity to engage in is an important aspect of motivation, as is the type of goals individuals have when they engage in tasks. Performance goals involve a focus on how well an individual will do at a task in relation to others, with an emphasis on appearing competent. In contrast, mastery goals emphasize the mechanics of learning to do the task well, and gaining skill and competence (Dweck, 1986). Elliot and colleagues have further proposed breaking down performance goals into performance-approach and performance-avoidance goals, involving appearing competent in front of others or not appearing incompetent, respectively (Elliot, McGregor, & Gable, 1999). Mastery goals have been associated with greater intrinsic interest in the task (e.g., Harackiewicz et al., 1997), greater persistence and effort (e.g., Elliot et al, 1999) and greater likelihood of choosing challenging tasks over excessively difficult or trivially easy tasks (Atkinson & Raphelson, 1956). Nevertheless, adopting mastery goals has not been consistently related to better performance, perhaps because individuals who are interested in a subject are more likely to seek out related interesting information that is not instrumental to better performance, and thus aren't as focused (Elliot et al., 1999). Rather, performance-approach goals seem to be most consistently related to better performance.

The next step after selecting goals is to work toward attaining them. How successful individuals are at achieving their goals may depend on their "motivational skills," which comprise the resources people have available to dedicate towards goal pursuit and task persistence (Heggestad & Kanfer, 1997). Unlike personality characteristics, which are presumably immutable, these skills may be developed over time. A related idea is the concept of active self, a resource used to inhibit actions that are tempting or habitual, initiate action, and make deliberate, conscious choices (Baumeister, Muraven, & Tice, 2000). According to this theory, self-control is a severely limited resource, and can be depleted by even short periods of engagement (i.e., <5 minutes). For example, when participants were asked to eat radishes and refrain from eating readily accessible cookies they persisted less on subsequent unsolvable geometry problems (Baumeister, Bratslavsky, Muraven, & Tice, 1998). This temporary depletion is associated with subsequent poorer performance on a wide range of tasks that require active self, from persisting on unsolvable problems to continuing to hold handgrip bars together (Baumeister et al., 1998, 2000; Muraven, Tice, & Baumeister, 1998). Such rapid depletion of self-control is problematic in that many everyday activities seem to require the engagement of an active self. Muraven & Baumeister (2000) suggest that a solution to this problem is to consider the active self as analogous to a muscle, in that it quickly tires from exertion, but can return to strength with rest. In addition, the frequent exercise of self-control can cause an increase in the amount of available resource and the length of time it can be engaged.

Whereas social psychology theories of motivation focus on situational factors such as how incentive structures influence motivation, personality theories look at the traits people have that influence their motivation. One classic distinction is the tendency towards approach (seeking rewards) vs. withdrawal (avoiding punishment). An influential personality theory dealing with approach and withdrawal is Jeffrey Gray's Behavioral Inhibition/Activation System (BIS/BAS), which posits that trait tendencies toward inhibition or activation are major determinants of behavior (Gray, 1970, 1981). The BAS system is related both to approach of rewarding stimuli, and active escape from negative stimuli. As such, it encompasses both positive affect in the sense of joy from receiving a reward, and relief at escaping a punishment. The BIS system is associated with a suspension of activity (freezing behavior), and encompasses both negative feeling of punishment, fear, and the loss of a potential reward, sadness/disappointment. The theory conceptualizes affect as a single bipolar dimension, running from negative to positive.

By contrast, Carver and colleagues (e.g., see Carver, Sutton & Scheier, 2000) view affect as two orthogonal dimensions, positive affect and negative affect. Low positive affect is characterized by sadness, and high positive affect by joy, whereas low negative affect is characterized by relief, and high negative affect by anxiety. Carver and White (1994) developed a scale to assess Gray's Behavioral Inhibition/Activation System, the BIS/BAS Scale. The difference between their scale and Gray's original concept is that BIS and BAS are treated as independent dimensions, consistent with the view that positive and negative affect are independent, thus, individuals can be high on both, low on both, or some combination of the two. The scale, which asks participants to endorse statements like, "It would excite me to win a contest" (BAS) and "I worry about making mistakes" (BIS), concentrates on emotional reactions to positive and negative life events. The idea is that emotional reactivity to events has effects on behavior, as individuals prioritize their time and effort into achieving goals that have the biggest impact on their happiness. Thus, an individual who worries a lot about making mistakes is more likely to be cautious in situations when failure is a possibility, even in situations where the potential rewards for taking action are high.

Studies examining the BIS and BAS systems have often relied on physiological data to test claims regarding the independence of the two systems, and their relationship to individual or group differences in affective style. A key component of this work is the assumption that approach and withdrawal are associated with greater relative activity in the left or right hemispheres of the brain, respectively (for review, *see* Davidson, 2003)—for example, EEG experiments that show that depressed patients have the same level of activation in right frontal cortex as controls, and what separates them is lower left frontal activity, characteristic of low positive affect (Henriques & Davidson, 1991). Anger is an important exception because it is negative, but it is also an approach emotion that prompts individuals to act (Lerner & Keltner, 2000, 2001). In an EEG experiment dealing with motivation, Sobotka and Davidson (1992) compared conditions in which participants had to withdraw their finger quickly from a button to avoid monetary penalties or had to press a button quickly to obtain monetary rewards. Right hemisphere activity was increased in the withdraw penalty condition, while left hemisphere activity increased in the approach reward condition. Further research has shown that differential activation of right and left hemispheres is a trait that underlies affective style from an early age. Specifically, children who have greater left-relative-to-right hemisphere activity at rest demonstrate a more positive affective style, while greater right than left have a more negative affective style (Fox et al., 1995). This tendency toward positive and negative style also correlated with social competence, as children with greater left frontal asymmetry displayed both more positive affect, and more socially competent behaviors in a free-play session.

Beyond emotional reactivity to events, it is possible to consider motivation as fitting into how individuals feel about themselves and their self-image. Higgins' (1999) self-discrepancy theory describes two self-guides that motivate people–the *ideal* self and the *ought* self. The ideal self possesses qualities that the individual personally aspires to, and is intrinsically motivated to attain, whereas the ought self is comprised of qualities that people feel obligated to attain. Discrepancies between the ideal self and the actual self produce disappointment and dejection, as individuals fail to achieve their personal goals. In contrast, differences between the ought self and the actual self produce anxiety, as individuals anticipate negative consequences (e.g., censure of others) for failing to achieve their obligatory goals.

The relative extent to which individuals focus on minimizing discrepancies between the actual self and these two self-guides (regulatory focus) can be measured as a stable personality trait (*see* Regulatory Focus Questionnaire, Higgins et al., 2001). Focus on the ideal versus the actual self is an indication of a promotion focus. Promotion-focused individuals direct more attention towards attaining rewards, thus they enjoy tasks that involve finding helpful items more than those that involve finding harmful items (Freitas & Higgins, 2002). Focus on the ought versus the actual self is an indication of a prevention focus. Prevention-focused individuals direct more attention towards avoiding negative outcomes, thus they enjoy tasks that involve finding (and avoiding) harmful items more than those that involve finding (and obtaining) helpful items. This concept of "regulatory fit" is one example of how personality can affect how motivated individuals are by different types of tasks.

Regulatory focus also has implications for how individuals set and pursue goals. Promotion-focused individuals tend to view goals as opportunities to achieve the best possible outcome, whereas the prevention-focused view goals as basic requirements to avoid a negative outcome. Thus, promotion-focused individuals may derive more utility from moving towards their goals than the prevention-focused. However, prevention-focused individuals tend to begin pursuit of their goals earlier (Freitas et al., 2002). Viewing a goal as a basic necessity rather than an intriguing possibility seems to urge earlier and more focused action.

Although regulatory focus can be measured as a personality trait, it can also be influenced in a state-like manner, by experimental manipulation of task conditions. Specifically, emphasizing that achievement of a goal is important to avoid negative outcomes primes a prevention focus, whereas concentrating on potential benefits primes a promotion focus. For example, participants in an experiment where the task is framed in a prevention-oriented way (write an essay in order to avoid being rejected for a fellowship and losing money) choose to begin writing the essay earlier than participants in a promotion-framed condition (write an essay to get accepted for a fellowship, and gain money) (Freitas et al., 2002).

In summary, social and personality psychology research demonstrates strong influences of a number of person and situation factors related to motivation, such as basic characteristics of affective style, and how motivational goals are framed or construed. More specifically, participants' behavior during laboratory tasks is affected by the incentives that are offered, individual differences in relative sensitivity towards reward versus punishment incentives, the types of goals set, and how progress towards those goals is monitored.

Neuroscience Perspectives

Neuroscientists have studied motivation by looking at how rewards and punishments are processed by the brain. Older views of motivation focused on "drive" models, in which the lack of an essential physiological variable is the source of drive in initiating actions to correct the deficiency. For example, a lack of food would cause foraging behavior to obtain food (Hull, 1943). The key idea was that the central nervous system implemented mechanisms for homeostasis—forces that maintain stable internal states—which correct any deviations in amount of nutrients, water, or changes in internal temperature from a set-point value. Moreover, according to the theory, drive-states also produce affective responses, with drive-reducing stimuli (e.g., presentation of a cheeseburger to a hungry person) being experienced as rewarding, and drive-increasing stimuli being experienced as aversive (e.g., presentation of salty foods when thirsty).

One of the further implications of drive theory is that stimulation that increases a behavior should be experienced as aversive, because by definition, it must have increased physiological drive. However, this inference was inconsistent with brain stimulation experiments, which showed that electrical stimulation of the same area that caused increased eating (the lateral hypothalamus) also appeared to be rewarding to the animal, as measured by tendency for the animal to also self-stimulate in that area (Valenstein, Cox, & Kakolewski, 1970). Another interesting finding from these experiments was that the response the animals gave to stimulation was not always eating—some animals drank, gnawed, or displayed sexual behavior, although which behavior was displayed was consistent within the same animal on different occasions. Thus, it seemed that the stimulation triggered a more general motivational state apart from filling (or increasing) a particular need.

In addition to providing data that conflicted with drive theory, the electrical stimulation studies also provided support for the next major theory of reinforcement, namely the idea that dopamine (DA) served as a "pleasure chemical" mediating the subjective experience of reward. In these experiments, electrical stimulation devices were implanted in limbic system structures, several in regions that are primary targets for dopamine, such as the nucleus accumbens (NAcc). These studies showed that animals and people with access to a self-stimulation apparatus often self-stimulated excessively, even to the exclusion of other motivated behaviors, including eating and sleeping (Heath, 1996; Olds, 1958). Researchers drew the conclusion that dopamine itself carried the hedonic signal, and thus by mimicking it (by directly stimulating dopamine targets), pleasurable states could be created.

However, this view was rejected by experiments in which DA was directly injected into the NAcc in conjunction with sucrose delivery (Wyvell & Berridge, 2000). In these studies, DA injection did not increase the consumption or apparent hedonic response to the sweet taste (as measured in animals by stereotypic indices of facial expression and reaction). Thus, DA did not increase liking of preferred foods (e.g., sweet tastes), which it should have if it amplified the foods' pleasure value. In addition, in studies with humans that had electrodes implanted in NAcc (for treatment of depression or debilitating pain), direct queries about their subjective experience typically yielded vague statements like, "I feel good" rather than descriptions of discrete episodes of pleasure following NAcc self-stimulation (Heath, 1996).

The hypothesis of dopamine as a pleasure chemical has been further challenged by more recent experimentation, leading to the development of several new theories of dopamine function. Below, these theories are briefly reviewed. Although they disagree with one another on the exact functional role of dopamine, they are united in rejecting the pleasure chemical hypothesis.

Modern Views of Dopamine Functions in Reward and Motivation

More recent studies of the role of dopamine in reward and motivation have relied upon direct single-cell recordings of neuronal activity within dopamine-releasing neurons in awake, behaving animals, such as primates. A key finding of this work is that dopamine neurons show phasic activity in response to reward-related information at the time of delivery for unpredicted rewards. Yet when rewards are predictable the dopamine response occurs to earlier reward-predictive cues and not to reward delivery itself (Schultz, 2002). These findings suggest that the dopamine system does not convey the hedonic aspects of reward, but rather acts as an informational signal about whether and when rewards will be delivered. This type of reward-prediction or reward-learning system could help to strengthen perceptual and response pathways that help maximize the obtainment of reward.

The theory of dopamine as central to reward learning has been challenged by experiments with animals where ascending DA pathways were nearly completely destroyed. Despite not having dopamine, these animals were able to learn an aversive reaction to a previously liked sweet taste when the sweet taste was paired with LiCl, a chemical that causes gastric distress (Berridge & Robinson, 1998). Later experiments showed that dopamine deficient mutant mice were able to learn the location of a food reward when given caffeine, but that the learning was not revealed until the mice were treated with L-DOPA, a DA precursor (Robinson, Sandstrom, Denenberg, & Palmiter, 2005). Together these results suggest that DA is not necessary to learn hedonic and behavioral responses to reward-related stimuli. Yet at the same time, the latter result provokes questions about why behaviors required to acquire reward are impaired in the absence of DA.

Berridge and colleagues, who emphasize that DA serves a motivational rather than hedonic function in the processing of reward, have put forth a response to this puzzle. Specifically, Berridge's view is that above and beyond informational aspects of reward (what is it? where is it?), and the hedonic or "liking" aspects of a reward, the system must produce a motivational state to induce a change in behavior (Berridge & Robinson, 1998). Berridge and Robinson attribute reward "wanting" or incentive salience to the action of the nucleus accumbens, which they suggest is the brain structure responsible for translating pure factual information about an available reward into a more affective desire to have it. This step could then feasibly lead to changes in behavior necessary to obtain the reward.

Support for the role of NAcc in wanting rather than liking has come from experiments demonstrating a clear dissociation of DA manipulations on hedonic responses vs. goal-directed instrumental behavior. Specifically, as described above, NAcc DA depletion in rats has no effect on sucrose consumption or facial displays of liking, but does dramatically decrease behavioral responses (e.g., lever pressing or maze running) required to obtain the reward (Berridge & Robinson, 1998). Moreover, Berridge suggests a reinterpretation of the NAcc self-stimulation studies, arguing that animals and humans are driven to self-stimulate in these situations because DA release can increase incentive salience in a general way, leading the world to look like a more attractive place (i.e., full of more valued rewards) rather than directly inducing a hedonic "pleasure" response (Berridge & Robinson, 1998).

Berridge is very clear in dissociating incentive salience "wanting" from the colloquial use of wanting, "cognitive wanting" (Berridge, 2004). Whereas incentive salience is not necessarily conscious, cognitive wanting is a declarative representation of a goal, based on previous experience with it, or an internal estimation of its value. In cognitive wanting, an understanding of the causal relationship between an individual's actions and the likelihood of obtaining the reward is necessary to facilitate those actions that lead to reward obtainment. In other words, a decision to pursue a reward is based on an expectation of how much the reward will be liked. Incentive salience, by contrast, can be triggered by mere perception of a stimulus, and can lead to irrational wanting of something that is not expected to be liked, such as a drug of abuse (Berridge, 2004).

The incentive salience model of addiction states that drugs that cause DA release create a false association of that drug (or drug paraphernalia) with an abnormally high "wanting" even when the drug user is long past the point when the drug use is actively enjoyed. Thus, drug users desperately "want" something they do not "like." This effect is especially pernicious because of how long it lasts. Whereas tolerance effects (fewer DA receptors as a result of repeated overabundance of DA in the synapse) disappear in a matter of weeks, sensitization effects can last for years, leading to a risk of relapse when encountering drug-related stimuli (Robinson & Berridge, 2000).

Thus, according to Berridge, motivation is not a drive-reducing process that attempts to correct deficits in reward signaled by the level of DA release. Instead, it is a process by which incentive salience becomes attached to stimuli, such that these stimuli attract approach-related behaviors that vary in strength in proportion to the estimated magnitude of salience. In contrast, the experience of actual liking, in Berridge's view, is mediated by an opiate system in the NAcc shell. This view is supported by findings that injection of opiates there increases positive taste reactions to sweet foods and promotes eating, while opiate blockers decrease positive taste reaction and decrease eating (Peciña & Berridge, 2000). Nevertheless, it is important to note that in a normally functioning system, wanting and liking of rewards go together, such that appropriate actions are implemented to gain a liked reward.

Panksepp has a similar view of DA function, in that it is primarily involved in instrumental behaviors needed to obtain rewards, rather than subserving the pleasure or liking of the reward itself. But unlike Berridge, Panksepp's view is not that NAcc mediates wanting, but rather "flexible seeking behavior" (Ikemoto & Panksepp, 1999). Thus, contrary to what would be predicted by the incentive salience hypothesis, when comparing two groups of hungry rats in a novel environment, the group that is dopamine-depleted will actually eat more than the control group (Koob et al., 1978). Panksepp explains these results by characterizing DA as a driver of exploration, and reward seeking. Thus, for normal rats in a novel environment, exploratory behaviors will be facilitated even at the expense of eating. In contrast, dopamine deficient rats feel no compulsion to explore and instead just eat their fill. Further, recordings of dopamine levels in the nucleus accumbens show that DA is especially high in thwarting paradigms, such as when a male rat is prevented from accessing a female rat. In these situations, the reward value of the female rat should be constant, but what is maximized is seeking behavior, as the test rat tries a large variety of behaviors in order to get to the female.

Another important distinction between Berridge and Panksepp is that Berridge does not distinguish between dopamine effects in the nucleus accumbens and dopamine in the dorsal striatum (i.e., the caudate and putamen structures of the basal ganglia), whereas Panksepp sees the two as very different. In Panksepp's model, dopamine projections from the ventral tegmental area (VTA) to the NAcc underlie flexible seeking behaviors, whereas dopamine projections from the substantia nigra to dorsal striatum mediate already learned behaviors (habits). Thus, when animals have already learned self-administration of a drug like heroine, or ethanol, selective dopamine depletion in the NAcc does not disrupt this behavior. Rather, because the drug use has become a habit, the nigro-striatal pathway is able to maintain the behavior. This is in contrast to the incentive salience hypothesis, which should predict that self-administration would be affected by the inability to attach salience to drug-related stimuli, and initiate approach.

Another closely related idea about dopamine's function has been advanced by Salamone (Salamone & Correa, 2002). Like Berridge and Panksepp, Salamone rejects the theory of dopamine as pleasure chemical, and instead believes DA has a central role in motivation. Further, he takes Berridge's concept of "wanting" and deconstructs it into two distinct aspects: directional (desire for a specific reward) and activational (willingness to initiate goal-directed behavior towards obtaining a reward). In Salamone's view, dopamine is critical for the activational aspect of wanting, but does not influence the appetitive value of rewards. He cites as evidence a number of studies in which dopamine antagonists impair instrumental responses for reward (e.g., lever pressing) but do not disrupt reward choice, or simple approach and consumption of food rewards.

Indeed, several studies cited by Salamone (e.g., Caul & Brindle, 2001) demonstrate that even fairly high doses of a DA antagonist like haloperidol will not disrupt performance on reward tasks that are not demanding. Thus, the appetite for the reward is not diminished, but rather the willingness to work for the reward. What DA reduction does, then, is alter the balance between costs and payoff, such that the costs seem to loom larger, making the animal less likely to put forth the effort to get the reward.

The role of dopamine in reward processing remains controversial, with researchers arguing for reward learning (Schultz), incentive salience (Berridge), flexible seeking (Panksepp), reward-directed action (Salamone), as well as other hypotheses. Nevertheless, there does seem to be a basic agreement that dopamine is intimately involved in motivation. For a normal response to new rewards in the environment to be accomplished, dopamine transmission is essential. However, what the animal or person does with reward information to change behavior is also critical if rewards are to be obtained rather than merely discovered or coveted.

Neural Circuitry of Reward Motivation

Dopamine neurons project to number of brain regions, not only to the ventral (NAcc) and dorsal (caudate and putamen) components of the basal ganglia, but also to medial and lateral frontal cortical regions. These projection sites appear to be involved with further processing of reward-related information. For example, it has been theorized that the basal ganglia is a site where rewards are linked to goal-directed motor actions (Kawagoe et al., 1998). Experiments in macaques recording the activity of neurons in the caudate during a saccade task have shown that introducing rewards can radically alter their activity (i.e. firing rates). At baseline, caudate neurons are selectively sensitive to stimuli in a particular location, usually the contralateral half of the visual field. However, when rewards were associated with only the nonpreferred direction, these neurons changed their preference, and responded more vigorously to the rewarded direction. This switch in preference was associated with faster and more accurate saccade responses to the rewarded direction (Kawagoe et al., 1998).

Reward signaling has also been investigated in humans, where several studies have used gambling-type tasks with differential reward payoffs to identify the basic neural circuitry involved in signaling reward. For example, Delgado et al. (2000, 2003) used such a paradigm, in which participants were asked to guess whether the value of a face-down card was higher or lower than five, with monetary rewards provided for correct guesses. Imaging results showed that in the caudate nucleus, a sustained hemodynamic response was observed on reward trials, whereas penalty trials led to a sharp drop-off in activation, with the basic valence effect also showing scaling according to the magnitude of the reward or punishment. This reward signaling is potentially useful for flagging certain actions or decisions as being associated with a positive outcome.

Reward representation refers to maintaining information about the type and current value of a reward (which may be context dependent), as opposed to merely signaling the presence or absence of a reward and/or its magnitude. Studies of reward representation have suggested the importance of areas within prefrontal cortex (PFC), specifically orbitofrontal cortex (OFC), for this function. A basic requirement of reward representation is being able to represent valence distinctions, as also occurs within basal ganglia structures. Consistent with this assumption, OFC regions appear to be anatomically dissociable in regards to valence effects. O'Doherty et al. (2001) found that medial areas of OFC activated to reward, whereas lateral areas responded to punishment during a reversal learning task with monetary reinforcers. Since this original finding, this basic pattern of spatial dissociation has been replicated a number of times (Kringelbach & Rolls, 2004).

Brain regions that represent reward value need to be able to integrate different categories of reinforcement into a unified representation of value (Montague & Berns, 2002; Montague & Cohen, 2006). Thus, when faced with a choice between rewards that are hard to compare, for example, an ice cream cone vs. an hour of reading a good book, individuals are able to consider their internal state (hunger, fatigue), general preferences for those activities, costs of the two options, and so forth, and generate a value for each. Representing the value of rewards in a way that would make them directly comparable would allow resources to be directed toward achieving the most attractive reward. The OFC appears well positioned for this function, as it shows responses to stimuli that appear to code them in terms of both their hedonic properties, which might be context-dependent, and their value to the individual in relationship to other potential reward-related stimuli.

These two properties of OFC function can be seen in terms of its response to primary sensory reinforcers, such as taste and smell. The role of the OFC in these domains is well established, given its strong anatomical linkage to primary gustatory and olfactory cortex (Rolls, 2001). Indeed, caudolateral OFC has been termed the secondary taste cortex (Rolls, 2001). In contrast to primary taste cortex, which responds consistently to taste sensation, secondary taste cortex ceases responding to tastes when the animal has been fed to satiety. Thus, Rolls (2005) states that secondary taste cortex represents the reward value of taste. However, as with abstract rewards, there is evidence that both positive and negative stimuli are represented in the OFC, as in O'Doherty et al. (2001), where positive taste (glucose) and negative (saltwater) were found to activate different parts of secondary taste cortex. The OFC response is also strongly sensitive to the organism's stable reward preferences. In free choice situations, OFC shows greater activity for the most preferred reward (Tremblay & Schultz, 1999). More recently, it has been shown that reward value in terms of choice preference appears to be directly coded in the level of activity within single OFC neurons (Padoa-Schioppa & Assad, 2006). In humans, a food stimulus once considered rewarding may also be represented as punishment if the individual is over-satiated (Small, Zatorre, Dagher, Evans, & Jones-Gotman, 2001). This result was replicated using liquids (tomato juice and chocolate milk) contrasted with a flavorless solution approximating saliva. The liquids which were ingested to satiety failed to activate OFC, whereas the liquid which was not continued to activate OFC (Kringelbach et al., 2003).

Although the primary focus on the prefrontal cortex with regard to reward-related processing has been on the OFC, some studies have revealed activation in the lateral prefrontal cortex (LPFC) as well. For example, Leon and Shadlen (1999) investigated reward incentive effects during cognitive task performance within area 46 in macaques, a region analogous to the human dorsolateral PFC. The traditionally assumed function of this region is to maintain goal-related information in working memory over short delays. Neurons in area 46 respond maximally to stimuli in a specific part of the visual field, maintaining this position over a delay so the animal can remember the location of a stimulus after is has disappeared. They found that the stimulus-specific neuronal activity was enhanced when an incentive cue indicated the possibility of a large reward for correct performance. This increased activity may have been indicative of a stronger representation of the location of the stimulus, which would have been advantageous in performing the task.

Watanabe et al. (2002) contend that the LPFC may in fact be a site of integration of cognitive and motivational operations that contribute to goal-directed behavior. Using single-cell recording in monkeys during a go/no-go task, they found a subset of neurons that discriminated between rewarded and unrewarded trials. Specifically, they increased firing in reward trials compared to unrewarded trials during the delay phase of the task. Higher activity was associated with more preferred rewards such as cabbage. The authors also looked at baseline activity and found significant differences, such that tonic activity levels were highest during blocks of trials in which the preferred reward was available (highest for cabbage). They theorize that this baseline activity in the different blocks may reflect continuous monitoring of the motivational context, and may be an index of motivational level. Regarding anatomy, the authors suggest a dorsal/ventral split, with dorsal activity reflecting the continuous monitoring of motivational context, and the ventral and arcuate areas tracking the motivational value of the presence or absence of individual rewards.

Although the precise functions of the dopamine system, the nucleus accumbens, the caudate nucleus, and the OFC are still under investigation, there is general agreement that these regions make up a reward processing system that handles reward information. In summary, when a reward is available in the environment, an animal must make a series of judgments about how intrinsically attractive is that reward, its relative value, which is dependent upon both stable preferences and transient fluctuations in internal state, and how much it costs to obtain the reward, in terms of required effort. If the reward is deemed worth the effort required, instrumental actions must be initiated to get the reward.

Neural Circuitry of Punishment Motivation

A great deal of research has been conducted to understand the neural circuitry of rewards and their relationship to instrumental behavior. Less is understood about how animals and humans process punishments or other aversive outcomes. One possibility is that punishments are coded within the same dopaminergic system as rewards. For example, Panksepp has commented that his theory of nucleus accumbens dopamine function, in which dopamine facilitates flexible approach toward rewards, can be equally applied to punishment, in that avoiding an aversive outcome entails approach toward safety, an incentive (Ikemoto & Panksepp, 1999). In Schultz's theory of reward learning, he notes that missed rewards depress dopamine firing, but given that DA cells have a resting firing rate of around five action potentials per second, there is not a large range to decrease to incorporate negative events (rather than just the absence of positive events) (Grace & Bunney, 1984). Indeed, Mirenowicz and Schultz (1996) showed that aversive stimuli closely matched to appetitive stimuli (saltwater and juice, respectively, that were calibrated to prompt the same behavioral performance) failed to activate dopamine neurons. Given this limitation, some researchers have proposed that incentive processing is a rectified system, whereby positive and negative incentives are processed by separate neural mechanisms.

The leading candidate for a neural counterpart mechanism to dopamine, in conveying punishment-related incentive signals, is the serotonin system. Serotonin (5-HT) projects widely throughout the nervous system, and has a myriad of functions (for a review of serotonin and behavior, see Lucki, 1998). The dorsal and medial raphe nuclei in the midbrain are the source of the serotonin that projects to the forebrain. One particular pathway that has received attention in motivation research is the projection of serotonergic cells from the dorsal raphe to the ventral striatum, also a target of dopamine cells. Due to this association, researchers (e.g., Daw et al., 2002; Deakin, 1983) hypothesized that serotonin and dopamine are opponent systems. Whereas dopamine facilitates approach behaviors towards positive stimuli, serotonin inhibits these behaviors, and mediates avoidance in the presence of aversive stimuli. Depending on which influence was stronger, the net result would be either approach or withdrawal. Daw et al. (2002) in particular extend the reward learning model of dopamine function to suggest that serotonin neurons' activity signals an error in the prediction of punishment, just as DA neurons signal an error in the prediction of reward.

In addition to questions regarding whether there is a distinct signaling mechanism that conveys information about punishments and aversive outcomes, it is also of interest to determine whether there are independent neural circuits for further processing this information, or if the neural circuitry of punishment overlaps with that for reward. There is some evidence for distinct components within the neural circuitry of punishment, such as the anterior insula and amygdala (Seymour et al., 2007). In animals, activity in the basolateral and central nuclei of the amygdala, respectively, has been associated with learning to predict and avoid aversive outcomes, such as an air puff to the face (Everitt et al., 2003; Paton et al., 2006). In humans, research has shown that amygdala damage is associated with the inability to learn associations between visual stimuli and negative outcomes such as a loud noise (Bechara et al., 1999), or an electric shock (LaBar et al., 1995). Insula activity has been observed in neuro-imaging paradigms that involve emotionally unpleasant (e.g., Coan et al., 2006), disgusting (e.g., Calder et al., 2007), or unfair outcomes (Sanfey et al., 2003).

In addition to its potentially unique components, the neural circuitry of punishment also appears to partially overlap with reward circuitry, since both seem to involve basal ganglia structures. For example, as discussed above, many investigators suggest that the nucleus accumbens plays a bivalent role in motivation, responding to both rewarding and aversive events (Salamone, 1994). Similarly, human neuro-imaging studies have observed nucleus accumbens direct activation during anticipation of not only rewarding but also aversive outcomes (Jensen et al., 2003). Nevertheless, some evidence suggests those positive and negative valences are represented in distinct anatomical subregions within the nucleus accumbens (Reynolds & Berridge, 2002).

The prefrontal cortex also appears to play a common role in processing both reward and punishment information. Single-cell recording experiments in primate LPFC, similar to those previously discussed examining effects of reward, have contrasted the effects of rewarding and aversive stimuli (Kobayashi et al., 2006). The goal of these studies has been to determine whether LPFC neurons also respond to punishments, such as an air puff to the eye, as potential outcomes for poor performance in tasks. Monkeys were to remember the location of a target over a 1- or 2-second delay, and then saccade to its location, which was the same task performed in the Leon and Shadlen (1999) study. Results showed that although the majority of neurons were sensitive to rewards only, a smaller, separate group was found that responded preferentially to punishments. Specifically, these neurons significantly increased or decreased their rate of responding during the delay period after cues indicating the possibility of an air puff. Unlike Leon and Shadlen (1999), they did not manipulate incentive magnitude, and thus it is not known whether larger punishments would be associated with greater changes in firing rate.

Roesch and Olson (2004) used a similar task, except that both reward and penalty were possible on each trial. Additionally, they focused on the OFC as well as a posterior region of LPFC (the premotor cortex). Each trial featured two locations, one of which displayed a cue indicating that a reward would be available if the monkey performed a saccade to that location, and the other indicating that a penalty would result if the monkey moved its eyes to that location. Punishments in this task were time-outs of 1 to 8 seconds, an aversive outcome because the thirsty monkeys were unable to perform trials to get liquid rewards during this time. Results showed a dissociation in the activity between OFC and premotor cortex. OFC neurons responded to the overall value of the cues present on each trial (i.e., the value of the reward minus the value of the punishment), whereas premotor cortex responded vigorously to both types of incentive relative to the neutral condition. Premotor neurons responded maximally to the high-reward, and high-penalty conditions. The authors suggest that these results stem from a distinction of the function of these two areas, such that OFC represents the combined value of potential incentives, whereas the premotor cortex represents the motivational impact of the incentives (i.e., that there is something of value at stake, whether positive or negative). However, it is still possible that distinct OFC subregions code for negative and positive value, as appears to occur within other structures such as the nucleus accumbens. Human neuroimaging studies provide evidence consistent with this type of organizational structure, with rewarding stimuli activating medial OFC, and aversive stimuli activating lateral OFC regions (O'Doherty et al., 2001).

Human neuroimaging studies examining the motivational effects of punishments are still sparse (partly because of increased ethical concerns), but a few imaging studies have been conducted during cognitive task performance. Small et al. (2005) conducted a spatial attention task using both positive and negative monetary incentives. In a replication of a previous study, they found that the amount of spatial biasing (directing attention to a cued location to be faster when the target appears) was correlated with activity in the posterior cingulate cortex (Small et al., 2003). Further, activity in PCC was enhanced by both positive and negative incentives, but the effects these had may have been mediated by separate limbic structures. The possibility of winning was associated with activity in OFC, while losing activated dorsal anterior cingulate cortex and insula. Small et al. (2005) make the point that all three of these regions are connected with PCC in the macaque, and if this is also the case in humans, they may mediate the increased activity in PCC they observed. Given the connectivity of PCC with limbic structures and its association with direction of attention, Small et al. (2005) have proposed that posterior cingulate cortex is a site of integration of motivation and top-down attentional control.

Taylor et al. (2006) manipulated whether errors produced monetary penalties or not during performance of an Eriksen flanker task. They also observed that potential for penalty led to increased activity in the ACC but in a rostral rather than dorsal region. Because there was no matched reward condition in this study, it was not possible to determine whether the effect in ACC was valence-selective.

In summary, neuroscience experiments have mapped out a network of brain regions that signal reward and punishment in the environment and represent those incentives such that they could be used as information to guide action choices and effort levels. However, this careful body of work has less to say about what happens to that information, and how it is integrated into ongoing processing that directly affects behavior.

Cognitive Neuroscience Perspectives

Both top-down social/personality theories and bottom-up neurobiological theories of motivation contribute to a greater understanding of motivation. Social and personality researchers focus on how the goals and attributions of individuals, as well as the social context, influence motivation to engage in particular tasks. Additionally, there is a strong emphasis on individual differences, such as the role of reward and punishment sensitivity as stable trait characteristics. Neuroscience-based theories focus on the neural systems responsible for generating a desire for rewards and translating this into simple instrumental actions like lever pressing to obtain rewards. The gap between these two is a process-level account of how people are able to channel their motivation to gain desired rewards into improved performance in more complex behavioral tasks. Cognitive neuroscience approaches may fill this gap by illuminating the neural mechanisms that translate wanting into getting. Specifically, cognitive neuroscience studies involve directly examining human brain activity during performance of cognitive tasks. Motivation can be studied in this context by examining how task-related brain activity is affected both by changes in motivational state (i.e., experimental manipulations), and motivation-related individual differences. Two possibilities are: *(1)* effects on high-level cognitive control structures and *(2)* direct effects on activity in task-related regions.

Cognitive Control Processes in the Brain

In any cognitive task, there are brain regions that make their contribution not by facilitating task-specific components of processing, but rather in the higher-level coordination and regulation of task elements. These processes of high-level coordination and regulation are collectively referred to as cognitive control functions. Cognitive control refers to the process by which cognition and behavior is directed towards the fulfillment of internally generated goals (Miller & Cohen, 2001). Many of the processes by which control is exerted during tasks, including sustained attention, task set maintenance and task preparation, and interference prevention, have been localized to the prefrontal cortex, particularly lateral prefrontal cortex. Miller and Cohen (2001) argue that the central theme to these functions is that the prefrontal cortex is able maintain information about goals, and how to achieve them, in a manner that is less susceptible to interference.

Motivational incentives may enable improved performance by influencing cognitive control functions. Specifically, task performance may improve with tighter control over cognitive processing and monitoring of behavioral performance, such that negative events like interference or momentary lapses of attention are prevented. Similarly, errors will be less frequent when participants successfully maintain task instructions in a highly accessible form that ensures maintenance of attentional focus and suppression of interference from distracting events. These operations are potentially costly, in terms of subjective mental effort and metabolic demands, but the possibility of reward may shift the cost–benefit relationship, leading to increased activity in cognitive control structures that enhance performance.

Locke and Braver (2008) looked at the effects of motivational incentives on brain activity in a continuous performance task (AX-CPT) that serves as a probe of cognitive control function. In previous neuroimaging studies of the AX-CPT, task performance is reliably associated with increases in lateral PFC activity associated with the active maintenance of task-relevant goals (Miller & Cohen, 2001). Locke and Braver (2008) found that under reward incentive conditions (good performance was associated with monetary bonuses), there was a sustained increase in activity within brain regions associated with cognitive control function. These included right lateral PFC, right parietal cortex, right inferior frontal cortex (BA 45), and left dorsal anterior cingulate (BA 32).

Right lateral prefrontal cortex (RLPFC) and right parietal cortex are thought to be the core components of a brain network that mediates sustained attention (Pardo et al., 1991; Posner & Petersen, 1990). Right inferior frontal cortex (RIFC) has been implicated in cognitive inhibition, or the suppression of irrelevant thoughts, memories, or responses (Aron, Robbins, & Poldrack, 2004). And finally, anterior cingulate activity, particularly dorsal anterior cingulate, has been implicated in the detection and resolution of response interference or conflict. Together, the increased tonic activity in these regions may reflect the greater level of cognitive control exerted by participants in the reward-incentive condition in order to improve performance and maximize reward attainment.

A theory of how individuals increase cognitive control processes to maintain or improve performance has been described in a review of the neurobiology of attention (Sarter et al., 2006). In response to theories of attention that focus solely on task demands, the authors counter that motivation to maintain attention is key, particularly in situations where there is a challenge, such as a distracter, a dual task, or general fatigue. In this theory, a decline in performance or increased response conflict is detected by the anterior cingulate, which then can call for increased attentional effort, depending on the costs and benefits of continuing versus disengaging. A right-lateralized frontoparietal network, the anterior attention network, can then exert top-down influences via cholinergic projections to sensory systems to reduce processing of distracters.

Pochon et al. (2002) demonstrated an increase in activity in regions associated with executive control in humans in an fMRI study. Comparing activity in an N-back task when reward was offered relative to activity when no reward was offered yielded an increase in activity in dorsolateral prefrontal cortex (DLPFC), an area involved in maintaining a task set. In their study, both increases in task demand and reward increased activity in DLPFC. A similar pattern of findings was observed in Taylor et al. (2004). These studies suggest that reward may act as a kind of call for increased effort in a similar way as a change in actual task demands. In both instances, more effort is required to maintain or improve behavior.

Interestingly, Pochon et al. (2002) also reported the recruitment of lateral frontopolar regions in response to reward. Frontopolar cortex is a region that has been hypothesized to be involved in goal–subgoal coordination, that is, maintaining an overarching goal while simultaneously performing a task with its own local set of subgoals (Braver & Bongiolatti, 2002; Koechlin, 1999). According to Pochon et al. (2002), reward, or what is at stake at any given moment, may be represented in the form of a primary, high-order goal in the frontopolar region. This goal may then coordinate the processing and management of subgoals that enable optimal completion of the task.

Gilbert and Fiez (2004) aimed to investigate the integration of reward and task in lateral PFC during working memory performance. They predicted that ventral PFC would show cue-related responses only when these indicated the potential for reward. Instead, the results indicated that following reward cues there was a significant decrease in ventral PFC activation during the delay period, while DLPFC activity increased. They attributed this result to an oppositional relationship between emotional/motivation regions and cognitive regions, such that a decrease in the former is necessary to obtain optimal performance in the latter (Gilbert & Fiez, 2004).

Task-Relevant Activity in the Brain

The second pathway for reward to have an impact on task performance would be through reward acting on the existing architecture that supports task-specific processing. When comparing regions that would normally be used to do a task under no-incentive versus incentive conditions, it may be possible to see enhanced activity in specific task-related brain regions, as a key means of improving performance. For example, Taylor et al. (2004) demonstrated an increase in activation in task-relevant regions in an object working memory study. In their study, they were able to decompose the trials into discrete phases, and during the delay phase they found increased activity in the premotor superior frontal sulcus and the intraparietal sulcus. These areas, involved in short-term storage of visuospatial information, also increased in response to working memory load. This suggests, as in Pochon et al. (2002), reward may elicit enhanced processing in a similar manner as an increase in task demands.

Although Pochon et al. (2002) showed prefrontal regions operating in a cognitive control capacity, such regions also project to task-relevant areas. Visual attention, for example, the way objects are selected for attention from an array of competing objects, may depend on a top-down biasing signal from PFC (Desimone & Duncan, 1995). During tasks with visual stimuli, it may be possible to see increased activity in visual processing areas as a result of top-down influences from prefrontal cortex. This may aid individuals in focusing attention on stimulus objects, or their particularly relevant features.

Even without reward incentives, participants may selectively enhance task-specific processing regions in accordance with task demands. An example of such a situation can be seen in the classic PET study of Corbetta et al. (1991), in which participants were asked to alternate attentional focus to specific stimulus dimensions (e.g., color, motion, shape) during a same–different matching task. Each dimension was associated with increased activity in the corresponding relevant regions of extrastriate cortex (e.g., attention to motion was associated with increase in area MT). In contrast, when participants divided their attention across a number of different features, activity was not greater than in extrastriate cortex compared to a passive viewing condition. Each of the selective attention conditions were also associated with activity in attentional regions such as the pulvinar, a finding the authors attribute to its role in directing visual attention.

In the study by Small et al. (2005), discussed above, that used rewards and punishments during a spatial attention task, the authors suggest that incentives modulated structures directly involved with spatial representation. Specifically, the posterior cingulate cortex, a region that had previously been observed to show correlations between activation level and degree of spatial biasing (increased speed of target detection when previously spatially cued to the relevant location), also showed further modulation by incentives. Small et al. (2005) hypothesized that posterior cingulate cortex is a site of integration of motivation and spatial attention control.

A study by Krawczyk et al. (2007) investigated activity related to perception of scenes (parahippocampus activity), and perception of faces (fusiform gyrus activity). Trials presented both scenes and faces, with participants instructed to attend to one category and ignore the other. Results showed that activity in the relevant region increased for the attended category, and decreased for the ignored category, relative to a passive viewing condition. These effects were magnified by the opportunity to earn points that could be exchanged for money. Together, these results suggest that attention can enhance activity in perceptual regions relevant to a given task, and that reward can amplify these attentional effects.

Effects of Individual Differences

The majority of studies in cognitive neuroscience focus on what is common across individuals. A cognitive process under study is investigated by running a number of participants and taking a group average. The key interest is in identifying variance that can be explained by the experimental manipulations, whereas differences among the participants are treated as error variance (Thompson-Schill et al., 2005). This approach works well in identifying regions associated with a core set of functions necessary for a task, because presumably these regions will be the same for every participant as they perform the task. What may be missing is activity in regions that aren't necessary but are optional resources individuals may bring to bear to perform well at a task. This kind of activity may be particularly relevant to motivation, as increases in motivation should be associated with greater use of resources to acquire rewards or avoid punishments. The use of individual differences measures, therefore, may be useful in identifying such regions (Locke & Braver, 2008).

Studies of how stable personality traits modulate brain activity are still in their infancy. Nevertheless, this work has already begun to reveal that motivation-related individual differences may influence how cognitive tasks are performed. Several studies have focused on the relationship between BAS and brain activation, mostly using EEG methods (e.g., Sutton & Davidson, 1997). More recently, fMRI studies of working memory and cognitive control have examined how task-related activity may be modulated by individual differences in BAS. Gray and Braver (2002) demonstrated that BAS scores were related to activity in the ACC during a working memory task. Specifically, higher BAS scores were associated with decreased activity in the caudal "cognitive" ACC, despite no differences in performance between high and low BAS participants. A follow-up study (Gray et al., 2005) indicated that this same pattern described a number of components of the cognitive control network (e.g., lateral PFC, parietal cortex). Moreover, a multiple regression analysis indicated that the effects of BAS on cognitive control-related brain activity were independent of individual differences in fluid intelligence (which also showed separate modulatory effects on activation). This suggests that BAS is associated with greater cognitive efficiency. Greater efficiency in cognitive control structures could be potentially useful in complex tasks like social interaction, which more extraverted individuals excel at, and which require the ability to multi-task (Lieberman & Rosenthal, 2001).

Effects of BAS have also been found on brain activity during more motivationally relevant tasks. Beaver et al. (2006) monitored brain activity while participants looked at pictures of appetizing food (contrasted with bland foods). They found BAS correlated with activity in a network of reward-sensitive regions, including ventral striatum, amygdala, and orbitofrontal cortex. Specifically, activity in these areas was correlated with the BAS drive scale, which measures the motivational, as opposed to seeking or hedonic aspects of reward. The BAS drive scale includes statements like, "When I want something I usually go all-out to get it." The authors argue that this subscale is especially sensitive because it most closely related to motivation per se, the drive to attain rewards in the environment, rather than to seek out new opportunities (fun-seeking), or the tendency to react strongly to receipt of rewards (reward sensitivity).

Other personality dimensions may also be relevant to motivation and reward processing. Pailing and Segalowitz (2004) examined how motivation and personality influenced behavioral performance and the neural response to error commission. Previous research on error detection and the anterior cingulate had shown that a specific brain wave form, the error-related negativity or ERN, could be observed following errors on a task. Pailing & Segalowitz found that although performance did not significantly differ in the different motivation conditions, the magnitude of the ERN was associated with the amount of reward available, such that making a mistake on a trial where a larger reward could have been achieved was associated with a larger ERN. ERN magnitude was also associated with scores on personality measures such as the conscientiousness subscale of the IPIP-NEO. As conscientiousness scores increased, differences in ERN amplitude across conditions decreased, perhaps because highly conscientious participants were equally concerned about making errors in all conditions. Conversely, Luu, Collins, and Tucker (2000) found that ERN amplitude was negatively correlated with trait differences in negative affect (NA), such that high NA individuals showed smaller ERNs. Interestingly, the effect interacted with time on task, such that high NA individuals initially showed greater amplitude ERN responses in the first task block, but in later task blocks showed a significantly decreased ERN response. The authors interpreted this pattern as suggesting that experience with a high degree of errors (which were induced by task demands and a quick response deadline) led the high NA individuals to eventually disengage from the task, in terms of attentional effort expended.

Differences in promotion and prevention regulatory focus were the subject of an fMRI study looking for neural correlates of these personality characteristics. In the study, participants made good/bad judgments about words while in the scanner. Results were that participants showed greater activity in amygdala, anterior cingulate, and extrastriate cortex to stimuli that were congruent with their regulatory focus, positive stimuli for the promotion focused, negative stimuli for the prevention focused. These results are consistent with the idea that regulatory focus influences how attention is allocated to incoming stimuli, such that more motivationally relevant stimuli garner more attention (Cunningham, Raye, & Johnson, 2005).

The studies outlined above show that some of the trait differences in personality have provided explanatory power in recent neuroscience studies using neuroimaging and EEG methods. They suggest that it may be possible, as well as desirable, to include measures that assess motivation-related concepts and traits in neuroscience experiments, particularly those that provide incentives for performance.

In summary, cognitive neuroscience experiments have provided evidence of how the presence of rewards and punishments in the environment may influence cognitive control and task-relevant regions of the brain to facilitate performance. By marshalling cognitive control processes to sustain attention, and maintain goals while suppressing interference, individuals can focus on the task at hand. By increasing activity in task-related regions, individuals can enhance detection and maintenance of relevant information, which can lead to faster and more accurate execution of responses. As demonstrated in classic experiments like Corbetta et al. (1991), since replicated a number of times, these processes can be engaged in response to tasks without any incentives. However, the recent research in motivation and the brain shows that incentives may engage these systems to a greater extent, as individuals attempt to maximize rewards (and/or minimize punishments) by doing well at a task. Finally, the engagement of at least a subset of the neural circuitry modulated by reward (and punishment) appears to be scaled further by stable individual variation in motivationally-relevant personality traits.

Convergence in Motivation Research

The previous sections have outlined the views of different groups of researchers working on motivation-related questions from different perspectives. Social and personality research has focused on how the framing and prioritization of goals, along with personality differences, influence the motivational structure of a situation. Neuroscience research has focused on theories that describe how the brain processes rewards and punishments. Cognitive neuroscience research has begun to show how both reward and punishment information, and motivation-related individual differences, can interact with regions in the brain responsible for cognitive control and task processing. Given the diverse methodologies and theoretical frameworks of these areas of research, drawing connections between them can seem like a daunting task. However, in studying motivation from multiple angles, it appears that there may be some convergence in the lines of research in what they reveal about motivation.

The hallmark of motivation is that it directs action toward a goal, whether intrinsically generated, or externally imposed. In the cognitive neuroscience literature, research indicates that areas of lateral prefrontal cortex are responsible for maintaining goals, and biasing ongoing processing towards the furtherance of goals. Deviations from goal pursuit are inhibited, and regions like anterior cingulate detect conflict to call for more control if needed. This process relates to Carver et al.'s (2000) feedback loop model of motivation in that a goal value is set, and progress towards that goal is closely monitored. However, what Carver et al. (2000) add to the standard cognitive model is that progress or lack thereof leads to positive or negative affect, respectively. It may be possible to incorporate affect into cognitive control in a region like anterior cingulate, which has both cognitive and affective functions, thus serving as a potential convergence zone.

Taylor et al. (2006) proposed a similar idea, namely that rostral ACC activity is reflective of the affective significance of monetary loss. It is possible to imagine that setbacks in progress towards the goal may have cognitive consequences (call for more top-down control) as well as affective consequences (negative feelings associated with setbacks). Similarly, Small et al. (2005) have looked at posterior cingulate cortex, and in noting that it is active in top-down control of attention, but also has rich connections to affective and motivational regions such as orbitofrontal cortex, have proposed that it may have a role in integrating emotion and attention.

The concept of the active self (Baumeister et al., 1998, 2000), which is responsible for inhibiting habitual responses or actions that may be tempting but are contextually inappropriate, bears more than passing resemblance to cognitive psychologists' view of inhibitory control. Importantly, however, in the active self literature, engaging this inhibitory mechanism for even very short amounts of time can have devastating effects on behavior by reducing the resources available for other physically or cognitively effortful tasks. This same dynamic does not appear to be present during cognitive situations involving inhibition. Performance can be maintained in typical cognitive inhibition tasks like the Stroop or a go/no-go across hundreds of trials.

The discrepancy between active self and cognitive inhibition paradigms may be that in the former, but not the latter, participants' persistence at a task is the relevant dependent variable. In this type of scenario, the choice may really be between staying in the laboratory versus leaving to do any number of other things. In certain situations there are conflicts between goals. In a laboratory setting, a participant might have the intention to do well on a task, but he or she also might want to leave early and get a cup of coffee. There may be something special about conflicts that pit two motivationally relevant choices against one another. This type of conflict may activate ACC in the same way that response conflict on a go/no-go task does, but may have a more emotional component as the different options are weighed against the costs, rather than comparing two motivationally irrelevant response choices. Nevertheless, it will also be important to figure out what the source of resource depletion is, and if it is truly a metabolic phenomenon (Gailliot et al., 2007). In this respect, it is interesting that the anterior cingulate has been implicated in metabolic regulation (Teves et al., 2004).

For larger goals to have influence on behavior, it is often beneficial to break them down into manageable smaller goals. Frontopolar cortex has been implicated in subgoal processing (Braver & Bongiolatti, 2002). This is similar to Gollwitzer's (1999) conception of goal intentions and implementation intentions. It is all very well to have a larger goal in mind, but without breaking it down into the necessary elements that direct behavior in specific ways, it is harder to execute. Implementation intentions are like subgoals, in that they specify how the larger goal should be pursued, but they are also special for another reason. The specific intention is paired with an external cue, such that it is executed when the external cue is encountered in the environment. This takes the pressure off the individual to internally cue the behavior. This is similar to research on prospective memory, while not reviewed here, involves the active maintenance of an intention (pick up dry cleaning) that is made easier if it is paired with a highly salient cue (when you see the red barn from the highway, take the exit for the dry cleaners) (Einstein & McDaniel, 2005). Thus, it may be that frontopolar cortex is engaged under conditions for which implementation intentions are not externally available, and must instead be endogenously generated based on goal intentions.

Another cognitive theory related to reliance on external cues is the dual mechanisms of control theory (Braver et al., 2007, 2009). Proactive control involves utilizing actively maintained context information to prime response selection. Thus, a proactive control strategy would keep the intention of going to the dry cleaners active all day until you arrived there. In contrast, reactive control involves utilizing context information only when needed. Thus, until the red barn was spotted from the highway, the intention would not be activated. Proactive control is generally very successful, but reactive control is less costly because the information only needs to be retrieved when triggered by an imperative event. Thus, the effect of a shift in motivational priority might have the effect of shifting individuals from a reactive cognitive control strategy into a more proactive one.

Other social psychology perspectives emphasize not just how goals are pursued, but how the types of goals people tend to adopt influence their behavior. One theory of goal choice is the research on performance versus mastery goals (Dweck, 1986). Results from many studies show that mastery goals tend to be associated with more positive outcomes. Mastery goals emphasize learning to do a task well rather than comparing your performance to others, and may be more effective exactly because they involve focusing attention on aspects of the task that can be improved. An individual's own behavior is under his or her control whereas others' is not. In addition, "Do better than person X" is more of a goal intention, whereas learning to do the task may be more of an implementation strategy. Thus, other-focused goals may be more cognitively effortful, and require the involvement of higher-order control mechanisms such as the frontopolar cortex.

Although goal choice is important, in laboratory settings the goal is often provided by the experimenter explicitly. When researchers provide incentives, the expectation is that these rewards or punishments (what's at stake) would be actively maintained by participants as they do the task. Evidence reviewed above showed that OFC, and LPFC maintain this kind of information about motivational context with reference to the desires of the individual. This kind of information could be the raw material for goals, as once the individual decides that a reward is worth pursuing (the receipt of the reward is of higher value than the effort required to achieve it), its attainment becomes a goal that is maintained. Given that at the outset of experiments, participants do not have direct experience of how much joy they will get out of a reward, these representations are predictions to some extent. Moreover, even after repeated experience with rewards, the hedonic qualities of these rewards may change (or be state-dependent), and thus it is important to have a maintained representation that captures the most recent experience rather than a more generic long-term memory trace or future prediction. The research on affective forecasting intersects here because it states that individuals decide on what goals to adopt based on their expected amount of positive affect from achieving them. Interestingly, this research has also shown that people are not particularly good at predicting how happy, or upset, they will be in future when they receive a reward or punishment. Thus, the biased predictions of participants may tend to enhance the effects of incentives, as individuals usually exaggerate how happy they will be on receiving a reward, and exaggerate how unhappy they will feel when penalized.

The way that people adopt, pursue, monitor, and react to achieving or failing to achieve goals may strongly relate to individual differences. People differ in how reactive they are to rewards and punishments, and what kinds of goals they feel are important to pursue. These differences may relate to physical differences in the brains of individuals. One theory that connects individual differences in personality to the physical structure of the brain has been conceptualized by Depue and Collins (1999). Their theory equates the functionality of the dopamine circuitry in VTA and NAcc with behavioral approach, such that high levels of DA facilitate approach behaviors aimed at bringing the organism in contact with reward stimuli. In this theory, genetic differences in the number of dopaminergic synapses at birth predisposes individuals with higher numbers of these synapses to explore more for rewards. Contact with more rewards then makes frequent use of the dopaminergic reward system, which leads to less pruning of these synapses, and thus to stable individual differences in the function of the dopamine system. People with greater amounts of dopaminergic activity may then be more reactive to cues of potential rewards, akin to the personality difference reflected in Jeffrey Gray's BAS. Linking this idea to Berridge's work, Depue & Collins state that greater dopaminergic activity may lead to greater incentive salience being attached to objects in the environment. That information could be stored in memory, and retrieved by the prefrontal cortex to motivate behavior even in absence of cues in the environment.

Although it has been less studied, it may be that processing of aversive stimuli and penalties may be subject to individual differences in the structure of the serotonergic system. Recent work in behavioral genetics has looked at variants of a gene that affects a serotonin transporter, 5-HTT. This transporter is situated on the presynaptic neuron, and is responsible for the reuptake of serotonin from the synapse (Hariri & Holmes, 2006). Individuals who have one or two copies of the "short" form of the allele reuptake serotonin less efficiently. Recent studies have examined the effect of this genetic variation on behavior. Individuals with short form 5-HTT are more likely to have higher levels of trait anxiety and harm avoidance (Lesch et al., 1996), have less capacity to cope with stress (Caspi et al., 2003), and show greater amygdala activity in response to fearful and angry faces (Hariri et al., 2002). Together, these results suggest that individuals who have the short allele version of the 5-HTT polymorphism may be more sensitive to cues indicating the possibility of punishments or other aversive outcomes.

In summary, the process by which people set and pursue goals can be studied both by asking about what those goals are, how people feel about them, and what behaviors they exhibit from a macro-level in the social-personality realm, or at the micro-level in cognitive and affective neuroscience research. Some of the concepts generated are similar, and may rely on the same underlying brain structures. More power to interpret the brain activity arising from those structures may be forthcoming if efforts are made to understand motivation on multiple levels.

Future Directions for Research

There are several key questions deserving of future research efforts that might be considerably more tractable using a multilevel approach. The first question is how motivation is created in individuals, in order to inspire greater effort to achieve a particular goal. One contributing set of factors is the incentive structures that are used. Research on animals often uses primary rewards, food or drink, whereas research on humans has used a number of different secondary or abstract rewards such as cash, or points. As described above, there is relatively less research on punishments, and less still that directly compares matched rewards and punishments.

Another set of factors is individual differences in participants in reward and punishment sensitivity, some of which are traceable to differences in brain anatomy and physiology. A combination of self-report or physiological measures on incentive-related variables and neuroimaging that can noninvasively reveal differences in the structure and function of relevant brain regions has the potential to quantify differences among participants.

To answer questions about how incentives and individual differences affect the creation of motivation, a good first step would be to have a picture of what reward and punishment motivational states look like in terms of their effects on behavior and brain activity. Understanding simple motivational states will then allow research to advance to look at how variations in incentives or the type of participants tested affects motivation. This research requires expertise in both psychometrics and neuroimaging, and is a prime candidate for interdisciplinary cooperation.

A second question for future research concerns how motivation, once created, affects cognition and action. How does having greater motivation facilitate the pursuit of goals? Looking at how the brain responds to incentives has yielded an important distinction between valence-dependent activity, which reflects the value of the incentives, and activity generated equally by both negative and positive incentives (valence-independent activity) (Roesch & Olson, 2004). On the valence-dependent side, research on reward processing has shown the complexity of reward representations. Specifically, rewards are evaluated for their magnitude, and their value to the individual in general (trait preferences, e.g., for certain foods) and at the time they are offered (state preferences, e.g., whether the person is hungry or not).

On the valence-independent side, there are prefrontal cortex neurons that respond just as strongly to high positive as high negative incentives, suggesting a translation has taken place that has stamped particular actions as important. Once a particular task has been deemed important, the next step is to translate desire to achieve a goal into action items. This process may be critical, and may be where some people who seem very motivated and/or report a high value on incentives fail. Recognizing importance per se is not sufficient, as shown by Gollwitzer's work contrasting goal intentions, which are vague goals like "do better in school," and implementation intentions, that explicate the goal with specific steps which are linked to external cues ("When I get home from school each day, I will spend an hour reading.").

A final question relates to how these two systems—that extract the value of different incentives, and then translate this information into specific action goals—interact to facilitate goal selection, and set effort levels for goal pursuit. To study this question, one strategy would be to design experiments where participants are presented with tasks that vary according to both the potential incentives possible (e.g., manipulate incentive valence and incentive value), and the actions or task dimensions to which they are attached. Ideally, these manipulations would be factorial (i.e., independent of each other). Given a task with a well-known behavioral and brain activity profile, it would be possible to see how the presence of incentives affects how people engage in the task, and whether these modulatory influences interact with either (or both) incentive manipulations or task goal manipulations.

The major questions outlined in this section about the creation of motivation, and the downstream effects of motivation, necessarily interact. For example, given that there are important individual differences in how participants react to the presence of reward and penalty cues, information from the valence-dependent system will be different for a participant who is indifferent to incentives than it will be for a participant strongly affected by incentives. In terms of channeling motivational drive into appropriate action, there may also be important differences in the capacity to break larger goals down into smaller, intermediate steps, perhaps related to working memory capacity or fluid intelligence.

Conclusion

Studying motivation, from its creation to how it affects brain activity and cognition, to ultimate changes in behavior, is an endeavor that requires consideration of multiple levels of analysis. In particular, motivation can be thought of as reflecting a process that sets the priority value of different behavioral goals, and hence the effort that will be expended to achieve them. This process affects goals and behaviors that span the range from primary drives and behaviors, such as eating and drinking, to future plans that may take years to complete (i.e., obtaining a Ph. D.). Thus, the factors that affect these goal priority mechanisms must involve computations that span the range from low-level assessments of physiological state (e.g., hunger), to complex and abstract evaluations of the cognitive demands and contextual constraints (e.g., social) of the current situation, as well as any stable trait-like differences in reactivity to these different variables. Consequently, a better understanding of the mechanisms of motivation and their influence on information processing may provide a crucial leverage point for researchers interested in self-control from various perspectives and domains of inquiry.

This chapter has reviewed work on, and theories of, motivation arising from both the neuroscience and social/personality literatures. We have further argued that a cognitive neuroscience approach may be particularly suited to helping close the gap between these two perspectives by providing hypotheses about the contributions of specific brain regions to cognitive processes that underlie the performance of tasks under incentive conditions, and how these are affected by both situational (i.e., task-related) and personality variables. It is our belief that future developments in the cognitive neuroscience of motivation will contribute strongly to a better understanding of the nature of self-control, by shedding light on the core processes that modulate the selection and pursuit of behavioral goals—the achievement of which is the hallmark of successful self-regulation.

References

Aron, A. R., Robbins, T. W., & Poldrack, R. A. Inhibition and the right inferior frontal cortex. Trends Cogn Sci 2004; 8: 170–177.

Atkinson, J., & Raphelson, A. Individual differences in motivation and behavior in particular situations. J Pers 1956; 24: 349–363.

Bandura, A. *Self-efficacy: The exercise of control.* New York: Freeman, 1997.

Baumeister, R. F., Bratslavsky, E., Muraven, M., & Tice, D. M. Ego depletion: Is the active self a limited resource? J Pers Soc Psychol 1998; 74: 1252–1265.

Baumeister, R. F., Muraven, M., & Tice, D. M. Ego depletion: A resource model of volition, self-regulation, and controlled processing. Soc Cogn 2000; 18: 130–150.

Beaver, J., Lawrence, A., van Ditzhuijzen, J., Davis, M., Woods, A., & Calder, A. Individual differences in reward drive predict neural responses to images of food. J Neurosci 2006; 10: 5160–5166.

Bechara, A., Damasio, H., Damasio, A., & Lee, G. P. Different contributions of the human amygdala and ventromedial prefrontal cortex to decision-making. J Neurosci 1999; 19: 5473–5481.

Berridge, K. C. Pleasures of the brain. Brain Cogn 2003; 52: 106–128.

Berridge, K. C. Motivation concepts in behavioral neuroscience. Physiol Behav 2004; 81: 179–209.

Berridge, K. C., & Robinson, T. E. What is the role of dopamine in reward: Hedonic impact, reward learning, or incentive salience? Brain Res Rev 1998; 28: 309–369.

Braver, T. S., & Bongiolatti, S. The role of frontopolar cortex in subgoal processing during working memory. Neuroimage 2002; 15: 523–536.

Braver, T. S., Gray, J. R., & Burgess, G. C. Explaining the many varieties of working memory variation: Dual mechanisms of cognitive control. In: Conway, A., Jarrold, C., Kane, M., Miyake, A., & Towse, J. (Eds.), *Variation in working memory.* Oxford: Oxford University Press, 2007.

Braver, T. S., Paxton, J. L., Locke, H. S., & Barch, D. M. Flexible neural mechanisms of cognitive control within human prefrontal cortex. PNAS 2009; 106: 7351–7356.

Calder, A. J., Beaver, J. D., Davis, M. H., van Ditzhuijzen, J., Keane, J., & Lawrence, A. Disgust sensitivity predicts the insula and pallidal response to pictures of disgusting foods. Eur J Neurosci 2007; 25: 3422–3428.

Cameron, J. Negative effects of reward on intrinsic motivation—a limited phenomenon: Comment on Deci, Koestner, and Ryan. Rev Edu Res 2001; 71: 29–42.

Cameron, J., & Pierce, W. D. Reinforcement, reward, and intrinsic motivation: A meta-analysis. Rev Edu Res 1994; 64: 363–423.

Carver, C. S., & White, T. L. Behavioral inhibition, behavioral activation, and affective responses to impending reward and punishment: The BIS/BAS scales. J Pers Soc Psychol 1994; 67: 319–333.

Carver, C., Sutton, S., & Scheier, M. Action, emotion, and personality: Emerging conceptual integration. Pers Soc Psychol Bull 2000; 26: 741–751.

Caspi, A., Sugden, K., Moffitt, T. E., et al. Influence of life stress on depression: Moderation by a polymorphism in the 5-HTT gene. Science 2003; 301: 386–389.

Caul, W. F., & Brindle, N. A. Schedule-dependent effects of haloperidol and amphetamine: Multiple-schedule task shows within-subject effects. Pharmacol Biochem Behav 2001; 68: 53–63.

Coan, J. A., Schaefer, H. S., & Davidson, R. J. Lending a hand: Social regulation of the neural response to threat. Psychol Sci 2006; 17: 1032–1039.

Corbetta, M., Miezin, F. M., Dobmeyer, S., Shulman, G. L., & Petersen, S. E. Selective and divided attention during visual discriminations of shape, color, and speed: Functional anatomy by positron emission tomography. J Neurosci 1991; 11: 2383–2402.

Cunningham, W. A., Raye, C. L., & Johnson, M. K. Neural correlates of evaluation associated with promotion and prevention regulatory focus. Cog Affect Behav Neurosci 2005; 5: 202–211.

Davidson, R. Affective neuroscience and psychophysiology: Toward a synthesis. Psychophysiology 2003; 40: 655–665.

Daw, N. D., Kakade, S., & Dayan, P. Opponent interactions between serotonin and dopamine. Neural Netw 2002; 15: 603–616.

Deakin, J. Roles of brain serotonergic neurons in escape, avoidance and other behaviors. J Psychopharmacol 1983; 43: 563–577.

Deci, E., & Ryan, R. *Intrinsic motivation and self-determination in human behavior.* New York: Plenum, 1985.

Delgado, M., Locke, H., Stenger, V., & Fiez, J. Dorsal striatum responses to reward and punishment: Effects of valence and magnitude manipulations. Cogn Affect Behav Neurosci 2003; 3: 27–38.

Delgado, M., Nystrom, L., Fissell C., Noll, D., & Fiez, J. Tracking the hemodynamic responses to reward and punishment in the striatum. J Neurophysiology 2000; 84: 3072–3077.

Depue, R., & Collins, P. Neurobiology of the structure of personality: Dopamine, facilitation of incentive motivation, and extraversion. Behav Brain Sci 1999; 22: 491–517.

Desimone, R., & Duncan, J. Neural mechanisms of selective visual attention. Ann Rev Neurosci 1995; 18: 193–222.

Dweck, C. Motivational processes affecting learning. Am Psychol 1986; 41: 1040–1048.

Einstein, G., & McDaniel, M. Prospective memory: Multiple retrieval processes. Curr Dir Psychol Sci 2005; 14: 286–290.

Elliot, A., McGregor, H., & Gable, S. Achievement goals, study strategies, and exam performance: A mediational analysis. J Educ Psychol 1999; 3: 549–563.

Elliott, R., Friston, K. J., & Dolan, R. J. Dissociable neural responses in human reward systems. J Neurosci 2000; 20: 6159–6165.

Everitt, B. J., Cardinal, R. N., Parkinson, J. A., & Robbins, T. W. Appetitive behavior: Impact of amygdala-dependent mechanisms of emotional learning. Ann NY Acad Sci 2003; 985: 233–250.

Fox, N., Rubin, K., Calkins, S., et al. Frontal activation asymmetry and social competence at four years of age. Child Dev 1995; 66: 1770–1784.

Freitas, A., & Higgins, T. Enjoying goal-directed action: The role of regulatory fit. Psychol Sci 2002; 13: 1–6.

Freitas, A., Liberman, N., Salovey, P., & Higgins, T. When to begin? Regulatory focus and initiating goal pursuit. Pers Soc Psychol Bull 2002; 28: 121–130.

Gailliot, M. T., Baumeister, R. F., DeWall, C. N., et al. Self-control relies on glucose as a limited energy source: Willpower is more than a metaphor. J Pers Soc Psychol 2007; 92: 325–336.

Gilbert, A. M., & Fiez, J. A. Integrating rewards and cognition in the frontal cortex. Cogn Affect Behav Neurosci 2004; 4: 540–552.

Gilbert, D., & Wilson, T. Miswanting: Some problems in the forecasting of future affective states. In: Forgas, J. P. (Ed.), *Feeling and thinking: The role of affect in social cognition*. New York, NY: Cambridge University Press, 2000: pp.178–197.

Gollwitzer, P. Strong effects of simple plans. Am Psychol 1999; 54: 493–503.

Grace, A., & Bunney, B. The control of firing pattern in nigral dopamine neurons: Single spike firing. J Neurosci 1984; 11: 2866–2876.

Gray, J. A. The psychophysiological basis of introversion-extraversion. Behav Res Ther 1970; 8: 249–266.

Gray, J. A. A critique of Eysenck's theory of personality. In: Eysenck, H. J. (Ed.), *A model for personality*. Berlin: Springer-Verlag, 1981: pp. 246–276.

Gray, J. R., & Braver, T. S. Personality predicts working-memory-related activation in the caudal anterior cingulate cortex. Cogn Affect Behav Neurosci 2002; 2: 64–75.

Harackiewicz, J. Predictors and consequences of achievement goals in the college classroom: Maintaining interest and making the grade. J Pers Soc Psychol 1997; 73: 1284–1295.

Hariri, A. R., & Holmes, A. Genetics of emotional regulation: The role of the serotonin transporter in neural function. Trends Cogn Sci 2006; 10: 182–191.

Hariri, A. R., Mattay, V. S., Tessitore, A., et al. Serotonin transporter genetic variation and the response of the human amygdala. Science 2002; 297: 400–403.

Heath, R. G. *Exploring the mind-brain relationship*. Baton Rouge, LA: Moran Printing, 1996.

Heggestad, E., & Kanfer, R. Individual differences in trait motivation: Development of the motivational trait questionnaire. Int J Educ Res 2000; 33: 751–776.

Henriques, J., & Davidson, R. Left frontal hypoactivation in depression. J Abnorm Psychol 1991; 100: 535–545.

Higgins, T. Promotion and prevention as a motivational duality: Implications for evaluative processes. In: Chaiken, S., & Trope, Y. (Eds.), *Dual-process theories in social psychology*. New York, NY: Guilford Press, 1999: pp. 503–525.

Higgins, E. T., Friedman, R. S., Harlow, R. E., Idson, L. C., Ayduk, O. N., & Taylor, A. Achievement orientations from subjective histories of success: Promotion pride versus prevention pride. Eur J Soc Psychol 2001; 31: 3–23.

Hull, C. L. *Principles of behavior, an introduction to behavior theory*. New York, NY: D. Appleton-Century, 1943.

Ikemoto, S., & Panksepp, J. The role of nucleus accumbens dopamine in motivated behavior: A unifying interpretation with special reference to reward-seeking. Brain Res Rev 1999; 31: 6–41.

Jensen, J., McIntosh, A. R., Crawley, A. P., Mikulis, D. J., Remington, G., & Kapur, S. Direct activation of the ventral striatum in anticipation of aversive stimuli. Neuron 2003; 40: 1251–1257.

Kawagoe, R., Takikawa, Y., & Hikosaka, O. Expectation of reward modulates cognitive signals in the basal ganglia. Nat Neurosci 1998; 1: 411–416.

Kobayashi, S., Nomoto, K., Watanabe, M., Hikosaka, O., Schultz, W., & Sakagami, M. Influences of rewarding and aversive outcomes on activity in macaque lateral prefrontal cortex. Neuron 2006; 51: 861–870.

Koechlin, E., Basso, G., Pietrini, P., Panzer, S., & Grafman, J. The role of the anterior prefrontal cortex in human cognition. Nature 1999; 399: 148–151.

Koob, G. F., Riley, S. J., Smith, S. C., & Robbins, T. W. Effects of 6-hydroxydopamine lesions of the nucleus accumbens septi and olfactory tubercle on feeding, locomotor activity, and amphetamine anorexia in the rat. J Comp Physiol Psychol 1978; 92: 917–927.

Krawczyk, D., Gazzaley, A., & D'Esposito, M. Reward modulation of prefrontal and visual association cortex during an incentive working memory task. Brain Res 2007; 1141: 168–177.

Kringelbach, M. L., O'Doherty, J., Rolls, E. T., & Andrews, C. Activation of the human orbitofrontal cortex to a liquid food stimulus is correlated with its subjective pleasantness. Cereb Cortex 2003; 13: 1064–1071.

Kringelbach, M. L., & Rolls, E. T. The functional neuroanatomy of the human orbitofrontal cortex: Evidence from neuroimaging and neuropsychology. Prog Neurobiol 2004; 72: 341–372.

LaBar, K., LeDoux, J., Spencer, D., & Phelps, E. Impaired fear conditioning following unilateral temporal lobectomy in humans. J Neurosci 1995; 15: 6846–6855.

Leon, M. I., & Shadlen, M. N. Effect of expected reward magnitude on the response of neurons in the dorsolateral prefrontal cortex of the macaque. Neuron 1999; 24: 415–425.

Lerner, J., & Keltner, D. Beyond valence: Toward a model of emotion-specific influences on judgment and choice. Cogn Emot 2000; 14: 473–493.

Lerner, J., & Keltner, D. Fear, anger, and risk. J Pers Soc Psychol 2001; 81: 146–159.

Lesch, K., Bengel, D., Heils, A., et al. Association of anxiety-related traits with a polymorphism in the serotonin transporter gene regulatory region. Science 1996; 274: 1527–1531.

Lieberman, M. D., & Rosenthal, R. Why introverts can't always tell who likes them: Multitasking and nonverbal decoding. J Pers Soc Psychol 2001; 80: 294–310.

Locke, H. S., & Braver, T. S. Motivational influences on cognitive control: Behavior, brain activation, and individual differences. Cogn Affect Behav Neurosci 2008 Mar; 8(1): 99–112.

Lucki, I. The spectrum of behaviors influenced by serotonin. Biol Psychiatry 1998; 44: 151–162.

Luu, P., Collins, P., & Tucker, D. Mood, personality, and self-monitoring: Negative affect and emotionality in relation to frontal lobe mechanisms of error monitoring. J Exp Psychol Gen 2000; 129: 43–60.

Markman, A. B., Maddox, W. T., & Baldwin, G. C. The implications of advances in research on motivation for cognitive models. J Exp Theor Artif Intell 2005; 17: 371–384.

Miller, E. K., & Cohen, J. D. An integrative theory of prefrontal cortex function. Ann Rev Neurosci 2001; 24: 167–202.

Mirenowicz, J., & Schultz, W. Preferential activation of midbrain dopamine neurons by appetitive rather than aversive stimuli. Nature 1996; 379: 449–451.

Montague, P. R., & Berns, G. S. Neural economics and the biological substrates of valuation. Neuron 2002; 36: 265–284.

Muraven, M., Tice, D. M., & Baumeister, R. F. Self-control as limited resource: Regulatory depletion patterns. J Pers Soc Psychol 1998; 74: 774–789.

Muraven, M., & Baumeister, R. F. Self-regulation and depletion of limited resources: Does self-control resemble a muscle? Psychol Bull 2000; 126: 247–259.

O'Doherty, J. P. Reward representations and reward-related learning in the human brain: Insights from neuroimaging. Curr Opin Neurobiol 2004; 14: 769–776.

O'Doherty, J., Kringelbach, M. L., Rolls, E. T., Hornak, J., & Andrews, C. Abstract reward and punishment representations in the human orbitofrontal cortex. Nat Neurosci 2001; 4: 95–102.

O'Doherty, J., Rolls, E. T., Francis, S., Bowtell, R., & McGlone, F. Representation of pleasant and aversive taste in the human brain. J Neurophysiol 2001; 85: 1315–1321.

Olds, J. Self stimulation of the brain. Science 1958; 127: 315–324.

Padoa-Schioppa, C., & Assad, J. A. Neurons in the orbitofrontal cortex encode economic value. Nature 2006; 441: 223–226.

Pailing, P. E., & Segalowitz, S. J. The error-related negativity as a state and trait measure:

Motivation, personality, and ERPs in response to errors. Psychophysiology 2004; 41: 84–95.

Pardo, J. V., Fox, P. T., & Raichle, M. E. Localization of a human system for sustained attention by positron emission tomography. Nature 1991; 349: 61–64.

Paton, J. J., Belova, M. A., Morrison, S. E., & Salzman, C. D. The primate amygdala represents the positive and negative value of visual stimuli during learning. Nature 2006; 439: 865–870.

Peciña, S., & Berridge, K. C. Opioid eating site in accumbens shell mediates food intake and hedonic 'liking': Map based on microinjection Fos plumes. Brain Res 2000; 863: 71–86.

Pochon, J. B., Levy, R., Fossati, P., et al. The neural system that bridges reward and cognition in humans: An fMRI study. Proc Nat Acad Sci USA 2002; 99: 5669–5674.

Posner M., & Petersen S. The attention system of the human brain. Ann Rev Neurosci 1990; 13: 25–42.

Reynolds, S. M., & Berridge, K. C. Positive and negative motivation in nucleus accumbens shell: Bivalent rostrocaudal gradients for GABA-elicited eating, taste "liking"/"disliking" reactions, place preference/avoidance, and fear. J Neurosci 2002; 22: 7308–7320.

Robinson, T. E., & Berridge, K. C. The psychology and neurobiology of addiction: An incentive-sensitization view. Addiction 2000; 95: S91–S117.

Robinson, S., Sandstrom, S. M., Denenberg, V. H., & Palmiter, R. D. Distinguishing whether dopamine regulates liking, wanting, and/or learning about rewards. Behav Neurosci 2005; 119: 5–15.

Roesch, M., & Olson, C. Neuronal activity related to reward value and motivation in primate frontal cortex. Science 2004; 304: 307–310.

Rolls, E. T. The rules of formation of the olfactory representations found in the orbitofrontal cortex olfactory areas in primates. Chem Senses 2001; 26: 595–604.

Rolls, E. T. Taste and related systems in primates including humans. Chem Senses 2005; 30(Suppl 1): i76–i77.

Salamone, J. D. Involvement of nucleus accumbens dopamine in appetitive and aversive motivation. Behav Brain Res 1994; 61: 117–133.

Salamone, J. D., & Correa, M. Motivational views of reinforcement: Implications for understanding the behavioral functions of nucleus accumbens dopamine. Behav Brain Res 2002; 137: 3–25.

Sanfey, A. G., Rilling, J. K., Aronson, J. A., Nystrom, L. E., & Cohen, J. D. The neural basis of economic decision-making in the ultimatum game. Science 2003; 300: 1755–1758.

Sarter, M., Gehring, W. J., & Kozak, R. More attention must be paid: The neurobiology of attentional effort. Brain Res Rev 2006; 51: 145–160.

Schultz, W. Getting formal with dopamine and reward. Neuron 2002; 36: 241–263.

Seymour, B., Singer, T., & Dolan, R. The neurobiology of punishment. Nat Rev Neurosci 2007; 8: 300–312.

Small, D., Gitelman, D., Gregory, M., Nobre A., Parrish, T., & Mesulam, M-M. The posterior cingulate and medial prefrontal cortex mediate the anticipatory allocation of spatial attention. Neuroimage 2003; 18: 633–641.

Small, D., Gitelman, D., Simmons, K., Bloise, S., Parrish, T., & Mesulam, M. Monetary incentives enhance processing in brain regions mediating top-down control of attention. Cereb Cortex 2005; 15: 1855–1865.

Small, D. M., Zatorre, R. J., Dagher, A., Evans, A. C., & Jones-Gotman, M. Changes in brain activity related to eating chocolate: From pleasure to aversion. Brain 2001; 124: 1720–1733.

Sobotka, S. S., Davidson, R. J., & Senulis, J. A. Anterior brain electrical asymmetries in response to reward and punishment. Electroencephalogr Clin Neurophysiol 1992; 83: 236–247.

Sutton, S., & Davidson, R. Prefrontal brain asymmetry: A biological substrate of the behavioral approach and inhibition systems. Psychol Sci 1997; 8: 204–210.

Taylor, S. F., Martis, B., Fitzgerald, K. D., et al. Medial frontal cortex activity and loss-related responses to errors. J Neurosci 2006; 26: 4063–4070.

Taylor, S. F., Welsh, R. C., Wager, T. D., Phan, K. L., Fitzgerald, K. D., & Gehring, W. J. A functional neuroimaging study of motivation and executive function. Neuroimage 2004; 21: 1045–1054.

Teves, D., Videen, T. O., Cryer, P. E., & Powers, W. J. Activation of human medial prefrontal cortex during autonomic responses to hypoglycemia. Proc Nat Acad Sci USA 2004; 101: 6217–6221.

Thompson-Schill, S., Braver, T., & Jonides, J. Individual differences. Cogn Affect Behav Neurosci 2005; 5: 115–116.

Tremblay, L., & Schultz, W. Relative reward preference in primate orbitofrontal cortex. Nature 1999; 398: 704–708.

Valenstein, E. S., Cox, V. C., & Kakolewski, J. W. Reexamination of the role of the hypothalamus in motivation. Psychol Rev 1970; 77(1): 16–31.

Watanabe, M., Hikosaka, K., Sakagami, M., & Shirakawa, S. Coding and monitoring of motivational context in the primate prefrontal cortex. J Neurosci 2002; 22: 2391–2400.

Wyvell, C. L., & Berridge, K. C. Intra-accumbens amphetamine increases the pure incentive salience of sucrose reward: Enhancement of reward 'wanting' without 'liking' or response reinforcement. J Neurosci 2000; 20: 8122–8130.

CHAPTER 8

The Common Neural Basis of Exerting Self-Control in Multiple Domains

Jessica R. Cohen and Matthew D. Lieberman

ABSTRACT

People regularly exert control over impulsive thoughts and behaviors in order to make appropriate decisions and take appropriate actions even when they are more difficult or less pleasant than alternative choices. A common theme in mental illnesses characterized by impulsivity, such as ADHD and substance abuse, is an impaired self-control mechanism. Therefore, understanding the mechanisms underlying an intact control mechanism can not only shed light on how healthy people exert self-control over their thoughts and behaviors, but help us to understand what is impaired in patient populations as well. The right ventrolateral prefrontal cortex (VLPFC) is a region in the brain that is commonly activated when people are exerting many different forms of self-control. It is noted that other prefrontal regions also consistently activated when one exerts self-control, such as the dorsolateral prefrontal cortex, anterior cingulate cortex, orbitofrontal cortex, and medial prefrontal cortex, may be recruited for other task demands and not self-control specifically. Although the right VLPFC has been linked to other functions as well, this review will focus on the hypothesized general role that it plays during acts of self-control.

There are infinite manners in which one can exert self-control. We limit our discussion to six forms of conscious, explicit control that are commonly addressed in the literature: motor response inhibition, suppressing risky behavior, delaying gratification, regulating emotion, memory inhibition, and thought suppression. First, we review the literature exploring the involvement of the right VLPFC in each type of self-control separately. Next, we explore the small amount of literature comparing different forms of self-control to each other and discuss the possibility that these forms of self-control are related constructs. We also discuss the anatomical positioning of the right VLPFC and point out that it is well suited to serving a key role in exerting self-control. Finally, we conclude that although more direct research must be conducted before firm conclusions can be made, there is evidence that the right VLPFC is utilized when exerting self-control regardless of the specific domain of control.

Keywords: Self-control, fMRI, right VLPFC

Self-control can be defined as "the overriding or inhibiting of automatic, habitual, or innate behaviors, urges, emotions, or desires that would otherwise interfere with goal directed behavior" (Muraven et al., 2006). Without self-control, capricious and enjoyable decisions would be made, statements uttered, and actions taken. For example, if one has a deadline at work for an unpleasant project, he or she may have the inclination to leave work and do something fun instead of taking the responsible, yet dull, path of meeting the deadline. Or, if one has a strong urge to disclose to one's boss his or her opinion of that horrendous project, the person may instead remain silent and sensibly agree to work late in order to competently finish the task at hand. Whereas it is sometimes more desirable to follow one's own whims, those actions could occur at the expense of practical and boring yet sensible decisions. Clearly it is important that some sort of internal control system be implemented in order to inhibit such impulses so that more appropriate decisions can be made and actions taken. When that control system is impaired problems occur, such as impulsive behavior in ADHD, gambling, poor financial decisions, substance abuse, etc. The negative behavioral and clinical manifestations of a lack of self-control underscore the importance of thoroughly understanding the basis of an intact control mechanism at many different levels of analysis, including cognitive, clinical, social, and neural. This review concentrates on the vast base of literature exploring the neural basis of self-control. This review demonstrates that the right ventrolateral prefrontal cortex (VLPFC) is a common neural region recruited for successful self-control in a variety of domains. The existence of this common neural mechanism implies that the different forms of self-control may be parts of a unitary concept.

It is important to point out that the right VLPFC has been associated with a number of diverse tasks and cognitive processes, such as executive self-control in a variety of domains (as discussed in this chapter), stimulus-driven, bottom-up attention and automatic alerting to unexpected, salient stimuli (Corbetta & Shulman, 2002), memory retrieval when one must differentiate between relevant and irrelevant aspects of a stimulus (Kostopoulos & Petrides, 2003), both object-oriented (Courtney et al., 1996) and spatially oriented (Rizzuto et al., 2005) working memory, and the interpretation of emotions (Kober et al., 2008). This chapter will focus on the association between the right VLPFC and self-control processes as one possible role of the right VLPFC. Although other right VLPFC roles are not discussed here, this review is not excluding the possibility that the right VLPFC plays a part in multiple functions. Moreover, other prefrontal areas, such as the dorsolateral prefrontal cortex (DLPFC), medial prefrontal cortex (mPFC), and anterior cingulate cortex (ACC) are often involved in tasks requiring control as well. However, they may be recruited for other task demands and are not thoroughly discussed here.

The extent and diversity of problems that can occur when self-control is impaired highlights the fact that various forms of self-control do exist. There is disagreement, however, as to whether the varieties of self-control and its assumed converse, impulsivity, are one construct (Monterosso & Ainslie, 1999) or many constructs (Evenden, 1999). Many personality and clinical psychologists have assumed that impulsivity can be decomposed into multiple independent parts, however there is disagreement as to the number and specifics of each and limited evidence to support those claims (Evenden, 1999). Although Ainslie, Evenden, and their colleagues have focused on temporal aspects of impulsivity, such as the inability to withhold responses or to wait for delayed rewards, and their research involves animals, their approaches to dissecting impulsivity and self-control can be expanded to other domains. A notable difference between self-control research with humans and animals is that humans have the advantage of being able to consciously and explicitly exert self-control, something that is much more difficult to observe or measure in animals, if it exists (Monterosso & Ainslie, 1999).

There are infinite manners by which to study self-control in humans. Control processes can either be unconscious, such as priming or speech control, or conscious and explicit. This chapter focuses on literature exploring a number of different forms of explicit self-control to determine if seemingly very different modes of control are in fact subserved by the same underlying process and therefore utilize similar neural mechanisms.

The specific forms of explicit self-control examined here are motor response inhibition, suppressing risky behavior, delaying gratification, regulating emotion, memory inhibition, and thought suppression. They are vastly different on the surface. However, it is possible that they are different manifestations of the same construct, namely the exertion of self-control over actions or cognitions while engaging in goal-directed behavior. If that is the case, the neural networks subserving each of these subtypes of self-control may be overlapping, if not identical. Although very little research has directly compared different forms of self-control to each other, there is evidence that the individual subtypes may rely on similar prefrontal networks. The right VLPFC, DLPFC, mPFC, and ACC are commonly activated in neuroimaging studies exploring the different self-control subtypes, although the right VLPFC is the region most consistently involved in such studies (Anderson et al., 2004; Aron & Poldrack, 2006; Elliott et al., 2000; Kalisch et al., 2005; Levesque et al., 2003; Matthews et al., 2004; McClure et al., 2004; Mitchell et al., 2007; Ochsner, 2004; Wyland et al., 2003). This review first focuses on the involvement of the right VLPFC in each variety of self-control separately. Next, it integrates the relevant literature and discusses the possibility that these forms of self-control are related constructs. Finally, it will conclude that although more direct research must be conducted before firm conclusions can be made, there is evidence that the right VLPFC is utilized when exerting self-control regardless of the specific domain of control. The commonly noted DLPFC and ACC activation, on the other hand, may subserve more general control mechanisms such as rule monitoring (Bunge, 2004) and conflict monitoring (Botvinick et al., 2004) respectively, whereas mPFC activity may be related to self-referential or emotional processing (Lieberman, 2007; Ochsner & Gross, 2005).

Motor Inhibition

Arguably one of the most thoroughly studied forms of self-control is motor response inhibition. Motor response inhibition can be simple, requiring the suppression of an action with no alternative response necessary, or more complex, requiring both the suppression of an action and the execution of a different action. For example, a baseball player must exert simple inhibitory control in order to stop an already-initiated swing when realizing the pitch being thrown will be a ball. If, however, one is driving and a child runs in front of the car, both inhibitory control must be exerted to remove pressure from the gas pedal and a different, unexpected response must be executed in order to slam on the brakes.

Both the go/no-go (Casey et al., 1997) and the stop-signal (Logan, 1994) tasks are simple motor inhibition paradigms that test one's ability to exert self-control by inhibiting a button press to a stimulus at the occurrence of a signal to immediately stop responding. The level of prepotency of responding can be manipulated by altering the proportion of stimuli not to be responded to; the fewer of those stimuli, the harder it is to inhibit a response. The dependent variable in the go/no-go task is either number of commission errors—responding to a no-go stimulus, or number of omission errors—not responding to a go stimulus. The dependent variable in the stop-signal task is stop-signal reaction time (SSRT), a measure of the time a participant needs to be able to inhibit his or her response. The main difference between the two tasks lies in the signal to stop: in the go/no-go task it is the stimulus itself (i.e., an "X" in a string of other letters), whereas in the stop-signal task it is a signal that occurs after the onset of the primary stimulus (i.e., a tone or a change in color of the primary stimulus). Successful performance on the stop-signal task reflects inhibitory self-control of an already-initiated response and may reflect a more pure form of response inhibition than that in the go/no-go task, which may more accurately reflect response selection. In the go/no-go paradigm participants are given the signal to inhibit their response before the response is actually initiated, since the stimulus itself is the signal to inhibit (Rubia et al., 2001). Importantly, however, imaging results with both tasks are very similar. The right VLPFC is consistently engaged in both go/no-go and stop-signal tasks when participants are inhibiting prepotent responses. It should be noted that other prefrontal and subcortical regions, namely the DLPFC (Garavan et al., 2002; Liddle et al., 2001; Rubia et al., 2001), ACC (Garavan et al., 2002; Rubia et al., 2001), and subthalamic nucleus (STN; Aron & Poldrack, 2006) are often found to be active during motor inhibition tasks. Although these regions are likely involved in cognitive control in addition to the right VLPFC, because this review focuses on the role of the right VLPFC in inhibitory self-control, they are not discussed further.

Early lesion work in rhesus monkeys demonstrated that the inferior frontal convexity, corresponding to the VLPFC in humans, is necessary for successful performance on go/no-go tasks, whereas the mPFC is not (Iversen & Mishkin, 1970). Single-cell recording in macaque monkeys has found neurons in the inferior DLPFC (analogous to the human VLPFC) that respond selectively either to go or to no-go stimuli (Sakagami & Niki, 1994).

Human neuroimaging studies have consistently found a similar reliance on right-lateralized VLPFC (or the inferior frontal cortex within the VLPFC; IFC) during successful no-go (Garavan et al., 2002; Garavan et al., 1999; Konishi et al., 1998; Liddle et al., 2001; Menon et al., 2001) and stop-signal (Aron & Poldrack, 2006; Chevrier et al., 2007; Rubia et al., 2003) performance (for a review, see Aron et al., 2004). In further support of the key role of the right IFC in successful stopping, it has been found that the right IFC and the STN, a subcortical region thought to be critical for successful stopping as well, were the only two regions correlated with SSRT; greater activity in each area was associated with faster SSRT and therefore better response inhibition (Aron & Poldrack, 2006).

Both human lesion studies and transcranial magnetic stimulation (TMS) studies have reinforced that not only is the right VLPFC utilized in successful motor response inhibition, but that it is necessary. One study examined stop-signal performance in patients with focal lesions in varying locations in the prefrontal cortex (Aron et al., 2003). The authors found that the extent of lesions in the right inferior frontal gyrus (IFG) was more correlated with slowing of SSRT than any other frontal lobe lesion. When regressing out effects resulting from right IFG damage no other frontal region damage correlated with SSRT, including the left IFG (Aron et al., 2003). Another study found that temporary disruption of the right IFG using TMS increased SSRT and decreased percent inhibition at a given stop-signal delay, regardless of what hand the subjects were using to respond (Chambers et al., 2006).

In conclusion, studies utilizing the go/no-go and stop-signal tasks consistently and fairly specifically implicate the right IFC in controlling simple motor response inhibition (Aron et al., 2004).

Reversal learning is a more complex form of motor inhibition that requires both the inhibition of a prepotent response and the substitution of that response with an alternate response that subjects had previously been instructed to avoid. Often, reward and punishment are used to develop prepotent responses and to signal when those responses must be overridden in favor of the opposite response (Clark et al., 2004).

Animal literature implicates the orbitofrontal cortex (OFC), a region within the larger VLPFC, as necessary for reversal learning. Lesioning the OFC consistently impairs reversal learning in a variety of species, including rats and primates (Clark et al., 2004). Early studies exploring focal frontal lesions in rhesus monkeys noted that the VLPFC/lateral OFC was crucial for reversal learning, whereas other frontal cortical regions such as the medial OFC, anterior OFC, and DLPFC were not (Butter, 1969; Dias et al., 1996; Iversen & Mishkin, 1970). Moreover, single unit recording from cells in both medial and lateral OFC in monkeys indicates that a population of cells responds to rewarded stimuli and that these same cells reverse their firing to the previously punished but newly rewarded stimulus after a reversal (Rolls et al., 1996; Thorpe et al., 1983).

Similar to the animal literature, human lesion studies consistently implicate the OFC as crucial for successful reversal learning (Fellows & Farah, 2003; Rolls et al., 1994). Because naturally occurring lesions in humans are not precise, a lateral/medial distinction cannot be made in these studies. Neuroimaging studies in healthy volunteers are beginning to elucidate the separable roles of specific sections of the OFC in reversal learning.

Tasks with healthy adults tend to use probabilistic reversal learning, meaning that subjects are given incorrect feedback on a certain percentage of responses, often 20%–30%, so as to increase the difficulty of the task and the number of reversal errors to be analyzed in event-related fMRI designs. Additionally, reversals occur after a range of correct responses in a row, for example, anywhere between 10 and 15, so the switch is not predictable (Cools et al., 2002). When comparing the final incorrect trial after a reversal, just before subjects reverse their response tendencies, to correct trials, trials where subjects did not subsequently change their response, or control tasks not requiring a decision to be made, the VLPFC/lateral OFC is consistently active (Cools et al., 2002; Kringelbach & Rolls, 2003; O'Doherty et al., 2003; Remijnse et al., 2005).

Crucially, in one study that further explored the presence of VLPFC activity during reversal learning, it was not more active for initial errors after a reversal (when the response was not subsequently changed) as compared to correct trials, or for probabilistic errors as compared to correct trials (Cools et al., 2002). This indicates that the main role of the VLPFC was to exert behavioral control over responses to inhibit the previously rewarded response so that a different response could be made. This purported VLPFC role is similar to that in go/no-go and stop-signal studies, where it has been implicated in inhibiting a prepotent response.

A theory that is largely supported by the aforementioned neuroimaging data is that the ventral PFC, and specifically the OFC, may be separated into functionally distinct areas, lateral and medial (Elliott et al., 2000). Given the diverse afferent and efferent connections between the OFC and other brain regions, including other prefrontal areas and limbic and subcortical areas, functional heterogeneity is not surprising. The medial OFC may keep track of reward contingencies in a dynamic environment, thus allowing for the realization of a necessary change in response if reward contingencies change, while the lateral OFC may exert behavioral control based on those contingencies (Elliott et al., 2000).

In conclusion, evidence is fairly consistent supporting a role for the right VLPFC, the IFG/lateral OFC in particular, in the behavioral inhibition of prepotent responses, whether the task requirement is simply to inhibit a response (Aron & Poldrack, 2006; Garavan et al., 2002; Rubia et al., 2003) or to inhibit a previously rewarded response in order to be able to make a different one (Cools et al., 2002; Elliott et al., 2000; Kringelbach & Rolls, 2003). When task demands include reward contingencies, as in most reversal learning paradigms, the mPFC is further recruited, particularly the medial OFC, which has been implicated both in processing reward-related information and in mediating emotion-related behavior, which may be relevant to reward monitoring as well (Elliott et al., 2000; O'Doherty et al., 2003).

Risk-Taking Behavior

When describing someone who engages in risky behavior, be it sensation-seeking such as sky diving, health-related such as drug use, or finance-related such as gambling, it seems intuitive to use adjectives and phrases such as "reckless", "impulsive", or "lacking self-control".

Although many self-report questionnaires have been developed to assess risky behavior, they are not ideal because subjects may not fill them out accurately due to lack of insight or self-presentational concerns (Lejuez et al., 2002). Thus, a handful of tasks have been designed to assess risky behavior in the laboratory while avoiding the pitfalls of self-report measures.

One of the earliest tasks designed to assess risky behavior was the Iowa Gambling Task (IGT; Bechara et al., 1994). Performance on the IGT is often impaired in patients with ventromedial PFC lesions, meaning that they tend to make risky choices that result in potentially higher gains in the short term, but a lower overall payoff (Bechara, 2004; Bechara et al., 1994; Bechara et al., 1998). However, there are interpretations of the processes involved in the IGT other than impaired self-control that can explain impaired performance on this task, such as learning outcome probabilities of each of the decks, developing a long-term strategy (Manes et al., 2002;

Wu et al., 2005), or reversal learning (Dunn et al., 2006; Fellows & Farah, 2005).

Many other studies have explored risky versus safe decision making in simpler gambling-related tasks with inconsistent results. In a recent meta-analysis, Krain and colleagues (2006) concluded that the lateral OFC and the mPFC are involved in risky decision-making. However, the involvement of these regions was task-general and was not related to whether the participants made risky or safe choices in the tasks. Some studies have examined risky versus safe choices more specifically. Although some have found that the right OFC/VLPFC was more active for risky as compared to safe trials (Cohen et al., 2005; Ernst et al., 2004; Eshel et al., 2007), a similar region has also been found to be more active for safe as compared to risky trials (Matthews et al., 2004). Moreover, lateral OFC activity has also been correlated positively with risk aversion (Tobler et al., 2007) and negatively with number of risky choices (Eshel et al., 2007), both relationships implying that greater lateral OFC activity is related to a tendency toward making safer choices.

The Balloon Analogue Risk Task (BART) is another procedure with which to explore risk-taking (Hunt et al., 2005; Lejuez et al., 2002). In this task, participants are shown a balloon and told to push one button to inflate it and another to end the trial. For each pump, a temporary bank of money is increased by a constant amount (e.g., five cents). With each additional pump, however, there is an increased chance that the balloon explodes. If the balloon explodes, the subject loses all the money in the temporary bank. If the subject chooses to end the trial before the balloon explodes, the money in the temporary bank becomes permanent winnings. The average number of pumps and the amount of money provided with each pump can be varied. In the first study using the BART, participants were healthy controls who varied on risk-taking tendencies as indexed by a variety of self-report measures (Lejuez et al., 2002). Number of pumps was correlated with a variety of risk-related concepts as indexed by the Barratt Impulsiveness Scale, the Eysenck Impulsiveness Scale, the Sensation Seeking Scale, and actual risky behavior including smoking, drinking, drug use, gambling, stealing, unprotected sex, and not using seatbelts. The correlation of performance on the BART with self-report scales was specific to risky behavior; it was not correlated with anxiety, depression, or empathy. This task is a good alternative to the gambling tasks because it is simple, provides immediate feedback, and, as sometimes occurs in the real world, risky behavior is rewarded up to a point before it is punished (Lejuez et al., 2002).

Whereas the BART has been related to impulsivity and presumably a lack of inhibitory control (Lejuez et al., 2002), its neural correlates have only begun to be explored. Preliminary data suggests that when suppressing prepotent responding on the BART, the right VLPFC is active (Stover et al., 2005, Cohen et al., 2009).

In conclusion, the risky decision-making literature provides support for the involvement of the lateral OFC and mPFC when confronted with risky choices (Krain et al., 2006). The involvement of the mPFC may result from the processing of reward-related information (Elliott et al., 2000). Although there is some support for the involvement of the right VLPFC in suppressing risky choices (Eshel et al., 2007; Matthews et al., 2004; Stover et al., 2005; Cohen et al., 2009; Tobler et al., 2007), the current literature is inconsistent. This phenomenon must be explored more thoroughly before any conclusions can be made regarding whether the suppression of a risky response is subserved in part by the same self-control mechanism that subserves successful motor response inhibition.

Temporal Discounting

Another focus of the literature exploring impulsivity regards temporal discounting. If given the choice of receiving $100 today or $110 tomorrow, many individuals choose to receive $100 today, even though it is a smaller reward. This may be viewed as impulsive behavior; people are sometimes unable to control their desire for an immediate payoff even though it would be beneficial to do so in the long run.

There have not been many studies exploring the neural basis of temporal discounting in humans. Research conducted with animals such as pigeons, rats, and primates, supports hyperbolic discounting models, in which the tendency to choose immediate rewards drops off steeply with time. Many animal studies have focused on the effects of lesions on temporal discounting behavior. When administering focal neural lesions to animals and then testing them on temporal discounting tasks, two regions consistently emerge that are associated with impulsive behavior when lesioned: the nucleus accumbens (NAcc) core and the OFC (Cardinal, 2006; Mobini et al., 2002). Moreover, single cell recordings in intact nidopallium caudolaterale (NCL) in pigeons, which corresponds to the human prefrontal cortex, have identified cells that fire during the delay between decision and reward when choosing the larger, delayed option. Activity in these cells was negatively correlated with length of delay, until a preference shift from the larger, later to the smaller, sooner reward. At that point, when the delay was zero for all decisions, cell activity remained constant (Kalenscher et al., 2005). Such cells, whose activity was negatively correlated with delay length, have also been identified in rhesus monkeys (Roesch & Olson, 2005). In both pigeons and monkeys, the same cells whose activity is negatively correlated with delay length fire more for greater reward magnitudes. In other words, OFC cells appear to code for overall subjective value of the rewards, incorporating both delay, which decreases subjective value, and reward, which increases subjective value (Kalenscher et al., 2005; Roesch & Olson, 2005). These cells are specific to the OFC, as cells in other prefrontal areas in the monkey, such as the DLPFC, frontal eye fields, supplementary eye fields, premotor area, and supplementary motor area did not code for delay length (Roesch & Olson, 2005).

In short, the animal literature has identified specific neural networks that seem to be involved in successful temporal discounting in a variety of species. This can provide clues as to which brain areas to study in humans in similar paradigms.

To date, there have been very few functional neuroimaging studies in healthy humans directly examining the neural systems underlying immediate as opposed to delayed rewards using choices between smaller rewards sooner or larger rewards later. In one such study, the authors hypothesized that a quasi-hyperbolic function, incorporating the two parameters ∂ (constant weighting of all delays) and ß (larger weight given to immediate outcomes), underlies temporal discounting (McClure et al., 2004). Other temporal discounting studies in humans have been more focused on exploring subjective value and other mental processes that will not be discussed here (*see* Kable & Glimcher 2007).

McClure and colleagues (2004) found that two dissociable neural systems were active during a temporal discounting task. The first was active for all trials where an immediate option was available (corresponding to ß). This network included limbic areas such as the ventral striatum, medial OFC, and mPFC. These areas have been associated with reward preference and visceral, emotional reactions. This limbic network can be seen as the neural parallel to Metcalfe and Mischel's (1999) "hot", emotional system. The second network was active during all decisions regardless of delay (corresponding to ∂), but more active for difficult than for easy decisions. Level of difficulty was defined by closer relative magnitudes of the two choices and greater variability in subject responses. This network included multiple PFC areas, including the right VLPFC, lateral OFC, and DLPFC. This lateral prefrontal network, which has been associated with response inhibition and rule representation, may be seen as the neural parallel to Metcalfe and Mischel's "cool", cognitive system. A dissociation between these two networks was seen when comparing their relative activation during trials where one option was immediate. When subjects chose the delayed option, the lateral prefrontal network was more active than the limbic network; when the immediate option was chosen, there was a trend toward the limbic network being more active than the lateral prefrontal network. While the use of a quasi-hyperbolic instead of a true hyperbolic function has been questioned (Ainslie & Monterosso, 2004), these results provide evidence that a right-lateralized prefrontal network, including the right VLPFC, is utilized when exerting self-control over temporal decision-making (McClure et al., 2004).

A similar paradigm was used in a neuroimaging study comparing methamphetamine abusers to healthy controls. The neural response to the task was qualitatively similar in the two groups even given the expected behavioral difference that methamphetamine abusers chose more impulsively than did healthy controls (Monterosso et al., 2007). Comparing hard choices to easy choices revealed significantly greater activation in the right VLPFC region found in other studies of self-control. Moreover, participants who chose the delayed option more had greater right VLPFC activity. (Monterosso et al., 2007).

While very little neuroimaging research has been conducted on temporal discounting, what does exist suggests that a network involving lateral prefrontal areas such as the VLPFC and DLPFC operates when deciding whether or not to delay gratification for a larger future payoff. This network is similar to that implicated in self-control and rule monitoring in both motor and behavioral economic domains (Aron & Poldrack, 2006; Cools et al., 2002; Garavan et al., 2002; Krain et al., 2006; Kringelbach & Rolls, 2003; O'Doherty et al., 2003; Rubia et al., 2003), and possibly includes more medial prefrontal areas including the OFC, implicated in emotion and reward processing and emotion regulation (Cools et al., 2002; Krain et al., 2006; O'Doherty et al., 2003).

Emotion Regulation

Whereas in many situations it can be adaptive to be in touch with and be able to express one's own emotions, there are some situations in which that is not appropriate. For example, a person who is a good sport may suppress feelings of disappointment and anger at not winning an award in order to congratulate and be happy for the person who beat him or her. Alternately, the winner may inhibit feelings of glee for achieving the award in order to be composed and supportive of the person he or she beat. In situations such as these, it is beneficial to be able to exert control over one's emotional state. Emotion regulation, which is the process by which people influence their emotional experience and expression (Gross, 1998), has been studied in order to understand the mechanisms behind self-control over affective processes. Successful emotion suppression requires that a person be able to exert self-control over his or her natural emotional response in order to dampen or strengthen it. It is thought that some mental disorders, such as anxiety and depression, may have their roots in the dysregulation of affect, thus there is much interest in discovering the mechanisms behind successful emotion regulation (Gross, 1998).

Although most research has been done on intentional emotion regulation, and that will be the focus in this chapter, unintentional emotion regulation can occur as well (Gross, 2002; Lieberman, 2007). For example, instructing participants to verbally label negative emotional stimuli appears to reduce negative emotion, even without a conscious attempt to do so. Some researchers studying emotion regulation are beginning to incorporate such designs into their research to explore unintentional emotion regulation (Hariri et al., 2000; Hariri et al., 2003; Lieberman et al., 2007).

The rapidly growing literature focusing on the neural correlates of emotion regulation fairly consistently implicates the right VLPFC/lateral OFC when suppressing as compared to maintaining negative emotions as varied as anxiety resulting from the anticipation of shocks, sadness, and viewing aversive images (Harenski & Hamann, 2006; Kalisch et al., 2005; Kim & Hamann, 2007; Levesque et al., 2003; Ochsner, 2004; Phan et al., 2005). Furthermore, the stimuli and emotion regulation strategies used vary greatly between studies, further emphasizing that the right VLPFC may play a role in exerting self-control in different contexts. Finally, an association between the magnitude of right VLPFC and right lateral OFC activity and self-reported decrease in negative emotion has been found, implying that this region is involved in controlling one's emotions (Ochsner et al., 2004; Phan et al., 2005). These results were specific to inhibiting negative emotion—when subjects were asked to decrease negative emotion, a more right-lateralized network involving the right lateral OFC was involved as compared to when they were asked to increase negative emotion. The left amygdala was more active when increasing as compared to decreasing negative emotion, upholding findings regarding its assumed role in the subjective experience of negative emotions (Ochsner et al., 2004). Multiple studies have found a negative correlation between right VLPFC activity and amygdala activity, implying that perhaps the right VLPFC has a role in suppressing the amygdala's natural response in negatively valenced situations (Hariri et al., 2000; Hariri et al., 2003; Lieberman et al., 2007; Phan et al., 2005).

The right VLPFC is consistently active when instructing subjects to suppress their emotions using a variety of stimuli and strategies. This suggests that inhibitory self-control, believed to be localized to the right VLPFC in a variety of self-control domains such as motor response inhibition, risk-taking, and temporal discounting, may be at work during emotion regulation as well.

Memory Inhibition

Although having good memory is adaptive, and forgetting information is often accompanied by negative consequences, there are situations in which it may benefit someone to forget something. For example, it is crucial to know one's own address and phone number. If people remembered every address and phone number they have ever had, however, it could make quickly recalling the current one more difficult or prone to error. In this case, it is adaptive to "forget" outdated information. Much research has been conducted exploring goal-oriented directed forgetting (MacLeod, 1998). The typical procedure displays a series of stimuli to participants, who are instructed to either remember all the stimuli or to forget some of them. Various stimuli have been used, such as strings of digits or consonants, individual words, word-pairs, and sentences. The cues to forget range from colored dots to the word "FORGET" being displayed. The recall and/or recognition of all stimuli is then tested, with "forget" stimuli consistently being remembered less than "remember" stimuli (MacLeod, 1998).

It has recently been asserted that the control of memory may involve a self-control process similar to behavioral inhibition as measured by stop-signal or go/no-go tasks (Levy & Anderson, 2002). In a memory-focused analogue to the go/no-go task, the think/no-think paradigm requires participants to suppress the memory for certain, previously learned unrelated word pair associations (no-think), while trying to remember others (think). The no-think condition results in successful directed forgetting (Anderson & Green, 2001).

Very little research exists to date exploring the neural basis of directed forgetting. The only neuroimaging study to explore directed forgetting utilized the think/no-think paradigm (Anderson et al., 2004). After the scanning session, participants were tested on cued recall of all the words, with an emphasis that they should recall think as well as no-think words. Behaviorally, the think/no-think manipulation worked. Participants recalled significantly fewer no-think words than think words in both recall tests (Anderson et al., 2004). When comparing the neural activity of no-think versus think trials, activity in the typical control network seen in studies of motor response inhibition was observed (Aron & Poldrack, 2006; Garavan et al., 2002; Rubia et al., 2003). Prefrontal activity included bilateral DLPFC and VLPFC (including the right inferior frontal gyrus), as well as the ACC, presupplementary motor area (preSMA), and dorsal premotor area. Supporting the theory that memories were actually suppressed during this task, bilateral hippocampal activation was less for the no-think trials than the think trials, possibly indicating that memory encoding was occurring during the think trials but not during the no-think trials. Correlational analyses were also conducted between activity during the scan and post-scanning recall. Activity in bilateral DLPFC and VLPFC was correlated with successful suppression—words that were not recalled were associated with greater lateral PFC activity during no-think trials. Hippocampal activity, on the other hand, was greater for think items that were later recalled as compared to those that were later forgotten (Anderson et al., 2004).

Although the functional MRI study using the think/no-think paradigm supported the hypothesis that inhibitory self-control, localized to the VLPFC in motor inhibition and other paradigms, is at work in directed forgetting (Anderson et al., 2004), more research must be conducted before firm conclusions may be drawn. Given that a wide range of prefrontal areas were recruited, it is possible that other control mechanisms were at work in addition to self-control during directed forgetting. For example, the DLPFC could have been recruited to monitor the think/no-think rule, while other regions may have been recruited to assist with differential rehearsal or other strategies used to control memory. It is interesting to note that although the think/no-think paradigm does not require motor control, a similar neural network to that seen during behavioral self-control was recruited, including the right VLPFC. This raises the possibility that the self-control network activated fairly consistently in a wide variety of forms of behavioral control such as those discussed above (including motor inhibition, risk-taking, temporal discounting, and emotion regulation), is also recruited for directed forgetting. However, given the dearth of research into this possibility to date, further research into the mechanisms behind and the neural basis of directed forgetting must be conducted before the role that inhibitory self-control plays in this phenomenon can be understood.

THOUGHT SUPPRESSION

Intuitively, it seems there can be advantages to being able to control one's own thoughts. For example, if a joke pops into someone's head at a funeral, it is preferable to be able to suppress thoughts about that joke so as not to smile or laugh inappropriately. In other situations, however, it may be disadvantageous to one's mental health to suppress certain thoughts, which can be an instance of an avoidant coping technique in mental disorders such as Post-Traumatic Stress Disorder (PTSD; Rassin et al., 2000). As a result of these somewhat contradictory effects of thought suppression, much research has been done in empirical settings to further understand this process.

The study of thought suppression has largely been conducted by Wegner and colleagues (for reviews, see Rassin et al., 2000; Wenzlaff & Wegner, 2000). In the classic "white bear" thought suppression experiment, two groups of participants were instructed to spend consecutive 5-minute periods either thinking about a white bear (the expression condition) or inhibiting thoughts about a white bear (the suppression condition) (Wegner et al., 1987). It was found that during suppression, thoughts of a white bear were reduced but not eliminated relative to expression. After suppression, however, a rebound effect occurred and there were increased thoughts about the target relative to control groups who did not have to previously suppress the thought. This pattern of results has been largely replicated (Wenzlaff & Wegner, 2000).

Parallels have been drawn between thought suppression and other domains of self-control (Wegner, 1992). Most similar is memory inhibition perhaps because of its shared reliance on cognitive, as opposed to behavioral control. The goal of both processes is to inhibit something from being consciously retrievable (Bjork, 1989; Rassin et al., 2000). Thought suppression shares some characteristics with other forms of self-control as well, such as emotion regulation and temporal discounting. When suppressing an emotion, subjects may rely on similar strategies, such as distraction, as when suppressing a thought (Gross, 2002). Subjects may also rely on distraction and attentional control when attempting to delay gratification (Mischel et al., 1989). Therefore, exploring the neural basis of thought suppression and how it may relate to the neural basis of other forms of self-control is a logical path to pursue.

There has been some research on the neural basis of thought suppression. Only two studies have directly explored the effects of thought suppression in a neutral, non-emotional setting (Mitchell et al., 2007; Wyland et al., 2003). The results of these two studies are not consistent. When contrasting thought suppression with free thought, Mitchell and colleagues found right VLPFC and right DLPFC to be active.

When contrasting clear-all thought with free thought, Wyland and colleagues found a large network of prefrontal areas to be active, including the bilateral insula, left IFC, and ACC. The lack of activity noted in the right VLPFC during self-control (the thought suppression condition) could be due to multiple reasons. First, it is possible that the right insular region they reported active in the clear-all thought versus free-think contrast overlaps with the right VLPFC noted in other self-control paradigms (Aron & Poldrack, 2006; Remijnse et al., 2005). Second, perhaps the right VLPFC is required for behavioral but not cognitive self-control. Both the current study and the memory-inhibition study also reported greater left VLPFC activity in the control versus no-control contrasts (Anderson et al., 2004; Wyland et al., 2003), although the other thought suppression study did not (Mitchell et al., 2007).

In support of the above theory of lateralization of the VLPFC, when comparing lateral and medial frontal EEG resting baseline activity in repressors as compared to nonrepressors, repressors have more left-lateralized activity than do nonrepressors, who show equivalent amounts of activity in both hemispheres (Tomarken & Davidson, 1994). Assuming that repressors are better able to suppress their thoughts, this finding could indicate that cognitive self-control may be localized to a left-lateralized frontal network. Additionally, when exploring thought suppression in an emotion regulation context, left lateral PFC has been found to be more active during thought suppression than during free-think conditions (Gillath et al., 2005; Kalisch et al., 2006). The authors hypothesized that the left VLPFC may have been recruited to produce distracting thoughts, a role consistent with the function attributed to Broca's area, found in the left VLPFC (Kalisch et al., 2006). This strategy may also have been used by subjects in the memory inhibition study (Anderson et al., 2004), thus providing an explanation for the left VLPFC activity noted during no-think as compared to think trials. This is in contrast to cognitive reappraisal, which recruits right anterolateral prefrontal regions (Kalisch et al., 2006). Although cognitive reappraisal is a cognitive self-control technique, it results in more directly observable/behavioral changes (i.e., changes in emotional intensity) than does thought suppression, thus this could explain why emotion regulation via cognitive reappraisal demonstrates right VLPFC activity, whereas emotion regulation via thought suppression does not. Clearly, however, more research must be conducted before any conclusions may be drawn from these data.

Synthesizing the Literature

Although the vast majority of studies exploring various forms of inhibitory self-control do not attempt to directly relate one variety to another, some research has been conducted with that goal in mind. Muraven and Baumeister in particular have conducted a series of studies relating multiple forms of self-control (for a review, see Muraven & Baumeister, 2000). They conducted a series of studies in which subjects were given a task involving any of a number of forms of self-control (such as emotion regulation, thought suppression, the stop-signal paradigm, the Stroop task, solving impossible anagrams, grasping a resistant handgrip for an extended period of time, or resisting sweets or alcohol) and were then tested on a second, completely different measure of self-control. Their performance was compared to a second group of participants who performed a task matched for effort and frustration as measured by self-report questionnaires that did not require self-control, such as solving mathematical problems or quickly typing a paragraph without feedback. The results consistently demonstrate that subjects who exerted self-control in an initial task were worse at the subsequent self-control task than those who initially performed a difficult initial task not requiring self-control, implying that not only is self-control a unified process, but that it is a limited resource that can be fatigued (Muraven & Baumeister, 2000; Muraven et al., 2006; Muraven et al., 1998).

The above line of research demonstrates that a wide range of measures of self-control all tap a common pool of resources, thus suggesting that they may all be subsumed by one general self-control process. In a different line of work that also attempts to equate different forms of self-control, a handful of studies have compared subject performance on two different forms in order to directly explore correlations in performance on the two measures, without regard for the effects of self-control fatigue. Overall, the studies have found that impairment in one domain of self-control is associated with impairment in another domain.

Much of this work focuses on motor response inhibition and its relation to other varieties of self-control, because it is arguably the simplest and the most completely studied form of self-control. A series of studies have directly compared the neural mechanisms utilized during stop-signal and go/no-go tasks to those utilized in other tasks that require cognitive control. Tasks used include the flanker task, which requires the suppression of irrelevant distracting information (Bunge et al., 2002; Wager et al., 2005), set shifting during the Wisconsin Card Sorting Test (Konishi et al., 1999), and an incompatible stimulus-response task that requires subjects to press left for a right arrow and vice versa (Wager et al., 2005). Conjunction analyses showed that right IFC and/or right anterior insula were active in adults across all tasks during the inhibition/conflict trials, the region commonly associated with response inhibition in go/no-go and stop-signal tasks (Bunge et al., 2002; Konishi et al., 1999; Wager et al., 2005).

Using a different tactic, some studies of response inhibition using the go/no-go or stop-signal paradigm focused on populations of subjects that have impairments in real world self-control, such as people with high levels of impulsivity (Logan et al., 1997), ADHD (Durston et al., 2006; Lijffijt et al., 2005; Logan et al., 2000), substance abuse problems (Fillmore & Rush, 2002; Monterosso et al., 2005), and obesity (Nederkoorn et al., 2006). Subjects in each of these impulsive populations performed worse at motor response inhibition (longer SSRTs or more errors in the go/no-go task) than healthy control subjects. Additionally, obese children also acted in a riskier manner than did healthy controls, suggesting that not only is impaired motor response inhibition associated with a lack of self-control in general, but impaired motor response inhibition may be associated with risky behavior as well (Nederkoorn et al., 2006). Moreover, children with ADHD have less of an increase in BOLD signal when comparing neural activity during no-go trials to that on go trials than do healthy control children in the right IFG (Durston et al., 2006).

Reversal learning has also been explored in populations at-risk for increased impulsivity, such as psychopaths, who performed worse on a reversal learning task than did healthy controls (Mitchell et al., 2002). The same psychopathic subjects performed worse on the IGT (i.e., chose more cards from the riskier, disadvantageous decks) than did those in the control group, suggesting a potential link between response inhibition as indexed by reversal learning and risky behavior (Mitchell et al., 2002).

Some studies have found a positive relationship between response inhibition and emotion regulation ability, as operationalized by either a greater ability to inhibit negative emotions during go/no-go blocks with negative feedback (Lewis et al., 2006) or less variability in anger ratings over a three- to four-day period (Hoeksma et al., 2004). In one study that invoked prepotent emotional and motor responses simultaneously (Berkman, Burklund, & Lieberman, 2009), intentional motor response inhibition was associated with decreased amygdala responses and the magnitude of these reductions was inversely associated with the strength of right VLPFC activity during response inhibition. In other words, intentionally inhibiting a motor response appears to have produced unintentional inhibition of affective responses as well, suggesting a common regulatory system for both.

Finally, memory inhibition has been theorized to be related to motor inhibition (Levy & Anderson, 2002). As mentioned in the memory inhibition section, the think/no-think paradigm was modeled after the go/no-go paradigm; both no-go and no-think conditions require overriding the natural tendency to either respond to a stimulus or to retrieve the previously studied target word associated with a cue word (Anderson & Green, 2001). Although the think/no-think paradigm has not been used in concert with the go/no-go paradigm, a similar right VLPFC region was active in an fMRI study utilizing the think/no-think paradigm that is active in motor inhibition tasks (Anderson et al., 2004). This suggests that a similar inhibitory process may underlie at least one aspect of successful memory inhibition.

There has been some research focusing on the link between risky behavior and other measures of impulsivity. As mentioned above, an association has been found between obesity in children, motor response inhibition, and risk-taking behavior (Nederkoorn et al., 2006). Risky behavior has also been found to be increased in other populations with a purported high level of impulsivity, such as psychopaths using the IGT (Mitchell et al., 2002) and those who abuse substances using two different risk-taking tasks, the IGT and the Cambridge Gamble Task (Monterosso et al., 2001).

There has also been research conducted relating risk-taking and emotion regulation. When inducing a negative mood in subjects and then asking them to play a lottery game, those who were told to use cognitive appraisal before making their decision acted in a less risky manner than those who were just asked to report their choice (Leith & Baumeister, 1996).

The decision to choose larger, delayed rewards over smaller, immediate rewards has also been linked to impulsivity and mood state. As described in the temporal discounting section, it has been proposed that the steepness of the temporal discounting curve may be an index of overall impulsivity (Monterosso & Ainslie, 1999). Additionally, a large literature exists exploring temporal discounting in populations purported to be impulsive, most specifically those who abuse substances; this literature consistently finds steeper discounting curves in those who abuse substances as compared to healthy controls (for a review, see Bickel & Marsch, 2001).

While there is a wide range of studies directly comparing more than one concept of self-control, there are still many unanswered questions. First, most studies only compare two of the six described varieties of conscious self-control. Second, the neural basis of self-control has not been explored while directly comparing multiple forms of self-control within a single population of subjects. Such a study design would be useful to note if there are coactivations or dissociations within the right VLPFC during acts of self-control. It is possible that the right VLPFC is involved in self-control generally, but that there are different subregions that underlie different forms of self-control. It is unclear from the existing literature, which uses different methods, different data acquisition tools, different data processing and analysis techniques, and, most importantly, different participants, whether there is a single right VLPFC region activated in all tasks requiring self-control or if there are unique subregions. The existence of behavioral associations between such a wide variety of self-control indices suggests that it would be fruitful to directly examine the extent of a shared neural basis of different forms of self-control.

THE VLPFC IS WELL-POSITIONED TO EXERT SELF-CONTROL

The right VLPFC is well suited to serving a key role in exerting self-control over actions. It has close anatomical associations with other control areas in the prefrontal cortex, such as the DLPFC, mPFC including the ACC, and the OFC (Miller & Cohen, 2001). Evidence that the right VLPFC is linked to the amygdala via the mPFC comes from affect labeling studies, which find increased right VLPFC activity associated with decreased amygdala activity (Hariri et al., 2000; Hariri et al., 2003; Lieberman et al., 2007), likely through mediation by the mPFC, which has dense reciprocal connections with both structures (Lieberman et al., 2007).

Recently, white matter connections identified using Diffusion Tensor Imaging (DTI) have been noted between the right VLPFC and the preSMA, an area thought to be involved in conflict detection (Aron et al., 2007). The preSMA may signal when there is a conflict between an intention (e.g., to resist drug use) and an impulse (e.g., craving)—in other words, when a temptation exists that may challenge one's self-control.

Crucially, the right VLPFC also has direct connections with motor output control areas of the basal ganglia such as the STN (Aron et al.). This may be the means through which the right VLPFC may send a signal to exert behavioral control and therefore underlie an act of self-control.

More detailed research on anatomical connections comes from studies of macaque monkeys. While it is not clear how similar human and monkey prefrontal cortices are, evidence exists that the cytoarchitecture of the macaque VLPFC (including the lateral OFC) is similar to that in humans (Ongur & Price, 2000; Petrides et al., 2005). Therefore, it is possible that similar anatomical connections exist. The monkey inferior arcuate sulcus, just anterior to the ventral premotor area, and its surrounding cortex may be the monkey homologue to the human VLPFC, as evidenced by a study comparing the cytoarchitecture of this region in monkeys and humans (Petrides et al., 2005). This region is directly connected in the monkey to many other cortical regions, such as the lateral and medial OFC, the dorsomedial PFC, the DLPFC, the ACC, the insula, the supplementary, premotor and primary motor areas, and areas of the superior temporal lobe (Deacon, 1992).

The lateral OFC receives sensory input from the primary taste cortex, visual areas via the inferior temporal cortex, and somatosensory areas such as the primary and secondary sensory cortices and the insula. It also has reciprocal connections with the amygdala, cingulate cortex, premotor areas, and DLPFC (Kringelbach & Rolls, 2004). Connections with some of those regions, specifically the ACC, DLPFC, preSMA and amygdala, have been noted in humans as well (Aron et al., 2007; Lieberman et al., 2007; Miller & Cohen, 2001). Lastly, the lateral OFC has efferent connections with the hypothalamus, periaqueductal gray area and striatum, the ventral caudate in particular (Kringelbach & Rolls, 2004). A recent DTI study of white matter tracts in humans has reinforced the finding of a direct connection between the lateral OFC and the ventral striatum in humans (Leh et al., 2007).

Conclusion

In conclusion, a fairly localized prefrontal network appears to underlie a variety of forms of self-control. Simple motor response inhibition, mostly assessed via go/no-go and stop-signal tasks, consistently activates the right IFC/VLPFC (Aron et al., 2004). This may be accompanied by DLPFC and ACC activity, which are most likely utilized generally for other cognitive processes, such as rule monitoring (Bunge, 2004) and performance/conflict monitoring (Botvinick et al., 2004) respectively. Support for this can be seen when noting the wide variety of tasks that have been associated with DLPFC or ACC activity, such as the Stroop task, the Wisconsin Card Sort Task, the flanker task, verb generation, and reward-related two-alternative forced-choice tasks, in addition to stop-signal and go/no-go tasks (for reviews, see Botvinick et al., 2004; Bunge, 2004; Ridderinkhof et al., 2004).

Most of the other tasks discussed that require self-control recruit a more diffuse prefrontal network. Reversal learning, for example, which requires the processing of reward contingencies in addition to self-control, recruits the reward-sensitive mPFC and medial OFC (Elliott et al., 2000). Risk-taking and temporal discounting similarly require the processing of relative rewards and recruit the mPFC in addition to the right VLPFC (Krain et al., 2006; McClure et al., 2004), although the specific nature of whether right VLPFC is utilized more when making risky or safe decisions is still under debate. Emotion regulation, which requires both self-control and emotion processing, recruits mPFC as well, which has been associated with emotion-related processing in addition to reward processing (Ochsner & Gross, 2005).

Finally, more cognitive forms of self-control, such as memory inhibition and thought suppression may require more monitoring of rules, which can explain the consistent DLPFC activation in these studies, as well as an increased amount of alternate processing needs, which can explain the diffuse and possibly left-lateralized PFC control network utilized during these tasks (Anderson et al., 2004; Mitchell et al., 2007;

Wyland et al., 2003). However, it is important to note that there are very few neuroimaging studies of memory inhibition and thought suppression, thus any conclusions that can be made are tentative.

In short, it is logical to conclude that the right VLPFC is an area commonly related to self-control, while other prefrontal regions may be recruited based on specific task demands during an act of self-control. It is important to note, however, that the right VLPFC is involved in other sorts of tasks, such as attention (Corbetta & Shulman 2002), memory (Courtney et al. 1996; Kostopoulos & Petrides 2003; Rizzuto et al. 2005), and emotion perception (Kober et al. 2008). Therefore, it cannot be concluded that the involvement of right VLPFC in a task means self-control is being exerted without examining theorized task demands (*see* Poldrack 2006).

Moreover, additional research directly comparing the neural networks recruited during different forms of self-control must be conducted before any conclusions can be drawn regarding the relative unity or segregation of right VLPFC activity. The research to date cannot resolve whether there is a single part of the right VLPFC that is involved in each of the aforementioned forms of self-control or if different subregions are recruited for different forms. Further research combining neuroimaging with multiple forms of self-control in the same population of subjects will help elucidate the specificity of right VLPFC activity as it relates to self-control, an important phenomenon to understand given the role of impaired self-control in a multitude of clinical problems, such as ADHD, substance abuse, gambling, and many others.

References

Ainslie, G., & Monterosso, J. Behavior. A marketplace in the brain? Science 2004; 306(5695): 421–423.

Anderson, M. C., & Green, C. Suppressing unwanted memories by executive control. Nature 2001; 410(6826): 366–369.

Anderson, M. C., Ochsner, K. N., Kuhl, B., et al. Neural systems underlying the suppression of unwanted memories. Science 2004; 303(5655): 232–235.

Aron, A. R., Behrens, T. E., Smith, S., Frank, M. J., & Poldrack, R. A. Triangulating a cognitive control network using diffusion-weighted magnetic resonance imaging (MRI) and functional MRI. J Neurosci 2007; 27(14): 3743–3752.

Aron, A. R., Fletcher, P. C., Bullmore, E. T., Sahakian, B. J., & Robbins, T. W. Stop-signal inhibition disrupted by damage to right inferior frontal gyrus in humans. Nat Neurosci 2003; 6(2): 115–116.

Aron, A. R., & Poldrack, R. A. Cortical and subcortical contributions to stop signal response inhibition: Role of the subthalamic nucleus. J Neurosci 2006; 26(9): 2424–2433.

Aron, A. R., Robbins, T. W., & Poldrack, R. A. Inhibition and the right inferior frontal cortex. Trends in Cogn Sci 2004; 8(4): 170–177.

Bechara, A. The role of emotion in decision-making: Evidence from neurological patients with orbitofrontal damage. Brain Cogn 2004; 55(1): 30–40.

Bechara, A., Damasio, A. R., Damasio, H., & Anderson, S. W. Insensitivity to future consequences following damage to human prefrontal cortex. Cognition 1994; 50(1–3): 7–15.

Bechara, A., Damasio, H., Tranel, D., & Anderson, S. W. Dissociation of working memory from decision making within the human prefrontal cortex. J Neurosci 1998; 18(1): 428–437.

Berkman, E. T., Burklund, L., & Lieberrman, M. D. Inhibitory spillover: Intentional motor inhibition produces incidental limbic inhibition via right inferior frontal cortex. Neuroimage 2009; 47(2): 705–712.

Bickel, W. K., & Marsch, L. A. Toward a behavioral economic understanding of drug dependence: Delay discounting processes. Addiction 2001; 96(1): 73–86.

Bjork, R. A. Retrieval inhibition as an adaptive mechanism in human memory. In: Roediger, H. L. III & Craik, F. I. M. (Eds.), *Varieties of memory and consciousness: Essays in honor of Endel Tulving*. Hillsdale, NJ: L. Erlbaum Associates, 1989.

Botvinick, M. M., Cohen, J. D., & Carter, C. S. Conflict monitoring and anterior cingulate cortex: An update. Trends Cogn Sci 2004; 8(12): 539–546.

Bunge, S. A. How we use rules to select actions: A review of evidence from cognitive neuro-

science. Cogn Aff Behav Neurosci 2004; 4(4): 564–579.
Bunge, S. A., Dudukovic, N. M., Thomason, M. E., Vaidya, C. J., & Gabrieli, J. D. Immature frontal lobe contributions to cognitive control in children: Evidence from fMRI. Neuron 2002; 33(2): 301–311.
Butter, C. M. Perseveration in extinction and in discrimination reversal tasks following selective frontal ablations in macaca mulatta. Physiol Behav 1969; 4(2): 163–&.
Cardinal, R. N. Neural systems implicated in delayed and probabilistic reinforcement. Neural Networks 2006; 19(8): 1277–1301.
Casey, B. J., Castellanos, F. X., Giedd, J. N., et al. Implication of right frontostriatal circuitry in response inhibition and attention-deficit/hyperactivity disorder. J Am Acad Child Adol Psychiatry 1997; 36(3): 374–383.
Chambers, C. D., Bellgrove, M. A., Stokes, M. G., et al. Executive "brake failure" following deactivation of human frontal lobe. J Cogn Neurosci 2006; 18(3): 444–455.
Chevrier, A. D., Noseworthy, M. D., & Schachar, R. Dissociation of response inhibition and performance monitoring in the stop signal task using event-related fMRI. Hum Brain Mapp 2007; 28(12): 1437–1458.
Clark, L., Cools, R., & Robbins, T. W. The neuropsychology of ventral prefrontal cortex: Decision-making and reversal learning. Brain Cogn 2004; 55(1): 41–53.
Cohen, J. R., & Poldrack, R. A. *The neural correlates of multiple forms of self-control.* San Francisco: Organization for Human Brain Mapping, 2009.
Cohen, M. X., Heller, A. S., & Ranganath, C. Functional connectivity with anterior cingulate and orbitofrontal cortices during decision-making. Brain Res: Cogn Brain Res 2005; 23(1): 61–70.
Cools, R., Clark, L., Owen, A. M., & Robbins, T. W. Defining the neural mechanisms of probabilistic reversal learning using event-related functional magnetic resonance imaging. J Neurosci 2002; 22(11): 4563–4567.
Corbetta, M., & Shulman, G. L. Control of goal-directed and stimulus-driven attention in the brain. Nat Rev Neurosci 2002; 3(3): 201–215.
Courtney, S. M., Ungerleider, L. G., Keil, K., & Haxby, J. V. Object and spatial visual working memory activate separate neural systems in human cortex. Cereb Cortex 1996; 6(1): 39–49.

Deacon, T. W. Cortical connections of the inferior arcuate sulcus cortex in the macaque brain. Brain Res 1992; 573(1): 8–26.
Dias, R., Robbins, T. W., & Roberts, A. C. Dissociation in prefrontal cortex of affective and attentional shifts. Nature 1996; 380(6569): 69–72.
Dunn, B. D., Dalgleish, T., & Lawrence, A. D. The somatic marker hypothesis: A critical evaluation. Neurosci Biobehav Rev 2006; 30(2): 239–271.
Durston, S., Mulder, M., Casey, B. J., Ziermans, T., & van Engeland, H. Activation in ventral prefrontal cortex is sensitive to genetic vulnerability for attention-deficit hyperactivity disorder. Biol Psychiatr 2006; 60(10): 1062–1070.
Elliott, R., Dolan, R. J., & Frith, C. D. Dissociable functions in the medial and lateral orbitofrontal cortex: Evidence from human neuroimaging studies. Cereb Cortex 2000; 10(3): 308–317.
Ernst, M., Nelson, E. E., McClure, E. B., Monk, C. S., Munson, S., Eshel, N., et al. Choice selection and reward anticipation: An fMRI study. Neuropsychologia 2004; 42(12): 1585–1597.
Eshel, N., Nelson, E. E., Blair, R. J., Pine, D. S., & Ernst, M. Neural substrates of choice selection in adults and adolescents: Development of the ventrolateral prefrontal and anterior cingulate cortices. Neuropsychologia 2007; 45(6): 1270–1279.
Evenden, J. L. Varieties of impulsivity. Psychopharmacology 1999; 146(4): 348–361.
Fellows, L. K., & Farah, M. J. Ventromedial frontal cortex mediates affective shifting in humans: Evidence from a reversal learning paradigm. Brain 2003; 126(Pt 8): 1830–1837.
Fellows, L. K., & Farah, M. J. Different underlying impairments in decision-making following ventromedial and dorsolateral frontal lobe damage in humans. Cereb Cortex 2005; 15(1): 58–63.
Fillmore, M. T., & Rush, C. R. Impaired inhibitory control of behavior in chronic cocaine users. Drug Alc Dependence 2002; 66(3): 265–273.
Garavan, H., Ross, T. J., Murphy, K., Roche, R. A., & Stein, E. A. Dissociable executive functions in the dynamic control of behavior: Inhibition, error detection, and correction. Neuroimage 2002; 17(4): 1820–1829.
Garavan, H., Ross, T. J., & Stein, E. A. Right hemispheric dominance of inhibitory control: An event-related functional MRI study. Proc Nat Acad Sci U S A 1999; 96(14): 8301–8306.

Gillath, O., Bunge, S. A., Shaver, P. R., Wendelken, C., & Mikulincer, M. Attachment-style differences in the ability to suppress negative thoughts: Exploring the neural correlates. Neuroimage 2005; 28(4): 835–847.

Gross, J. J. The emerging field of emotion regulation: An integrative review. Rev Gen Psychol 1998; 2(3): 271–299.

Gross, J. J. Emotion regulation: Affective, cognitive, and social consequences. Psychophysiology 2002; 39(3): 281–291.

Harenski, C. L., & Hamann, S. Neural correlates of regulating negative emotions related to moral violations. Neuroimage 2006; 30(1): 313–324.

Hariri, A. R., Bookheimer, S. Y., & Mazziotta, J. C. Modulating emotional responses: Effects of a neocortical network on the limbic system. Neuroreport 2000; 11(1): 43–48.

Hariri, A. R., Mattay, V. S., Tessitore, A., Fera, F., & Weinberger, D. R. Neocortical modulation of the amygdala response to fearful stimuli. Biol Psychiatr 2003; 53(6): 494–501.

Hoeksma, J. B., Oosterlaan, J., & Schipper, E. M. Emotion regulation and the dynamics of feelings: A conceptual and methodological framework. Child Develop 2004; 75(2): 354–360.

Hunt, M. K., Hopko, D. R., Bare, R., Lejuez, C. W., & Robinson, E. V. Construct validity of the balloon analog risk task (bart): Associations with psychopathy and impulsivity. Assessment 2005; 12(4): 416–428.

Iversen, S. D., & Mishkin, M. Perseverative interference in monkeys following selective lesions of inferior prefrontal convexity. Exp Brain Res 1970; 11(4): 376–386.

Kable, J. W., & Glimcher, P. W. The neural correlates of subjective value during intertemporal choice. Nat Neurosci 2007; 10(12): 1625–1633.

Kalenscher, T., Windmann, S., Diekamp, B., Rose, J., Gunturkun, O., & Colombo, M. Single units in the pigeon brain integrate reward amount and time-to-reward in an impulsive choice task. Curr Biol 2005; 15(7): 594–602.

Kalisch, R., Wiech, K., Critchley, H. D., Seymour, B., O'Doherty, J. P., Oakley, D. A., et al. Anxiety reduction through detachment: Subjective, physiological, and neural effects. J Cogn Neurosci 2005; 17(6): 874–883.

Kalisch, R., Wiech, K., Herrmann, K., & Dolan, R. J. Neural correlates of self-distraction from anxiety and a process model of cognitive emotion regulation. J Cogn Neurosci 2006; 18(8): 1266–1276.

Kim, S. H., & Hamann, S. Neural correlates of positive and negative emotion regulation. J Cogn Neurosci 2007; 19(5): 776–798.

Kober, H., Barrett, L. F., Joseph, J., Bliss-Moreau, E., Lindquist, K., & Wager, T. D. Functional grouping and cortical-subcortical interactions in emotion: A meta-analysis of neuroimaging studies. Neuroimage 2008; 42(2): 998–1031.

Konishi, S., Nakajima, K., Uchida, I., Kikyo, H., Kameyama, M., & Miyashita, Y. Common inhibitory mechanism in human inferior prefrontal cortex revealed by event-related functional MRI. Brain 1999; 122 (Pt 5): 981–991.

Konishi, S., Nakajima, K., Uchida, I., Sekihara, K., & Miyashita, Y. No-go dominant brain activity in human inferior prefrontal cortex revealed by functional magnetic resonance imaging. Eur J Neurosci 1998; 10(3): 1209–1213.

Kostopoulos, P., & Petrides, M. The mid-ventrolateral prefrontal cortex: Insights into its role in memory retrieval. Eur J Neurosci 2003; 17(7): 1489–1497.

Krain, A. L., Wilson, A. M., Arbuckle, R., Castellanos, F. X., & Milham, M. P. Distinct neural mechanisms of risk and ambiguity: A meta-analysis of decision-making. Neuroimage 2006; 32(1): 477–484.

Kringelbach, M. L., & Rolls, E. T. Neural correlates of rapid reversal learning in a simple model of human social interaction. Neuroimage 2003; 20(2): 1371–1383.

Kringelbach, M. L., & Rolls, E. T. The functional neuroanatomy of the human orbitofrontal cortex: Evidence from neuroimaging and neuropsychology. Progress Neurobiol 2004; 72(5): 341–372.

Leh, S. E., Ptito, A., Chakravarty, M. M., & Strafella, A. P. Fronto-striatal connections in the human brain: A probabilistic diffusion tractography study. Neurosci Lett 2007; 419(2): 113–118.

Leith, K. P., & Baumeister, R. F. Why do bad moods increase self-defeating behavior? Emotion, risk taking, and self-regulation. J Pers Soc Psychol 1996; 71(6): 1250–1267.

Lejuez, C. W., Read, J. P., Kahler, C. W., Richards, J. B., Ramsey, S. E., Stuart, G. L., et al. Evaluation of a behavioral measure of risk taking: The balloon analogue risk task (bart). J Exp Psychol Appl 2002; 8(2): 75–84.

Levesque, J., Eugene, F., Joanette, Y., Paquette, V., Mensour, B., Beaudoin, G., et al. Neural circuitry underlying voluntary suppression of sadness. Biol Psychiatry 2003; 53(6): 502–510.

Levy, B. J., & Anderson, M. C. Inhibitory processes and the control of memory retrieval. Trends Cogn Sci 2002; 6(7): 299–305.

Lewis, M. D., Lamm, C., Segalowitz, S. J., Stieben, J., & Zelazo, P. D. Neurophysiological correlates of emotion regulation in children and adolescents. J Cogn Neurosci 2006; 18(3): 430–443.

Liddle, P. F., Kiehl, K. A., & Smith, A. M. Event-related fMRI study of response inhibition. Hum Brain Mapp 2001; 12(2): 100–109.

Lieberman, M. D. Social cognitive neuroscience: A review of core processes. Ann Rev Psychol 2007; 58: 259–289.

Lieberman, M. D., Eisenberger, N. I., Crockett, M. J., Tom, S. M., Pfeifer, J. H., & Way, B. M. Putting feelings into words: Affect labeling disrupts amygdala activity to affective stimuli. Psychol Sci 2007; 18(5): 421–428.

Lijffijt, M., Kenemans, J. L., Verbaten, M. N., & van Engeland, H. A meta-analytic review of stopping performance in attention-deficit/hyperactivity disorder: Deficient inhibitory motor control? J Abn Psychol 2005; 114(2): 216–222.

Logan, G. D. On the ability to inhibit thought and action: A users' guide to the stop signal paradigm. In: Dagenbach, D. & Carr, T. H. (Eds.), *Inhibitory processes in attention, memory, and language.* San Diego, CA: Academic Press, 1994: pp. 189–240.

Logan, G. D., Schachar, R. J., & Tannock, R. Impulsivity and inhibitory control. Psychol Sci 1997; 8(1): 60–64.

Logan, G. D., Schachar, R. J., & Tannock, R. Executive control problems in childhood psychopathology: Stop signal studies of attention deficit hyperactivity disorder. In: Monsell, S. & Driver, J. (Eds.), *Control of cognitive processes: Attention and performance* (Vol. XVIII). Cambridge: MIT Press, 2000: pp. 653–677.

MacLeod, C. M. Directed forgetting. In: Golding, J. M. & MacLeod, C. M. (Eds.), *Intentional forgetting: Interdisciplinary approaches.* Mahwah, NJ: Lawrence Erlbaum Associates Publishers, 1998: pp. 1–57.

Manes, F., Sahakian, B. J., Clark, L., et al. Decision-making processes following damage to the prefrontal cortex. Brain 2002; 125(3): 624–639.

Matthews, S. C., Simmons, A. N., Lane, S. D., & Paulus, M. P. Selective activation of the nucleus accumbens during risk-taking decision making. Neuroreport 2004; 15(13): 2123–2127.

McClure, S. M., Laibson, D. I., Loewenstein, G., & Cohen, J. D. Separate neural systems value immediate and delayed monetary rewards. Science 2004; 306(5695): 503–507.

Menon, V., Adleman, N. E., White, C. D., Glover, G. H., & Reiss, A. L. Error-related brain activation during a go/nogo response inhibition task. Hum Brain Mapp 2001; 12(3): 131–143.

Metcalfe, J., & Mischel, W. A hot/cool-system analysis of delay of gratification: Dynamics of willpower. Psychol Rev 1999; 106(1): 3–19.

Miller, E. K., & Cohen, J. D. An integrative theory of prefrontal cortex function. Ann Rev Neurosci 2001; 24: 167–202.

Mischel, W., Shoda, Y., & Rodriguez, M. I. Delay of gratification in children. Science 1989; 244(4907): 933–938.

Mitchell, D. G., Colledge, E., Leonard, A., & Blair, R. J. Risky decisions and response reversal: Is there evidence of orbitofrontal cortex dysfunction in psychopathic individuals? Neuropsychologia 2002; 40(12): 2013–2022.

Mitchell, J. P., Heatherton, T. F., Kelley, W. M., Wyland, C. L., Wegner, D. M., & Neil Macrae, C. Separating sustained from transient aspects of cognitive control during thought suppression. Psychol Sci 2007; 18(4): 292–297.

Mobini, S., Body, S., Ho, M. Y., et al. Effects of lesions of the orbitofrontal cortex on sensitivity to delayed and probabilistic reinforcement. Psychopharmacology 2002; 160(3): 290–298.

Monterosso, J. R., & Ainslie, G. Beyond discounting: Possible experimental models of impulse control. Psychopharmacology 1999; 146(4): 339–347.

Monterosso, J. R., Ainslie, G., Xu, J., Cordova, X., Domier, C. P., & London, E. D. Frontoparietal cortical activity of methamphetamine-dependent and comparison subjects performing a delay discounting task. Hum Brain Mapp 2007; 28(5): 383–393.

Monterosso, J. R., Aron, A. R., Cordova, X., Xu, J., & London, E. D. Deficits in response inhibition associated with chronic methamphetamine abuse. Drug Alc Dependence 2005; 79(2): 273–277.

Monterosso, J. R., Ehrman, R., Napier, K. L., O'Brien, C. P., & Childress, A. R. Three decision-making tasks in cocaine-dependent patients: Do they measure the same construct? Addiction 2001; 96(12): 1825–1837.

Muraven, M., & Baumeister, R. F. Self-regulation and depletion of limited resources: Does self-control resemble a muscle? Psychol Bull 2000; 126(2): 247–259.

Muraven, M., Shmueli, D., & Burkley, E. Conserving self-control strength. J Pers Soc Psychol 2006; 91(3): 524–537.

Muraven, M., Tice, D. M., & Baumeister, R. F. Self-control as limited resource: Regulatory depletion patterns. J Pers Soc Psychol 1998; 74(3): 774–789.

Nederkoorn, C., Braet, C., Van Eijs, Y., Tanghe, A., & Jansen, A. Why obese children cannot resist food: The role of impulsivity. Eat Behav 2006; 7(4): 315–322.

O'Doherty, J., Critchley, H., Deichmann, R., & Dolan, R. J. Dissociating valence of outcome from behavioral control in human orbital and ventral prefrontal cortices. J Neurosci 2003; 23(21): 7931–7939.

Ochsner, K. N. Current directions in social cognitive neuroscience. Curr Opin Neurobiol 2004; 14(2): 254–258.

Ochsner, K. N., & Gross, J. J. The cognitive control of emotion. Trends Cogn Sci 2005; 9(5): 242–249.

Ochsner, K. N., Ray, R. D., Cooper, J. C., et al. For better or for worse: Neural systems supporting the cognitive down- and up-regulation of negative emotion. Neuroimage 2004; 23(2): 483–499.

Ongur, D., & Price, J. L. The organization of networks within the orbital and medial prefrontal cortex of rats, monkeys and humans. Cereb Cortex 2000; 10(3): 206–219.

Petrides, M., Cadoret, G., & Mackey, S. Orofacial somatomotor responses in the macaque monkey homologue of broca's area. Nature 2005; 435(7046): 1235–1238.

Phan, K. L., Fitzgerald, D. A., Nathan, P. J., Moore, G. J., Uhde, T. W., & Tancer, M. E. Neural substrates for voluntary suppression of negative affect: A functional magnetic resonance imaging study. Biol Psychiatry 2005; 57(3): 210–219.

Picton, T. W., Stuss, D. T., Alexander, M. P., Shallice, T., Binns, M. A., & Gillingham, S. Effects of focal frontal lesions on response inhibition. Cereb Cortex 2007; 17(4): 826–838.

Poldrack, R. A. Can cognitive processes be inferred from neuroimaging data? Trends Cogn Sci 2006; 10(2): 59–63.

Rassin, E., Merckelbach, H., & Muris, P. Paradoxical and less paradoxical effects of thought suppression: A critical review. Clin Psychol Rev 2000; 20(8): 973–995.

Remijnse, P. L., Nielen, M. M., Uylings, H. B., & Veltman, D. J. Neural correlates of a reversal learning task with an affectively neutral baseline: An event-related fMRI study. Neuroimage 2005; 26(2): 609–618.

Ridderinkhof, K. R., Ullsperger, M., Crone, E. A., & Nieuwenhuis, S. The role of the medial frontal cortex in cognitive control. Science 2004; 306(5695): 443–447.

Rizzuto, D. S., Mamelak, A. N., Sutherling, W. W., Fineman, I., & Andersen, R. A. Spatial selectivity in human ventrolateral prefrontal cortex. Nat Neurosci 2005; 8(4): 415–417.

Roesch, M. R., & Olson, C. R. Neuronal activity in primate orbitofrontal cortex reflects the value of time. J Neurophysiol 2005; 94(4): 2457–2471.

Rolls, E. T., Critchley, H. D., Mason, R., & Wakeman, E. A. Orbitofrontal cortex neurons: Role in olfactory and visual association learning. J Neurophysiol 1996; 75(5): 1970–1981.

Rolls, E. T., Hornak, J., Wade, D., & McGrath, J. Emotion-related learning in patients with social and emotional changes associated with frontal lobe damage. J Neurol Neurosurg Psychiatry 1994; 57(12): 1518–1524.

Rubia, K., Russell, T., Overmeyer, S., et al. Mapping motor inhibition: Conjunctive brain activations across different versions of go/no-go and stop tasks. Neuroimage 2001; 13(2): 250–261.

Rubia, K., Smith, A. B., Brammer, M. J., & Taylor, E. Right inferior prefrontal cortex mediates response inhibition while mesial prefrontal cortex is responsible for error detection. Neuroimage 2003; 20(1): 351–358.

Sakagami, M., & Niki, H. Encoding of behavioral significance of visual stimuli by primate prefrontal neurons: Relation to relevant task conditions. Exp Brain Res 1994; 97(3): 423–436.

Shafritz, K. M., Collins, S. H., & Blumberg, H. P. The interaction of emotional and cognitive neural systems in emotionally guided response inhibition. Neuroimage 2006; 31(1): 468–475.

Stover, E. R. S., Fox, C. R., C., T., & Poldrack, R. A. Risk and uncertainty in reward-based feedback learning: Neural activation varying with risk and expected value on the balloon analogue risk task. Washington, DC: Society for Neuroscience, 2005.

Thorpe, S. J., Rolls, E. T., & Maddison, S. The orbitofrontal cortex: Neuronal activity in the behaving monkey. Exp Brain Res 1983; 49(1): 93–115.

Tobler, P. N., O'Doherty, J. P., Dolan, R. J., & Schultz, W. Reward value coding distinct from risk attitude-related uncertainty coding in human reward systems. J Neurophysiol 2007; 97(2): 1621–1632.

Tomarken, A. J., & Davidson, R. J. Frontal brain activation in repressors and nonrepressors. J Abnorm Psychol 1994; 103(2): 339–349.

Verbruggen, F., & De Houwer, J. Do emotional stimuli interfere with response inhibition? Evidence from the stop signal paradigm. Cogn Emot 2007; 21(2): 391–403.

Wager, T. D., Sylvester, C. Y., Lacey, S. C., Nee, D. E., Franklin, M., & Jonides, J. Common and unique components of response inhibition revealed by fMRI. Neuroimage 2005; 27(2): 323–340.

Wegner, D. M. You can't always think what you want: Problems in the suppression of unwanted thoughts. In: Zanna, M. P. (Ed.), *Advances in experimental social psychology* (Vol. 25). San Diego: Academic Press, 1992: pp. 193–225.

Wegner, D. M., Schneider, D. J., Carter, S. R., 3rd, & White, T. L. Paradoxical effects of thought suppression. J Pers Soc Psychol 1987; 53(1): 5–13.

Wenzlaff, R. M., & Wegner, D. M. Thought suppression. Ann Rev Psychol 2000; 51: 59–91.

Wu, G., Zhang, J., & Gonzalez, R. Decision under risk. In: Koehler, D. & Harvey, N. (Eds.), *Handbook of judgment and decision making*. Oxford, UK: Blackwell, 2005: pp. 399–423.

Wyland, C. L., Kelley, W. M., Macrae, C. N., Gordon, H. L., & Heatherton, T. F. Neural correlates of thought suppression. Neuropsychologia 2003; 41(14): 1863–1867.

PART II

Mental

CHAPTER 9

Working Memory Capacity: Self-Control Is (in) the Goal

James M. Broadway, Thomas S. Redick, and Randall W. Engle

ABSTRACT

Self-control is defined in relation to current goals of an organism. Working memory capacity (WMC) is defined as a cognitive system for maintaining access to goal representations as needed. Self-control depends on cognitive control, which depends in large part on WMC. We discuss the proposal that WMC reflects the abilities to control attention and to control retrieval from long-term memory. From within this dual-component framework (Unsworth & Engle, 2007) we discuss research that has examined relations between WMC and some types of mental self-control failure like over-general autobiographical memories, intrusive thoughts, and mind-wandering. We also discuss research examining the relation between WMC and delay discounting, a popular experimental paradigm for assessing self-control (Rachlin, 2000). Evidence suggests that for some of these phenomena, WMC is a more primary factor than the associated clinical disorders. In other cases, WMC appears to be secondary to other factors such as intelligence. Across these mixed findings at least two generalities can be derived. The positive findings demonstrate that individual differences in WMC can be a confounding "third variable" for a proposed relation between, for example, depression and over-general autobiographical memories (Dalgleish et al., 2007). On the other hand, the negative findings illustrate that individual differences in WMC can obscure more primary influences in a situation like delay discounting (Shamosh et al., 2008). In either case it would be advisable for researchers to measure WMC as a participant factor, if only to control a major source of interindividual variability in their data. Overall, we hold to our position that WMC is critically important for maintaining good self-control in support of a wide variety of goals.

Keywords: Individual differences, working memory capacity, goal-directed behavior, over-general autobiographical memories, intrusive thoughts, mind-wandering, delay discounting, self-control

Working memory is defined as a system for maintaining access to goal-relevant information in support of ongoing complex behavior and cognition. Functional limits of the working memory system define its capacity and this differs between individuals. Self-control is defined in relation to goals and it is also a goal for its own sake. The working memory system has survival value for the self because it selectively processes and records information that is goal-relevant. Working memory capacity has been closely identified with the ability to control attention (Engle, 2002; Engle & Kane, 2004), an ability that would seem to be critical for self-control and self-regulation. Self-regulation is closely related to self-control, except that the "correct answer" is not clearly defined. Self-regulation and self-control must interact in important ways, and we suggest that working memory capacity supports both of these functions. In this paper we explore the implications of this line of reasoning and suggest how individual differences in working memory capacity might be related to individual differences in the ability to exercise self-control and to self-regulate.

A goal is a reference point around which behavior is organized. Self-control becomes relevant when a person has to make a choice between actions that lead to incompatible goals. Experimental psychologists study how people handle this problem in special situations where the correct choice is defined by the experimenter. When incorrect choices are reflexive, habitual, or salient, then an individual must resist these sources of interference to ultimately make correct choices. To the extent that he or she is successful, the person has maintained good self-control. Otherwise the person has ceded some degree of control to the interfering stimulus or habit.

We are interested in the mental processes that enable a person to deal with interference and distraction to avoid passing control outside the self. These processes depend on executive control, "the process by which the mind reprograms itself" (Logan, 2004, p. 227). In our discussion of how working memory might support self-control we emphasize the function of selection, "the very keel on which our mental ship is built" (James, 1890, p. 680). Working memory maintains access to relevant information and suppresses irrelevant information. We believe the selective function of working memory is important for mental control generally and self-control particularly when a person is tempted to pursue conflicting goals. First we discuss working memory capacity in terms of selective attention and memory processes, mainly in the contexts of laboratory situations in which the experimenter defines both the interference and the goal. Afterward we explore possible links to cognitive control problems that are examined out of concern for psychological health and well-being.

Working Memory Capacity: Background

The study of individual differences in working memory capacity was initiated with the development of the reading span task (Daneman & Carpenter, 1980), intended to measure the ability to simultaneously store and process information. Individuals read a series of sentences for comprehension and attempt to remember the final word of each sentence for later testing. Daneman and Carpenter showed that performance on the reading span task correlated with a measure of complex cognition (reading comprehension) but performance on a simple word span task did not. Numerous variations on the reading span procedure have been devised, referred to collectively as complex span tasks. In the operation span task (Turner & Engle, 1989; Unsworth et al., 2005), participants solve a series of simple math equations, with each equation followed by an unrelated item for later recall. Regardless of content domain of the memoranda or difficulty of the interleaved processing tasks, performance on complex span tasks has shown to be predictive of a wide range of higher- and lower-order abilities (Ackerman, Beier, & Boyle, 2005; Conway et al., 2002; Engle & Kane, 2004; Engle et al., 1999; Kane et al., 2007; Turner & Engle, 1989).

Engle et al. (1999) hypothesized that performance on simple span tasks reflects contributions from short-term memory, but performance on complex span tasks reflects contributions from short-term memory plus the control of attention. This notion is reasonable because in complex span tasks a person must frequently divert attention away from to-be-remembered items to do the processing task, and then back again to encode a new item for later recall. Engle et al. (1999) formed a latent variable to represent the ability to control attention by separating the variance unique to complex span tasks from the variance shared with simple span tasks. The residual variable representing control of attention was more strongly related to a latent variable for intelligence than was the variable for short-term storage. This supported the proposal that the ability to control attention determines how much a person can remember in a complex span task and is also responsible for relationships between such performance and complex cognition. This conclusion leads to the further hypothesis that individuals who perform differently on complex span tasks should also perform differently on tasks that do not require much remembering but make heavy demands on attending. The theory of working memory capacity defined as the executive control of attention has been presented fully in other publications (Engle & Kane, 2004; Kane, A.R.A. Conway, Hambrick, & Engle, 2007). We review a few important empirical studies comprising support for that view.

Working Memory Capacity and the Executive Control of Attention

Antisaccade

Kane et al. (2001) used an extreme-groups design to compare participants who had scored low on a complex memory span task (low spans) to those who had scored high (high spans) in the antisaccade procedure (Hallett, 1978). Participants must inhibit a reflexive-orienting response to an attention-capturing, sudden-onset stimulus, to quickly and accurately perform a simple task such as detecting a letter subsequently appearing in a different location. Low spans were slower and less accurate to detect letters appearing in a location opposite to a flashing stimulus than high spans were. High and low spans did not differ in detecting letters that appeared in the same location as a flashing stimulus (prosaccade condition). In Unsworth, Schrock, and Engle (2004), the experimental task entailed merely looking away from the flashing stimulus in antisaccade conditions, or toward it in prosaccade conditions. Eye tracking data showed that low spans were more likely to incorrectly look first toward, rather than away from, the abrupt-onset stimulus in antisaccade conditions. Low spans also were slower than high spans to initiate correct eye movements away from the stimulus. As in the study by Kane et al., low spans and high spans did not perform differently in prosaccade conditions, however, when looking toward the flash was the correct response.

These results illustrate how working memory capacity differentiates individuals in reference to specific goals. Working memory capacity differentiates individuals when attention-capture interferes with goal-attainment in the antisaccade conditions (detect the letter or look away from the flash). Working memory capacity does not differentiate individuals when attention-capture actually facilitates goal-attainment in prosaccade conditions (detect the letter or look toward the flash). These results indicate that working memory capacity is important for either inhibiting prepotent behaviors that are incorrect with respect to current goals, or for activating correct behaviors that are weakly supported in the current environment. Either approach leads to the conclusion that maintaining robust goal representations in working memory is decisively important for self-control in this sort of situation.

Stroop

In the studies by Kane and Engle (2003) high- and low-span participants were instructed to name the color in which a color-word appeared (Stroop, 1935). On incongruent trials, the color-word was different from the color in which it appeared. On congruent trials, the color-word and color matched. Measures of interference were derived by comparing response times and errors on incongruent trials to those on congruent trials. Low-span individuals showed more Stroop interference than high spans, suggesting that low-span individuals were less able to maintain the goal to name the color, particularly when that goal was only weakly supported by the environment. To more fully test this hypothesis, Kane and Engle (2003) varied the proportion of incongruent trials across blocks. They reasoned that the goal of naming the color would be easier for low spans to maintain in blocks in which incongruent trials were more frequent because such trials could serve to remind the goal (Logan & Zbrodoff, 1979). Therefore, differences in interference effects between span groups should be reduced. In contrast, responding to the incorrect stimulus dimension (i.e., the name of the word) would be coincident with a correct response on a majority of trials in blocks in which congruent trials were more frequent. In these blocks low spans should be more likely to lose the goal of color-naming and lapse into word-reading responses. As predicted the magnitude of Stroop interference was greatest during blocks in which only a small percentage of trials were incongruent and so too was the difference between high and low spans in interference. Similarly to the antisaccade studies, these data suggest that in the absence of environmental or contextual support, low working-memory-capacity individuals have difficulty executing novel behaviors when these conflict with habitual responses.

Dichotic Listening

Individual differences in working memory capacity predict the ability to block the capture of attention by strongly associated cues. A.R.A. Conway, Cowan, & Bunting (2001) instructed participants to repeat aloud a continuous message presented in one ear while ignoring the message presented in the other ear (Moray, 1959). During the course of the procedure, each participant was presented with his or her own first name in the unattended message. Of the high-span participants, only 20% reported hearing their own name in the unattended message, whereas 65% of the low spans reported doing so. High spans more effectively ignored the attention-capturing stimulus to concentrate their efforts on the goal of shadowing the attended message.

Comparison of Successive Visual Arrays

Control of attention allows high spans to restrict access to immediate memory, protecting to-be-remembered information from interference from irrelevant material (Engle, 2002). Some evidence consistent with this hypothesis was obtained by Vogel, McCollough, & Machizawa (2005). In the visual arrays task (Luck & Vogel, 1997) participants are briefly shown a target display containing a number of colored rectangles. After a delay period, participants view a probe display that might be identical to the first one or changed with respect to some attribute of one of the rectangles (e.g., its color or orientation). Cowan et al. (2005) used this task as a measure of working memory capacity and reported strong correlations with performance on complex span tasks.

Participants in the studies by Vogel et al. (2005) were cued to attend only the right or left side of each target array, knowing they would be probed for memory concerning the cued side only. There were either four items on the relevant side and no items on the irrelevant side, or else two items on each side of the target display. For individuals with low working memory capacity (measured in a separate arrays task), slow cortical potentials measured by EEG during the delay period between standard and test arrays were indistinguishable whether there were four items on the relevant side and none on the irrelevant side, or two items on the relevant side and two on the irrelevant side. For high working-memory-capacity individuals, slow cortical potentials during the delay period showed an orderly relationship to the number of items on the relevant side of the target display only. Behavioral results were correlated with the EEG results (Vogel et al., 2005).

WORKING MEMORY CAPACITY AND CONTROLLED RETRIEVAL

Individuals differing in working memory capacity also differ in the ability to selectively focus on goal-relevant information and ignore goal-irrelevant information, even in situations where memory demand is low. In many cases there is only one critical thing to remember, the goal of the task. The next section focuses on studies that have examined working memory capacity in terms of controlled, effortful retrieval, in contexts rich in interference.

Proactive Interference

One kind of interference in memory that has been extensively studied is called proactive interference (Wickens, Born, & Allen, 1963). This refers to reduced learning during the course of a memory experiment resulting from interference from earlier learned items. Recall declines when people are exposed to successive lists composed of words from the same category (e.g., farm animals). Individual differences in working memory capacity predict susceptibility to proactive interference (Kane & Engle, 2000; Rosen & Engle, 1998). Proactive interference can be demonstrated in scores from complex span tasks used to measure working memory capacity, and correlations with intellectual abilities are reduced when complex span lists are manipulated to reduce proactive interference (Bunting, 2006).

Low spans in the studies by Kane and Engle (2000) showed greater decrements in recall across lists than high working memory capacity individuals. Participants were also required to perform an attention-demanding, finger-tapping task during list encoding, recall, or both. This additional mental-motor load caused high-span individuals to show proactive interference effects equivalent to low spans regardless of when the load was imposed. This suggested that high working-memory-capacity individuals normally used executive control processes unavailable to low-capacity individuals to resist proactive interference, but such control processes were no longer available to high spans when under a load.

Rosen and Engle (1998) tested high and low spans with lists of paired-associates in A–B, A–C, A–B form. High spans were faster to reach criterion learning and produced fewer first-list intrusions than low spans when learning the second list (A–C) that shared cues with the first list (A–B). High spans were slower than other high spans in a control condition to relearn the A–B list when presented a second time after learning the A–C list. These results were proposed to reflect suppression of the earlier list by high spans (Rosen & Engle, 1998). Conversely, low-capacity individuals were faster to relearn the A–B list than their matched controls, suggesting an ironic benefit from not doing the mental work necessary to combat interference while learning the A–C list.

Fan Interference

Participants in Cantor and Engle (1993) were shown lists of sentences to study for later recognition memory testing in the fan paradigm (Anderson, 1974). Some sentences uniquely mapped persons to places—for example, "The artist is in the house." Learning sentences that together violated one-to-one mapping—for example, "The fireman is in the store; the fireman is in the zoo; the doctor is in the house"—created fan interference, which increases with increasing cue-overlap. Generally when people are shown the studied and new sentences for recognition, response times are slower and people make more errors as fan increases. In Cantor and Engle (1993) individual differences in fan-related slowing were strongly related to working memory capacity. So much so in fact, that the two measures redundantly predicted verbal aptitude in regression analyses.

WORKING MEMORY CAPACITY AS PRIMARY AND SECONDARY MEMORIES

Unsworth and Engle (2007; 2006b) proposed that individual differences in working memory capacity reflect the contributions of two processes: *(1)* active maintenance of information in an attention-like primary memory and *(2)* controlled, cue-driven retrieval from a secondary memory. A participant in a memory experiment is able to code his or her experiences as belonging to a global context (or experimental context), a list context, or an item-level context. Contextual features or attributes are associated in secondary memory with the attributes of the to-be-remembered list items. Contextual levels may be distinguished by their specificity along the temporal dimension. When it is time to retrieve the needed information, participants use retrieval cues to point with greater or lesser precision at the context in which the information was encoded. Similar proposals are found in many existing memory models (e.g., Atkinson & Shiffrin, 1968).

Over-general retrieval cues result in the inclusion of irrelevant information in the searched areas of secondary memory, resulting in more forgetting. Low-span individuals suffer more from proactive and other kinds of interference because they fail to constrain search of secondary memory to relevant information. Consistent with this idea, low spans tend to commit more errors than high spans by incorrectly recalling items from previous lists (Unsworth & Engle, 2006). Failing to access information specific to appropriate contexts is another potential source of variability in cognitive control that might have implications for self-control.

Working Memory Capacity, Self-Control, and Self-Regulation

We have sketched a picture in which individual differences in working memory capacity reflect individual differences in the interrelated abilities to selectively attend and remember information, and use that information effectively to achieve simple goals defined within controlled experimental environments. We believe that such abilities are critical for living successfully outside the laboratory as well. In this section we review three ways the construct of working memory capacity has been applied to questions about controlling the contents of the mind: *(1)* retrieving autobiographical memories, *(2)* suppressing unwanted thoughts, and *(3)* keeping the mind from wandering. We conclude by examining working memory capacity in a situation requiring bona fide self-control.

Working Memory Capacity and the Self-Memory System

The self is undoubtedly a powerful organizing framework for memories. Macrae and Roseveare (2002) showed that self-oriented memories were relatively immune to retrieval-induced forgetting (Anderson, Bjork, & Bjork, 1994) compared to memories oriented toward other people. If accessible, autobiographical memories can be used to support current goal seeking (Williams et al., 2007). Disorders of over-general autobiographical memory retrieval have been associated with clinical problems like depression and post-traumatic stress disorder (Brewin, 1998) as well as diminished working memory capacity and problem-solving ability (Williams et al., 2006). In the Autobiographical Memory Test (Williams & Broadbent, 1986), participants are shown lists of word cues and are asked to respond to each cue by reporting a specific memory (an event occurring at a particular place and time, lasting less than one day). Errors other than omissions are judged incorrect with respect to one of these two requirements. ("I always enjoyed going to the beach" is too general because it does not specify a particular beach and "I went to the beach all last year" is too general because it does not specify a particular time).

Williams et al. (2007) proposed a complicated system including affect-regulation mechanisms interacting with memory processes (*see also* Conway & Pleydell-Pearce, 2000) to explain the phenomenon of over-general autobiographical memories in depression. In its broad outlines, the theory borrows from general memory models (e.g., Atkinson & Shiffrin, 1968), so it is also similar in many respects to the working memory capacity frameworks presented by Unsworth and Engle (2007) as well as those that place more emphasis on executive control (Engle & Kane, 2004). The self-memory system is composed of two major components: a long-term memory called the autobiographical knowledge base and a working memory called the working self (Conway & Pleydell-Pearce, 2000). The autobiographical knowledge base is organized according to global, intermediate, and specific levels of description. Strategic, top-down search of autobiographical memory uses appropriately specific cues to sample information level by level. Direct spontaneous retrieval depends on associative processes (Conway & Pleydell-Pearce, 2000) that are subjected to inhibiting or filtering by the working self. The working self sets up retrieval plans to guide search of the autobiographical knowledge base, compares recovered candidate memories, and allows output of approved memories. These activities are all constrained by active goals of the self-system. Williams and colleagues (Williams, 2006; Williams et al. 2007) proposed several specific processes to explain the phenomenon of over-general autobiographical memories.

Through related processes of capture and rumination, intrusive thoughts gain control of the self-memory system. Through related processes of functional avoidance and mnemonic interlock (Williams, 1996) intermediate-level information tends to cue only other intermediate-level information, and strategic search of associative memory cannot proceed to access memories represented at specific-event levels. Some memories at the specific-event level are related to traumatic events, and the threats to self recorded in such memories may become over-generalized to other specific-event level memories. If this is the case, a depressed person will be unable to access specific-event level memories resulting from a generalized protective habit of avoiding threatening cognitions.

Evidence suggests that the over-general retrieval phenomenon is more directly related to working memory capacity than to affective disorders. This suggests further that some of the more specialized mechanisms proposed by Williams et al. (2007; Williams, 2006) may not be necessary to account for relevant data. For example, Dalgleish et al. (2007) found that reporting over-general autobiographical memories was strongly predicted by performance on a verbal fluency task, a test of executive function shown to be related to working memory capacity (Rosen & Engle, 1997). The relationship held even after removing the variance shared with depression, suggesting that the over-general autobiographical memory disorder is not an outcome of depression. Participants in Dalgleish et al. (2007) were also given a reversed version of the test, in which they were instructed to respond with general memories instead of specific ones. Results were likewise reversed: More depression and less working memory capacity were related to more overly specific memories.

Depressed participants in Dalgleish et al. (2007) erred by recalling too many specific autobiographical memories, but they should not have been able to do this at all according to the theory of over-general autobiographical memories outlined above. These results suggest a general observation: Working memory capacity will be helpful for producing a memory on demand, whether the goal is to produce a series of very general memories (e.g., name all the animals you can think of that start with the letter "f") or very specific ones (e.g., name all the cities you have visited this year and when).

Working Memory Capacity and Suppressing Unwanted Thoughts

Brewin and Beaton (2002) studied the ability to suppress unwanted thoughts in relation to working memory capacity and intelligence using the "White Bear" paradigm (Wegner et al., 1987). Participants were left alone in a room to continuously verbalize their thoughts over three consecutive sessions. In the first session, participants were just instructed to freely verbalize. Before the second session participants were instructed not to think about white bears and to report any such thoughts (suppression condition). Before the third session, participants were instructed to think about white bears and report the occurrence of such thoughts (expression condition). Working memory capacity and intelligence were negatively related to the number of reports of white bear thoughts in the suppression condition but not in the expression condition. Generalizing these results, Brewin and Smart (2005) found working memory capacity–related differences in intrusive thoughts of a more personally relevant nature, independent of mood.

Working Memory Capacity and Mind Wandering

An alternative approach to suppressing unwanted thoughts is to practice mindfulness (Sīlānanda, 2002). Trying to suppress unwanted thoughts can ironically cause them to persist (Wegner, 1997). Instead of trying to fight off intrusive thoughts, invite them to stay as your guest and they will lose their power to dominate the mind (Sīlānanda, 2002). The opposite of remaining mindful is to let one's mind wander. "Mind wandering represents a state of decoupled attention because, instead of processing information from the external environment, our attention is directed toward our own private thoughts and feelings" (Smallwood, Fishman, & Schooler, 2007). Kane et al. (2007) used an experience-sampling methodology to look at whether working memory capacity is related to frequency of task-unrelated thoughts. Reports of task-unrelated thoughts were moderated by working memory capacity when people were involved in challenging tasks compared to routine activities, with higher span people less likely than lower spans to let their thoughts wander off-task.

Mindfulness-based cognitive therapy (Williams et al., 2000) incorporates traditional Buddhist methods of training the mind to stay focused on moment-to-moment experiences and to avoid rumination and distractibility. Williams et al. (2000) assigned formerly depressed patients to either treatment-as-usual or mindfulness-based cognitive therapy, administering the autobiographical memory test before treatment and again after. Patients in the mindfulness group reported more specific autobiographical memories when tested the second time. The control group did not show any change. The groups did not differ in mood at either time, suggesting that performance on the autobiographical memory test may be more directly related to cognitive control abilities than to mood (see also, Brewin & Smart, 2005; Dalgleish et al., 2007).

Working Memory Capacity and Impulsive Decision Making

Delay discounting is a choice situation for studying self-control (Rachlin, 2000). Participants are offered hypothetical choices between two sums of money. The smaller sum is available immediately but the larger one is not available until after a specified delay. People consistently prefer the smaller-sooner reward to the larger-later one (Rachlin, 2000). A measure of the degree to which a person is impulsive and "myopic" regarding future consequences can be obtained by offering various sums at various delays and plotting the individual's discounting function (Rachlin, 2000; p. 10).

Working memory capacity has seemed like a plausible source of variation across individuals in delay discounting and related "gambling" tasks (Fellows & Farah, 2005; Frank & Claus, 2006). Like the flash in the antisaccade task (Unsworth, Schrock, & Engle, 2004) or the drink now as opposed to sobriety over time (Rachlin, 2000), the small but immediate reward may serve as a salient cue that requires executive control to resist (Stout et al., 2005). Results occasionally have suggested a role for working memory capacity in delay discounting (Hinson, Jameson, & Whitney, 2003; but *see* Franco-Watkins, Pashler, & Rickard, 2003). However, the weight of the evidence at present suggests that impulsive decision making is related more directly to intelligence than to working memory capacity (Finn & Hall, 2004; Shamosh et al., 2008; Whitney, Jameson, & Hinson, 2004).

Conclusion

We have presented results of experimental and correlational studies indicating that working memory capacity is most of all about controlling the contents of the mind by selectively attending and remembering goal-relevant information. We have suggested how these processes might be central to the general problems of self-control and self-regulation, and reviewed some recent applications of the working memory capacity construct to these problems.

Working memory capacity appears to be a more central factor than the associated clinical ailments for some phenomena (e.g., over-general autobiographical memories). In contrast, working memory capacity is apparently not central to the paradigmatic self-control problem of impulsive decision making. It is somewhat counterintuitive to observe that working memory capacity predicts counting (Tuholski, Engle, & Bayliss, 2001; Unsworth & Engle, 2008) but not delay discounting (Shamosh et al., 2008), but such anomalies can guide future research to establish boundary conditions for the working memory capacity construct (see also, Kane, Poole, Tuholski, & Engle, 2006). Meanwhile there remain strong theoretical reasons and a growing body of empirical findings to suggest that working memory capacity is important for maintaining self-control in support of a wide variety of goals.

References

Ackerman, P. L., Beier, M. E., & Boyle, M. O. Working memory and intelligence: The same or different constructs? Psycholog Bull 2005; 131: 30–60.

Anderson, J. R. Retrieval of propositional information from long-term memory. Cogn Psychol 1974; 3: 288–318.

Anderson, M. C., Bjork, R. A., & Bjork, E. L. Remembering can cause forgetting: Retrieval dynamics in long-term memory. J Ex Psychol: Learning, Memory, Cognition 1994; 20: 1063–1087.

Atkinson, R. C., & Shiffrin, R. M. Human memory: A proposed system and its control processes. In: Spence, K. W. & Spence, J. T. (Eds.). *The psychology of learning and motivation: Advances in research and theory*, Vol. 2, New York, NY: Academic Press, 1968: pp. 89–195.

Brewin, C. R. Intrusive autobiographical memories in depression and post-traumatic stress disorder. Appl Cogn Psychol 1998; 12: 359–370.

Brewin, C. R., & Beaton, A. Thought suppression, intelligence, and working memory capacity. Behav Res Ther 40: 923–930.

Brewin, C. R., Dalgleish, T., & Joseph, S. A dual representation theory of posttraumatic stress disorder. Psycholog Rev 1996; 103: 670–686.

Brewin, C. R., & Smart, L. Working memory capacity and suppression of intrusive thoughts. J Behav Ther Exp Psychiatry 2005; 36: 61–68.

Bunting, M. Proactive interference and item similarity in working memory. J Exp Psychol: Learning, Mem Cogn 2006; 32: 183–196.

Cantor, J. & Engle, R. W. Working memory capacity as long-term memory activation: An individual differences approach. J Exp Psychol: Learning Mem Cogn 1993; 19: 1101–1114.

Conway, A. R. A., Cowan, N., Bunting, M. F., Therriault, D. J., & Minkoff, S. R. B. A latent variable analysis of working memory capacity, short-term memory capacity, processing speed, and general fluid intelligence. Intelligence 2002; 30: 163–183.

Conway, A. R. A., Cowan, N., & Bunting, M. F. The cocktail party phenomenon revisited: The importance of working memory capacity. Psychonomic Bull Rev 2001; 8: 331–335.

Conway, M. A. & Pleydell-Pearce, C. W. The construction of autobiographical memories in the self-memory system. Psycholog Rev 2000; 107: 261–288.

Dalgleish, T., Williams, J. M. G., Golden, A. J., et al. Reduced specificity of autobiographical memory and depression: The role of the executive control. J Exp Psychol: Gen 2007; 136: 23–42.

Daneman, M., & Carpenter, P. A. Individual differences in working memory and reading. J Verbal Learning Verbal Behav 1980; 19: 459–466.

Engle, R. W. Working memory capacity as executive attention. Current Direct Psycholog Sci 2002; 11: 19–23.

Engle, R. W., Cantor, J., & Carullo, J. Individual differences in working memory and comprehension: A test of four hypotheses. J Exp Psychol: Learning Mem Cogn 1992; 18: 972–992.

Engle, R. W., & Kane, M. J. Executive attention, working memory capacity, and a two-factor theory of cognitive control. In: Ross, B. (ed.), *The Psychology of Learning and Motivation* (Vol. 44). New York, NY: Elsevier, 2004: pp. 145–199.

Engle, R. W., Tuholski, S. W., Laughlin, J. E., & Conway, A. R. A. Working memory, short-term memory and general fluid intelligence: A latent variable approach. J Exp Psychol, Gen 1999; 128: 309–331.

Fellows, L. K., & Farah, M. J. Different underlying impairments in decision-making following ventromedial and dorsolateral frontal lobe damage in humans. Cerebral Cortex 2005; 15: 58–63.

Finn, P. R., & Hall, J. Cognitive ability and risk for alcoholism: Short-term memory capacity and intelligence moderate personality risk for alcohol problems. J Abn Psychol 2004; 113: 569–581.

Franco-Watkins, A. M., Pashler, H., & Rickard, T. C. Does working memory load lead to greater impulsivity? Commentary on Hinson, Jameson, and Whitney. J Exp Psychol: Learning Memory Cogn 2003; 32: 443–447.

Frank, M. J., & Claus, E. D. Anatomy of a decision: Striato-orbitofrontal interactions in reinforcement learning, decision making, and reversal. Psychol Rev 2006; 113: 300–326.

Hallett, P. E. Primary and secondary saccades to goals defined by instructions. Vision Research 1978; 18: 1279–1296.

Hinson, J. M., Jameson, T. L., & Whitney, P. Impulsive decision making and working memory. J Exp Psychol: Learning Memory Cogn 2003; 29: 298–306.

James, W. The principles of psychology. New York, NY: Dover Publications, 1890, 1950.

Kane, M. J., Bleckley, M. K., Conway, A. R. A., & Engle, R. W. A controlled-attention view of working memory capacity. J Exp Psychol, Gen 2001; 130: 169–183.

Kane, M. J., Brown, L. H., McVay, J. C., Silvia, P. J., Myin-Germeys, I., & Kwapil, T. R. For whom the mind wanders, and when. Psycholog Sci 2007; 18: 614–621.

Kane, M. J., Conway, A. R. A., Hambrick, D. Z., & Engle, R. W. Variation in working memory capacity as variation in executive attention and control. In: Conway, A.R.A., Jarrold, C., Kane, M. J., Miyake, A., and Towse, J. N. (Eds.), Variation in Working Memory. New York, NY: Oxford University Press, 2007: pp. 21–48.

Kane, M. J., & Engle, R. W. Working memory capacity and the control of attention: The contributions of goal neglect, response competition, and task set to Stroop interference. J Expl Psychol: Gen 2003; 132: 47–70.

Kane, M. J., & Engle, R. W. Working-memory capacity, proactive interference, and divided attention: Limits on long-term memory retrieval. J Expl Psychol: Learning Memory Cogn 2000; 26: 336–358.

Kane, M. J., Hambrick, D. Z., Tuholski, S. W., Wilhelm, O., Payne, T. W., & Engle, R. W. The generality of working memory capacity: A latent-variable approach to verbal and visuospatial memory span and reasoning. J Expl Psychol: Gen 2004; 133: 189–217.

Kane, M. J., Poole, B. J., Tuholski, S. W., & Engle, R. W. Working memory capacity and the top-down control of visual search: Exploring the boundaries of "executive attention". J Expl Psychol: Gen 2006; 32: 749–777.

Logan, G. D. Cumulative progress in formal theories of attention. Ann Rev Psychol 2004; 55: 201–234.

Logan, G. D., & Zbrodoff, N. J. When it helps to be misled: Facilitative effects of increasing the frequency of conflicting stimuli in a Stroop-like task. Memory Cognition 1979; 7: 166–174.

Luck, S. J., & Vogel, E. K. The capacity of visual working memory for features and conjunctions. Nature 1997; 390: 279–281.

Macrae, C. N., & Roseveare, T. A. I was always on my mind: The self and temporary forgetting. Psychonomic Bull Rev 2002; 9: 611–614.

Moray, N. Attention in dichotic listening: Affective cues and the influence of instructions. Quarterly J Exp Psycholog 1959; 11: 59–60.

Rachlin, H. The science of self-control. Cambridge MA: Harvard University Press, 2000.

Rosen, V. M., & Engle, R. W. Working memory capacity and suppression. J Memory Language 1998; 39: 418–436.

Rosen, V. M., & Engle, R. W. The role of working memory capacity in retrieval. J Expl Psychol: Gen 1997; 126: 211–227.

Shamosh, N. A., DeYoung, C. G., Green, A. E., et al. Individual differences in delay discounting: Relation to intelligence, working memory, and anterior prefrontal cortex. Psycholog Sci 2008; 19: 904–911.

Sīlānanda, Ven. U. The four foundations of mindfulness. Boston MA: Wisdom Publications, 2002.

Smallwood, J., Fishman, D. J., & Schooler, J. W. Counting the cost of an absent mind: Mind wandering as an underrecognized influence on educational performance. Psychonomic Bull Rev 2007; 14: 230–236.

Stout, J. C., Rock, S. L., Campbell, M. C., Busemeyer, J. R., & Finn, P. R. Psychological processes underlying risky decisions in drug abusers. Psychol Add Behav 2005; 19: 148–157.

Stroop, J. R. Studies of interference in serial verbal reactions. J Expl Psychol: Gen 1935; 121: 15–23.

Tuholski, S. W., Engle, R. W., & Baylis, G. C. Individual differences in working memory capacity and enumeration. Memory Cogn 2001; 29: 484–492.

Turner, M. L., & Engle, R. W. Is working memory task dependent? J Mem Language 1989; 28: 127–154.

Unsworth, N., & Engle, R.W. A temporal-contextual retrieval account of complex span: An analysis of errors. J Mem Language 2006a; 54: 346–362.

Unsworth, N., & Engle, R.W. Simple and complex memory spans and their relation to fluid abilities: Evidence from list-length effects. J Mem Language 2006b; 54: 68–80.

Unsworth, N., & Engle, R.W. The nature of individual differences in working memory capacity: Active maintenance in primary memory and controlled search from secondary memory. Psycholog Rev 2007; 114: 104–132.

Unsworth, N., & Engle, R.W. Speed and accuracy of accessing information in working memory: An individual differences investigation of focus switching. J Expl Psychol: Learning Mem Cogn 2008; 34: 616–630.

Unsworth, N., Heitz, R.P., Schrock, J.C., & Engle, R.W. An automated version of the operation span task. Behav Res Meth 2005; 37: 498–505.

Unsworth, N., Schrock, J. C., & Engle, R. W. Working memory capacity and the antisaccade task: Individual differences in voluntary saccade control. J Expl Psychol: Learning Mem Cogn 2004; 30: 1302–1321.

Vogel, E. K., McCollough, A. W., & Machizawa, M. G. Neural measures reveal individual differences in controlling access to working memory. Nature 2005; 434: 500–503.

Wegner, D. M. When the antidote is the poison: Ironic mental control processes. Psycholog Sci 1997; 8: 148–150.

Wegner, D. M., Schneider, D. J., Carter, S. R., & White, T. L. Paradoxical effects of thought suppression. J Personal Soc Psychol 1987; 53: 5–13.

Whitney, P., Jameson, T., & Hinson, J. M. Impulsiveness and executive control of working memory. Personal Individual Diff 2004; 37: 417–428.

Wickens, D. D., Born, D. G., & Allen, C. K. Proactive inhibition and item similarity in short-term memory. J Verb Learning Verb Behav 1963; 2: 440–445.

Williams, J. M. G. Depression and the specificity of autobiographical memory. In: Rubin, D.C. (Ed.), *Remembering our past: Studies in autobiographical memory*. Cambridge, England: Cambridge University Press, 1996: pp. 244–267.

Williams, J. M. G. Capture and rumination, functional avoidance, and executive control (CaRFAX): Three processes that underlie overgeneral memory. Cogn Emotion 2006; 20: 548–568.

Williams, J. M. G., Barnhofer, T., Crane, C., Hermans, D., Raes, F., Watkins, E., & Dalgleish, T. Autobiographical memory specificity and emotional disorder. Psycholog Bull 2007; 133: 122–148.

Williams, J. M. G., & Broadbent, K. Autobiographical memory in suicide attempters. J Abn Psychol 1986; 95: 144–149.

Williams, J. M. G., Chan, S., Crane, C., Barnhofer, T., Eade, J., & Healy. Retrieval of autobiographical memories: The mechanisms and consequences of truncated search. Cogn Emotion 2006; 20: 351–382.

Williams, J. M. G., Teasdale, J. D., Segal, Z. V., & Soulsby, J. Mindfulness-based cognitive therapy reduces overgeneral autobiographical memory in formerly depressed patients. J Abn Psychol 2000; 109: 150–155.

CHAPTER 10

The Dynamic Control of Human Actions

Florian Waszak, Anne Springer, and Wolfgang Prinz

ABSTRACT

Human action serves two complementary purposes. On the one hand, actions are meant to achieve desired effects in the environment. On the other hand, people act as a consequence of external events, trying to accommodate to environmental demands. While the former type of action is usually referred to as "voluntary," "goal-directed," or "intention-based," the latter is often conceptualized as "response," "reaction," or "stimulus-based." At the same time, the concepts of intention- and stimulus-based action control are inseparably interwoven. Although intention-based actions by definition rely on intentions, the planning process also needs to consider stimulus information from the agent's actual environment. Similarly, although stimulus-based actions are triggered by external stimuli from the environment, stimulus information is not a sufficient condition for the execution of the action: to respond to the external information in the appropriate way, it rather needs to be complemented by an intentional set.

In this chapter, we address theoretical and experimental approaches to the cognitive underpinnings of action control. We outline current theories of human action control and review experimental paradigms addressing this issue by comparing intention-based and stimulus-based actions or by investigating the interference between both types of action control. Finally, we discuss the function of the *self* within the proposed cognitive framework. For this purpose, we link the action control theories under discussion to cross-cultural and social-psychological evidence suggesting that individuals differ in self-regulatory performance depending on their social orientation—that is, how they define the self in connection to other people.

We will claim that research on self-control can profit from cognitive research on action control because both fields deal with situations in which automatic behavioral tendencies need to be controlled and adjusted to the individual's goals and desires, for example, losing weight, abstaining from alcohol, cigarettes or drugs, or avoiding violent and aggressive reactions to others. In turn, to understand from a cognitive perspective why people fail to withstand predominant responses to the environment, and, on the other hand, how they can successfully regulate their behavior can be viewed as a substantial part of understanding self-control.

Keywords: Action control, voluntary action, intention, stimulus–response behavior, executive control, independent and interdependent self

The investigation of human action is a very thorny undertaking. Human behavior is extremely complex. People restlessly act. Human actions are rather a continuous stream of nested actions, with one action often depending on or preparing the other. This stream of action comprises segments on very different levels, from very simple, detached gestures like scratching one's head to very complex endeavours made up of a vast number of intermediate steps like finding a new job. It is not only that it is very difficult to segment the stream of actions for proper investigation, it is also that if unconstrained people show an immense inventiveness as concerns the actual realization of one and the same goal. Just think of the number of different ways in which people can realize goal as simple as getting oneself some new gloves. Internet shopping? Or do you prefer to knit them yourself? It is probably for these and other difficulties that human action has been somewhat neglected in cognitive sciences (Prinz & Hommel, 2002; Rosenbaum, 2005).

This neglect becomes especially apparent in the fact that, despite the cognitive revolution during the 1950s and 1960s and the enthusiastic willingness to vanquish the stimulus–response (S-R) psychology of the behaviorism, research on human action never managed to replace the classical behavioristic methodology. Stimulus-response experiments, tracing back human behavior to the given stimulus information, are still the prevailing method in research on human action. Evidently, with this tool of choice research on human action can hardly generate theories that measure up to the complexity of human behavior. The dominance of experiments based on stimulus-triggered responses is probably due to the fact that they have the essential advantage that the researcher knows more or less what the subject will do, and when. That is, the researcher can easily quantify the subject's behavior, for example, by measuring response times. By contrast, the quantification of actions that are triggered by some internal inducement is more difficult, as the researcher has no bearing on the what and when of really voluntary actions (unless he makes the subject perform a "voluntary" action by the presentation of an external cue).

However, some new paradigmatic and theoretic developments have recently emerged in the domain of action control, both stressing the importance of internal factors. We will sketch two of these developments in more detail. Section 1 deals with the relationship between intentional, top-down processes and automatic, bottom-up processes. It seems that top-down and bottom-up processes relate in a more interwoven way than assumed by earlier accounts of action control. The second development pertains to the fact that research has attended increasingly to actions based on the agent's intentions or goals as opposed to the "traditional" type of action that is performed in response to some stimulus information in the environment. Moreover, the question as to what are the functional and neurophysiological differences between these two types of action, goal-based and stimulus-based, has become a major topic within various fields of psychological research. Section 2 will address these topics. Section 3 will point to another line of research, social-psychological research on different self-construals. This research suggests that individuals differ in their cognitive control processes, which, in turn, may lead to stable individual differences in self-regulatory abilities.

Sensorimotor Action: Responding to the Environment

Humans can interact with their environment in two ways: They may either carry out actions to produce certain effects in the environment meant to meet the agent's goals or desires; or they may carry out actions to accommodate to environmental demands. The former type of action is usually called voluntary, volitional, or intentional. We will refer to it as goal-based, since it is selected on the basis of the agent's goals. The latter kind of action is usually called reaction. Section 1 will deal with this type of action. We will refer to it as stimulus-based, since it is selected with respect to a prior stimulus event.

As mentioned above, the main focus of research has been on actions performed in response to external stimuli. The history of reaction time experiments is almost as long as the history of experimental psychology itself.

Along with psychophysics—developed at the same time by pioneers of experimental psychology like Wilhelm Wundt and Gustav Theodor Fechner—reaction time studies were a method of choice in the early experimental laboratories and have been ever since. The pioneer in reaction time studies was the physiologist F.C. Donders (1868). Donders invented a method to estimate the duration of the mental processes taking place when the subject faces different tasks. Among other things, he measured the response times for simple responses, choice responses, and go/no-go responses. He demonstrated that simple reactions are faster than go/no-go reactions, and that choice reactions were slower than the other two types of reactions. He assumed that the time the subjects need for the different tasks depends on the number and types of mental stages involved in the task. The simple reaction-time (RT) task requires only perception and motor time. The go/no-go task requires the two stages of the simple RT task plus the time to discriminate the two stimuli (the go and the no-go stimulus), and so forth. By subtracting the RTs measured for the different tasks, Donders determined the processing time of the different stages. With this simple idea Donders initiated a research program that has been continued until today, the componential processing analysis of human task performance that has been advanced especially by Sternberg (1969).

The following 150 years witnessed an enormous number of studies investigating, as Donders did, stimulus-based actions. Just to name a few prominent examples: Hick (1952) showed that choice reaction times are proportional to $\log_n$, where n is the number of different possible stimuli ("Hick's Law"). Many studies have investigated RTs in different modalities. They found that RTs to auditory stimuli are faster than to visual stimuli (e.g., Galton, 1899; Welford, 1980) and that this is true for simple and complex responses (Sanders, 1998). Piéron (1920) demonstrated that simple reaction time decays as a hyperbolic function of luminance ("Piéron's law"). There are much more studies of this kind, relating RTs to gender, age, handedness, peripheral vs. focal vision, and the like.

The notion that people may carry out actions in response to external stimuli evokes the metaphor of humans as sophisticated machines that act only if exposed to some external impulse. Evidently, the strong versions of the behaviorism are very close to this view, assuming that all behavior can be explained without recourse to internal mental states. However, despite its methodological proximity to the behaviorism, psychological research during the second half of the last century realized that the strong behaviorism does not capture the complexity of human behavior. We will outline some of the more recent developments below. Nevertheless, the question as such is still pertinent: Who controls our actions, internal demands or the demands of our environment? Or at least, to which degree are our actions determined by external and internal factors?

We will try to outline some of the answers that psychological research has given to this question in the course of the last decades. The present chapter focuses on research on action control—that is, on research that focuses on the question of how concrete motor behavior is controlled. However, we would like to point out that research on action control and research on self-control partially share a common subject: Situations that are inherently conflicting are not only at the heart of research on self control, but represent also, as we will show in the first part of the present chapter, a major object of research on action control proper. This becomes especially apparent in the fact that experimental research on self-control and experimental research on action control use the same paradigms to operationalize conflict.

Dual Route Logic

More recently, theories resulting from RT studies became more sophisticated. The instance theory of automaticity (Logan, 1988), for example, is a full-fledged theory of skill acquisition. It holds that skill acquisition can be explained by the accumulation of examples or instances of the skill. People begin with some set of general algorithms to solve a particular task. However, whenever a particular skill is performed, a trace of the instance of that action is stored in memory.

When the task has to be performed again, a race ensues between the algorithm and retrieval of previously stored instances. Because memory retrieval is getting faster as more instances are stored, it soon begins to dominate the race over algorithmic processing. In other words, beginners rely on explicit strategies, whereas experts merely remember what they did before.

The instance theory of automaticity (Logan, 1988) already points to one of the most pertinent issues in recent research on stimulus based action control—the role that controlled processes, on the one hand, and automatic processes, on the other hand, play in the transformation of stimulus information into motor action (Shiffrin & Schneider, 1977). In everyday life, we usually react to the environment in a way that meets our purposes, indicating that stimulus–response (S-R) translation (the mechanisms that translates stimulus information into a corresponding motor response) is controlled by our intentions. However, sometimes we fail to carry out the intended action, performing, instead, some other unintended action. Very often, this unintended action is one that we are accustomed to perform in the given situation or that is otherwise compatible with the circumstances. This suggests that humans are liable to processes of S-R translation that do not obey the agent's intentions. In other words, over and above the controlled route of S-R translation there seems to be an automatic route of response translation. Some processes are thought to be innately automatic; others are thought to become automatic through practice (Spelke, Hirst, & Neisser, 1976). The standard account maintains that, in contrast to controlled S-R translation, automatized translation is initiated unintentionally upon stimulus presentation, is fast, and cannot be prevented or stopped.

As mentioned before, both research on action control as well as research on self-control deal with situations in which automatic tendencies need to be overridden. However, self-control is related to a much broader spectrum of "behavior," as it refers to the control of one's thoughts, emotions, and complex social behavior. The "behavior" that needs to be overridden is often linked to basic drives or urges. Self-control refers to tempering these urges or pursuing one's well-intentioned but effortful resolutions, like maintaining a diet or performing better in school. Successive self-control is linked to a multitude of desirable outcomes, as, for example, better mental health, reduced aggression, superior academic performance, reduced susceptibility to drugs, and so forth (*see* Duckworth, & Seligman, 2005; Tangney, Baumeister, & Boone, 2004). That is, in a certain sense, self-control is about how people counteract "behavioral" tendencies that are detrimental to their long-term goals, like losing weight or getting better grades. In research on action-control, on the other side, "conflict" refers to stimulus-driven response tendencies that are inappropriate in the given task context (i.e., to the facilitation of a specific motor output automatically triggered by the mere sight of a particular stimulus). Here, control refers to how people counteract behavioral tendencies that are detrimental for their imminent short-term goals, like responding to a particular stimulus despite the presence of a number of other potentially action-relevant distractor stimuli.

This is not to say that the two levels of control are not interrelated. In the contrary, long-term goal maintenance despite recurring temptations to diverge from the right path does not only necessitate indefatigably keeping up the motivation to achieve one's goals, but also the capability to handle all the enticements one is exposed to everyday; the intention to do some extra work in the evening can only be realized if one is able to resist turning on the television in passing. Consequently, to understand why people fail to withstand the response tendency triggered by an environmental stimulus constitutes an integral part of research on self-control.

Although genuine action errors occur rather seldom and are, therefore, rather difficult to explore, there is an increasing literature investigating different types of action errors and the situations in which they occur (e.g., Heckhausen & Beckmann, 1990; Reason, 1990). However, the better part of research in this context is not on real action slips but on small delays in the execution of an action when the subject is faced with ambivalent stimuli. One of the most famous examples is the so-called color-word Stroop

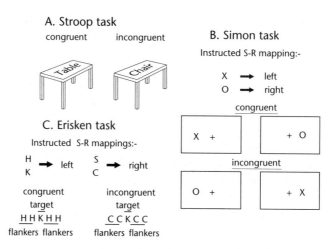

Figure 10–1. Illustration of the Stroop, Simon and Eriksen tasks. (a) The Stroop task is illustrated in its picture–word version. Both stimulus components refer to either the same object (congruent) or to different objects (incongruent). (b) In the Simon task, the subject is instructed to respond with a left or right keypress, respectively, to a nonspatial stimulus attribute (form in this case). The stimulus is presented to the left or the right of the fixation cross, such that the code of the correct action corresponds (congruent) or does not correspond to the spatial code of the stimulus (incongruent). (c) In the Eriksen task, the subject is instructed to react with a left or right keypress in response to one of four different letters presented in the middle of a string of letters (target). The remaining letters of the string (flankers) can map either to the same key as the target (congruent) or to the other key (incongruent).

task (Stroop, 1935; for a review, *see* MacLeod, 1991; for an illustration of the Stroop task in a version that uses pictures instead of colors, *see* Fig. 10–1a). In this task, subjects are required to name the ink color of words. Usually, performance is less error prone and faster if the meaning of the word is congruent with the to-be-named ink color (e.g., the word BLUE written in blue ink) than if the word is incongruent (e.g., the word GREEN written in blue ink). This has been taken as evidence for word reading being automatic: a skilled reader cannot avoid reading the word despite instructions to attend only to the color in which it is printed. When the color-word stimulus is incongruent, the automatic processing of the color word elicits a conflict between the intended action of naming the color and the highly automatized action of reading the word.

Similar conclusions have been drawn from the Simon effect, which refers to the common finding that in a task where stimulus location is irrelevant, RTs are nevertheless shorter when stimulus and response locations are congruent than when they are not (Simon & Rudell, 1967; *see* reviews by Lu & Proctor, 1995; Umiltà & Nicoletti, 1990). For example, Craft and Simon (1970) presented subjects with a red or a green stimulus that could appear either to the left or to the right of the fixation point (*see* Fig. 10–1b for an illustration with two forms instead of two colors). Subjects were to respond with a left-key press to the green stimulus and a right-key press to the red stimulus. Notice that the side on which the stimulus was presented is completely irrelevant to the task. Nevertheless, RTs were significantly faster when the green light was presented to the left field and when the red light was presented to the right field.

Several explanations to account for the "Simon effect" have been put forward (for a review, *see* Hommel & Prinz, 1997). The general consensus seems to be that the location of the stimulus automatically generates a spatial code, even though the target of the task is a non-spatial dimension, such as color (Umiltà & Nicoletti, 1990; Wallace, 1971). If, so the reasoning goes, this spatial code overlaps with the relevant response codes derived from the target dimension, the stimulus will automatically prime the feature-overlapping response, whether this response is correct or not (dimensional-overlap model, Kornblum, Hasbroucq, & Osman, 1990). If the codes are congruent, the same motor code is activated via two routes, the intended route controlling the target color task and the automatic route triggered by the spatial stimulus code. The two routes working together result in a speed-up of response selection. If, however, the two codes are incongruent, two responses are facilitated at the same time giving rise to a time-consuming conflict that the system needs to resolve before it can settle down to a unique response.

Another source of evidence for response translation not entirely taking place in the intended way are flanker experiments. In Eriksen and Eriksen's (1974) classic flanker interference paradigm, the subjects are required to make a speeded discrimination between, for example, two letters. Subjects are presented with a string of five letters. The target stimulus is located at the center of the string. The remaining stimuli left and right of the target are congruent or incongruent to the target, that is, they are either identical to the current target or they represent the alternate target. The usual outcome of this type of experiment is that trials with congruent flankers yield shorter RTs than trials with incongruent flankers (e.g., Eriksen & Hoffman, 1973; Eriksen & Eriksen, 1974).

That this is not a mere effect of stimulus similarity is demonstrated by the fact that the effect is also found when flankers and target are always dissimilar but map either to the same response (congruent) or to a different response (incongruent; e.g. Miller, 1991; see Fig. 10–1c). It seems that, although irrelevant to the task (i.e., not intended to be translated into a response), flankers are automatically processed to a certain degree, probably up to response-related stages (e.g., Coles, 1989).

The results from research on task switching can be interpreted in a similar vein. In the task switching paradigm, subjects are required to alternate between performing two different tasks—usually choice RT tasks. The central finding is that switching between competing tasks produce substantial performance costs in terms of RTs and error rates. In the present context, one common finding that refers to cross talk between the two tasks is of special interest: When the two tasks used in the experiment are both mapped to the same response keys, trials are either response congruent or incongruent. Sudevan and Taylor (1987), to name but one example, showed that subjects' performance is better on congruent than on incongruent trials (see Rogers & Monsell, 1995). The existence of *cross-talk* demonstrates that it is not only the S-R mapping of the currently valid task that is applied upon stimulus presentation. Rather, both valid and currently invalid mapping are applied concurrently, yielding a conflict the resolution of which is time consuming.

Another example of automatically applied S-R mappings has been investigated by Waszak and colleagues (Waszak, Hommel, & Allport, 2003, 2004, 2005; Waszak & Hommel, 2007), who explored the influence of long-term stimulus-response (S-R) priming in task-switching (see Wylie & Allport, 2000). For example, Waszak and colleagues (2003) made participants orally name either the word- or the picture-constituent of incongruent picture-word conjunctions (e.g., the picture of a TABLE with the word "chair" superimposed on it, see Fig. 10–1a). Subjects had to switch tasks every second or third trial, depending on the particular experiment. Within the word-reading task, participants could encounter two types of stimuli: (1) picture–word stimuli that had never been presented in the context of picture-naming (unprimed stimuli), or (2) picture–word stimuli that they had picture-named previously (primed items). Waszak et al. (2003) showed that word-reading RTs in response to primed stimuli were much slower than to unprimed stimuli, even when more than 200 trials intervened between the priming event (the occurrence of a particular item in picture-naming) and the probe event (the occurrence of a the same item in word-reading). To explain this item-specific priming effect, they suggested that when subjects perform a particular action in response to a particular stimulus, they encode the underlying stimulus-, response-, and task-related codes into an integrated S-R episode, or event code (Allport, 1987; Hommel, 1998; Hommel, Pösse, & Waszak, 2000; Logan, 1988, Neill, 1997). If, so the reasoning goes, the stimulus of the encoded S-R episode is encountered again, the entire episode is automatically retrieved. This, in turn, makes switching between arbitrary tasks using the same stimuli more difficult, because it elicits a time-consuming conflict between the current task goal and the automatically retrieved response and task-related codes; another instance of unintended, automatic stimulus–response translation.

To make a long story short, there is a multitude of experimental evidence demonstrating that S-R translation doesn't evolve in a strictly intended manner. Instead, it seems that stimuli are, to a certain degree, automatically translated into a motor response code, giving rise to the conflict phenomena in Stroop, Simon, Eriksen, and task-switching experiments. To account for these findings, various dual routes models have been suggested, all of which consider stimulus-based action control to be based on a sort of competition between intentional or controlled processes, on the one hand, and automatic processes, on the other.

At the beginning of Section 1, we asked what controls our actions, internal or external demands? In light of the experiments outlined above, the answer would be: Our actions are not entirely controlled by external demands, as the strong behaviorism assumes, but by our intention to act. However, so the reasoning goes, our actions are not entirely controlled by our intentions either, since they can be thwarted by external impulses taking the automatic route.

Evidently, the dual route logic does not only encompass the control of rather simple sensorimotor behavior tasks tested in Stroop or Eriksen experiments. It can also be applied to the control of conflict in more complex behavior. The most typical example of a complex conflict in which a low-level goal or habit counteracts a high-level goal or intention is dieting. In terms of the dual route logic, the tasty looking cake with whipped cream constitutes a distractor that automatically triggers the craving to eat. This gives the dieter a hard time to stick to his intention to maintain the diet.

As mentioned above, although action and self-control concern two different levels of behavior, they are interrelated in that successful self-control depends on the ability to control one's actions. Moreover, there is evidence that the two levels of control, action and self-control, draw on the same mechanisms. Muraven and Baumeister (2000) propose that self-control is a limited resource (self-control strength); exerting self-control consumes self-control strength and, as a consequence, reduces the amount of strength that is available for later acts of self-control. Moreover, Muraven and Baumeister suggest that all kinds of self-control (cognitive, emotional, and physical) draw upon the same limited resource. It is important to note that the self-control strength model distinguishes between effort and self-control. It claims that self-control is involved only if the task requires overriding a response (Muraven, Tice, & Baumeister, 1998).

Support for this idea comes, for example, from a recent study by Inzlicht and Gutsell (2007). These authors had participants watch an emotional movie; participants were either to watch the movie normally or they were to suppress their emotions while watching. Thereafter, the participants completed a Stroop task. Compared with the control group, participants who suppressed their emotions performed worse on the Stroop task. Moreover, Inzlicht and Gutsell also recorded electroencephalographic activity. They observed that the behavioral deficit of the experimental group was mediated by weaker error-related negativity (ERN) signals. The ERN is a negative voltage deflection that peaks about 80 miliseconds after response onset. The ERN has been shown to originate from the dorsal anterior cingulate cortex (DACC) (van Veen & Carter, 2002). It is considered to reflect preconscious processes related to an impending error (Nieuwenhuis et al., 2001) and can be considered to index the activity of the error monitoring system. The study from Inzlicht and Gutsell, therefore, shows that when people exert self-control on their emotional response to an arousing movie, their neural mechanisms for monitoring conflicts and errors in sensorimotor tasks (like the Stroop task) are weakened, resulting in a reduced ability to monitor losses of control. This study shows the close relation between mechanisms of self-control and of action control.

Prepared Reflex Logic

Hommel (2000a) noted recently that conceiving S-R translation as a two-lane route, on one of which the subject's "will" picks the intended response upon stimulus presentation, and on the other of which the competing stimulation automatically translates into the corresponding action code, may be misleading. As Hommel showed, in most if not all demonstrations of automatic S-R translation, the "unintended" activation of the "wrong" response is in one way or the other related to the subject's intentions.

Take the Simon effect, which refers to the finding that in a task where stimulus location is irrelevant (e.g., respond to green with a right-key press; respond to red with a left-key press), RTs are nevertheless shorter when stimulus and response locations (left vs. right) are congruent. Grice, Boroughs, and Canham (1984) presented subjects with two stimuli on each trial, one on each side of fixation. One of these stimuli was the target and the other a neutral distractor. The symmetrical display should have cued both left- and right-response codes, such that no Simon effect should have been observed. Yet, Grice and colleagues found a standard Simon effect. Evidently, one can speak of "automatic" response translation in the sense that the subject did not voluntarily choose to translate the side of target presentation into the corresponding response code, (making, as a consequence, congruent trials (e.g., left response to left target) faster than incongruent trials). However, the results show that it is not each and every stimulus that is translated into a response but only the particular stimulus the subject intends to respond to.

In a similar vein, the Eriksen flanker effect seems to depend on the flanker taking part in the (instructed) task set. Cohen and Shoup (1997) showed that a flanker effect is obtained only if target and flankers are defined on the same dimension (e.g., color). Thus, flanker stimuli may be irrelevant in the current trial in that the correct response does by no means depend on the flankers; however, to be "automatically" translated into a response, they need to be valid targets (appearing at the wrong location).

A similar conclusion can be drawn from a series of experiments on the Stroop effect done by Besner and colleagues (e.g., Besner, Stoltz & Boutilier, 1997; Besner & Stoltz, 1999; Besner, 2001; Risko, Stolz, & Besner, 2005). Besner, Stoltz, and Boutilier (1997) presented subjects with color-word Stroop stimuli with either all letters colored or a single letter colored. They found that Stroop interference was reduced or even eliminated when only one letter was colored. They argued that "automatic" word reading occurs only when the subject adopts a mental set that includes lexical-semantic processing. However, coloring a single letter causes the subject to adopt a mental set that restrains subjects from treating the stimulus as a word, dissuading subjects from reading it.

A very similar picture arises as for the automatic retrieval of stimulus-task bindings slowing down the subject's performance when required to accomplish a different, competing task (Waszak et al., 2003; 2004; 2005). As mentioned above, Waszak and colleagues showed that it is harder to read the word of a picture-word Stroop stimulus (e.g., the picture of a TABLE with the word "chair" superimposed on it), when the subject previously encountered the same stimulus in the picture-naming task. They suggest that the task/response that has been associated to the stimulus during previous S-R events (naming the picture, in this case) are reactivated upon stimulus presentation (in the word-reading task, in this case), slowing down the response.

This reactivation is automatic in the sense that the subject does not intend to retrieve a task set that is detrimental to the execution of the currently appropriate task. However, in similar experiments Waszak and Hommel (2007) show that the "wrong" S-R event is reactivated only if the task set for the "wrong" task remains active across the switch of the task and, therefore, competes with the activations now needed for the new task. The subjects intend to perform the new task (and most of the time finally do so), but the intention of the previous trials with the other tasks still lingers in the system, making the activation of the new task harder. Allport and colleagues (1994) called this effect Task Set Inertia (TSI). It is due to this left-over activity of the competing task set that subjects are vulnerable to interference by the competing task. In this state, between-tasks interference is further enhanced if the current stimulus has been primed in the competing task context. In other words, the "automatic" retrieval of the competing stimulus-task binding is preconditioned on the presence of some top-down bias in favor of that stimulus-task binding. This represents another example of automatic processing being contingent on controlled processing.

To sum up, there is evidence pointing out that the relationship between intentional and automatic process might not be that of a two-lane route. Instead, intentional processes seem to set the stage for automatic processes: the automatic translation of a stimulus into a motor action is most of the time based on some intentional state of the subjects implemented before the presentation of the stimulus (for a review, *see* Hommel, 2000). According to Hommel (2000) the idea that stimuli may automatically trigger an action upon stimulus presentation based on instructions organized before the S-R event has been put forward for the fist time by Exner (1879). The term "prepared reflex" has been coined by Woodworth (1938), who concluded from his introspective studies on speeded reactions that "[…] no new will impulse is needed after the entrance of the stimulus in order that the reaction shall follow" and that "[t]he only voluntary act is the preparation" (Woodworth, 1938, p. 305).

Automatic response activation by instructed S-R mappings

Evidently, from a prepared reflex perspective, research that focuses on the question of how humans carry out certain reactions—that is, how they "rattle off" a motor action in response a stimulus (the reflex part), looks right through the more interesting question as to how the system gets ready to do so (the "prepared" part).

From this point of view, some of the most intriguing steps enabling the subject to accomplish the task at hand are already completed the moment the investigator leaves the experimental cabin. These steps take place during the instructions, before the assessment of the subject's behavior begins.

The power of verbal instructions is obvious. Humans are able to perform any voluntary motor response to any environmental stimulus (i.e., even if stimulus and response are in no way compatible or otherwise linked) such that the specific S-R mapping has to be set-up just before the experiment. Humans merely need to be instructed verbally. How important these arbitrary visuomotor mappings (e.g.,Wise & Murray, 2000) are for the understanding of human behavior becomes evident from the tremendous difference between humans and monkeys. In order to make a monkey produce a certain arbitrary visuomotor translation, the investigator has to shape his behavior by operant conditioning for days or even weeks by way of rewarding the monkey each time he performed a correct response.

There are a number of studies investigating whether instructions influence the codes used in S-R translation. One of them is the study by Hommel (1993). Hommel made subjects perform a standard Simon task in which they had to press left and right keys in response to the pitch of a tone. The tones were presented either through a left or right loudspeaker. However, each keypress switched on a light on the opposite side of the keypress. One group of subjects was instructed to "press the left/right key in response to the low/high pitch." Another group performed the very same task, but was instructed to "flash the right/left light in response to the low/high pitch." The standard Simon effect was obtained for the group instructed to press the key, (that is, faster RTs when the location of the loudspeaker and the key press corresponded). Importantly, the group instructed to flash the light showed exactly the opposite result—that is, faster RTs when the location of the loudspeaker and light corresponded. Hommel suggested that the light group had coded their responses in terms of light location. Because these were opposite to

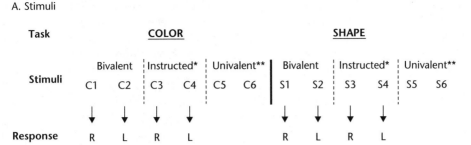

Figure 10–2. Illustration of the study from Waszak et al. (2008). Panel A show the S-R assignment for the six colors (C) and the six shapes (S). Subjects switched between color- and shape-naming. On each trial, subjects were presented with a color–shape combination. Combining the colors and shapes from the illustrated sets generated three conditions that differed with respect to the irrelevant, distractor dimension, as shown in panel B for the color task. Distractor bivalent: The distractor of the given trial is also used as a target in the other task. Both stimulus dimensions are valid targets. Distractor univalent: The distractor of the given trial is never used as a target in the other task. The distracter dimension of these stimuli had no valid S-R mapping. Distractor instructed: The distractor of a given trial is never used as a target in the other task, as for the univalent stimuli; however, the distracter is associated with the competing task because, during the instructions, subjects had included a valid S-R mapping into the task set.

the response keys, the effect was reversed. In other words, the effect of automatic response translation was a rather direct function of task instructions, demonstrating its dependency on intentional processes. (A detailed discussion of the influence of the subject's interpretation, of stimulus and response coding, of the current task goals and preparation, and of task-related strategies on the Simon effect can be found in Hommel [2000b]). Similarly, Wenke and Frensch (2005), who manipulated the response instructions in a dual-task paradigm, showed that response instructions directly influence the way responses are coded for motor control.

However, more recently, some studies have more directly investigated the prime example for a prepared reflex in the sense defined above—S-R translations of mappings the subjects have no prior experience with—that is, translations of instructed (but not applied) S-R mappings. Waszak, Wenke, and Brass (2008; see also Wenke, Gaschler, & Nattkemper, 2005) report a task-switching experiment assessing the cross-talk of arbitrary S-R mappings as a part of the instructed task representation, on the one hand, and the cross-talk of repetitively applied mappings, on the other hand. The subjects' task was either to respond to the color or to the shape of the stimulus by performing a left or a right key press. On each trial, a cue indicated which task to perform next. Waszak and colleagues used six shapes and six colors allocated to three different stimulus subsets (see Fig. 10–2). By combining the colors and shapes from the three sets they generated three stimulus conditions that differed with respect to the irrelevant distractor dimension [color or shape], which had to be ignored in the given trial. In the bivalent stimulus condition, the distractor of the given trial was also used as a target in the other task. In the univalent stimulus condition, the distractor of the given trial was never used as a target in the other task. In the instructed stimulus condition, the distractor of a given trial was never used as a target in the other task, exactly as for univalent stimuli. However, in contrast to univalent stimuli, the distractor of instructed stimuli was associated with the competing task because, during the instructions, subjects had included a valid S-R mapping into the task set. Not surprisingly, bivalent stimuli showed considerably slower RTs than univalent stimuli, once more demonstrating that the system needs to overcome bottom-up interference caused by the distractor-triggered activation of the competing task (*see* above). Importantly, instructed stimuli were also much slower than univalent stimuli, demonstrating that (distractor) stimuli automatically activate task or response codes with which they are associated by mere instruction (and not by S-R traces accumulated in memory each time the mapping is applied in the course of the experiment, as is the case for bivalent stimuli). These findings show that taking part in the intentional set may not only be a necessary condition for automatic stimulus processing to occur, but also a sufficient condition.

Interim Conclusion

Humans are very efficient in responding in a prespecified way to external stimulation. They can easily perform a certain motor action when a particular stimulus appears. However, sometimes they perform an unintended action, committing an *error*. Moreover, external stimulation that counteracts the intended stimulus-response translation can *interfere* with the intended action, slowing performance down without causing an overt action error. These observations have brought about dual route models according to which human performance is controlled by two mechanisms of response translation, a slow, conscious, and intentional mechanism that serves the subjects to perform the correct actions and a fast, unconscious, and automatic mechanism that may thwart the subjects' intention if intentional and automatic route do not activate the same response. However, as we tried to show in the second part of section 1, automatic processes may not be as unintended as they appear to be. There is ample evidence that the *intentional* aspects of motor control precedes the stimulus that triggers the execution of the motor action, rather than intervenes between stimulus and response, and that *automatic* processes are conditioned on these intentional control states, rather than being completely independent from them. This prepared reflex logic (*see* Hommel, 2000) is suggested by the fact that automatic response translations as assessed in interference paradigms such as the Simon, Stroop, or Eriksen paradigms, are always in one way or the other depending on the mental set of the subject. Moreover, other studies show that establishing a link between a stimulus and a response by mere instruction is sufficient to observe effects of response or task activation upon presentation of the stimuli, a very convincing case of automatic processes for which intentional processes set the stage.

Hence, as concerns the question as to what controls our actions, internal or external demands, the prepared reflex logic has a simple answer: The conceptual separation between internal and external demands is not sustainable; automatic and controlled, internal and external intertwine.

Comparable ideas have been discussed in the literature. For example, Allport's (1980) selection-for-action account argues that actions are controlled by processes that guide the selection of environmental objects as triggers of or targets for action. Anderson's ACT-R (Anderson & Lebiere, 1998) and Meyer and Kieras' (1997) EPIC model, two full-range cognitive architectures, incorporate similar assumptions.

Moreover, most theories on prospective memory rely on similar mechanisms. Most of the time an action follows the intention to act immediately (or the triggering stimulus follows the intention promptly). However, a delay between intention and triggering condition/action is not uncommon. Sometimes we are simply unable to act on an intention on the spot. We cannot put a letter into the mailbox before we actually are at the mailbox. Research on prospective memory (Meacham & Leiman, 1982) investigates the question how intention results in future behavior. (Other authors have preferred the term "delayed intentions" [Ellis, 1996]). In a prospective memory task, subjects form the intention to act in the future and then engage in other activities "waiting" for the suitable moment to act. Gollwitzer and colleagues (Gollwitzer, 1999; Brandstätter, Lengfelder, & Gollwitzer, 2001) have argued that what they call "implementation intentions" have the format "if X happens, then I will do Y". They assume that cue X is capable of triggering the postponed action automatically, at least if the cue is specified concretely enough. Bargh and Gollwitzer (1994) compare implementation intentions to habits. When the plan is sound, all you need to carry out the action is the perception of the specified cue.

The studies outlined above suggest that automatic and controlled processes are interrelated in that controlled processes set the stage for automatic processes. However, there is also evidence that behavior that is traditionally labelled "controlled" is subject to influences from "automatic" processes. One of the most prominent examples stems from recent research in experimental social psychology: It has been shown that incidentally activated knowledge can influence complex social behavior. For example, Bargh, Chen, and Burrows (1996) covertly primed participants with trait knowledge about rudeness, stereotypes of the elderly, and stereotypes of African Americans. The latter two were meant to prime the traits *slowness* and *hostility*, respectively. This was done by means of a "Scrambled Sentence Test", presented to participants as a test of language ability. Participants are presented with five "scrambled" words and are required to construct as quickly as possible a grammatically correct sentence. To prime rudeness, for example, some of the five word lists contained an adjective or verb semantically related to the trait rude (e.g., intrude, obnoxious, etc). Bargh, Chen, and Burrows showed that participants primed with trait information were more likely than nonprimed participants to show behavior related to the primed trait—for example, to interrupt another person (rudeness), to walk slowly down the hallway (slowness), or express hostility after being provoked (hostility). This was the case, although none of the participants was aware of a connection between the priming manipulation and their behavior.

Similarily, Dijksterhuis and van Knippenberg (1998) primed participants with the traits "intelligence" or "stupidity." Thereafter, they made participants complete an allegedly unrelated knowledge test. As expected, participants primed with the trait "intelligence" showed better performance than control participants, whereas participants primed with "stupidity" showed worse performance than control participants. These studies show that perception can automatically activate behavioral representations, which, in turn, can guide actual behavior in complex social contexts that is typically thought to be conscious, intentional, and controlled.

IDEOMOTOR ACTION: MAKING THE ENVIRONMENT RESPOND

Up to now we focused on stimulus-based actions—that is, actions that are carried out in response to some external event (although, as we showed, they are nevertheless contingent on the subjects' intentions). However, as stated in the introduction, stimulus-based actions constitute only one part of the spectrum of human behavior. The other part constitutes the main characteristic of human flexibility—namely, the capacity to act—decoupled from the environment, in a way that is not stimulus driven, but that, in the contrary, is meant to shape the external world according to intentions. Psychological research has only begun to explore this type of action. The common-sense concept of these internally generated, voluntary actions (Jahanshahi & Frith, 1998) typically refers to rather complex behavior in which actions are the means for satisfying long-term desires or needs. That is, the key feature of the commonsense concept of voluntary actions is that they are goal-directed, or, more precisely, produced by an internally generated, goal-directed mental state or thought. Evidently, this type of action is also based on the subject's intentions (to achieve a certain goal) and elsewhere we referred to it as intention-based action (*see* Waszak et al., 2005; Keller et al., 2006; Herwig, Prinz, & Waszak, 2007). However, to discriminate this goal-directed type of intention from the type of intention discussed in Section 1 (which refers to the "controlled route" of S-R translation in terms of the dual route logic and to the "preparation part" of S-R translation in terms of the prepared reflex logic), we refer to it in the following as goal-based.

Most scientific definitions of voluntary actions agree with this definition. A movement is considered a voluntary action if its aim is to produce some internally pre-specified effect. Moreover, action goals—as anticipated representations of intended effects of actions—are also believed to play a core role in the acquisition and control of intentional actions (*see* Hommel 2003; Hommel et al., 2001; Prinz, 1997). In the ideomotor theory of action control, the intended action effect is considered also to be the cause of the action. The ideomotor principle can be traced back to the middle of the 19th century, when Lotze (1852) and Harleß (1861) suggested a model of how humans, given a particular goal, select a suitable action. According to this model, an actor acquires bi-directional associative links between the code of a particular action and the codes of the action's sensory consequences. These associations provide the actor with the information regarding the means by which a given sensory consequence can be produced. Hence, the actor can use this information to produce desired effects.

Numerous experimental results support this framework. Elsner and Hommel (2001) showed that subjects acquire action-effect links and that these links are bidirectional. They made subjects first work through an acquisition phase (*see* Fig. 10–3), in which a self-selected key press was always followed by a certain tone (e.g., left key press followed by a high pitch tone;

A. Acquisition phase

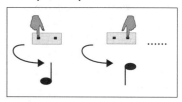

B. Test phase

consistent   inconsistent

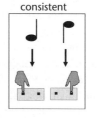 

Figure 10–3. Illustration of the study from Elsner and Hommel (2001).

right key press followed by a low pitch tone). In a subsequent test phase, they presented the same tones as target stimuli in response to which subjects had to perform a speeded-choice response. For half of the subjects, the S-R mapping of the test phase corresponded to the action-effect mapping of the acquisition phase; for the other half, it was reversed. Elsner and Hommel (2001) showed that, in the test phase, subjects' RTs were faster when responding to tones that their actions had produced in the acquisition phase than when responding to tones that had been produced by the alternative actions. This result suggests that associations between the key presses and the subsequent tones acquired during the acquisition phase influence the activation of the choice responses in the test phase. In other words, the perception of a learned sensory effect activates "backwardly" the action it is associated with, exactly as the ideomotor principle predicts. The same conclusions can be drawn from studies with visual effects (Zießler, 1998) and electrocutaneous action effects (Beckers, De Houwer, & Eelen, 2002). That these bilateral associations are indeed functional in action control is demonstrated by a study from Kunde, Hoffmann, and Zellmann (2002; *see also* Kunde, Koch, &

Hoffmann, 2004). Kunde et al. made subjects prepare a certain action A that consistently led to a tone of certain pitch. However, before the execution of the action A, subjects were to carry out another action B, which resulted in a tone of either the same or a different pitch. The authors found that action B was faster when it resulted in the same effect as a concurrently prepared (but not yet executed) action A, suggesting that the anticipation of the effect associated with action A must play a role in the preparation of the action.

Comparing Sensorimotor and Ideomotor Action

Evidently, one of the most interesting questions in this context is how the control of stimulus-based actions on the one side and of goal-based actions on the other are related. A first step toward an answer of this question is to investigate how the two types of action are controlled in the brain.

The direct comparison of stimulus-based and goal-based actions is a challenge, because they differ in some central aspects. For example, stimulus-based actions are *preceded* by some stimulus event in the environment, whereas goal-based actions are *followed* by a change in the environment. Moreover, the timing is usually different. Whereas stimulus-based actions are triggered by the stimulus, goal-based actions are not.

To solve these problems, Waszak et al. (2005) devised a new paradigm allowing for the direct comparison of the functional signatures of goal-based and stimulus-based actions by examining both behavioral and electrophysiological measures. In this paradigm, subjects were asked to make left or right key presses at the midpoint between adjacent items in evenly timed sequences of visually presented pacing signals. In each block, subjects were presented with 35 stimuli (Xs presented to the left or right of fixation). The interstimulus interval (ISI) was constant at 1200 milliseconds. Subjects were to perform key presses that bisect the intervals between two consecutive stimuli. Thus, the sequence of stimuli was interwoven with a sequence of actions (SASASA). Each subject took part in two experimental conditions. Under the stimulus-based condition, subjects were instructed to react to the position of the previous stimulus with the appropriate (i.e., spatially compatible) key press. Under the goal-based condition, the subject's action (which key was pressed) determined the position of the subsequent stimulus. This paradigm allowed Waszak and colleagues to directly compare stimulus-based and goal-based actions, although the sensorimotor context of the actions, their kinematics and timing, as well as the entropy of the sequences is utterly the same for the two types of movement.

It is only in the stimulus-based condition that Waszak et al. (2005; *see also* Keller et al., 2006) observed S-ERPs (stimulus-locked event-related potentials) which reflect processes of stimulus-driven stimulus-response translation (and which probably take place in parietal and lateral premotor areas). This demonstrates that stimuli were processed right up until response preparation only, or to a larger degree, if the stimuli were relevant for the upcoming action. In the response-locked ERPs (i.e., event-related potentials averaged with respect to the action), Waszak et al. (2005) found a complementary pattern of activity—specifically, ERPs that indicate pronounced motor preparation in frontal motor areas before movements in the goal-based condition than in the stimulus-based condition. Intriguingly, the response-locked LRP (lateralized readiness potential; increased cortical activity contralateral to a forthcoming response) remained essentially invariant under the stimulus-based and goal-based conditions. The LRP is thought to reflect specific motor preparation (Gratton et al., 1988; De Jong et al., 1988). Hence, the specific programming of the actions did not differ between the two conditions. Waszak et al.'s results, thus, support theories suggesting that stimulus-based and goal-based actions are controlled by different neural structures that have a final common pathway (Passingham, 1997; *see also* Sherrington, 1906).

The behavioral results of studies by Waszak et al. (2005) and Keller et al. (2006) also point to functional differences in the control of stimulus-based and goal-based actions. Specifically, in both of these studies, temporal attraction effects in the timing of actions were observed: key presses, which generally preceded the true ISI (inter-stimulus interval) bisection point by about 80 milliseconds on average, occurred about 50 milliseconds earlier in the stimulus-based condition than in the goal-based condition. Thus, goal-based actions occurred relatively close in time to their anticipated effects, whereas stimulus-based actions occurred closer to their triggering stimuli. Waszak and colleagues considered these effects on the actual timing of actions to be a "behavioral analog" of the perceptual effects found by Haggard and colleagues (Haggard, Clark & Kalogeras, 2002; Haggard et al., 2002). Haggard et al. made subjects judge the time when they pressed a key or when a tone was presented by indicating the position of a clock hand (Libet's clock procedure, Libet et al., 1983). They found that perceptual onset times of actions and their ensuing effects, on the one hand, and of stimuli and subsequent actions in response to them, on the other hand, attracted each other in time. Haggard et al. suggested that action-effect binding effects (and perceptual attraction between the two events) occur whenever the system detects a match between an external event and a predicted action effect. The same may hold true for stimulus-response binding in that it takes place whenever an action is contingent on a prior sensory event.

It seems, thus, that stimulus-based and goal-based actions are controlled by different neural substrates and that the activity of these substrates goes along with two different types of perceptuomotor integration. As for goal-based actions, the action's motor code is bound to the ensuing effect of the action. As for stimulus-based actions, the action's motor code is bound to the triggering stimulus. Herwig, Prinz, and Waszak (2007) recently explored this notion. They reasoned, that, if the framework concerning the difference in perceptuomotor integration (stimulus–response vs. action-effect) between the two kinds of action holds true, action-effect integration as demonstrated by Elsner and Hommel (2001; *see* above) should occur only if the subjects operate in the goal-based "mode" of action.

Herwig et al. (2007) replicated the experiments by Elsner and Hommel (2001) with two different types of acquisition phase: The first type—the goal-based acquisition—corresponded to the one used by Elsner and Hommel. In this group, subjects freely selected between two possible actions. In the second group—the stimulus-based acquisition group—the actions were triggered by external stimulus events. However, in all groups, the actions were contingently followed by certain effect tones. Thus, the only difference between the groups was whether the actions were selected in a goal- or in a stimulus-based mode of movement. As in the study by Elsner and Hommel (2001), Herwig et al. then presented the same tones as target stimuli in response to which subjects had to perform a speeded choice response (test phase). As in the study from Elsner and Hommel, for half the subjects, the S-R mapping of the test phase corresponded to the action-effect mapping of the acquisition phase, for the other half it was reversed.

Herwig et al. (2007) found that in the goal-based acquisition group, subjects responding with "compatible" actions (i.e., actions that preceded the target tones in the acquisition phase) were faster than subjects responding with the "incompatible" actions, replicating the finding of Elsner and Hommel (2001). More importantly, in the stimulus-based acquisition group there was no compatibility effect whatsoever ($ß < 0.05$). In other words, action-effect (i.e., ideomotor) learning takes place in goal-based actions only or at least to a larger degree.

Herwig et al. (2007) took all these results to suggest that in a given stimulus-action-effect sequence, the codes of two events that belong to the same processing episode are bound together. In the goal-based mode of movement, action and ensuing effect are associated. In the stimulus-based mode, however, stimulus and ensuing action are linked. They suggest that these two types of perceptuomotor integration reflect the way the two modes work in. In the stimulus-based mode, the subjects pass on control to the stimuli in that the system merely acts upon presentation of a particular stimulus in a prespecified way (the prepared reflex outlined in the first part of this chapter; Exner, 1879; Hommel, 2000). In this mode, actions are selected with respect to their sensory antecedents. In the goal-based mode, by contrast, actions are guided by the ideomotor principle—that is, they are selected with respect to their sensory consequences.

We have reviewed experimental and theoretical approaches to stimulus-based and goal-based action control. Against this background, we address in the following the role of the self from another point of view, with respect to individual differences in self-regulatory processes.

Self-control has been conceived of as a master function that integrates the self's manifold processes and activities (e.g., Baumeister, 1998; Higgins, 1990). According to Baumeister (1998, 2002) the self is based on three principles. First, basic awareness of the self enables the individual to develop self-knowledge or self-esteem (*reflexive consciousness of the self*). Second, self-consciousness draws on social interaction with others (*self as an interpersonal being*). Third, the self has an executive function, which comprises both externally oriented acts, like choice and volition, and internally oriented processes of self-control, like alternating or overriding one's response tendencies, including thoughts, emotions, and actions (*executive function of the self*).

In the present context, the self's executive function is most important. For example, how and by which cognitive processes may the self control externally evoked (but rather unwanted) behavioral tendencies, like daydreaming instead of working, or drinking and eating too much? The last part of our chapter discusses this question based on cross-cultural and social-psychological research showing that a person's type of self-definition (derived from the social context) impacts self-regulatory abilities. This suggests individuals to differ in their general self-regulatory strength and even in their basic cognitive control processes, such as explored in Stroop interference or Task Switching paradigms outlined in the section on *Dual Route Logic*.

INDIVIDUAL DIFFERENCES IN ACTION CONTROL: THE ROLE OF THE SELF

During the last decade, social psychology has been increasingly stimulated by a notion that originates from cross-cultural psychology—that is the concepts of independent and interdependent self-construal (for review, *see* Cross, Bacon & Morris, 2000; Nisbett et al., 2001; Markus & Kitayama, 1991, 1998). This notion holds that depending on their cultural affiliation, individuals will define the self as an autonomous entity comprising a unique configuration of traits (European American cultural context; independent self-construal), or rather as an entity that incorporates elements of the social world, such as close relationships, important roles, and group memberships (Eastern, Southeast Asian cultural context; interdependent self-construal).

Cross-cultural psychologists have argued that self-control is more central to collectivistic cultures in order to maintain or create ingroup harmony, respectively, whereas individualistic cultures encounter a wider range of acceptable behaviors, which should result in less motivation to self-control and, therefore, less practice in self-control activities (for review, *see* Markus & Kitayama, 1991, 1998; Matsumoto, 1990, 1992, 2006; Matsumoto et al., 1998; Ekman, 1999). As various studies have shown, emotional experience (particularly intensity and expression of emotions) differs depending on the individual's cultural affiliation. For example, although in the United States the public expression of anger is accepted as a means to achieve personal rights, it is despised in Japan as a childish loss of self-control (e.g., Markus & Kitayama, 1991). It has been proposed that Asians are generally educated to hide their emotional state and control their facial expressions, particularly when emotional expression will threaten ingroup harmony (e.g., Yuki, Maddux & Masuda, 2007). Accordingly, to interpret the emotional state of others, they concentrate on facial regions that are hard to control—namely, the eyes—as cues to emotional states. In contrast, West-Europeans do not tend to cover their emotions, and in fact, quite overtly express their feelings. Correspondingly, they are focused on the mouth as the most expressive face region (Yuki et al., 2007). Furthermore, although members of individualistic cultures express feelings more openly, they presume that others are not feeling emotions as intensely as they overtly show them. In contrast, collectivistic persons more readily "read into" others' expressions and assume others to feel more intensely than they overtly allow themselves to show (Matsumoto et al., 2002; Matsumoto, Kasri, & Kooken, 1999).

Recently, a person's cultural belonging and the associated type of self-definition (i.e., independent vs. interdependent self-construal) have been demonstrated to affect a general ability for self-control. According to Baumeister and colleagues (Baumeister, Heatherton, & Tice, 1994; Baumeister et al., 1998; Baumeister, 2002) all efforts of self-control draw upon the same limited resource, which may lead to rapid depletion of the resource (*ego depletion*; cf. section *Dual Route Logic*). Although self-control abilities appear to be limited, they can be restored and improved such as by high motivation to self-control (Muraven, 1998) or practice at self-control (Muraven, Baumeister, & Tice, 1999). However, a fundamental need to be socially accepted has been proposed as the most basic driving force behind self-regulatory efforts, which, in fact, enable various benefits from others, like shared resources, help, or affection (e.g., Baumeister & Leary, 1995).

Accordingly, Seeley and Gardner (2003) assumed that Asians (interdependent self-construal) are more strongly motivated to engage in self-regulation in everyday social interaction, and therefore achieve, over time, greater self-regulatory strength relative to U.S. Americans (independent self-construal). Initially, the authors asked their research participants to speak into a tape recorder anything that came to their mind for 5 minutes. In the experimental condition, participants were instructed to think of a white bear, and then to suppress any thoughts of that bear (i.e., self-control requirement). In contrast, participants in the control condition were instructed to think about anything they wanted, including (but not limited to) a white bear (i.e., no self-control requirements) (Wegner et al., 1987). All participants had to knock on a desk each time they thought of a white bear during the 5-minute-session. Following this initial task, the participants were asked to squeeze a handgrip exerciser for as long as possible, drawing on self-regulatory strength for exercising hand force.

As predicted, U.S. American participants were found to replicate prior findings of self-control depletion—they showed less hand force in the suppression than in the control condition. On the other hand, Asian participants, associated with higher practice in self-control, failed to show effects of self-depletion—that is, they did not differ in the two conditions. Importantly, the authors found the same pattern of results when they classified the participants according to their chronic self-construal via Singelis' Self-Construal Scale (Singelis, 1994); or as high versus low in chronic social orientation by use of Snyder's (1974) Self-Monitoring Scale (Seeley & Gardner, 2003; Study 2). These findings support the notion that people with interdependent self-definition (e.g., collectivistic cultural background; high social need to belong) have greater self-regulatory abilities and less depletion after regulatory efforts relative to those with an independent view of the self (e.g., individualistic cultural background; lower social orientation). Thus, the results suggest that repeated exercise of self-control may strengthen self-regulatory abilities over time, and that social orientation may crucially encourage this exercise. An illustrative example may be acts of self-improvement like starting a diet or quitting smoking, which indeed are often motivated by the wish to please others, and often more successful when undertaken in a group than attempted alone (e.g., fitness partners meetings, anti-smoking seminars; McAuley et al., 2000).

Evidence for individual variations in self-regulatory performance has also been provided by research on gender differences. For example, men, more likely than women, have been shown to engage in aggressive behavior towards strangers (for review, *see* Baumeister & Sommer, 1997; Cross & Madson, 1997). These and other findings have been explained on the basis that women tend to an interdependent self-construal and seek connectedness with others, while men prefer independent self-definitions and being separate from others (Cross & Madson, 1997; Gabriel & Gardner, 1999). In turn, women more strongly control physically aggressive reactions that threaten their social relationships, whereas the fear of damaging social relationships will be less likely to stop men from being aggressive. As Cross and Madson (1997) argue, women further tend to frame aggression as a loss of self-control, whereas men see it as a means to control others who are threatening their superiority. Thus, aggressive acts appear to have more practical use for men (associated with independent self-construal) than for women (associated with interdependent self-construal).[1]

More recently, the cognitive mechanisms that may underlie differences between the independent and interdependent self have been addressed, although focusing on differences in action and cognition (Hannover & Kühnen, 2002, 2004; Hannover et al., 2005). For example, why are persons with interdependent (relative to independent) self-construal more likely to seek stronger physical closeness to an interaction partner (Holland et al., 2004), mimic another person's behavior (van Baaren et al., 2003), remember contextual information (Kühnen & Oyserman, 2002; Masuda & Nisbett, 2001) and have a more field-dependent perception (Chua, Boland & Nisbett, 2005; Ishii, Reyes, & Kitayama, 2003)?

As outlined in the next subsection, these and similar findings have been accounted for on the basis that persons with independent and interdependent self differ in their basic cognitive procedures for the control of own response tendencies (Hannover et al., 2005). In this context, the self is defined as a knowledge structure comprising both independent and interdependent aspects, which can differ in chronic but also in situational accessibility (e.g., Hannover, 1997; Roeder & Hannover, 2002). Accessibility describes the ease with which information can be retrieved from memory, and the fact that it is stronger the more recently (situational accessibility) and the more frequently (chronic accessibility) the information has been used (for review, see Bargh, 1997). In turn, the relative accessibility of both kinds of self-knowledge is determined by a person's cultural context, and can further be manipulated by situational cues (i.e., via semantic priming techniques). Irrespective of whether the chronic self is more independent or more interdependent, individuals will define their selves as more independent whenever the actual context triggers independent self-knowledge, but as more interdependent whenever it activates interdependent self-knowledge (e.g., Trafimow, Triandis, & Goto, 1991). More specifically, it is claimed that a person will interpret incoming information against the background of those aspects of knowledge—either independent or interdependent—that are highly accessible within a given situation. In other words, the particular content of accessible self-knowledge determines the semantic interpretation of incoming information (*semantic mechanism of the self*; Kühnen, Hannover, & Schubert, 2001).

Furthermore, it has been shown that whether people define their self as independent or interdependent affects the individual's thinking on the level of cognitive procedures (*procedural mechanism of the self*; Hannover et al., 2005; Kühnen et al., 2001). For example, Kühnen and colleagues (2001) used a priming paradigm by Gardner, Gabriel, and Lee (1999) to activate an independent or an interdependent view on the self. Their participants were asked to read a short paragraph about a trip to a city, and instructed to mark all pronouns in the text. In the independence priming condition these pronouns referred to the self as autonomous entity (e.g., I, me, mine), whereas in the interdependence priming condition, the pronouns were related to the self in relation to others (e.g., we, us, our). The authors found interdependence-primed participants to be more strongly field-dependent than independence-primed participants, as indicated by the Embedded Figures Test (Witkin, 1950; 1969), in which participants have to visually identify simple figures embedded in complex geometrical shapes. Using the same priming technique, Kühnen and Oyserman (2002) found interdependence-primed participants outperformed independence-primed participants in remembering the spatial locations of particular objects. Since the participants did not know in advance that they would have to memorize the objects' locations, the better performance of the interdependence-primed participants suggests that they had spontaneously applied a field-dependent type of encoding processes.

These and similar findings (for a review, see Hannover et al., 2005) support the notion that depending on which type of self-knowledge is accessible, individuals tend to process stimuli more or less strongly affected by the context in which these stimuli appear. In the context-independent mode of thinking, individuals will interpret incoming information—may it be about the self, about other people, or other stimuli—as if it is unaffected by the context in which it occurs. In the context-dependent processing mode, on the other hand, individuals interpret information in its respective context—that is, consider contextual influences more strongly (*context-independent* vs. *context-dependent processing mode*; Kühnen et al., 2001; Hannover & Kühnen, 2002, 2004).

In brief, it is argued that different cultural upbringing results in different types of self-construal (*independent* vs. *interdependent self*). Furthermore, a person's view of the self can be manipulated by the current context, and the mechanism by which the type of self-construal affects individual experience is two-fold. First, incoming information is assimilated towards the different contents of accessible independent or interdependent self-knowledge. Second, the activation of either one or the other type of self-knowledge coincides with the activation of different processing modes—that is, a context-independent or context-dependent mode of thinking. This interconnection between self-knowledge and processing mode is explained on the basis that independent and interdependent aspects differ not only in their contents but also in their mental structure. While independent aspects (i.e., individual traits) are stored independently of knowledge about other persons or social contexts, interdependent aspects (i.e., social characteristics) contain representations of the self in relation to others and social contexts (e.g., Hannover, 1997, 2002; Rhee et al., 1995). As a result, access to and retrieval of mainly independent aspects should foster context-independent processing, whereas access to mainly interdependent aspects should substantiate a context-dependent mode of processing.

However, both types of self-knowledge appear to induce not only a general style of thinking. As we will show in the following section, they have also been found to differentially influence the particular cognitive mechanisms applied for the control of one's responses, as suggested by studies in which Stroop interference or Task Switching paradigms are applied to participants for whom either independent or interdependent aspects of self-knowledge have been temporarily made accessible.

Impact of the Self on Action Control Mechanisms

In a recent series of studies, Springer (2005) and Hannover et al. (2005) proposed that context-independent and context-dependent modes of thinking can also be differentiated in terms of basic cognitive control processes. More specifically, the authors assumed that context-independent information processing bears strongly on selective attention and inhibition—that is, on selectively focusing on a particular (commonly response-relevant) stimulus and on actively suppressing (commonly response-irrelevant) information. These processes assure context-independent processing whenever a stimulus configuration automatically activates several behavioural response tendencies at once (cf. section, *Dual Route Logic*). Similarly, *not* selectively focusing on focal stimulus information and *not* inhibiting contextual information that is irrelevant to the intended response promotes context-dependent information processing. This rather *wide* attentive focus including focal as well as contextual cues should facilitate performance whenever contextual information is in fact *needed* for appropriate task responding (such as, for example, in the study by Kühnen & Oyserman [2002] outlined in the previous section). Furthermore, context-independence should facilitate switching between tasks, because task switching involves the configuration of the cognitive elements required to perform an upcoming task while inhibiting interfering influences from the previously performed one (e.g., Rogers & Monsell, 1995) as outlined in the section titles *Dual Route Logic*.

In sum, Springer and colleagues (Hannover et al., 2005; Springer, 2005) assumed that access to independent self-knowledge promotes three different functions of cognitive control that enable context-independent information processing—namely, selective attention to response-relevant information, inhibition of response-irrelevant information, and task switching. By the same token, access to interdependent self-knowledge should promote cognitive control functions that support context-dependent processing—that is, a broad focus of attention on both response-relevant and response-irrelevant (contextual) stimuli, and less inhibition of contextual stimuli.

To test the selective attention hypothesis, Springer (2005) used a Stroop-like reaction time task in which participants had to respond to a particular dimension of a stimulus (response-relevant information), while ignoring other response-irrelevant (contextual) aspects of that same stimulus triggering a conflicting response tendency (*see* Fig. 10–1a for an example of a Stroop stimulus). Interference effects should occur to the extent that the person does not selectively attend to the response-relevant stimulus aspect. Accordingly, individuals with access to interdependent (relative to independent) self-knowledge should produce stronger interference effects. Specifically, Springer asked her participants to decide whether a word presented on a computer screen appeared above or below a reference line ("yes" or "no" response). Whereas the contents of some of the words were semantically neutral as regards the spatial position (e.g., "also," "much"), others were congruent or incongruent with the word's spatial position (e.g., "deep," "atop"). The word's spatial position was the response-relevant dimension, while the word's semantic meaning represented the contextual dimension. In incongruent trials, the word's position and semantic meaning were conflicting, whereas congruent trials did not trigger conflicting response tendencies but the same type of response. Reaction times in incongruent and congruent trials, respectively, compared to neutral trials indicate interference and facilitation effects of the response-irrelevant, contextual stimulus dimension.

In order to activate either independent or interdependent self-knowledge before the experimental task, the author made use of the priming paradigm by Gardner et al. (1999) as outlined above. As expected, context-effects (i.e., significantly faster responses to congruent relative to neutral and incongruent stimuli) were found for interdependence-primed participants, whereas no such effects occurred for the group of independence-primed participants. More specifically, both types of self-knowledge appeared to differentially influence selective attention. Participants to whom independent self-knowledge was made accessible managed to more strongly focus their attention on the response-relevant information, while participants with highly accessible interdependent self-knowledge, engaged in context-dependent processing, (i.e., they attended to both the word's spatial position and the word's semantic meaning, resulting in RT effects of the semantic meaning in congruent trials when compared to neutral and incongruent trials).

To measure the inhibition of response-irrelevant information (as the second cognitive control function that should differ depending on the relative accessibility of self-knowledge), Springer (2005) used a negative priming paradigm (e.g., Tipper, 1985). Note that the function of focusing one's attention on response-relevant information is not identical to the active inhibition of contextual, response-irrelevant information. Whereas focusing prevents contextual information from being encoded, inhibition implies that the person actively suppresses a reactive tendency, such that its reactivation is more difficult when it is needed at a later moment. The prolonged reaction times associated with this reactivation process are referred to as negative priming effects (Neill, 1997; Tipper, 1985). Springer assumed persons with easy access to independent self-knowledge to strongly inhibit contextual information and thus produce strong negative priming effects. To test this hypothesis, the author again used the Stroop-like task described above. However, she included pairs of incongruent trials related such that the semantic stimulus information that was response-irrelevant and hence to be inhibited in trial n-1 (e.g., the word "below" presented *above* the reference line) became response-relevant in the subsequent trial n (a word presented *below* the reference line, such that the correct response was "below"). As expected, negative priming effects were observed after a priming of independent self-knowledge, but not after priming of interdependent self-knowledge. Thus, it appears that compared to interdependence primed participants, participants with high access to independent self-knowledge had a tendency to actively inhibit response-irrelevant, contextual information, as a result of which their reaction times were prolonged when the inhibited dimension became subsequently response-relevant.

In another study, Springer (2005) explored the task switching performance of interdependence and independence primed participants. She hypothesized that persons with easy access to independent as compared to interdependent self-knowledge would be more efficient in abandoning a completed task—that is, in efficiently controlling interfering response tendencies triggered by that recently activated task, and would thus switch more rapidly toward a new task. After being primed with either independent or interdependent self-knowledge, participants performed two RT tasks, switching task on every second trial (alternating runs paradigm; Rogers & Monsell, 1995) (cf. section *Dual Route Logic*). Springer measured switch costs—the RT difference between trials in which the same task repeated (e.g., RT in task A when the previous task was also A) and trials following a task switch (e.g., RT in task A when the previous task was B). As expected, independence primed participants were found to be substantially faster in switching between the tasks relative to interdependence primed participants.

Drawing on differences in the chronic accessibility of the two types of self-knowledge, the experimental findings reported above could be replicated for participants who had been classified as either independent or interdependent by means of Singelis' (1994) Self-Construal Scale in advance of the experimental sessions (Springer, 2005).

Together, the results support the notion that different kinds of self-knowledge promote differences in cognitive control functions that are commonly conceived of as universal. Specifically, access to independent self-knowledge promotes differences in cognitive functions that foster context-independent information processing (i.e., selective attention to response-relevant information) inhibition of response-irrelevant information, and task switching. In contrast, access to interdependent self-knowledge fosters a wide focus of attention, i.e. a tendency to process response-relevant and response-irrelevant information in parallel, resulting in a context-dependent processing mode. This is evidence for the executive function of the self—that is, internally oriented processes of self-control, such as alternating or overriding predominant response tendencies in order to achieve goal-directed outcomes (Baumeister, 1998, 2002). Moreover, it suggests that, depending on the type of self-construal, individuals differ in basic cognitive control processes. In fact, compared to those with independent self and low social orientation, people with interdependent self and high social orientation may engage more strongly in self-control activities, like working instead of daydreaming, or eating and drinking appropriately, since they are driven by high social sensitivity and a strong motivation to fit in with others (e.g., Matsumoto, 1990; 1992; 2006; Seeley & Gardner, 2003).

Conclusion

We have tried to show that internally generated mental states play a core role in the control of not only goal-based actions but also of stimulus-based actions. As for goal-based actions, the mental state is an internal anticipation of a desired effect which automatically activates the motor code necessary to produce this effect (ideomotor principle). Notice that the notion of intentions, as used in the present chapter, does by no means imply consciousness. Purposeful behavior can easily go without conscious experience; otherwise we would need to suspect the mouse purposely stealing the cheese from the kitchen table to enjoy conscious experience (but maybe we should). Similarly, Frith (1992) has argued that schizophrenic delusions reflect a tendency to attribute own voluntary actions to external factors, implying that one does not need to be conscious of the goals one is pursuing.

As for stimulus-based actions, on the other hand, the mental state can be conceived of as an internal anticipation of the possible stimulus conditions to which the agents want to react along with the corresponding motor response (prepared reflex logic). From this point of view, the basis of human action is always the agent's intention to interact with the environment, either by consigning control to the environment or by manipulating it (controlling it, so to say).

In the case of stimulus-based actions, the ability to anticipate possible stimulus conditions and to decide in advance how to react in response to the different possibilities goes to the expense of a certain proneness to interference from unintended S-R translation which are "automatically" triggered if the stimulus constellation is ambiguous with respect to the intentions. This notion is reminiscent of the flexibility-stability dilemma (Goschke, 2000): On the one hand, humans need to limit the ability of external stimuli to evoke actions irrelevant to current goals, but on the other hand, in face of the uncertainty of the environment, it is maladaptive to be too locked into one particular S-R condition.

In the third part of the chapter, we reviewed cross-cultural and social-psychological research suggesting that the psychological mechanisms underlying self-control are affected by the individual's (socially shaped) way of reflecting upon the self. People with interdependent self (i.e., high social orientation) may be better practiced in self-control activities, which are in fact indispensable for getting along with others and satisfying a high need to belong to the social group. Accordingly, those with interdependent self may build better self-regulatory abilities over time, relative to those with rather independent self-definition (i.e., lower social orientation; e.g., Matsumoto, 1990, 1992; Matsumoto et al., 1999, 2002; Seeley & Gardner, 2003).

Furthermore, both types of self-knowledge appear to impact the implicit cognitive processes for the control of own response tendencies (i.e., selective attention, inhibition, and task switching), which can be provoked by current external stimuli, like sweets, alcohol, or drugs, but are incompatible with one's long-term goals and desires, like losing weight or abstaining from drugs. This suggests individuals do differ in their cognitive control processes that allow modifying reactive tendencies, and in turn, controlling one's thoughts, emotions, and actions (*executive function of the self*; Baumeister, 1998). Moreover, it suggests the individual's view of the self to be a crucial mediator for differences in action control.

Note

1 Baumeister and Sommer (1997), for example, argue that both men and women have a comparable "need to belong," but differ in their social orientation. Whereas women are orientated toward close relationships, men orientate toward broader social structures to seek social connection, for example, by competing for a good position in a status hierarchy. In turn, men may engage more in physical aggression toward strangers because it seems more effective for them than for women. However, toward family members or relationship partners, women appear to physically aggress as much as men (e.g., Breslin, Riggs, O'Leary & Arias, 1990) and are less likely to feel guilty over marital aggression (Harris, 1994). Thus, gender differences in aggression are mainly accounted for by aggression toward strangers.

References

Allport, D. A. Patterns and actions: Cognitive mechanisms are content-specific. In: Claxton, G. (Ed.), *Cognitive psychology*. London: Routledge, 1980: pp. 26–63.

Allport, A. Selection for action: Some behavioral and neurophysiological considerations of attention and action. In: Heuer, H., & Sanders, A. F. (Eds.), *Perspectives on perception and action*. Hillsdale, NJ: Lawrence Erlbaum, 1987: pp. 395–419.

Allport, D. A., Styles, E. A., & Hsieh, S. Shifting intentional set: Exploring the dynamic control of tasks. In: Umiltà, C., & Moscovitch, M. (Eds.), *Attention and performance XV: Conscious and nonconscious information processing*. Cambridge, MA: MIT Press, Bradford Books, 1994: pp. 421–452.

Anderson, J. R., & Lebiere, C. *The atomic components of thought*. Mahwah, NJ: Lawrence Erlbaum Associates, 1998.

Bargh, J. A. The automaticity of everyday life. In: Wyer, R. S., Jr. (Ed.), *The automaticity of everyday life: Advances in social cognition*, Vol. 10. Mahwah, NJ: Erlbaum, 1997: pp. 1–61.

Bargh, J. A., & Gollwitzer, P. M. Environmental control of goal-directed action: Automatic and strategic contingencies between situations and behavior. In: Spaulding, W. D. (Ed.), *Integrative views of motivation, cognition, and emotion. Nebraska symposium on motivation*, Vol. 41. Lincoln, NE: University of Nebraska Press, 1994: pp. 71–124.

Bargh, J. A., Chen, M., & Burrows, L. The Automaticity of Social Behaviour: Direct effects of trait concept and stereotype activation on action. J Pers Soc Psychol 1996; 71: 230–244.

Baumeister, R. The self. In: Gilbert, D., Fiske, S., & Lindzey, G. (Eds.), *Handbook of social psychology*. New York, NY: Oxford University Press, 1998: pp. 680–740.

Baumeister, R. Ego depletion and self-control failure: An energy model of the self's executive function. Self Identity 2002; 1: 129–136.

Baumeister, R. F., Bratlavsky, E., Muraven, M., & Tice, D. M. Ego depletion: Is the active self a limited resource? J Pers Soc Psychol 1998; 74: 1252–1265.

Baumeister, R. F., Heatherton, T. F., & Tice, D. M. *Losing control: How and why people fail at self-regulation*. New York, NY: Academic Press, 1994.

Baumeister, R. F., & Leary, M. R. The need to belong: Desire for interpersonal attachments as a fundamental human motivation. Psychol Bull 1995; 117: 497–529.

Baumeister, R. F., & Sommer, K. L. What do men want? Gender differences and two spheres of belongingness: Comment on Cross and Madson. Psychol Bull 1997; 122: 83–84.

Beckers, T., De Houwer, J., & Eelen, P. Automatic integration of non-perceptual action effect features: The case of the associative affective Simon effect. Psychol Res 2002; 66: 166–173.

Besner, D. The myth of ballistic processing: Evidence from Stroop's paradigm. Psychon Bull Rev 2001; 8(2): 324–330.

Besner, D., & Stoltz, J. A. Unconsciously controlled processing: The Stroop effect reconsidered. Psychon Bull Rev 1999; 6: 449–455.

Besner, D., Stoltz, J. A., & Boutilier, C. The Stroop effect and the myth of automaticity. Psychon Bull Rev 1997; 4: 221–225.

Brandstätter, V., Lengfelder, A., & Gollwitzer, P. M. Implementation intentions and efficient action initiation. J Pers Soc Psychol 2001; 81: 946–960.

Breslin, F. C., Riggs, D. S., O'Leary, K. D., & Arias, I. Family precursors: Expected and actual consequences of dating aggression. J Interpers Violence 1990; 5: 247–258.

Chua, H. F., Boland, J. E., & Nisbett, R. E. Cultural variation in eye movements during scene perception. Proc Natl Acad Sci USA 2005; 102: 12,629–12,633.

Cohen, A, & Shoup, R. Perceptual dimensional constraints on response selection processes. Cogn Psychol 1997; 32: 128–181.

Coles, M. G. H. Modern mind-brain reading: Psychophysiology, physiology, and cognition. Psychophysiology 1989; 26: 251–269.

Craft, J. L., & Simon, J. R. Processing symbolic information from a visual display: Interference from an irrelevant directional cue. J Exp Psychol 1970; 83: 415–420.

Cross, S. E., Bacon, P. L. & Morris, M. L. The relational-interdependent self-construal and relationships. J Pers Soc Psychol 2000; 66: 268–275.

Cross, S. E., & Madson, L. Models of the self: Self-construals and gender. Psychol Bull 1997; 122: 5–37.

Gabriel, S. & Gardner, W. Are there "his" and "her" types of interdependence? Gender differences in collective versus relational interdependence. J Pers Soc Psychol 1999; 77: 642–655.

De Jong, R., Wierda, M., Mulder, G., & Mulder, L. J. M. Use of partial stimulus information in response processing. J Exp Psychol Hum Percept Perform 1988; 14: 682–692.

Dijksterhuis, A. & van Knippenberg, A. Automatic social behavior or how to win a game of trivial pursuit. J Pers Soc Psychol 1998; 74: 865–877

Donders, F. C. On the speed of mental processes. Translated by W. G. Koster, 1969. Acta Psychol 1868; 30: 412–431.

Duckworth, A. L., & Seligman, M. E. P. Self-discipline outdoes IQ in predicting academic performance of adolescents. Psychol Sci 2005; 16: 939–944.

Ekman, P. Basic Emotions. In: Dalgleish, T. & Power, M. (Eds.), *Handbook Cogn Emotion*. New York, NY: John Wiley, 1999: pp. 301–320.

Ellis, J. Prospective memory or the realization of delayed intentions: A conceptual framework for research. In: Brandimonte, M., Einstein, G. O., & McDaniel, M. A. (Eds.), *Prospective memory: Theory and applications*. Mahwah,

Elsner, B., & Hommel, B. Effect anticipation and action control. J Exp Psychol Hum Percept Perform 2001; 27(1): 229–240.

Eriksen, C. W., & Hoffman, J. E. The extent of processing of noise elements during selective encoding from visual displays. Percept Psychophys 1973; 14: 155–160.

Eriksen, B. A., & Eriksen, C. W. Effects of noise letters upon the identification of a target letter in a nonsearch task. Percept Psychophys 1974; 16: 143–149.

Exner, S. Physiologie der Grosshirnrinde. In: Hermann, L. (Ed.), Handbuch der Physiologie, Vol. 2. Leipzig, Germany: Vogel, 1879: pp. 189–350.

Frith, C. D. *The cognitive neuropsychology of schizophrenia.* Hove: Erlbaum, 1992.

Galton, F. On instruments for (1) testing perception of differences of tint and for (2) determining reaction time. J Anthropol Inst 1899; 19: 27–29.

Gardner, W., Gabriel, S., & Lee, A. "I" value freedom but "we" value relationships: Self-construal priming mirrors cultural differences in judgment. Psychol Sci 1999; 10: 321–326.

Gollwitzer, P. M. Implementation intentions: Strong effects of simple plans. Am Psychol 1999; 54: 493–503.

Goschke, T. Intentional reconfiguration and involuntary persistence in task-set switching. In: Monsell, S., & Driver, J. (Eds.), *Control of cognitive processes: Attention and performance* (Vol XVIII). Cambridge, MA: MIT Press, 2000: pp. 331–355.

Gratton, G., Coles, M. G. H., Sirevaag, E. J., Eriksen, C. W., & Donchin, E. Pre- and post-stimulus activation of response channels: A psychophysiological analysis. J Exp Psychol Hum Percept Perform 1988; 14: 331–344.

Grice, G. R., Boroughs, J. M., & Canham, L. Temporal dynamics of associative interference and facilitation produced by visual context. Percept Psychophysics 1984; 36: 499–507.

Haggard, P., Clark, S., & Kalogeras, J. Voluntary action and conscious awareness. Nat Neurosci 2002; 5: 382–385.

Haggard, P., Aschersleben, G., Gehrke, J. & Prinz, W. Action, binding and awareness. In: Prinz, W.& Hommel, B. (Eds.), *Common mechanisms in perception and action: Attention and Performance,* Vol. XIX Oxford: Oxford University Press, 2002: pp. 266–285.

Hannover, B. Das dynamische Selbst. Zur Kontextabhängigkeit selbstbezogenen Wissens [The dynamic self. The context-dependency of self-related knowledge.]. Bern, Switzerland: Huber, 1997.

Hannover, B. One man's poison ivy is another man's spinach: What self-clarity is in independent self-construal, a lack of context-dependency is in interdependent self-construal. Revue Internationale De Psychologie Sociale/Internat Rev Soc Psychol 2002; 15: 65–88.

Hannover, B., & Kühnen, U. Der Einfluss independenter und interdependenter Selbstkonstruktionen auf die Informationsverarbeitung im sozialen Kontext [The influence of independent and interdependent self-construals on information processing in the social context]. Psychol Rundsch 2002; 53: 61–76.

Hannover, B., & Kühnen, U. Culture, context, and cognition: The Semantic Procedural Interface model of the self. Eur Rev Soc Psychol 2004; 15: 297–333.

Hannover, B., Pöhlmann, C., Springer, A., & Roeder, U. Implications of independent versus interdependent self-knowledge for motivated social cognition: The Semantic Procedural Interface Model of the self. Self Identity 2005; 4: 159–175.

Hannover, B., Pöhlmann, C., Roeder, U., Springer, A., & Kühnen, U. Eine erweiterte Version des Semantisch-Prozeduralen Interface-Modells des Selbst: Funktion des mentalen Interface und Implikationen des Modells für motivierte Prozesse [An Extended Version of the Semantic Procedural Interface Model of the Self: A Closer Look at the Mental Interface and the Model's Implications for Motivated Social Cognition]. Psychol Rundsch 2005; 56: 99–112.

Harleß, E. Der Apparat des Willens. Zeitschrift fuer Philosophie und philosophische Kritik 1861; 38: 50–73.

Harris, M. B. Gender of subject and target as mediators of aggression. J Appl Soc Psychol 1994; 24: 453–471.

Heckhausen, H., & Beckmann, J. Intentional action and action slips. Psychol Rev 1990; 97: 36–48.

Herwig, A., Prinz, W., & Waszak, F. (2007). Two modes of sensorimotor integration in intention-based and stimulus-based action. Q J Exp Psychol A 2007; 60(11): 1540–1554.

Hick, W. E. On the rate of gain of information. Q J Exp Psychol 1952; 4: 11–26.

Higgins, E. T. Personality, social psychology, and person-situation relations: Standards and knowledge activation as a common language. In: Pervin, L. A. (Ed.), *Handbook of personality: Theory and research.* New York: Guilford, 1990: pp. 301–338.

Holland, R., Roeder, U., van-Baaren, R., Brandt, A., & Hannover, B. Don't stand so close to me: The effects of self-construal on interpersonal closeness. Psychol Sci 2004; 15: 237–242.

Hommel, B. Inverting the Simon effect by intention. Psychol Res 1993; 55: 270–279.

Hommel, B., & Prinz, W. *Theoretical issues in stimulus-response compatibility.* Amsterdam: North Holland: Elsevier, 1997.

Hommel, B. Event files: Evidence for automatic integration of stimulus–response episodes. Visual Cognition 1998; 5: 183–216.

Hommel, B. The prepared reflex: Automaticity and control in stimulus-response translation. In: Monsell, S., & Driver, J. (Eds.), *Control of cognitive processes: Attention and performance XVIII.* Cambridge, MA: MIT Press. 2000a: pp. 247–273.

Hommel, B. Intentional control of automatic stimulus-response translation. In: Rossetti, Y., & Revonsuo, A. (Eds.), *Interaction between dissociable conscious and nonconscious processes.* Amsterdam: John Benjamins Publishing Company, 2000b: pp. 223–244.

Hommel, B. Acquisition and control of voluntary action. In: Maasen, S., Prinz, W., & Roth, G. (Eds.), *Voluntary action: Brains, minds, and sociality.* Oxford: Oxford University Press, 2003: pp. 34–48.

Hommel, B., Müsseler, J., Aschersleben, G., & Prinz, W. The theory of event coding (TEC). A framework for perception and action planning. Behav Brain Sci 2001; 24: 849–937.

Hommel, B., Pösse, B., & Waszak, F. Contextualization in perception and action. Psychol Belg 2000; 40: 227–245.

Inzlicht, M., & Gutsell, J. N. Running on empty: Neural signals for self-control failure. Psychol Sci 2007; 18(11): 933–937.

Ishii, K., Reyes, J. A., & Kitayama, S. Spontaneous attention to word content versus emotional tone: Differences among three cultures. Psychol Sci 2003; 14(1): 39–46

Jahanshahi, M., & Frith, C. D. Willed action and its impairments. Cogn Neuropsychol 1998; 15(6–8): 483–533.

Keller, P., Wascher, E., Prinz, W., Waszak, F., Koch, I., & Rosenbaum, D. Differences between intention-based and stimulus-based actions. J Psychophysiol 2006; 20(1): 9–20

Kornblum, S., Hasbroucq, T., & Osman, A. Dimensional overlap: Cognitive basis for stimulus-response compatibility—a model and taxonomy. Psychol Rev 1990; 97: 253–270.

Kühnen, U., Hannover, B., & Schubert, B. A Semantic-Procedural Interface Model of the Self: The role of self-knowledge for context-dependent versus context-independent modes of thinking. J Pers Soc Psychol 2001; 80: 397–409.

Kühnen, U., & Oyserman, D. Thinking about the self influences thinking in general: Procedural consequences of self-construal activation. J Exp Soc Psychol 2002; 38: 492–499.

Kunde, W., Koch, I., & Hoffmann, J. Anticipated action effects affect the selection, initiation, and execution of actions. Q J Exp Psychol A 2004; 57(1): 87–106.

Kunde, W., Hoffmann, J., & Zellmann, P. The impact of anticipated action effects on action planning. Acta Psychol 2002; 109: 137–155.

Libet, B., Gleason, C. A., Wright, E. W., & Pearl, D. K. Time of conscious intention to act in relation to onset of cerebral activity (readiness potential). The unconsciuosly initiation of a freely voluntary act. Brain 1983; 106: 623–642.

Logan, G. D. Toward an instance theory of automatization. Psychol Rev 1988; 95: 492–527.

Lu, C. H., & Proctor, R. W. The influence of irrelevant location information on performance: A review of the Simon and spatial Stroop effects. Psychon Bull Rev 1995; 2: 174–207.

Lotze, R. H. *Medicinische Psychologie oder die Physiologie der Seele.* Leipzig: Weidmann'sche Buchhandlung, 1852.

MacLeod, C. M. Half a century of research on the Stroop effect: An integrative review. Psychol Bull 1991; 109: 163–203.

Markus, H., & Kitayama, S. Culture and the self: Implications for cognition, emotion, and motivation. Psychol Rev 1991; 98: 224–253.

Markus, H., & Kitayama, S. The cultural psychology of personality. J Cross Cult Psychol 1998; 29(1): 63–87.

Masuda, T., & Nisbett, R. E. Attending holistically versus analytically: Comparing the context sensitivity of Japanese and Americans. J Pers Soc Psychol 2001; 81: 922–934.

Matsumoto, D. Cultural similarities and differences in display rules. Motiv Emot 1990; 14: 195–214.

Matsumoto, D. American-Japanese cultural differences in the recognition of universal facial expressions. J Cross Cult Psychol 1992; 23: 72–84.

Matsumoto, D. Are cultural differences in emotion regulation mediated by personality traits? J Cross Cult Psychol 2006; 37: 421–437.

Matsumoto, D., Takeuchi, S., Andayani, S., Kouznetsova, N., & Krupp, D. The contribution of individualism vs. collectivism to cross-national differences in display rules. Asian J Soc Psychol 1998; 1: 147–165.

Matsumoto, D., Kasri, F., & Kooken, K. American-Japanese cultural differences in judgments of expression intensity and subjective experience. Cogn Emot 1999; 13: 201–218.

Matsumoto, D., Consolacion, T., Yamada, H., Suzuki, R., Franklin, B., Paul, S., Ray, R., & Uchida, H. American-Japanese cultural differences in judgments of emotional expressions of different intensities. Cogn Emot 2002; 16(6): 721–747.

McAuley, E., Blissmer, B., Katula, J., & Duncan, T. E. Exercise environment, self-efficacy, and affective responses to acute exercise in older adults. Psychol Health 2000; 15: 341–355.

Meacham, J. A., & Leiman, B. Remembering to perform future actions. In: Neisser, U. (Ed.), *Memory observed: Remembering in natural contexts*. San Francisco, CA: W. H. Freeman and Company, 1982: pp. 327–336.

Meyer, D. E., & Kieras, D. E. A computational theory of executive cognitive processes and multiple-task performance: Part 1. Basic Mechanisms. Psychol Rev 1997; 104: 3–65.

Miller, J. The flanker compatibility effect as a function of visual angle, attentional focus, visual transients, and perceptual load: A search for boundary conditions. Percept Psychophys 1991; 49: 270–288.

Muraven, M. Mechanisms of self-control failure: Motivation and limited resources. Disseration Abstracts International, 59(5-B), (UMI No. 95022254). 1998.

Muraven, M., & Baumeister, R. F. Self-regulation and depletion of limited resources. Does self-control resemble a muscle? Psychol Bull 2000; 126: 247–259.

Muraven, M., Baumeister, R. F., & Tice, D. M. Longitudinal improvement of self-regulation through practice: Building self-control strength through repeated exercise. J Soc Psychol 1999; 139: 446–457.

Muraven, M., Tice, D. M., & Baumeister, R. F. Self-control as a limited resource: Regulatory depletion patterns. J Pers Soc Psychol 1998; 74: 774–789.

Neill, W. T. Episodic retrieval in negative priming and repetition priming. J Exp Psychol Learn Mem Cogn 1997; 23: 1291–3105.

Nieuwenhuis, S., Ridderinkhof, K. R., Blom, J., Band, G. P. H., & Kok, A. Error-related brain potentials are differentially related to awareness of response errors: Evidence from an antisaccade task. Psychophysiology 2001; 38: 752–760.

Nisbett, R. E., Peng, K., Choi, I., & Norenzayan, A. Culture and systems of thought. Psychol Rev 2001; 108: 291–310.

Passingham, R. E. Functional organisation of the motor system. In: Frackowiak, R. S. J., Friston, K. J., Frith, C. D., Dolan, R. J., & Mazziotta, J. C. (Eds.), *Human brain function*. San Diego, CA: Academic Press, 1997: pp. 243–274.

Piéron, H. Nouvelles recherches sur l'analyse du temps de latence sensorielle et sur la loi qui relie ce temps a l'intensité de l'excitation. Annee Psychol 1920; 22: 58–142.

Prinz, W. Perception and action planning. Eur J Cogn Psychol 1997; 9: 129–154.

Prinz, W., & Hommel, B. (Eds.). *Common mechanisms in perception and action: Attention and performance XIX*. Oxford: Oxford University Press, 2002.

Reason, J. T. *Human error*. New York: Cambridge University Press, 1990.

Rhee, E., Uleman, J. S., Lee, H. K., & Roman, R. J. Spontaneous self-descriptions and ethnic identities in individualistic and collectivistic cultures. J Pers Soc Psychol 1995; 69: 142–152.

Risko, E. F., Stolz, J. A., & Besner, D. Basic processes in reading: Is visual word recognition obligatory? Psychon Bull Rev 2005; 12(1): 119–124.

Roeder, U., & Hannover, B. Kontextabhängigkeit als Dimension der Selbstkonstruktion: Entwicklung und Validierung der Dortmunder Kontextabhängigkeits-Skala (DKS) [The measurement of context-dependency of self-construals]. Zeitschrift für Differentielle und Diagnostische Psychologie 2002; 23: 339–352.

Rogers, R. D., & Monsell, S. Costs of a Predictable switch between Simple Cognitive Tasks. J Exp Psychol Gen 1995; 124, 2: 207-231.

Rosenbaum, D. A. The Cinderella of psychology—the neglect of motor control in the science of mental life and behavior. Am Psychol 2005; 60: 308-317.

Sanders, A. F. *Elements of human performance: Reaction processes and attention in human skill*. Mahwah, New Jersey: Lawrence Erlbaum Associates, Publishers, 1998: 575 pages.

Seeley, E. A., & Gardner, W. L. The "Selfless" and Self-Regulation: The role of chronic other-orientation in averting self-regulatory depletion. Self Identity 2003; 2: 103-117.

Sherrington, C. S. *Integrative action of the nervous system*. New Haven, CT: Yale University Press, 1906.

Shiffrin, R. M., & Schneider, W. Controlled and automatic human information-processing: II. Perceptual learning, automatic attending, and a general theory. Psychol Rev 1977; 84: 127-190.

Simon, J. R., & Rudell, A. P. Auditory S-R compatibility: The effect of an irrelevant cue on information processing. J Appl Psychol 1967; 51: 300-304.

Singelis, T. M. The measurement of independent and interdependent self-construals. Pers Soc Psychol Bull 1994; 20: 580-591.

Snyder, M. Self-monitoring of expressive behavior. J Pers Soc Psychol 1974; 30: 562-537.

Spelke, E., Hirst, W., & Neisser, U. Skills of divided attention. Cognition 1976; 4: 215-230.

Springer, A. Wie das Selbst das Denken steuert. Der Einfluss independenten und interdependenten Selbstwissens auf exekutive Funktionen zur Steuerung der Informationsverarbeitung [How the self affects thinking. The influence of independent and interdependent self-knowledge on executive functions for the cognitive control of information processing]. Hamburg, Germany: Kovac, 2005.

Sternberg, S. The discovery of processing stages: Extensions of Donders' method. Acta Psychologica 1969; 30: 276-315.

Stroop, J. R. Studies of interference in serial verbal reactions. J Exp Psychol 1935; 28: 643-662.

Sudevan, P., & Taylor, D. A. The cuing and priming of cognitive operations. J Exp Psychol Hum Percept Perform 1987; 13: 89-103.

Tangney, J. P., Baumeister, R. F., & Boone, A. L. High self-control predicts good adjustment, less pathology, better grades, and interpersonal success. J Personal 2004; 72: 271-322.

Tipper, S. P. The negative priming effects: Inhibitory effects of ignored primes. Q J Exp Psychol 1985; 37A: 571-590.

Trafimow, D., Triandis, H. C., & Goto, S. G. Some tests of the distinction between the private self and the collective self. J Pers Soc Psychol 1991; 60: 649-655.

Umiltà, C., & Nicoletti, R. Spatial stimulus-response compatibility. In: Proctor, R. W., & Reeve, T. G. (Eds.), *Stimulus-response compatibility: An integrated perspective. Advances in psychology*, Vol. 65. Amsterdam, The Netherlands: North-Holland, 1990: pp. 89-116.

van Baaren, R. B., Maddux, W. W., Chartrand, T. L., Bouter, de C., & van Knippenberg, A. It takes two to mimic: Behavioral consequences of self-construals. J Pers Soc Psychol 2003; 84: 1093-1102.

van Veen, V., & Carter, C. S. The anterior cingulate as a conflict monitor: fMRI and ERP studies. Physiol Behav 2002; 77: 477-482.

Wallace, R. J. S-R Compatibility and idea of a response code. J Exp Psychol 1971; 88: 354-363.

Waszak, F., & Hommel, B. The costs and benefits of cross-task priming. Mem Cogn 2007; 35(5): 1175-1186.

Waszak, F., Hommel, B., & Allport, A. Task-switching and long-term priming: Role of episodic stimulus-task bindings in task-shift costs. Cogn Psychol 2003; 46(4): 361-413.

Waszak, F., Hommel, B., & Allport, A. Semantic generalization of stimulus-task bindings. Psychon Bull Rev 2004; 11(6): 1027-1033.

Waszak, F., Hommel, B., & Allport, D. A. Interaction of task readiness and automatic retrieval in task switching: Negative priming and competitor priming. Mem Cogn 2005; 33(4): 595-610.

Waszak, F., Wenke, D., & Brass, M. Cross-talk of instructed and applied arbitrary visuomotor mappings. Acta Psychologica 2008; 127: 30-35.

Waszak, F., Wascher, E., Keller, P., Koch, I., Aschersleben, G., & Rosenbaum, D. A. Intention-based and stimulus-based mechanisms in action selection. Exp Brain Res 2005; 162: 346-356.

Wegner, D. M., Schneider, D. J., Carter, S. R., & White, T. L. Paradoxical effects of thought suppression. J Pers Soc Psychol 1987; 55: 5-13.

Wenke, D., & Frensch, P. A. The influence of task instruction on action coding: Constraint setting or direct coding? J Exp Psychol Hum Percept Perform 2005; 31: 803-819.

Wenke, D., Gaschler, R., & Nattkemper, D. Instruction-induced feature binding. Psychol Res/Psychol Forsch 2007; 71: 92–106.

Welford, A. T. Choice reaction time: Basic concepts. In: Welford, A. T. (Ed.), *Reaction times*. New York, NY: Academic Press, 1980: pp. 73–128.

Wise, S. P., & Murray, E. A. Arbitrary associations between antecedents and actions. Trends Neurosci 2000; 23: 271–276.

Witkin, H. A. Individual differences in the case of perception of embedded figures. J Personal 1950; 19: 1–15.

Witkin, H. A. *Embedded figures test*. Palo Alto, CA: Consulting Psychologists Press, 1969.

Woodworth, R. S. *Experimental psychology*. New York, NY: Holt, Rinehart and Winston, 1938.

Wylie, G., & Allport, A. Task switching and the measurement of "switch costs." Psychol Res 2000; 63: 212–233.

Yuki, M., Maddux, W. W., & Masuda, T. Are the windows to the soul the same in the East and West? Cultural differences in using the eyes and mouth as cues to recognize emotions in Japan and the United States. J Exp Soc Psychol 2007; 43: 303–311.

Ziessler, M. Response-effect learning as a major component of implicit serial learning. J Exp Psychol Learn Mem Cogn 1998; 24: 962–978.

CHAPTER 11

Task Switching: Mechanisms Underlying Rigid vs. Flexible Self-Control

Nachshon Meiran

ABSTRACT

This chapter reviews the historical and current literature on task switching, focusing primarily on cognitive-behavioral studies on healthy human subjects. It outlines what I see to be widely accepted conclusions. These include the notion that tasks have mental representations ("task sets") and that a change in this representation results in slowing (although the exact reasons for the slowing are debated). Following Ach (2006/1910), the chapter divides the processes that are currently mentioned in the literature into those making an inner obstacle against a task switch (thus causing rigidity) and those that enable a task switch (thus supporting flexibility). It also discusses some major controversies in the field and suggest that many of these controversies are more apparent than real by pointing out the many issues where a broad consensus exists.

Keywords: Task switching, flexibility, literature review

Typical examples of situations involving self-control are characterized by their "don't" feature, such as being offered a bowl of delicious ice cream although you are on a diet (e.g., Vohs & Heatherton, 2000). Nonetheless, "do" situations also involve self-control. An example is your boss entering your office asking you to do something although you are deeply immersed in working on your favorite project. Here, self-control is needed both to interrupt the ongoing activity and to initiate the alternative activity. Research shows that complying with "do" commands is more difficult (Logan & Burkel, 1986) and is developmentally delayed (Kochanska et al., 2001) as compared with "don't" commands. One reason for this may be that self-control (also termed "cognitive control" and "executive functioning," *see* Miller & Cohen, 2001; Wood & Grafman, 2003) comprises several abilities. This is evident in low interindividual correlations between various control functions (e.g., Miyake et al., 2000) and somewhat differential functional anatomy (e.g., Bunge, 2004; Collette & Van der Linden, 2002).

At least five relatively independent domains of executive functioning have been identified, including working memory (WM) updating (e.g., Kessler & Meiran, 2006; Morris & Jones, 1990), behavioral inhibition (Logan, 1994; Friedman & Miyake, 2004), online performance monitoring (e.g., Botvinick et al., 2001), multitask coordination (e.g., Logan & Gordon, 2001; Meyer & Kieras, 1997a, 1997b; Pashler, 1994), and mental set shifting. Emotion regulation (Ochsner & Gross, 2005) may represent a sixth dimension in this taxonomy. The present chapter addresses the most widely used approach to the study of mental set shifting (or cognitive flexibility), which is the task switching paradigm (Logan 2003; Monsell, 2003, for recent reviews). Given the recent explosion of knowledge on this topic, the present review is inherently limited and it focuses on cognitive behavioral studies performed on healthy young human adults.

It is now widely claimed that self-control has developed in the Homo Sapiens to match the cognitive challenges of the complex social interactions that characterize our species (e.g., Adolphs, 2003; Barkley, 2001; Baumeister, 2005). Although the role of inhibition in social interaction has been widely acknowledged (e.g., Barkley, 2001), the role of cognitive flexibility in the social domain has been less explored. Nonetheless, theoretical analysis suggests that cognitive flexibility features a key role in social interactions. First, Baumeister noted that a crucial feature of Homo Sapiens is collaborative interaction involving division of labor and collaborative problem solving. To negotiate the division of labor while working on the solution, individuals must understand the intentions of their collaborators as well as the collaborator's potential contribution to the solution. In addition, there is a need for flexible shifts of attention between perspectives (one's own perspective and the collaborator's perspective), between focusing on the problem and focusing on understanding the collaborator's approach, and so forth. Supporting evidence comes from a study by Bonino and Cattelino (1990). They showed that children who were characterized as flexible according to their performance on the Wisconsin Card Sorting Test (arguably, the predecessor of the modern task-switching paradigm) were better able to negotiate collaborative problem solving than their less flexible counterparts. Second, the complex division of labor that characterizes modern social structures further demands individuals to constantly change their perspective as they change their social roles and their relative position in the social hierarchy. Third, this complex structure of labor division puts individuals in a position in which multiple long-term tasks are carried out in parallel, such as parenting, earning an income, and working toward an academic degree. These ongoing tasks are broken down into manageable subtasks and these subtasks become constantly intertwined. As a result, everyday life is often characterized by frequent switches between the subtasks, such as switching between answering a phone call from your child to continuing work on your Ph.D. dissertation. Finally, collaborative interaction often involves emotion regulation to maintain the socially appropriate stance. Furthermore, some emotion regulation strategies, such as reappraising a situation, involve a shift in mental perspective, and as such are functionally analogous to task switching (Shepps & Meiran, 2007). It should be noted that generating collaboration often involves reappraising the situation and accepting the collaborator's perspective.

The Task Switching Paradigm

Most of the current chapter is devoted to a review of the cognitive literature on task switching. This research has focused almost exclusively on the physical domain. It has helped to reveal the cognitive and neural mechanisms supporting rigidity and flexibility. I hope that the present review will promote the much needed work of understanding the role of human cognitive rigidity and flexibility in the social domain.

Historic Foundations

This section mentions only four papers, which I view as cornerstones in this literature. Ach (2006/1910) noted that the will's duty is to overcome inner obstacles that prevent it from achieving its goals. According to him, these inner obstacles are mostly habits. He used a paradigm that is surprisingly similar to modern task-switching paradigms. His paradigm consisted of two phases. In Phase 1, subjects learned to perform a given task on a set of stimuli. In Phase 2, subjects were required to perform another task on the same stimuli, and reaction time (RT) was measured to the nearest millisecond (!). Ach showed that the task switch resulted in considerable slowing in Phase 2 reactions, a phenomenon that today we call "switch cost." He interpreted the switch cost as reflecting the effort of the will in overcoming the habit formed during Phase 1. Ach also measured the number of Phase 1 practice repetitions required to form a habit that was sufficiently strong to overcome the will during Phase 2. Overcoming the will was reflected in the erroneous execution of the task practiced in Phase 1 during Phase 2. This measure was taken as an index of what he called the "*associative equivalent* of the will," which is an indirect index of will power. Note that many current works view switch cost as an index for lack of control. In contrast, Ach's view of switch cost is of a measure of will *effort*, not of will *power*. Peculiarly, his measure of will power has not to my knowledge been applied in later studies.

Jersild (1927) is often cited as the inventor of the modern task-switching paradigm (although what is the first study to my knowledge was conducted by Jones, 1915, cited by Bernstein, 1924). He asked his subjects to alternate between two familiar tasks that were performed on the same set of stimuli, such as adding a digit and subtracting another digit. (e.g., switching between a "+3" operation and a "−1" operation.). Note that Ach's (2006/1910) and Jersid's paradigms have in common the feature that the execution of one task creates a tendency (or habit) to perform Task A that forms the "inner obstacle" while performing Task B. Jersild's experiments resulted in three important findings. First, switch costs were robust when the stimuli were *bivalent* (affording two tasks, e.g., a digit when the tasks are "+1" and "−1"). Second, there were no switch costs when the stimuli were *univalent* (uniquely affording one task, e.g., words requiring an "opposite" task, e.g., white → black, and digits requiring a numerical task such as "+3"). The first two findings show that the habit to perform a given task tends to bind with the stimuli (Waszak et al., 2003). Finally, the costs were markedly reduced when there was an external reminder regarding which task was required (the stimuli for one task were on the right and those for the other task were on the left). The last effect shows that switch costs are partly determined by task-goal (un)certainty. All these effects were replicated in later studies.

Biederman (1972) provided crucial evidence that subjects can (at least partially) ignore information that was deemed irrelevant by the current task instructions. For example, if instructions indicate that color is relevant and size is irrelevant, subjects can partly ignore the size information. Shaffer (1965) developed the task cuing paradigm in which the tasks change randomly and each trial begins with a task cue telling which task to execute. Shaffer was also the first researcher who compared three conditions in the same experiment. These include a condition with a single task, and two conditions from the blocks that involved task switching: *task-switch trials*, in which the task has just switched, and *task-repetition trials*, in which the current task was the same as the previously executed task. This design allows one to separate two costs associated with switching tasks, later discussed by Braver et al. (2003); Fagot (1994), Kray and Lindenberger (2000), Koch et al. (2005), Los (1996), Meiran et al. (2000) and Rubin and Meiran (2005), among others. The first is "switch cost" and it refers to the difference between task-switch and task-repetition trials. The second is "mixing cost" and it refers to the cost associated with being in a situation involving potential switching. Mixing cost is defined as the performance difference between task-repetition trials (taken from blocks in which the tasks switch) and single-task blocks.

Task Switching Paradigms

Although researchers often refer to *the* task switching paradigm, there are, in fact, many different paradigms, and it still remains to be shown whether these paradigms tap the exact same abilities. This issue is particularly important for researchers who wish to incorporate task-switching into their studies as measures of cognitive flexibility. Based on the following sections, my advice is to use, if possible, more than one task-switching paradigm to ensure that one measures flexibility as opposed to task-specific processes. The currently used task-switching paradigms differ from one another in at least four respects: *(a)* the tasks; *(b)* whether the tasks involve a single step or multiple steps; *(c)* how the tasks are instructed; and *(d)* when they are instructed.

Task Differences

Task switching paradigms involve two or more tasks. In terms of responses, the tasks may require manual responses (often key presses, e.g., Rogers & Monsell, 1995), vocal responses (e.g., Allport et al., 1994; Jersild, 1927), or eye movements (e.g., Hunt & Klein, 2002). In terms of memory access, the tasks may require perceptual classification (such as color decision, e.g., Fagot, 1994; Hartley et al., 1990), semantic retrieval, as in making an odd–even judgment on digits or vowel–consonant judgment on letters (Rogers & Monsell, 1995), spatial location judgments (de Jong, 1995; Meiran, 1996; Shaffer, 1965), or episodic memory retrieval (Mayr & Kliegl, 2000). In terms of decision type, the tasks may require classification (e.g., Rogers & Monsell, 1995), odd-item out decisions (Mayr & Keele, 2000), or same–different judgments (Meiran & Marciano, 2002). Only a few studies have tried to examine how the choice of tasks influences task-switching performance. The few studies that looked into this issue found a remarkable degree of task specificity. For example, Yehene and Meiran (2007) showed that most of the reliable individual differences variance in switch costs is task specific (.64 to .85) and in another switch-related effect (the congruency effect, explained below), the entire reliable variance is task specific. Meiran et al. (2002) and Meiran and Marciano (2002) used the same stimuli with manual responses but using classification and same-different judgments, respectively. The results of the two studies differed substantially. Specifically, although preparation drastically reduced switch costs in classification tasks, it did not reduce the costs at all with same-different judgments. Hunt and Klein (2002) showed that switch costs were eliminated for eye movement responses, whereas they are not eliminated for key-press and vocal responses. Mayr and Kliegl (2000) showed that semantic memory tasks and episodic memory tasks produce very different switch cost profiles.

Single Step vs. Multistep Tasks

Although nearly all the studies studied single-step tasks, in which each task constitutes a single stimulus and a single response and S-R sequence, some studies focused on multistep tasks. Luria et al., (2006), Luria and Meiran (2003, 2006) and Schneider and Logan (2006) studied the effects of changing the order of tasks when the tasks themselves remain the same. Hayes et al. (1998) studied the influence of changing response sequence. Both of these changes incurred switch cost in spite of the fact that the subtasks were kept constant across the order switch. Namely, every trial required both color classification and letter classification, only in different orders.

Methods for Task Instruction

There are three principled methods to instruct subjects about the task change: from memory, by means of an instructional cue, or by the subjects' choice. Instruction method appears to have an important influence on the processes contributing to the observed switch costs (e.g., Altmann, 2007).

Instructions from Memory

Instruction from memory was the method chosen by Jersild (1927). In his experiments, subjects either executed each of two tasks (Task A and Task B) in isolation or alternated between them (ABAB...). Such a method requires keeping in memory the task sequence and monitoring the progress along that sequence (*see* Rubinstein et al., 2001). The method has been criticized on the basis of lacking a proper baseline. To remedy this problem, Rogers and Monsell (1995) introduced the *alternating runs paradigm*, in which subjects alternated between runs of trials involving the same task. When the run length is 1, the method is equivalent to Jersild's. When it is 2, the sequence is AA-BB-..., and so forth.

There are now additional techniques to instruct the tasks from memory. For example, subjects may receive a cue indicating the task sequence in the next two (Sohn & Carlson, 2000) or more trials (Gopher et al., 2000).

Capitalizing on the advantage of this technique, Gopher et al. showed that the initial slowing in the beginning of the run is found also for task repetitions, indicating a "restart cost" (*see also* Allport & Wylie, 2000). A similar approach was developed by Logan (2004) to assess working memory (WM) capacity for task sequences. Logan found that when subjects are asked to memorize a long series of tasks, they form task chunks that are analogous in many respects to the item chunks in short-term memory tasks (Miller, 1956). Moreover, Logan reported that the beginning of a "task-chunk" is indicated by response slowing. Subjects can also be told to SWITCH tasks or STAY on a task (Forstmann et al., 2005), so that the next task is based on their memory of the preceding task, or receive a cue in the beginning of a run of trials, forcing them to maintain the task goal in memory throughout the run (Altmann & Gray, 2002). Additionally, subjects may learn a task sequence implicitly, so that their performance is assisted by the (implicit) memory of the task sequence (e.g., Gotler et al., 2003; Heuer et al., 2001; Koch, 2001).

Instructions in Each Trial

The first author to have used this method was Shaffer (1965), as described already. The advantage of this method is that it enables tight control over task preparation time. This advantage has been used by Hartley et al. (1990), Meiran (1996), Shaffer (1965), and Sudevan and Taylor (1987) to study task preparation effects by varying the interval between the task instructions and the target stimulus. A methodological issue is that the task set adopted in the preceding trial decays over time (Allport et al., 1994; Meiran et al., 2000). Thus, task preparation time and the time allowed for the previous task set to decay are potentially confounded. Solutions to this problem were offered by Meiran (1996) and Meiran et al. (2000). Another problem is that task repetitions are associated with a repetition of the task cue, and this may contribute to switch effects. Solutions to this problem were developed by Arrington et al. (2007), Logan and Bundesen (2003), Mayr and Kliegl (2003), and Monsel and Mizon (2006). It is still debated whether the cue repetition effect is a purely perceptual phenomenon unrelated to task control (Logan & Bundesen, 2003; Schneider & Logan, 2005) or representing a component control process (Arrington et al., 2007; Mayr & Kliegl, 2003, *see also* Gade & Koch, 2007).

There are numerous studies comparing the alternating runs paradigm in which task instructions come from memory with the cuing paradigm described here. Rogers and Monsell (1995) looked at position in run effects. Specifically, they had runs of four trials (AAAA-BBBB-…) so that performance in the 1st through 4th positions could be compared. Their results used longer runs to show that the first trial in the run (which is the switch trial) is associated with poorer performance than the remaining trials in the run (repeat trials), which show similar level of performance. This "position-in-run" effect depends on the paradigm. When the tasks are ordered randomly and instructed by means of an external cue, position in run leads to response facilitation even for repeat trials (Meiran et al., 2000). Monsell et al. (2003), who made a direct comparison between these two techniques, suggested that the trend for speeding observed in the cuing paradigm results from a gradual increase in task commitment. In the alternating runs paradigm, in which the task order is known in advance, full commitment is achieved immediately.

Self-Selected Tasks

There are three procedures in which subjects choose which task to execute. In one procedure (Arrington & Logan, 2004, 2005), subjects are told to switch between tasks under the constraint that the number of task switches will be roughly equal to that of task repetitions. This procedure yields switch effects of usual size and apparently the difference relative to the experimenter-instructed approaches described above does not lie there (Mayr & Bell, 2006). It lies in the frequency in which subjects switch, for which the common finding shows that subjects prefer to stay on the task, and switch on less than 50% of the trials, contrary to instructions. Although it is tempting to interpret this tendency to stay on a task as reflecting autonomous choice, it in fact is affected by stimulus factors such as stimulus repetition (Mayr & Bell, 2006). Another approach was used by Forstmann et al. (2006), who gave their subjects more than just two tasks and instructed them to either STAY or SWITCH. Because there were more than two optional tasks to switch to, the switch condition involved a task choice. A recently introduced third method involves letting subjects to choose freely the task to execute without any constraint. The surprising finding is that nearly all the subjects choose to switch tasks even when switching is from an easy task to a more difficult one (Kessler et al., 2009).

Mental Sets and Task Sets

The concept of mental set has a long and perhaps notorious history in psychology. Gibson's (1941) classical review of the early works was highly critical of the concept and argued that it is poorly defined. A much improved definition of mental set is afforded by indices taken from the task switching paradigm. It should be kept in mind, however, that the greater precision came at the cost of a potential loss of generality, because what the task switching paradigm presumably involves is a "task set" which may be a narrower term than "mental set."

The term "task set" refers to the active mental representations that afford the chosen cognitive activity. As elaborated beautifully by Prinz (1997), even the simple RT task, that arguably is the simplest cognitive task, requires a task set. The task requires making a predetermined response (e.g., press a key) to any stimulus that is presented without making any judgment concerning the stimulus. According to Prinz, even this incredibly simple task requires the intention (or "task set") to make the required response, without which no response would be made.

What does a task set include? It includes *(a)* the goal state; *(b)* the selection of task- relevant information by attentional mechanisms, including the relevant stimulus information and the relevant feedback information; *(c)* activated task-relevant semantic information (e.g., when the task requires odd-even judgment, the relevant numerical information needs to be active); *(d)* activated response information, such as the readiness to respond with the right and the left hand, and what each response means in the context of the given task (e.g., "a right key press means 'odd' "); *(e)* the activated response rules, such as "if ODD, press LEFT;" *(f)* the order of actions and action interrelatedness information if the task is one involving multiple actions (e.g., Luria & Meiran, 2003, 2006; Schneider & Logan, 2006), and so forth. All these could be described system parameters that change when the task changes (Logan & Gordon, 2001; Meiran, 2000a; Meiran et al., 2008).

As a rule, when any of these parameters changes, responding is slowed. This has been shown for changes involving the stimulus dimension (Meiran & Marciano, 2002; Ward, 1982), decision rule (Allport et al., 1994; Schneider & Logan, 2007), response modality (Philipp & Koch, 2005), subtask order in multistep tasks (Luria & Meiran, 2003; Luria et al., 2006), a change in the meaning associated with a given key press (Brass et al., 2003; Meiran, 2000b; 2005; Meiran & Marciano, 2002), and so forth.

Once we know that a parameter change results in slowing we can begin asking, how are task sets mentally represented? One possibility is that task sets are represented in a unified format, which is analogous to a computer file that contains all the parameters and is retrieved as a unit. Alternatively, task sets may be viewed as *ad hoc* assemblies of parameters (distributed representation) that change asynchronously. There is no definitive answer to this question. One approach to answer this question is to compare conditions involving multiple parameter changes to conditions with a single parameter change. Some studies report that changing multiple parameters results in greater slowing as compared with changing a single parameter (e.g., Arrington et al., 2003; Steinhauser & R. Hübner, 2005; R. Hübner et al., 2001). Such a result suggests that each parameter change is associated with independent slowing—that slowing is additive. It implies that task sets have distributed representations (supporting positions like those of Logan & Gordon, 2001 and Meiran, 2000a). Other studies find that a change in two parameters results in equal slowing as a change in one parameter (e.g., Allport et al., 1994), a result that supports the unified representation idea. Hahn et al. (2003) and R. Hübner et al. (2001) observed additive slowing in some conditions but in other conditions a double parameter switch incurred as much slowing as did a single parameter switch. What seems to explain this discrepancy between the two sets of conditions is subjects' strategic tendency to form a unified representation. This choice of strategy is apparently made when the parameters are coupled in the experiment (a certain relevant stimulus dimension always went together with a specific judgment). I (Meiran, 2000b) looked at the time course of preparation as a function of the parameter change. In this study, it was shown that a change in the direction of attention to the relevant stimulus features occurred in anticipation of the target, but a change in the meaning of the responses occurred after or during response. This finding supports the distributed representation notion.

Regardless of these not yet fully resolved issues, most researchers seem to regard switch-related slowing as a marker for a mental set change. (The only notable exceptions are the theories of Logan & Bunsesen, 2003, and Schneider & Logan, 2005, and related empirical works). They disagree on whether the set change reflects goal-related "top-down" processes (e.g., Rogers & Monsell, 1995; Rubinstein et al., 2001), more reflexive "bottom-up" processes (e.g., Allport et al., 1994; Allport & Wylie, 2000) or a combination of both (Koch & Allport, 2006; Meiran, 1996, 2000a, 2000b; Meiran et al., 2000; Sohn & Anderson, 2001; Yeung & Monsell 2003b).

This dispute had and still has an immense influence on the field. I believe that the disagreement is more apparent than real. For example, those who argue that the control operating in task switching is reflexive still acknowledge the fact that performance on this paradigm involves cognitive control. In fact, their idea is that task sets persist beyond the time in which they were relevant (Allport et al., 1994) or get automatically retrieved by the stimuli to which they applied beforehand (Allport & Wylie, 2000). Similarly, those researchers who argue that the task-set change is goal-directed (e.g., Rogers & Monsell, 1995; Rubinstein et al., 2001) currently acknowledge the fact that task sets have inertia (*see* especially Yeung & Monsell, 2003a, 2003b), which may be regarded as a more passive form of control.

What Makes a Cognitive Task?

Perhaps a more basic question to ask before characterizing task sets is, "What makes a cognitive task?" Recent studies have shown that the definition of a cognitive task depends, to a large extent, on what subjects subjectively perceive to be a task. This answer is very much in line with the conclusions above that it is a change in the mental representation (of the task parameters) that results in slowing.

Support comes from studies on multi-step tasks, showing switch effects when the task pair changes (from, say "respond to color → respond to letter" to "respond to letter → respond to color"), but the tasks remain the same. Note that the perceptual grouping of the letter and color tasks into a task pair dictates switch effects and, hence, how the tasks are represented mentally (Lien & Ruthruff, 2004; Luria & Meiran, 2003, 2006). Corroborating evidence comes also from studies on single-step tasks showing that the same transition incurs a task switch cost when the instructions refer to tasks but not when they refer to individual stimulus–response pairings (Dreisbach et al., 2006, 2007). Finally, Yehene et al. (2005) reported a neurological case, AF, who was asked to perform a standard task switching paradigm involving SHAPE and SIZE tasks. This patient showed task mixing costs in spite of the fact that she had stopped switching and performed only the SIZE task. Similar effects were found in that study among a group of control participants who were instructed to be ready for a task switch upon a given instruction, which never occurred.

It has long been acknowledged that goals and tasks are arranged in hierarchies (e.g., the seminal paper by Norman & Shallice, 1986). Recent works using task-switching paradigms provide support for this position. This evidence comes from two sources. One is from studies in which the tasks involve multiple steps (Luria &

Meiran, 2003, 2006; Luria et al., 2006; Schneider & Logan, 2006). The other comes from studies involving switching between multiple tasks that form a hierarchy (Kleisorge & Heuer, 1999). The signature for hierarchical control in multi step size is the sequence initiation time, seen in slower responding in the first trial of the sequence (*see also* Logan, 2004). The signature of hierarchical control within a multi-layered task array is the propagation of the switch-signal in the hierarchy. This propagation results in a somewhat paradoxical phenomenon of easier switching when multiple task elements are switched (for detail, *see* Kleisorge & Heuer, 1999).

Processes

Once we have some grasp concerning what makes a task and what is a task set, we can proceed in asking what processes are involved in switching tasks and task sets (and, by extension, which processes contribute to mental rigidity and flexibility). Based on Ach's (2006/1910) conception, I divide the processes into two broad classes. The first class includes those processes that create the inner obstacle (and contribute to cognitive rigidity). The second class includes processes that ensure successful goal achievement (and contribute to cognitive flexibility). A similar hybrid account for task switching performance appears in many recent theories (e.g., Gilbert & Shallice, 2002; Koch & Allport, 2006; Meiran, 2000; Meiran et al., 2000; Meiran & Daichman, 2005; Sohn & Anderson, 2001; Sumner & Ahmed, 2006; Yeung & Monsell, 2003a, 2003b).

Inner Obstacles and Contributions to Rigidity

Task-Set Inertia

One process that enhances the inner obstacle is the inertia of the mental sets (Allport et al., 1994; *see* Sumner & Ahmed, 2006, for a discussion of the various potential forms of the task-set inertia). Presumably, tasks that demand more intensive control efforts form more durable memory traces, a fact that makes switching *away* from these tasks more difficult. For example, when subjects switch between tasks of unequal familiarity, such as color naming and word reading of Stroop stimuli, the less habitual task (color naming) requires a greater degree of top-down control (e.g., MacDonald et al., 2000, for evidence). Consequently, switching away to the easier task results in a seemingly paradoxical increase in the switch cost observed in the *easier* task, a phenomenon called "switch asymmetry". This phenomenon is also observed when bilingual subjects switch between a dominant and a nondominant language (Meuter & Allport, 1999). However, there are studies reporting the opposite, more intuitively predicted trend whereby switch costs are larger for the more difficult task (e.g., Rubintein et al., 2001). This latter finding is referred to as "reversed asymmetry." Yeung and Monsell (2003a) found switch asymmetry when there was a high degree of response conflict and reversed asymmetry with lower response conflict.

When switching from Task A to Task B, two control operations are required. One is to inhibit Task set A, which has become irrelevant. The other is to activate Task set B, which has become relevant. The task-set inertia process discussed above is usually interpreted as a carryover of *activation*. Another form of inner obstacle is lingering inhibition. Lingering inhibition has been studied in two paradigms. Mayr and Keele (2000) introduced the backward inhibition paradigm. According to them, task transition involves the suppression of the abandoned task, which enables one to go on to the next task in the sequence. Moreover, this task suppression tends to persist. Accordingly, they designed a three-task paradigm, with Tasks A, B, and C. This enabled them to compare two kinds of task sequences, both involving an immediate task switch. In one sequence, Task A was performed after having just been abandoned. This was the A → B → A sequence. In the control condition, Task A was performed after having been abandoned a longer time before, a C → B → A sequence. The major finding was that performance was poorer when the task had just been abandoned as compared to when it had been abandoned a longer time before. The effect was labeled "backward inhibition" and it has been replicated in a variety of paradigms since then (e.g., Arbuthnott & Frank, 2000; Schuch & Koch, 2003).

In the experiments of Masson et al. (2003), subjects reacted to pairs of stimuli. The first stimulus required naming in which color was written either an incompatible color word or a row of Xs. The second stimulus was verbal and required word reading. The results show that word reading was slowed when preceded by incompatible color naming as compared to neutral (Xs) color naming. A similar effect was found for other reading tasks as well. This result shows that when subjects executed the color naming response, they needed to block the word reading processing pathway. It also shows that this inhibition persisted beyond the time when it was needed, indicating task-set inertia.

Stimulus-Set Binding

Yet another form of inner obstacle is *stimulus-set binding*, according to which task sets bind with the stimuli on which the task was executed. Note that unlike the task-set inertia idea, assuming that the task set persists in an active state, the stimulus-set binding idea is that the set gets automatically retrieved when the stimuli are re-encountered. Potentially, this is a useful process because in most cases, stimuli consistently require a given task. However, rarely, a given stimulus might require a new task, making it likely that the wrong task set will be retrieved. Accordingly, switch costs are increased for stimuli that have been previously associated with the alternative task set (Allport & Wylie, 2000; Gilbert & Shallice, 2002; Waszak et al., 2003).

Retroactive Adjustments

The final form of inner obstacle concerns the fine tuning of the control parameters that presumably takes place in order to continuously optimize performance in a given task. This process creates an inner obstacle because when there is a task switch from Task A to Task B, the cognitive system has just become better tuned to execute Task A, and this fine tuning impairs performance on the following Task B (Meiran, 1996, 2000a, 2000b). Essentially, this form of inner obstacle resembles (or may even be identical with) negative transfer effects discussed in the learning literature. It is well-documented that subjects may switch to a more cautious strategy after making an error or after encountering an error-prone condition (e.g., Brown et al., 2007; Goschke, 2000). Similarly, subjects pay more attention to a given stimulus aspect after that stimulus was relevant in the preceding trial (M. Hubener et al., 2004). In task switching, it appears that the adjustment applies mostly to the meaning associated with each key press. Specifically, in many task switching experiments, a given key press is associated with two meanings, one for each task. For example, in switching between COLOR and SHAPE, pressing the left key may indicate both CIRCLE and RED, depending on the task. Results suggest that the association between the given meaning (e.g., CIRCLE) and the key press (left key) is strengthened after executing the SHAPE task. As a result, there is a performance cost when the task switches and the left key is used to indicate the other meaning (RED) (Brass et al., 2003; Meiran, 2000a; Schuch & Koch, 2004).

Based on this rationale, one would predict that switch costs would be lessened if there were less opportunity for retroactive adjustment. Indeed, switch costs are actually eliminated when the pre-switch response is inhibited (e.g., Philipp et al. 2007; Schuch & Koch, 2003). Similarly, when the preswitch trial involves the erroneous execution of the wrong task, switch costs turn into switch gains, because what is nominally a task switch is actually a task repetition (Steinhauser & R. Hübner, 2006). Finally, one would predict that a greater degree of retroactive adjustment would result in enlarged switch costs. This prediction was borne out by Sumner and Ahmed (2006).

Conclusions

When people engage in a cognitive activity such as task execution, they adopt mental sets. These sets involve the needed operations but also the inhibition of the no longer needed processes. The sets are memory representations, and as such, they persist in time beyond the point in which they were relevant. Moreover, the sets are automatically retrieved when the target stimuli associated with them are re-presented. Finally, mental sets dictate the type of online fine tuning of the system that is normally enacted to ensure continuous improvement in performance. This fine tuning becomes counterproductive when the activity changes. All these factors form inner obstacles and are therefore causes for behavioral and cognitive rigidity. The role of the intentional control processes described in the following section is to combat the inner obstacles in order to ensure flexible and intentional processing.

Control Processes and Contributions to Flexibility

According to the present definition, a control process is any process that helps to overcome the inner obstacles in the service of flexible goal directed behavior. No assumption is made that control processes need to reflect the action of an autonomous agent or that control processes need to be conscious (e.g., *see* Gotler et al., 2003; Meiran et al., 2002, showing evidence for nonconscious control). Finally, no assumption is made that control processes are more endogenous than the processes contributing to the inner obstacle. Here, I refer to Rogers and Monsell's (1995) influential division of processes into "endogenous" (which according to them means "intentional") and "exogenous." In fact, some of the inner obstacles just described stem from within (from subjects' memories) and are, in that sense, "endogenous." Concomitantly, some control processes are invoked by an external stimulus, such as a task cue, and should therefore be regarded as "exogenous."

Three control processes feature a major role in the current literature on task switching. These include (*a*) deciding which task to execute and maintaining goal representation in memory; (*b*) inhibiting the alternative tasks and filtering out task irrelevant information; and (*c*) performance monitoring. I will discuss these processes in turn.

Task Decision and Goal Maintenance

Before executing a task, subjects must know what the required task is. The information regarding task identity may either be retrieved from memory (Logan, 2004; Rubinstein et al., 2001) or become available via the processing of a task cue. These processes have been described as "goal setting" (Rubinstein et al., 2001) or "task decision" (Fagot, 1994). An additional process is holding the task identity in some form of WM. Below, I review some supporting evidence.

The first is Jersild's (1927) task-cuing effect described in Historical Foundations (smaller switch costs when the tasks are cued), which indicates that the retrieval of the next goal affects performance. Rubinstein et al. (2001) further showed that the task-cuing effect is additive with (and hence, independent from) the effect of task complexity. This result led the authors to suggest two serially ordered executive processes: goal setting and task-rule implementation. Support for the idea that task goals are held in some WM comes from studies showing that loading this system impairs task-switching performance at least in some cases (Baddeley et al., 2001; Bryck & Mayr, 2005; Emerson & Miyake, 2003).

The second piece of evidence comes from Jersild (1927), who showed that when each task is associated with a distinct stimulus set, switching costs are absent. One could interpret this finding as evidence that when the stimulus sets are disjoint (and clearly distinguishable, *see* Sumner & Ahmed, 2006) there is no need to make a top-down task decision. The role of unambiguous stimulus-task association has been elegantly demonstrated by Allport et al. (1994, Experiment 4). In this study, the authors used disjoint (univalent) stimuli: groups of digits (e.g., 333) and Stroop stimuli (e.g., the word "RED" written in blue ink), each affording only one task in the given context. As found by Jersild, there were no switch costs. However, when the task performed on each subset of the stimuli changed (from counting the digits to saying the digit value, for example) switch costs were found. Note that the stimuli remained univalent even after the task change because each stimulus category was uniquely associated with one task. Arguably, the switch costs were caused by the created need to recall which exact task to execute on the given stimulus type.

Following Braver et al. (2003), Fagot (1994), and Los (1996), among others, Rubin and Meiran

(2005) distinguished between switching cost and mixing cost (*see* the description of Shaffer, 1965, in Historical Foundations). They showed that stimulus bivalence (whether the stimulus affords the competing task) affected mixing costs and not switching costs and suggested that, when the tasks change unpredictably, mixing costs represent mostly task decision difficulty. Bryck and Mayr (2005) provided additional evidence supporting this assumption.

Further evidence for task decision comes from the increase in RT with an increasing number of tasks (Biederman, 1973; Dixon, 1981; Dixon & Just, 1986; Meiran et al., 2002). Interestingly, all these studies found the effect in conditions with little time to prepare and not when subjects were given time to prepare, suggesting that the choice among the potential tasks is made in anticipation of the task. Similarly, RT was affected by explicit block-wide task expectancy (Dreisbach et al., 2002; Ruthruff et al., 2001; Sohn & Carlson, 2000) and by implicit task expectancy. The last form of expectancy is created in experiments in which subjects are led to believe that the task sequence is random, whereas in fact it consists of a repetitive pattern. Replacing the repetitive pattern by a new pattern resulted in response slowing. This result indicates that the subjects have learned the repetitive task pattern and made use of it. The effects of expectancy are equivalent for switch trials and task repetition trials (Gotler et al., 2003; Heuer et al., 2001; Koch, 2001; *see also* Koch, 2005), presumably because of the need to make a task decision in both cases in the absence of an instructed task sequence.

Although the studies, reviewed above, point to the importance of having a vivid WM representation of the task goal, Altmann and Gray (2002; 2008) show evidence that forgetting this goal may also be functional. Specifically, they were interested in the effect of the position in run on RT. A run is defined as a series of trials in which the task is repeated in a context in which tasks may switch. Specifically, the first position in the run is a switch trial, because the previous trial involved another task. All the remaining positions in the run involve task repetition. Altmann and Gray showed that RT slightly but consistently increased over the run. Based on this evidence and a formal model they argued that the "within-run slowing" is evidence for the forgetting of the goal that, according to them, is functional because it allows for the smooth encoding of the next goal. It should be pointed out that although the within-run slowing was replicated in Altmann's lab under a variety of conditions (e.g., Altmann, 2002), and a similar (albeit nonsignificant) trend was found by Rogers and Monsell (1995) among others, many other studies consistently find within-run speeding, especially when the task order is random (Meiran et al., 2000; Meiran & Marciano, 2002; Monsell et al., 2003; Sumner & Ahmed, 2006; *see also* Tornay & Milàn, 2001, for the same, albeit nonsignificant trend). A potential difference between the two types of studies is the presentation of a task cue in the beginning of the run (Altmann's studies) or in every trial (the remaining studies; *see* Altmann, 2002).

Inhibition

Task switching requires the interruption of one task in favor of the alternative task. The interruption aspect is likely inhibitory. The fact that task switching involves inhibition is supported by two pieces of evidence that were already reviewed, including Mayr and Keele's (2000) backward inhibition effect and Masson et al.'s (2003) effects concerning pathway inhibition.

Additional evidence for the involvement of inhibition in task switching comes from Logan and Burkell (1986), who studied inhibition within the framework of the stop-signal paradigm (Logan, 1994, for review). In the stop-signal paradigm subjects are first pretrained on a task to create a strong tendency to execute this task. Afterward, they are required to withhold task execution on a certain (low) proportion of the trials, and their inhibitory abilities are measured. The Logan-Burkel paradigm requires that instead of withholding responses (as in the standard stop-signal paradigm), subjects execute another task. In that respect, this paradigm resembles the task-switching paradigm. Logan and Burkel's results indicate that inhibition was less effective (and more demanding) in this stop-switch paradigm as compared to the standard stop-signal paradigm in which no task switching was required. Nonetheless, the difference was not large: about 40 milliseconds.

Finally, the fact that task switching involves inhibition is supported by three additional facts. One is the fact that the Stroop task, commonly taken as a measure of inhibition, and task switching activate similar regions in the prefrontal cortex, including the posterior lateral prefrontal cortex (Derfusss et al., 2005) and the anterior cingulate gyrus (Dosenbach et al., 2006). Second, switch costs show a substantial individual differences correlation with the ability to suppress dominant responses and ignore interference (Friedman & Miyake, 2004). Finally, Yeung et al. (2006; *see also* Wylie et al., 2004) asked subjects to switch between a color task and a face task, capitalizing on the distinct brain topography of these two tasks. Importantly, they showed that individual differences in switch costs were positively correlated with individual differences in the degree to which the currently irrelevant brain region was active. Namely, when performing the FACE task, for example, brain regions associated with the COLOR task were active, and the degree to which they were active determined the size of the behavioral switch costs observed in the FACE task. This evidence clearly associated switch costs with the failure to inhibit the irrelevant task. It should be mentioned that the contribution of inhibition to switch effects remains somewhat controversial (e.g., Lien et al., 2006).

Information Filtering

Although the term "inhibition" usually refers to the entire task, filtering refers to the selection of specific task related information such as word information when performing color naming (Masson et al., 2003). Hübner et al. (2003) additionally showed that information that was relevant for the task that has been executed in the preceding trial becomes inhibited after a task switch.

A well established finding in the task switching literature is the task-rule congruency effect, which indicates imperfect filtering of the information related to the currently irrelevant task. Sudevan and Taylor (1987) were the first ones to document this effect. They studied switching between two numerical classification tasks. They asked subjects to press the left key in response to odd numbers (in an odd-even classification) and in response to numbers higher than 5 (in a high-low classification). Right hand responses were required for even numbers and for numbers smaller than 5. As a consequence, there were numbers that required a left-key press in both tasks (e.g., "7," which is both ODD and HIGH). These are congruent targets. Other targets required different responses in each task. For example, "8" required a left-hand response in the High-Low task (because it is high) and a right-hand response in the odd–even task (because it is even). These were *incongruent* targets. More generally, congruency relates to whether the target stimulus requires the same response according to both rules. The fact that congruence effects are highly replicable indicates filtering failure. They show that the currently irrelevant task rule is operative and affects response choice. It should be mentioned here that Meiran and Daichman (2005) have recently suggested that the congruency effect found in errors represents the erroneous execution of the wrong task. In contrast, Meiran and Kessler (2008) show that the RT congruency effect reflects the activation of overlearned stimulus category codes.

Biederman's (1972) study, described in Historical Foundations, shows that subjects are only partly successful in filtering out information that is relevant for one task rule but is irrelevant for the task rule that is currently required. Additional evidence that information filtering is used for task control comes from studies comparing univalent stimuli with bivalent stimuli. These studies show smaller switch costs for univalent stimuli (e.g., Mayr, 2001; Meiran, 2000b). A similar role of information filtering has been noted with respect to the responses. When the responses used in the two (or more) tasks overlap, each response becomes associated with multiple meanings. For example, when switching between an ODD–EVEN task and a HIGH–LOW task, a given key press might be used to indicate ODD when the first task is required and HIGH when the second task is required. One way in which control could be achieved is by selectively attending to one response meaning and temporarily ignoring the other response meaning. To study this information filtering function, researchers compared univalent response setups (in which the responses for the two tasks are disjointed) with bivalent response setups (in which the responses for the two tasks overlap). Further evidence suggests that the link between the motor act and its symbolic meaning is adjusted by each task execution (Meiran, 2000b, Philipp et al., 2007) or even by the mere activation of the relevant response representation (R. Hübner & Druey, 2006).

Monitoring

By monitoring, I refer to ongoing recording of changes in control demands and consequent behavioral adjustments. A relevant piece of evidence is the increase in switch costs following incongruent trials (Goschke, 2000). Brown et al. (2007) have recently replicated and extended this finding. Importantly, their formal model attributes this change to monitoring effects. Indirect evidence for the involvement of monitoring comes from brain imaging studies in which the dorsal anterior cingulate gyrus is often implicated (especially Dosenbach et al., 2006), because this region is believed to involve monitoring in a variety of other paradigms (for review, Botvinick et al., 2001).

Preparation

I mention "preparation" here although it is a process that is not described at the same level of analysis as task decision, inhibition, and monitoring. Preparation reflects the changing in advance of any of those, plus additional, non switch-related task aspects, such as phasic alertness (Posner & Bois, 1971) and stimulus timing (Los & Van Der Heuvel, 2001; Meiran et al., 2000). In fact, evidence for anticipatory change has been found regarding task identity (Gade & Koch, 2007; Meiran & Daichman, 2005; Sohn & Anderson, 2001; Sohn & Carlson, 2000), the retrieval of the stimulus-response mapping from episodic memory (Mayr & Kliegl, 2000; see also Lien et al., 2005, A-L. Cohen et al., in press), the direction of attention to the relevant stimulus dimension (Meiran, 2000b), the blocking of irrelevant information (M. Hübner et al., 2004) and the order of sub-tasks making a multi-step task (Luria & Meiran, 2003, 2006; Luria et al., 2006).

This anticipatory preparation was considered to be a hallmark of executive functioning by some authors (Allport et al., 1994; Rogers & Monsell, 1995; Meiran, 1996). One surprising finding that keeps intriguing researchers is the persistence of switch cost even after ample preparation time. For example, Meiran and Chorev (2005) found that switch costs were only slightly and non-significantly smaller after 10 seconds of advance preparation as compared to 1.4 seconds. There are some notable exceptions to this rule, showing that switch costs can be eliminated by advance preparation. These include using nonoverlapping responses for the two tasks (Meiran, 2000b), switching between two eye-movement tasks: a pro-saccade task (orient towards the stimulus) and an anti-saccade task (orient away from the stimulus, Hunt & Klein, 2002), presenting the task cue only briefly (Verbruggen et al., 2007, but see Gotler & Meiran, 2001, for a different result with a similar procedure), and avoiding a response in the pre-switch trial (e.g., Schuch & Koch, 2003).

There are numerous hypotheses regarding the nature of this "residual switch cost". Most of these hypotheses share the idea that some, but not all of the task set is prepared in advance. The theories differ as to when the remaining preparation or adjustment is being made. Advance preparation theories suggest that preparation (if used) precedes task execution processes. Rogers and Monsell (1995), who were the first to show the residual switch cost, suggested that preparation is postponed until the target stimulus is presented ("stimulus-cued reconfiguration"). de Jong (2000, see also Nieuwenhuis & Monsell, 2002; Brown et al., 2006) argued that the participants prepare fully but do so only on a subportion of the trials. Lien et al. (2005), who observed preparation effects for some but not all the responses, argued that subjects prepare some response rules but not others.

Another idea is that the residual switch cost results from the persistent nature of the inner obstacles. Allport et al. (1994), for example, suggested that preparation barely affects the switch cost and the cost is mainly due to task set inertia. Mayr and Keele (2000; *see also* Arbuthnott & Frank, 2000; Schuch & Koch, 2003) suggested that the residual cost results from a carryover of task inhibition that took place in previous trials. I suggested that the residual cost is a form of negative transfer because of retroactive adjustment (Meiran 1996, 2000a, 2000b; *see above*).

Luria et al. (2006) contrasted these two accounts using multistep tasks that required subjects to make three responses to each stimulus according to the three dimensions of the stimulus. They found switch costs in the second response in the response triplet, suggesting that the task preparation was not completed after the stimulus was presented because it persisted beyond the first response. This result is incompatible with the postponed preparation idea of de Jong (2000) and Rogers and Monsell (1995) but is in line with the idea concerning a carryover of inner obstacles.

Conclusion

There are numerous processes that contribute to cognitive flexibility. They include the decision of which activity to execute, the vivid representation of goals, the inhibition of previous goals, and the filtering of no-longer relevant information. The goals are usually arranged in hierarchies so that more global goals (such as completing writing this chapter) are subdivided into smaller goals (such as completing writing this section). Flexible performance is ensured by online monitoring and consequent behavioral adjustments. Finally, many of these processes can be carried out in preparation for the activity. This preparation, although useful, is rarely complete, and in most circumstances, the inner obstacles influence (but do not dictate) behavior, at least until the first execution of the next task or activity.

References

Ach, N. On volition (T. Herz, Trans.). (Original work published 1910) Available at: University of Konstanz, Cognitive Psychology Web site: http://www.uni-konstanz.de/kogpsych/ach.htm. 2006.

Adolphs, A. Cognitive neuroscience of human social behavior. Nat Rev Neurosci 2003; 4: 165–178.

Allport, A., Styles, E. A., & Hsieh, S. Shifting intentional set: Exploring the dynamic control of tasks. In: Umiltà, C., & Moscovitch, M. (Eds.), *Attention and performance XV: Consciousand unconscious processing*. Cambridge, MA: MIT Press, 1994: pp. 421–452.

Allport, A., & Wylie, G. "Task-switching," stimulus-response bindings and negative priming. In: Monsell, S., & Driver, J. (Eds.), *Attention and performance XVIII: Control of cognitive processes*. Cambridge, MA: MIT Press, 2000: pp. 35–70.

Altmann, E. M. Functional decay of memory for tasks. Psychol Res 2002; 66: 287–297.

Altmann, E. M. Comparing switch costs: Alternating runs and explicit cuing. J Exp Psychol Learn Mem Cogn 2007; 33: 475–483.

Altmann, E. M., & Gray, W. D. Forgetting to remember: The functional relationship of decay and interference. Psychol Sci 2002; 13: 27–33.

Altmann, E. M., & Gray, W. D. An integrated model of cognitive control in task switching. Psychol Rev 2008; 115: 602–639.

Arbuthnott, K. D., & Frank, J. Executive control in set switching: Residual switch cost and task-set inhibition. Can J Exp Psychol 2000; 54: 33–41.

Arrington, C. M., Altmann, E. M., & Carr, T. H. Tasks of a feather flock together: Similarity effects in task switching. Mem Cogn 2003; 31: 781–789.

Arrington, C. M., & Logan, G. D. The cost of a voluntary task switch. Psychol Sci 2004; 15: 610–615.

Arrington, C. M., & Logan, G. D. Voluntary task switching: Chasing the elusive homunculus. J Exp Psychol Learn Mem Cogn 2005; 31: 683–702.

Arrington, C. M., Logan, G. D., & Schneider, D. W. Separating cue encoding from target processing in the explicit task cuing procedure. Are there "true" task switch effects? J Exp Psychol Learn Mem Cogn 2007; 33: 484–502.

Baddeley, A. D., Chincotta, D., & Adlam, A. Working memory and the control of action: Evidence from task switching. J Exp Psychol Gen 2001; 130: 641–657.

Barkley, R. A. The executive functions and self regulation: An evolutionary neuropsychological perspective. Neuropsychol Rev 2001; 11: 1–29.

Baumeister, R. F. *The cultural animal: Human nature, meaning, and social life*. New York: Oxford University Press, 2005.

Bernstein, E. Quickness and intelligence. Br J Psychol Monogr Suppl 1924; 7(3): VI–55.

Biederman, I. Human performance in contingent information processing tasks. J Exp Psychol 1972; 93: 219–238.

Biederman, I. Mental set and mental arithmetic. Mem Cogn 1973; 1: 383–386.

Bonino, S., & Cattleino, E. The relationship between cognitive abilities and social abilities in childhood: A research on flexibility in thinking and cooperation with peers. Int J Behav Develop 1999; 23: 19–36.

Botvinick, M. M., Braver, T. S., Carter, C. S., Barch, D. M., & Cohen, J. D. Evaluating the demand for control: Anterior cingulate cortex and crosstalk monitoring. Psychol Rev 2001; 108: 624–652.

Brass, M., Ruge, H., Meiran, N., et al. When the same response has different meaning: Recoding the response meaning in the lateral prefrontal cortex. NeuroImage 2003; 20: 1026–1031.

Braver, T. S., Reynolds, J. R., & Donaldson, D. I. Neural mechanisms of transient and sustained cognitive control during task switching. Neuron 2003; 39: 713–726.

Brown, J. W., Reynolds, J. R., & Braver, T. S. A computational model of fractionated conflict-control mechanisms in task switching. Cogn Psychol 2007; 55: 37–85.

Brown, S., Ratcliff, R., & Smith, P.L. A critical test of the failure to engage theory of task switching. Psychon Bull Rev 2006; 13: 152–159.

Bryck, R. L., & Mayr, U. The role of verbalization during task set selection: Switching or serial order control. Mem Cognit 2005; 33: 611–633.

Bunge, S. A. How we use rules to select actions: A review of evidence from cognitive neuroscience. Cogn Aff Behav Neurosci 2004; 4: 564–579.

Cohen, A-L, Bayer, U. C., Jaudas, A., & Gollwitzer, P. M. Self regulatory strategy and self control: Implementation intention modulates task switching and Simon task performance. Psychol Res in press.

Collette, F., & van der Linden, M. Brain imaging of the central executive component of working memory. Neurosci Behav Rev 2002; 26: 105–125.

De Jong, R. Strategical determinants of compatibility effects with task uncertainty. Acta Psychol 1995; 88: 187–207.

De Jong, R. An intention-activation account of residual switch costs. In: Monsell, S., & Driver, J. (Eds.), *Attention and performance XVIII: Control of cognitive processes*. Cambridge, MA: MIT Press, 2000: pp. 357–376.

Derrfuss, J., Brass, M., Neumann, J., & von Cramon, D. Y. Involvement of the inferior frontal junction in cognitive control: meta-analyses of switching and Stroop studies. Hum Brain Mapp 2005; 25: 22–34.

Dixon, P. Algorithms and selective attention. Mem Cogn 1981; 9: 177–184.

Dixon, P., & Just, M. A. A chronometric analysis of strategy preparation in choice reactions. Mem Cogn 1986; 14: 488–500.

Dosenbach, N. U. F., Visscher, K. M., Palmer, E. D., et al. A core system for the implementation of task sets. Neuron 2006; 50: 799–812.

Dreisbach, G., Haider, H., & Kluwe, R. H. Preparatory processes in the task-switching paradigm: Evidence from the use of probability cues. J Exp Psychol Learn Mem Cogn 2002; 28: 468–483.

Dreisbach, G., Goschke, T., & Haider, H. Implicit task sets in task switching? J Exp Psychol Learn Mem Cogn 2006; 32: 1221–1233.

Dreisbach, G., Goschke, T., & Haider, H. The role of rules and stimulus-response mappings in the task-switching paradigm. Psychol Res 2007; 71: 383–392.

Emerson, M. J., & Miyake, A. The role of inner speech in task switching: A dual-task investigation. J Mem Lang 2003; 48: 148–168.

Fagot, C. *Chronometric investigations of task switching*. Unpublished Ph.D. thesis, University of California, San Diego, 1994.

Forstmann, B. U., Brass, M., Koch, I., & von Cramon, D. Y. Internally generated and directly cued task sets: an investigation with fMRI. Neuropsychologia 2005; 43: 943–952.

Forstmann, B., Brass, M., Koch, I., & von Cramon, D. Y. Voluntary selection of task-sets revealed by functional magnetic resonance imaging. J Cogn Neurosci 2006; 18: 388–398.

Friedman, N., & Miyake, A. The relations among inhibition and interference control functions. J Exp Psychol Gen 2004; 133: 101–135.

Gade, M., & Koch, I. Cue-task associations in task switching. Q J Exp Psychol 2007; 60: 762–769.

Gibson, J. J. A critical review of the concept of set in contemporary experimental psychology. Psychol Bull 1941; 38: 781–817.

Gilbert, S. J., & Shallice, T. Task switching: A PDP model. Cogn Psychol 2002; 44: 297–337.

Gopher, D., Armony, L., & Greenshpan, Y. Switching tasks and attention policies. J Exp Psychol Gen 2000; 129: 308–339.

Goschke, T. Intentional reconfiguration and involuntary persistence in task-set switching. In: Monsell, S., & Driver, J. (Eds.), *Attention and performance XVIII: Control of cognitive processes*. Cambridge, MA: MIT Press, 2000: pp. 331–355.

Gotler, A., & Meiran, N. Cognitive processes underlying a frontally-mediated component of task-switching. Brain Cogn 2001; 47: 142–146.

Gotler, A., Meiran, N., & Tzelgov, J. Nonintentional task-set activation: Evidence from implicit task sequence learning. Psychon Bull Rev 2003; 10: 890–896.

Hahn, S., Andersen, G. J., & Kramer, A. F. Multidimensional set switching. Psychon Bull Rev 2003; 10: 503–509.

Hartley, A. A., Kieley, J. M., & Slabach, E. H. Age differences and similarities in the effects of cues and prompts. J Exp Psychol Hum Percept Perform 1990; 16: 523–538.

Hayes, A. E., Davidson, M. C., Keele, S. W., & Rafal, R. D. Toward a functional analysis of the basal ganglia. J Cogn Neurosci 1998; 10: 178–198.

Heuer, H., Schmidtke, V., & Kleinsorge, T. Implicit learning of sequences of tasks. J Exp Psychol Learn Mem Cogn 2001; 27: 967–983.

Hübner, R., & Dreuy, M. Response execution, selection, or activation: What is sufficient for response-related repetition effects under task shifting? Psychol Res 2006; 70: 245–261.

Hübner, R., Futterer, T., & Steinhauser, M. On attentional control as a source of residual shift costs: Evidence from two-component task shifts. J Exp Psychol Learn Mem Cogn 2001; 27: 640–653.

Hübner, M., Dreisbach, G., Haider, H., & Kluwe, R. H. Backward inhibition as a means of sequential task-set Control: Evidence for reduction of task competition. J Exp Psychol Learn Mem Cogn 2003; 29: 289–297.

Hunt, A. R., & Klein, R. M. Eliminating the cost of task set reconfiguration. Mem Cogn 2002; 30: 529–539.

Jersild, A.T. Mental set and shift. Arch Psychol 1927; 14: 81.

Kessler, Y., & Meiran, N. All updateable objects in working memory are updated whenever any of them is modified: Evidence from the memory updating paradigm. J Exp Psychol Learn Mem Cogn 2006; 32: 570–585.

Kessler, Y., Shencar, Y., & Meiran, N. Choosing to switch: Spontaneous task switching despite associated behavioral costs. Acta Psychol 2009; 131: 120–128.

Kleinsorge, T., & Heuer, H. Hierarchical switching in a multi-dimensional task space. Psychol Res 1999; 62: 300–312.

Koch, I. Automatic and intentional activation of task sets. J Exp Psychol Learn Mem Cogn 2001; 27: 1474–1486.

Koch, I. Sequential task predictability in task switching. Psychon Bull Rev 2005; 12: 107–112.

Koch, I., & Allport, A. Cue-based preparation and stimulus-based priming of tasks in task switching. Mem Cogn 2006; 34: 433–444.

Koch, I., Prinz, W., & Allport, A. Involuntary retrieval in alphabet-arithmetic tasks: Task-mixing and task-switching costs. Psychol Res 2005; 69: 252–261.

Kochanska, G, Coy, K. C., & Murray, K. T. The development of self regulation in the first four years of life. Child Develop 2001; 72: 1091–1111.

Kray, J., & Lindenberger, U. Adult age differences in task-switching. Psychol Aging 2000; 15: 126–147.

Lien, M.-C., & Ruthruff, E. Task switching in a hierarchical task structure: Evidence for the fragility of the task repetition benefit. J Exp Psychol Learn Mem Cogn 2004; 30: 697–713.

Lien, M.-C., Ruthruff, E., & Kuhns, D. On the difficulty of task switching: Assessing the role of task set inhibition. Psychon Bull Rev 2006; 13: 530–535.

Lien, M.-C., Ruthruff, E., Remington, R. W., & Johnston, J. C. On the limits of advance preparation for a task switch: Do people prepare all the task some of the time or some of the task all of the time. J Exp Psychol Hum Percept Perform 2005; 31: 299–315.

Logan, G. D. On the ability to inhibit thought and action: A users' guide to the stop signal paradigm. In: Dagenbach, D., & Carr, T. H. (Eds.), *Inhibitory processes in attention, memory, and language*. San Diego, CA: Academic Press, 1994: pp. 189–239.

Logan, G. D. Executive control of thought and action: In search of the wild homunculus. Current Direct Psychol Sci 2003; 12: 45–48.

Logan, G. D. Working memory, task switching, and executive control in the task span procedure. J Exp Psychol Gen 2004; 133: 218–236.

Logan, G. D., & Bundesen, C. Clever homunculus: Is there an endogenous act of control in the explicit task-cuing procedure? J Exp Psychol Hum Percept Perform 2003; 29: 575–599.

Logan, G. D., & Burkell, J. Dependence and independence in responding to double stimulation: A comparison of stop, change, and dual-task paradigms. J Exp Psychol Hum Percept Perform 1986; 12: 549–563.

Logan, G. D., & Gordon, R. D. Executive control of visual attention in dual-task situations. Psychol Rev 2001; 108: 393–434.

Los, S. A. On the origin of mixing costs: Exploring information processing in pure and mixed blocks of trials. Acta Psychol 1996; 94: 145–188.

Los, S. A., & Van Den Heuvel, C. E. Intentional and unintentional contributions to nonspecific preparation during reaction time foreperiods. J Exp Psychol Hum Percept Perform 2001; 27: 370–386.

Luria, R., & Meiran, N. Online order control in the psychological refractory period paradigm. J Exp Psychol Hum Percept Perform 2003; 29: 556–574.

Luria, R., & Meiran, N. Dual route for subtask order control: Evidence from the Psychological Refractory Period paradigm. Q J Exp Psychol Section A, 2006; 59: 720–744.

Luria, R., Meiran, N., & Dekel-Cohen, C. Stimulus cued completion of reconfiguration and retroactive adjustment as causes for the residual switching cost in multi-step tasks. Eur J Cogn Psychol 2006; 18: 652–668.

MacDonald III, A.W., Cohen, J.D., Stenger, V.A., & Carter, C.S. Dissociating the role of the dorsolateral prefrontal and anterior cingulated cortex in cognitive control. Science 2000; 288: 1835–1838.

Masson, M. E. J., Bub, D. N., Woodward, T. S., & Chan, J. C. K. Modulation of word-reading processes in task switching. J Exp Psychol Gen 2003; 132: 400–418.

Mayr, U. Age differences in the selection of mental sets: The role of inhibition, stimulus ambiguity, and response-set overlap. Psychol Aging 2001; 16: 96–109.

Mayr, U., & Bell, T. On how to be unpredictable? Evidence from the voluntary task switching paradigm. Psychol Sci 2006; 17: 774–780.

Mayr, U., & Keele, S. W. Changing internal constraints on action: The role of backward inhibition. J Exp Psychol Gen 2000; 129: 4–26.

Mayr, U., & Kliegl, R. Task-set switching and long-term memory retrieval. J Exp Psychol Learn Mem Cogn 2000; 26: 1124–1140.

Mayr, U., & Kliegl, R. Differential effects of cue changes and task changes on task-set selection costs. J Exp Psychol Learn Mem Cogn 2003; 29: 362–372.

Meiran, N. Reconfiguration of processing mode prior to task performance. J Exp Psychol Learn Mem Cogn 1996; 22: 1423–1442.

Meiran, N. Modeling cognitive control in task-switching. Psychol Res 2000a; 63: 234–249.

Meiran, N. The reconfiguration of the stimulus task-set and the response task set during task switching. In: Monsell, S., & Driver, J. (Eds.), *Attention and performance XVIII: Control of cognitive processes.* Cambridge, MA: MIT Press, 2000b: pp. 377–400.

Meiran, N. Task rule congruency and Simon-like effects in switching between spatial tasks. Q J Exp Psychol Section A, 2005; 58A: 1023–1041.

Meiran, N., & Chorev, Z. Phasic alertness and the residual task-switching cost. Exp Psychol 2005; 52: 109–124.

Meiran, N., Chorev, Z., & Sapir, A. Component processes in task switching. Cogn Psychol 2000; 41: 211–253.

Meiran, N., & Daichman, A. Advance task preparation reduces task error rate in the cueing task-switching paradigm. Mem Cogn 2005; 33, 1272–1288.

Meiran, N., Hommel, B., Bibi, U., & Lev, I. Consciousness and control in task switching. Conscious Cogn 2002; 11: 10–33.

Meiran, N., & Kessler, Y. The task rule congruency effect in task switching reflects activated long term memory. J Exp Psychol Hum Percept Perform 2008; 34: 137–157.

Meiran, N., & Marciano, H. Limitations in advance task preparation: Switching the relevant stimulus dimension in speeded same-different comparisons. Mem Cogn 2002; 30: 540–550.

Meiran, N., Kessler, Y., & Adi-Japha, E. Control by Action Representation and Input Selection (CARIS): A theoretical framework for task switching. Psychol Res 2008; 72: 473–500.

Meuter, R. F., & Allport, A. Bilingual language switching in naming: asymmetrical costs of language selection. J Mem Lang 1999; 40: 25–40.

Meyer, D. E., & Kieras, D. E. A computational theory of executive cognitive processes and multiple-task performance: Part 1. Basic mechanisms. Psychol Rev 1997a; 104: 3–65.

Meyer, D. E., & Kieras, D. E. A computational theory of executive cognitive processes and multiple-task performance: Part 2. Accounts of psychological refractory-period phenomena. Psychol Rev 1997b; 104: 749–791.

Miller, E. K., & Cohen, J. D. An integrative theory of prefrontal cortex function. Ann Rev Neurosci 2001; 24: 167–202.

Miller, G. A. The magical number seven, plus or minus two: Some limits on our capacity to process information. Psychol Rev 1956; 63: 81–97.

Miyake, A., Friedman, N. P., Emerson, M. J., Witzki, A. H., Howerter, A., & Wager, T. D. The unity and diversity of executive functions and their contributions to complex "frontal lobe" tasks: A latent variable analysis. Cogn Psychol 2000; 41: 49–100.

Monsell, S. Task switching. Trends Cogn Sci 2003; 7: 134–140.

Monsell, S., & Mizon, G. A. Can the task cuing paradigm measure an "endogenous" task set reconfiguration process? J Exp Psychol Hum Percept Perform 2006; 32: 493–516.

Monsell, S., Sumner, P., & Waters, H. Task-set reconfiguration with predictable and unpredictable task switches. Mem Cogn 2003; 31: 327–342.

Morris, N., & Jones, D. M. Memory updating in working memory: The role of the central executive. Br J Psychol 1990; 81: 111–121.

Muraven, M. R., & Baumeister, R. F. Self-regulation and depletion of limited resources: Does self-control resemble a muscle? Psychol Bull 2000; 126: 247–259.

Nieuwenhuis, S., & Monsell, S. Residual costs in task switching: Testing the failure-to-engage hypothesis. Psychon Bull Rev 2002; 9: 86–92.

Norman, D. A., & Shallice, T. Attention to action: Willed and automatic control of behavior. In: Davidson, R. J., Schwartz, G. E., & Shapiro, D. (Eds.), *Consciousness and self-regulation: Advances in research and theory*, Vol. 4. New York: Plenum, 1986: pp. 1–18.

Ochsner, K. N., & Gross, J. J. The cognitive control of emotion. Trends Cogn Sci 2005; 9: 242–249.

Pashler, H. Dual-task interference in simple tasks: Data and theory. Psychol Bull 1994; 116: 220–244.

Philipp, A. M., Jolicœur, P., Falkenstein, M., & Koch, I. Response selection and response execution in task switching: Evidence from a go-signal paradigm. J Exp Psychol Learn Mem Cogn 2007; 33: 1062–1075.

Philipp, A. M., & Koch, I. Switching of response modalities. Q J Exp Psychol A Hum Exp Psychol 2005; 58A: 1325–1338.

Posner, M. I., & Boies, S. Components of attention. Psychol Rev 1971; 78: 391–048.

Prinz, W. Why Donders has led us astray? In: Hommel, B., & Prinz, W. (Eds.), Theoretical issues in stimulus response compatibility. *Advances in Psychology*. US: Elsevier Science, 1997: pp. 247–267.

Rogers, R. D., & Monsell, S. The cost of a predictable switch between simple cognitive tasks. J Exp Psychol Gen 1995; 124: 207–231.

Rubin, O., & Meiran, N. On the origins of the task mixing cost in the cuing task switching paradigm. J Exp Psychol Learn Mem Cogn 2005; 31: 1477–1491.

Rubinstein, J. S., Meyer, D. E., & Evans, J. E. Executive control of cognitive processes in task switching. J Exp Psychol Hum Percept Perform 2001; 27: 763–797.

Ruthruff, E., Remington, R. W., & Johnston, J. C. Switching between simple cognitive tasks: The interaction of top-down and bottom-up factors. J Exp Psychol Learn Mem Cogn 2001; 27: 1404–1419.

Schneider, D. W., & Logan, G. D. Modeling task switching without switching tasks: A short-term memory priming account of explicitly cued performance. J Exp Psychol Gen 2005; 134: 343–367.

Schneider, D. W., & Logan, G. D. Hierarchical control of cognitive processes: Switching tasks in sequences. J Exp Psychol Gen 2006; 135: 623–640.

Schneider, D. W., & Logan, G. D. Defining task-set reconfiguration: The case of reference point switching. Psychon Bull Rev 2007; 14: 118–125.

Schuch, S., & Koch, I. The role of response selection for inhibition of task sets in task shifting. J Exp Psychol Hum Percept Perform 2003; 29: 92–105.

Schuch, S., & Koch, I. The cost of changing the representation of action. J Exp Psychol Hum Percept Perform 2004; 30: 566–582.

Shaffer, L.H. Choice reaction with variable S-R mapping. J Exp Psychol 1965; 70: 284–288.

Sohn, M. H., & Anderson, J. R. Task preparation and task repetition: Two-component model of task switching. J Exp Psychol Gen 2001; 130: 764–778.

Sohn, M. H., & Carlson, R. A. Effects of repetition and foreknowledge in task-set reconfiguration.

J Exp Psychol Learn Mem Cogn 2000; 26: 1445–1460.

Steinhauser, M., & Hübner, R. Mixing costs in task-shifting reflect sequential processing stages in a multi-component task. Mem Cogn 2005; 33: 1484–1494.

Steinhauser, M., & Hübner, R. Response-based strengthening in task-shifting: Evidence from shift effects produced by errors. J Exp Psychol Hum Percept Perform 2006; 32: 517–534.

Sudevan, P., & Taylor, D. A. The cueing and priming of cognitive operations. J Exp Psychol Hum Percept Perform 1987; 13: 89–103.

Sumner, P., & Ahmed, L. Task switching: The effect of task recency with dual and single affordance stimuli. Q J Exp Psychol 2006; 59: 1255–1276.

Tornay, F. J., & Milàn, E. G. A more complete task-set reconfiguration in random than in predictable task switch. Q J Exp Psychol 2001; 54: 785–803.

Verbruggen, F., Liefooghe, B., Vandierendonck, A., & Demanet, J. Short cue presentations encourage advance task preparation: A recipe to diminish the residual task switching cost. J Exp Psychol Learn Mem Cogn 2007; 33: 342–356.

Vohs, K. D., & Heatherton, T. F. Self-regulatory failure: A resource depletion approach. Psychol Sci 2000; 11: 249–254.

Ward, L. M. Determinants of attention to local and global features of visual forms. J Exp Psychol Hum Percept Perform 1982; 8: 562–581.

Waszak, F., Hommel, B., & Allport, A. Task-switching and long-term priming: Role of episodic stumulus-task bindings in task-shift costs. Cogn Psychol 2003; 46: 361–413.

Wood, J. N., & Grafman, J. Human prefrontal cortex: Processing and representational perspectives. Nat Rev Neurosci 2003; 4: 139–147.

Wylie, G. R., Javitt, D. C., & Foxe, J. J. Don't think of a white bear: an fMRI investigation of the effects of sequential instructional sets on cortical activity in a task-switching paradigm. Hum Brain Mapp 2004; 21: 279–297.

Yehene, E., & Meiran, N. Is there a general task switching ability? Acta Psychol 2007; 126: 169–195.

Yehene, E., Meiran, N., & Soroker, N. Task alternation cost without task alternation: Measuring intentionality. Neuropsychologia 2005; 43: 1858–1869.

Yeung, N., & Monsell, S. Switching between tasks of unequal familiarity: The role of stimulus-attribute and response-set selection. J Exp Psychol Hum Percept Perform 2003a; 29: 455–469.

Yeung, N., & Monsell, S. The effects of recent practice on task switching. J Exp Psychol Hum Percept Perform 2003b; 29: 919–936.

Yeung, N., Nystrom, L. E., Aronson, J. A., & Cohen, J. D. Between-task competition and cognitive control in task switching. J Neurosci 2006; 26: 1429–1438.

CHAPTER 12

Unconscious Influences of Attitudes and Challenges to Self-Control

Deborah L. Hall and B. Keith Payne

ABSTRACT

Recent developments in social psychology have brought to light important questions concerning the nature of implicit cognitions about race and the implications for self-control. If people are unaware of their own prejudices, how can they keep them in check? This chapter distinguishes the unconscious influence of people's attitudes and beliefs from attitudes and beliefs that are themselves hidden from consciousness. It reviews research that provides little evidence of unconscious racial attitudes, but much stronger evidence that racial attitudes influence people's judgments and behavior in numerous unconscious ways. It also discusses the unique challenges to self-control that unconscious influences present and then concludes by highlighting strategies for preventing prejudice by limiting the unconscious influence of people's attitudes about race.

Keywords: Unconscious, attitudes, implicit, prejudice, self-control

---

Remember the last time you were stuck in traffic, only to find that after an eternity of inching, there was nothing but a fender-bender neatly off to the side of the road to explain the delay? It is maddening to realize that the whole delay resulted from one driver after the other slowing down to take a good look. This realization usually occurs precisely as you are craning your neck, slowing down to take a good look. It is hard not to be captivated by the details of people's misfortune—just ask the folks who read gossip columns, watch horror films, or study social psychology.

Social psychologists have always been excited by findings that are nothing but bad news for our cherished ideas about how humans operate. Classic research on cognitive dissonance challenged the view of humans as rational decision-makers by showing that people would sometimes perform acrobatic contortions of self-justification rather than accept threatening inconsistencies (Festinger, 1957). Milgram's (1963) studies of obedience revealed how chillingly far people would go in response to simple commands by a person wearing a lab coat. More recently, social psychologists have told us that feeling in control of our actions does not make it so, and that much of our behavior can rattle off pretty well without us (Bargh & Chartrand, 1999; Wegner, 2002). After decades of optimism about the steady decline of prejudice, they have told us that rumors of racism's demise have been greatly exaggerated and that valuing equality does not guarantee a conscience clear of bias (Devine, 1989; Greenwald & Banaji, 1995). And, if you are thinking that you cannot recall a single instance in which these findings would apply to you, social psychologists are here to tell you that is exactly what they would expect someone in your position to say (Pronin, Gilovich, & Ross, 2004).

The excitement is understandable. Ideas that change the way we think about being human are the signposts that let us know psychological science is getting somewhere. Still, that progress leaves a lot of misfortune on the side of the road. These kinds of findings raise important questions about agency, responsibility, and self-control. In this essay, we focus on the questions raised by research on implicit social cognition, especially as they pertain to stereotypes and attitudes about race. A common worry is that if biases reside outside of consciousness, even enlightened and well-intentioned people will be hard pressed to control them. "Never forget," warned Baudelaire, "when you hear the progress of enlightenment vaunted, that the devil's best trick is to persuade you that he doesn't exist," (1869/1997). How can people control these devilish biases if they do not know they exist? Does implicit social cognition research show that people hold attitudes and stereotypes of which they are unconscious? And does it cause problems for practical issues of self-control or moral issues of responsibility?

A careful look at the research literature on implicit cognition shows good evidence for some versions of unconsciousness but not others. We distinguish unconscious *influences* of attitudes and beliefs from attitudes and beliefs that are themselves hidden from consciousness because the literature has very different things to say about them. For instance, it seems clear that people can be, and often are, influenced by attitudes and beliefs without being aware of it. But hard evidence that people have attitudes and beliefs that they do not know about, or cannot know about when they try, is difficult to find.

At this point, some readers will breathe a sigh of relief because if the research doesn't prove that people have widespread unconscious prejudices, then maybe it does not threaten comfortable ideas about self-control and responsibility after all. Other readers will feel a little disappointed at the news because if the research does not prove that people have widespread unconscious prejudices, then maybe the science is not as groundbreaking as was once thought. But it may be too soon for sighs of relief or disappointment. The varieties of unconsciousness that are well-supported by research pose plenty of problems for people trying to keep their biases under control. In our view, both unconscious attitudes and unconscious influences pose potential problems for controlling actions; but the evidence for unconscious influence is much more substantial than for unconscious attitudes. Ultimately, a lot of people may be rubbernecking in the wrong direction because it is unconscious *influence* that really threatens our ideas about responsibility and self-control.

Does Research Support the Link between Implicit Measures and Unconscious Beliefs?

The development of indirect techniques for measuring attitudes about race and stereotypes has been driven by a remarkable surge of interest in implicit social cognition (Greenwald & Banaji, 1995; Petty, Fazio, & Brinol, 2008). A common interpretation of these measures is that they reveal unconscious attitudes and beliefs—mental content that people are not aware they possess. This interpretation is so widespread that it is rarely justified or empirically tested. Instead, when an implicit test is used to measure some concept, not only the concept itself, but almost anything associated with that test is often assumed to be unconscious. This interpretation, however, has not gone without criticism. Several authors have noted that the evidence that these tests reveal unconscious attitudes and stereotypes is weak (e.g., Fazio & Olson, 2003; Gawronski, Hofmann, & Wilbur, 2006). Nevertheless, when implicit or indirect tests show different patterns from explicit or direct measures, it is typical to read explanations based on unconscious versus conscious attitudes. In this chapter, we draw on these previous papers as well as on observations from our own research to critically evaluate notions of unconscious attitudes and unconscious influence.

Why Do Implicit and Explicit Attitude Tests Diverge?

The key evidence used to support the argument that implicit tests reveal unconscious attitudes comes from dissociations between implicit and explicit tests. That is, when the results of implicit and explicit measures of the same construct fail to correlate, the conscious-unconscious distinction is immediately invoked. One of the most informative findings comes from a meta-analysis conducted by Hofmann et al. (2005). Looking at 126 studies in which participants' attitudes were measured by the Implicit Association Test (IAT; Greenwald, McGhee, & Schwartz, 1998) and via self-report, they found that the average correlation between implicit and explicit attitudes was $r = .24$. Reviews of implicit cognition research using affective priming and other methods have found similarly low correlations (e.g., Fazio & Olson, 2003).

This pattern of dissociation is especially common in studies of racial attitudes, to which we now turn our focus. The most common interpretation is that people report their consciously held, egalitarian attitudes towards racial minorities on direct measures, whereas indirect measures reveal the less-positive racial attitudes that linger in their unconscious. In essence, explicit measures reveal a vaunted enlightenment, but implicit measures do not fall for the mind's devilish tricks.

The lack of correspondence between implicit and explicit tests, however, may result from other factors that are unrelated to unconscious attitudes. For example, research subjects may be motivated to present themselves in a positive, unbiased light. Concerns with self-presentation, therefore, may influence the degree to which participants provide honest responses on self-report measures of racial bias, and this may ultimately attenuate the relationship between implicit and explicit measures. Another factor driving the weak correlation may be the low reliability of some indirect measures. Furthermore, when subjects are asked to report their attitudes or feelings, they may be perfectly aware that they tend to get squeamish around certain sorts of people, but they might not feel as though that squeamishness reflects their true attitude. Instead, they may consider it something else—a bad habit left over from their less-enlightened days or a primitive gut reaction they do not believe maps on to their "true attitude." Finally, methodological differences between implicit and explicit tests that encourage different mental processes—what we refer to as a lack of 'structural fit'—may also mask higher implicit-explicit correlations. We next look more carefully at each of these issues.

The Motive to Self-Present

Tolerance for the expression of discriminatory and racially insensitive attitudes has been diminishing. As a result, many people are reluctant to report their true feelings about other races when doing so may paint them in a prejudicial light. The correlation between the implicit and explicit racial attitudes these people report should be low to the extent that: *(1)* people deliberately report racial attitudes that deviate from their true beliefs on explicit measures, and *(2)* they are unable to control the responses reflecting their true racial attitudes on implicit measures. People who are less concerned about appearing prejudiced, on the other hand, should have no reason to alter their responses on explicit measures of racial attitudes. For these people, there should be a much stronger positive correlation between implicit and explicit racial attitudes.

Research has consistently shown that people's willingness to report their genuine attitudes depends on how concerned they are about acting without prejudice. Fazio et al. (1995) examined the relationship between the racial attitudes participants explicitly reported on the Modern Racism Scale (McConahay, 1986), a self-report measure of opinions on racially charged political issues, and the attitudes revealed using a priming measure of participants' affective associations with Black versus White faces. They found that the strength of the correlation between the implicit and explicit measures differed depending on how motivated participants were to control potentially prejudiced responses. Participants who were highly motivated to suppress reactions that might reflect racial bias on their part showed the typical dissociation between directly and indirectly measured racial attitudes. The racial attitudes they reported on the Modern Racism Scale were weakly (negatively) correlated with the amount of racial bias they demonstrated in their affective reactions to Black and White faces. For participants who were low in the motivation to control prejudiced responses, a stronger positive correlation between implicit and explicit measures emerged.

We obtained similar results in our own research on racially biased perceptual errors (Payne, 2001). Imagine performing the following task: An image appears on a computer screen in front of you and your job is to identify whether the image is a picture of a gun or a tool. Just prior to the appearance of the gun or tool, you see a photograph of either a Black or a White person's face on the computer. You are instructed to ignore the photograph of the face and to classify the image of the gun or tool as quickly and as accurately as you can. The presentation of either a Black or White face followed by the appearance of either a gun or a tool appears a number of times in succession, and on each trial, your job is always to correctly identify the guns and tools. We have asked participants to do just this and have found a consistent pattern of results. Participants are more likely to misidentify a harmless tool as a gun when primed with a Black face than a White face (Payne, 2001). The stronger the association people have between "Black" and "gun," the more pronounced the bias. The weapon identification task can thus be used as an implicit measure of people's racially biased attitudes toward Blacks.

Paralleling the findings of Fazio et al. (1995), we have discovered that the strength of the correlation between racial bias on the weapon identification task and participants' self-reported racial attitudes varies based on their motivation to control prejudiced responses. In one study (Payne, 2001), we had participants complete the Modern Racism Scale before performing the weapon identification task. We also measured participants' motivation to control their prejudice using Dunton and Fazio's (1997) Motivation to Control Prejudiced Reactions Scale. The degree of correspondence between race bias on the weapon identification task and the racial attitudes reported on the Modern Racism Scale depended on participants' motivation to control prejudice. For participants low in the motivation to control prejudiced responses, there was a positive relationship between implicit and explicit racial bias against Blacks. Greater implicit bias on the weapon identification task was associated with more negative attitudes towards Blacks on the Modern Racism Scale. The relationship between implicitly and explicitly measured racial bias evaporated for participants reporting a stronger motivation to control prejudiced responses. That is, participants' responses on the Modern Racism Scale were uncorrelated with their performance on the weapon identification task.

Another line of support for the role of self-presentation motives comes from a study by Nier (2005) in which participants' racial attitudes were assessed with a direct and an indirect measure. Using the "bogus pipeline" manipulation, Nier led half of his participants to believe that attempts at misrepresenting their true racial attitudes would be detected by the experimenter. This procedure reduced the motive to present oneself in a positive light, as any deviations from the truth would ultimately be exposed. Nier's results mirrored our own with respect to the motivation to control prejudice. For participants who could act on a self-presentation motive without fear of being detected, the correlation between implicit and explicit racial attitudes was nonsignficant. But for participants who could not act on this motive, there was a significant positive relationship between the implicit and explicit measures of racial attitudes.

We have obtained comparable results using an indirect attitude measure called the affect misattribution procedure, or AMP, for short, to assess implicit racial bias (Payne et al., 2005). The AMP is based on the idea that people's evaluation of neutral, ambiguous stimuli can reveal something about their underlying attitudes. In a typical study using the AMP, participants view pairs of pictures on a computer screen. The first picture in each pair is an affective prime—an image that invokes a pleasant or an unpleasant reaction, depending on one's attitude. The second image that appears on the computer screen is a Chinese symbol, or pictograph, that participants are asked to rate. These pictographs are generally unfamiliar to participants in the United States and are pretested to be neutral in valence. The key dependent measure is participants' ratings of the pleasantness of the Chinese pictographs. The logic behind the AMP is that participants' reactions to the affective primes will be misattributed to their evaluations of the neutral Chinese pictographs. Participants' evaluations of the pictographs can thus provide an indirect measure of participants' emotional responses to the primes.

We have found that the strength of the correlation between self-reported racial attitudes and race bias measured by the AMP varies depending on participants' self-presentational concerns. For example, we recruited a sample of Black and White college students and asked them to report how "warm and favorable" versus "cold and unfavorable" they felt towards Blacks and Whites. This served as our explicit measure of racial attitudes. We used two measures, Plant and Devine's (1998) Internal and External Motivation to Respond Without Prejudice Scale and Dunton and Fazio's (1997) Motivation to Control Prejudiced Responses Scale, to assess participants' motivational concerns. Finally, we administered a version of the AMP that contained photographs of the faces of young Black and White males as the affective primes to measure implicit racial attitudes.

Both the implicit and explicit measures of racial attitudes revealed a significant own-race preference. White participants evaluated the pictographs paired with White faces more favorably than the pictographs paired with Black faces and provided explicit reports of more warm and favorable feelings towards Whites than Blacks. Black participants did just the opposite. They evaluated the pictographs paired with Black faces more favorably than those paired with White faces, and their feelings towards Blacks were warmer and more favorable than their feelings towards Whites. Importantly, the implicit-explicit correlation was moderated by participants' general motivation to control prejudiced responses. There was a strong positive correlation between the implicit and explicit racial attitudes of participants who were low in the motivation to control their prejudice, however, the implicit and explicit racial attitudes of participants who were high in the motivation to control prejudice were virtually uncorrelated (Payne et al., 2005).

These patterns suggest that when subjects complete an explicit measure of their racial attitudes, many are faced with a self-control dilemma. When motivations to be unprejudiced clash with automatic affective responses, they have to find some way to resolve that conflict. Research suggests that the discrepancy between motivations and affective responses may itself initiate the conflict resolution process. For example, Moskowitz and colleagues (1999) suggested that for subjects motivated to control prejudice, processing stereotype-related information activates not only prejudiced responses, but also egalitarian goals. This finding is related to a more general phenomena, in which tempting stimuli elicit not only automatic or impulsive inclinations, but also counteractive control motives (Trope & Fishbach, 2000; *see also* Amodio et al., 2004). Because responses on explicit tests are easier to monitor and control than implicit tests, these control efforts can be expected to mainly influence self-reports on explicit tests.

Taken together, these findings indicate that motivational concerns are a major factor determining the relationship between directly and indirectly measured racial attitudes (*see also* Akrami & Ekehammer, 2005; Dunton & Fazio, 1997; Gawronski, Geschke, & Banse, 2003; Hofmann, Gschwendner, & Schmitt, 2005; Nosek, 2005). The more lack of correspondence between implicit and explicit tests can be explained by self-presentation motives, the less reason there is to assume that biases driving responses on indirect measures are hidden from consciousness. People can and do report their automatic racial attitudes when they are properly motivated to do so.

Reliability of Indirect Measures

The low correlation between direct and indirect measures of racial attitudes may also be a consequence of the low internal reliability of many indirect measures, particularly measures that use reaction times as the key dependent variable. Compared to direct attitude measures, indirect measures that use reaction times tend to be higher in measurement error (Cunningham, Preacher, & Banaji, 2001). This is not surprising, given the multiple factors that can affect response times on even simple tasks and the high variability inherent in response-time measures (Lane et al., 2007). The measurement error resulting from the use of indirect measures may attenuate the correlation between direct and indirect tests because reliability sets the upper limit on correlations. The highest possible correlation between two tests depends on the product of their reliabilities. Two tests each with reliabilities of .50, for example, cannot correlate higher than .25. Cunningham, Preacher, and Banaji (2001) used latent variable analysis to control for measurement error in implicit tests. In doing so, they obtained a much higher correlation between implicit and explicit racial attitude tests than has typically been found with other statistical procedures.

One way to control for measurement error is to use low-reliability tests, estimate the amount of measurement error, and then estimate what the correlations between tests would be in the absence of measurement error using sophisticated statistical procedures. The success of this approach hinges on the ability to accurately estimate the amount of measurement error. It requires multiple tests for each construct and relatively large sample sizes. Another way to surmount the problem of measurement error is to use indirect measures with higher reliability. The AMP, which does not rely on reaction times, circumvents the issue of high measurement error. Across several studies, the AMP has demonstrated sufficiently high reliability, with an internal consistency that is usually above .80. As a result, we have been able to detect strong positive correlations between implicit and explicit racial attitudes when using the AMP as our indirect measure.

Self-presentation and reliability have been discussed by several authors as reasons for divergence between implicit and explicit tests. To these well-established factors, we add two more that both have to do with what the results of implicit attitude tests really mean. The first issue is how research subjects construe what their "attitude" is. The second has to do with how well our measurement techniques capture the most important distinctions between implicit and explicit tests.

What Is Your Real Attitude, Anyway?

Any attitude test is an attempt to operationally define a particular attitude. That includes identifying where a person stands on some dimension, such as from "like" to "dislike," but it also includes some assumptions about what an attitude is to begin with. Some attitude tests, such as a "feeling thermometer," allow subjects to define their attitudes largely for themselves by simply asking them about their feelings. Other attitude tests, such as the Modern Racism Scale, define the attitude for participants by asking about specific policy views, such as whether or not they support busing policies as a means for achieving racial integration. The assumption behind the measure is that subjects' racial attitudes can be inferred from those policy preferences. Implicit attitude tests also include some assumptions, often tacit, about what an attitude is. Some of those assumptions have been controversial, as researchers disagree about whether the kinds of behaviors measured by particular implicit tests should be considered indicators of attitudes or something else (Arkes & Tetlock, 2004; Blanton & Jaccard, 2006).

If attitude researchers have differing ideas about what an attitude is and whether any given test captures it, it is a good bet that ordinary attitude-holders have an even more diverse set of ideas about what constitutes their attitudes. Oftentimes when participants (and researchers) take an implicit test, they are surprised by what the test reveals about their beliefs and preferences. One interpretation of this feeling is that the attitude is unconscious. "If I am surprised by my own test results, then maybe I don't know what my attitude is." From this perspective, taking an implicit test can be like finding out from an X-ray that you swallowed a marble: "How did that get in there?!" Another interpretation of the surprise is that people may not consider the associations, feelings, or experiences driving their responses on implicit tests to reflect their "real attitude." In this case, the first response is "Oh no—a foreign object?!" but then, "Oh, wait, you just mean that marble I swallowed last Tuesday." The trick is knowing when the researcher and the subject are talking about the same marble.

An intriguing study has recently made headway on this problem. Ranganath, Smith, and Nosek (2008) gave research subjects three reaction-time-based implicit measures of attitudes toward gay people. They also asked subjects to rate their attitudes toward gay people using scales that distinguished between "gut reactions" and "actual feelings." The gut reactions subjects reported were more negative than their actual feelings. And, importantly, those gut reactions corresponded well with the implicit measures, whereas ratings of actual feelings did not. This pattern suggests that when studies of stereotypes and intergroup bias ask people to report their attitudes, the low correlations frequently found with implicit measures may stem in part from the fact that subjects do not consider their "gut reactions" to be their "actual feelings." It also suggests that subjects have some awareness of the attitudes revealed by implicit tests because when asked the "right" questions, they can report them in a way that matches their responses on implicit tests.

This study illustrates the importance of carefully considering what subjects think words like "attitudes" and "feelings" mean. It is difficult to know what to make of results when the researcher and the subject are not defining these words in the same way. The more general version of this argument is that when measuring implicit and explicit responses, it is important for the tests to be controlled in such a way that the only key difference between them is along the dimension of interest—be it consciousness versus unconsciousness, automatic versus controlled responses, etc. If the key difference is confounded with other inconsistencies between the two tests, such as differing meanings of "attitude" or different construals of the attitude object, it is hard to know what the results mean. Below we discuss some of the problems with using implicit and explicit tests that differ in ways beyond the implicit-explicit distinction and describe a potential solution.

Structural Fit: Equating Implicit and Explicit Tests on Extraneous Differences

The obvious difference between implicit and explicit tests is that implicit tests are supposed to be implicit—that is, they are supposed to measure automatic or unconscious processes—and explicit tests are supposed to be explicit measures of consciously controlled processes. But take a minute to think about all of the other ways that implicit and explicit tests tend to differ. In other words, the weak correlation between implicit and explicit racial attitudes may often be driven by differences in the way direct and indirect measures are typically structured. Take, for example, the low correlation that many researchers have found between racial attitudes reported on the Modern Racism Scale (MRS) and performance on the IAT. The difference between these measures that is usually of greatest theoretical interest is that one test, the MRS, asks participants to directly report their attitudes, whereas the other test, the IAT, provides an indirect index of participants' racial attitudes. However, these two measures differ in a number of other ways that are entirely independent of the direct-indirect (or, implicit-explicit) distinction. For example, the MRS asks participants to indicate how much they agree or disagree with verbal statements that capture attitudes about racially charged political issues. Completing the MRS requires that participants read complex propositional statements and translate their attitude into a response that falls along a bipolar (disagree-agree) continuum. Participants taking the IAT are asked to indicate whether words or pictures representing different categories are "good" or "bad" based on a decision rule that is established at the beginning of the test. As this comparison illustrates, although the Modern Racism Scale and the IAT differ with respect to how directly they measure racial bias, they also differ in ways that are unrelated to the implicit-explicit distinction. Without controlling for these incidental differences, conclusions about the underlying cause of the low correlation cannot necessarily be drawn.

Using the AMP as an indirect measure of racial attitudes, we have found that the strength of its relationship with direct attitude measures depends heavily on how closely matched the structural components of the two tests are. In a recent series of studies, subjects performed the AMP, with photos of Black and White male faces as the affective primes that preceded the presentation of the neutral Chinese pictographs. Participants were instructed to ignore the pictures of the faces, and to rate the pleasantness of each Chinese pictograph on a 4-point scale. In a second version of the AMP, subjects saw the same face-pictograph pairs, but were instructed to ignore the Chinese pictographs and rate the pleasantness of the Black and White faces on the same 4-point scale. The standard version of the AMP, in which participants rated the Chinese pictographs, served as an implicit or indirect measure of participants' attitudes towards Blacks and Whites. The modified version, in which participants provided their direct evaluations of the Black and White faces, served as an explicit measure. Unlike other studies comparing implicit and explicit tests, however, these measures were equated on their other structural features. Subjects were presented with the same items, given matched tasks, and responded using the same scales. Correlations between implicit and explicit attitudes were strong using this method, in contrast to the many weak correlations that have been reported comparing reaction time measures to self-report tests.

In another study, we examined the relationships between several different implicit and explicit measures of racial attitudes that varied in their degree of structural fit. We included two

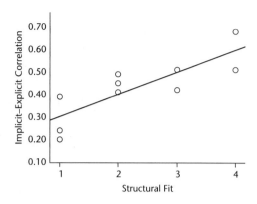

Figure 12–1. The size of the implicit-explicit correlation as a function of structural fit (reverse rank order). Adapted from Payne, Burkley, & Stokes (2008, Study 2).

versions of the AMP—one that used Black and White faces as primes as described above, and one that used verbal group labels (i.e., African Americans, Blacks, European Americans, Whites). Subjects rated the pictographs in some blocks, and they rated the primes in other blocks. In addition to these implicit and explicit tests, subjects completed three traditional self-report measures of racial attitudes, including a "feeling thermometer," the Modern Racism Scale, and the Attitudes Toward Blacks Scale (Brigham, 1993). To compare how well these tests cohered, we first rank ordered them by how well their structural features were matched. As summarized in Figure 12–1, we found that the magnitude of the correlation between each of the direct measures and scores on the indirect AMP varied as a function of structural fit. Figure 12–1 shows the strength of the relationship between structural fit (reverse-rank ordered so that higher numbers reflect greater structural fit) and the size of the implicit-explicit correlation. Structural fit was correlated at a strikingly high 0.90 with the size of implicit-explicit correlations. When the tests had poorly matched structures, as in virtually all implicit attitude studies, the implicit-explicit correlations were very low. When the tests were structurally equated, implicit and explicit responses were strongly positively correlated, painting a much more consistent picture of people's attitudes about race.

Summary

It appears that the verdict on unconscious race bias and the existence of attitudes that are hidden from consciousness altogether may have been issued too soon. With the development and growing popularity of implicit attitude measures came the assumption that these instruments were tapping into sentiments that people harbor unknowingly. But closer inspection of the key evidence supporting this claim—the disconnect between the results of implicit and explicit tests—reveals a number of factors unrelated to the conscious/unconscious distinction that help to explain this finding (*see* Gawronski et al., 2006, for a similar conclusion). Self-presentational concerns may lead some participants to alter their responses on self-report measures of prejudice to portray themselves in a more egalitarian light. Furthermore, differences in the indicators that researchers and participants are using to determine what a "true attitude" is, the failure to equate implicit and explicit tests that are being compared on all but one critical dimension, and the low reliability of implicit measures that rely on response latencies can also attenuate the correlation between implicit and explicit attitudes about race. In the section above, we reviewed research demonstrating that when each of these factors is taken into account, the correspondence between implicit and explicit attitudes increases substantially. The stronger correlation has important implications for the hidden bias debate; it suggests that people do, in fact, have access to the mental content they reveal on implicit tests.

If unconscious attitudes about race—previously seen as the major obstacle for controlling one's prejudice—are not as widespread as was once thought, where does this leave people who are worried about keeping their biases in check? Still in plenty of trouble, we argue. A major assumption has been that consciously held attitudes are subject to voluntary control. That is, to the extent that we are aware of our feelings towards others, we should also be aware of how these feelings shade our interactions. As a result, we are (and should be held) responsible for the behavior we perform that stems from these attitudes. The flipside of this logic is that unconscious attitudes shade our interactions in ways that we cannot control. It is difficult to control the impact of feelings we do not even know we have. The idea of unconscious attitudes has thus given rise to considerable debate concerning how accountable we should be for the behaviors they produce.

We argue that an attitude need not be unconscious to influence our thoughts and behaviors without our awareness. There is plenty of research indicating that attitudes of which we are perfectly conscious can and do influence our thoughts and behaviors in countless *unconscious* ways. The section below briefly reviews some of the seminal research on unconscious influence. We then turn to several findings from our own research that highlight the unconscious influence of attitudes about race. As our work demonstrates, we may be painfully aware of our prejudices towards others but blissfully *un*aware of how these prejudices come to affect our judgments and behavior. And as it turns out, unconscious influence may be every bit as fascinating or threatening, depending on how you view it.

UNCONSCIOUS INFLUENCE OF ATTITUDES

The notion of unconscious influence is hardly new to psychology. The story of Clever Hans should have a familiar ring for anyone with at least a semester of general psychology under his or her belt. Clever Hans was a student of Wilhelm von Osten, a German school teacher during the early 1900s. Under the tutelage of von Osten, Hans had acquired an array of skills; he could multiply fractions, perform long division, construct grammatically correct sentences, and even identify the tones in a musical scale. What had earned Clever Hans his nickname was the fact that he was a horse. Von Osten would present Hans with a question or problem to solve, and Hans would convey an answer by stomping his hoof a certain number of times (Pfungst, 1911).

The pair drew audiences from throughout Germany. For some, witnessing Clever Hans perform his feats was enough to convince them that he was a remarkable horse. Others remained skeptical, questioning just how clever Hans really was. Detractors assumed that Hans' abilities were part of an elaborate hoax orchestrated by von Osten. In response to mounting controversy, a team of scientists set out to investigate whether Clever Hans was, in fact, a sham. The scientists discovered that Hans' hoof stamps were driven by subtle, nonverbal cues that the person questioning him would emit in anticipation of the correct answer. These cues appeared to be unintentional and involuntary—even the team of scientists had emitted them in the early stages of their investigation. Much to von Osten's dismay, Clever Hans' extraordinary skill was not the ability to read and do math, but the ability to pick up on nonverbal cues.

We tell the story of Clever Hans because it demonstrates how our attitudes—and expectations—can leak out without our awareness. More contemporary research by Rosenthal and Jacobson (1966, 1968) showed that it was not only horses that could pick up on these cues, but that children could, too. In a now-classic study, the researchers informed a sample of kindergarten teachers that several of their pupils would "bloom" intellectually throughout the coming year. Although they had been chosen at random, the children who were identified as potential "bloomers" did, in fact, demonstrate superior academic ability by year's end. The schoolteachers had been emitting subtle cues based on their expectations that the bloomers would outperform their less-gifted peers, and it was the teachers' expectations of success that caused the actual surge in the bloomers' intellectual growth.

Other seminal findings on unconscious influence come from research by Nisbett and Wilson (Nisbett & Wilson, 1977; Wilson & Nisbett, 1978) demonstrating that people lack awareness of their higher-order cognitive processes. That is, people may know exactly *how* they feel about various stimuli (e.g., objects, people, social groups), but their explanations for *why* they feel a certain way aren't based on any actual introspection into their thought processes. Instead, people generate explanations for their feelings and decisions based on *a priori* theories that seem plausible (e.g., choosing a college because it is nationally ranked)

or salient features of their environment (e.g., attributing a bad mood to rainy weather). In one of their most well-known illustrations of this phenomenon, Nisbett and Wilson (1978) placed four identical pairs of stockings on a table in a department store and asked shoppers passing by to identify the pair of the highest quality. Despite the fact that the stockings were virtually identical, the shoppers' preferences showed evidence of a strong positioning effect. The further to the right the stockings appeared in the line-up, the greater the likelihood that shoppers identified it as the best in quality, with 40% of shoppers choosing the pair positioned on the far right. When asked to explain how they had reached their decision, no shoppers indicated that they had been influenced by the placement of the stockings on the display. Rather, they cited differences in knit, weave, sheerness, and elasticity—explanations that seemed plausible, but reflected a lack of awareness of the actual mental processes that had led to their decision.

A related notion is that of mental contamination, a term Wilson and Brekke (1994) have used to describe unwanted influences on people's thoughts, emotions, and behaviors. According to Wilson and Brekke, mental contamination can occur for two reasons: people are unaware of the unwanted influence altogether; or, they are aware that the influence may be taking place, but are unable to control it. For example, employers may want to hire the best person for the job, but they may not realize how an applicant's race, gender, or degree of physical attractiveness may be affecting their decision. Alternatively, employers may be aware of the potential for such factors to shade their evaluations of an applicant, but may still be unable to prevent the "contamination" from taking place. Research from our own lab suggests that in either case, unconscious influence seems to be the source of the problem.

Knowing When We Are Biased

Do people lack insight into the mental processes that result in patterns of racial bias? One way that we have addressed this question has been to look at the correlation between people's subjective experience of bias—whether or not they *think* they have been influenced by racial cues—and the degree to which they actually demonstrate it. If people are aware of the influence of their attitudes about race on their judgments, evaluations, and behavior, we would expect to find a positive relationship between people's perception and actual demonstration of bias.

We investigated this relationship in a study in which participants performed a simple memory task (Payne, Jacoby, & Lambert, 2003). We gave participants a list of names that were stereotypical of either Black or White males (e.g., Jamal vs. Gregory). Each name on the list was paired with one of two occupations: basketball player (an occupation that is more stereotypical for Black males) or politician (an occupation that is more stereotypical for White males). Thus, participants memorized a list of name-occupation pairs, some of which were consistent with prevailing stereotypes and some that were not. Later, we asked participants to do two things: recall the occupation that had been paired with each name and indicate how confident they were that each of their answers was correct.

This memory task allowed us to separate the automatic and controlled processes that were driving participants' responses. Once again, imagine yourself as a participant in this study. You are prompted with the name "Keyshawn" and asked to remember whether you'd seen this name paired with the occupation "politician" or "basketball player" a few minutes back. One of two things may happen. You may have a conscious recollection of the occupation that was actually paired with this name. In this case, the answer you choose based on your recollection results from the controlled use of memory and will in most instances be correct. However, when prompted with the name Keyshawn, the stereotype-congruent occupation may also come to mind automatically. This automatic bias may occur regardless of whether the stereotypical occupation is the correct or incorrect response.

We were interested in how both of these processes might relate to participants' subjective experience. We used participants' confidence in their responses as an indicator of the degree of insight they had into the accuracy of their memory and their degree of automatic bias. Although participants showed good insight into their ability to recall the accurate profession overall, as evidenced by the positive correlation between confidence and accuracy, they showed poor insight into their degree of automatic bias. That is, their degree of confidence and demonstration of bias were virtually unrelated. Another way of looking at this is that when participants were correct, they knew they were correct. When participants were wrong, they were more likely to misremember that stereotypically White names had been paired with "politician" and that stereotypically Black names had been paired with "basketball player" than to misremember the opposite pairing. And, they were just as likely to show this bias when they reported perfect confidence in their memories as when they reported having no confidence at all.

We have observed a similar effect using both the weapon identification task and the AMP. In a study using the weapon identification task, we asked participants to indicate how accurate they felt they were in their classification of the weapons and household tools (Payne, Lambert, & Jacoby, 2002). Recall that in this paradigm, participants are asked to classify images they view on a computer screen as either a "gun" or a "tool" and that these images are paired with a photograph of either a Black or White male face. This time, however, we varied the amount of time participants had to make the classifications, with deadlines ranging from 200 to 700 milliseconds after the presentation of the photos. The variation in deadlines allowed us to examine more closely the automatic and controlled processes driving participants' responses. With a longer deadline, participants would have enough time to discriminate guns from tools on the basis of the objective features of each image. In other words, they would be able to think carefully about whether they were viewing a picture of a gun or a tool. Participants' ability to exert cognitive control in these instances would result in a high rate of accurate responding. With a much shorter deadline, participants would have to rely, at least in part, on automatic processing. In these instances, participants' responses would be much more likely to reflect an accessibility bias. To the extent that participants associate guns more strongly with Black versus White males, they should show a greater tendency to make stereotype-congruent identification errors.

As predicted, accuracy rates were highest when participants were given 700 milliseconds to respond and lowest when they had only 200 milliseconds. When participants did make errors, their errors were consistent with prevailing stereotypes. They were more likely to misidentify tools as guns after being primed with a Black face, and were more likely to misidentify guns as tools after being primed with a White face. Participants' confidence in their accuracy and their actual rate of accuracy were positively correlated, but this was driven entirely by the positive correlation between participants' confidence and cognitive control. There was no relationship between confidence and the amount of automatic bias they demonstrated.

These findings highlight a major obstacle we encounter when we try to control prejudice. It's hard to know how to control unwanted influence if we cannot tell when it is at work. In support of this, we have investigated two specific strategies for reducing bias: *(1)* warning people about the potential for bias and *(2)* giving people the opportunity to refrain from making potentially biased responses altogether. We have discovered that both strategies are ineffective precisely because they rely on people's ability to detect bias as it is happening. Consider a study in which participants performing the weapon identification task were randomly assigned to one of three conditions (Payne, Lambert, & Jacoby, 2002). In the first condition, participants were told to try their best not to let the pictures of the Black and White faces influence their identification of the weapons or tools. In the second condition, participants were actually encouraged to stereotype, and were instructed to use the Black and White faces as cues to help them identify the weapons and tools. In the third condition, participants were given no explicit instructions regarding the pictures of the Black and White faces, and thus served as a "no goal" control. We found that participants who were instructed to use race as a cue did, in fact, report significantly stronger intentions to do so than participants in the other two conditions. In spite of this, the amount of bias that participants actually did demonstrate did not vary across conditions. Participants mistook tools for guns with greater frequency after being primed with a Black face and mistook guns for tools with greater frequency after being primed with a White face, regardless of whether they had been trying to use or refrain from using race as a cue. It appeared that giving participants any explicit warning about race, whether it was to use or to avoid using race as a cue, seemed to activate stereotypes about race and increase the magnitude of race bias.

A similar finding has emerged in our research using the AMP. That the AMP provides evidence of unconscious influence should come as no surprise. The success of the procedure as an indirect measure of attitudes relies on the unintentional transfer of people's affective reactions to various primes. In a sense, it is a measure that is defined by unconscious influence. Interestingly, this misattribution of affect occurs even after participants are told that the primes may be influencing their evaluations of the neutral Chinese pictographs. Participants who are warned about the potential for bias and instructed not to let their reactions to pleasant and unpleasant photos (Payne et al., 2005; Experiments 1 & 2) or Black and White faces (Payne et al., 2005; Experiment 5) influence their ratings of the pictographs show just as much bias as participants who are not.

We have also looked at another strategy for controlling bias. Imagine yourself as a contestant on your favorite game show. The answers you provide to various questions (or the questions you provide to various answers, if *Jeopardy!* was the show that came to mind) will fall into one of two categories: correct and incorrect. But what if you could refrain from providing an answer altogether if you were not sure your answer was correct? Using the paradigms we have discussed, we have given participants an opportunity to "take a pass" when they feel that their responses may be biased. For instance, in a study high in external validity for anyone who has forgotten an acquaintance's name at a social gathering and faced the dilemma of whether to hazard a guess or avoid incrimination by keeping one's mouth shut, we had participants memorize a list of names and professions that were stereotypically Black or White, but this time gave participants the opportunity to "pass" on certain trials if they were unsure of the correct answer (Payne, Jacoby, & Lambert, 2003). We wanted to know whether people would hold their tongues, so to speak, when their recall of the correct answer was shaky. To examine this, we varied the format of the items on the recall test so that each participant received a combination of forced and free response trials. On the forced-response trials, participants had to provide an answer. If they had trouble remembering the correct occupation for that name, they had to make a guess. On the free-response trials, participants had the option to "pass" if they couldn't recall the correct occupation. Thus, participants could reduce the number of errors they made on the free response trials by cutting out the guesswork. We found that providing participants with the "free response" option did little to reduce their stereotype-congruent errors in recall. Because participants did not know *when* their responses were being influenced by stereotypes about race, the opportunity to hold their tongues did not afford them better control over their bias.

Yet another challenge that unconscious influence poses for the control of prejudice comes from research by Irene Blair, Jennifer Eberhardt, and their colleagues (Blair, Judd, & Chapleau, 2004; Eberhardt et al., 2006) showing that despite our best attempts to curtail unwanted influence, our attitudes about race may bias our judgments in new and unexpected ways. Their work suggests that controlling the influence of attitudes about race may be like using a bucket to stop rain as it falls through a leaky roof—placing the bucket under one hole does not prevent water from dripping down through another. The research by Blair and Eberhadt comes in the wake of the acknowledgement that the color of a defendant's skin has an impact on the length and severity of the sentence he or she receives. Numerous states have established sentencing guidelines that require legal decisions to be made on the basis of objective criteria, and these guidelines have been enacted in part to address well-documented racial disparities in the criminal justice system (*see* Tonry, 1995). Even with these guidelines, subtle forms of race bias continue to shape legal outcomes. In their investigation of the criminal sentences assigned to a random sample of inmates in the state of Florida, Blair, Judd, and Chapleau (2004) found that independent of racial group membership, there was a significant relationship between the severity of the punishment received and the degree to which a defendant's facial features were stereotypically Afro-centric. Although there were no significant differences in the sentences that Black versus White defendants received once prior criminal records and the severity of the crime were statistically controlled, defendants with more stereotypically Afro-centric facial features (as determined by nose width, lip thickness, and hair texture) received longer sentences than defendants with less Afro-centric features. This difference remained even after criminal history and the severity of the offense were taken into account. Defendants with more stereotypically Afro-centric features received harsher sentences, regardless of whether they were Black or White.

Eberhart and colleagues (2006) observed a similar effect in cases involving the death penalty. Specifically in capital sentencing cases involving a Black defendant and a White victim, they found that defendants with highly Afro-centric facial features were twice as likely to receive the death penalty as defendants with less Afro-centric features. This disparity held even after aggravating and mitigating circumstances, severity of the murder, and the socioeconomic status of both the victim and defendant were controlled. These studies suggest that although sentencing guidelines can help reduce the likelihood of discrimination based on a defendant's skin color, race-related bias may still influence legal outcomes in less obvious ways.

The difficulty in overcoming unconscious influences of racial bias has parallels in studies of the effects of social context on person perception. Trope (1986) has distinguished between influences of context on identification of behaviors versus influences of context on adjustment of judgments for situational demands. When context is used to adjust judgments, perceivers make deliberate inferences, for example, that fearful actions in a frightening situation do not necessarily mean that the actor is fearful, just that he is responsive to the situation. In contrast to this type of deliberate inference, context also influences people automatically by biasing their interpretations of exactly what the act is to begin with. So a quick movement that happens immediately after a loud noise might be interpreted as a fearful startle, whereas the same movement without the noise might be interpreted as excitement. Trope and Alfieri (1997) found that influences of context on inferential adjustments depended on processing resources, but influences of identification did not. Instead, influences of context on identification depended on how ambiguous the action was, but not on cognitive resources. The important point for the present purpose is that deliberately thinking about the judgment was not sufficient to overcome the influences of context on the identification of behavior. The reason is that the process of interpreting the behavior was presumably influenced unconsciously, and only the end product of that process was available to introspection.

These studies illustrate how in some cases, by the time a person tries to inspect their own judgments, it may be too late. In some cases, unconscious influences on the interpretation of ambiguous events may constrain the kinds of explicit inferences that are made. Uhlmann and Cohen (2005; Experiment 1) asked participants to evaluate one of two candidates for the job of police chief. One candidate was presented as being "street smart" but lacking in formal education, whereas the other candidate was presented as being well-educated but lacking in "street smarts." Participants were also led to believe the candidate was either male or female. The key dependent measures were how likely participants would be to hire the candidate as police chief and how important various attributes, including being street smart and being well-educated, were for the job. Uhlmann and Cohen found that participants were more likely to hire the male candidate than the female candidate—a finding that fits with traditional gender stereotypes linking men with both law enforcement and positions of leadership. What is fascinating is that participants' ratings of the importance of being street smart versus being educated fluctuated in such a way that supported their preference for a male candidate (*see also* Norton, Vandello, & Darley, 2004). Participants who evaluated a male candidate with street smarts rated this attribute as more important than educational background, whereas those who evaluated a male candidate with a formal education rated education as more important than street smarts. In short, they selectively valued or devalued the attributes to justify their preferences for a male candidate, and these mental gymnastics allowed participants to maintain a belief in their objectivity. In fact, self-perceived objectivity was actually correlated with actual amount of bias. Those participants who felt their hiring decision was based on "rational," "objective," and "logical" decision-making were the ones who showed the most bias in the value they placed on job criteria.

Summary

This section reviewed research demonstrating the pervasiveness of unconscious influence. From intelligent horses to controlling the influence of stereotypes and attitudes about race, unconscious influence has been shown to be difficult to keep at bay. Across several studies, we have found that people are not particularly good at detecting bias as it is happening. This is evidenced by the striking disconnect between people's subjective experience of bias (whether or not they think they have been biased) and their actual demonstration of it. The inability to detect bias in real time renders certain strategies for controlling prejudice ineffective. For example, warning people about the potential for unwanted bias and providing them with the opportunity to refrain from making potentially biased judgments do little to reduce the bias from taking place. In some instances, these strategies actually increase the likelihood of bias. In others, bias may be curbed on one level, but it may continue to "leak out" on other levels. Thus, the unconscious influence of stereotypes and attitudes about race provides a major challenge for self-control.

STRATEGIES FOR SELF-CONTROL

The research we have reviewed thus far has converged on two main points. First, implicit biases can often be reported accurately on explicit measures under the right conditions. The conditions are "right" when (a) subjects are motivated to report their attitudes; (b) the tests are reliable enough to detect them; (c) experimenters and subjects have the same psychological constructs in mind; and (d) implicit and explicit tests do not differ in ways that confound the implicit/explicit distinction. These findings suggest that the mental content assessed by implicit tests is not necessarily inaccessible to consciousness. The second point is that even consciously reportable attitudes and beliefs can influence behaviors without our awareness. When people attempt to control their behaviors, such unconscious influence presents just as many challenges. People are not particularly good at detecting when they are versus when they are not being biased, and as a result, attempts at controlling one's prejudice may be unsuccessful.

How, then, can prejudice be kept in check? One frequently advocated solution to the problem of hidden bias is consciousness-raising. The idea is that by taking an implicit test, people can discover their unconscious attitudes and begin to address them. This is similar to the logic underlying the classic psychoanalytic notion that by gaining insight into our unconscious mind, we can exercise control over it. But if unconscious attitudes and beliefs are not at the heart of the problem, then consciousness-raising may not be the best solution. Our emphasis in this chapter has been on the pervasiveness of unconscious influence, rather than on attitudes that are themselves hidden from consciousness, and we believe this analysis points to a different set of strategies for self-control. In this final section, we describe some strategies that may be particularly helpful for reducing unconscious influence.

Solution 1: Limiting the Potential for Bias

Warning people about the potential for bias does little to prevent the bias from occurring. Nevertheless, warnings may provide an indirect benefit for people struggling to control unconscious influence. If people are aware that they are entering situations where the potential for bias is high, they can take measures to shield their mental processes from the unwanted influence ahead of time (Wilson & Brekke, 1994). Consider a hiring practice that many symphony orchestras have adopted over the years. Historically, orchestras in the United States and Europe were comprised almost entirely of men, a trend that was due in large part to negative stereotypes about women's musical ability. To ensure that the very best musicians were being hired for the job, many orchestras implemented the practice of "blind auditions." Musicians were required to audition from behind a screen that masked their physical appearance from the judges evaluating their performance. Doing so made it impossible for judges to evaluate them on the basis of anything other than the sound being produced. Research has shown that as the popularity of this procedure increased, the gender composition of symphony orchestras began to change. Today, the orchestras that hold blind auditions are the ones with the highest proportion of female musicians (Goldin & Rouse, 2000).

Comparable practices are routinely employed in other areas as well. Blind and double-blind experimental designs are now standard in many scientific fields. Papers submitted to academic journals frequently undergo a process of blind peer review. Teachers often grade assignments blind to students' names to ensure that the evaluations they make are based on merit alone, and not on attitudes they may have about particular students. These measures help eliminate unconscious influence by placing a barrier between the person making the evaluative judgment and all nonessential features of the target. In other words, decision makers are still making a subjective judgment, but situational constraints have been put in place that make it impossible for the mental processes leading up to the judgment to be influenced by extraneous factors.

There are undoubtedly times when this strategy may be difficult or impossible to employ. It would be hard to prevent jurors from observing the race or gender of a defendant sitting in a courtroom or medical professionals from noticing these characteristics in the patients they treat. In these situations, it may be still be possible to limit the potential for bias by relying on statistical prediction rules. Statistical prediction rules and their application in both legal and medical domains have been part of a larger debate concerning clinical versus actuarial judgment (Dawes, Faust, & Meehl, 1989). A judgment that is "clinical" is based on human reasoning, whereas "actuarial" judgments are automated and based purely on statistical modeling (such as probability and linear regression models). Actuarial judgments are automated in the sense that a specific piece of information or combination of data will always yield the same result.

In an investigation of over 100 studies directly comparing clinical and actuarial predictions, Paul Meehl (1986; Dawes, Faust, & Meehl, 1989) found that actuarial judgments were consistently more reliable and accurate. In virtually all of the studies, actuarially based predictions were more accurate than the predictions made by experienced professionals in the relevant fields—even when the people making the predictions were given an informational advantage. Statistical prediction rules, for example, provide more accurate predictions of academic success than admissions officers are able to make (e.g., Dawes, Swets, & Monahan, 2000), are better at predicting whether violent criminals will commit future acts of violence (Monahan, 1995), and lead to more accurate diagnosing by mental and physical health practitioners (e.g., Goldberg, 1968). So what are the implications for self-control? If people relinquish their control over the decision-making process and instead rely on statistical models, there will be less opportunity for unconscious influence to interfere and people may ultimately be happier with the results.

As a strategy for combating unconscious influence, putting on literal or statistical blinders may be particularly effective for several reasons. First, it circumvents the stage in mental processing when people may be especially susceptible to influences they are not aware of. That is not to say that evaluating the quality of a musical performance, deciding whether to accept a manuscript for publication, or choosing what criteria to include in a prediction equation is not subjective. But people *can* be confident that factors like race and gender that are irrelevant to the evaluation at hand will not be influencing their decisions. The second key element of this strategy is that it is proactive. Because constraints are put into place ahead of time, successful self-control does not hinge on people's ability to detect bias as it is happening. Research in other domains has shown that proactive forms of self-control can be much more effective than strategies that are put into place after problems arise. Aspinwall and Taylor (1997), for example, have argued that implementing proactive strategies for coping with potential stressors in one's life can not only lessen the impact of stress when it does occur, but it can also reduce the likelihood that the stress will occur in the first place. Similarly, strategies for eliminating or at least limiting the potential for unwanted bias that are implemented proactively should help surmount the problem of unconscious influence.

Solution 2: Proactive Control Over Automatic Responses

Warning people about the potential for bias, in itself, doesn't prevent unconscious influence from happening. Ironically, giving people the explicit goal of not being biased—telling them *not* to use race as a cue—can actually increase the likelihood that the bias will occur. Recall the weapon identification study described earlier (Payne, Lambert, & Jacoby, 2002) where we gave participants the explicit goal of not being influenced by race, encouraged the use of race as a cue, or gave participants no explicit goal at all. The participants who'd been told to avoid using race as a cue showed just as much bias as the participants who'd actually been encouraged to rely on racial stereotypes. Similarly, MacCrae et al. (1994) have found that trying not to think in stereotypical terms can backfire, leading to an ironic surge in stereotypical thinking later on. In their research, participants who were given explicit instructions to avoid thinking about a social target in stereotypical terms showed an increase in stereotypical thinking when they stopped trying to suppress. This rebound effect had significant implications; participants experiencing a post-suppression surge in stereotype-related thoughts were more likely to act on their prejudices in a later phase of the study. Additional support comes from research by Trawalter and Richeson (2005) documenting executive impairment following cross-racial interactions. Building on their finding that people show cognitive impairment after interacting with a member of a different race (particularly when people have a negative attitude about the other race), Trawalter and Richeson examined the impact of different interaction goals. They found that participants who went into a cross-racial interaction with the goal of avoiding the expression of prejudice showed more subsequent cognitive impairment than participants who entered the situation with the goal of having a positive interracial exchange.

Although trying not to be biased may backfire in many contexts, there may be ways to proactively exert control over automatic responses before we are in their grasp. Specifically, we have found that race bias can be reduced by asking participants to commit to implementation intentions that activate counter-stereotypical thoughts (Stewart & Payne, 2008). Implementation intentions are plans that link a behavioral opportunity to a specific response. They take the form of "if, then" guidelines that use cues in a person's environment to dictate specific goal-directed behavior (Gollwitzer & Brandstatter, 1997). For example, if you have borrowed this book from the library, you probably have the broad intention to return it on time. An efficient way to help you achieve success would be to devise an implementation intention to carry out once you have finished the book; *if* you see it sitting on your desk, *then* you will put it in your briefcase or bag so you can drop it off.

In an initial study using implementation intentions, we had participants perform the weapon identification task. Rather than give them the fairly abstract goal to avoid being biased, we randomly assigned them to commit to one of three implementation intentions. In the first condition, we gave participants an implementation intention we believed would override the activation of the stereotype linking Black males with threat. We told them that during the weapon identification task, every time they saw a Black face, they should immediately think "safe." We had participants commit to this implementation intention by saying to themselves "whenever I see a Black face on the screen, I will think the word 'safe.'" Participants in the other two conditions were told to think either "quick" or "accurate" whenever a Black face appeared on the screen, and they were also asked to commit to these intentions. These were used as "dummy" intentions because the task already required subjects to respond quickly and accurately. We chose "quick" and "accurate" so that participants in these control conditions would carry out an implementation intention that was relevant to the task at hand, but unrelated to stereotypes about Blacks.

Participants were more likely to mistake tools for guns when primed with a Black face, and were more likely to mistake guns for tools when primed with a White face. However, this effect was further modified by participants' implementation intention (*see* Fig. 12-2). Only participants in the "think quick" and "think accurate" conditions showed the typical pattern of race bias. Participants in the "think safe" condition were statistically no more likely to mistake tools for guns after seeing Black faces than White faces. The implementation intention reduced racial bias even though subjects responded just as quickly as in the other conditions. Moreover, the intention took effect within the first several trials, and remained effective throughout the entire task. Even under conditions of fast and demanding responses, implementation intentions appear to be highly effective at reducing automatic bias.

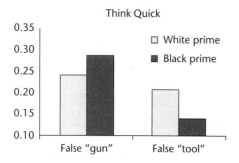
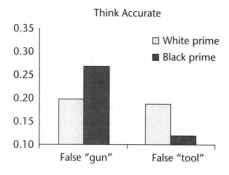
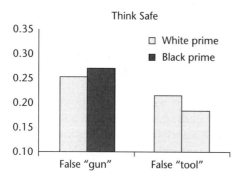

Figure 12–2. False gun and false tool responses in the weapon identification task as a function of three specific intentions. The intention to think "safe" in response to Black faces eliminated the race bias even under fast responding. Adapted from Stewart & Payne (2008).

Solution 3: Creating Cognitive Connoisseurs

A final approach is to strengthen people's awareness of the influences on their thoughts and behaviors over time. Discussions of consciousness and unconsciousness often have a static tone. A belief or process is said to be conscious or unconscious, without considering

that what is unconscious at one moment (e.g., the exact position of your left thumb) may be conscious at another moment (e.g., now). This example hinges only on directing your focus, but in other cases, bringing mental processes into consciousness may require training and practice. Consider, for example, how experienced meditators learn to attend precisely to their breathing and body position, or how wine connoisseurs learn to become aware of many aspects of wine that they formerly could not discern. The last approach we discuss for reducing unconscious influence is to become a *cognitive connoisseur*. That is, one way to overcome unconscious influences is to learn, with practice, to become aware of subtle influences on thought and behavior that may not have been perceived before.

One way to accomplish this goal may be through meta-cognitive training. The objective of meta-cognitive training is to teach people how to monitor their own thought processes. We know of no specific applications of meta-cognitive training within the context of prejudice reduction, but programs designed to teach people to become better "thinkers" have been successful in other domains. For example, meta-cognitive training has been used within the context of computer-based learning to help students develop critical problem-solving skills. In a study by Teong (2003), 11- and 12-year-olds with low scores on a math achievement exam participated in a training program that encouraged them to use meta-cognition to solve math word problems. After being presented with a math problem, students were prompted to ask themselves out loud if they understood what the problem was asking. They were also instructed to think of the possible strategies they could use to help solve the problem, to ask themselves if they were on the right track after choosing and implementing a strategy, and, after obtaining a solution, to ask themselves if their answer made sense. Compared to students in a control condition who had been given the same math problems without the meta-cognitive prompts, students who received the meta-cognitive training showed greater improvement in their math skills over time.

We believe this study points to a crucial first step that must be taken by aspiring cognitive connoisseurs. That is, the first step in a meta-cognitive training regimen may be as simple as getting people to stop and think about all potential sources of influence before they make a judgment or decision. Think about the countless evaluations you make in a day. Chances are, in only a handful of the cases did you ask yourself why you may have felt or acted a certain way, and, in still fewer cases did you stop to think about potential sources of influence beforehand. When people do stop to consider these things (or are prompted to by an experimenter), their analysis probably is not based on a careful consideration of *all* possible sources of influence, but rather on the sources that seem most plausible or are the most salient. Asking people to do something as simple as listing all possible factors that *could* exert an influence in their heads—even the things they are certain they will not be swayed by—may increase their awareness of influences that might otherwise go undetected.

These considerations point to the direction of attention as an important factor determining what mental contents are conscious at any given time. A related approach has been taken in studies of emotion awareness, a topic that could be instructive for understanding awareness of attitudes. Lane's (2008) model of emotional awareness argues that the ability to become aware of emotional processes is a cognitive skill like any other. Emotional awareness, in this view, can occur at different levels of sophistication, from simple and vague feelings (e.g., I feel bad) to discrete emotions (e.g., I feel angry), and then to blends of emotions and blends of blends (e.g., I feel slightly jealous, with a dollop of resentment, balanced by a hint of admiration). Lane's model suggests emotional awareness follows a developmental trajectory toward greater sophistication across development. But emotion awareness also differs across individuals at any given developmental stage, and critically, it can be improved by training (Subic-Wrana et al., 2005).

Although meta-cognitive training is an approach that has yet to be applied to the domain of stereotypes and prejudice, it could have important implications. The programs that have been used to improve learning and memory demonstrate that people can sharpen their insight into their mental processes with training, and that cognitive processes that are at one time inscrutable may become easily monitored and more effectively controlled with practice. Although this strategy is one that requires a considerable investment of time and effort, it may hold the greatest promise for overcoming unconscious influences. Whereas most of the solutions we have discussed are enacted on a situation-to-situation basis, the benefits of sharpened meta-cognitive awareness would be experienced across situations. As a result, meta-cognitive training remains an important avenue for future research.

Conclusion

We reviewed findings from our own and others' research that call into question the assumption that indirect or implicit measures tap into unconscious thought. Despite the typically low correlation between the attitudes that people explicitly report and those that are revealed on implicit measures, there is relatively little evidence to suggest that this divergence is driven by the existence of attitudes that people harbor subconsciously. Current methodologies fail to provide a critical test of unconscious mental content. Moreover, when factors that are extraneous to the conscious/unconscious distinction are taken into account, the correlation between implicitly and explicitly measured attitudes increases substantially. The motive to conceal attitudes that are viewed as socially unacceptable, the low reliability of many implicit measures, differences in what people construe as their "true" attitudes, and poor structural fit between implicit and explicit measures are all factors that can attenuate the correlation. The amplification of the implicit-explicit correlation that occurs when these factors are accounted for suggests that attitudes that are measured implicitly are not necessarily hidden from consciousness.

Although there is not strong evidence of attitudes and beliefs that linger in the reaches of our unconscious, there is plenty of research indicating that our attitudes and beliefs influence us in countless unconscious ways. Our research has shown that people are not particularly good at telling when they are versus when they are not being influenced by their biases on a moment-to-moment basis. This disconnect between people's subjective experience of bias and their actual demonstration of it makes it hard to keep one's biases in check. As a result, warning people about the potential for bias or giving them opportunities to refrain from making potentially biased judgments does little to prevent the unwanted influence from occurring.

There may be ways, however, to constrain unconscious influences that deserve further attention. Whereas general consciousness-raising does not seem to solve the problem of unconscious influence, strategies that limit the potential for bias and meta-cognitive training may prove to be more effective. By replacing the emphasis on static unconscious attitudes and beliefs with an emphasis on unconscious influences as processes that change over time, debates over unconscious bias can be cast in a new light. There is room for excitement about the fascinating new science of unconscious influence, alongside a more nuanced understanding of what we may and may not be able to achieve, even as cognitive connoisseurs.

REFERENCES

Akrami, N., & Ekehammar, B. The association between implicit and explicit prejudice: The moderating role of motivation to control prejudiced reactions. Scand J Psychol 2005; 46: 361-366.

Amodio, D., Harmon-Jones, E., Devine, P., Curtin, J., Hartley, S., & Covert, A. Neural signals for the detection of unintentional race bias. Psychol Sci 2004; 15: 88-93.

Arkes, H. R., & Tetlock, P. E. Attributions of implicit prejudice, or 'Would Jesse Jackson 'fail' the Implicit Association Test?' Psychol Inq 2004; 15: 257-278.

Baudelaire, C. *The Parisian prowler: Le spleen de Paris: Petits poemes en prose* (E. K. Kaplan, Trans.). Athens, GA: University of Georgia Press, 1997. (Original work published in 1869.)

Blair, I. V., Judd, C. M., & Chapleau, K. M. The influence of Afrocentric facial features in criminal sentencing. Psychol Sci 2004; 15: 674-679.

Blanton, H., & Jaccard, J. Arbitrary metrics in psychology. Am Psychol 2006; 61: 27-41.

Cunningham, W. A., Preacher, K. J., & Banaji, M. R. Implicit attitude measures: Consistency, stability, and convergent validity. Psychol Sci 2001; 12: 163-170.

Dawes, R. M., Faust, D., & Meehl, P. E. Clinical versus actuarial judgment. Science 1989; **243,** 1668-1674.

Devine, P. G. Stereotypes and prejudice: Their automatic and controlled components. J Pers Soc Psychol 1989; 56: 5-18.

Dunton, B. C., & Fazio, R. H. An individual difference measure of motivation to control prejudiced reactions. Pers Soc Psychol Bull 1997; 23: 316-326.

Eberhardt, J. L., Davies, P. G., Purdie-Vaughns, V. J., & Johnson, S. L. Looking deathworthy: Perceived stereotypicality of Black defendants predicts capital-sentencing outcomes. Psychol Sci 2006; 17: 383-386.

Fazio, R. H., Jackson, J. R., Dunton, B. C., & Williams, C. J. Variability in automatic activation as an unobtrusive measure of racial attitudes: A bona fide pipeline? J Pers Soc Psychol 1995; 69: 1013-1027.

Fazio, R. H., & Olson, M. A. Implicit measures in social cognition research: Their meaning and use. Ann Rev Psychol 2003; 54: 297-327.

Festinger, L. *A theory of cognitive dissonance.* Evanston, IL: Row, Peterson, 1957.

Gawronski, B., Geschke, D., & Banse, R. Implicit bias in impression formation: Associations influence the construal of individuating information. Eur J Soc Psychol 2003; 33: 573-589.

Gawronski, B., Hofmann, W., & Wilbur, C. J. Are "implicit" attitudes unconscious? Conscious Cogn 2006; 15: 485-499.

Greenwald, A. G., & Banaji, M. R. Implicit social cognition: Attitudes, self-esteem, and stereotypes. Psychol Rev 1995; 102: 4-27.

Greenwald, A. G., McGhee, D. E., & Schwartz, J. K. L. Measuring individual differences in implicit cognition: The Implicit Association Test. J Pers Soc Psychol, 1998; 74: 1464-1480.

Hofmann, W., Gawronski, B., Gschwendner, T., Le, H., & Schmitt, M. A meta-analysis on the correlation between the Implicit Association Test and explicit self-report measures. Pers Soc Psychol Bull 2005; 31: 1369-1385.

Hofmann, W., Gschwendner, T., & Schmitt, M. On implicit–explicit consistency: The moderating role of individual differences in awareness and adjustment. Euro J Pers 2005; 19: 25–49.

Homer *The Odyssey* (W. H. D. Rouse, Trans.). New York, NY: Signet Classics, 1999.

Jennings, J. M., & Jacoby, L. J. Improving memory in older adults: Training recollection. Neuropsychol Rehabil 2003; 13: 417–440.

Karpinski, A., & Hilton, J. L. Attitudes and the Implicit Association Test. J Pers Soc Psychol 2001; 81: 774–788.

Lane, K. A., Banaji, M. R., Nosek, B. A., & Greenwald, A. G. Understanding and using the Implicit Association Test: IV: Procedures and validity. In: Wittenbrink B. & Schwarz, N. (Eds.), *Implicit measures of attitudes: Procedures and controversies*. New York, NY: Guilford Press, 2007: pp. *59–102*.

Macrae, C. N., Bodenhausen, G. V., Milne, A. B., & Jetten, J. Out of mind but back in sight: Stereotypes on the rebound. J Pers Soc Psychol 1994; 67: 808–817.

McConahay, J. B. Modern racism, ambivalence, and the modern racism scale. In: Dovidio, J. F., & Gaertner, S. L. (Eds.), *Prejudice, discrimination, and racism*. New York, NY: Academic Press, 1986: pp. 91–126.

Milgram, S. Behavioral study of obedience. J Abnorm Soc Psychol 1963; 67: 371–378.

Lane, R. D. Neural substrates of implicit and explicit emotional processes: A unifying framework for psychosomatic medicine. Psychosom Med 2008; 70: 214–231.

Moskowitz, G. B., Gollwitzer, P. M., Wasel, W., & Schaal, B. Preconscious control of stereotype activation through chronic egalitarian goals. J Pers Soc Psychol 1999; 77: 167–184.

Nier, J. A. How dissociated are implicit and explicit racial attitudes?: A bogus pipeline approach. Group Process Intergroup Relat 2005; 8: 39–52.

Nisbett, R. E., & Wilson, T. D. Telling more than we can know: Verbal reports on mental processes. Psychol Rev 1977; 84: 231–259.

Norton, M. I., Vandello, J. A., & Darley, J. M. Casuistry and social category bias. J Pers Soc Psychol 2004; 87: 817–831.

Nosek, B. A. Moderators of the relationship between implicit and explicit evaluation. J Exp Psychol Gen 2005; 134: 565–584.

Payne, B. K. Prejudice and perception: The role of automatic and controlled processes in misperceiving a weapon. J Pers Soc Psychol 2001; 81: 181–192.

Payne, B. K., Burkley, M. A., & Stokes, M. B. Why do implicit and explicit attitude tests diverge? The role of structural fit. J Pers Soc Psychol 2008; 94: 16–31.

Payne, B. K., Cheng, C. M., Govorun, O., & Stewart, B. D. An inkblot for attitudes: Affect misattribution as implicit measurement. J Pers Soc Psychol 2005; 89: 277–293.

Payne, B. K., Jacoby, L. L., & Lambert, A. J. Memory monitoring and the control of stereotype distortion. J Exp Soc Psychol 2004; 40: 52–64.

Payne, B. K., Lambert, A. J., & Jacoby, L. L. Best laid plans: Effects of goals on accessibility bias and cognitive control in race-based misperceptions of weapons. J Exp Soc Psychol 2002; 38: 384–396.

Petty, R. E., Fazio, R. H., & Briñol, P. *Attitudes: Insights from the new wave of implicit measures*. Mahwah, NJ: Erlbaum, 2008.

Plant, E. A., & Devine, P. G. Internal and external motivation to respond without prejudice. J Pers Soc Psychol 1988; 75: 811–832.

Pronin, E., Gilovich, T. D., & Ross, L. Objectivity in the eye of the beholder: Divergent perceptions of bias in self versus others. Psychol Rev 2004; 111: 781–799.

Ranganath, K., Smith, C., & Nosek, B. Distinguishing automatic and controlled components of attitudes from direct and indirect measurement methods. J Exp Soc Psychol 2008; 44: 386–396.

Rosenthal, R., & Jacobson, L. Teachers' expectancies: Determinants of pupils' IQ gains. Psychol Rep 1966; 19: 115–118.

Rosenthal, R., & Jacobson, L. *Pygmalion in the classroom: Teacher expectation and pupils' intellectual development*. New York, NY: Holt, Rinehart & Winston, 1968.

Stewart, B. D., & Payne, B. K. Bringing automatic stereotyping under control: Implementation intentions as an efficient means of thought control. Pers Soc Psychol Bull 2008; 34: 1332–1345.

Subic-Wrana, C., Bruder, S., Thomas, W., Lane, R., & Köhle, K. Emotional awareness deficits in inpatients of a psychosomatic ward: A comparison of two different measures of alexithymia. Psychosom Med 2005; 67: 483–489.

Teong, S. K. The effect of metacognitive training on mathematical word-problem solving. J Comput Assist Learn 2003; 19: 46–55.

Tonry, M. *Malign neglect: Race, crime, and punishment in America.* New York, NY: Oxford University Press, 1995.

Trawalter, S., & Richeson, J. A. Regulatory focus and executive function after interracial interactions. J Exp Soc Psychol 2006; 42: 406–412.

Trope, Y. Identification and inferential processes in dispositional attribution. Psychol Rev 1986; 93: 239–257.

Trope, Y., & Alfieri, T. Effortfulness and flexibility of dispositional judgment processes. J Pers Soc Psychol 1997; 73: 662–674.

Trope, Y., & Fishbach, A. Counteractive self-control in overcoming temptation. J Pers Soc Psychol 2000; 79: 493–506.

Uhlmann, E. L., & Cohen, G. L. Constructed criteria: Redefining merit to justify discrimination. Psychol Sci 2005; 16: 474–480.

Wegner, D. M. *The illusion of conscious will.* Cambridge, MA: MIT Press, 2002.

Wilson, T. D., & Brekke, N. C. Mental contamination and mental correction: Unwanted influences on judgments and evaluations. Psychol Bull 1994; 116: 117–142.

Wilson, T. D., & Nisbett, R. E. The accuracy of verbal reports about the effects of stimuli on evaluations and behavior. Soc Psychol 1978; 41, 118–131.

CHAPTER 13

Self-Control Over Automatic Associations

Karen Gonsalkorale, Jeffrey W. Sherman, and Thomas J. Allen

ABSTRACT

Processes that permit control over automatic impulses are critical to a range of goal-directed behaviors. This chapter examines the role of self-control in implicit attitudes. It is widely assumed that implicit attitude measures reflect the automatic activation of stored associations, whose expression cannot be altered by controlled processes. We review research from the Quad model (Sherman et al., 2008) to highlight the importance of two self-control processes in determining the influence of automatically activated associations. The findings of this research indicate that processes relating to detecting appropriate responses and overcoming associations contribute to performance on implicit attitude measures. These two processes work together to enable self-control of automatic associations; one process detects that control is needed, and the other process overcomes the associations to permit correct behavior. Implications for understanding self-control dilemmas are discussed.

Keywords: Implicit attitudes, automatic associations, self-regulation, detection, overcoming bias

It is Friday morning and John is excited because he is finally going on the road trip that he has been planning for months. He loads his bags into the car, completes a safety check, reviews the road atlas for the umpteenth time, and pulls out of the driveway. With his favorite music blaring and the traffic flowing smoothly, John is feeling happy and relaxed. A few minutes later, he makes a wrong turn and finds himself heading towards the office instead of his intended destination. Realizing his mistake, John curses himself for being on "auto-pilot."

In this example, John's automatic habit has prevented him from successfully executing his trip (at least temporarily). What did John need to do to achieve his goal? First, he had to be able to identify the correct route to his vacation. Having virtually committed the route to memory, John more than satisfied this first condition. Given that he knew the required route, John then needed to overcome the automatic response that compelled him to follow his commute to work. On this second count, John failed.

We argue that these processes—detecting appropriate responses and regulating automatic habits—are critically important across a wide range of goal-directed behaviors. Our approach is not limited to instances in which self-control resolves conflicts between lower-level and higher-order goals (e.g., satisfying a sugar craving versus staying on a diet). Rather, we propose that self-control processes also will be crucial whenever a situation is characterized by competition between automatic impulses and responses that promote goal attainment. In the example above, John has no goal to go to work; he simply has an automatic habit[1] that temporarily disrupts his vacation plans. Many psychological phenomena have the same basic structure. For example, in the Stroop task (Stroop, 1935), the automatic habit to read the word must be overcome to report the color of the word accurately. In implicit measures of attitudes, automatic associations with targets (e.g., associations between Blacks and negativity) must be overridden to perform the task accurately when the task requires an association-incompatible response. Though participants generally seek to perform these tasks correctly, rarely do they have a goal to implement habitual responses or activate automatic associations in the course of performing them. Thus, although the automatic and controlled processes produce competing responses, this conflict typically does not arise from competing goals.

Contribution to Understanding Self-Control Dilemmas

We explore self-control issues by investigating the processes that contribute to performance on tasks that place automatic and controlled processes in opposition to one another. Specifically, in our research, we have applied the Quadruple Process Model (Quad model; Conrey et al., 2005; Sherman et al., 2008) to dissociate the processes that influence responses on such tasks. As described below, this model assesses the extent to which individuals detect correct responses and overcome automatic associations. The model also estimates the degree to which automatic associations are activated while performing the task and the influence of response biases. On the Stroop task, for example, the Quad model is able to assess the relative influence of processes relating to the automatic habit to read, accuracy in identifying the color of the words, regulation of the automatic habit, and response biases, as contributors to task performance.

Our primary level of analysis is the mind. We are interested in understanding the cognitive, affective, and motivational processes that enable individuals to control their attitudes and behavior in social life. A primary goal of our research is to separate the multiple automatic and controlled processes that underlie people's attitudes and behaviors. This approach has important linkages to levels of analysis at the brain and at society. Given that automaticity and control are associated with distinct regions in the brain, a complete account of the processes involved in self-control requires consideration of the neural systems that underlie them. Conversely, neuroscientific approaches may benefit from mapping a range of automatic and controlled processes (observed behaviorally) onto specific brain systems. Our level of analysis also has implications for societal-level analyses. The automatic and controlled processes occurring within individuals may influence the effectiveness of society's efforts at controlling its citizens. For instance, anti-discrimination laws may only be effective if individuals are willing and able to control their automatic stereotypes and prejudices. Reverse effects may occur also; society may facilitate or constrain the extent to which automatic and controlled processes affect thought and behavior. If stereotypes are salient within a society, for example, individuals may require stronger self-control to combat the effects of the stereotypic associations that they hold. The relationships between the three levels of analysis highlight the importance of a multi-disciplinary approach to the issue of self-control.

Although our approach is applicable to a wide range of situations, in this chapter we will review research findings in the domain of implicit attitudes,[2] which has been the primary focus of our work thus far. It is widely assumed that implicit attitude measures reflect the unintended, automatic activation of stored associations, whose expression cannot be altered or inhibited by controlled processes (e.g., Bargh, 1999; Devine, 1989; Fazio et al., 1995; Greenwald, McGhee, & Schwartz, 1998). Thus, self-control issues have been seen as largely irrelevant to understanding responses on implicit measures. In contrast, we propose that both automatic and controlled processes underlie implicit task performance, and that these processes can be independently measured using the Quad model.

Consider the Stroop task again. A young child who knows colors but does not know how to read will likely perform very well on the task, making few errors. An adult with full reading ability may achieve the same level of success. However, these performances would be based on very different underlying processes. In the case of the adult, to perform the task accurately, the automatic habit to read the word must be overcome to report the color of the word accurately. In contrast, the child has no automatic habit to overcome on incompatible trials (e.g., the word "blue" written in red ink). The same logic applies to implicit measures of attitudes, many of which have the same compatibility structure as the Stroop task. The identical responses of two individuals on an Implicit Association Test (IAT; Greenwald et al., 1998), for example, may reflect moderately biased associations (e.g., between Blacks and negativity) in one case, but strong associations that are successfully overcome in the other.

Thus, investigating self-control within implicit attitude measures is important for gaining a more complete understanding of what these measures assess and how they should be conceptualized. For example, common interpretations of implicit measures may underestimate not only the extent of controlled processing, but also the extent of automatic processing, because a strong ability to overcome automatic bias may mask the true extent of that bias. Another implication is that self-control in implicit task performance may be partly responsible for a host of effects that are typically attributed to the operation of automatic processes. Findings that scores on implicit attitude measures vary across individuals and are responsive to experimental interventions (*see* Blair, 2002) have often been interpreted as evidence that automatic associations differ among individuals and can be readily changed. However, if controlled processes also are responsible for variability and malleability in implicit task performance, then the implications of such results would be very different. For example, the results may not indicate the ease with which implicit associations may be changed but rather may reflect a greater role for intentions and motivations than has been previously assumed. As a final example, better understanding of the role of self-control in implicit attitudes may help to better identify means for changing those attitudes. If an implicit bias stems from biased automatic associations, then a strategy that directly influences those associations may be most effective. In contrast, if the bias stems from deficits of self-control, then interventions that improve self-control may be most effective.

Importantly, the implications of our research extend well beyond the exertion of self-control during implicit task performance. Because implicit measures and tasks like the Stroop create self-control needs that mirror those encountered in everyday life, exploring the processes required to successfully perform these tasks can enhance understanding of many real-world self-control dilemmas. In particular, this approach can shed light on any situation in which a goal may be thwarted in favor of an automatic response. Although we do not view conflict between lower-level and higher-order goals as a necessary feature of self-control dilemmas, our approach may nevertheless yield insight into such situations. The ability to recruit self-control processes to perform a task effectively will likely predict success at mediating between immediate and longer-term goals. For example, how well a person who wants to quit smoking is able to overcome positive associations with cigarettes on an implicit attitude measure might predict whether she later smokes a cigarette to satisfy her nicotine craving (lower goal), or behaves in line with her desire to stop smoking (higher goal). Thus, our work contributes to understanding how individuals behave in situations that require control over impulses, both at the task level and at the broader goal level.

Control and Automaticity in Social Behavior

The Quad model was developed, in part, by considering the processes that appear across a wide spectrum of dual-process models of social and cognitive psychology (e.g., Chaiken & Trope, 1999; Sherman, 2006). By definition, all four processes of the Quad model are never found within any particular dual-process model. Rather, the goal of dual-process models is to assess the extent to which a judgment or behavior reflects one type of automatic processing and one type of controlled processing. The Quad model incorporates the processes that are most commonly identified across the various dual-process models. These processes have been shown to be fundamental and ubiquitous components of judgment and behavior.

Although they are not always explicitly presented as such, dual-process models almost always are relevant to questions of self-control. They are concerned with delineating the circumstances under which judgment and behavior are driven by controlled, intended processes versus automatic, unintended processes. In examining these questions, dual-process models have generally been concerned with one of two different types of control. In some models, control is characterized by stimulus detection processes that attempt to provide an accurate depiction of the environment. For example, in dual-process models of persuasion, the controlled process is involved in discrimination between strong and weak arguments (e.g., Chaiken, 1980; Fazio, 1990; Petty & Cacioppo, 1981). In models of impression formation, the controlled process entails attention to and integration of target behaviors, providing an individuated (and presumably accurate) impression of the person (e.g., Brewer, 1988; Fiske & Neuberg, 1990). In Jacoby's Process Dissociation models (Jacoby, 1991; Lindsay & Jacoby, 1994; Payne, 2001), control represents an ability to determine and provide a correct response.

However, in other dual-process models, control is characterized by self-regulatory processes that attempt to inhibit unwanted or inappropriate information. For example, in Devine's (1989) model of stereotyping, control must be exerted to overcome the automatic influence of stereotypes. In Wegner's (1994) model of thought suppression, control must be exerted to inhibit unwanted thoughts. In many models of social judgment, self-regulatory control is exerted when people try to correct their judgments for subjectively expected biases (e.g., Martin, 1986; Wegener & Petty, 1997). These types of dual-process models have been more explicitly recognized as pertaining to self-control.

Both detection and regulation processes are controlled processes in that they require intention and cognitive resources, and can be terminated at will (e.g., Bargh, 1994). However, they have different functions and have very different influences on attitudes and behavior. It is clear that, on many occasions, they operate simultaneously. For example, a police officer's decision whether to shoot a Black man who may or may not have a gun may depend both on his ability to discriminate whether the man has a gun and, when there is no gun, his ability to overcome an automatic bias to associate Blacks with guns and to shoot. Thus, we believe that there is much to be gained by distinguishing between these types of control and measuring their contributions to behavior independently.

Dual-process models also have generally been concerned with one of two different types of automaticity. Most commonly, automaticity is represented as simple associations that are triggered by the environment without the perceiver's awareness or intent. Stereotypes play this role in dual-process models of impression formation (e.g., Brewer, 1988; Fiske & Neuberg, 1990). In models of persuasion (e.g., Chaiken, 1980; Petty & Cacioppo, 1981) and judgment (e.g., Epstein, 1991; Sloman, 1996), heuristics function in much the same way. This is the kind of automaticity that implicit attitude measures are intended to assess (e.g., Devine, 1989; Fazio et al., 1995; Greenwald et al., 1998).

In other dual-process models, however, automatic processes influence behavior only when control fails. For example, Jacoby's Process Dissociation model of memory (Jacoby, 1991) proposes that, when controlled attempts at recollection fail, people may instead rely on automatically generated feelings of familiarity to identify the stimulus as old. Others have portrayed the influence of implicit stereotypes in (mis)identifying weapons as operating in this manner (e.g., Payne, 2001). Another example is the implicit preference shown for items on the right side of a display when conscious introspection provides no rational basis for this preference (Nisbett & Wilson, 1977).

Although both types of automatic processes may operate without conscious intention, awareness, or the use of cognitive resources, clearly they are different kinds of processes. For example, a police officer's decision to shoot might be influenced by automatically activated associations between Blacks and guns. In the absence of such associations, however, the officer's decision might still be influenced by a secondary automatic bias to presume danger in the absence of clear evidence to the contrary, and guess that the person is holding a gun. We believe it is important to distinguish between these types of automatic processes and to measure their contributions to behavior separately.

THE QUAD MODEL

The Quad model is a multinomial model (*see* Batchelder & Riefer, 1999) designed to estimate the independent contributions of each of the four processes described above to a given behavior. According to the model, responses on implicit measures of bias reflect the operation of four qualitatively distinct processes: Activation of Associations (AC), Detection (D), Overcoming Bias (OB), and Guessing (G). The AC parameter refers to the degree to which biased associations are automatically activated when responding to a stimulus. All else being equal, the stronger the associations, the more likely they are to be activated and to influence behavior. The D parameter reflects a relatively controlled process that discriminates between appropriate and inappropriate responses. Sometimes, the activated associations conflict with the detected correct response. For example, on incompatible trials of the Stroop task or incompatible trials of implicit attitude measures (e.g., a Black face prime followed by a positive target word), automatic associations or habits conflict with detected correct responses. In such cases, the Quad model proposes that an overcoming bias process resolves the conflict. As such, the OB parameter refers to self-regulatory efforts that prevent automatically activated associations from influencing behavior when they conflict with detected correct responses. Finally, the G parameter reflects general response tendencies that may occur when individuals have no associations that direct behavior, and they are unable to detect the appropriate response. Guessing can be random, but it may also reflect a systematic tendency to prefer a particular response. For example, pressing the "unpleasant" key in response to a target face in the IAT (Greenwald et al., 1998) could be considered a socially undesirable response. To avoid that possibility, participants may adopt a conscious guessing strategy to respond with the positive rather than the negative key. Thus, guessing can be relatively automatic or controlled. The Quad model employs multinomial modeling to estimate the influence of each of these processes within a single task (for a review, *see* Batchelder & Riefer, 1999).

The structure of the Quad model is depicted as a processing tree in Figure 13–1. In the tree, each path represents a likelihood. Processing parameters with lines leading to them are conditional upon all preceding parameters. For instance, Overcoming Bias (OB) is conditional upon both Activation of Associations (AC) and Detection (D). Similarly, Guessing (G) is conditional upon the lack of Activation of Associations (1 – AC) and the lack of Detection (1 – D). Note that these conditional relationships do not imply a serial order in the onset and conclusion of the different processes. Rather, these relationships are mathematical descriptions of the manner in which the parameters interact to produce behavior. Thus, attempts to detect a correct response (D) and attempts to overcome automatic biases (OB) may occur simultaneously. However, in determining a response on a trial of a given task, the influence of attempts to overcome bias will be seen only in cases in which detection is successful.

The conditional relationships described by the model form a system of equations that predict the number of correct and incorrect responses in different conditions (e.g., compatible and incompatible trials). We will illustrate with reference to the Black-White IAT (Greenwald et al., 1998), one of the most frequently used implicit measures of attitudes toward Blacks and Whites. On each trial of the IAT, participants are presented in the middle of

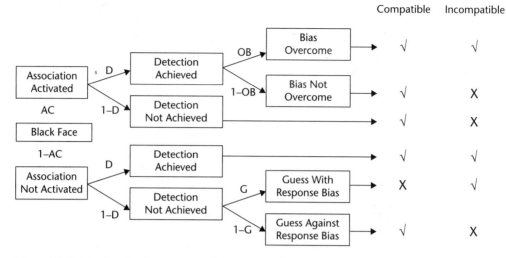

Figure 13–1. The Quadruple Process Model (Quad Model). Each path represents a likelihood. Parameters with lines leading to them are conditional upon all preceding parameters. The table on the right side of the figure depicts correct (√) and incorrect (X) responses as a function of process pattern and trial type. In this particular figure, the response bias refers to guessing with the positive key.

a computer screen with a stimulus from one of four categories: Black faces, White faces, pleasant words, and unpleasant words. Participants are asked to indicate, as quickly and accurately as possible, to which category the stimulus belongs by pressing the appropriate key, according to labels at the top of the screen. In the "compatible" block, participants are instructed to press one key in response to Black faces and unpleasant words, and the other key in response to White faces and pleasant words. The keys used to categorize Black and White faces are switched in the "incompatible" block, such that the Black and pleasant categories are assigned to one key, and the White and unpleasant categories to the other key. Participants who respond more quickly in the compatible block compared to the incompatible block are thought to have implicit negative associations toward Blacks relative to Whites.

According to the Quad model, a Black face stimulus in an incompatible block of a Black–White IAT will be assigned to the correct side of the screen with the probability: AC × D × OB + (1 − AC) × D + (1 − AC) × (1 − D) × G. This equation sums the three possible paths by which a correct answer can be returned in this case. The first part of the equation, AC × D × OB, is the likelihood that the association is activated *and* that the correct answer can be detected *and* that the association is overcome in favor of the detected response. The second part of the equation, (1 − AC) × D, is the likelihood that the association is not activated *and* that the correct response can be detected. Finally, (1 − AC) × (1 − D) × G, is the likelihood that the association is not activated *and* the correct answer cannot be detected *and* that the participant guesses by pressing the positive ("pleasant") key. Because the "pleasant" and "Black" categories share the same response key in the incompatible block, pressing the positive key in response to a Black face stimulus will return the correct answer. The respective equations for each item category (e.g., Black faces, White faces, positive words, and negative words in both compatible and incompatible blocks) are then used to predict the observed proportion of errors in a given data set. The model's predictions are then compared to the actual data to determine the model's ability to account for the data. A χ^2-estimate is computed for the difference between the predicted and observed errors. To best approximate the model to the data, the four parameter values are changed through maximum likelihood estimation until they produce a minimum possible value of the χ2. The final parameter values that result from this process are interpreted as relative levels of the four processes. For a complete description of data analysis within the Quad model, *see* Conrey et al. (2005).

Behavioral and Neurological Evidence for the Validity of the Quad Model's Parameters

Numerous research findings have established the validity of the Quad model. Conrey et al. (2005) demonstrated that the Quad model accurately describes performance on two of the most widely used implicit measures of attitudes, the IAT and the sequential priming task (e.g., Payne, 2001). Because Detection and Overcoming Bias are relatively controlled processes, fewer cognitive resources should decrease the extent of D and OB. One experiment showed that limiting participants' ability to engage in controlled processing by forcing them to respond quickly reduced estimates for D and OB, but did not affect Activation of Associations or Guessing. Thus, capacity constraints influence only the relatively controlled processes. In another experiment, manipulating the base rate of correct responses requiring a right- or left-handed button press shifted response bias (G) in the direction of the base rate, but did not affect any of the other parameters. These findings indicate that the four processes of the Quad model contribute to performance on implicit attitude measures, respond to experimental manipulations in a predictable manner, and are empirically separable.

Other research findings provide further support for the construct validity of the Quad model's parameters. For example, reaction time bias on the IAT is positively correlated with AC, but negatively correlated with OB (Conrey et al., 2005). This finding suggests that greater association activation increases implicit racial bias, but the ability to inhibit automatic associations attenuates this bias. Furthermore, a neuro-imaging study of IAT performance (Beer et al., 2008) showed that AC was correlated with activity in the amygdala and insula, which are involved in emotional processing and arousal (Phan et al., 2006; Phelps et al., 2000). This finding is consistent with the depiction of AC as measuring association activation. At the same time, on trials in which automatic associations and controlled processes compete to determine performance (i.e., incompatible trials), D was associated with activation in both the dorsal anterior cingulate cortex (dACC) and the dorsolateral prefrontal cortex (DLPFC). Whereas activity in the dACC has been related to detecting conflict between competing behavioral responses (e.g., Botvinick et al., 1999), activity in the DLPFC has been linked to inhibitory control over pre-potent responses (e.g., Chee et al., 2000; Taylor et al., 1998). Thus, when automatic and controlled processes compete to direct behavior, the D parameter predicts brain activity associated with detecting appropriate behavior among competing responses and inhibiting inappropriate automatic reactions. This is consistent with the Quad model's depiction of D as a controlled process that selects appropriate behavior and feeds into efforts to overcome inappropriate automatic influences.[3]

The next section reviews research in which we applied the Quad model to examine the processes underlying implicit attitudes. This research illustrates the benefits of considering self-control issues when exploring implicit attitudes.

APPLYING THE QUAD MODEL: DETECTION AND OVERCOMING BIAS IN IMPLICIT ATTITUDES

We have conducted a series of studies to demonstrate that performance on implicit attitude measures can vary and be changed through a number of different mechanisms. In these studies, we used the Quad model to analyze new and published data on implicit attitude variability and malleability. These findings highlight the utility of using the Quad model to gain deeper insight into the different processes responsible for implicit task performance. As we will describe below, some of these insights were masked in previous research by data analytic approaches that could not separate the influences of Detection, Overcoming Bias, and Association Activation.

Public versus Private Contexts

In one of the earliest demonstrations of how the Quad model can shed new light on existing data, Conrey et al. (2005) re-analyzed a study by Lambert et al. (2003) on the effects of public versus private contexts on implicit attitudes. Lambert et al. (2003) found that bias on the Weapons Identification Task (Payne, 2001) was amplified when participants believed that their performance would be made public. Two competing explanations have been advanced for this effect. According to the habit-strengthening account, public contexts lead to enhanced bias by increasing the influence of dominant responses. In contrast, the impairment of control account proposes that anticipation of a public setting creates a cognitive load that weakens the ability to control responses.

Conrey et al.'s (2005) Quad model re-analysis of the Lambert et al. (2003) data indicated that the public context led to an increase in Activation of Associations and a decrease in Detection. These findings are consistent with both the habit strengthening and the impairment of control explanations. Moreover, the Quad model analysis revealed an increase in Overcoming Bias in the public context, a finding that was not predicted by either account, and could not be detected in the original analyses based on Process Dissociation (Jacoby, 1991). This finding indicates that public accountability does not impair all aspects of self-control. Although it does reduce the ability to determine correct responses, it also increases people's ability to overcome unwanted bias. More broadly, this finding is important because it shows the value of separating different types of control that may be influenced in opposite ways by the same context.

Individual Differences in Motivation to Respond Without Prejudice

We also have applied the Quad model to understand motivation-based individual differences in the expression of racial bias. Research suggests that individuals who are either internally or externally motivated to respond without prejudice show lower levels of bias on explicit measures compared to individuals who are not motivated to control prejudice (Plant & Devine, 1998). However, only those individuals who are internally but not externally motivated (high IMS/low EMS participants) are able to respond without bias on implicit prejudice measures (Amodio, Devine, & Harmon-Jones, 2008; Amodio et al., 2003; Devine et al., 2002). By contrast, individuals motivated by internal and external reasons (high IMS/high EMS participants) and those who are not internally motivated (low IMS participants) exhibit bias on implicit measures.

Although there is substantial evidence that high IMS/low EMS individuals are effective in regulating race bias on implicit measures, relatively little is known about how they achieve non-prejudiced responding. Recently, Amodio et al. (2008) found that high IMS/low EMS individuals showed heightened electrophysiological responses corresponding to conflict detection following stereotypical errors on a priming measure of implicit stereotypes. Hence, high IMS/low EMS participants showed heightened conflict monitoring when their responses were discrepant with the goal to be non-prejudiced. This suggests that these individuals are particularly adept at detecting competing appropriate and inappropriate responses. If this is the case, then these same individuals should demonstrate higher estimates of the Quad model's Detection parameter than other participants. To test this hypothesis, we (Sherman et al., 2008) re-analyzed the accuracy data from Amodio et al. (2008).

In addition to analyzing the Detection parameter, we examined the possibility that high IMS/low EMS individuals also have less biased associations automatically activated and/or greater ability to overcome those associations than other individuals. Just as enhanced detection would produce less implicit racial bias among high IMS/low EMS individuals, so, too, would reduced activation of biased associations or a greater ability to regulate those associations. Based on an application of Jacoby's (1991) PD procedure, Amodio et al. (2008) reported no differences in automatic activations among individuals with different motivations.

Our Quad model re-analysis of the data indicated that, compared to the other participants, the high IMS/low EMS participants were more able to detect appropriate and inappropriate responses on the weapons identification task. In addition, these individuals exhibited less activation of stereotypic associations compared to individuals who were not internally motivated to respond without prejudice. We replicated these findings in a follow-up study using a different measure of implicit attitudes. Specifically, we found that high IMS/low EMS participants showed less implicit racial bias on the IAT (Greenwald et al., 1998), higher estimates of Detection, and lower levels of Association Activation, compared to other individuals. Importantly, there was no evidence of differences in Overcoming Bias as a function of different motivations in either study.

These findings are consistent with Amodio et al.'s conclusion that high IMS/low EMS participants are more effective in controlling race bias because they have a more finely tuned conflict detection system. For conflict detection to occur, the correct and incorrect responses must be identified. Our results also suggest that high IMS/low EMS individuals may have less biased automatic associations to overcome compared to other individuals. The finding that the OB parameter did not differentiate among individuals with different motivations would appear to indicate that motivation-based individual differences in implicit bias may have little to do with inhibition processes. As such, it seems that high IMS/low EMS individuals may not be especially good at regulating, per se. Rather, it appears that they are particularly able to detect when regulation is required (i.e., when there are conflicting responses and a danger of responding inappropriately; Amodio et al., 2008; Monteith et al., 2002), thereby increasing the likelihood of behaving appropriately.

Training to Negate Biased Associations

The findings described in the previous section may shed light on how certain prejudice reduction strategies work. If high IMS/low EMS individuals show less implicit bias because of enhanced behavioral monitoring and lower association activation, then prejudice reduction strategies that improve awareness of appropriate and inappropriate responses should also reduce implicit bias via the same mechanisms. To test this possibility, we trained participants to negate anti-Black and pro-White associations before performing the IAT (Sherman et al., 2008). On each trial of the training task a Black or White face was presented together with a positive or negative word. Participants in the negate associations condition were instructed to press the "YES" key whenever they saw a Black face with a positive word below it or a White face with a negative word below it, and to press the "NO" key whenever a Black face appeared with a negative word or a White face appeared with a positive word. Participants in the maintain associations condition were given the opposite instructions. This type of training has been shown to reduce implicit stereotyping in previous research (Kawakami et al., 2000). We hypothesized that the attention and effort required to successfully execute the training task would enhance Detection and reduce Association Activation. Based on our finding that high IMS/low EMS individuals did not show greater levels of Overcoming Bias compared to other individuals, we did not expect the training to improve OB.

Results supported our hypotheses. Replicating Kawakami et al.'s (2000) findings, the participants who had been trained to negate negative associations with Blacks and positive associations with Whites showed significantly less IAT bias than the participants who were trained to maintain these associations. Analysis using the Quad model showed that the negation training not only weakened participants' automatically activated associations (AC) but also improved their ability to determine the correct response (D). The finding of enhanced detection is consistent with the idea that training enables individuals to develop cues for recognizing and then controlling non-prejudiced responses (Monteith et al., 2002). This suggests that people can be trained to engage self-control in a manner similar to individuals who are internally motivated to be non-prejudiced (and, presumably, train themselves).

In other studies, we have identified important roles for Overcoming Bias in accounting for variability in implicit attitudes. As a measure of the extent to which activated associations can be overcome when they conflict with appropriate behavior, we reasoned that OB should be particularly sensitive to variations in self-regulatory ability. To test this hypothesis, we investigated populations that are known to differ in self-regulatory ability in two studies. The first study examined the processes underlying the effects of alcohol intoxication on implicit racial bias.

Alcohol Intoxication

Research indicates that alcohol impairs cognitive and motor performance by reducing the ability to regulate prepotent responses. For example, intoxicated individuals are less able to inhibit distracting thoughts and restrain inappropriate responses on cognitive tasks (Easdon & Vogel-Sprott, 2000). Applying these findings to the domain of social attitudes, Bartholow, Dickter, and Sestir (2006) hypothesized that alcohol increases stereotypic responding by impairing self-regulatory ability. To explore this possibility, they modified a priming measure of implicit racial stereotyping (e.g., Dovidio, Evans, & Tyler, 1986) such that it included "go" trials and "stop" trials. The primes were Black or White faces and houses (the control primes), and the target words consisted of adjectives that can be used to describe people or houses (e.g., *carpeted, furnished*). Half of the person adjectives were stereotypic of Blacks (e.g., *athletic, lazy*) and half were stereotypic of Whites (e.g., *intelligent, boring*). For the "go" trials, participants were instructed to indicate whether the adjective could ever be used to describe the picture of the person or house that preceded it. Participants were instructed to withhold responses on the "stop" trials. Errors on the stop trials served as the behavioral index of regulation failure (Logan & Cowan, 1984). Results indicated that alcohol affected the pattern of errors on the stop trials, such that the high-dose group committed significantly more errors on stereotype-consistent trials than on stereotype-inconsistent trials, whereas the placebo group did not show this stereotyping effect. Moreover, electrophysiological data indicated that the alcohol-based impairment of inhibition of stereotype-based responses was mediated by a reduction in the negative slow wave (NSW) component of the event-related potential. The NSW has been associated with the engagement of cognitive control processes sub-serving inhibition (West & Alain, 1999). At the same time, alcohol had no effect on a neurological marker of stereotype activation (the P300). These behavioral and neurological data suggest that alcohol impairs the ability to regulate the expression of stereotypic associations.

Applying the Quad model to Bartholow et al.'s priming data, we found that Overcoming Bias was the only parameter that differed across alcohol consumption conditions (Sherman et al., 2008). This finding suggests that alcohol intoxication interferes with people's ability to regulate automatically activated associations. There were no significant differences in Association Activation or Detection as a function of alcohol consumption. Thus, the Quad model findings provide converging evidence that the effects of alcohol on race-based responding are specific to inhibition failure. The Quad model offers a non-invasive means of detecting such effects with behavioral data alone.

Aging

In our second study that focused on the Overcoming Bias parameter, we explored the effects of aging on implicit racial bias (Gonsalkorale, Sherman, & Klauer, 2009). Large national surveys have consistently shown that older White Americans tend to express more racial prejudice than their younger counterparts (e.g., Wilson, 1996). Recent research suggests that age-related differences in racial bias extend to the implicit level, with one large study (Nosek, Banaji, & Greenwald, 2002) reporting a positive correlation between age and implicit racial bias. These findings often are interpreted as evidence that older people's racial associations are more biased than those of younger adults, reflecting generational changes in societal attitudes.

An alternative explanation for age differences in prejudice is that deficits in self-regulatory ability alter the attitudinal expression of older adults. Given that the ability to inhibit automatically activated stereotypes enables people to behave non-prejudicially (Bartholow et al., 2006; Devine, 1989; Moskowitz, Salomon, & Taylor, 1999), and that inhibitory functioning declines with age (Connelly, Hasher, & Zacks, 1991; Hasher & Zacks, 1988), losses in inhibitory ability may increase stereotyping and prejudice during old age, even if the underlying associations are of equivalent (or even declining) strength across the life span. Consistent with this possibility, von Hippel, Silver, and Lynch (2000) found that losses in inhibitory functioning mediated increases in explicit racial stereotyping among the elderly. This research also indicated that, contrary to popular wisdom, older adults reported strong desires to control their prejudices, suggesting that they were willing but not able to control their biases.

We conducted a study to examine whether inhibitory processes can account for age differences in racial bias on an implicit measure. Race IAT data were collected from White participants who visited the IAT demonstration Web site (http://implicit.harvard.edu/; Nosek et al., 2002). We modeled the data as a function of participant age, which ranged from 11 to 94. The results suggested that age-related differences in IAT bias arose from differences in the ability of older and younger adults to regulate automatically activated associations. Despite showing stronger IAT effects, the older adults demonstrated less activation of biased associations and a greater likelihood of detecting the correct response than the younger adults. However, as predicted, Overcoming Bias decreased with age. It appears that, despite weaker activation of associations and greater detection of correct responses, the older adults exhibited stronger implicit bias behaviorally because they were less able to inhibit their activated associations. These findings suggest that age differences in implicit racial bias may be caused by age-related losses in regulatory functions.

Predicting the Quality of Intergroup Interactions

The findings reviewed above indicate that Detection and Overcoming Bias are important underlying processing components of performance on implicit measures of racial bias. We believe that the processes that direct performance on these immediate response tasks are likely to predict success at resolving impulse regulation conflicts in broader domain relevant contexts. To illustrate this relationship, we will now describe a study in which we examined the ability of the Quad model's parameters to predict behavior in an intergroup interaction (Gonsalkorale, von Hippel, Sherman, & Klauer, 2009).

The goal of this study was to test hypotheses regarding the processes underlying the relationship between implicit attitudes and behavior in intergroup interactions. According to one account, implicit race bias predicts unfriendly behavior in cross-race interactions because biased automatic associations drive prejudice-consistent behavior in attitude-relevant situations (e.g., Dovidio, Kawakami, & Gaertner, 2002). In previous research, correlations between scores on implicit attitude measures and interaction behavior (e.g., Dovidio et al., 2002; Dovidio et al., 1997; Fazio et al., 2005; McConnell & Leibold, 2001) have been taken as evidence that automatic associations direct behavior. However, implicit attitude measures are not pure reflections of the automatic associations that are hypothesized to drive behavior in intergroup settings (Amodio et al., 2004; Bartholow et al., 2006; Conrey et al., 2005; Payne, 2001; Sherman, 2009; Sherman et al., 2008). Thus, correlations between scores on these measures and behavior in the presence of outgroup members do not necessarily indicate the influence of those associations. In contrast, we tested this idea directly by examining whether Activation of Associations predicts poorer interaction quality.

An alternative account proposes that people may be able to prevent their automatic biases from influencing behavior by regulating their behavior when interacting with outgroup members (e.g., Richeson & Shelton, 2007). Consistent with this possibility, one study (Shelton et al., 2005) found that Whites who were high in implicit racial bias were evaluated more favorably than their low-bias counterparts because the former were perceived to be more engaged in the interaction. In our study, we wanted to examine whether the immediate regulation of automatic associations, as reflected in responses on implicit measures, is sufficient to influence interaction behavior. This possibility, which has not been considered in previous research, is important, as it may signal an "upstream," early cognitive process that attenuates the influence of automatic associations and facilitates smooth intergroup interactions, independently of the ability to control behavior during the course of an interaction. If overcoming associations contributes to smooth intergroup interactions, Overcoming Bias should predict better interaction quality.

To examine these issues, we asked non-Muslim Caucasian participants to interact with an experimental confederate who appeared to be and was described as Muslim. Following the interaction, the confederate rated how much he liked the participants, whereas the participants completed a Go/No-Go Task (GNAT; Nosek & Banaji, 2001) measuring implicit bias toward Muslims. The GNAT is a variant of the IAT that assesses attitudes toward a single target group (e.g., Muslims) rather than relative evaluations of two groups (e.g., Muslims versus non-Muslims). Participants who showed more negative attitudes toward Muslims on the GNAT were evaluated less positively by the Muslim confederate. We applied the Quad model to the GNAT data to examine the extent to which different processes may contribute to the quality of the interaction. The confederate's ratings of how much he liked the participants were predicted by an interaction between automatic negative associations and ability to overcome bias. Specifically, when the strength of participants' negative associations with Muslims was low, participants' level of overcoming bias was unrelated to the confederate's ratings. In contrast, the ability to regulate automatic negative associations predicted greater liking when those associations were strong. Thus, among participants with strong anti-Muslim associations, the ability to recruit self-control to perform the GNAT effectively predicted success at regulating behavior during the intergroup interaction. These findings are the first to show that process estimates derived from the Quad model are related to self-control in a broader behavioral context that plays out over extended time.

This study also illustrates how the Quad model may enhance interpretability of data from implicit attitude measures. We found that participants who exhibited greater bias against Muslims on the GNAT received less positive ratings from the Muslim confederate. Taken on its own, this result might be interpreted in a variety of ways. For example, approaches that treat implicit attitude measures as pure reflections of automatic associations would conclude that stronger associations predict disliking. The negative relationship between GNAT bias scores and likeability might also be used to refute the importance of self-regulation, as those who are presumed to regulate the most (i.e., those with higher implicit-measure bias scores; Shelton et al., 2005) were liked the least. In contrast, our Quad model findings indicate that biased associations alone do not jeopardize the quality of an intergroup interaction. The modelling further demonstrates that regulation of associations plays an important role when people have strong automatic associations. In the absence of the Quad model findings, the data from the implicit measure would lead to very different conclusions. Providing a means to tease apart multiple possible interpretations of effects involving implicit attitudes is one of the strengths of the Quad model.

Summary and Implications

Our findings highlight the importance of self-control processes in determining the influence of automatically activated associations. Across multiple studies, we have found that processes relating to both detecting appropriate responses and overcoming associations contribute to performance on measures of implicit knowledge. These two processes work together to permit self-control of automatic associations; one process detects that control is needed, and the other process overcomes the associations to permit correct behavior.

The idea that different types of controlled processes play a role in implicit task performance has important implications for understanding the nature of implicit attitudes. If implicit measures are presumed to reflect only the automatic activation of associations, then malleable performance on such measures must, by definition, be taken as evidence that the associations activated in performing a given task have been altered. That is, implicit attitude malleability must reflect either changes in the nature of the underlying associations or changes in the particular associations that are temporarily accessible. However, our research suggests that such conclusions are likely to significantly overestimate the extent to which activated associations can be altered. In some cases, malleability effects will be due, at least in part, and maybe entirely, to response processes that have nothing to do with the underlying associations, per se. As such, though implicit attitude malleability is certainly cause for optimism about people's ability to avoid automatic stereotyping and prejudice effects (e.g., Blair, 2002), caution is warranted in concluding that associative knowledge is easily altered.

Implications for Treating Implicit Measures as "Process Pure"

There is now considerable evidence that implicit measures engage multiple processes, both automatic and controlled, as underlying associations are translated into behavioral responses on the tasks. Our research shows the consequences of this task complexity for interpreting implicit attitude effects. Specifically, our findings highlight the danger of assuming a one-to-one correspondence between performance on implicit measures and the extent of automatic association activation. It follows that implicit measures should not be assumed to provide estimates of processes (e.g., Fazio et al., 1995), representations (Greenwald et al., 1998; Wilson et al., 2000), or systems (Rydell & McConnell, 2006; Strack & Deutch, 2004) that are independent from intention and control. In the studies described earlier, we demonstrated these conclusions in the domains of implicit attitude variability and malleability. However, our findings have implications for other implicit attitude effects, as well. For example, dissociations between implicit and explicit measures cannot be assumed to reflect the separate and independent contributions of automatic and controlled processes, representations, or systems to performance on the two tasks. One difficulty for dual representation and dual system models is the frequent lack of correlations among different implicit measures of the same attitude (e.g., Sherman, 2006). If all implicit measures are tapping the same automatic process, representation, or system, then different implicit measures should correlate more highly than they often do. However, from the current perspective, performance on the different measures may reflect a variety of differences in the processes recruited in performing the tasks. The key differences between any two measures (implicit or explicit) may have to do with the nature of the associations activated by the different tasks, the nature of response biases in performing the tasks, or the nature of more controlled detection or self-regulation processes engaged while performing the tasks. Finally, the same considerations surround interpretations of the relationships between implicit measures and behavior. It is not necessarily the case that correlations between an implicit measure and a behavior reflect the operation of automatic processes, representations, or systems; other components of task performance may also (or instead) be responsible for the relationship.

Implications for Attitude Change Strategies

An important question in the minds of people who are interested in promoting egalitarianism is how to design successful prejudice-reduction strategies. If interventions designed to change implicit attitudes do not always lead to changes in the activation of associations, then how useful are they? In our view, changing underlying associations is but one method of changing people's implicit attitudinal and behavioral responses. Our research suggests that different attitude-change strategies may be best suited to changing different kinds of implicit attitudes. For example, if the attitudinal bias stems not from biased associations but from an inability to monitor ongoing behavior for appropriateness (as among high IMS/high EMS participants), then interventions designed to enhance the detection of conflicting responses might be advised. However, if the attitudinal bias stems from a deficit in self-regulation (as among older adults), then the best intervention might be one that serves to strengthen the ability to overcome unwanted associations. The current research demonstrates how such an attitude-intervention matching process may be achieved, by identifying the bases for individual differences in implicit attitudes and the bases of the effects of interventions on implicit attitude change.

Future Directions: Predicting Success in Achieving Broader Goals

At the beginning of this chapter, we argued that application of the Quad model to immediate response tasks (e.g., implicit measures) holds promise for increasing understanding of broader goal-directed behavior. The ability to recruit and apply self-control processes on immediate response tasks is likely to predict success at resolving conflicts within the same domain between low-level, narrow goals and high-level, global goals. The findings from the intergroup interaction study provide the first indication of such a relationship between task-level control and the regulation of broader domain-relevant behaviors and goals.

Our ongoing research aims to provide further evidence for this application of the model by examining other relationships between process estimates derived from task performance and real-world behavior. For example, research has shown that smokers have less negative implicit attitudes about cigarettes than do ex-smokers and non-smokers (Sherman et al., 2003). We are applying the Quad model to try to understand the reasons for this effect, and to help generate effective interventions to help people quit smoking. For example, it may be that smokers have less negative automatic associations with cigarettes than do non-smokers. Alternatively, it may be that smokers are less able to determine appropriate smoking-related behaviors than non-smokers, or are less able to regulate the expression of their more favorable associations. By understanding how these groups differ on these processes, we can better understand why some people start smoking and others do not, why some people are able to quit smoking and others cannot, and what specific processes might need to be addressed in interventions aimed at reducing smoking.

As a general model of impulse control, the Quad model is relevant to a range of self-control dilemmas that are characterized by competing goals. Thus, the model may be able to predict whether dieters will choose healthy foods in the wake of tempting alternatives, whether recovering gambling addicts will be enticed by the lure of a casino, when people will be able to control affective reactions such as anger or happiness that may interfere with important decisions, and so on. In these scenarios and many others, automatic response tendencies that satisfy lower goals also have the potential to thwart higher order goals in a manner described by the Quad model. It is our hope that the model's broad applicability will lead to enhanced understanding of self-control issues in many different domains of judgment and behavior.

Notes

1 Note that this habit could reflect automatic goal activation occurring at an unconscious level. Our main point here is that the approach described in this chapter need not involve the activation or application of goals (either consciously or unconsciously).

2 In this chapter we focus on applying the Quad model to implicit measures, which we argue reflect the joint operation of automatic associations and controlled processes. However, there are many instances in which the association, impulse, or habit may not be automatic, per se.

The Quad model is relevant to these situations, as long as the association or impulse conflicts with a desired or intended response (thus producing incompatible responses).

3 For methodological reasons having to do with the different trials used to derive estimates of overcoming bias (OB) and brain activity, we were unable to associate that parameter with specific brain activity.

REFERENCES

Amodio, D. M., Devine, P. D., & Harmon-Jones, E. Individual differences in the regulation of intergroup bias: The role of conflict monitoring and neural signals for control. J Pers Soc Psychol 2008; 94: 60–74.

Amodio, D. M., Harmon-Jones, E., & Devine, P. G. Individual differences in the activation and control of affective race bias as assessed by startle eyeblink response and self-report. J Pers Soc Psychol 2003; 84: 738–753.

Amodio, D. M., Harmon-Jones, E., Devine, P. G., Curtin, J. J., Hartley, S. L., & Covert, A. E. Neural signals for the detection of unintentional race bias. Psychol Sci 2004; 15: 88–93.

Bargh, J. A. The four horsemen of automaticity: Awareness, intention, efficiency, and control in social cognition. In: Wyer, R. S., & Srull, T. K. (Eds.), Handbook of social cognition, Vol. 1: Basic processes, 2nd ed. Hillsdale, NJ: Erlbaum, 1994: pp. 1–40.

Bartholow, B. D., Dickter, C. L., & Sestir, M. A. Stereotype activation and control of race bias: Cognitive control of inhibition and its impairment by alcohol. J Pers Soc Psychol 2006; 90: 272–287.

Batchelder, W. H., & Riefer, D. M. Theoretical and empirical review of multinomial process tree modeling. Psychon Bull Rev 1999; 6: 57–86.

Beer, J. S., Stallen, M., Lombardo, M. V., Gonsalkorale, K., Cunningham, W. A., & Sherman, J. W. The Quadruple Process Model approach to examining the neural underpinnings of prejudice.NeuroImage 2008; 42: 775–783.

Blair, I. V. The malleability of automatic stereotypes and prejudice. Pers Soc Psychol Rev 2002; 6: 242–261.

Botvinick, M., Nystrom, L. E., Fissell, K., Carter, C. S., & Cohen, J. D. Conflict monitoring versus selection-for-action in anterior cingulate cortex. Nature 1999; 402: 179–181.

Brewer, M. B. A dual process model of impression formation. In Srull, T. K., & Wyer, R. S. (Eds.), Advances in social cognition, Vol. 1. Hillsdale, NJ: Erlbaum, 1988: pp. 1–36.

Chaiken, S. Heuristic versus systematic information processing and the use of source versus message cues in persuasion. J Pers Soc Psychol 1980; 39: 752–766.

Chaiken, S., & Trope, Y. (Eds.). Dual-process theories in social psychology. New York, NY: Guilford Press, 1999.

Chee, M. W. L., Sriram, N., Soon, C. S., & Lee, K. M. Dorsolateral prefrontal cortex and the implicit association of concepts and attributes. Neuroreport 2000; 11: 135–140.

Connelly, S. L., Hasher, L., & Zacks, R. T. Age and reading: The impact of distraction. Psychol Aging 1991; 6: 533–541.

Conrey, F. R., Sherman, J. W., Gawronski, B., Hugenberg, K., & Groom, C. J. Separating multiple processes in implicit social cognition: The Quad Model of implicit task performance. J Pers Soc Psychol 2005; 89: 469–487.

Devine, P. G. Stereotypes and prejudice: Their automatic and controlled components. J Pers Soc Psychol 1989; 56: 5–18.

Devine, P. G., Plant, E. A., Amodio, D. M., Harmon-Jones, E., & Vance, S. L. The regulation of explicit and implicit race bias: The role of motivations to respond without prejudice. J Pers Soc Psychol 2002; 82: 835–848.

Dovidio, J. F., Evans, N., & Tyler, R. B. Racial stereotypes: The contents of their cognitive representations. J Exp Soc Psychol 1986; 22: 22–37.

Dovidio, J. F., Kawakami, K., & Gaertner, S. L. Implicit and explicit prejudice and interracial interaction. J Pers Soc Psychol 2002; 82: 62–68.

Dovidio, J. F., Kawakami, K., Johnson, C., Johnson, B., & Howard, A. On the nature of prejudice: Automatic and controlled processes. J Exp Soc Psychol 1997; 33: 510–540.

Easdon, C. M., & Vogel-Sprott, M. Alcohol and behavioral control: Impaired response inhibition and flexibility in social drinkers. Exp ClinPsychopharm 2000; 8: 387–394.

Epstein, S. Cognitive-experiential self theory: An integrative theory of personality. In Curtis, R. (Ed.), The self with others: Convergences in psychoanalytical, social, and personality psychology. New York, NY: Guilford, 1991: pp. 111–137.

Fazio, R. H. Multiple processes by which attitudes guide behavior: The MODE model as an integrative framework. Adv Exp Soc Psychol 1990; 23: 75–109.

Fazio, R. H., Jackson, J. R., Dunton, B. C., & Williams, C. J. Variability in automatic activation as an unobstrusive measure of racial attitudes: A bona fide pipeline? J Pers Soc Psychol 1995; 69: 1013–1027.

Fiske, S. T., & Neuberg, S. L. A continuum of impression formation, from category-based to individuating processes: Influences of information and motivation on attention and interpretation. Adv Exp Soc Psychol 1990; 23: 1–74.

Greenwald, A. G., McGhee, D. E., & Schwartz, J. L. K. Measuring individual differences in implicit cognition: The Implicit Association Test. J Pers Soc Psychol 1998; 74: 1464–1480.

Gonsalkorale, K., Sherman, J. W., & Klauer, K. C. Aging and prejudice: Diminished regulation of automatic race bias among older adults. J Exp Soc Psychol 2009; 45: 410–414.

Gonsalkorale, K., von Hippel, W., Sherman, J. W., & Klauer, K. C. Bias and regulation of bias in intergroup interactions: Implicit attitudes toward Muslims and interaction quality. J Exp Soc Psychol 2009; 45: 161–166.

Hasher, L., & Zacks, R. T. Working memory, comprehension, and aging: A review and a new view. In Bower, G. H. (Ed.), *The psychology of learning and motivation: Advances in research and theory*, Vol. 22. San Diego, CA: Academic Press, 1998: pp. 193–225.

Jacoby, L. L. A process dissociation framework: Separating automatic from intentional uses of memory. J Mem Lang 1991; 30: 513–541.

Kawakami, K., Dovidio, J. F., Moll, J., Hermsen, S., & Russin, A. Just say no (to stereotyping): Effects of training in the negation of stereotypic associations on stereotype activation. J Pers Soc Psychol 2000; 78: 871–888.

Lambert, A. J., Payne, B., Jacoby, L. L., Shaffer, L. M., Chasteen, A. L., & Khan, S. R. Stereotypes as dominant responses: On the "social facilitation" of prejudice in anticipated public contexts. J Pers Soc Psychol 2003; 84: 277–295.

Lindsay, D. S., & Jacoby, L. L. Stroop process-dissociations: The relationship between facilitation and interference. J Exp Psychol Hum Percept Perform 1994; 20: 219–234.

Logan, G. D., & Cowan, W. B. On the ability to inhibit thought and action: A theory of an act of control. Psychol Rev 1984; 91: 295–327.

Martin, L. L. Set/reset: Use and disuse of concepts in impression formation. J Pers Soc Psychol 1986; 51: 493–504.

McConnell, A. R., & Leibold, J. M. Relations among the Implicit Association Test, discriminatory behavior, and explicit measures of racial attitudes. J Exp Soc Psychol 2001; 37: 435–442.

Monteith, M. J., Ashburn-Nardo, L., Voils, C. I., & Czopp, A. M. Putting the brakes on prejudice: On the development and operation of cues for control. J Pers Soc Psychol 2002; 83: 1029–1050.

Moskowitz, G. B., Salomon, A. R., & Taylor, C. M. Preconsciously controlling stereotyping: Implicitly activated egalitarian goals prevent the activation of stereotypes. Soc Cogn 2000; 18: 151–177.

Nisbett, R. E., & Wilson, T. D. Telling more than we can know: Verbal reports on mental processes. Psychol Rev 1977; 84: 231–259.

Nosek, B. A., & Banaji, M. R. The Go/No-go Association Task. Soc Cogn 2001; 19: 625–666.

Nosek, B. A., Banaji, M., & Greenwald, A. G. Harvesting implicit group attitudes and beliefs from a demonstration web site. Group Dynam Theory Res Pract 2002; 6: 101–115.

Payne, B. K. Prejudice and perception: The role of automatic and controlled processes in misperceiving a weapon. J Pers Soc Psychol 2001; 81: 181–192.

Payne, B. K., Lambert, A. J., & Jacoby, L. L. Best laid plans: Effects of goals on accessibility bias and cognitive control in race-based misperceptions of weapons. J Exp Soc Psychol 2002; 38: 384–396.

Petty, R. E., & Cacioppo, J. T. *Attitudes and persuasion: Classic and contemporary approaches*. Dubuque, IA: Brown, 1981.

Phan, K. L., Wager, T., Taylor, S. F., & Liberzon, I. Functional neuroanatomy of emotion: A meta-analysis of emotion activation studies in PET and fMRI. Neuroimage 2002; 16: 331–348.

Phelps, E. A., O'Connor, K. J., Cunningham, W. A., et al. Performance on indirect measures of race evaluation predicts amygdala activation. J Cogn Neurosci 2000; 12: 729–738.

Plant, E. A., & Devine, P. G. Internal and external motivation to respond without prejudice. J Pers Soc Psychol 1998; 75: 811–832.

Richeson, J. A., & Shelton, J. N. Negotiating interracial interactions: Costs, consequences, and possibilities. Curr Direct Psychol Sci 2007; 16: 316–320.

Rydell, R. J., & McConnell, A. R. Understanding implicit and explicit attitude change: A systems of reasoning analysis. J Pers Soc Psychol 2006; 91: 995–1008.

Shelton, J. N., Richeson, J. A., Salvatore, J., & Trawalter, S. Ironic effects of racial bias during interracial interactions. Psychol Sci 2005; 16: 397–402.

Sherman, J. W. On building a better process model: It's not only how many, but which ones and by which means? Psychol Inq 2006; 17: 173–184.

Sherman, J. W. Controlled influences on implicit measures: Confronting the myth of process-purity and taming the cognitive monster. In: Petty, R. E., Fazio, R. H., & Briñol, P. (Eds.), *Attitudes: Insights from the new wave of implicit measures*. Hillsdale, NJ: Erlbaum, 2009: pp. 391–426.

Sherman, J. W., Gawronski, B., Gonsalkorale, K., Hugenberg, K., Allen, T. J., & Groom, C. J. The self-regulation of automatic associations and behavioral impulses. Psychol Rev 2008; 115: 314–335.

Sherman, S. J., Rose, J. S., Koch, K., Presson, C. C., & Chassin, L. Implicit and explicit attitudes toward cigarette smoking: The effects of context and motivation. J Soc Clin Psychol 2003; 22: 13–39.

Sloman, S. A. The empirical case for two systems of reasoning. Psychol Bull 1996; 119: 3–22.

Strack, F., & Deutsch, R. Reflective and impulsive determinants of human behavior. Pers Soc Psychol Rev 2004; 8: 220–247.

Stroop, J. R. Studies of interference in serial verbal reactions. J Exp Psychol 1935; 18: 643–662.

Taylor, S. F., Kornblum, S., Lauber, E. J., Minoshima, S., & Koeppe, R. A. Isolation of specific interference processing in the Stroop Task: PET Activation studies. J Neuroimaging 1998; 6: 81–92.

von Hippel, W., Silver, L. A., & Lynch, M. E. Stereotyping against your will: The role of inhibitory ability in stereotyping and prejudice among the elderly. Pers Soc Psychol Bull 2000; 26: 523–532.

Wegener, D. T., & Petty, R. E. The flexible correction model: The role of naïve theories of bias in bias correction. Adv Exp Soc Psychol 1997; 29: 141–208.

Wegner, D. M. Ironic processes in mental control. Psychol Rev 1994; 101: 34–52.

West, R., & Alain, C. Event-related neural activity associated with the Stroop task. Cogn Brain Res 1999; 8: 157–164.

Wilson, T. D., Lindsey, S., & Schooler, T. Y. A model of dual attitudes. Psychol Rev 2000; 107: 101–126.

CHAPTER 14

Perish the Forethought: Premeditation Engenders Misperceptions of Personal Control

Carey K. Morewedge, Kurt Gray, and Daniel M. Wegner

ABSTRACT

People are normally encouraged to engage in premeditation—to think about the potential consequences of their behavior before acting. Indeed, planning, considering, and studying can be important precursors to decision-making, and often seem essential for effective action. This view of premeditation is shared by most humans, a kind of universal ideal, and it carries an additional interesting implication: Even the *hint* that premeditation occurred can serve as a potent cue indicating voluntary action, both to actors and observers. In legal and moral contexts, for example, actors are seen as especially culpable for the consequences of their actions if those consequences were premeditated, whether or not the premeditation influenced the decision. In this chapter, we review evidence indicating that even irrelevant premeditation can lead people to believe that an action's consequences were under personal control. We present research exploring how various forms of premeditation—including foresight, effortful forethought, wishful thinking, and the consideration of multiple possible outcomes of action—may lead actors to prefer and to feel responsible for action outcomes even when this premeditation has no causal relation to the outcomes.

Keywords: Premeditation, rational action, actors' perceptions, decision-making, complete control, random control, no control, priority, prior knowledge, delayed knowledge, consistency, intention, choice blindness, exclusivity, situational constraints, external influences, obedience experiments, cognitive dissonance, facilitated communication, controlled effort, Eureka error, meta-desires, counterfactual blame, dispositionalism, unconscious deliberation

"Except only the defendant's intention to produce a given result, no other consideration has affected our feeling that it is or is not just to hold him responsible for the result as its foreseeability."
—Edgerton, (1929, p. 1134). *Legal Cause.* Harper and James. (cited by Hart & Honoré, 1959/2002, pp. 254)

It feels necessary to think carefully before making important decisions. Whether buying a house, picking a spouse, or deciding to have children, it seems unwise to make a decision quickly, or to simply pick the first option we considered. Even relatively inconsequential choices are preceded by significant forethought. When buying a new digital camera, we might read *Consumer*

Reports, compare features such as lenses and weight, and try to imagine which would be better suited for our next family gathering or vacation. We engage in such forethought because premeditated decisions—choices guided by prior conscious deliberation of alternatives and their consequences—are considered superior to decisions made on the fly or in its absence.[1] Our parents, teachers, and peers continually advise us to "Look before you leap" and "Think before you speak," and when we make poor choices, they are often attributed to errors committed during premeditation ("What on Earth was I thinking?"). Premeditated decisions appear to determine the behavior most under our control, and behavior we control is presumed to be better than behavior we do not. Indeed, both the legal system and society at large view the presence of premeditation as the most important indicator of rational action (Denno, 2003).

Premeditation plays an important role in classic self-control dilemmas such as deciding whether to spend or to save, eat or diet, and engage in risky or safe behavior. Its presence suggests that not only were we aware of the consequences of our behavior, but we had the power to act rationally and control that behavior. Consequently, we blame ourselves more when lapses in self-control were premeditated than when they were not premeditated (Klimchuck, 1994). Buying a new car impulsively may seem rash, but doing so after considering whether to put the money aside for your child's college education seems both rash and shameful. Deciding to order steak tartare seems like a mild indulgence, whereas ordering that steak after considering healthier options seems both indulgent and irresponsible. This is also true when viewing the actions of others, as we sympathize less with those who are injured if they had consciously considered (and then ignored) ways to protect themselves in advance (e.g., wearing a helmet). No matter what the domain, it appears that the presence of premeditation is an important indicator of the extent to which people possess self-control.

Determining whether an act was premeditated is considered to be one of the principle tasks, if not the primary task, of modern legal systems (Denno, 2003). When a person is tried for a crime, juries are asked to determine the fulfillment of two criteria: *Actus reus* and *mens rea*. *Actus reus*, or "guilty act," refers to whether the defendant actually committed a crime. *Mens rea*, or "guilty mind," refers to whether the defendant intentionally committed that crime—usually synonymous with whether it was premeditated. In the legal setting, premeditation suggests that the defendant both foresaw the consequences of her action and had sufficient personal control to have prevented the criminal act. Although a defendant may still be held accountable for an unintended guilty action, judging that a crime was premeditated dramatically increases ascriptions of blame and punishment. Murder and manslaughter are both convictions for the same action (i.e., killing a person), but the differences in attributions of intention result in markedly different sentences: The recommended sentence for first-degree murder (i.e., a premeditated killing) is life in prison, whereas the sentence for voluntary manslaughter (i.e., an unpremeditated killing) is 10 years in prison (United States Sentencing Commission, 2004). The difference in punishment for these otherwise identical crimes highlights the importance society ascribes to premeditation when determining whether actions were under personal control.

Despite frequently making premeditated decisions and making inferences about which decisions made by other people were premeditated, we know little about what premeditation specifically entails. We may recognize at an intuitive level when an action is premeditated (killing your boss after days of scheming), or not premeditated (reflexively kicking your leg when the doctor taps your knee), but what constitutes premeditation? One obvious criterion is that you have to think about *something* in advance for it to be a premeditated outcome, but will any thought do? This chapter examines the nature of premeditation—how it is perceived and whether it is a reliable indicator of self-control. We explore the five requirements for an outcome to look and feel like it was produced by a premeditated decision, and investigate the efficacy of premeditation. As phenomenal experience and

Ideal Premeditation

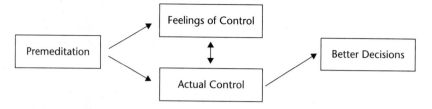

Actual Premeditation

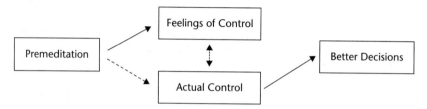

Figure 14–1. The diagram above depicts ideal premeditation, as conceived of by society and the law. It assumes that premeditation influences both feelings of control and actual control to a similar degree, resulting in better decisions. An alternative, actual premeditation, is presented in the diagram beneath it. It suggests that premeditation influences feelings of control, but may not influence one's actual control over decision or lead to better decisions.

the law suggest, premeditation may influence both our feelings of control and actual control over our actions and the outcomes they produce (illustrated in the top half of Fig. 14–1).

Alternatively, in spite of the importance placed upon premeditation, it may often be incidental. In other words, premeditation may fail to influence our decision making, even when we feel certain that it determined the outcome we chose. Like superstitious fans who believe they can influence the outcome of a football game by crossing their fingers or wearing the jersey of their favorite player, premeditation may be an ineffectual ritual. We suggest that although premeditation may increase feelings of self-control, it may not serve as a reliable indicator of actual control. Conscious thought may often only be tenuously linked to our actual behavior, and in such situations, the presence of premeditation may trick us into feeling that we controlled an outcome (as illustrated by the bottom half of Fig. 14–1). Consequently, if premeditation leads to the experience of self-control without providing authentic control, then the increased guilt and self-loathing we experience after succumbing to the vices we tried to resist may be largely unwarranted.

Actors' Perceptions

Each of us has experienced what it feels like to have successfully made a decision—we desired some outcome and persevered until it was produced. We also know what it feels like to have done something without thinking about it in advance or consciously making a decision ("How did I end up with this bowl of chocolate pudding and a ladle?"). What distinguishes between such "premeditated" decisions and "unmeditated" behavior—what leads us to feel that outcomes were intentionally chosen or unintentionally produced? Drawing from Apparent Mental Causation theory (Wegner, 2002, 2003; Wegner & Wheatley, 1999) and Rational Choice theory (e.g., von Neumann & Morgenstern, 1944), we suggest that there are five basic requirements that reflect the general structure of premeditated decisions.

Premeditated decisions begin with the identification of a problem or an unfulfilled goal, whether minor ("What should I eat for lunch?") or major ("Should I have surgery or try physical therapy?"). Once a problem or goal has been identified, information is gathered about the alternatives that are available ("Peanut butter and jelly again, or pizza?") such as each alternative's positive and negative consequences. After identifying an alternative as satisfactory or the best available (Schwartz et al., 2002; Simon, 1957), controlled effort is applied in an attempt to bring about the outcome one desires (e.g., spreading preserves, putting a plate in the microwave, or consulting a second physician). Whether evaluating the choice of a meal or medical procedure, we may thus consider the fulfillment of five criteria to determine if an outcome was premeditated. Premeditation appears to be thought that (1) is used to identify a preferred option from a set consisting of *multiple alternatives*, (2) occurs *prior* to the outcome produced, (3) is *consistent* with the outcome produced, (4) is *exclusively* related to the outcome, and (5) leads to the application of *effort* to bring about the outcome.

Although thought that fulfills each of these criteria will lead us to feel that an outcome was premeditated and under our control, all of these need not necessarily be fulfilled for us to feel that we are in control. Rather than a strict checklist, these factors appear to represent a set of rough guidelines. The presence of some of these factors in situations in which we lack control may erroneously suggest that we possessed control over an outcome that we did not. In other words, purely incidental thoughts may lead us to experience the feeling of control, fooling us into believing those outcomes were produced by our premeditation.

Requirement 1: Multiple Potential Courses of Action

The first guideline indicating that an outcome was premeditated and under our control is the existence of multiple alternatives, whether they are real or illusory. This can include having considered two or more alternatives, having considered whether to produce a single outcome, or having the *potential* to have considered alternatives. When deciding what to eat for dinner, leftover pizza may be the only food left in your refrigerator. You may have eaten it 99% of the time in similar past instances, but you may still feel like you have control over your dinner selection because you know that you *could* order delivery from a neighborhood restaurant. You may be just as likely to eat the leftover pizza at home as you would be if captors gave it to you while in solitary confinement, but in the former case you would attribute pizza-eating to your decision, whereas the lack of alternatives in solitary confinement (among other factors) would make you feel little control over the contents of your dinner. In many cases we may select the option that is dictated by our habits or culture (e.g., to leave a tip), and only perceive that option to have been intentionally chosen if we also considered its alternatives (Fiske, 1989).

The availability of multiple possible outcomes may be a legitimate requirement for a premeditated decision, yet there are instances in which alternatives to the outcome produced were not realistic or feasible, and still their presence makes us feel like we were in control. One outcome may be so preferable that its alternatives would never be selected (e.g., "Hmm…looks like leftover pizza or dog food." and "Honey, would you prefer to watch *Dr. Strangelove* or a documentary about root canals tonight?"). Desirable alternatives may exist but may not be feasible economically ("Should we buy the Ford or the Maserati?"), socially ("Should we have beef brisket or eat the neighbors?"), or temporally ("Sorry, I'm married."). And we may only realize that alternatives existed after an outcome was selected ("They had chocolate *and* strawberry?"). In short, there are a host of situations in life where actors may erroneously believe that alternatives were available at the time of a decision.

Conversely, we often feel no control over our premeditated decisions when alternatives to the course of action we select are ignored. Deciding not to act or simply discontinuing a current course of action (i.e., maintaining the status quo), are often not perceived to be decisions. Avoiding or delaying a decision, or asking someone else to make a decision for us, are also not usually perceived to be decisions (Anderson, 2003). Perhaps this is why we generally consider ourselves and others less responsible for outcomes that result from a decision not to act (i.e., acts of omission) than for identical outcomes produced by an action (i.e., acts of commission; Spranca, Minsk, & Baron, 1991).

As we believe that a decision was premeditated when alternatives to the chosen outcome were unfeasible and that no decision was made when nonobvious alternatives were present, the *mere presence* of alternatives may engender the feeling of control—whether we could have chosen those alternatives or not. A series of art selection experiments (Morewedge, Wegner, & Vosgerau, 2009) explored this possibility by testing whether we are more likely to consider outcomes to be under our control when alternative outcomes are present at the time of selection, but cannot actually be selected.

Participants randomly "selected" one of two works of art to see for 20s on each trial in a thirty-trial experiment by pressing a key on the right or left of their keyboard. The term "selected" is used because participants had no real control; there was no consistent relation between the button they pushed and the artwork they saw. Sometimes the right and left keys corresponded to the work on the right and left, respectively. Other times those keys corresponded to the works on the left and right, respectively. Thus, in a sense they were randomly hitting keys, and the computer was randomly presenting photographs and paintings. Importantly, participants were informed of the lack of consistent correspondence between the keys and artworks at the beginning of the experiment.

Before randomly selecting a photograph or painting, participants saw thumbnail-sized previews of both, one, or none of the artworks they could see in that trial (*see* Fig. 14–2). At the end of each trial, participants reported the extent to which they felt they controlled the outcome of the selection, and whether the artwork selected was the artwork they intended to see. These reports were averaged to create an index of participants' perceived control over the selection of outcomes. Although people often claim to control outcomes that are purely random (Langer, 1975), notice that these selections were equally random, regardless of the number of previews participants saw in the beginning of each trial.

If participants reported feeling more control over the selection of artwork in trials when they saw more rather than fewer previews, it would appear that participants used the *mere presence* of alternatives to infer the extent to which they controlled the artwork selected. They did just that. Participants were more likely to report feeling control over the works selected when both of the artworks were previewed than when only one work was previewed. And they were more likely to report feeling control over works selected when only one work was previewed than when neither work was previewed—even when the work previewed was inconsistent with the work selected. In short, the greater the number of previews seen before a selection was displayed, the more control participants felt over the selection, whether or not the preview matched the randomly selected outcome (Morewedge et al., 2009, Experiment 1).

A second experiment manipulated the number of alternatives previewed and whether participants could actually control the selection (Morewedge et al., 2009, Experiment 2). As before, participants saw two, one, or no previews of the works of art they could see in each trial (within subjects). Control over selections varied between subjects: Participants in a *true control* condition could control the painting they selected (the left and right keys consistently referred to the left and right artworks on all trials). Participants in a *random control* condition randomly selected paintings as described in the previous experiment, and participants in a *random assignment* condition simply pressed a spacebar to have the computer program randomly select a work for them.

If participants were simply confused about the extent to which they controlled the selection process in the previous experiment, we would expect that those in the *true control* and *random control* conditions would report similar feelings of control, but both groups would report feeling more control than participants in the *random assignment* condition (whose lack of control should be fairly clear). Participants did not appear to be confused, as they were sensitive to the amount of actual control they possessed in the experiment. Those with *true control* reported feeling greater control than participants with *random control*, and both groups of participants reported feeling more control than participants with no control—participants in the *random assignment* condition.

As in the first experiment, those in the *true control* and *random control* conditions reported feeling more control over selections when they saw more rather than fewer previews. Interestingly, participants in the *random assignment* condition reported feeling no more control over outcomes whether they were presented with more or fewer previews. The appearance that we have a choice of alternatives seems necessary for premeditation to feel effective—it is not enough to simply see alternative potential outcomes. Important to note is that the *perception* of possible alternatives influences judgments of control, independent of a person's actual ability to control the alternative selected. What follows is an examination of additional indicators of premeditation, and whether those indicators can lead to specious inferences of personal control.

Requirement 2: Temporal Priority

A second requirement for premeditated outcomes is that they were considered *before* the action producing those outcomes was executed. Thinking about outcomes after executing the critical action can hardly be considered a premeditated decision, since that thought could have had no causal impact on the action or its consequences (c.f., Radin, 2006). This priority principal is demonstrated by an experiment (Wegner & Wheatley, 1999) in which each participant and a confederate placed their hands upon a *planchette*—an object used to spell out messages and select responses on a Ouija board—to move a computer mouse beneath the planchette across a computer display. The display consisted of objects such as a house, duck, and rowboat. The participant and confederate moved the planchette together to select an object while listening to audio tracks (played over headphones). Unbeknownst to participants, on critical trials a certain object was preselected, and the confederate subtly pushed the planchette so that it rested upon the preselected object. On these critical trials, the participants would hear the name of the preselected object over headphones. Whether the participant heard the name of the preselected object *before* or *after* the selection was varied within subjects. Participants reported feeling more control over the selection of the object if it was named before than after the selection. In other words, only thoughts occurring *prior* to an outcome were perceived to produce it.

Perhaps prior knowledge of our alternatives is also crucial for an outcome to be considered the result of one's premeditated decision. Another art selection experiment examined this possibility (Morewedge et al., 2009). Participants in a *prior knowledge* condition and a *delayed knowledge* condition randomly selected works of art to see in a paradigm similar to the first art selection experiment. In the prior knowledge condition, participants saw two, one, or no thumbnail previews of works of art *before* they randomly selected a work to see for 20s. In the delayed knowledge condition, participants saw two, one, or no thumbnail post-views of works of art *after* they randomly selected a work, but before they saw the work they selected. Thus, both groups of participants received the same information about the options available in each trial, but only those in the prior knowledge condition knew those options before making a selection.

As in the other experiment (where all participants had prior knowledge), participants in the prior knowledge condition reported feeling more control over their selection when they saw more rather than fewer previews. Participants in the delayed knowledge condition, however, reported feeling equivalent control over selections irrespective of the number of post-views they saw in trials, and also reported feeling less control than participants in the prior knowledge condition. Although participants in prior and delayed knowledge conditions had the same amount of control over the works of art selected in this experiment and saw the same number of pre- and post-views, only participants who saw the alternatives in advance felt they personally controlled outcomes. In other words, even though all works were randomly selected, participants felt control over selections only when the priority requirement was fulfilled. Thus, people appear to consider outcomes to be result of their decisions only when their alternatives were considered beforehand, even when that consideration was purely incidental.

Requirement 3: Consistency

For an outcome to be considered the product of a premeditated decision, it must presumably be consistent with the thoughts preceding it. If we think about eating hamburgers all day and find turkey on our dinner plate, we may question how it got there. We expect outcomes to conform to our thoughts and use the consistency between the two to determine which outcomes we controlled. In a laboratory experiments, people examine their thoughts just before an outcome to determine if they consciously willed it (Wegner, 2002, 2003). In a legal domain, people examine the thoughts of defendants to assess how much punishment they deserve for the crime they committed. If a killer said, "I thought about how I could kill Roger with a blow to the head," she would presumably be incarcerated for a long time. On the other hand, if she said, "I thought about how I could hit a nail and discovered that I had hit Roger instead," the inconsistency between her thought and the outcome suggests that she did not plan to kill Roger or control the action that did (Hart & Honoré, 1959/2002).

Even when outcomes are produced by chance, consistency between intention and outcome may give rise to the feeling of personal control. The repeated production of intended events—such as successfully flipping a coin so it turns up heads five times in a row—may give rise to the feeling that we controlled those events through skill rather than chance (Langer, 1975, 1983; Langer & Roth, 1975). One implication of the consistency requirement is that people should take more credit for their triumphs (which they are likely to imagine) than their failures (which they would prefer to ignore), a tendency that has been demonstrated to be robust (for a review, see Miller & Ross, 1975). Indeed, this tendency is so robust that people even take credit for the outcomes they *merely intended* to produce. People praise themselves for simply intending to help the needy, for example, whether or not they actually did (Kruger & Gilovich, 2004). Depressed people appear to be one of the few exceptions to this pervasive tendency (Alloy & Abramson, 1979), as they claim less responsibility for successes than do nondepressed people.

Not only can the consistency principle lead us to feel control over the things we did not cause, it can lead even those of us who are sane to believe we possess supernatural powers. In one study, participants were asked to stick pins into the head of a voodoo doll made to resemble a confederate of the experimenter. Participants were more likely to report feeling personally responsible for the confederate suddenly experiencing a headache if they were previously led to dislike the confederate (i.e., when he showed up late to the study wearing a trucker hat emblazoned with the motto, "Stupid People Shouldn't Breed"), than if they were not previously led to dislike the confederate (i.e., when he showed up on time and was dressed inoffensively; Pronin et al., 2006).

It seems that thoughts only need to appear consistent with outcomes for us to claim those outcomes were intentionally produced. Indeed, decision makers can be led to believe they chose an option they actually rejected. Such *choice blindness* is typically produced by having people pick between two similar alternatives and then tricking them into believing they selected the unchosen alternative. In one demonstration, men were shown pairs of cards depicting similar women and were asked to choose which woman was more attractive (Johansson et al., 2005). They then explained why they preferred the woman they chose. On choice blindness trials, the experimenter (who was a magician) used sleight of hand to switch the cards and showed each man his unchosen alternative—acting as if he had preferred the woman he said was less attractive. Thirty percent of the men completely missed the switch! Even more surprising was their readiness to give reasons for having "chosen" the woman they had actually rejected:

"She's radiant. I would rather have approached her at a bar than the other one" (Johansson et al., 2006, p.118). Participants apparently forgot which woman they chose, and inferred from the similarity between their choice and the outcome that they preferred the less attractive woman. In hindsight we seem to overestimate the consistency between our original intentions and the outcomes of our decisions (Pieters, Baumgartner, & Bagozzi, 2006).

Sometimes rather than forgetting which option we chose, we entirely forget all of our options. Chance and Norton (2007) showed research participants pairs of colors and asked them to identify which color they preferred in each pair. Although the behavior of participants demonstrated they had seen the color pairs before—they performed above chance when later distinguishing between old and new pairs—on average participants falsely reported having never seen 25% of the original color pairs. Interestingly, participants were especially likely to forget the original color pairs that were most difficult to choose from.

People may forget which alternatives they originally chose or were available, but there are likely to be limits to their forgetfulness. With an art selection experiment, we investigated the extent to which participants used false feelings of consistency to infer that they made a decision. Using a paradigm similar to the previously described art selection experiments, participants randomly selected one of two works of art to see by pressing one of two keys after seeing two, one, or no previews of the works. In this experiment, however, previews could either see *accurate* previews of the works of art (as before), *inaccurate* previews of the works of art, or accurate previews of the works of art that were *distorted* beyond recognition (*see* Fig. 14–2).

As in previous experiments, participants felt more control when they saw a greater number of *accurate* previews. More important, feelings of control were not affected by the number of *inaccurate* and *distorted* previews that participants saw (Morewedge et al., 2009, Experiment 4). When the previews represented possible outcomes, participants felt more responsible with more previews, but when previews provided

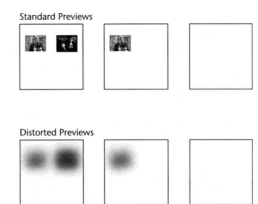

Figure 14–2. Examples of previews and postviews depicting two, one, and no works of art that were used in art selection experiments appear in the top row. Examples of distorted previews appear in the bottom row. Each preview was 100 pixels wide and presented 25 pixels from the center of a 17" monitor with a resolution of 1028 × 768.

information that was irrelevant or uninformative, perceived control was unrelated to the number of previews they saw. It appears that outcomes need not be consistent with the most preferred choice. Rather, the outcome of a decision must be consistent with at least some of the *possible* alternatives. These findings suggest that when examining the consistency between our intentions and outcomes, we reconstruct rather than recall the outcome we originally intended to produce.

Requirement 4: Exclusivity

For an outcome to feel like the result of a premeditated decision, it must appear to have been produced exclusively by one's intentions. If Bobby's mother tells him to eat his peas and Bobby decides to eat them, who gets credit for the pea-eating? Bobby's pea-loving mother. Whether situational constraints are strong or subtle, we are quick to deny responsibility for undesirable outcomes produced by our behavior when it can be attributed to external influences. Obedience experiments by Milgram (1974) demonstrate that the perception of responsibility for nefarious outcomes—those endangering the life of another human being—may be abnegated when we receive direction from an authority figure. The suggestions of an authority led teachers in Milgram's experiments to believe that they were forced to, rather than decided to, expose a learner to levels of electric shock that could prove fatal, even though teachers had full control over the fate of learners and had the option to disobey the authority figure.

Like the other requirements for premeditation, it appears that perceived rather than actual exclusivity is crucial when determining personal control over outcomes. We can be easily led to ignore real exclusivity: Cognitive dissonance experiments show that we are quick to attribute the discomfort arising from our decisions to placebos such as a sugar pill (Harmon-Jones & Mills, 1999; Zanna & Cooper, 1974). Conversely, we sometimes claim exclusivity for outcomes caused by external forces. For example, we take full responsibility for and generate reasons why we produced actions that were performed because they were suggested by a hypnotist (Wheatley & Haidt, 2005), or why we made decisions that were due to irrelevant contextual influences such as the order in which options were presented (Nisbett & Wilson, 1977).

The tendency to confuse exclusivity has important implications, as in the case of facilitated communication—a technique in which a facilitator (usually a therapist) lays their hands atop the hands of a patient (e.g., an autistic child), and interprets their slight movements as meaningful responses to questions posed to the patient. Although it sounds like a good idea in theory, in practice facilitators answer the questions themselves and attribute their answers to the patients (Wegner, Fuller, & Sparrow, 2003).

Requirement 5: Effort

Once an outcome has been chosen, the translation of that thought into action requires controlled effort (James, 1890/1955), which appears to be the final requirement for an outcome to be perceived as premeditated. As a result, outcomes chosen out of habit or that are easy to perform feel less controlled than those that are unusual or difficult (Fiske, 1989).

Perhaps this is because effort is usually exerted during the process of decision-making, as evidenced by a decreased ability to perform effortful tasks after having made a decision. Like the performance of tasks requiring self-control, making judgments and decisions appears to deplete a limited resource that some have suggested is will-power (Masicampo & Baumeister, 2008; Vohs & Faber, 2007). In perhaps the best illustration, participants were placed in a room for five minutes with a plate of freshly baked chocolate chip cookies and a plate of radishes. Participants instructed to eat only radishes quit a subsequent (impossible) task more quickly than controls and those allowed to eat only cookies (Baumeister et al., 1998). Presumably, suppressing one's desire to eat the cookies taxed participants' ability to persist in the subsequent task.

Expending effort while an outcome is produced, however, may mistakenly lead us to feel responsible for producing it. These *Eureka errors* are particularly likely when we expend more mental or physical effort during the generation of a solution than during the presentation of a solution. In one study, participants were paired with a partner and asked to solve anagrams together (Preston & Wegner, 2007). When participants were asked to squeeze a handgrip while thinking of solutions, they claimed credit for the solution, regardless of whether they or their partner solved it. Similar effects were found in other experiments; participants were more likely to falsely claim credit for solutions to anagrams when the anagrams were displayed in a difficult-to-read font than an easier-to-read font. Participants were thus likely to feel responsible for an outcome when they engaged in effortful thinking during its production.

Indeed, merely engaging in effortful thinking about *anything* before producing an outcome may lead us to claim more control over that outcome, even when it was randomly produced. In another art selection task, participants saw thumbnail previews of the two works of art that could be selected in each trial. On two-thirds of trials, participants performed one of two other tasks before selecting artwork. Half of the time, participants answered a relevant question—what they were thinking and feeling.

Half of the time, participants answered irrelevant questions such as, "*What word do you associate with tube socks?*" and "*What's a fair price for a banana cream pie?*"

Although participants reported more control over trials in which they performed an additional task, the results of this study indicate that the content of the task did not matter. Relative to no-task trials, participants reported feeling greater, but equal amounts of control over the trials in which they answered relevant and irrelevant questions. Regardless of whether their thoughts were relevant to the task at hand, participants felt more control over outcomes after having thought about something—anything—before the outcome was produced (Morewedge et al., 2009, Experiment 5).

Summary

These five sections outline the necessary components of premeditation, which give rise to the experience of control over the outcomes of decisions, and suggest that their presence often fools us into believing we controlled them. Multiple alternatives need merely be present, not possible, to make outcomes feel like the product of our decisions. These alternatives must appear in advance of the decision, regardless of whether they can actually be chosen, to make the outcome feel like it was under our control. Even when we have no real choice, we feel that we determined outcomes when they appear consistent with our intentions and free of external influence. Finally, any kind of thinking before a decision, whether related or not, appears to lead to the perception that an outcome was the result of a premeditated decision—that our thoughts and actions controlled it. Although premeditation may indicate that we considered our options before making a decision, we appear to have difficulty determining when it did and did not lead to the production of the outcomes we intended.

OBSERVERS' PERCEPTIONS

When determining whether an outcome was the result of a premeditated decision, actors have the privilege of knowing the thoughts and alternatives they considered prior to its production. Observers have a more difficult task, as they must infer which thoughts and alternatives actors considered. Although actors and observers differ in the way they ultimately attribute responsibility for outcomes (Gilbert & Malone, 1995; Jones & Harris, 1967; Kelley, 1973; Ross & Nisbett, 1991), observers who believe they possess knowledge of the prior thoughts and alternatives available to actors judge premeditation according to criteria similar to the criteria that actors use. Observers examine whether actors had multiple options to consider or choose from, could have foreseen and desired to produce the outcome, whether actors' intentions appeared to be the sole cause of the outcome, and the effort actors expended toward creating the outcome produced (Pizarro, Uhlmann, & Bloom, 2003; Shaver, 1985; Weiner, 1995). Like actors, however, observers may merely examine whether these requirements appear to have been satisfied. Observers who do not believe they possess such knowledge judge premeditation by making more general inferences about the actor and the apparent goal of the outcome produced. In both cases, however, these judgments are made under uncertainty. Observers' judgments may thus substitute their own intuitions and emotions for evidence when none is available (Haidt, 2001; Kahneman & Frederick, 2002; Kahneman, Schkade, & Sunstein, 1998).

Furthermore, observers may be motivated to infer that the requirements have been satisfied to validate their feelings and intuitions that someone should be praised or blamed for the outcome (Alicke, 2000; Haidt, 2001; Gray & Wegner, in press; Kahneman, Schakde, & Sunstein, 1998). Indeed, such intuitions are unlikely to be corrected, as people generally believe their judgments are veridical assessments of the world (Nisbett & Wilson, 1977; Pronin, Gilovich & Ross, 2004). Although observers may examine whether actions meet the five requirements actors use to assess premeditation, observers' examination of these requirements is often cursory, merely a ritual used to justify the ascription of praise or blame.

REQUIREMENT 1: MULTIPLE POTENTIAL COURSES OF ACTION

Observers are most likely to consider an outcome the result of a premeditated decision and thus controlled when actors appear to have had multiple alternatives from which to choose. In some way, observers are more likely than actors to believe that alternatives were possible, as they tend to perceive that actors could have produced an alternative outcome even when actors were unaware that alternative existed. Wells and Gavanski (1989), for example, asked people to assess an employer's responsibility for indirectly killing his employee by ordering an entrée containing an ingredient to which she was highly allergic. In one version of the scenario, the employer selected between two entrées for the employee, both containing the lethal ingredient. In the other version, only one of the two entrées contained the lethal ingredient. Although the employer was unaware of her allergy in both scenarios, he was seen as having greater control over her life and death when only one entrée was fatal than when both were fatal. Of course, the employer had no real control over her fate in either scenario—in both cases he did not know that either entrée would kill his employee. Yet, observers inferred that because the employer could have chosen a nonfatal entrée, the employer was more responsible for choosing an entrée that killed her.

Requirement 2: Priority

Retrospectively discerning whether outcomes were foreseeable at the time of judgment is a challenging task for observers. A *curse of knowledge*—knowing with certainty the events that occurred—hinders observers' ability to imagine actors' thoughts when faced with a decision (Camerer, Loewenstein, & Weber, 1989). Knowledgable observers may have difficulty conceiving of a situation in which actors were not aware of all of the consequences of the outcome chosen. Children have particular difficulty accounting for knowledge that they possess when predicting naïve actors' behavior (Wimmer & Perner, 1983). Although adults are influenced to a lesser degree by prior knowledge than children, they are still likely to be initially egocentric (Epley, Morewedge, & Keysar, 2004). The certainty hindsight provides also makes the foreseeability of an outcome seem greater, even influencing expert judgment (Fischhoff, 1975). It is difficult for observers to discern actors' thoughts at the time of judgment and ignore knowledge of the outcomes of those actions.

Requirement 3: Consistency

Consistency between an actor's prior thoughts or desires and an outcome lead observers to consider actions premeditated, whether or not the actor's thoughts and desires could influence the outcome. Woolfolk, Doris, and Darley (2006) showed that actors are held more culpable for actions they intended, even if their intentions did not influence the action. Observers read a vignette in which a protagonist was given a drug by terrorists that made him powerless to resist their command to kill a man, and then judged how responsible he was for that killing. Half of participants were informed that the protagonist had previously thought about killing the man and wanted him dead, whereas the other half were informed that the protagonist had never thought about killing the man and wanted him to live. In both cases, the protagonist had no control over the outcome he produced (the man's death), but observers ascribed more blame to the target when the outcome was consistent than inconsistent with his previous desires.

Consistency between the outcome that occurred and the outcome an actor intended to produce, however, does not appear to be sufficient for an act to be considered the result of a premeditated decision. Observers are sensitive to the consistency between the action intended to produce an outcome and the action that actually produced it. A person intending to break into his uncle's house and kill him, for example, is considered less culpable for the death of his uncle if he accidentally ran into and killed the uncle while driving to the house, than if he killed his uncle in the house as he originally planned (Pizarro, Uleman, & Bloom, 2003). Although the actor produced the outcome he desired—the death of his uncle—observers only considered the outcome to have been premeditated when the sequence of events producing his uncle's death were consistent with the sequence he planned to execute.

When observers are unsure of actors' prior intentions, they may base judgments of premeditation on the consistency between outcomes and their lay theories of actors' general inclinations or *meta-desires*. People are considered less blameworthy for moral transgressions that appear to have been committed impulsively than moral transgressions that appear to involve premeditation ("She doesn't really dislike me—she only chewed me out because she was drunk"), because most people have a meta-desire to avoid doing bad. Conversely, people are not considered less praiseworthy for good behaviors committed impulsively than good behaviors that appear premeditated, because most people have a meta-desire to do good (Pizarro, Bloom, & Salovey, 2003). Interestingly, this suggests that actors who are generally perceived to desire to do evil are not excused from impulsive immoral acts because those acts are considered consistent with their meta-desires.

Finally, observers occasionally engage in attribute substitution (Kahneman & Frederick, 2002), overextending consistency to include unintended side effects of the outcome produced. In the Wells and Gavanski (1989) case mentioned before, observers overextended the consistency between an employer's intention (i.e., to select an entrée for his employee) and action (i.e., ordering an entrée for his employee) to include an unintentional side effect (i.e., his employee's death). People are held similarly accountable by the law for unintended consequences of their actions, if the actions themselves were premeditated (Hart & Honoré, 1959/2002). In a celebrated case, a man and an exotic dancer conspired to rob a rich client. The dancer entertained the client in his car until her accomplice opened the car door and pointed a gun at the client. Although her accomplice had no desire to shoot the client, he slipped on a patch of ice and pulled the trigger. The client was killed by the stray bullet and the accomplice was convicted of felony murder because he was engaging in an intentional illegal act (Alicke, 2000). Observers are especially likely to attribute unintended side effects to actors' intentions when the outcome and original intention are similar. A person is considered more culpable for having caused a car accident while speeding, for example, if he was rushing home to hide a vial of cocaine than to hide an anniversary gift. In this case, criminal intentions to hide an illegal substance presumably were generalized to include any potentially harmful outcome (Alicke, 1992).

Requirement 4: Exclusivity

Under ordinary circumstances, for a variety of reasons, observers tend to assume that actors' behavior was due to their intentions and dispositions (Gilbert & Malone, 1995; Ichheiser, 1949; Jones & Davis, 1965; Jones & Harris, 1967; c.f., Krull, 1993). For example, observers are generally less sensitive than actors to the presence of external influences when inferring what caused actors to behave and choose as they do. A seminal demonstration by Jones and Harris (1967) asked participants to read an essay supporting or opposing Fidel Castro's governance of Cuba and then infer the essay writer's opinion. Some readers were told that writers were free to choose a position to endorse, whereas other readers were told that writers were assigned to endorse a position. Despite knowing that essay writers were forced to support or oppose Castro in the latter condition, essay readers inferred that the writers' attitudes toward Castro were reflected by the position participants (unwillingly) espoused.

One notable exception to this correspondence bias is observers' tendency to attribute counterfactual blame (Miller, Visser, & Staub, 2005). When the constraints of the situation are so severe that there are no possible alternative courses of action available, observers' dispositionalism is sometimes so strong that they assume actors' intentions were contrary to their behavior. Miller and colleagues (2005) showed a person taking a test with or without strict supervision to observers and then asked observers to assess the test taker's character. When the test taker was under strict supervision, observers considered her less trustworthy than when the test taker was under no supervision. Observers assumed that she would have cheated in the latter case if there were weaker situational constraints. In short, observers appear to consider exclusivity quite differently than do actors. They generally perceive actors to have controlled actions that were performed as a result of substantial coercion and believe they can infer actors' intentions even when actors are forced to perform the actions they do.

Requirement 5: Effort

Although observers often discount the influence of the situation in attributing responsibility, certain outcomes appear premeditated because of the substantial effort required to enact them. In any situation, there is usually a course of action that is normative, and when a counter-normative or unusual action is observed, it is thought to have been consciously chosen. For example, most people buy groceries at the supermarket, so learning that a person only purchases groceries from a farm co-op may lead you to believe that she really cares about where her broccoli comes from. Similarly, most people drive at or above the speed-limit, so we may infer that a person driving more slowly than the limit intends to do so, assuming we do not simply infer that the driver is incompetent (Morewedge, Preston, & Wegner, 2007). Choosing such nondominant alternatives is often called making the "hard choice" (Fiske, 1989), a term implying that the decision maker was not only aware of the alternative but possessed both the cognitive resources and sufficient control over her situation to reject the modal response (Gilbert, 1991; Shiv & Fedorikhin, 1999).

Summary

Observers appear to evaluate the same requirements actors do when discerning whether an outcome was due to a premeditated decision, and like actors, they often evaluate those requirements in a cursory manner. Observers' inability to know actors' desires, whether alternatives were recognized or considered, and the difficulty of disentangling the person from the situation, leads observers to more often attribute premeditation of and control over outcomes to actors than actors attribute to themselves. This is exacerbated by observers' tendency to generalize actors' intentions and a failure to discount knowledge gained from hindsight.

Effects of Premeditation

The evidence presented suggests that the presence of premeditation does not serve as a reliable index of personal control. If premeditation is present before outcomes that we did not cause and absent before outcomes that we did, it is questionable whether premeditation deserves to be considered an indicator of intentional action. In a broad sense, judgments and decisions are influenced by information that is cognitively accessible when they are made (Kahneman & Frederick, 2002; Morewedge, Gilbert, & Wilson, 2005; Morewedge, Holtzman, & Epley, 2007; Tversky & Kahneman, 1973; Winkielman & Schwarz, 2001), so it would be surprising if information considered during premeditation had no effect on outcomes produced by the premeditator. Given the serious social and legal consequences of considering an outcome to have been caused by a premeditated decision (Denno, 2003; Federal Sentencing Guidelines, 2004; Kadish & Schulhofer, 1995; Weiner, 1995), it is natural to wonder: What are the effects of premeditation on decision making?

It is clear that merely thinking about a desired action makes us more likely to produce it, particularly when that action is easy to imagine (Levav & Fitzsimmons, 2006). Indeed, we are more likely to perform pro-social actions and less likely to perform antisocial actions if we think about them in advance (Sherman, 1980). Registered voters asked if they intended to vote in an election were 25% more likely to vote in that election than voters who were not asked (Greenwald et al., 1987). M.B.A. students led to think about flossing their teeth more frequently flossed than those led to think about reading for pleasure (Levav & Fitzsimmons, 2006). Perhaps most surprising, participants in a nationally representative sample of more than 40,000 American households were 35% more likely to make a purchase costing thousands of dollars—a *car*—within 6 months of the survey if they were asked if they intended to buy one in the near or distant future (Morwitz, Johnson, & Schmittlien, 1993).

One way premeditation may affect the production of an outcome is through the formation of implementation intentions, which cue goal-directed responses when presented with specific situations (Gollwitzer, 1999). For example, you may wish to finish some home renovations by the end of the week, and thus plan to decline the after-work invitations of your colleagues so you can go home early to renovate. By deciding beforehand how to respond when a relevant situation presents itself, you can work towards your goals and avoid getting sidelined by distractions and procrastination (Dellarosa & Bourne, 1984). Furthermore, preemptively deciding how to behave may lead to the eventual automatization of the intended goal-directed behavior (Bargh, 1997; Gollwitzer, 1993, 1996).

Although premeditation can help us work toward goals by increasing the probability that an intended outcome will be produced, the effects of premeditation are not always beneficial. Wilson and Schooler (1991) found that novice jam-tasters asked to rate jams after introspecting about their opinions were less accurate in their taste ratings than novice jam-tasters in a control condition. Similarly, students who introspected about their artistic preferences before choosing one of two posters to adorn their dorm room were less likely than controls to choose a high-quality art poster rather than a lower-quality humorous poster (e.g., *Starry Night* by Van Gogh rather than *Garfield* by Jim Davis). Furthermore, those introspectors reported feeling less satisfied with their decision 3 weeks later than did controls, suggesting that premeditation can impair the stability of preferences used to inform judgments and decisions (Wilson et al., 1993).

Dijksterhuis (2004) examined effect of premeditation on the quality of decisions and found its usefulness depends on the complexity of the decision and the kind of deliberation used to make it. He suggests that conscious deliberation improves the quality of simple decisions such as which shampoo or oven mitts to buy, as it is possible to simultaneously compare a few dimensions or features consciously. Dijksterhuis also suggests, however, that conscious deliberation reduces the quality of complex decisions such as which person to marry or house to buy, because we cannot simultaneously compare a large number of dimensions or features consciously. Instead, we end up overweighting the few features or dimensions we can consider and underweighting unconsidered features when making complex decisions.

To test this detrimental effect of conscious deliberation, he asked participants to choose the best apartment out of a set of four, which varied across 12 attributes (e.g., amount of sunlight, attractiveness of the location, size). One apartment was clearly better than the other three because it possessed more positive and fewer negative attributes, although the descriptions were complex enough that comparing between apartments was difficult. Once all of the features of the apartments had been presented, participants indicated the apartment they preferred either *(a)* immediately; *(b)* after consciously deliberating about the choice for five minutes; or *(c)* after performing an unrelated task for five minutes. Conscious deliberation did not improve the quality of decisions, as participants who spent five minutes deliberating before choosing were as likely to choose the best apartment as participants who chose immediately. More important, participants who performed an unrelated task for 5 minutes before choosing were most likely to pick the best apartment, a finding Dijksterhuis and colleagues (2005; Dijksterhuis & Nordgen, 2006) attribute to the superiority of unconscious deliberation. They propose that unconscious deliberation possesses a greater capacity to simultaneously compare complex multidimensional stimuli and more accurately weight their attributes when making judgments and decisions (c.f., Payne, Samper, Bettman, & Luce, 2008).

In sum, the effects of premeditation appear to depend on the decision at hand. Premeditation increases the likelihood that actors will produce the premeditated outcomes and desired goal-directed behavior, which can be good or bad (e.g., committing one's life to justice vs. committing murder). Whether premeditation improves or impairs decisions appears to depend on the alternatives considered. Premeditation appears to improve the quality of decisions involving alternatives that are relatively simple, but sometimes impairs the quality of decisions involving alternatives that are more complex.

Implications for Self-Control

The findings presented in this chapter suggest that premeditation may not reliably indicate when we possess control over a decision or outcome. Disturbingly, these findings not only imply that premeditation may not deserve the importance afforded it by society and the law, but also that we should reconsider and carefully examine the extent to which our conscious thoughts actually enable us to exert self-control. If the link between premeditation and control is specious for unimportant decisions (such as art selection), it is possible that conscious forethought also has little influence on more important decisions, such as those made when we are confronted with the choice to consume or abstain from the consumption of unhealthy foods, alcohol, cigarettes, and drugs.

We may lament "Why didn't I think this through?" after succumbing to temptation, but it may be that even with significant premeditation we would have awoken to find our shirts covered in powdered sugar and donut crumbs, our breath reeking of whisky, the only difference being the greater sting of guilt caused by considering our gluttonous behavior before we engaged in it. Indeed, much of our behavior is affected by unconscious goals and motivations (Dijksterhuis, Chartrand, & Aarts, 2007; Förester, Liberman, & Friedman, 2007). On the other hand, when we are able to resist our impulses and make the appropriate choice, having consciously rebuffed temptation may make us feel empowered. The experience of conscious conflict may have the additional benefit of making us more satisfied with our choice to behave well (Brehm, 1956; Festinger, 1957).

We are not, however, advocating the embrace of the belief that we are all self-indulgent automatons, powerless to avoid our impulses to eat fattening foods and wash them down with stiff drinks, particularly as advocating such a belief may have profoundly negative consequences (Vohs & Schooler, 2008). Rather, evidence suggests that the conscious struggle we engage in when faced with self-control dilemmas does actually have some influence over our behavior. We appear to have a limited resource of self-control that we can allocate toward specific tasks (Baumeister et al., 1998; Wegner, 1994). The key puzzle appears to be determining what role conscious thought plays in the theater of self-control. Perhaps conscious thought determines the impulses that we should regulate with this limited resource and the impulses that are not worth attempting to control. We may thus decide to apply our limited resource of self-control to the dilemmas we consider most important, or to prevent the lapses that would reflect most poorly upon ourselves, just as people are able to selectively inhibit the forms of prejudice that are frowned upon by their society (Franco & Maass, 1999). Alternatively, conscious thought may allow us to create automatic behavioral scripts—implementation intentions—that are carried out when we must exert self-control (Gollwitzer, 1999). Ironically, then, premeditation may influence decisions indirectly by automatizing the behaviors we consciously intend to produce.

Conclusion

Actors, observers, society, and the law place faith in premeditation as a principle indicator of personal control. The findings presented in this chapter illustrate many instances in which these faith-based judgments are misleading. Actors have difficulty discerning the efficacy of their premeditation, and rely on the apparent rather than the actual fulfillment of the five criteria we reviewed: *(1)* having considered or the ability or consider *multiple alternatives*; *(2)* if the meditation they engaged in occurred *prior* to the action or outcome produced; *(3)* if outcomes were *consistent* with the alternatives they considered; *(4)* if outcomes were *exclusively* due to internal causation on the part of the actor;

and (5) if outcomes occurred as a result of the actor's controlled *effort*.

Observers more superficially assess the fulfillment of these five requirements to determine whether outcomes were caused by an actor's premeditation, and exhibit a general tendency to over-attribute outcomes to actors' premeditated decisions. The cursory examination of alternatives, priority, consistency, exclusivity, and effort by observers is largely due to their lack of first hand knowledge, which is often inferred from their own intuitions, emotions, and beliefs about actors' meta-desires.

Although premeditation increases the likelihood that we will produce the outcome we intend to perform, its benefits are questionable. Given the serious nature of its social and legal consequences, it may thus be time to re-examine the privileged position of this often ineffectual behavior. Premeditation appears to often be no more than ritual; incidental cognition that engenders false feelings of personal control. In spite of this, there may be hope for self-control if we can premeditate far enough in advance to form automatic action plans. Nevertheless, these findings suggest it may be time to seriously question whether premeditation deserves the importance granted to it by society and the law, and whether it should continue to serve as a primary indicator of personal control.

Note

1. In this chapter, the terms premeditation, deliberation, and making decisions are distinct. We define premeditation as conscious thought corresponding to the performance of an action or decision before it is produced or determined. We define deliberation as conscious or unconscious consideration of a stimulus, potential action, etc., that may occur before, during, or after the performance of an action or decision. We define making decisions as the process of selecting a course of action to pursue or stimulus when faced with multiple alternatives, whether those alternatives are multiple courses of action and stimuli, or those alternatives involve either selecting an action or stimulus or not selecting that particular action or stimulus.

References

Alicke, M. D. Culpable causation. J Pers Soc Psychol 1992; 63(3): 368–378.

Alicke, M. D. Culpable control and the psychology of blame. Psychol Bull 2000; 126(4): 556–574.

Alloy, L. B., & Abramson, L. Y. Judgment of contingency in depressed and nondepressed students: sadder but wiser? J Exper Psychol Gen 1979; 108: 441–485.

Anderson, C. J. The psychology of doing nothing: Forms of decision avoidance result from reason and emotion. Psychol Bull 2003; 129: 139–167.

Baumeister, R. F., Bratslavsky, E. Muraven, M., & Tice, D. M. Ego depletion: Is the active self a limited resource? J Pers Soc Psychol 1998; 74, 1252–1265.

Baumeister, R. F., & Heatherton, T. F. Self-regulation failure: An overview. Psychol Inq 1996; 7: 1–15.

Bem, D. J., & McConnell, H. K. Testing the self-perception explanation of dissonance phenomena: On the salience of premanipulation attitudes. J Pers Soc Psychol 1970; 14: 23–31.

Brehm, J.W. Postdecision changes in the desirability of alternatives. J Abnorm Soc Psychol 1956; 52: 384–389.

Camerer, C., Loewenstein, G., & Weber, M. The curse of knowledge in economic settings: An experimental analysis. J Polit Econ 1989; 97: 1232–1254.

Chance, Z., & Norton, M. I. Decision amnesia: Why taking your time leads to forgetting. Unpublished manuscript, Harvard Business School, 2007.

Denno, D. W. A mind to blame: New views on involuntary acts. Behav Sci Law 2003; 21: 608–618.

Dijksterhuis, A. Think different: The merits of unconscious thought in preference development and decision making. J Pers Soc Psychol 2004; 87: 586–598.

Dijksterhuis, A., Chartrand, T., & Aarts, H. Effects of priming and perception on social behavior and goal pursuit. In: Bargh, J. A. (Ed.), *Social psychology and the unconscious: The automaticity of higher mental processes*. New York, NY: Psychology Press, 2007: pp. 51–132.

Dijksterhuis, A., & Nordgen, L. F. A theory of unconscious thought. Perspect Psychol Sci 2006; 1: 95–109.

Epley, N., Morewedge, C. K., & Keysar, B. Perspective taking in children and adults:

Equivalent egocentrism but differential correction. J Exp Soc Psychol 2004; 40(6): 760–768.

United States Sentencing Commission. *Federal sentencing guidelines*, 2004. Available at: http://www.ussc.gov/2004guid/2a1_1.htm (accessed December 8, 2007).

Feinberg, T. E. *Altered egos: How the brain creates the self.* New York, NY: Oxford University Press, 2001.

Festinger, L. *A theory of cognitive dissonance.* Stanford, CA: Stanford University Press, 1957.

Fischhoff, B. Hindsight does not equal foresight: The effect of outcome knowledge on judgment under uncertainty. J Exp Psychol Hum Percept Perform 1975; 1: 288–299.

Fiske, S. Examining the role of intent: Toward understanding its role in stereotyping and prejudice. In: Uleman, J. S., & Bargh, J. A. (Eds.), *Unintended thought.* New York, NY: Guilford Press, 1989: pp. 253–286.

Förster, J., Liberman, N., & Friedman, R. S. Seven principles of goal activation: A systematic approach to distinguishing goal priming from priming of non-goal constructs. Pers Soc Psychol Rev 2007; 11: 211–233.

Franco, F. M., & Maass, A. Intentional control over prejudice: When the choice of the measure matters. Euro J Soc Psychol 1999; 29: 469–477.

Gilbert, D. T. How mental systems believe. Am Psychol 1991; 46: 107–119.

Gilbert, D. T., & Malone, P. S. The correspondence bias. Psychol Bull 1995; 117: 21–38.

Goethals, G. R., & Reckman, R. F. The perception of consistency in attitudes. J Exp Soc Psychol 1973; 9: 491–501.

Gollwitzer, P. M. Implementation intentions: Strong effects of simple plans. Am Psychol 1999; 54: 493–503.

Gray, K., & Wegner, D. M. Blaming God for our pain: Human suffering and the divine mind. Pers Soc Psychol Rev in press.

Greenwald, A.G., Carnot, C.G., Beach, R., & Young, B. Increasing voting behavior by asking people if they expect to vote. J App Psychol 1987; 72: 315–318.

Haidt, J. The emotional dog and its rational tail: A social intuitionist approach to moral judgment. Psychol Rev 2001; 108(4): 814–834.

Harmon-Jones, E., & Mills, J. (Eds.). *Cognitive dissonance: Progress on a pivotal theory in social psychology.* Washington, DC: American Psychological Association, 1999.

Hart, H. L. A., & Honoré, T. *Causation and the law.* New York, NY: Oxford University Press, 1959/2002.

Ichheiser, G. *Misunderstandings in human Relations: A study in false social perception.* Chicago, IL: University of Chicago Press, 1949.

James, W. *The principles of psychology.* Mineola, New York: Dover, 1890/1955.

Johansson, P., Hall, L., Sikstrom, S., & Olsson, A. Failure to detect mismatches between intention and outcome in a simple decision task. Science 2005; 310(7): 116–119.

Jones, E. E., & Davis, K. E. From acts to dispositions: the attribution process in person perception. In: Berkowitz, L. (Ed.), *Advances in experimental social psychology,* Vol. 2. New York, NY: Academic Press, 1965; pp. 219–266.

Jones, E. E., & Harris, V. A. The attribution of attitudes. J Exp Soc Psychol 1967; 3: 1–24.

Kadish, S. H., & Schulhofer, S. J. *Criminal law and its process: Cases and materials.* New York, NY: Aspen Law & Business, 1995.

Kahneman, D., & Frederick, S. Representativeness revisited: Attribute substitution in intuitive judgment. In: Gilovich, T., Griffin, D., & Kahneman, D. (Eds.), *Heuristics & biases: The psychology of intuitive judgment.* New York, NY: Cambridge University Press, 2002: pp. 49–81.

Kahneman, D., Schkade, D., & Sunstein, C. R. Shared outrage and erratic awards: The psychology of punitive damages. J Risk Uncertaint 1998; 16: 49–86.

Kelley, H. H. The processes of causal attribution. Am Psychol 1973; 28: 107–128.

Klimchuck, D. Review: Outrage, self-control, and culpability. U Toronto Law J 1994; 44: 441–468.

Kruger, J., & Gilovich, T. Actions and intentions in self-assessments: The road to self-enhancement is paved with good intentions. Pers Soc Psychol Bull 2004; 30: 328–339.

Krull, D. S. Does the grist change the mill? The effect of the perceiver's interferential goal in the process of social inference. Pers Soc Psychol Bull 1993; 19: 340–348.

Langer, E. J. The illusion of control. J Pers Soc Psychol 1975; 32: 311–328.

Langer, E. J. *The psychology of control.* Beverly Hills, CA: Sage, 1983.

Langer, E. J. & Roth, J. Heads I win, tails it's chance: The illusion of control as a function of the sequence of outcomes in a purely chance task. J Pers Soc Psychol 1975; 32: 951–955.

Masicampo, E. J., & Baumeister, R. F. Toward a physiology of dual-process reasoning and judgment: Lemonade, willpower, and expensive rule-based analysis. Psychol Sci 2008; 19: 255–260.

McFarland, C., Ross, M., & DeCourville, N. Women's theories of menstruation and biases in recall of menstrual symptoms. J Pers Soc Psychol 1989; 57: 522–531.

Milgram, S. *Obedience to authority: An experimental view.* New York, NY: Harpercollins, 1974.

Miller, D. T., & Ross, M. Self-serving biases in the attribution of causality: Fact or fiction? Psychol Bull 1975; 82: 213–225.

Miller, D. T., Visser, P. S., & Staub, B. D. How surveillance begets perceptions of dishonesty: The case of the counterfactual sinner. J Pers Soc Psychol 2005; 89: 117–128.

Morewedge, C. K., Gilbert, D. T., & Wilson, T. D. The least likely of times: How remembering the past biases forecasts of the future. Psychol Sci 2005; 16(8): 626–630.

Morewedge, C. K., Holtzman, L., & Epley, N. Unfixed resources: Perceived costs, consumption, and the accessible account effect. J Consum Res 2007; 34: 459–467.

Morewedge, C. K., Preston, J., & Wegner, D. M. Timescale bias in the attribution of mind. J Pers Soc Psychol 2007; 93: 1–11.

Morewedge, C. K., Wegner, D. M., & Vosgerau, J. *The premeditation ritual: Going through the motions of thinking ahead.* Manuscript in preparation, Carnegie Mellon University, 2009.

Morwitz, V.G., Johnson, E.J., & Schmittlein, D. Does measuring intent change behavior? J Consum Res 1993; 20: 46–61.

Nisbett, R. E., & Wilson, T. D. Telling more than we can know: Verbal reports on mental processes. Psychol Rev 1977; 84(3): 231–259.

Payne, J. W., Samper, A., Bettman, J. R., & Luce, M. F. Unconscious thought in complex decision making. Psychol Sci 2008; 19(11): 1118–1123.

Pieters, R., Baumgartner, H., & Bagozzi, R. Biased memory for prior decision making: Evidence from a longitudinal field study. Organ Behav Hum Decis Process 2006; 99: 34–48.

Pizarro, D. A., Uhlmann, E., & Bloom, P. Causal deviance and the attribution of moral responsibility. J Exp Soc Psychol 2003; 39: 653–660.

Pizarro, D. A., Uhlmann, E., & Salovey, P. Asymmetry in judgments of moral blame and praise: The role of perceived metadesires. Psychol Sci 2003; 14(3): 267–272.

Preston, J., & Wegner, D. M. The eureka error: J Pers Soc Psychol 2007; 92: 575–584.

Pronin, E., Gilovich, T., & Ross, L. Objectivity in the eye of the beholder: Divergent perceptions of bias in self versus others. Psychol Rev 2004; 111(3): 781–799.

Pronin, E., Wegner, D. M., McCarthy, K., & Rodriguez, S. Everyday magical powers: The role of apparent mental causation in the overestimation of personal influence. J Pers Soc Psychol 2006; 91: 218–231.

Radin, D. I. Psychophysical evidence of possible retrocausal effects in humans. In: Sheehan, D. P. (Ed.), *Frontiers of time, retrocausation—experiment and theory.* New York: Springer-Verlag, 2006: pp. 193–213.

Ross, L. D., & Nisbett, R. E. *The person and the situation: Perspectives of social psychology.* New York, NY: McGraw-Hill, 1991.

Schacter, D. L., & Addis, D. R. The cognitive neuroscience of constructive memory: Remembering the past and imagining the future. Philos Trans R Soc Lond B Biol Sci 2007; 362: 773–786.

Schwartz, B., Ward, A., Monterosso, J., Lyubomirsky, S., White, K., & Lehman, D. R. Maximizing versus satisficing: Happiness is a matter of choice. J Pers Soc Psychol 2003; 83: 1178–1197.

Shaver, K. G. *The attribution of blame: Causality, responsibility, and blameworthiness.* New York, NY: Springer-Verlag, 1985.

Shiv, B., & Fedorikhin, A. Heart and mind in conflict: Interplay of affect and cognition in consumer decision making. J Consum Res 1999; 26: 278–282.

Simon, H. A. *Models of man, social and rational: Mathematical essays on rational human behavior.* New York, NY: Wiley, 1957.

Spranca, M., Minsk, E., & Baron, J. Omission and commission in judgment and choice. J Exp Soc Psychol 1991; 27(1): 76–105.

Tversky, A., & Kahneman, D. Availability: A heuristic for judging frequency and probability. Cog Psychol 1973; 5: 207–232.

Vohs, K. D., & Faber, R. J. Spent resources: Self-regulatory resource availability affects impulse buying. J Consum Res 2007; 33: 537–547.

Vohs, K. D., & Schooler, J. W. The value of believing in free will: Encouraging a belief in determinism increases cheating. Psychol Sci 2008; 19: 49–54.

von Neumann, J., & Morgenstern, O. *Theory of games and economic behavior.* Princeton: Princeton University Press, 1944.

Wegner, D. M. Ironic processes of mental control. Psychol Rev 1994; 101: 34–52.

Wegner, D. M. *The illusion of conscious will.* Cambridge, MA: MIT Press, 2002.

Wegner, D. M. The mind's best trick: How we experience conscious will. Trends Cogn Sci 2003; 7: 65–69.

Wegner, D. M., Fuller, V. A., & Sparrow, B. Clever hands: Uncontrolled intelligence in facilitated communication. J Pers Soc Psychol 2003; 85: 5–19.

Wegner, D. M., & Wheatley, T. P. Apparent mental causation: Sources of the experience of will. Am Psychol 1999; 54: 480–492.

Wheatley, T., & Haidt, J. Hypnotic disgust makes moral judgments more severe. Psychol Sci 2005; 16: 780–784.

Weiner, B. *Judgments of responsibility: A foundation for a theory of social conduct.* New York, NY: Guilford Press, 1995.

Wells, G. L., & Gavinsky, I. Mental simulation of causality. J Pers Soc Psychol 1989; 56: 161–169.

Wimmer, H., & Perner, J. Beliefs about beliefs: Representation and constraining function of wrong beliefs in young children's understanding of deception. Cognition 1983; 13: 103–128.

Winkielman, P., & Schwarz, N. How pleasant was your childhood? Beliefs about memory shape inferences from experienced difficulty of recall. Psychol Sci 2001; 12: 176–179.

Woolfolk, R. L., Doris, J. M., & Darley, J. M. Identification, situational constraint and social cognition: Studies in the attribution of moral responsibility. Cognition 2006; 100: 283–301.

Zanna, M. P., & Cooper, J. Dissonance and the pill: An attribution approach to studying the arousal properties of dissonance. J Pers Soc Psychol 1974; 29: 703–709.

CHAPTER 15

The Power of Planning: Self-Control by Effective Goal-striving

Peter M. Gollwitzer, Caterina Gawrilow, and Gabriele Oettingen

ABSTRACT

As highlighted by Kurt Lewin, goal attainment is not yet secured solely by forming strong commitments to highly desirable and feasible goals. There is always the subsequent issue of implementing a set goal, and one wonders what people can do to enhance their chances of being successful at this second phase of goal pursuit. A promising answer seems to be the following: People may plan out in advance how they want to solve the problems of goal implementation. But what are these problems? There are at least four problems that stand out. These problems include getting started with goal striving, staying on track, calling a halt, and not overextending oneself. We will describe research showing that making if-then plans (i.e., form implementation intentions) on how to deal with these problems indeed facilitates solving the crucial problems of goal implementation. Thereafter, we will ask whether implementation intentions foster goal attainment even under conditions that are commonly viewed as not amenable to self-regulation attempts, such as succeeding on an intelligence test or overcoming spider phobia. Finally, we will report research showing that implementation intentions can even foster goal-striving in those samples (e.g., children with ADHD) that are known to suffer from impaired action control.

Keywords: Implementation intentions, Goal intentions, Medial/lateral pre-frontal cortex, Action initiation, Goal shielding, Disengagement, Overextension, Academic test performance, Negotiation performance, Winning competitions, Overcoming habitual responses, Simon effect, Spider phobia, Weapon identification task, Behavior change interventions, Children with ADHD, Response inhibition, Delay of gratification, Set-shifting, Multi-tasking

Research on self-regulation and self-control has defined its object of interest by emphasizing different phenomena and processes. The many targeted phenomena include overriding unwanted thoughts (e.g., related to distractions, temptations, stereotyping, self-inflation), feelings (e.g., anger, disgust, fear, sadness, prejudice) and behaviors (e.g., aggressive, immoral, risky, health-damaging, underachieving). The various processes that are assumed to promote self-regulation and self-control pertain to fostering the wanted over the unwanted by cognitively inhibiting the unwanted and/or activating the wanted, or by modifying one's current or anticipated emotions so that the wanted can be executed more easily, and the unwanted can be more easily halted or prevented. Often it is assumed that effective self-regulation and self-control requires a switch; for instance, a switch from a hot mode of information processing to a cool mode, from a low-level to a high-level construal, a short-term to a long-term time perspective, from impulsive to reflective action control, from habitual bottom-up direct action control by present stimuli to top-down control by the desired end states specified in respective goals, or from low-priority/low-importance goals to high-priority/high-importance goals. It is this latter process of achieving self-regulation and self-control by striving for goals that is focused on in the present chapter. We will argue that goal attainment is facilitated when people plan out their goal-striving in advance. More specifically, we suggest that effectively regulating one's goal-striving by making if–then plans (i.e., form implementation intentions) is a reliable and powerful way to achieving self-control.

IMPLEMENTATION INTENTIONS: PLANNING OUT GOAL-STRIVING IN ADVANCE

To form an implementation intention (Gollwitzer, 1993, 1999), one needs to identify a future goal-relevant situational cue (i.e., the *if-component*) and a related planned response to that cue (i.e., the *then-component*). Whereas a goal intention specifies the desired event in the form of "I intend to perform Behavior X/to reach Outcome X" (e.g., to exercise regularly/ to get an A in Introductory Psychology), an implementation intention specifies both an anticipated goal-relevant situation and a proper goal-directed response. Thus, an implementation intention that serves the goal intention to "get an A in the social psychology class" would follow the form "**If** Situation Y arises (e.g., when my roommates will be asking me to go out tonight), **then** I will perform Behavior Z (e.g., will say that I will be joining them next week when my exam is over)."

There is added benefit of an implementation intention: a meta-analysis by Gollwitzer and Sheeran (2006) involving over 8,000 participants in 94 independent studies reported an effect size of $d = 0.65$. This medium-to-large effect size (Cohen, 1992) represents the additional facilitation of goal achievement by implementation intentions compared to goal intentions alone. As goal intentions by themselves already have a facilitating effect on behavior enactment (Webb & Sheeran, 2006), the size of this effect is remarkable.

How Do Implementation Intention Effects Come About?

The mental links created by implementation intentions facilitate goal attainment on the basis of psychological processes that relate to both the anticipated situation (the "if" part of the plan) and the intended behavior (the "then" part of the plan). Because forming an implementation intention implies the selection of a critical future situation, the mental representation of this situation becomes highly activated, and hence more accessible (Gollwitzer, 1999). This heightened accessibility of the "if" part of the plan was observed in several studies (e.g., Aarts, Dijksterhuis, & Midden, 1999; Parks-Stamm, Gollwitzer, & Oettingen, 2007; Webb & Sheeran, 2007, 2008) and means that people are in a good position to identify and take notice of the critical situation when they subsequently encounter it (e.g., Webb & Sheeran, 2004).

Studies also indicate that implementation intentions forge a strong association between the specified opportunity and the specified response (Webb & Sheeran, 2007, 2008). The upshot of these strong links is that the initiation of the goal-directed response specified in the if–then plan becomes automated—that is, exhibits features of automaticity including immediacy, efficiency, and redundancy of conscious intent. The idea is that people do not have to deliberate anymore about when and how they should act when they have formed an implementation intention—unlike people who have formed mere goal intentions. Evidence that if–then planners act quickly (Gollwitzer & Brandstätter, 1997, Experiment 3), deal effectively with cognitive demands (Brandstätter, Lengfelder, & Gollwitzer, 2001), and do not need to consciously intend to act at the critical moment (Bayer, Achtziger, Gollwitzer, & Moskowitz, 2009; Sheeran, Webb, & Gollwitzer, 2005, Study 2) is consistent with this idea.

These component processes of implementation intentions (enhanced cue accessibility, automatization of responding) mean that if-then planning enables people to see and seize good opportunities to move towards their goals. Fashioning an if-then plan thus *strategically automates* goal-striving (Gollwitzer & Schaal, 1998) because people delegate control of goal-directed behaviors to pre-selected situational cues with the explicit purpose of reaching their goals, that is, automatic action initiation originates in a conscious act of will (if–then planning).

Implementation Intentions and Overcoming Problems of Goal-Striving

Given these special features of action control by implementation intentions, one wonders whether people benefit from forming implementation intentions when goals geared at showing a high amount of self-control or self-discipline are at stake. Let us discuss this question by addressing the four central problems of goal realization.

Getting started

Numerous studies suggest that problems of *getting started* on one's goals can be solved effectively by forming implementation intentions. For example, Gollwitzer and Brandstätter (1997, Study 2) analyzed a goal intention (i.e., writing a promised report about how one spent Christmas Eve) that had to be performed at a time (i.e., during the subsequent Christmas holiday) where people are commonly busy with other things. Still, research participants who had furnished their goal intention with an implementation intention that specified when, where, and how one wanted to get started on this project were about three times as likely to actually keep their promise to write the report than mere goal intention participants. Similarly, Oettingen, Hönig, and Gollwitzer (2000, Study 3) observed that implementation intentions helped college students to act on their goal to regularly practice solving certain math problems (e.g., at 10 a.m. in the morning of every Wednesday over the next 4 weeks).

Other studies have examined the ability of implementation intentions to foster goal-striving that is somewhat unpleasant to perform and thus are associated with an initial reluctance to act. For instance, the goal to perform regular breast examinations (Orbell, Hodgkins, & Sheeran, 1997) or cervical cancer screenings (Sheeran & Orbell, 2000), resume functional activity after joint replacement surgery (Orbell & Sheeran, 2000), eat a low-fat diet (Armitage, 2004), recycle (Holland, Aarts, & Langendam, 2006), and engage in physical exercise (Milne, Orbell, & Sheeran, 2002), were all more readily acted upon when people had furnished these goals with implementation intentions.

Staying on Track

Many goals cannot be accomplished by simple discrete one-shot actions but require that people keep striving for the goal over an extended period of time. Such staying on track may get very difficult when certain internal (e.g., being anxious, tired, overburdened) or external stimuli (e.g., temptations, distractions) are not conducive to goal realization but instead generate interferences that could potentially derail the ongoing goal pursuit. Implementation intentions can facilitate the shielding of such goal pursuits from the negative influences of interferences from outside the person by suppressing these negative influences (Gollwitzer & Schaal, 1998). For example, if a person wants to avoid being unfriendly to a friend who is known to make outrageous requests, she can protect herself from showing the unwanted unfriendly response by forming suppression-oriented implementation intentions. Such suppression-oriented implementation intentions may take various forms: "And if my friend approaches me with an outrageous request, then I will not respond in an unfriendly manner!" or "…, then I will respond in a friendly manner!" or "…, then I'll ignore it!"

But suppression-oriented implementation intentions can also be used to shield ongoing goal pursuits from disruptive inner states. Achtziger, Gollwitzer, and Sheeran (2008, Study 1) report a field experiment concerned with dieting in which goal shielding was supported by suppression implementation intentions geared at controlling potentially interfering inner states (i.e., cravings for junk food). An alternative way of using implementation intentions to protect ongoing goal-striving from getting derailed by adverse inner states (e.g., inappropriate moods, ego-depletion, irritation) is forming implementation intentions geared at stabilizing the ongoing goal pursuit at hand (Bayer, Gollwitzer, & Achtziger, in press). Using again the example of a person who is approached by her friend with an outrageous request, let us assume that this person is also tired or irritated and thus particularly likely to respond in an unfriendly manner. If this person has stipulated in advance in an implementation intention what she will converse about with her friend, the critical interaction may simply run off as planned, and being tired or irritated should fail to affect the person's goal to relate to her friend in a civilized manner.

Calling a Halt

The self-regulatory problem of calling a halt to a faulty goal pursuit can also be ameliorated by forming implementation intentions. People often fail to readily disengage from chosen means and goals that turn out to be faulty because of a strong self-justification motive (Brockner, 1992). Such escalation phenomena (also referred to as "throwing good money after bad") can be controlled effectively, however, by the use of implementation intentions that specify exactly when and how to consider a switch to a different means or a different goal. For instance, Henderson, Gollwitzer, and Oettingen (2007) asked participants who had chosen a certain strategy for a given task goal to either form an implementation intention that specified a complex reflection response ("If I receive disappointing feedback, then I'll think about how things have been going with my strategy!") or a more simple action response ("If I receive disappointing feedback, then I'll switch my strategy!"), or merely set the goal to always use the best strategy available. Henderson et al. observed that action implementation intentions facilitated disengagement as a response to experienced failure no matter whether there were signs that things were picking up or that they would continue to stay bleak. Reflection implementation intention participants, on the other hand, integrated information about recent improvement in forming their disengagement decision (i.e., they were less willing to disengage when things were picking up). This study shows that implementation intentions can be used to control the costly escalation of behavioral commitment commonly observed when people experience failure with a chosen strategy of goal-striving. Using reflection implementation intentions (as compared to action implementation intentions) even allows for flexible disengagement in the sense that recent turns to the better are respected in one's decision to switch (or not) to a different goal-striving strategy.

Not Overextending Oneself

The assumption that implementation intentions subject behavior to the direct control of situational cues (Gollwitzer, 1993) implies that the self is not implicated when behavior is controlled via implementation intentions. As a consequence, the self should not become depleted (Muraven & Baumeister, 2000) when task performance is regulated by implementation intentions, and thus for individuals using implementation intentions, not over-extending themselves should become easier. Indeed, using different ego-depletion paradigms, research participants who had used implementation intentions to self-regulate in a first task do not show reduced self-regulatory capacity in a subsequent task. Whether the initial self-regulation task was controlling emotions while watching a humorous movie (Gollwitzer & Bayer, 2000), or performing a Stroop task (Webb & Sheeran, 2003, Study 1), implementation intentions successfully preserved self-regulatory resources as demonstrated by greater persistence on subsequent difficult tasks (i.e., solving difficult anagrams).

WHEN THE GOING GETS TOUGH: IMPLEMENTATION INTENTION EFFECTS IN THE FACE OF HARDSHIPS

In the rest of the chapter we will explore whether implementation intention unveil their beneficial effects even under conditions where goal-striving becomes tough. This question and the respective line of research have been stimulated by Aristotle's concept of *akrasia* (lack of will power). We felt that any strategy of goal-striving that psychology claims to contribute to people's self-control has to prove itself under conditions where people commonly fail to demonstrate willpower. Such conditions are manifold and thus we concentrated on the following four: *(a)* situations in which a person's knowledge and skills constrain performance (such as taking academic tests); *(b)* situations in which an opponent's behavior limit one's performance (such as sports competitions or negotiation settings); *(c)* situations in which the wanted behavioral (e.g., no littering), emotional (e.g., no fear), or cognitive (e.g., no stereotyping) response runs into conflict with habits favoring an antagonistic response; and *(d)* situations in which individuals who are known to have problems with action control, such as children with ADHD, have to tackle the typical problems burdening goal-striving (e.g., shielding ongoing goal-striving from distractions).

Academic Test Performance

Performance on academic tests (math tests, general intelligence tests) is by definition determined primarily by a person's knowledge, analytic capability, and cognitive skills. To increase test scores by willpower, a person thus may want to focus on motivational issues such as staying concentrated on the various test items throughout the test or by reducing worry cognitions (e.g., Did I find the right answer on the last item?) and self-doubts (e.g., Do I have the skills to find the right solution for the item at hand?).

Taking a Math Test

Bayer and Gollwitzer (2007, Study 1) asked female high school students to complete a math test (composed by high school math teachers) under two different instructions. Half of the participants were asked to form the mere achievement goal intention: "I will correctly solve as many tasks as possible!" The other half of the participants had to furnish this goal intention with the following self-efficacy-strengthening implementation intention "And if I start a new task, then I will tell myself: I can solve this task!" We observed that participants in the implementation intention group showed a better performance in the math test (in terms of number of tasks solved correctly) than participants in the mere goal intention condition, indicating that self-efficacy-strengthening implementation intentions facilitate successful goal-striving in a challenging achievement situation.

Implementation intentions are usually constructed by specifying a situational cue in the if-part and linking it to goal-directed cognitive or behavioral responses in the then-part. In the present study, a critical situational cue (i.e., starting a new test item) in the if-part was linked to a motivational response (i.e., a private self-efficacy strengthening statement) in the then-part. Interestingly, this pre-programmed, inner self-motivating speech sufficed to produce better test performance. This suggests that implementation intentions can be used to ameliorate also motivational problems of goal implementation (such as self-doubts in the face of challenging test items) and thus increase a person's willpower (i.e., the potential to exert self-control).

The present manipulation to increase willpower was particularly parsimonious, as we only had participants asked to form a plan in respect to when they will have to execute an inner self-efficacy strengthening statement. Still, these findings leave open a pressing question: Does this inner speech need to take the format of an implementation intention? Maybe it suffices that participants simply form an additional goal intention geared towards holding up self-efficacy, such as "And I will tell myself: I can solve these problems!" To explore this possibility, Bayer and Gollwitzer (2007) conducted a follow-up study in which participants had to take an intelligence test; this study included a further control condition (i.e., a self-efficacy-strengthening goal intention condition).

Taking an Intelligence Test

Bayer and Gollwitzer (2007, Study 2) asked college students to perform the Raven's Advanced Progressive Matrices intelligence test. They found that students who had been asked to form a self-efficacy-enhancing implementation intention showed higher test performance than participants with a mere goal intention to perform well. This replication of the implementation intention effect observed with high school students working on a math test is particularly noteworthy as the well-established Raven's Advanced Progressive Matrices intelligence test (Raven, 2000) was used. The tasks on the test get increasingly harder requiring greater cognitive capacity to encode, analyze, and solve them correctly; the test is considered to allow for reliable assessments of a person's analytical reasoning capabilities.

In this study, it was also examined whether adding a self-efficacy-strengthening goal intention ("And I will tell myself: I can do these problems!") to the achievement goal intention ("I will correctly solve as many tasks as possible!") improves participants' test performance. As it turned out, test performance improved only when participants were instructed to form additional self-efficacy strengthening implementation intentions. This finding is important for several reasons. First, many of the field and laboratory studies investigating the benefits of implementation intentions (e.g., on health behaviors, job safety, environment protection; see meta-analysis by Gollwitzer & Sheeran, 2006) do not use an additional condition that spells out the "then" part of the implementation intention in terms of a further goal intention (for an exception, see Oettingen, Hönig, & Gollwitzer, 2001). Therefore, in these studies the benefits of implementation intentions as compared to mere goal intentions could potentially be based on having access to additional information on how to act. With the present study, we can rule out this alternative account as the use of the strategy of strengthening one's self-efficacy in terms of forming a mere goal intention did not lead to higher test scores. Only when this strategy was suggested to participants in the format of an if–then plan, positive effects on test performance emerged.

Finally, the observed differences between self-efficacy-strengthening implementation intentions and self-efficacy-strengthening goal intentions further support the assumption (Gollwitzer, 1999) that implementation intentions—by specifying situational cues—recruit different action control processes than goal intentions. Whereas the latter are said to instigate effortful, conscious processes, the former trigger automatic processes. Assuming that performing the Raven test is quite demanding and thus burdens cognitive capacities, it is not surprising that only self-efficacy-strengthening implementation managed to be effective.

Dealing With Opponents

Often our performances are restrained by others who are competing with us for positive outcomes. Typical examples are sports competitions where athletes try to triumph over their opponents or negotiations in which a common good has to be shared between two opposing parties. In such situations, exerting willpower means to effectively protect one's goal-striving from unwanted influences of the competitive situation. In the following, we report two studies showing that implementation intentions can be used to bolster such willpower.

Winning Tennis Competitions

Studies on sports competitions discovered that negative inner states (e.g., performance anxiety, Covassin & Pero, 2004; anger, Wughalter & Gondola, 1991; feelings of stress, Hanegby & Tenenbaum, 2001) hamper the quality of athletic performance. This is particularly true for sports that involve a direct competition with an opponent as it is the case with tennis. Accordingly, Achtziger, Gollwitzer, and Sheeran (2008, Study 2) wondered whether specifying these negative inner states as critical cues in the "if" component of implementation intentions and a goal-shielding response in the "then" component, supports staying on track under such adverse inner states. The specified goal-shielding responses were taken from research that analyzed strategies of improving one's performance during a tennis match (e.g., ignoring the noise made by the audience, focusing one's attention on the ball, engaging in relaxation behavior; see Anshel & Anderson, 2002).

Achtziger et al. manipulated participants' goal intention to perform well in a tennis match rather than simply measuring it. Accordingly, we had a no-goal control condition and a mere goal intention condition in which tennis players (taken from German tennis leagues) were assigned the following goal the day before a critical match: "I will play each ball with utmost concentration and effort to win the match!" In the implementation intention condition, participants were also assigned this goal, and in addition, asked to form four if–then plans that specified internal states in the "if" components (e.g., performance anxiety, physiological arousal) and staying-on-track responses in the "then" components (e.g., ignoring the adverse inner state). It was observed that self- and other-ratings (by trainers and teammates) of physical fitness and performance were higher for implementation intention participants as compared to both goal intention and no-goal (control) participants.

This field experiment is one of the few implementation intention studies (e.g., Murgraff et al., 1996) in which participants could choose their implementation intentions in an individualized manner and make use of more than just one implementation intention. Participants selected the four, most personally relevant, negative inner states as "if" components of their implementation intentions and the four goal-directed responses that they assessed as being suited best for shielding their goal-striving as "then" components. These findings speak to the idea that implementation intention inductions can easily be tailored to one's particular self-regulatory problems at hand. To facilitate goal-striving and increase rates of goal attainment in competitive situations, people apparently can also form multiple if–then plans, geared at alleviating the particular inner states they find most detrimental to their performance, and linking them to goal-directed responses they perceive as most useful (instrumental).

Moreover, in the present study the goal-directed responses specified in the then-component of the implementation intentions studied were coping responses linked to various handicaps and weaknesses originating inside the person and not to challenges arising from outside the person. One therefore may wonder whether if–then plans that link coping responses to disruptive external events will also be effective in shielding an ongoing goal pursuit. Research by Gollwitzer and Schaal (1998) on resistance to temptations provides an affirmative answer to this question, given that certain coping responses are specified in the "then" component (e.g., ignore responses). Participants who had to perform tedious arithmetic problems for a period of 15 minutes were more successful in doing so despite the presentation of various interspersed attractive video clips, when participants had formed implementation intentions that specified "attractive video clips" in the "if" component and an "ignore" response in the "then" component.

Reflecting on the pros and cons of forming implementation intentions that link a coping response to negative external events versus detrimental inner states, the following should be kept in mind: specifying inner states has the advantage that these detrimental states could function as a summary label for all of those negative external events that might compromise goal-striving—even those one is not aware of or may fail to anticipate. Accordingly, whenever a person is not in a good position to know about and anticipate critical events, specifying detrimental inner states seems to be the safer strategy to shield one's goal-striving. For instance, patients with panic attacks are usually not aware of which kinds of external events trigger the attack and whether these events are to be expected in a forthcoming external context (e.g., Hinton, Nathan, & Bird, 2002). For these patients it would make sense to specify inner states (i.e., upcoming anxiety) as "if" components to control their negative emotions.

Prevailing in Loss Framed Negotiations

Negotiations particularly lend themselves well to investigate the power of implementation intentions: Negotiations are cognitively very demanding tasks in which a large amount of information has to be processed online, and the course of events is hard to predict as one is not performing a task alone but conjointly with an opponent. Thus, negotiations can be understood as the prototype of a complex situation in which striving for desired goals can easily become derailed. Therefore, analyzing whether the beneficial effects of implementation intentions found in previous research also hold true in negotiations is of great interest to assess whether needed willpower accrues from forming if–then planning.

In their negotiation research, Trötschel and Gollwitzer (2007) explored whether the self-regulation strategy of forming implementation intentions allows negotiators to arrive at high performance levels in finding agreements even if they have to operate under the adverse conditions of a loss frame (i.e., the negotiation outcomes are loss-framed and thus a resistance to concession making is induced; e.g., Bottom & Studt, 1993; Olekalns, 1997). In one of their experiments, pairs of negotiators were assigned the role of representatives of two neighboring countries (i.e., the blue and the orange nations) and asked to negotiate the distribution of a disputed island (i.e., its regions, villages, and towns). One group of pairs of negotiators was asked to form the mere pro-social goal of "I want to cooperate with my counterpart!", and a second group to furnish this goal with a respective implementation intention: "And if I receive a proposal on how to share the island, then I will make a cooperative counterproposal!" Both groups were then subjected to a frame manipulation, whereby both members of the pair received a loss-frame manipulation (i.e., each region's value was expressed in points that are lost when the region is given away). In addition, two control conditions were established: A first control condition contained pairs of negotiators who were not assigned pro-social goals and asked to negotiate under a loss frame; the second control condition also consisted of pairs of negotiators who were not assigned pro-social goals, but these pairs of negotiators were asked to negotiate under a gain frame (i.e., each region's value was expressed in points that are won when the region is kept). These two control conditions were used to establish the negative influence of loss versus gain frames on joint profits. In addition, the loss-frame control condition served as a comparison group for the two critical experimental groups (i.e., the pro-social goal group and the pro-social goal plus implementation intention group).

When looking at the agreements achieved (i.e., level of joint outcomes), it was observed that pairs of loss-frame negotiators with a pro-social goal intention managed to somewhat reduce the resistance to concession making arising from the loss-frame negotiation context, but that only negotiators who furnished their pro-social goal intentions with respective implementation intentions were successful in completely abolishing the negative impact of the loss-frame negotiation context (i.e., showed a negotiation performance that was not different from that of gain-frame negotiators). In addition, action control via implementation intentions was found to be very efficient (i.e., implementation intentions abolished the negative effects of loss framing by leaving the negotiators' cognitive capacity in tact); negotiators who had formed implementation intentions were more likely to use the cognitively demanding integrative negotiation strategy of logrolling (i.e., making greater concessions on low rather than high priority issues).

In a follow-up experiment, the effects of pro-social goals and respective implementation intentions on the course of the negotiation were analyzed. The analyses on the course of the negotiation indicated that loss-frame pairs of negotiators who had furnished their pro-social goals with corresponding implementation intentions revealed a steeper progress in finding agreements than loss-frame pairs of negotiators without pro-social if–then plans; actually, at the end of the negotiation implementation intention, participants had achieved negotiation agreements that were comparable to those of gain-frame pairs of negotiators. Furthermore, implementation intentions were again strongly associated with using the integrative negotiation strategy of logrolling. Apparently, having one's negotiation behavior controlled by implementation intentions saved cognitive resources that could be used to successfully discover integrative solutions.

Overcoming Habitual Responses

The self-regulation of one's goal-striving becomes difficult when habitual responses conflict with initiating and executing the needed goal-directed responses instrumental to goal attainment. In such cases, having willpower means to assert one's will of attaining the chosen goal against one's "bad" habits. Accordingly, we wondered whether the self-regulation strategy of forming if–then plans can help people to let their goals win out over their habits.

Behavioral Responses

By assuming that action control by implementation intentions is immediate and efficient, and adopting a simple horse race model of action control, people can be expected to be in a position to break habitualized responses by forming implementation intentions (i.e., if–then plans that spell out a response that is contrary to the habitualized response to the critical situation). Such studies have been conducted successfully in the field (Holland, Aarts, & Langendam, 2006), but also in the laboratory (Cohen et al., 2008, Study 2).

Holland, Aarts, and Langendam (2006) addressed whether implementation intentions could help break unwanted habits (and replace them with new wanted behaviors) in a field experiment conducted in an institution. The goal of the researchers was to increase the use of recycling bins for plastic cups and paper, and reduce the bad habit of throwing out these recyclable items in personal wastebaskets. Participants were randomly assigned to one of six conditions: a no-treatment control condition, a control condition with a behavior report questionnaire, a facility condition where each participant received her own recycle bin, a combined facility and questionnaire condition, and two implementation intention conditions: one with a personal facility, and one without. Recycling behavior was substantially improved in the facility as well as in the implementation intentions conditions in Week 1 and Week 2, and still two months after the manipulation. In addition, the correlation between past and future behavior was strong in the control conditions, whereas these correlations were nonsignificant and close to zero in the implementation intention conditions. Apparently, implementation intentions effectively broke old habits by facilitating new recycling behavior. This shows that even strongly habitualized behaviors can be replaced by newly planned goal-directed behaviors via implementation intentions.

Cohen et al. (2008, Study 2) explored the suppression of habitual responses in a more controlled laboratory experiment using the Simon task. In this paradigm, participants are asked to respond to a non-spatial aspect of a stimulus (i.e., whether a presented tone is high or low) by pressing a left or right key, and to ignore the location of the stimulus (i.e., whether it is presented on one's left or on the right side). The difficulty of this task is in ignoring the spatial location (left or right) of the tone in one's classification response (i.e., pressing a left or right response key; Simon, 1990). The cost in reaction times is seen when the location of the tone (e.g., right) and required key press (e.g., left) are incongruent, as people habitually respond to stimuli presented at the right or left side with the corresponding hand. Cohen et al. (2008, Study 2) found that implementation intentions eliminated the Simon effect for the stimulus that was specified in the "if" component of the implementation intention. Reaction times for this stimulus did not differ between the congruent and incongruent trials (i.e., they were fast throughout).

Emotional Responses

Recent research has also explored whether adding implementation intentions to emotion-regulation goals would make these goals more effective (Schweiger et al., 2009). In one study, participants were exposed to a series of pictures used to elicit disgust. When participants formed a response-focused implementation intention ("If I see disgusting scenes, then I'll stay calm and relaxed."), they exhibited a reduction in arousal compared to a control group. As anticipated, participants who operated on mere goals to not get disgusted could not willfully reduce their arousal to the disgusting pictures. A second study analyzed the control of spider fear in spider phobics. Both participants with response-focused implementation intentions ("If a see a spider, then I will stay calm and relaxed.") and antecedent-focused implementation intentions ("If I see a spider, then I'll ignore it.") experienced less negative affect in the face of spider pictures than a no self-regulation control group; again, mere goal intentions to not get frightened failed to achieve this effect. Moreover, spider phobics using implementation intentions even managed to control their fear to the low level observed with a sample of participants who were pre-selected on the basis of having no fear of spiders at all.

In a final study using dense-array electroencephalography (EEG) to assess event-related potentials (ERPs), the effectiveness of ignore-implementation intentions for the control of spider fear in spider phobics was replicated. More important, participants who added ignore-implementation intentions to their goal intentions to not get frightened showed a lower positivity in the P1 (an ERP assessed around 120 ms after stimulus presentation in the occipital and parietal brain areas) when detecting spider pictures as compared to mere goal and control participants; no such difference was found for pleasant or neutral control pictures. Indeed, previous research has shown that the P1 can discriminate high-arousing negative stimuli from neutral and positive stimuli (Smith et al., 2003), as well as spiders from nonthreatening animals such as butterflies (Carretié et al., 2003). In line with these findings, participants in our study without any emotion-regulation goal intention (control condition) or those with a mere goal intention showed the typical positivity of the P1 at about 120 ms when detecting threatening stimuli. In contrast, forming an implementation intention led to a down-modulation of this component, resulting in a significantly lower positivity of the P1. This attests to the specificity of the implementation intention effect, and again supports our assumption that forming implementation intentions leads to strategic automation of the goal-directed responses specified in their "then" part, as conscious efforts to inhibit the activation of the mental representation of a presented stimulus are commonly assumed to show their effects later than 300 ms after stimulus presentation (overview by Bargh & Chartrand, 2000).

Cognitive Responses

Automatic cognitive biases such as stereotyping represent another type of habitualized response that can be in opposition to one's goals. Although one may have the goal to be egalitarian, automatic stereotyping happens quickly and unintentionally; some attempts to control automatic stereotyping have even resulted in backfire effects. Extending earlier work by Gollwitzer and Schaal (1997), Stewart and Payne (2007) examined whether implementation intentions designed to counter automatic stereotypes (e.g., "When I see a Black face, I will then think 'safe.'") could reduce stereotyping towards a category of individuals (versus a single exemplar). They used the Process Dissociation Procedure (PDP; Jacoby, 1991) to estimate whether the reduction in automatic stereotyping came about by reducing automatic stereotyping, increasing control, or a combination of these two processes. It was found that implementation intentions reduced stereotyping in a weapon identification task (Studies 1 and 2) and an IAT task (Study 3) by reducing automatic effects of the stereotype (without increasing conscious control). This reduction in automatic race bias held for even new members of the category (Study 2). These studies suggest that implementation intentions are an efficient way to overcome automatic stereotyping.

Implications: Implementation Intentions in the Brain

The reported research findings on the control of habitual responses by implementation intentions implies that action control by if–then plans turns top-down control by goals into bottom-up control by the situational cues specified in the "if" component of an implementation intention. This implication was recently tested in an MRI study by Gilbert et al. (2009). The study draws on the gateway hypothesis of rostral prefrontal cortex (area 10) function by Burgess, Simons, Dumontheil, and Gilbert (2005) suggesting a distinction between action control that is primarily triggered by low level stimulus input, and action control that is primarily guided by higher-level goal representations.

Looking at the results of a host of studies using different kinds of executive function tasks, Burgess et al. observed in a meta-analysis that stimulus-driven, bottom-up action control is associated with medial area 10 activity, whereas goal-driven, top-down action control is associated with lateral area 10 activity. Accordingly, Gilbert et al. (2009) postulated that action control by implementation intentions should by characterized by medial area 10 activity, whereas action control by mere goals should be associated with lateral area 10 activity.

To test this hypothesis, a prospective memory (PM) paradigm was used. Prospective memory tasks require participants to perform an ongoing task (e.g., a lexical decision task, a classification task), but remember to also perform an additional response (i.e., the PM response, e.g., pressing the space bar) whenever a particular stimulus is presented within the ongoing task (e.g., a particular word, a particular constellation of the stimuli to be classified). In the Gilbert et al. study, each participant had to perform two different prospective memory tasks, one with a goal intention to perform the PM responses and the other with an implementation intention to perform these responses. As it turned out, implementation intentions facilitated the performance of PM responses as compared to mere goal intentions, and this gain in performance did not lead to any additional costs in performing the ongoing task. Even more importantly, PM performance based on a goal intention were accompanied by greater lateral area 10 activity, whereas PM performances based on implementation intentions were associated with greater activity in the medial area 10. Moreover, the difference in brain activity associated with correctly responding to PM targets under goal vs. implementation intentions correlated strongly and significantly with the behavioral difference as a consequence of acting on the basis of goal versus implementation intentions. The fact that acting on implementation intentions is associated with medial area 10 activity, whereas acting on goal intentions is associated with lateral area 10 activity, adds further support to the theory that by forming implementation intentions, people can switch from goal-striving that is guided by conscious top-down control to direct, stimulus-triggered goal-striving (Gollwitzer, 1999).

Critical Samples: Children with Attention-Deficit/Hyperactivity-Disorder (ADHD)

In the past, implementation intention research with clinical samples has been conducted to test process hypotheses on how implementation intentions achieve their effects. For instance, the hypothesis that implementation intentions lead to efficient action control (i.e., action control by implementation intentions does not suffer from cognitive load) was tested by assessing whether heroine addicts during withdrawal, patients suffering from schizophrenia (Brandstätter, Lengfelder, & Gollwitzer, 2001, Studies 1 and 2), and patients with a frontal lobe damage (Lengfelder & Gollwitzer, 2001) also benefit from forming implementation intentions. The present line of research on children with ADHD instead serves the purpose of testing the power of implementations in terms of improving goal-striving even under adverse conditions. Children with ADHD are known to be inattentive, hyperactive, and particularly impulsive (APA, 1994). They show impairments in focusing, sustaining, and switching attention, as well as inappropriate motor activity, and limited inhibitory control of responses (e.g., Barkley, 1997; Nigg, 2001). ADHD emerges in the preschool years and affects 3% to 5% of school-aged children.

Approximately two-thirds (50%–70%) of individuals diagnosed with ADHD exhibit comorbid clinical problems related to learning ability, social adjustment and functioning, and/or emotional well-being. ADHD-related symptoms, as well as comorbid symptoms, lead to several difficulties in everyday life. Children with ADHD show moderate to large deficits in academic achievement (Frazier et al., 2007) and experience interpersonal problems (Hoza et al., 2005). Consequently, in comparison to children without ADHD, children with ADHD tend to receive poorer grades in school, more frequently need to repeat a school year, often require tutoring or placement in special classes, and show reading disabilities. Furthermore, in comparison to children without ADHD, children with ADHD are less popular with their peers, more frequently experience rejection, and are less likely to have dyadic friendships.

ADHD not only leads to difficulties in everyday life, but also to impaired performance on different neuropsychological tasks measuring inhibitory control (Halperin & Schulz, 2006), such as the Go/NoGo task (Rubia et al., 2001), the Continuous Performance Test (CPT; Rosvold et al., 1956), the Stop Signal Test (SST; Logan & Cowan, 1984), and the Stroop Test (Stroop, 1935). For example, in a Go/NoGo task, a response must be either executed or inhibited in response to the presentation of a Go or a NoGo signal, respectively. In most studies using the Go/NoGo paradigm, the inhibitory effect is enhanced with a predominant Go response resulting from frequent Go trials and infrequent NoGo trials. Generally, children with ADHD have longer response times to Go stimuli (i.e., targets) and make more errors when presented with NoGo stimuli (i.e., nontargets).The basic paradigm in a CPT is typically a Go/NoGo task in which the participant responds to the presentation of a target and withholds the response to the presentation of a non-target. Importantly, CPT derived measures predict the presence of most ADHD symptoms (Epstein, Erkanli, & Conners, 2003): Children with ADHD miss more targets and show more false alarms to nontargets than children without ADHD. Unlike the Go/NoGo task and the CPT, the SST requires children to inhibit a motor response while the action is being executed. Furthermore, research illustrates that children with ADHD have slower stop signal reaction times and significantly slower reaction times in Go trials than comparison children without ADHD on the SST (e.g., Oosterlaan, Logan, & Sergeant, 1998). Finally, in a Stroop Interference Test (Stroop, 1935) participants have to name the color (e.g., green) of a noncongruent color word (e.g., red). Participants must inhibit the automatic response (i.e., reading) in favor of the non-automatic response (i.e., naming the color). In a meta-analysis, Homack and Riccio (2004) showed that children with ADHD consistently exhibit poorer performance on Stroop Tasks when compared to children without ADHD. In sum, these findings support the idea that behavioral inhibition is a core deficit in children with ADHD.

Children with ADHD also have difficulties on tasks requiring cognitive flexibility (i.e., shifting a cognitive set). For example, on the Wisconsin Card Sorting Test (WCST; Berg, 1948), children with ADHD tend to preserve instead of respond flexibly (i.e., they stick to unsuccessful hypotheses instead of adapting their strategy). Accordingly, this tendency appears to be another characteristic feature of ADHD, explaining these children's poorer performances on various dependent variables measured with the WCST. Moreover, children with ADHD also have difficulties in multitasking. The Six Elements Test, for example, which measures complex planning behavior in the sense of multitasking situations, requires the simultaneous execution of six different tasks (SET; Burgess, 2000). Teenagers with ADHD show significantly worse results on the SET than children without ADHD (Clarke, Prior, & Kinsella, 2000). In sum, these findings seem to suggest that children with ADHD have a host of difficulties in regulating their own behavior.

Implementation Intentions Facilitate Response Inhibition in Children with ADHD

Our initial research investigated whether implementation intentions can support children with ADHD on a Go/NoGo task. The task required children to both classify randomized stimuli presented on a computer screen, as well as inhibit classification in response to a NoGo signal (Gawrilow & Gollwitzer, 2008). Specifically, children had to respond as quickly as possible to pictures of vehicles or animals by pressing a respective vehicle or animal key (i.e., Go trial). On one third of the trials, however, an audible NoGo signal presented before the stimulus announced to the children that they should inhibit their response on that trial (i.e., NoGo trial). In two experiments, children with ADHD were assigned to either a goal intention group ("I will not press a key for pictures that have a sound!") or an implementation intention group ("And if I hear a tone, then I will not press any key!"). Both the response times for the Go trials and the number of successful stops on the NoGo trials were measured.

In the first study, the performances of children with and without ADHD were compared. Children with ADHD who furnished a suppression goal with implementation intentions improved inhibition of an unwanted response on a Go/NoGo task to the same level observed in children without ADHD. The second study compared the performances of children with ADHD with and without psychostimulant medication. In this study, a combination of implementation intentions and psychostimulant medication resulted in the highest level of suppression performance in children with ADHD (Gawrilow & Gollwitzer, 2008, Studies 1 and 2).

In a recent follow-up study, Gawrilow, Oettingen, and Gollwitzer (2009a, Study 1) asked children with and without ADHD to complete a Number-Stroop task. All children saw rows of single-digit numbers containing numbers from 1 to 9 on a computer screen. The amount of numbers on each trial was varied from two to seven. The task required children to either name the number in a naming task (e.g., press key "one" to a row of four ones) or count the amount of numbers in a counting task (e.g., press key "four" to a row of four ones). All children began with the (easier) naming task, followed by the (more difficult) counting task. Error rates and response times were measured as dependent variables.

To establish the three experimental conditions, prior to the counting task one third of the children received only a goal intention ("I will count the numbers."), while the remaining children additionally received one of two different implementation intentions: a task-facilitating implementation intention (e.g., "As soon as a new row of numbers is presented, I will concentrate on the amount of numbers.") or a distraction-inhibiting implementation intention (e.g., "As soon as a new row of numbers is presented, I will ignore the type of the numbers.").

Whereas children with and without ADHD both profited from if–then plans, children with and without ADHD had an increase in errors on the counting task. This occurred despite having the explicit goal to count the numbers. Only when given an additional implementation intention did they manage to nearly remain at the same error level in the counting task as in the naming task. In line with research by Mischel and colleagues (e.g., Patterson & Mischel, 1976) both children with and without ADHD profited more from distraction-inhibiting than from task-facilitating plans. Only in the distraction-inhibiting implementation intention condition were the errors made during the naming task comparable to the amount of errors children made during the counting task. Thus, ignoring the distraction (i.e., type of numbers) is a more effective strategy than concentrating on the main task (i.e., counting the numbers).

Implementation Intentions Facilitate Delay of Gratification Performance in Children with ADHD

Go/NoGo and delay of gratification paradigms have one fundamental characteristic in common: In both tasks, performance requires controlling a prepotent response, whether it is producing a behavioral response in the Go/NoGo task (i.e., pressing the key although the NoGo signal indicates that the response has to be inhibited) or attending to the immediate reward in the delay of gratification task (i.e., grabbing the immediate reward although the delayed reward is more valuable). Not surprisingly then, performance on both tasks have been linked to activity in similar areas in the prefrontal cortex (Eigsti et al. 2006; Mischel & Ayduk, 2002). In the same vein, stop and delay of gratification paradigms have both been used to measure impulsivity, even though these two types of tasks may be suited particularly well for assessing one rather than the other aspect of impulsivity as described in the literature. These aspects include *(a)* the ability to collect and evaluate information before reaching decisions; *(b)* the ability to choose larger delayed rewards over small immediate rewards; and *(c)* the ability to suppress motor responses that have been rendered prepotent (Chamberlain & Sahakian, 2007). In sum, Go/NoGo task and delay of gratification paradigms seem to measure aspects of cognitive functioning that pertain to impulsivity.

The aim of the following two studies was to explore whether if–then plans help children with ADHD in a delay of gratification paradigm (Gawrilow, Gollwitzer, & Oettingen, 2009). Specifically, children with and without ADHD were confronted with a computerized delay of gratification game, modeled after paradigms used in experiments by Mischel and colleagues (Mischel, Shoda, & Rodriguez, 1989) and by Sonuga-Barke and colleagues (Sonuga-Barke, Taylor, Sembi, & Smith, 1992). For this task, children had to decide between an immediate and less valuable (red pictures showing vehicles or animals with a value of one point) or a delayed and more valuable gratification (blue pictures showing vehicles or animals with a value of three points) presented on a computer screen. As an incentive, children could exchange their points for money at the end of the experiment.

Participants in the first study were children with ADHD in a German medical center specializing in the treatment of children with ADHD. Children were randomly assigned to three conditions: One-third of the children received a neutral sentence ("Red pictures are one point, blue pictures are three points") and one-third received a sentence with a goal intention ("I will earn as many points as possible"). The remaining children received the goal intention and an additional implementation intention ("If a red picture appears, then I will wait for the blue one."). As compared to the neutral sentence (control) group, the children benefited (i.e., made more points) from the implementation intention but not from the mere goal intention.

In a second study we invited not only children with ADHD, but we compared children with ADHD to children without any known psychological disturbances. Again, children were randomly assigned to the same three conditions (neutral vs. goal intention instruction vs. goal intention plus implementation intention instruction) as in the previous study. Both children with and without ADHD benefited from having formed implementation intentions. Specifically, children who had formed an if–then plan, in contrast to those with a neutral statement or mere goal intention, managed to delay gratifications better during the game and therefore ultimately earned more money at the end of the game. Consistent with the previous study, goal intentions were not superior to neutral instructions in children with ADHD; thus formulating the goal to obtain more points and therefore a bigger reward was not helpful for delaying gratifications in children with ADHD. Furthermore, implementation intentions enabled children with ADHD to wait significantly more often in the second half compared to the first half of the game. Thus, children with ADHD can use making if–then plans as a strategy to sustain their waiting behavior for delayed rewards over a longer period of time although impulsivity is one of the main symptoms of ADHD.

Implementation Intentions Facilitate Set-Shifting and Multitasking in Children with ADHD

Children with ADHD not only show deficits in inhibitory control as measured by Go/NoGo and delay of gratification paradigms, but also struggle on other neuropsychological tasks (Halperin & Schulz, 2006). We therefore investigated the effect of implementation intentions on executive functioning in two more studies: in tasks assessing set-shifting and in tasks that require multitasking.

In the *set-shifting* study, we presented children with ADHD with a slightly modified version of a WCST (Gawrilow, Oettingen, & Gollwitzer, 2009a, Study 2). At the outset, the children received several stimulus cards and a stack of additional cards. The children were then asked to match each of the additional cards to one of the stimulus cards, but received no instruction (i.e., rule) on how to match the cards; they did however receive immediate feedback from the experimenter about the correctness of their choice. Following a child's 10th correct match, the experimenter changed the rule. Prior to the task, one-third of the children were randomly assigned to a neutral instruction condition ("The additional cards need to be matched to the cards on the table."), one-third to a goal intention condition ("I will match as many cards as possible with the correct rule."), and the remaining children to an implementation intention condition ("And if my rule is wrong, then I will try another rule immediately."). The implementation intention group ended up with a significantly lower level of preservation errors in relation to the percentage of total errors than both the goal intention and control instruction group. This significant main effect implies that children with ADHD can benefit from implementation intentions in shifting cognitive sets as measured with a modified WCST.

In the *multitasking* study (Gawrilow, Oettingen, & Gollwitzer, 2009b), we used a multitasking paradigm modeled after the Six Elements Test (SET). Children received three different tasks that were associated with different colors (i.e., green, blue, & red) and every task consisted of two parts. Children were instructed to work on two green tasks (consisting of counting and calculating items), on two blue tasks (consisting of naming items), and two red tasks (consisting of sorting items). In our modified version, children worked on each of the six tasks at least twice over a period of 10 minutes without working on a task with the same color consecutively. Children with and without ADHD were randomly assigned to a neutral instruction ("There are a lot of tasks and it is not allowed to work on two tasks with the same color consecutively."), goal intention instruction ("I will try to solve at least 10 items of every color alternating."), or an additional implementation intention ("And as soon as I have solved 10 items of a color, then I will switch to another color."). Children with ADHD achieved a higher score as compared to the control group only with implementation intentions; for children without ADHD, this was true already with goal intentions.

Conclusion and Outlook

We have argued that making if–then plans is an effective strategy to make goal-striving effective so that getting started, staying on track, calling a halt, and not overextending oneself when striving for one's goals become more manageable. We have then reported evidence that if–then planning holds up its promise even when goal-striving is challenged by the lack of relevant skills, competitive opponents, habitual antagonistic responses, or a psychological disorder handicapping action control in general (i.e., ADHD). Given this track record, if–then planning qualifies as an effective goal-striving strategy to be taught to people in interventions to facilitate their everyday goal-striving. Such interventions need to create conditions where people are particularly motivated and capable to make if–then plans and where the effects of if–then plans are known to be particularly strong. This has recently been attempted in intervention studies that combined mental contrasting (MC; i.e., considering obstacles to a vividly imagined desired future goal state, Oettingen, Pak, & Schnetter, 2001) and implementation intentions into one intervention (MCII-intervention) to promote exercising in middle-aged women (Stadler, Oettingen, & Gollwitzer, 2009) and coping with the stressors of everyday life in college students (Oettingen et al., 2009).

References

Aarts, H., Dijksterhuis, A. P., & Midden, C. To plan or not to plan? Goal achievement of interrupting the performance of mundane behaviors. Eur J Soc Psychol 1999; 29: 971–979.

Achtziger, A., Gollwitzer, P. M., & Sheeran, P. Implementation intentions and shielding goal striving from unwanted thoughts and feelings. Pers Soc Psychol Bull, 2008; 34: 381–393.

Armitage, C. J. Evidence that implementation intentions reduce dietary fat intake: A randomized trial. Health Psychol 2004; 23: 319–323.

American Psychiatric Association. *Diagnostic and statistical manual of mental disorders*, 4th ed. Washington, DC: Author, 1994.

Bargh, J. A., & Chartrand, T. L. The mind in the middle. A practical guide to priming and automaticity research. In: Reis, H. T. & Judd, C. M. (Eds.), *Handbook of research methods in social and personality psychology*. Cambridge, UK: Cambridge University Press, 2000: pp. 253–285.

Barkley, R. A. Behavioral inhibition, sustained attention, and executive functions: Constructing a unifying theory of ADHD. Psychol Bull 1997; 121: 65–94.

Bayer, U. C., Achtziger, A., Gollwitzer, P. M., & Moskowitz, G. Responding to subliminal cues: Do if-then plans cause action preparation and initiation without conscious intent? Soc Cogn 2009; 27: 183–201.

Bayer, U. C., Gollwitzer, P. M., & Achtziger, A. Staying on track: Planned goal striving is protected from disruptive internal states. J Exp Soc Psychol, in press.

Bayer, U. C., & Gollwitzer, P. M. Boosting scholastic test scores by willpower: The role of implementation intentions. Self Identity 2007; 6: 1–19.

Berg, E. A. A simple objective test for measuring flexibility in thinking. J Gen Psychol 1948; 39: 15–22.

Bottom, W. P., & Studt, A. Framing effects and the distributive aspect of integrative bargaining. Organ Behav Hum Decis Process 1993; 56: 459–474.

Brandstätter, V., Lengfelder, A., & Gollwitzer, P. M. Implementation intentions and efficient action initiation. J Pers Soc Psychol 2001; 81: 946–960.

Brockner, J. The escalation of commitment to a failing course of action: Toward theoretical progress. Acad Manage Rev 1992; 17: 39–61.

Burgess, P. W. Strategy application disorder: The role of the frontal lobes in human multitasking. Psychol Res 2000; 63: 279–288.

Burgess, P. W., Simons, J. S., Dumontheil, I., & Gilbert, S. J. The gateway hypothesis of rostral PFC function. In: Duncan, J., Phillips, L., & McLeod, P. (Eds.), *Measuring the mind: Speed control and age*. Oxford, UK: Oxford University Press, 2005: pp. 215–246.

Carretié, L., Hinojosa, J. A., Martín-Loeches, M., Mercado, F., & Tapia, M. Automatic attention to emotional stimuli: Neural correlates. Hum Brain Mapp 2004; 22: 290–299.

Chamberlain, S. R., & Sahakian, B. J. The neuropsychiatry of impulsivity. Curr Opin Psychiatr 2007; 20: 255–261.

Clarke, C., Prior, M., & Kinsella, G. Do executive function deficits differentiate between adolescents with ADHD and Oppositional Defiant/Conduct Disorder? A neuropsychological study using the Six elements test and Hayling sentence completion test. J Abnorm Child Psychol 2000; 28: 403–414.

Cohen, A.-L., Bayer, U. C., Jaudas, A., & Gollwitzer, P. M. Self-regulatory strategy and executive control: Implementation intentions modulate task switching and Simon task performance. Psychol Res 2008; 72: 12–26.

Cohen, J. A power primer. Psychol Bull 1992; 112: 155–159.

Covassin, T., & Pero, S. The relationship between self-confidence, mood state, and anxiety among collegiate tennis players. J Sport Behav 2004; 27: 230–242.

Eigsti, I., Zayas, V., Mischel, W., et al. Predictive cognitive control from preschool to late adolescence and young adulthood. Psychol Sci 2006; 17: 478–484.

Epstein, J. N., Erkanli, A., & Conners, C. K. Relations between continuous performance test performance measures and ADHD behaviors. J Abnorm Child Psychol 2003; 31: 543–554.

Frazier, T. W., Youngstrom, E. A., Glutting, J. J., & Watkins, M. W. ADHD and achievement: Meta-analysis of the child, adolescent, and adult literatures and a concomitant study with college students. J Learn Disab 2007; 40: 49–65.

Gawrilow, C., & Gollwitzer, P. M. Implementation intentions facilitate response inhibition in children with ADHD. Cognit Ther Res 2008; 32: 261–280.

Gawrilow, C., Gollwitzer, P. M., & Oettingen, G. *If-then plans benefit delay of gratification performance in children with and without ADHD*. Cognitive Therapy and Research, in press.

Gawrilow, C., Oettingen, G., & Gollwitzer, P. *If-then plans benefit executive functions in children with ADHD*. Manuscript submitted for publication, 2009a.

Gawrilow, C., Oettingen, G., & Gollwitzer, P. M. *Implementation intentions support multitasking in children with ADHD*. Manuscript in preparation, 2009b.

Gilbert, S. J., Gollwitzer, P. M., Cohen, A.-L., Oettingen, G., & Burgess, P. W. Separable brain systems supporting cued versus self-initiated realization of delayed intentions. J Exp Psychol Learn Mem Cogn 2009; 35: 905–915.

Gollwitzer, P. M. Goal achievement: The role of intentions. Eur Rev Soc Psychol 1993; 4: 141–185.

Gollwitzer, P. M. Implementation intentions. Strong effects of simple plans. Am Psychol 1999; 54: 493–503.

Gollwitzer, P. M., & Bayer, U. C. *Becoming a better person without changing the self*. Paper presented at the Self and Identity Pre-conference of the Annual Meeting of the Society of Experimental Social Psychology, Atlanta, Georgia, October 2000.

Gollwitzer, P. M., & Brandstätter, V. Implementation intentions and effective goal pursuit. J Pers Soc Psychol 1997; 73: 186–199.

Gollwitzer, P. M., & Schaal, B. Metacognition in action: The importance of implementation intentions. Pers Soc Psychol Rev 1998; 2: 124–136.

Gollwitzer, P. M., & Sheeran, P. Implementation intentions and goal achievement: A meta-analysis of effects and processes. Adv Exp Soc Psychol 2006; 38: 69–119.

Halperin, J. M., & Schulz, K. P. Revisiting the role of the prefrontal cortex in the pathophysiology of ADHD. Psychol Bull 2006; 132: 560–581.

Hanegby, R., & Tenenbaum, G. Blame it on the racket: Norm-breaking behaviors among junior tennis players. Psychol Sport Exerc 2001; 2: 117–134.

Henderson, M. D., Gollwitzer, P. M., & Oettingen, G. Implementation intentions and disengagement from a failing course of action. J Behav Decis Mak 2007; 20: 81–102.

Hinton, D., Nathan, M., & Bird, B. Panic probes and the identification of panic: A historical and cross-cultural perspective. Cul Med Psychiatry 2002; 26: 137–153.

Holland, R. W., Aarts, H., & Langendam, D. Breaking and creating habits on the working floor: A field experiment on the power of implementation intentions. J Exp Soc Psychol 2006; 42: 776–783.

Homack, S., & Riccio, C. A. A meta-analysis of the sensitivity and specificity of the Stroop Color and Word Test with children. Arch Clin Neuropsychol 2004; 19: 725–743.

Hoza, B., Mrug, S., Gerdes, A. C., et al. What aspects of peer relationships are impaired in children with attention-deficit/hyperactivity disorder? J Consult Clin Psychol 2005; 73: 411–423.

Lengfelder, A., & Gollwitzer, P. M. Reflective and reflexive action control in patients with frontal brain lesions. Neuropsychology 2001; 15: 80–100.

Logan, G. D., & Cowan, W. B. On the ability to inhibit thought and action: A theory of an act of control. Psychol Rev 1984; 91: 295–327.

Milne, S., Orbell, S., & Sheeran, P. Combining motivational and volitional interventions to promote exercise participation: Protection motivation theory and implementation intentions. Br J Health Psychol 2002; 7: 163–184.

Mischel, W., & Ayduk, O. Self-regulation in a cognitive-affective personality system: Attentional control in the service of the self. Self Identity 2002; 1: 113–120.

Mischel, W., Shoda, Y., & Rodriguez, M. L. Delay of gratification in children. Science 1989; 244: 933–938.

Muraven, M., & Baumeister, R. F. Self-regulation and depletion of limited resources: Does self-control resemble a muscle? Psychol Bull 2000; 126: 247–259.

Murgraff, V., White, D., & Phillips, K. Moderating binge drinking: It is possible to change behavior if you plan it in advance. Alcohol Alcohol 1996; 6: 577–582.

Oettingen, G., Barry, H., Guttenberg, K., & Gollwitzer, P. M. Self-regulation of time management: Mental contrasting with implementation intentions. Manuscript submitted for publication, 2009.

Oettingen, G., Hönig, G., & Gollwitzer, P. M. Effective self-regulation of goal attainment. Int J Educ Res 2000; 33: 705–732.

Oettingen, G., Pak, H., & Schnetter, K. Self-regulation of goal-setting: Turning free fantasies about the future into binding goals. J Pers Soc Psychol 2001; 80: 736–753.

Olekalns, M. Situational cues as moderators of the frame-outcome relationship. Br J Soc Psychol 1997; 36: 191–209.

Orbell, S., & Sheeran, P. Motivational and volitional processes in action initiation: A field study of the role of implementation intentions. J App Soc Psychol 2000; 30: 780–797.

Orbell, S., Hodgkins, S., & Sheeran, P. Implementation intentions and the theory of planned behavior. Pers Soc Psychol Bull 1997; 23: 945–954.

Oosterlaan, J., Logan, G. D., & Sergeant, J. A. Response inhibition in ADHD, CD, comorbid ADHD + CD, anxious and normal children: a meta-analysis of studies with the stop task. J Child Psychol Psychiatr 1998; 39: 411–425.

Parks-Stamm, E. J., Gollwitzer, P. M., & Oettingen, G. Action control by implementation intentions: Effective cue detection and efficient response initiation. Soc Cogn 2007; 25: 248–266.

Patterson, C. J., & Mischel, W. Effects of temptation-inhibiting and task-facilitating plans of self-control. J Pers Soc Psychol 1976; 33: 209–217.

Raven, J. C. The Raven's progressive matrices: Change and stability over culture and time. Cogn Psychol, 2000; 41: 1–48.

Rosvold, H. E., Mirsky, A. F., Sarason, I., Bransome, E. D., & Beck, L. H. A continuous performance test of brain damage. J Consult Psychol 1956; 20: 343–350.

Rubia, K., Russell, T., Overmeyer, S., et al. Mapping motor inhibition: conjunctive brain activations across different versions of Go/NoGo and stop tasks. NeuroImage 2001; 13: 250–261.

Schweiger Gallo, I., Keil, A., McCulloch, K. C., Rockstroh, B., & Gollwitzer, P. M. Strategic automatization of emotion control. J Pers Soc Psychol 2009; 96: 11–31.

Sheeran, P., & Orbell, S. Using implementation intentions to increase attendance for cervical cancer screening. Health Psychol 2000; 19: 283–289.

Sheeran, P., Webb, T. L., & Gollwitzer, P. M. The interplay between goal intentions and implementation intentions. Pers Soc Psychol Bull 2005; 31: 87–98.

Simon, J. R. The effects of an irrelevant directional cue on human information processing. In: Proctor, R. W. & Reeve, T. G. (Eds.), *Stimulus-response compatibility: An integrative perspective*. Amsterdam: North-Holland, 1990: pp. 31–86.

Sonuga-Barke, E. J. S., Taylor, E., Sembi, S., & Smith, J. Hyperactivity and delay aversion—I. The effect of delay on choice. J Child Psychol Psychiatry 1992; 33: 387–398.

Stadler, G., Oettingen, G., & Gollwitzer, P. M. Effects of a self-regulation intervention on women's physical activity. J Prevent Med 2009; 36: 29–34.

Stroop, J. R. Studies of interference in serial verbal reactions. J Exp Psychol 1935; 28: 643–662.

Trötschel, R., & Gollwitzer, P. M. Implementation intentions and the willful pursuit of goals in negotiations. J Exp Soc Psychol 2007; 43: 519–598.

Webb, T. L., & Sheeran, P. Can implementation intentions help to overcome ego-depletion? J Exp Soc Psychol 2003; 39: 279–286.

Webb, T. L., & Sheeran, P. Identifying good opportunities to act: Implementation intentions and cue discrimination. Eur J Soc Psychol 2004; 34: 407–419.

Webb, T. L., & Sheeran, P. Does changing behavioral intentions engender behavior change? A meta-analysis of the experimental evidence. Psychol Bull 2006; 132: 249–268.

Webb, T. L., & Sheeran, P. How do implementation intentions promote goal attainment? A test of component processes. J Exp Soc Psychol 2007; 43: 295–302.

Webb, T. L., & Sheeran, P. Mechanisms of implementation intention effects: The role of goal intentions, self-efficacy, and accessibility of plan components. Br J Soc Psychol 2008; 47: 373–395.

Wughalter, E. H., & Gondola, J. C. Mood states of professional female tennis players. Percept Motor Skills 1991; 73: 187–190.

CHAPTER 16

Unpacking the Self-Control Dilemma and Its Modes of Resolution

Arie W. Kruglanski and Catalina Kőpetz

ABSTRACT

The problem of self-control has been an old preoccupation since the time of Greek philosophers. In modern psychology, self-control and related concepts, such as conscientiousness, ego resilience, willpower, or the human agency, have been of longstanding interest to theorists yielding to invaluable insights into people's abilities and difficulties to cope with self-control concerns. However, an overarching conception that would guide our understanding of self-control phenomena is still in great need. The purpose of the present chapter is to organize the most recent theorizing and empirical research on self-control and to sketch the contours of such a framework around two main issues: 1) the essential "ingredients" of the self-control problem (saliency of seemingly incompatible objectives, and their relative value); 2) the basic ways of responding to the self-control problem (goal-choice, and multifinality quest).

Keywords: Self-control, goal pursuit, goal-saliency, goal-value, goal-choice, multifinality

INTRODUCTION

The exercise of self-control has been long hailed as a supreme human facility, indispensable for the worthy life. Already the Greek philosophers touted willpower as a major virtue, and viewed yielding to temptations as a deplorable weakness.[1] In the Judeo-Christian tradition, the archetypal story of Adam and Eve and their lamentable fall from grace is attributed to their succumbing to temptation (eating from the tree of knowledge), accorded the dubious glory of defining the "original sin." Within the Eastern philosophies too, temptations have been treated as hindrances to the good life, depending as it does on the transcendence of the self and its ephemeral desires (Flew, 1999).

These ancient attitudes are ubiquitously represented in modern thought. In Freud's portrayal (Freud, 1962, 1990), the aim of psychoanalysis is to overcome id-based "temptations" through powers of the realistic and socially aware Ego, aiming to safeguard worthier and culturally more significant concerns. The Protestant Ethic (Weber, 1930) highlights the Calvinist value of constant labor and touts as morally superior the foregoing of momentary pleasures for the sake of more profound objectives (e.g., one's salvation, the accumulation of wealth, or both).

Self-control and related concepts, such as conscientiousness, ego resilience, willpower, or the human agency have been of longstanding interest to psychological theorists (Freud, 1962, 1990; James, 1890; Mischel, 1974, 1996; Bem

& Allen, 1974; McCrae & Costa, 1999; Block & Block, 1980; Block, 2002; Carver & Scheier, 1982; Gollwitzer & Bargh, 1996). Research on these issues has yielded a wealth of intriguing insights into ways people cope with self-control concerns, their strategies of overcoming temptations, and the personality traits that describe those individuals who are capable or incapable of successful self-control. These efforts, although of considerable value and importance, have been relatively scattered and fragmented; they lack an overarching conception of self-control in which known facts about the phenomenon may be ordered and understood. The purpose of the present chapter is to sketch the contours of such a framework. We begin by considering the essential meaning of the self-control concept, and proceed by identifying what appear to be the essential ingredients of the self-control problem: *(1)* saliency of seemingly incompatible objectives and *(2)* their relative value. We then consider the basic ways of responding to the self-control problem, namely via *(1)* goal-choice and *(2)* multifinality quest, and sketch out a likely sequence of psychological events involved in confronting the self-control challenge.

What is Self-Control? Matters of Definition

In the relevant research literature, self-control has been typically defined as the sacrifice of immediate, low priority/importance goals in favor of long term, high priority/importance objectives. In Mischel's (Mischel, 1974) delay of gratification paradigm, the self-control issue arises from a conflict between immediately available small rewards and delayed (though qualitatively similar) larger rewards (e.g., fewer versus more numerous candies). As Mischel and Ayduck (2004) noted, the self-control problem entails an "essentially simple 'less now' versus 'more later' dilemma" (p. 118). Similarly, Fishbach & Trope (2005, p. 256, emphasis added) proposed that the self-control dilemma concerns the case wherein "*short-term* outcomes...tempt people to act against their *long-term* interests." And Fujita, Trope, Liberman and Levin-Sagi (2006, p. 351, emphasis added) remarked that self-control has been thought of as decisions and actions "in accordance with...*long term* rather than...*short term* outcomes."

In these and other similar definitions, it is not quite clear whether the essence of the self-control dilemma is assumed to reside in the distinction between *long term* versus *short term* goals, or in between *low priority/importance* versus *high priority/importance goals*. Conceptually, these two dimensions are different and separable. Thus, it is possible to envisage instances wherein the high-priority goal was proximal and immediate, whereas the low-priority goal was distal and long term: saving one's child from drowning may represent an immediate goal of incomparably greater priority than a long-range goal of advancing one's career by arriving in time for a job interview, for example. Would pursuit of the short-term goal in such circumstances be perceived as a loss of self-control? Intuitively, it would not.

To determine whether the temporal dimension (of immediate versus long-term goals) or the priority dimension (of low- versus high-priority goals) matters more to perceptions of self-control, we presented participants (University of Maryland undergraduates, $N = 42$) with vignettes depicting an actor engaged in pursuit of a long-term goal (maintaining a healthy diet) while confronting a salient short-term goal (indulging a high-calorie, tasty food) in three experimental conditions.

In one condition (*dieting*), the actor chose the dieting goal over the eating goal. In the second condition (*justified consumption*), the actor's consumption of the high-calorie food was portrayed as enacted deliberately to generate sufficient energy to study for an impeding exam. In the third condition (*unjustified consumption*), the actor's consumption of the highly caloric food was not accompanied by an additional justification. Following their perusal of these vignettes, participants were asked to determine the degree to which the actor *(1)* was acting in accordance with her or his more important goal and *(2)* believed that she or he was exercising self-control. We found, unsurprisingly, that participants believed that the actor was acting in accordance with her more important goal in both the *dieting* and the *justified consumption* condition ($M = 7.40$ vs. $M = 7.00$, $t < 1$) even though in the first case she chose her long-term goal (of health/appearance) whereas in the second condition she chose to pursue the short-term goal (of energy generation). Of greater interest, our participants also believed that the target exercised an equal degree of self-control in both these cases ($M = 3.00$ vs. $M = 3.13$, $t < 1$).

By contrast, in the third, *unjustified consumption*, condition where the actor chose to consume the highly caloric food, without an added justification for this act, participants believed less than their counterparts in the *dieting* and *justified consumption* conditions that she/he acted in line with her more important goal ($M = 4.93$ vs. $M = 7.40$, $t(42) = 3.04$, $p < 0.05$ and $M = 4.93$ vs. $M = 7.00$, $t(42) = 2.55$, $p < 0.05$, for comparisons of the former with each of the latter conditions). Of greatest interest, participants in the *unjustified consumption* condition considered that she also lost her self-control as compared with the *dieting* condition ($M = 4.93$ vs. $M = 3.00$, $t = 2.21$, $p < 0.05$) and the *justified consumption* condition ($M = 4.93$ vs. $M = 3.13$, $t = 2.05$, $p < 0.05$).

Although based solely on one scenario with very specific constraints, these findings suggest that for our participants, self-control meant the choice of a higher priority/importance over a lower priority/importance goal, rather than the choice of a distal/long-term goal over a proximal/short-term goal.

Essential and Effective Goal Hierarchies

Consistent with the "lay theory" suggested by the foregoing data, in the present paper we address *self-control* in terms of an *essential* hierarchy of goal importance on which individuals' objectives may be ordered. Thus, rather than stressing the short- versus long-term nature of the competing goals, we consider self-control as a choice in favor of a more important over a less important objective when the two are in conflict. The notion of an *essential hierarchy* of goals requires further comment. After all, recent research on goal-priming phenomena (for reviews, see Fischbach & Ferguson, 2007; Morsella, Bargh & Gollwitzer, 2009; Moskowitz & Grant, 2009; Kruglanski & Kőpetz, 2009a, 2009b) suggests that goals may be pervasively activated by various environmental features. In that sense, the *effective* goal hierarchy guiding current actions may differ from one moment to the next pertaining as it does to a subset of goals activated during a particular instant. If so, one could argue that at any given moment, one's choice of an action was in the service of what at that particular instant constituted one's most important goal. This would divest from all meaning the notion of self-control understood here as the problem of choosing between a less and a more important objective.

From this perspective, the notion of an *essential goal hierarchy* serves the indispensable function of a referential standard in relation to which the experience of self-control is determined.

The Immediacy Factor

Although the essence of the self-control issue may revolve around a conflict between goals of different degrees of importance, the *immediacy* (or *proximity*) factor is crucial to defining the dilemma. Typically, *temptations* (a piece of cake for a dieter, a cigarette for a "would be" quitter) are closer temporally and oftentimes spatially than overarching goals higher in one's essential goal hierarchy. This is likely to create a psychological proximity (Liberman, Trope, & Stephan, 2007; Trope & Liberman, 2003) and result in increased saliency or degree of activation that the immediate goal may enjoy (Higgins, 1996). Indeed, it is the psychological salience of the proximal and less valuable goal in comparison with the distal and hence pallid albeit more significant objective that renders temptations so hard to resist. Along these lines, it has been suggested that given the considerable pull exercised by the proximal, and hence salient, temptation, the individual will need a certain degree of mental resources to extricate herself or himself from its "magnetic field" to evaluate the current goal in comparison with other more distal goals which it may threaten to undermine (Baumeister, Bratslavsky, Muraven,

& Tice, 1998; Muraven, Tice, & Baumeister, 1998; Gailliot et al., 2007; Vohs & Heatherton, 2000). Presumably, if the more and less important goals were equal in their degree of activation, the difference in their relative importance would be readily apparent, and most people would choose the more important over the less important goal.

In this vein, Papies, Stroebe, and Aarts (2007) have found that for restrained eaters, the biased attention to tasty food items prompted by food pre-exposure tended to disappear after participants were primed with the diet construct. Also consistent with the implied importance of goal saliency, Mann and Ward (2004) showed how in situations of self-control dilemma, salient cues may direct individuals' behavior, especially when their cognitive resources are drained. In their study, chronic dieters were exposed to a highly caloric food (milkshake) that they would normally avoid. The experimenters manipulated cognitive load and orthogonally altered the saliency of a particular external cue. In one experimental condition, participants' dieting goal (of high importance) was made salient by asking them to complete a common measure of diet habits and by showing in the lab a series of external cues relevant to dieting, such as a scale, diet books, and a recipe for milkshake. No such cues were present in the alternative condition, referred to as the shake-saliency condition. Participants were then left alone in the room with a milkshake for 7 minutes. The quantity of milkshake consumed during this interval represented the main dependent variable.

The results showed that participants in the diet-saliency condition drank less milkshake than did participants in the control condition. This effect was qualified by the load manipulation. In the absence of load, participants in the diet-saliency differed only slightly in the amount of milkshake they consumed compared to participants in the milkshake saliency conditions. Presumably, in this condition where participants' cognitive resources were unaffected by the load manipulation, they were capable to overcome the attentional drawing power of the eating temptation and to behave more in line with their overarching goal (dieting). However, under cognitive load, participants in the diet-saliency condition consumed significantly less milkshake than participants in the milkshake-saliency condition. These findings are consistent with data suggesting that under cognitive load, or other similar manipulations known to induce a need for cognitive closure (e.g., time pressure, noise, fatigue) individuals may "seize" and "freeze" their judgments and attendant actions on judgments suggested by the most salient or accessible cues (Ford & Kruglanski, 1995; Kruglanski, 2004; Thompson et al., 1994).

What these suggest then is that the saliency (proximity, activation) of goals involved in the self-control dilemma draws attention to these goals' subjective value or importance. If an immediate, and hence salient, goal (e.g., the temptation) was assessed as of sufficient value (i.e., appeal) to warrant goal consistent action and if the individual lacked the epistemic motivation or cognitive resources to contemplate the potential impact this may have for further goals (cf. Kruglanski, 1989; 2004) he or she might end up pursuing that particular goal without further ado (hence, succumbing to the temptation). However, in the presence of sufficient processing resources, the individual might well contemplate such impact and if it turned out to entail the sacrifice of other more valuable objectives, the individual might well refrain from pursuit of the immediate goal despite the initial impulse.

In summary, we have viewed the self-control problem as one of sacrificing a more important objective for a less important objective. Although on the face of it, this may appear as an "irrational" choice, it may be psychologically valid nonetheless. Specifically, at the moment of choice, the individual's attention may be drawn toward the psychologically proximal and salient goal, and away from the distal objective. This may augment the experienced value of the proximal goal: in the moment it may appear considerably more valuable than would be warranted otherwise. In other words, in the individual's *effective* goal hierarchy (that which is momentarily active) the proximal goal may appear considerably more valuable than it is in the person's *essential* goal hierarchy representing her or his general ordering of goals' importance in "neutral" states devoid of strong proximal goal stimuli. Issues of relative saliency and value of goals involved in the self-control dilemma have been emphasized in researchers' discussions of various self-control strategies. We now turn to a brief review of this work.

Strategies of Self-Control

I. Manipulating the Relative Saliency of Goals and Temptations

Much recent work has recognized the impact that saliency may have on one's capability to exercise self-control, and has identified mechanisms and strategies whereby self-control can be enhanced to protect people's important goals (Ariely & Wertenbroch, 2002; Fishbach, Friedman, & Kruglanski, 2003; Trope & Fishbach, 2000; Metcalfe & Mischel, 1999; Freitas, Liberman, & Higgins, 2002, Gollwitzer, 1990). Many such strategies are directly aimed either to increase the saliency of the high-priority goal or to decrease the saliency of the temptations.

Issues of saliency and immediacy figure prominently in Metcalfe and Mischel's (1999) dual process model involving two interacting systems—a "cool" cognitive system and a "hot" emotional system. The "cool" system is referred to as a "know" cognitive system consisting of interconnected cognitive representations and responsible for generating rational, reflective and strategic behavior. According to the authors, it is the "cool" system that allows persons to keep a goal in mind (maintaining its prominence or saliency) and monitor the progress toward its attainment. By contrast, the "hot" or the "go" system "specializes" in quick and emotional processing, and responds impulsively and reflexively, on the basis of unconditional and conditional trigger features. In their empirical research, Metcalfe and Mischel (1999) showed that the ability to "balance" these two systems is critical for successful self-control.

In this vein, Metcalfe & Mischel (1999) found that participants' ability to overcome immediate temptations and to delay gratification of higher-priority goals depended on their ability to transform a hot, motivating reward into its cool, "informational" representation. For example, children were able to delay their gratification for 18 minutes when they pretended that the rewards they were confronting were not real but rather constituted pictures (e.g., children waiting for pretzels were instructed to "put a frame around them in their head"). By contrast, participants were able to wait for less than 6 minutes when they pretended that the real rewards were present in front of them. In present terms, translation into the "cool" system may decrease the *psychological proximity* of the tempting stimulus, hence lower its saliency and allow the striving for the alternative goal to be maintained.

Activation of the High Importance Goals by Forming Implementation Intentions

Saliency concerns are also involved in the formation of "implementation intentions," another technique relevant to safeguarding the pursuit of important goals. This strategy has been articulated and extensively researched by Gollwitzer and his colleagues (Gollwitzer, 1993, 1999; Gollwitzer, Bayer, & McCullough, 2005). According to these investigators, successful pursuit of important goals (hence, successful self-control) may be enhanced by forming "if-then" associations between specific features of situations that an individual expects to enter, and the appropriate intention to execute the goal-directed response. The saliency issue enters here in that the relevant situational features are thought to automatically activate, and hence render more salient, the intention to execute the goal correspondent behavior (for a review, *see* Gollwitzer et al., 2005).

For example, a possible implementation intention that would promote the self-control of eating behavior might associate a suitable situational context (e.g., placing a dinner order at a restaurant) with the intention to make the appropriate choice (e.g., ask for a vegetarian dish). In finding oneself in the designated setting the appropriate choice is made salient and the correspondent behavior may be enacted. In support of this analysis, a series of studies found that experimentally prompted formation of implementation intentions versus that of simple goal intentions (i.e., an intention to pursue a certain goal) was more effective in initiating goal pursuit (Gollwitzer & Brandstatter, 1997) and increasing commitment to goal attainment (Brandstatter, Lengfelder, & Gollwitzer, 2001; Oettingen, Honig, & Gollwitzer, 2000).

Additional studies have shown that forming implementation intentions increased goal attainment rates concerning important yet unpleasant or difficult to pursue goals such as submitting oneself to regular breast examinations (e. g., Orbell, Hodgkins, & Sheeran, 1997), carrying out cervical cancer screenings (Sheeran & Orbel, 2000), or engaging in physical exercise (Milne, Orbell, & Sheeran, 2002). Forming implementation intentions was also found to increase individuals' ability to resist tempting distractions. In a study relevant to this issue, Gollwitzer and Schaal (1998) asked participants to perform a series of arithmetic problems while distracting clips from award-winning commercials were interspersed at random intervals on the monitor screens. Implementation intentions were found to be more effective in protecting participants from these distractions than mere goal intentions. In summary, the formation of implementation intentions may be said to increase one's ability to exercise self-control by enhancing the *saliency* of the goal-appropriate behavior represented by the intention to implement it.

Activation of Goals by Temptations

Whereas work by Gollwitzer and his colleagues addresses the possibility of increasing the saliency of goal-appropriate behaviors by situational features, research by Fishbach, Friedman, and Kruglanski (2003) addresses the possibility that high importance goals and their correspondent behaviors may be activated by the tempting stimuli themselves. In other words, individuals may exercise self-control by forming strong associations between temptations and the high-importance goals with which they are in conflict. In line with this reasoning, Fishbach et al. (2003) found that successful self-regulators in a domain are more likely to form such temptation-goal associations than unsuccessful self-regulators.

In one study (Fishbach, Friedman, & Kruglanski, 2003, Study 3), University of Maryland undergraduates who (by their own admission) were either successful or unsuccessful academically, performed a lexical decision task after first being exposed to a subliminal prime. On some trials, the primes related to temptations to avoid studying such as "television," "procrastinate," "phone" and "internet," and target words related to the goal of studying; for example, "study," "grades," "homework," and "graduate." On other trials, the study words were the primes and the temptation-words, the lexical targets. The results showed that for successful students, temptation words activated study words to a significantly greater extent than vice versa, and goal words resulted in an inhibition of temptation-related targets. By contrast, for unsuccessful students, study words activated temptation words to a greater extent than vice versa. Such asymmetric pattern of associations between short-term objectives (temptations) and the higher-value goals with which they interfere allows participants to exercise self-control through activation of the latter goals and the enactment of goal-consistent behaviors.

Explicit versus Implicit Strategies

Self-Control as a Goal

Fishbach and Shah (2006) explored the possibility that saying "no" to temptations may be facilitated by a *general predisposition* toward approaching goals and avoiding temptations. Using a joystick paradigm as a way of operationally defining basic approach and avoidance responses (Solarz, 1960), Fishbach and Shah (2006) showed that individuals were relatively faster to pull the joystick (representing an approach response) to goal-related concepts and were relatively faster to push the joystick (representing the avoidance response) to temptation-related concepts (Study 1). Such tendencies were stronger for individuals who held an overarching goal (e.g., dieting) (Study 2) and when the temptations were perceived to be particularly attractive.

Specifically, the more attractive the temptations appeared to be, the faster effective self-regulators were in pulling the joystick toward themselves when presented with concepts related to a high-importance goal, and to push the joystick away from themselves when presented with concepts related to temptations (Study 3). Such predispositions to automatically approach high-importance goals and to avoid temptations were significantly related to individuals' intentions to behave consistently with their overarching goals, as well as to their actual behavior.

These findings suggest that self-control may constitute an important goal *in and of itself* that may be chronically active, or may be activated by contextual cues. That is, beyond the desire to protect specific high-priority goals (e.g., dieting) from specific temptations, some individuals (or most individuals in some situations) may be motivated to protect (what they perceive as) more important goals from interference by momentarily salient, and hence "tempting" low importance goals. A general goal of self-control suggests the forming of an intention to *avoid* stimuli classified as temptations and to *approach* stimuli classified as (high-priority) goals. The presence of such general self-control goal implies that individuals who subscribe to it may invest their energies in an *assessment process* designed to determine the relative importance of various goals affording their classification into the "temptation" and "high-priority goal" categories. Furthermore, individuals with a high commitment to the self-control goal might feel frustrated and upset should the results of their assessment be inconclusive (e.g., because one's various goals were close in value and hence refractory to prioritization).

To summarize, extant research findings suggest that self-control dilemmas typically arise in situations where a high-priority goal has a low activation level and a low-priority goal (the temptation) has a high activation level. Focusing attention on immediately accessible, yet low-priority enticements may render them hard to "resist" in the absence of comparison with the high-priority goals that are hard to access. Considerable cognitive and motivational resources might be needed to activate those latter goals. Forecasting the difficulty, some individuals may *automatize* the process by forming strong associative links between specific temptations and the high priority goals these threaten to undermine. Automatization may reduce the need to invest exorbitant ego resources in the exercise of self-control, thus increasing the likelihood of successful control. Individuals may also adopt a general self-control goal and form a generalized intention to resist temptations wherever those appear. Such goal may require the investment of resources in an assessment process designed to determine the value for the individual of various goals. A chronic self-control goal of this kind may result in a prior classification of a given stimulus as a "goal" or a "temptation." These possibilities could be profitably investigated in subsequent research.

Varying the Relatives Value of "Goals" versus "Temptations"

Activating a goal allows a comparison of its overall value with that of the "temptation." To the extent that the goal was seen as considerably more valuable than the temptation it would be pursued, and the temptation sacrificed. Thus, given an above-threshold level of activation needed to carry out a value comparison, the choice of the goal over the temptation may depend on the perceived value-differential between them. One strategy of self-control could consist, therefore, of augmenting the value attached to the goal, and/or reducing the value attached to the temptation, thus increasing the discrepancy between the two.

In this vein, Trope and Fishbach (2000) hypothesized that in anticipation of the self-control dilemma, people may bolster the value of the "goal" by linking it to a reward contingent on goal-appropriate behavior. In one study, these investigators found that participants for whom health was important were willing to award themselves a bonus contingent on completion of a medical test. Although participants could have received the bonus without taking the test, they chose to make the bonus conditional on the test when the test results were deemed of considerable importance and when they expected the test to be rather unpleasant (Study 2).

In the same vein, Muraven and Slessareva (2003) showed that people may be able to overcome resource depletion and exercise self-control if their action was perceived as instrumental to an additional goal. Specifically, in one study, participants initially went through a resource depletion phase where they performed a thought-suppression task. Subsequently, as a measure of self-control (resistance to frustration) they were required to perform an unsolvable problem-solving task, and the time they spent trying to solve the task was measured. Some participants were led to believe that the task might help to answer important questions about memory that could potentially lead to better treatment for Alzheimer's disease; others were given the same descriptions without any mention of its benefits to Alzheimer's treatment. The results showed that participants whose resources were depleted by the thought-suppression task quit working on the task much sooner when the task was not associated with any additional benefits than when the task was believed to be instrumental in developing new treatment for Alzheimer's disease.

In addition to bolstering self-control by associating pursuit of the high-priority goal by associating it with additional goals or benefits, one may strategically downgrade the value of a temptation (e.g., eating high-glucose foods) by imposing a *penalty* contingent on succumbing to the temptation. In a study relevant to this point, Trope and Fishbach (2000, Study 1) offered participants the opportunity to test the influence of glucose intake on their cognitive functioning. The feedback from the test was described to be very useful, but it required abstinence from food containing glucose for either a short (6 hours) or a long period of time (3 days). Before deciding whether to take the test, participants were asked to specify the amount of money they would be willing to pay as a penalty for failing to take the test. The results showed that participants who expected the temptation to be strong during a long period of abstinence versus weak during a short period of abstinence imposed on themselves a larger penalty for failure to comply with the test requirements.

Trope and Fishbach's (2000) results suggest that boosting the value of the high-priority goal and downgrading the value of the "temptation" can be accomplished by making activities that serve these goals *negatively or positively multifinal* (Kruglanski et al., 2002)—that is, attached to additional (positive or negative) objectives. For example, the self-imposition of penalties can be seen as representing negative multifinality; it makes the temptation-driven behavior instrumental to a negative consequence. Negative multifinality may be involved also in de-valuing a temptation, that is, in remembering the negative consequences of pursuing the temptation driven behavior. Attaching self-imposed rewards to resisting to a temptation represents positive multifinality; it creates a situation wherein resistance is instrumental to other important goals as well (the self-imposed rewards). Positive multifinality is involved also in bolstering the more important goal, that is, in recalling the additional positive consequences of engaging in a behavior instrumental to that goal.

Fujita, Trope, Liberman, and Levin-Sagi (2006) proposed that self-control may be enhanced when people are able to see "the forest beyond the trees," or, in their terms, to make decisions and act in accordance with global, high-level construals of the situation rather than with local, low-level construals. To test this notion, they procedurally primed participants in one condition to use high-level construals by considering questions related to *why* they engaged in a particular action (e.g., in attempts to maintain good health) or, in another condition, to use low-level construals by considering questions about *how* they engaged in the same action. The results showed that considering *why* versus *how* they engaged in a particular action resulted in better self-control expressed in a reduced preference for immediate over delayed outcomes (higher delay of gratification).

From the present perspective, having participants think about *why* they engaged in an action may remind them about the ultimate value or desirability of pursuing it. On the other hand, responding to questions related to *how* they engaged in an action may have prompted participants to consider a means to pursue a goal, and be primarily concerned with the goal's attainability. It is thus possible that a focus on value, implicit in the "why" question (and awareness of the value sacrifice that succumbing to a temptation would exact) may have been responsible for participants' better ability to delay gratification in the Fujita et al. (2006) experiments.

In summary then, beyond self-control strategies related to appropriately altering the *relative saliency* of the (proximal) temptation and the (distal) goal, there exist various strategies related to bolstering the value of the goal and reducing the value of the temptation.

Multifinality Quest as Response to the Self-Control Dilemma

Implicit in the self-control literature discussed above was the notion that resolution of the self-control dilemma involves *goal choice*. Specifically, an individual's options were seen to consist of either resisting the temptation, and safeguarding one's high-priority goal, or succumbing to the temptation, hence sacrificing the goal. Yet, a very different option might exist that does not frame the situation in zero sum terms, and allows one to "eat the cake and have it too." Such option is to identify a means that satisfies all the co-active goals at once. In fact, such a solution is related to a common manufacturing practice these days of equipping one's products with multiple features responsive to consumers' varied desires. The food industry, for example, is producing a large array of diet foods that promise to deliver taste while reducing one's caloric intake, allowing one to maintain (or acquire) a slim figure at the same time. Similarly, the fitness industry has come up with energy-enhancing, muscle-building products that also promise to restore one's mineral balance without compromising on taste. The electronics industry has managed to create sundry products equipped with a variety of features, cell phones that serve as cameras, computer terminals, personal calendars, and music boxes (e.g., Apple's iPhone) "all in one."

Given the unpalatable possibility of having to give up on some of their goals, people may seek alternative strategies to deal with multiple goal situations. For example, Dhar and his colleagues (Dhar & Simonson, 1999; Khan & Dhar, 2007) suggested that when confronted with two goals such as pleasure and good health, people may attempt to *balance* their choice of means to satisfy the goals (i.e., order a tasty but unhealthy appetizer followed by a healthy entrée). In a similar vein, research on the compromise effect (Simonson, 1989) demonstrated a general preference for choice alternatives that partially meet several goals at once rather than ones that fully meet a single goal.

The multifunctional nature of consumer products is explicit and overt. It figures prominently in the products' advertising campaigns and constitutes a selling point deigned to increase their appeal. But recent research in social cognition has demonstrated that numerous goal-driven effects can be implicit and unconscious. Can the quest for multifinality be unconscious as well? Apparently so. In fact, much of contemporary research on unconscious goal priming can be considered to involve multifinal strivings. Thus, in a classic study, Bargh et al. (2001) engaged participants with the activity of fishing in a small lake for the goal of profit (G1) while priming them subliminally with a cooperation or competition goal (G2). In this situation participants were more cooperative or competitive in accordance with the priming, but presumably they were also pursuing their focal goal of profit maximization.

Hassin and Bargh (2004) asked participants to perform the Wisconsin Card Sorting Task (assumed to assess flexibility) after being primed with a high-performance goal or a neutral goal. Participants were more flexible in the experimental vs. the control condition. In other words, participants *were doing what the experimenter asked them to do*, but also they did it in a way that corresponded to their primed goal. In a yet another study, Chen, Lee-Chai & Bargh (2001) primed participants with the concept of "power" assumed to activate in individuals with a communal orientation the goal of social responsibility, and in individuals with an exchange orientation the goal of taking care of one's own needs. Participants were then instructed to perform exercises that took varied amounts of time. Both the communal and the exchange participants performed their focal goal of doing the exercises, but they did it in different ways corresponding to their divergent goals. Specifically, communal individuals performed the task in a way that made it easier for other, alleged, participants by taking upon themselves to perform the more time consuming exercises leaving the briefer ones for their partners. The exchange individuals made the opposite choice and hence made the task easier for themselves.

To summarize then, in classic goal-priming studies, participants seem to be pursuing two goals at the same time: They comply with the experimental instructions, defining the focal, explicit goal, and they do so in a way that satisfies the second, primed goal as well. Of interest, participants in goal-priming studies are typically unaware that their behavior was *multifinal*, and they are typically unconscious of the primed goal that was having demonstrable impact on their behavior.

Multifinality in the Choice Paradigm

The multifinality interpretation of goal-priming effects implies that the observed goal-consistent behavior is performed because it promises to afford the greatest value given the individual's multiple objectives, rather than because it is simply activated by the primed goal, a notion implicit in the "perception-action expressway" (Dijksterhuis & Bargh, 2001). It follows that multifinality considerations should influence preferences also in explicit choice contexts where several different objects or behavioral options are all activated to a more or less equal degree.

To investigate these issues we revisited the classic study by Wilson and Nisbett (1978) in which passersby at a department store were asked to select among four identical pairs of nylon stockings or four identical nightgowns of the highest quality. A strong position effect obtained such that the two rightward objects in the array were heavily over chosen.

The highlight of this research, of central interest to Wilson and Nisbett, was that participants seemed entirely unaware of having been systematically biased in their choices, and justified them entirely in terms of the choice-objects' quality. But why did the rightward bias occur in the first place? The notion of *multifinality* offers a possible insight into this phenomenon. From that perspective, participants in the Wilson and Nisbett (1978) study, in addition to the goal explicitly assigned them by the experimenter, that of *making a reasonable choice*, also harbored a goal of *reaching quick closure* after inspecting the entire array of stockings. Whereas the former goal was highly explicit and conscious, the latter may have constituted a background goal of which the participants were not explicitly aware.

Consider, additionally, that English is written and read from left to right and, as a consequence, the scanning habit within the American culture shows a left to right directionality (Maass & Russo, 2003). If so, *both goals* above (making a reasonable choice and reaching quick closure) would have been satisfied by the rightward objects in the array (more so than by the leftward objects), the last ones to be inspected following an initial, necessary, sweep. In the present terminology then, the rightward objects would have been more multifinal than their preceding, left lying alternatives. In this sense, multifinality might have constituted a major reason why Wilson and Nisbett's (1978) participants ended up over-choosing the rightward objects and exhibiting the enigmatic position effect of which they were eminently unaware.

To investigate this possibility, Chun, Kruglanski, Sleeth-Keppler & Friedman (2005, Study 1) conceptually replicated Wilson and Nisbett's (1978) study with one difference: they kept the participants' conscious goal constant while systematically manipulating the putative unconscious goal of closure. In a variant of Wilson and Nisbett's (1978) procedure, participants were given the conscious and explicit goal of choosing among four pairs of (actually identical) athletic socks the pair of the best quality. To manipulate the unconscious goal, participants in one condition were placed under time pressure to heighten their need for closure (Kruglanski 2004; Kruglanski & Webster 1996; Webster & Kruglanski 1998). In a second condition, no time pressure was applied and participants were given *accuracy* instructions to reduce their need for closure, or introduce the need to avoid closure (Kruglanski & Webster 1996; Kruglanski et al., 2006). Consistent with expectations, the authors replicated the Wilson & Nisbett (1978) effect in time-pressure condition and eliminated it in the accuracy condition.

In subsequent studies, Chun et al. (2005) furnished additional evidence for multifinality quest using the background goals of identifying versus disidentifying with one's university and identifying versus disidentifying with the U.S. (for review, *see* Kruglanski & Kőpetz, 2009a, 2009b). These investigators found that when identification with the university was the background goal, participants tended to view a swatch of fabric colored red which constitutes one of the UMD colors as superior in quality to a different swatch of (actually the same) fabric colored purple, which is a non-university color. This result was replicated with Winthrop University students who under the identification condition tended to view a garnet-colored paper (constituting a Winthrop color) as superior in quality to a blue-colored paper. Finally, participants who viewed Coke as a more American drink than Pepsi tended to select a drink bearing the Coke label when induced to identify with the U.S. as superior in taste than the same drink bearing the Pepsi label.

Multifinality constraints on means generation

The quest for multifinal means seems to benefit the process of self-control as it appears to maximize value and avoid a difficult and costly goal choice. Nonetheless, identifying multifinal means has a downside as well, related to the reduced number of means to the focal goal that one might end up as a consequence of the effort to attain multifinality. Kőpetz, Fishbach & Kruglanski (2008) recently carried out a number of experiments addressing these phenomena.

In one of our studies we activated participants' alternative goals by asking them in one condition to list three activities they had planned for the rest of that day (operationally defining the *uncompleted goals* condition). In another condition participants listed three activities that they had already accomplished that day (defining the *completed goals* condition). We assumed that the latter goals would have lost their driving potential, or in Lewinian terms, would have their "tension-system" drained. Participants were then asked to choose the foods they desired for lunch from a list of 20 foods. These foods were pre-tested such that 10 of them were generally readily available at the food court where we ran the study (e.g., Chinese food, tacos, fries) whereas the remaining 10 were foods considered to be unavailable at the food court, though available at other campus locations (e.g., salmon, macaroni and cheese, crab cakes). We assumed that "easy-to-get" foods are multifinal in that choosing them will help participants to fulfill their focal goal (of having lunch) and also to save time for alternative goals that they had planned for the day. That is, in fact, what happened.

In a second study (Study 3), we found that participants were more selective with regard to the foods they considered for lunch when they were also concerned with "keeping a healthy diet." Provided with a list of foods containing both high and low-caloric foods and asked to choose the foods they wanted for lunch, our participants restricted their choice predominantly to the low-caloric foods. However, when the goal of eating became particularly important (i.e., participants felt hungrier) as a result of priming with eating-related words, participants inhibited the "healthy diet" goal such that no selectivity in their food choice transpired anymore.

To summarize, the quest for multifinal means appears to represent a basic response to multiple goal environments. Marketing researchers (Thompson, Hamilton, & Rust, 2005) have found that when consumers face a choice between different models of a product, a majority tends to choose the model with the most features. Our own research suggests that the preference for multifinal means may occur implicitly and without individuals' awareness of the goals that affect their choices. One way of understanding this tendency is in terms of value maximization. Simply put, individuals may be loath to give up on any of their goals. Hence, in the presence of multiple activated goals, people's initial impulse might be to explore whether a choice between them can be avoided and whether a clever solution may be found that avoids any kind of loss.

GENERAL DISCUSSION

In conducting their everyday affairs, people typically are reminded of multiple goals they would like to pursue. A laptop on which one attempts to write a chapter may call to mind the many emails that one owes one's friends, the books that one wishes to purchase, and the holiday trips one would like to take. As much recent research suggests, (*see* Morsella, Bargh, & Gollwitzer, 2009, for reviews) such goal reminders, even though they necessitate an increment in the goals' activation, need not necessarily reach conscious awareness for them to affect behavior. Often, the activated goals seem incompatible with other objectives. This defines the self-control dilemma (Trope & Fishbach, 2000) to which individuals need to find a response. In the preceding pages we discussed this common predicament and unpacked its various components.

The self-control dilemma typically refers to the case where a salient low-priority goal evokes action tendencies which if realized would jeopardize the attainment of a less salient yet a higher-priority objective in an *essential goal hierarchy* to which the individual may subscribe. From this perspective, the essence of the dilemma concerns the values of the conflicted goals. Pursuit of the less-valuable goal runs counter to the precept of value maximization and hence entails an experienced loss of self-control. Similarly, the decision to refrain from such a pursuit produces a sense of control.

The foregoing analysis suggests that the self-control experience entails a *value comparison* between the goal pursued and that sacrificed. Occasionally, such experience may arise after the fact, that is, after the goal-driven action had already been undertaken. This might occur, for instance, if the nonchosen goal with which the chosen one needs to be compared was inhibited because of commitment to the latter goal (Shah, Friedman, & Kruglanski, 2002). Also, an asymmetry may well exist between cases wherein the low-priority goal (or the "temptation") was succumbed to or resisted: Decision to pursue the low-priority goal may involve an inhibition of the higher-priority goal with which it is in conflict, hence entail a temporary unawareness of its existence. In contrast, the decision to resist the low-priority goal is likely to be made in the context of acute awareness of the higher-priority goal which pursuit it may undermine. At any rate, awareness of having sacrificed a more for a less valuable objective may give rise to a sense of control loss and to feelings of regret.

It is of interest to speculate about the sequence of events that may take place in circumstances of a self-control dilemma. First, a low importance goal may get activated by features of the external environment or by one's own stream of associations. An assessment may then take place to determine whether the goal's value is of a sufficient magnitude to warrant appropriate (goal-consistent) action. At that phase too, the individual might consider whether the goal interferes with other objectives and what is the relative value of the goal as compared to those alternatives. Whether the assessment will be extensive or brief would depend on the individual's resources. When such resources are sufficient, the assessment phase may be extensive and detailed. In the absence of sufficient resources, it may be brief and limited, such that the individual may not even consider any alternative objectives and proceed "impulsively" with pursuit of the activated goal. Prior formation of linkages between various goals may automatize the process (Fishbach et al., 2003) and liberate it from dependence on extensive cognitive and motivational resources.

Where in course of the assessment process several differently-valued goals are activated, three principal outcomes are possible: *(1)* pursuit of the most highly valued goal, affording a sense of self-control; *(2)* pursuit of a less-valued goal, leading to perceived loss of control on further reflection about the sacrificed goals, and *(3)* identification of multifinal means affording the satisfaction of one's currently active objectives without the need to incur any sacrifices. The first two outcomes represent *goal choice* wherein some value at least is foregone; the last outcome avoids choice altogether, and represents a "clever," value-maximizing, solution to the self-control dilemma.

The actual process of goal choice (outcomes 1 and 2 above) is insufficiently well understood at this time. Plausibly, it is a dynamic process wherein the individual vacillates between different choice options, being pulled in different directions and finally "falling" into the force field exerted by a given goal and committing to it while inhibiting its alternatives. The latter process entails a kind of complexity that might be best understood, perhaps, via a nonlinear modeling approach to self-control phenomena.

The phenomenon of multifinality quest (outcome 3 above) also requires further elaboration and probing. The way we discussed it implies a mechanism of parallel constraint satisfaction wherein the same means or activity concomitantly mediates several objectives (Kunda & Thagard, 1996). However, a sequential multifinality quest may often take place wherein people pursue a "go now, pay later" strategy of currently succumbing to a temptation (e.g., a highly caloric food item), and deciding to compensate for it by subsequent choice (e.g., appropriately reducing one's caloric intake on subsequent occasions, or performing an arduous physical exercise). Possibly, the foregoing sequential strategy may allow one to avoid choice where choice is particularly difficult to make (e.g., where the competing goals are nearly equal in value), and no concomitantly multifinal means seems available. These matters could well be explored in future self regulatory research.

NOTE

1 According to Plato, a major virtue is one of temperance, "the ordering or controlling of certain pleasures and desires this is... implied in the saying of being his own master... in the human soul there is a better and also a worse principle and when the better has the worse under control the man is said to be master of himself and this is a term of praise.

REFERENCES

Ariely, D., & Wertenbroch, K. Procrastination, deadlines, and performance: Self-control by precommitment. Psychol Sci 2002; 13(3): 219–224.

Bargh, J. A., Gollwitzer, P. M., Lee-Chai, A., Barndollar, K., & Troetschel, R. The automated will: Nonconscious activation and pursuit of behavioral goals. J Pers Soc Psychol 2001; 81(6): 1014–1027.

Baumeister, R. F., Bratslavsky, E., Muraven, M., & Tice, D. M. Ego-depletion: Is the active self a limited resource? J Pers Soc Psychol 1998; 74(5): 296–309.

Bem, D. J., & Allen, A. On predicting some of the people some of the time. The search for cross-situational consistencies in behavior. Psychol Rev 1974; 81: 506–520.

Block, J. *Personality as an affect-processing system: Toward an integrative theory.* Mahwah, NJ: Lawrence Erlbaum Associates, 2002.

Block, J., & Block, J. J. The role of ego-control and ego-resiliency in the organization of behavior. In: Collins, W. A. (Ed.), *The Minnesota symposium on child psychology*, Vol. 13. Hillsdale, NJ: Erlbaum, 1980: pp. 39–101.

Brandstatter, V., Lengfelder, A., & Gollwitzer, P. M. Implementation intentions and efficient action initiation. J Pers Soc Psychol 2001; 81: 946–960.

Carver, C. S., & Scheier, M. F. Control-theory: A useful conceptual framework for personality, social, clinical, and health psychology. Psychol Bull 1982; 92: 111–135.

Chen, S., Lee-Chai, A. Y., & Bargh, J. A. Relationship orientation as a moderator of the effects of social power. J Pers Soc Psychol 2001; 80(2): 173–187.

Chun, W., Kruglanski, A. W., Sleeth-Keppler, D., & Friedman, R. *On the psychology of quasi-rational decisions: The multifinality principle in choice without awareness.* Manuscript submitted for publication, 2005.

Dhar, R., & Simonson, I. Making complementary choices in consumption episodes: Highlighting versus balancing. J Mark Res Soc 1999; 36(1): 29–44.

Dijksterhuis, A., & Bargh, J. A. The perception-behavior expressway: Automatic effects of social perception on social behavior. In: Zanna, M. P.

(Ed.), *Advances in experimental social psychology*. San Diego, CA: Academic Press, 2001; 33: 1–40.

Fishbach, A., & Ferguson, M. J. The goal construct in social psychology. In: Kruglanski, A. W., & Higgins, E. T. (Eds.), *Social psychology: Handbook of basic principles*. New York: Guilford Press, 2007: pp. 490–515.

Fishbach, A., & Shah, J. Y. Self-control in action: Implicit dispositions toward goals and away from temptations. J Pers Soc Psychol 2006; 90(5): 820–832.

Fishbach, A., & Trope, Y. The substitutability of external control and self-control. J Exp Soc Psychol 2005; 41(3): 256–270.

Fishbach, A., Friedman, R. S., & Kruglanski, A. W. Leading us not unto temptation: Momentary allurements elicit overriding goal activation. J Pers Soc Psychol 2003; 84(2): 296–309.

Fitzsimons, G. M., & Bargh, J. A. Automatic Self-regulation. In: Baumeister, R. F. & Vohs, K. D. (Eds.), *Handbook of self-regulation: Research, theory, and applications*. New York, NY: Guilford Press, 2004: pp. 151–170.

Flew, A. *A dictionary of philosophy*, 2nd ed. Gramercy, 1999.

Ford, T. E., & Kruglanski, A. W. Effects of epistemic motivations on the use of accessible constructs in social judgment. Pers Soc Psychol Bull 1995; 21(9): 950–962.

Freitas, A. L., Liberman, N., Higgins, E. T. Regulatory fit and resisting temptation during goal pursuit. J Exp Soc Psychol 2002; 38: 291–298.

Freud, S. *The ego and the id*. New York, NY: W. W. Norton & Company, 1962.

Freud, S. *Beyond the pleasure principle*. New York, NY: W. W. Norton & Company, 1990.

Fujita, K., Trope, Y., Liberman, N., & Levin-Sagi, M. Construal levels and self-control. J Pers Soc Psychol 2006; 90(3): 351–367.

Gailliot, M. T., Baumeister, R. F., DeWall, C. N., et al. Self-control relies on glucose as a limited energy source: Willpower is more than a metaphor. J Pers Soc Psychol 2007; 92(2): 325–336.

Gollwitzer, P. M. Action phases and mind-sets. In: Higgins, E. T., & Sorrentino, R. M. (Eds.), *Handbook of motivation and cognition: Foundations of social behavior*, Vol. 2. New York, NY: Guilford Press, 1990: pp. 53–92.

Gollwitzer, P. M. Implementation intentions: Strong effects of simple plans. Am Psychol 1999; 54(7): 493–503.

Gollwitzer, P. M., & Schaal, B. Metacognition in action: The importance of implementation intentions. Pers Soc Psychol Rev 1998; 2: 124–136.

Gollwitzer, P. M., Bayer, U. C., & McCulloch, K. C. The control of the unwanted. In: Hassin, R. R., Uleman, J., & Bargh, J. A. (Eds.), *The new unconscious*. New York, NY: Oxford University Press, 2005: pp. 485–515.

Gollwitzer, P. M. Goal achievement: The role of intentions. Eur Rev Soc Psychol 1993; 4: 141–185.

Gollwitzer, P. M., & Bargh, J. A. (Eds.). *The psychology of action: Linking cognition and motivation to behavior*. New York, NY: Guilford Press, 1996.

Gollwitzer, P. M., & Brandstatter, V. Implementation intentions and effective goal pursuit. J Pers Soc Psychol 1997; 73: 186–199.

Gross, J. J. Emotion regulation: Past, present, future. Cogn Emot 1999; 13: 551–573.

Hassin, R. R., & Bargh, J. A. *Flexible automaticity: Evidence from automatic goal pursuit*. Manuscript under review, 2004.

James, W. *Principles of psychology*. New York, NY: Holt, 1890.

Khan, U., & Dhar, R. Where there is a way, is there a will? The effect of future choices on self-control. J Exp Psychol Gen 2007; 136(2): 277–288.

Kőpetz, C., Fishbach, A., & Kruglanski, A. W. Taking alternative goals into account: Multifinality constraints on means generation. Manuscript submitted for publication, 2008.

Kruglanski, A. W., & Kőpetz, C. E The role of goal-systems in self-regulation. In: Bargh, J., Gollwitzer, P., & Morsella, E. (Eds.), *The psychology of action: Vol 2: The mechanisms of human action*. New York: Guilford, 2009a: pp. 350–367.

Kruglanski, A. W., & Kőpetz, C. E. What is so special (and non-special) about goals? A view from the cognitive perspective. In: Moscovitz, G., & Grant, H. (Eds.), *Goals*. New York: Guilford Press, 2009b: pp. 27–55.

Kruglanski, A. W. *Lay epistemic and human knowledge: Cognitive and motivational bases*. New York: Plenum, 1989.

Kruglanski, A. W. *The psychology of closed mindedness*. New York: Psychology Press, 2004.

Kruglanski, A. W., & Webster, D. M. Motivated closing of the mind: "Seizing" and "freezing." Psychol Rev 1996; 103(2): 263–283.

Kruglanski, A. W., Pierro, A., Mannetti, L., & Grada, E. De Groups as epistemic providers: Need for closure and the unfolding of groupcentrism. Psychol Rev 2006; 113(1): 84–100.

Kunda, Z., & Thagard, P. Forming impressions from stereotypes, traits, and behaviors: A parallel-constraint-satisfaction theory. Psychol Rev 1996; 103(2), 284–308.

Maass, A., & Russo, A. Directional bias in the mental representation of spatial events: Nature or culture? Psychol Sci 2003; 14: 296–301.

Mann, T., & Ward, A. To eat or not to eat: Implications of the attentional myopia model for restrained eaters. J Abnorm Psychol 2004; 113(1): 90–98.

McCrae, R. R., & Costa, P. T. A five-factor theory of personality. In: Pervin, L. A., & John, O. P. (Eds.), *Handbook of personality: Theory and research*, 2nd ed. New York, NY: Guilford Press, 1999: pp. 139–153.

Metcalfe, J., & Mischel, W. A hot/cool-system analysis of delay of gratification: Dynamics of willpower. Psychol Rev 1999; 106(1): 3–19.

Milne, S., Orbell, S., & Sheeran, P. Combining motivational and volitional interventions to promote exercise participation: Protection motivation theory and implementation intentions. Br J Health Psychol 2002; 7: 163–184.

Mischel, W. Processes in delay of gratification. In: Berkowitz, L. (Ed.), *Advances in experimental social psychology*, Vol. 7. New York, NY: Academic Press, 1974: pp. 249–292.

Mischel, W. From good intentions to willpower. In: Gollwitzer, P. M., & Bargh, J. A. (Eds.), *The psychology of action: Linking cognition and motivation to behavior*. New York, NY: Guilford Press, 1996: pp. 197–218.

Mischel, W., & Ayduk, O. Willpower in a cognitive-affective processing system: The dynamics of delay of gratification. In: Baumeister, R. F., & Vohs, K. D. (Eds.), *Handbook of self-regulation: Research, theory, and applications*. New York, NY: Guilford Press, 2004: pp. 99–129.

Morsella, E., Bargh, J. A., & Gollwitzer, P. M. (Eds.). *The psychology of action (Vol 2): The mechanisms of human action*. New York, NY: Guilford Press, 2009.

Moskowitz, G. B., & Grant, H. (Eds.). *Goals*. New York, NY: Guilford Press, 2009.

Muraven, M., & Slessareva, E. Mechanisms of self-control failure: Motivation and limited resources. Pers Soc Psychol Bull 2003; 29: 894–906.

Muraven, M., Tice, D. M., & Baumeister, R. F. Self-control as a limited resource: Regulatory depletion patterns. J Pers Soc Psychol 1998; 74(3): 774–789.

Oettingen, G., Honig, G., & Gollwitzer, P. M. Effective self-regulation of goal-attainment. Int J Educ Res 2000; 33: 705–732.

Orbell, S., Hodgkins, S., & Sheeran, P. Implementation intentions and the theory of planned behavior. Pers Soc Psychol Bull 1997; 23: 945–954.

Papies, E. K., Stroebe, W., & Aarts, H. *On the allure of forbidden food: Self-regulatory cognition in chronic dieters*. Paper presented at the Annual Meeting of the Society for Experimental Social Psychology, Chicago, IL, October 2007.

Richards, J. M., & Gross, J. J. Emotion regulation and memory: The cognitive costs of keeping one's cool. J Pers Soc Psychol 2000; 79: 410–424.

Shah, J. Y., Friedman, R., & Kruglanski, A. W. Forgetting all else: On the Antecedence and consequences of goal shielding. J Pers Soc Psychol 2002; 83: 1261–1280.

Sheeran, P., & Orbell, S. Using implementation intentions to increase attendance for cervical cancer screening. Health Psychol 2000; 19: 283–289.

Solarz, A. K. Latency of instrumental responses as a function of compatibility with the meaning of eliciting verbal signs. J Exp Psychol 1960; 59: 239–245.

Thompson, D. V., Hamilton, R. W., & Rust, R. T. Feature fatigue: When product capabilities become too much of a good thing. J Market Res 2005; 42(4): 431–442.

Thompson, M. M., Roman, R. J., Moskowitz, G. B., & Chaiken, S. Accuracy motivation attenuates cover priming: The systematic reprocessing of social information. J Pers Soc Psychol 1994; 66(3): 447–489.

Trope, Y., & Fishbach, A. Counteractive self-control in overcoming temptation. J Pers Soc Psychol 2000; 79(4): 493–506.

Vohs, K. D., & Heatherton, T. F. Self-regulatory failure: A resource-depletion approach. Psychol Sci 2000; 11(3): 249–254.

Weber, M. *The Protestant ethic and the spirit of capitalism*. London: G. Allen & Unwin, Ltd., 1930.

Webster, D. M., & Kruglanski, A. W. Cognitive and social consequences of the motivation for closure. Eur Rev Psychol 1998; 8: 133–173.

Wilson, T. D., & Nisbett, R. E. The accuracy of verbal reports about the effects of stimuli on evaluations and behavior. Soc Psychol 1978; 41(2): 118–131.

CHAPTER 17

Conflict and Control at Different Levels of Self-Regulation

Abigail A. Scholer and E. Tory Higgins

ABSTRACT

Traditionally, self-control conflicts have been defined as conflicts between some immediate, short-term gratification versus some delayed, long-term gain. Although this is certainly a self-control issue, we argue that a focus on this definition of self-control has obscured the broader self-control issue: self-control is about resolving and managing conflict. In this chapter, we take a broad view of defining self-control within the self-regulatory system by considering how conflicts and control are represented within a self-regulatory hierarchy. Specifically, we suggest that self-control involves managing conflicts at multiple levels: managing conflicts between and within the levels of behaviors, tactics, strategies, and goals. In particular, we suggest that self-control conflicts can exist both between and within multiple levels in a hierarchy. Vertical conflicts occur *between* levels in a self-regulatory hierarchy (e.g., between higher-order and lower-order concerns, between goal orientations and strategies, between strategies and tactics). Horizontal conflicts occur *within* levels in a self-regulatory hierarchy (e.g., between goals, between strategies, between tactics, between behaviors). We review evidence that individuals exert self-control both horizontally and vertically at the goal and strategic levels of the self-regulatory hierarchy. We end by discussing the possibility that conflict representations are malleable. Representing the *same* conflict in different ways (e.g., vertically vs. horizontally) may have significant implications for interventions.

Keywords: Self-control, conflict, self-regulatory hierarchy, regulatory focus, regulatory fit

---

We begin by considering two classic examples of self-control problems. In other words, we begin by considering sex and marshmallows. Most contemporary approaches to thinking about self-control problems derive, either explicitly or implicitly, from the Freudian distinction between the Id as an unconscious entity abiding by the pleasure principle and the Super-ego as the lofty agent of internalized prohibitions and oughts. Self-control is the purview of the Ego (abiding by the reality principle), which must balance the unmitigated impulses of the Id with the demands and prohibitions of the Super-ego. The traditional interpretation of this conflict is that self-control is about an impulse to be repressed, a temptation to be resisted, a response to be inhibited. The Id represents lower-order, impulsive, and base desires; the Super-ego represents higher-order, abstract, and valued standards. Thus, the traditional view of self-control suggests that self-control is about impulse control at the behavioral level; self-control is exerted when higher-order standards prevail over lower-order desires (cf. Baumeister, Schmeichel, & Vohs, 2007; Mischel, Cantor, & Feldman, 1996).

This view of self-control has also been the dominant interpretation of Walter Mischel's classic self-control (delay-of-gratification) studies (e.g., Mischel & Ebbesen, 1970; Mischel & Baker, 1975; Moore, Mischel & Zeiss, 1976). A child has a choice between one treat (e.g., one marshmallow) immediately versus two treats (two marshmallows) later; the ability to delay gratification by waiting for two marshmallows is considered to be a mark of effective self-control. The choice to forgo immediate, short-term gratification (the pleasure of eating one marshmallow) for future greater gain (two marshmallows) is seen as the triumph of a higher-order (long-term) goal over a lower-order (short-term) impulse.

There is no doubt that one aspect of self-control is about exerting control at the behavioral level to overcome some relatively low-level desire in the service of a higher-order goal. Many approaches have argued that this is the central self-control concern (cf. Baumeister et al., 2007). We argue, however, that a focus on this definition of self-control has obscured the broader self-control issue: *self-control is about resolving and managing conflict*. A re-analysis of the classic self-control situations supports the notion that there are other ways in which to conceptualize these conflicts.

Freud conceived of the Ego as a harmonizing agent, bringing the relative unreasonableness of both the Id and the Super-ego into line (Freud, 1923/1961). Although it is not surprising that in contemporary discussion of self-control conflicts the Id has been replaced with lower-order concerns, there is nothing inherent about self-control conflicts that necessitates that the conflict must arise from a lower-order impulse. Furthermore, although the Super-ego, by definition, is "above" the Ego, the Ego also has conflicts with the Super-ego about its overly severe demands. Thus, it was not the case in Freud's conception that the Super-ego was unequivocally good and the Id unequivocally bad, with the Super-ego's interests always being preferred to the Id's interests. The Ego had to resolve the conflict *between* the Id *and* the Super-ego; the Ego was not just the agent of the Super-ego. Both higher-order and lower-order concerns can be valid—and *both* can be too extreme. A critical aspect of self-control is managing the potential conflict that can exist between them as competing interests.

It is also possible to think more broadly about the self-control conflict in the classic delay-of-gratification paradigm. Traditionally, the conflict has been defined as one in which self-control is exerted if immediate, short-term gratification is sacrificed for a later, greater gain (e.g., Trope & Fisbach, 2000), often seen to be a reflection of the preferential activation of the cool (maximize gain) versus hot (consume now) system (Metcalfe & Mischel, 1999). However, the conflict could also be viewed instead as a classic double approach-avoidance conflict (Lewin, 1935; Miller, 1944). There are, in fact, trade-offs to both options: eating one marshmallow immediately versus waiting for two. By eating a single marshmallow now, I give up the pleasure of a second marshmallow, but I also give up the pain of waiting. By waiting for two marshmallows, I get the pleasure of a second marshmallow (maximizing outcomes), but I also accept the pain of waiting. Re-conceptualized as a double approach-avoidance conflict, the situation represents difficult trade-offs with no clear dominant alternative. Choosing to wait does not have to be interpreted as the triumph over an impulse; choosing to wait is just one way to resolve the conflict.

Thus, we argue that what is central to self-control is the management and resolution of conflicts. Additionally, we suggest that self-control is broader than control of conflicts at the behavioral level. Even if we ultimately care about behavioral outcomes, there are multiple levels in the self-regulatory hierarchy where conflicts can exist and where control can be targeted. Self-control conflicts also extend beyond tensions between "higher-order" (e.g., Super-ego) and "lower-order" (e.g., Id) concerns. Even if conflicts often involve higher-order (long-term/abstract) concerns versus lower-order (short-term/concrete) concerns, these represent only a subset of all of the possible self-control conflicts.

We argue that self-control conflicts can exist both between and within multiple levels in a self-regulatory hierarchy. The classic self-control conflict between higher-order and lower-order concerns has often been represented as a vertical conflict between a higher-order goal (e.g., academic success) and a lower-order means of behaving—the temptation—that is incompatible with (i.e., impedes) attaining that goal (e.g., partying all night) (cf. Fujita, Trope, Liberman, & Levin-Sagi, 2006). In our approach, this is just one example of a *vertical* self-control conflict (conflict between levels of the self-control hierarchy).[1] Vertical conflicts are not restricted to incompatibilities between higher-order versus lower-order concerns. There can also be conflicts between a type of goal orientation and a type of strategy that is a non-fit with that type of goal orientation, such as pursuing a goal with a promotion focus using a vigilant strategy that is a non-fit with a promotion focus, or conflicts between strategies and tactics, such as implementing an eager strategy using a conservative tactic that is inappropriate for a given context. Conflicts can also exist *within levels* in a self-control hierarchy, which we refer to as *horizontal* self-control conflicts. A horizontal self-control conflict is any conflict that exists within a level in a self-control hierarchy—between goals, between strategies, between tactics, and between behaviors. It is also possible that the *same* conflict can be represented as *either* vertical or horizontal. A vertical conflict between the goal of achieving academic success versus the temptation of partying with friends could be reframed as a horizontal conflict between the two higher-order goals of connecting with friends versus investing in scholarship. The fact that the same conflict may be represented in different ways could have significant implications for interventions and treatments of conflict, a point we return to at the end of the chapter.

This approach makes clear that self-control involves managing conflicts at multiple levels: managing conflicts between and within the levels of behaviors, tactics, strategies, and goals. Individuals experience conflicts between managing multiple goals (Shah, Friedman, & Kruglanski, 2002). Individuals can experience conflict over the kinds of strategies they employ or protect (Higgins, 2000; Showers & Cantor, 1985). Individuals can experience conflicts about whether or not a given tactic serves an underlying strategy (Scholer, Stroessner, & Higgins, 2008). Success and failure can be measured not only in terms of control over one's behavior, but also in terms of control over one's goals, strategic orientation, or tactics. In this chapter, we take a broad view of defining self-control within the self-regulatory system by considering how conflicts and control are represented within a self-regulatory hierarchy.

HIERARCHIES OF SELF-REGULATION

Various theoretical approaches have emphasized in different ways the importance of differentiating among levels of self-regulation—between goals and subgoals (Miller, Galanter, & Pribram, 1960), between principles, programs, and sequences of movement (Carver & Scheier, 1998), between low and high levels of action identification (Vallacher & Wegner, 1985; 1987), between life-task goals, strategies, and plans or tactics (Cantor & Kihlstrom, 1987), between self-regulatory systems and strategies (Higgins, 1997; Higgins et al., 1994), and between temperaments, motive dispositions, goals, and behaviors (Elliot, 2006; Elliot & Church, 1997; Gable, 2006; Pervin, 1989; 2001). Although these approaches differ in their preferred terminology and in the number of distinctions they wish to make, they all emphasize the importance of recognizing that the levels of self-regulation are independent (at any lower level in the hierarchy, there are multiple means that can serve a higher level) and are defined by different concerns (e.g., defining goals, determining strategies, engaging in behaviors).

Although there are many different ways that we could explore these hierarchical approaches, we restrict our discussion to a simplified hierarchy composed of goals, strategies, tactics, and behaviors. These levels, commonly reflected in different self-regulatory hierarchies, allow for interesting comparisons of how control processes operate at different levels while setting useful constraints on our exploration. We begin by defining these levels, and then emphasize how control can be targeted at the conflicts that can occur both within and between the goal and strategic levels.

Goals serve as the end-states, standards or references points that guide behavior (*see* Kruglanski, 1996). Goals specify the domain of regulation ("physical fitness"), the motivational concerns of the individual (e.g., accomplishment, security), as well as telling us whether an individual is regulating in relation to a desired end-state (e.g., a goal to be fit and svelte) or undesired end-state (e.g., a goal to avoid being fat). Horizontal goal conflicts can occur between motivational concerns (e.g., between pursuing one's *aspiration* to be a rock star versus upholding one's *duty* to provide for one's new spouse), between reference points (e.g., between aiming for an A versus avoiding an F), or between life domains (e.g., performing well in *school* versus getting along with siblings at *home*).

Some of the ways in which individuals can target control at the level of goals are by focusing on high-level versus low-level construals (Fujita et al., 2006), linking goals to important motivational concerns (Elliot & Sheldon, 1999), choosing approach versus avoidance goals (Elliot, 2006; Elliot, Sheldon, & Church, 1997), or by shielding competing goals (e.g., Shah & Kruglanski, 2002). Such control has important implications for goal commitment and persistence—and ultimately, behavior. However, goals do not specify the strategic, tactical, or behavioral ways in which the goal's directional motivation may be playing out.

Strategies are the links between goals and tactics or behavior, reflecting the general plans or means for goal pursuit. For instance, there are different ways both of approaching desired end-states and of avoiding undesired end-states. Regarding the most common case of approaching desired end-states, individuals can employ, for example, eager approach strategies (moving toward the desired end-state by approaching matches to it) or vigilant approach strategies (moving toward the desired end-state by avoiding mismatches to it). Though less commonly discussed, there are also different ways of avoiding undesired end-states. To move away from an undesired end-state at the system level, one can either strategically approach mismatches or strategically avoid matches to it (see Higgins et al., 1994). Thus, strategies differ from goals because they are about the general means or process, rather than the endpoints. However, strategies differ from tactics or behavior because they are about the broad-level descriptions of the means ("Be eager!") rather than the more specific tactical instantiations (e.g., "Be risky—be willing to make a mistake to seize an opportunity for advancement.") or behavioral instantiations (e.g., "Say 'Yes!' when Sally asks you if you want to try that aerobic hula-hoop class.") of those means. At the level of strategies, horizontal conflicts can occur because of the trade-offs associated with different strategies. For example, employing an eager strategy leads to increased speed, whereas employing a vigilant strategy leads to increased accuracy (Förster, Higgins, & Bianco, 2003). Given that both speed and accuracy can be critical, the trade-offs associated with different strategies can create conflict.

The same strategy may be served by multiple tactics; tactics are the instantiation of a strategy in a given context and are about the means or process at a more concrete, in-context level (Cantor & Kihlstrom, 1987; Higgins, 1997). For example, one can protect and maintain a vigilant strategy by engaging in either risky or conservative tactics (Scholer et al., 2008). In the signal detection sense, these risky or conservative tactics are reflected in the bias for the acceptance threshold that one adopts. A bias toward a lenient or liberal criterion for acceptance is a tactic that maximizes hits (even at the cost of false alarms)—a so-called "risky" bias. In contrast, a bias towards a strict criterion for acceptance is a tactic that maximizes correct rejections (even at the cost of misses)—a so-called "conservative" bias (Swets, 1973).

Depending on the nature of the situation, either risky tactics or conservative tactics may better serve the underlying vigilant strategy (Scholer et al., 2008; Scholer & Higgins, 2008).

For example, when all is well, playing it safe tactically by adopting a conservative bias best serves a vigilant strategy because it minimizes the possibility of mistakes. However, when the context is negative or threatening, making a mistake (i.e., missing a negative signal) undermines strategic vigilance. In this context, strategic vigilance is served by doing anything necessary, including being tactically risky, to get back to safety. As with strategies, the use of a given tactic has both benefits (e.g., "If I'm risky, I won't miss any negative signals!") and costs (e.g., "If I'm risky, I may misclassify some positive signals as negative.") Thus, conflicts can occur horizontally resulting from the trade-offs between tactics; conflicts may also occur vertically between strategies and tactics. For instance, a vertical conflict can occur when an individual employs a conservative tactic that interferes with an eager strategy.

It is important to note that our notion of tactics is not simply synonymous with behavior. A risky tactic, for example, may result in different kinds of behaviors depending on what is being demanded or afforded in a given context. A risky tactic may be reflected in the behavior of adopting a liberal threshold for acceptance when a recognition judgment is demanded, but it may also be reflected in the behavioral preference for a risky choice over a sure thing when a gambling decision is demanded (e.g., Kahneman & Tversky, 1979). Thus, although the tactical level reflects a more concrete instantiation of the strategic level, even the tactical level is reflected more concretely in different specific behaviors.

In the remainder of this chapter, we consider in more detail how individuals exert control at different levels of the self-regulatory hierarchy in the service of effective self-regulation. We restrict our exploration to how individuals exert control at the goal and strategic levels of the self-regulatory hierarchy. Although we believe that the tactical and behavioral levels are also interesting and important, empirical research on conflicts and control targeted at the levels of goals and strategies is much more extensive at present than research on conflicts and control at the tactical and behavioral levels.

Indeed, researchers have not always measured or operationalized variables in ways that allow clear distinctions to be made between tactical versus behavioral control or even tactical versus strategic control. Consequently, given the nature of the existing literature, a discussion of tactical or behavioral control has the potential to muddy rather than clarify the waters. We do believe, however, that the waters can and should be clarified and that such clarifications represent exciting avenues for future research—a point we return to at the conclusion of the chapter.

Although we limit our primary discussion to control targeted at the levels of goals and strategies, there are still many rich questions and issues to explore: How do people target control to make a particular goal itself more effective? How do individuals protect, maintain, and enhance self-control strategies? We discuss different ways in which control is exerted both horizontally and vertically at these levels, how control at these levels affects behavior, and the implications of this approach for developing more effective interventions.

Goals and Control

There is no doubt that the construct of goals is central to issues of self-control. Much attention in the self-control literature has focused on how and when the processes of goal *pursuit* go awry or work effectively, but the goal-*setting* process is also critical (cf. Janis & Mann, 1977; Heckhausen & Gollwitzer, 1987). The types of goals that individuals pursue and the ways in which individuals juggle and manage multiple goals can have a profound impact on how self-control processes unfold. In particular, some types of goals may result in better outcomes than others because of the reference point guiding regulation (e.g., Carver & Scheier, 1990; Higgins, 1997), the motivational concerns they reflect (Deci & Ryan, 1985; Sheldon & Elliot, 1998; 1999), or their relation to other accessible goals (Shah, Friedman, & Kruglanski, 2002; Shah & Kruglanski, 2002). This section reviews various ways that control can be exerted at the level of goals to produce more effective self-regulation. In particular, we focus on the types of horizontal conflicts that have often been overlooked by traditional conceptualizations of self-control conflicts.

We want to begin by clarifying how we envision the difference between control exerted at the level of goals versus control at the level of behavior. One way that we have found helpful for differentiating what it means to target control at different levels of the self-regulatory hierarchy is to consider how success in exerting control can be measured or evaluated at each level. At the level of behavior, success is measured by whether the target behavior was achieved or not: Did I finish that chapter on time? Did I eat that brownie? At the level of goals, success is measured by whether the goals themselves are regulated effectively: Did I pursue an approach or avoidance goal? Did I pursue a self-concordant or externally controlled goal? Of course, ultimately success is measured by whether the behavior is regulated—whether the goal is obtained or not. However, given that control exerted at the level of goals can have a significant impact on behavior, it is important not only to consider behavioral successes and failures, but also to consider self-control successes and failures at other levels in the hierarchy. Doing so makes clear just how many different ways in which individuals can more effectively engage in self-control.

We also want to make clear that when we suggest that control can be exerted at different levels in the self-regulatory hierarchy, we do not mean that this control is necessarily conscious. Goals can be activated outside of awareness and pursued in a manner that resembles conscious goal pursuit (e.g., Chartrand & Bargh, 2002; Fitzsimons & Bargh, 2003; Hassin, Bargh, & Zimerman, 2009; see also Moskowitz, 2002; cf. Oettingen et al., 2006). Individuals can exert control and can engage in self-regulation even when they don't intend to; control does not need to be intended, within awareness, or demanding of cognitive resources to be effective. Recent evidence also suggests that nonconscious goal pursuit, like conscious goal pursuit, allows for the flexibility that may be critical for effective self-control (Hassin et al.). However, as we will discuss later, it may sometimes be helpful to bring some of these processes into awareness for engaging in effective interventions.

Managing Conflicts between Multiple Goals

The pursuit of even a single goal is challenging. Donald, on a diet once again, faces a temptation as he strolls by the big plate of cookies next to the office water cooler. "Damn those office birthday parties," he thinks. "There goes my diet!" Walking by the tempting cookies next to the water cooler, in and of itself, would be a self-control triumph for our dear Donald. Alas, Donald has many other goals to juggle also (getting that report to his boss on time, controlling his temper when his wife forgets their anniversary, expressing appreciation to his co-worker Betty for taking the time to bake the cookies that so vex him), and the presence of these other goals can reduce Donald's commitment to his diet goal. At the level of goals, conflicts often exist between the pursuit of many varied goals. Indeed, one of the primary challenges (and opportunities) for exerting control at the level of goals is in the management of multiple goals. Research suggests that the activation or accessibility of alternative goals can have a variety of impacts on the pursuit of the focal goal, even when individuals are not consciously aware that an alternative goal has been activated (Shah & Kruglanski, 2002).

If the alternate goal is unrelated to the focal goal, its activation can pull resources away from the focal goal, reducing commitment, persistence, the development of novel means, and the emotional intensity of success or failure feedback related to the focal goal (Shah & Kruglanski, 2002). In other words, if Donald is reminded of an upcoming deadline just as he approaches the water cooler, the activation of this alternative unrelated goal may make it more difficult for Donald to refuse the temptation of Betty's cookies. In contrast, if an alternative goal is facilitatively related to the focal goal, it can increase persistence, performance, and the emotional intensity of success or failure feedback.

The activation of the focal goal can also shield individuals from other goals that lure away self-regulatory attention and resources (Shah, Friedman, & Kruglanski, 2002). When a focal goal is activated (e.g., "reading more books"), it inhibits the accessibility of alternative subordinate goals that are substitutable for the superordinate end (e.g., "reading more fiction books" and "reading more non-fiction books" both serve the superordinate end of reading more books). In contrast, alternative goals that *facilitate* progress towards the superordinate end, but are not substitutable, are not inhibited to the same degree by the activation of the focal goal (Shah, Friedman, & Kruglanski, 2002). Not all individuals in all situations engage in the same degree of automatic inhibition of these alternative goals. Individuals who are more strongly committed to the focal goal or who are high in need for closure (Webster & Kruglanski, 1994) are more likely to inhibit alternative goals. Additionally, the regulatory concern underlying the goal (whether goals are perceived as prevention goals whose attainment is necessary or promotion goals whose attainment is an aspiration) can also impact the extent to which intergoal inhibition occurs (Shah et al., 2002). The more goals are viewed as duties/responsibilities, the more likely they are to inhibit other goals. The more goals are viewed as ideals, the *less* likely they are to inhibit other goals.

These findings suggest that control can be targeted at horizontal conflicts at the level of goals by modifying the relations between goals, by targeting the regulatory concern underlying a given goal, or by targeting the likelihood of alternative goal activation. For instance, alternative goals that could draw resources away from some focal goal could be reframed in ways that make it less likely that they will pull resources away from the primary pursuit. Control can also be exerted more actively and directly by increasing the possibility that facilitative, non-substitutable alternative goals get activated in conjunction with the focal goal to bolster self-regulatory resources.

Although it is likely that much of the control exerted in the management of multiple goals occurs automatically, individuals may also be able to consciously and intentionally modify and manipulate intergoal relationships. Eventually, such intentional control may become over-learned and applied automatically (cf. Shah et al., 2002). Specifically, goals may be linked to certain situations (e.g., Bargh et al., 2001) or social relationships (Fitzsimons & Bargh, 2003; Shah, 2003) in ways that automatically facilitate effective and flexible (Hassin et al., 2009) juggling of multiple pursuits.

Regulatory Reference and Control

Control at the level of goals can be exerted not only in the management of multiple goals but also in the management of what types of individual goals are adopted. One of the most fundamental distinctions that has been made at the level of goals is whether individuals are regulating in relation to desirable end-states or reference points or undesirable end-states or reference points (e.g., Carver & Scheier, 1981, 1990, 1998; Elliot & Thrash, 2002; Freud; 1920/1952; Gray, 1982; Lang; 1995; Lewin, 1935; Miller, 1944; Mowrer, 1960). Whether individuals are regulating in relation to desired or undesired end-states corresponds to the idea of *regulatory reference* (Higgins, 1997) and reflects the standard that individuals are using to evaluate performance for progress. For example, individuals can have the goal to finish the marathon (a desired end-state) or can have the goal to avoid dropping out of the marathon (an undesired end-state). When individuals are regulating in relation to a desired end-state, they either attempt to minimize the discrepancy between the current state and that desired end-state (Carver & Scheier, 1981, 1990) or attempt to approach matches or avoid mismatches to that desired end-state (Higgins, 1997). Similarly, when individuals are regulating in relation to an undesired end-state, they either attempt to amplify the discrepancy between the current state and the undesired end-state (Carver & Scheier, 1981, 1990) or attempt to avoid matches and approach mismatches to the undesired end-state (Higgins, 1997). Consequently, one type of horizontal conflict that can exist at the level of goals is the tension between adopting a desired end-state versus an undesired end-state as a regulatory reference point. What are the implications for self-control of having these different reference points?

There is mounting evidence to support the idea that all else being equal, individuals who pursue approach goals are better off than individuals who pursue avoidance goals. Individuals who pursue approach goals rather than avoidance goals have higher subjective well-being (Elliot, Sheldon, & Church, 1997) and report fewer physical illness symptoms (Elliot & Sheldon, 1998). A number of suggestions have been posited to explain why undesired end-states as standards might be more problematic than desired end-states as standards. Avoidance goals might be more likely to elicit anxiety or to increase feelings of threat as they remind an individual of the undesired state to be avoided, undermining effective self-regulation (Elliot & Sheldon, 1997; Elliot & Church, 1997; Hembree, 1988; Speilberger & Vagg, 1995). Carver and Scheier (1981) have also suggested that avoidance goals may be partly problematic in that they only specify what to move away *from* but not what to move *toward* (cf. Skinner, 1971; *see also* Higgins, 1997, for a discussion of this issue). It should be noted, however, that the problems with avoidance goals may depend, in part, on the strategies that individuals employ in reference to them—specifically, whether they are avoiding matches or approaching mismatches to the undesired end-state.

Avoidance goals not only lead to reduced well-being, but have also been linked to specific problems in self-regulatory processes. Avoidance goals have been shown to lead to greater disorganization (Elliot & McGregor, 2001), procrastination (Howell & Watson, 2007), and to poorer performance (Elliot & McGregor, 2001). Avoidance goals also lead to perceptions of poorer progress relative to approach goals (Elliot, Sheldon, & Church, 1997), decreased feelings of competence, and increased perceptions that goals are externally or internally pressured or demanded, the concept of *controllingness* (Elliot & Sheldon, 1997). Thus, at the level of goals, it seems that one way that individuals can target control is by adopting approach, rather than avoidance goals.

However, the story at the level of goals is not all about approach and avoidance. Although there is certainly evidence that individuals should choose their reference points wisely, there are a number of other concerns, orthogonal to approach and avoidance, that also have an important impact on regulation at the level of goals, such as whether individuals are pursuing learning or mastery goals versus performance goals (Dweck, 1999; Elliot & McGregor, 2001), whether goals reflect extrinsic or intrinsic concerns (Deci & Ryan, 1985; Kasser & Ryan, 1996), whether goals are authentic to enduring interests and values (Elliot & Sheldon, 1999), or whether goals reflect promotion focus or prevention focus concerns (Higgins, 1997). For example, within the achievement goal domain, Elliot & McGregor (2001) have shown that not all avoidance goals are the same in terms of their negative effect on self-regulation. Although both mastery-avoidance and performance-avoidance goals lead to disorganization, increased test anxiety and emotionality, only performance avoidance goals actually lead to poorer performance and more health center visits (Elliot & McGregor; Elliot & Church, 1997). This kind of empirical data suggests the importance of considering how control can be targeted at multiple concerns at the level of goals to impact self-regulation. The next sections focus on how control in relation to self-concordance and regulatory focus concerns affects self-control processes at the level of goals.

Self-Concordance and Control

Self-concordance is the extent to which goals authentically reflect an individual's enduring interests and values (Sheldon & Elliot, 1998, 1999). Not all goals are self-concordant; some goals are *controlled* by external constraints and pressures, by internally demanding introjects, or by fleeting and impulsive desires. When individuals pursue goals that are "authentic," arising from intrinsic or identified motivation (Deci & Ryan, 1985), the goals have an internal perceived locus of causality and the individual is believed to "feel ownership as he or she pursues the goal" (Sheldon & Elliot, 1999, p. 483).

The self-concordance model has focused on the impact of *pursuing* self-concordant goals on subjective well-being and regulatory processes, rather than on the goal selection process that leads individuals to select self-concordant versus controlled goals. In particular, the model suggests that because self-concordant goals represent enduring values and interests, they are more likely to be pursued persistently over time, especially when the inevitable obstacles and challenges arise (Sheldon & Elliot, 1999).

Indeed, self-concordant goals have been shown to result in higher goal attainment than goals that are externally controlled or goals that reflect introjected demands (Sheldon & Elliot, 1998, 1999). The higher goal attainment achieved with self-concordant goals appears to be mediated by sustained effort throughout the goal pursuit process. Although initial investments of effort do not differ between self-concordant versus controlled goals, in longitudinal datasets self-concordance has been linked to continued effort, independent of the effects of self-efficacy, implementation intentions, avoidance goals (Sheldon & Elliot, 1999), or intrinsic/extrinsic goal content (Sheldon et al., 2004).

Not only are self-concordant goals pursued more effectively but, when attained, they also lead to higher subjective well-being (Sheldon et al., 2004) and increased experiences of competence, autonomy, and relatedness (Sheldon & Elliot, 1999). Furthermore, because individuals who pursue self-concordant goals put more effort into pursuing them and are more likely to achieve their goals, they are likely to have higher self-efficacy, leading to increased self-regulatory effectiveness in general (Bandura, 1977; 1997).

It is interesting to consider again the classic self-control conflict between a higher-order goal versus a lower-order temptation in the context of our discussion of regulatory reference and self-concordance. For an individual whose higher-order concern is her social relationships, the self-control conflict between investing in one's close relationships versus staying at home to study may be more successfully resolved if the "higher-order" concern (the priority of connectedness to close others) is represented as a high-level construal (e.g., Fujita et al., 2006).

However, the conflict can also be represented as a horizontal conflict at the level of goals that can be resolved in a number of other ways as well. For example, the literature we have reviewed suggests that it may also be more successfully resolved if the goal for connectedness is represented as an approach goal, or if the goal for connectedness is seen as autonomous, rather than arising from external or introjected demands.

The evidence we have reviewed thus far, describing both the approach/avoidance distinction and the self-concordant/controlled distinction, suggests one way to think about control at the level of goals: certain types of goals are better than others for effective self-regulatory control. Approach goals and self-concordant goals not only lead people to feel greater subjective well-being, but do so because they facilitate more effective goal pursuit. Generally speaking, approach goals are better than avoidance goals, and self-concordant goals are better than controlled goals. In the next section, however, we would like to explore the idea that control at the level of goals is not just about whether a certain type of goal is better *in general* for effective goal pursuit, but is also about whether some types of goals are better than others given certain self-regulatory demands or given certain desired outcomes.

Regulatory Focus and Control

Building on earlier distinctions (e.g., Bowlby, 1969, 1973; Higgins, 1987; Mowrer, 1960), regulatory focus theory distinguishes between two co-existing regulatory systems or orientations that serve critically important but different survival needs (Higgins, 1997). Individuals can regulate in the promotion or the prevention system, a distinction orthogonal to whether individuals are regulating in relation to desired or undesired end-states. The promotion orientation regulates nurturance needs and is concerned with growth, advancement, and accomplishment. Individuals in a promotion focus are striving towards ideals, wishes, and aspirations and are particularly sensitive to the presence and absence of positive outcomes (gains and non-gains). In contrast, the prevention orientation regulates security needs. Individuals in a prevention focus are concerned with safety and responsibility and with meeting their oughts, duties, and responsibilities, and they are particularly sensitive to the absence and presence of negative outcomes (non-losses and losses). Regulatory focus theory is orthogonal to the earlier distinction between approaching desired end-states and avoiding undesired end-states because at the level of goals, promotion and prevention orientations *both* involve approaching desired end-states (e.g., approaching accomplishment or safety, respectively) and avoiding undesired end-states (e.g., avoiding nonfulfillment or danger, respectively).

Self-control involves unavoidable trade-offs (Higgins, 1998); this is particularly evident with regards to the promotion and prevention systems. Because the underlying concerns of the promotion and prevention systems differ, each system is uniquely equipped to deal effectively with different kinds of self-regulatory challenges. Neither system is better than the other; they simply have different strengths and result in different kinds of vulnerabilities (Higgins, 1997). Thus, another type of horizontal conflict that can exist at the level of goals is between promotion versus prevention concerns. This section explores how control can be targeted in the pursuit of promotion versus prevention goals to both influence self-regulatory effectiveness on different kinds of tasks and to more effectively regulate emotions.

Regulatory Focus and Task Demands

Imagine that Donald's boss has just told him that he has 24 hours to come up with a new slogan for the company's flagship product. The situation presents a number of commonly faced self-regulatory challenges (i.e., Donald has to find a way to initiate the task, sustain engagement in the task, and perform well) that may involve conflicts at the level of goals. Based on the literature review from the prior sections, Donald might do well to adopt an approach goal ("I'm going to come up with the best slogan this company has ever seen!") rather than an avoidance goal ("I'm going to make sure I don't come up with a slogan that stinks."). Donald might do even better if the goal is self-concordant rather than controlled ("I want to do well on this assignment because it serves my enduring value of creativity." *vs.* "I want to do well on this assignment because my boss told me I had to."). Beyond either of these considerations, however, there is evidence that Donald would also be served well by adopting a promotion-focused, rather than prevention-focused, goal (Crowe & Higgins, 1997; Friedman & Förster, 2001), as promotion-focused concerns are more relevant to creativity. Friedman & Förster (2001) have found that individuals in a promotion focus exhibit more creative insight and creative generation than individuals in a prevention focus, in part because being in a promotion focus encourages eager strategies and leads to a memory search for more novel responses.

In contrast, if Donald's boss assigned him the task of checking the page proofs of the product manual, Donald might do better with a prevention-focused goal (Förster et al., 2003). Individuals in a prevention focus are vigilant against mistakes; they have a greater motivational investment in being accurate than individuals in a promotion focus. When a task emphasizes accuracy, an individual in a prevention focus should perform better than an individual in a promotion focus. Indeed, Förster et al. have found that individuals in a prevention focus perform more accurately on a proofreading task than individuals in a promotion focus; furthermore, as prevention-focused participants approach task completion, their greater accuracy than promotion-focused participants increases.

These data suggest another important way in which control can be targeted at the level of goals: whether individuals adopt a promotion- or prevention-focused orientation. Although individuals differ in the chronic strength of the promotion or prevention systems (Higgins et al., 2001; Higgins, Shah, & Friedman, 1997), a promotion or prevention orientation can also be temporarily or situationally primed, increasing the possibility of regulatory flexibility in the use of these systems for different kinds of task demands. If individuals are aware of which system is better suited for a task, they may be able to intentionally increase the strength of one system to more effectively exert self-control. The association between a given system (e.g., promotion) and a given task demand (e.g., being creative) may also become automatically linked, such that being in given situation automatically activates promotion or prevention concerns, leading to more effective self-regulatory control.

Regulatory Focus and Emotions

Regulatory focus concerns can impact the effectiveness of self-regulatory processes resulting from a match between the underlying goal orientation and the demands of the task. Additionally, whether an individual is in a promotion or prevention focus also has important implications for how self-regulatory success or failure is *affectively* experienced. Success in a promotion focus reflects the presence of a gain (a positive outcome) and results in happiness and satisfaction, cheerfulness-related emotions. Success in a prevention focus, in contrast, reflects the absence of a loss (absence of a negative outcome) and results in peacefulness and calm, quiescence-related emotions. Similarly, failure in a promotion focus reflects a non-gain (the absence of a positive outcome) and results in dejection-related emotions like sadness and disappointment, failure in a prevention focus reflects a loss (the presence of a negative outcome) and results in agitation-related emotions like anxiety and worry (Higgins, 1997; Shah & Higgins, 2001). In support of these predictions, several studies have found that priming ideal (promotion) discrepancies leads to increases in dejection whereas priming ought (prevention) discrepancies leads to increases in agitation (Higgins et al., 1986; Strauman & Higgins, 1987). Furthermore, individuals in a promotion-focus are faster at appraising how cheerful or dejected a given object makes them feel, whereas individuals in a prevention focus are faster at appraising how quiescent or agitated an object makes them feel (Shah & Higgins, 2001).

Both prevention- and promotion-focused individuals experience contentment and a sense of well-being when they successfully attain a goal. However, the contentment achieved by the success of a prevention-focused goal is very different than the contentment achieved by the success of a promotion-focused goal. Consequently, one major implication of the relationship between regulatory focus concerns and emotional responses to success and failure is that individuals who have a goal to be happy (in the sense of joyful and ebullient) should set promotion goals. Setting promotion goals, however, will make individuals vulnerable to sadness and depression if they fail to meet them.

Self-system theory (SST) is a recently developed structured psychotherapy to treat the depression that is associated with individuals who have chronic promotion goals and are failing (Vieth et al., 2003). SST is based on the hypothesis that "chronic or catastrophic failure to meet promotion goals is a contributory causal factor in the onset and maintenance of depressive episodes for individuals with a promotion focus" (Vieth et al., 2003, p. 249). As part of the educational and goal-setting aspects of the theory, therapist and client discuss the implications of the different concerns of the promotion and prevention systems. Interestingly, the authors describe a case study in which the client was under the mistaken impression that attaining prevention goals would lead to the happiness and satisfaction that she dearly wanted. Part of the usefulness of the therapy for her (and for other clients) appears to be learning that succeeding or failing at promotion versus prevention goals has distinct emotional consequences.

All of this suggests that being able to target control at the system level (i.e., adopting promotion- versus prevention-related goals) could lead to more effective self-control because of its emotion-regulation impact. Individuals often have beliefs or hopes about how achieving a particular goal will make them feel (e.g., "If I can go to the gym three times this week, I'll feel really happy."). Successful goal attainment is one way in which individuals boost future self-control successes; self-efficacy and motivational engagement are likely to increase (Bandura 1977, 1997; Veith et al., 2003). However, to the extent that individuals experience successful self-control as being about *both* achieving the desired end-state (e.g., going to the gym three times a week) *and* achieving the desired affective state (e.g., happiness), the impact of self-control successes may be undermined if individuals expect promotion-related emotions from prevention successes (and vice versa). This is a form of emotion regulation fundamentally different than emotion-regulation strategies that are targeted at altering a given emotional response (e.g., Gross, 1998); it is about exerting emotional control at the level of the goals one adopts.

Another way in which individuals appear to engage in controlling emotions at the level of goals was reported by Shah et al. (2003) in their work on goal shielding theory. In general, depressed participants show decreased goal-shielding (decreased inhibition of alternative goals); they are less effective at exerting control at the level of goals. However, what if the focal goal has the potential of alleviating the depression if it were attained? Shah et al. found that when participants were depressed, they did show greater inter-goal inhibition for one type of focal goal—*ideal* goals—the goals that would result in happiness and satisfaction if obtained. Similarly, when participants were anxious, they selectively showed greater inter-goal inhibition when the focal goal was an *ought* goal. In other words, participants were more likely to shield and protect a goal that could alleviate their emotional stress if it were attained—emotional control enacted at the level of goals.

Goals and Control: A Brief Summary

The research reviewed in this section suggests that there are numerous horizontal conflicts that can exist at the level of goals. Targeting control at these conflicts is one significant way to increase the effectiveness of self-control. Through the management of multiple goals, the adoption of specific types of goals (e.g., approach goals or self-concordant goals) or motivational orientations underlying goals (e.g., promotion or prevention), individuals can engage in more effective self-control. Control at the level of goals is just one way that self-regulatory control can play out in the hierarchy; goals are linked to specific tactics and behaviors through the strategies that individuals adopt. In the following section, we discuss some of the ways in which *strategic* control is targeted and managed—both horizontally (conflicts between strategies) and vertically (conflicts between goals and strategies).

STRATEGIES AND CONTROL

Strategies are the links between goals and behavior, reflecting the plans or means that individuals use to pursue their goals. Conflicts can occur horizontally between competing strategies, such as competing strategies for coping with stress or competing strategies for attaining a goal. To deal with such conflicts, individuals can target control at the strategic level to employ a particular strategy (e.g., problem-focused coping) that is effective for a specific situation (e.g, a controllable stressor) or a particular strategy (e.g., finding friends who listen but do not encourage) that is effective for their specific orientation to a task (e.g., defensive pessimists preparing for an exam; Cantor, 1986a, 1986b, 1994; Showers, 1992). As discussed in more detail later in this section, conflicts can also occur vertically between strategies and goal orientations (e.g., when a promotion-focused employee is assigned to use a vigilant strategy).

Individuals can experience horizontal conflicts between competing strategies when selecting the optimal approach for coping with a particular type of stressor. One way in which control can be exerted is by selecting strategies that are compatible with the type of stressor or self-regulatory challenge that is being confronted (Aldwin, 1994; Cantor, 1994; Carver & Scheier, 1994; Linville & Clark, 1989; Taylor & Aspinwall, 1996). The distinction most often discussed in the stress and coping literature is between strategies that are directly aimed at altering the stressful situation (problem-focused strategies) versus strategies that are aimed at managing emotional responses (emotion-focused strategies) (e.g., Carver, Scheier, & Weintraub, 1989; Lazarus & Folkman, 1984). Such strategies can be employed in the management of more extreme stressors, but they are also relevant to a discussion of more mundane, everyday stressors and challenges that are commonly considered in the domain of self-regulation.

Although problem-focused strategies are generally associated with less emotional distress than emotion-focused strategies (e.g., Penley, Tomaka, & Wiebe, 2002), the effectiveness of the strategy seems to also depend on the controllability or mutability of the stressor or challenge. Indeed, better adjustment seems to be associated with coping strategies that match the controllability of the stressor (e.g., Cheng; 2003; Compas, Malcarne, & Fondacro, 1988). For example, Compas et al. (1988) found that children who generated problem-focused strategies in response to controllable stressors did better in terms of emotional and behavioral adjustment than those who generated problem-focused strategies in response to uncontrollable stressors. In general, individuals who are able to flexibly adopt different strategies for the demands of different situations experience less distress and more effective regulation (Cheng, 2003; Chiu et al., 1995).

The compatibility between a given strategy and a self-regulatory challenge is not the only type of "fit" that is important and that can be targeted for control. Another important way in which individuals can adopt more effective strategies is elucidated by the principles of regulatory fit (Higgins, 2000). The remainder of this section reviews evidence from the regulatory fit and defensive pessimism literatures that suggests how strategic control can be exercised in the pursuit of effective resolution of both horizontal conflicts within the strategic level and vertical conflicts between goals and strategies.

Regulatory Fit

Regulatory fit theory (Higgins, 2000) posits that individuals derive value from using strategic means that fit their underlying regulatory orientations. As we have discussed, one way to differentiate between types of strategic means is to recognize that people may use either approach strategic means or avoidance strategic means. Within the context of regulatory focus theory, approach strategic means (eager means) "fit" a promotion focus whereas avoidance strategic means (vigilant means) "fit" a prevention focus. When individuals experience regulatory fit by using strategic means that sustain their underlying orientation, they "feel right" about what they are doing (Higgins, 2000) and also experience increased engagement (Higgins, 2006). Regulatory fit has been shown to not only affect the value of the goal pursuit activity but also to affect the value of subsequent object appraisals (Higgins et al., 2003) and to lead to better task performance (Shah, Higgins, & Friedman, 1998). Although regulatory fit theory is not restricted to fit between regulatory focus orientations and means (e.g., Avnet & Higgins, 2003; Bianco, Higgins, & Klem, 2003), we focus on research that has explored the implications of pursuing promotion or prevention goals using eager or vigilant strategies.

Both horizontal and vertical conflicts may arise in the context of regulatory fit theory. When an individual is deciding between strategies (e.g., between an eager vs. vigilant strategy), the conflict is horizontal. However, when an individual employs a strategy that is a nonfit with their goal orientation, the conflict is vertical. Individuals may experience vertical conflicts when situational constraints require that they use a non-fitting strategy (e.g., a prevention-focused worker instructed to use an eager strategy by their manager). Individuals may themselves select non-fitting strategies. Some individuals may be more skilled than others at discerning whether a given strategy is appropriate for their current orientation (cf. Chiu et al., 1995). Some individuals may also experience a chronic vertical conflict resulting from receiving nonfitting self-regulatory messages while growing up, such as learning from Dad to have a promotion orientation toward goals while learning from Mom to use vigilant strategies in goal pursuit. Regarding this latter possibility, as individuals grow up they are exposed to many different messages from parents, siblings, teachers, peer groups, the media, and so on. People are not only socialized for particular goal orientations (e.g., caretaker interactions communicating that what matters is fulfilling hopes and aspirations), but are also socialized for particular strategies, tactics, and behaviors.

Consequently, someone could experience a chronic vertical conflict by learning a particular kind of goal orientation (e.g., promotion) from one socialization agent but being trained to use a nonfitting goal pursuit strategy (e.g., vigilance) by a different socialization agent. In sum, it is clear that strategic conflicts can occur at multiple stages of self-regulation—when individuals select among alternative strategies (potential for horizontal conflicts) and when individuals pursue a goal using a nonfitting strategy (vertical conflict). In this section, we review some of the implications of being in a state of fit versus nonfit, where that state could relate to either a horizontal or a vertical conflict.

One type of self-regulatory goal that individuals often adopt is to perform well on some endeavor. Shah, Higgins, & Friedman (1998) found that performance on an anagram task is better under conditions of fit versus nonfit. For participants with a chronic promotion focus, strategic approach framing led to better performance. In contrast, for participants with a chronic prevention focus, strategic avoidance framing led to better performance. Not only do individuals perform tasks more effectively under fit, but Freitas & Higgins (2002) further found that under fit individuals reported greater task enjoyment and greater subjective perceptions of success.

These studies also highlight an important distinction between the benefits and costs of approach and avoidance at the system level versus the benefits and costs of approach and avoidance at the strategic level. Although numerous studies have found support for the idea that approach goals result in better outcomes than avoidance goals (e.g., Elliot & Harackiewicz, 1996; Elliot & Sheldon, 1997, 1998; Elliot, Sheldon, & Church, 1997), this may be primarily with regards to approach and avoidance at the level of *goals*. At the strategic level neither approach nor avoidance strategies per se produce better outcomes overall (in terms of performance or subjective well-being). Rather, value at the strategic level may be derived more from a fit between the strategic means and one's regulatory orientation, rather than from a main effect of approach versus avoidance. In other words, successful resolution of the vertical conflict may be particularly critical.

This distinction between the *effects of avoidance at the level of goals* versus the *effects of avoidance at the level of strategies* provides a clear demonstration of why it is useful to distinguish between levels of self-regulation. If these levels were not independent, it would be hard to reconcile why avoidance sometimes leads to unhappiness (e.g., Elliot, Sheldon, & Church, 1997) and sometimes leads to enjoyment (Freitas & Higgins, 2002). However, when one considers the hierarchy of self-regulation that distinguishes avoidance goals from avoidance strategies, then the apparent contradiction in these findings is resolved.

Spiegel, Grant-Pillow, and Higgins (2004) extended the test of regulatory fit theory to the regulation of real-life health behaviors. In one study, for example, participants were told that they were participating in a two-session experiment to track the nutritional habits of college students. In the initial session, all participants were given health messages that advocated pursuit of the same desired end-state—eating more fruits and vegetables. The key manipulations took place as part of the messages that participants received. Although all participants were given the same message ("eat more fruits and vegetables"), a promotion versus prevention focus was manipulated through the concerns (accomplishments vs. safety, respectively) that were highlighted within the messages. Additionally, within each regulatory focus condition, participants were asked either to imagine the benefits they would get if they complied with the health message (strategic approach) or the costs they would incur if they didn't comply with the health message (strategic avoidance). Participants were asked to keep a nutritional log for the following week; the critical dependent variable was whether participants in conditions of fit would consume more fruits and vegetables than participants in conditions of non-fit. Indeed, participants in conditions of fit ate more fruits and vegetables in the week following the first session than participants in conditions of nonfit.

Grant et al. (2008) further investigated the impact of regulatory fit on the regulation of daily life problems. In a daily diary study, participants reported the "most upsetting or bothersome incident" that happened to them each day. Participants also completed a daily measure of distress as well as a strategic coping measure that assessed the use of eager strategies and vigilant strategies for coping. Prior to beginning the diary portion of the study, all participants completed the regulatory focus questionnaire (Higgins et al., 2001), a chronic measure of regulatory focus that assesses the extent to which participants believe that they have achieved success within the promotion and prevention systems.

Using the principles of regulatory fit theory, Grant et al. (2008) predicted that regulatory fit would increase a strategy's effectiveness by influencing the extent to which individuals would "feel right" about whatever coping strategies they used, thereby directly reducing the experience of distress. Specifically, they predicted that on days when participants used more coping strategies that fit their underlying orientations, they would experience less distress. Grant et al. also predicted that there would be a significant impact of nonfit: on days when participants used more coping strategies that did not fit their underlying orientations, they would experience *more* distress.

At the level of goals, all participants were avoiding an undesired end-state (daily stressors). Support for the regulatory fit predictions would provide evidence that there are different strategic ways that individuals can avoid undesired end-states. Both predictions were supported. Promotion-focused individuals who used eager (approach) coping strategies to deal with undesired end-states experienced *less* distress than promotion-focused individuals who used vigilant (avoidant) coping strategies. Similarly, prevention-focused individuals who used vigilant strategies reported less distress than prevention-focused individuals who used eager coping strategies. It is important to note that there was no main effect of promotion or prevention pride on distress; chronic regulatory orientation did not affect reactivity to stress. What was critical was the successful resolution of conflicts within the level of strategies and between the levels of goals and strategies in the way in which individuals coped with daily stressors. Although there was some evidence in this study that approach strategies generally led to less distress than avoidance strategies, it was the fit or nonfit between participants' underlying orientations and the strategy employed that was especially important for well-being.

Defensive Pessimism

Research on defensive pessimism also highlights individual differences in strategic control. Cantor and colleagues (Norem & Cantor, 1986a, 1986b; Showers, 1992) have identified different strategies that people use to cope with task pursuit when the possibility of failure exists. This work highlights not just differences in *how* people regulate strategically, but also highlights differences in the temporal unfolding of strategic regulation. Work on defensive pessimism suggests that some individuals actually have better outcomes when they adopt a negative, rather than a positive, outlook for anticipated events (Norem & Cantor, 1986a, 1986b; Showers, 1992). Defensive pessimists "expect the worst" when entering a new situation, despite the fact that they generally do not perform differently than those with a more optimistic outlook (Cantor & Norem, 1989; Cantor et al., 1987). In contrast to "true" pessimism, defensive pessimism serves two goals—a self-protective goal of preparation for possible failure in the future and a motivational goal of increasing effort in the present to prevent negative possibilities. Defensive pessimists experience potential conflict between their preferred vigilant strategy and a more eager and optimistic strategy that would be warranted by their past successes. Indeed, Norem & Cantor (1986b) found that simply pointing out the inconsistency between their current less-positive expectations and their more positive past performance disrupted their ability to harness vigilance in their preferred way. When the strategic coping mechanisms of defensive pessimists were disrupted in this manner, they performed more poorly.

In another research program, Showers (1992) has demonstrated that it is not only anticipated performance failure but more general anticipation of negative outcomes that appears to be important to defensive pessimists' strategic response to an upcoming task. For these studies, participants were selected based on chronic defensive pessimism or optimism scores in social situations. When participants arrived for the study, they were told that they would be having a "get acquainted" conversation with another participant, a social situation that has the potential to go badly. Prior to the conversation, some participants filled out a questionnaire that highlighted the possibilities for positive outcomes in the upcoming discussion and other participants filled out a questionnaire highlighting negative outcome possibilities. Showers (1992) found that defensive pessimists in the "negative possibilities" condition exhibited more positive behaviors during the social interaction than did defensive pessimists in the "positive possibilities" condition—they talked more, exerted more effort, and the conversations were rated more positively by the confederates with whom they were interacting. Another study found that defensive pessimists in the negative (versus positive) possibilities condition reported feeling more prepared for the interaction. It appears that the act of reflecting on possible negative outcomes is a critical component of the defensive pessimists' way of coping with possible failure. In contrast, such negative reflection can interfere with the performance of those individuals who are optimists (*see* Norem & Illingworth, 1993).

Flexible Control of Strategies

Exploring the nature of strategic control highlights the importance of flexibility in self-regulation. To exert control at the strategic level most effectively, strategies must be adapted to the changing demands of self-regulatory challenges and particular situations. Furthermore, strategies that "fit" an individual's underlying motivational orientation lead to increased engagement and self-regulatory payoffs; for example, there is recent evidence that regulating in a state of fit makes individuals less depleted (Hong & Lee, 2008) and better able to confront subsequent regulatory challenges. In other words, successful resolution of horizontal conflicts between strategies and vertical conflicts between goals and strategies may facilitate successful resolution of subsequent conflicts.

As we have reviewed, however, individuals do not always successfully manage conflicts between strategies or between strategies and goals. Situational constraints or cognitive or motivational limitations may make it difficult for individuals to always select (consciously or unconsciously) the most effective strategy. Precisely because the levels of the self-regulatory hierarchy are dissociable, individuals pursuing promotion goals will not always use eager strategies (e.g., Grant et al., 2008). In the final section, we discuss possibilities for intervention given our conception of control within the self-regulatory hierarchy. How can the principles of the self-regulatory hierarchy be harnessed to more creatively exert self-control? How and at what levels is control most effectively exercised?

Targeting Control in the Hierarchy

In this chapter we have explored how the construct of control operates within the self-regulatory hierarchy. In particular, we have focused on self-control conflicts at the levels of goals and strategies. We have examined how control operates at the level of goals (resolving conflicts between goals), at the level of strategies (resolving conflicts between strategies) and at their vertical intersection (making one's strategies "fit" one's goals). Although control can also be directly targeted at the levels of tactics or behavior that serve these goals and strategies, we have outlined how the concept of self-control can be conceived of more broadly to include both horizontal and vertical control directed at higher levels in the self-regulatory hierarchy. In doing so, we have expanded the "classic" conceptualization of self-control as the triumph of higher-order goals over lower-order temptations to self-control as the triumphal resolution of both horizontal and vertical conflicts at all levels of the self-regulatory hierarchy.

Thinking about control in this way highlights an obvious, but important point: at any lower level of the hierarchy, multiple ways exist to serve an upper level. Strategies serve underlying goals; tactics and behaviors serve underlying strategies. Thus, there is great flexibility in how control can be targeted. Furthermore, it may be possible to use the independence of the levels to target control at the level that is most malleable in a given context or individual. For example, if an individual is strongly committed to a particular strategy, it may be most effective to work on changing or modifying maladaptive tactics that serve that strategy rather than attempting to alter the strategy directly (cf. Vieth et al., 2003). By focusing on the underlying motivational forces that drive a particular behavior, it is possible to consider ways to utilize the motivational orientation to one's advantage.

As an illustration, the Grant et al. (2008) diary study suggests that interventions that identify the best way to frame a coping strategy, given an underlying motivational orientation, may be particularly effective. For example, people suffering from clinical depression tend to have a dominant promotion focus, and those suffering from clinical anxiety tend to have a dominant prevention focus (Strauman, 1992). If new (and generally positive) tactics for coping with the client's everyday problems are being proposed (e.g., by the therapist), then it would make sense to frame a new intervention tactic as serving an eager (approach) strategy for the depressed client and as serving a vigilant (avoidance) strategy for the anxious client—thereby intensifying the attractiveness of the tactical activity to the client and increasing the likelihood that he/she will engage in it. For example, choosing to exercise more, which could benefit clients with either depression or anxiety, could be framed differentially. In other words, the tactical direction of the therapist to these two clients could be the same (e.g., "begin an exercise regimen") but the framing of the tactic could be tailored to reflect the eager approach strategy or vigilant avoidance strategy that would be most motivating given the client's underlying regulatory focus. By helping the client to successfully resolve a vertical conflict, the therapist could help increase the client's self-control.

It may also be beneficial to target interventions at different levels simultaneously to take advantage of motivational relations between and among levels. For example, the low engagement strength of depressed clients could be temporarily reduced by intervening both at the goal level and at the strategic level. By both inducing a prevention focus and inducing strategic vigilance together, the resulting regulatory fit could increase strength of engagement that could then be used productively to remove a depressed client from the debilitating symptom of "no interest in anything." Intervening just at the goal level and trying to induce a prevention focus in depressed clients might not work because their routinized tendency for strategic eagerness would be a nonfit with a prevention focus (a difficult vertical conflict), thereby maintaining the low engagement strength. Similarly, intervening just at the strategic level by inducing strategic vigilance would be a nonfit with the underlying promotion focus. Only by simultaneously targeting both the goal level and the strategy level would the intervention be effective (see also Vieth et al., 2003).

The current approach also suggests the possibility that the same self-control conflict can be envisioned either as a horizontal conflict or as a vertical conflict. Indeed, we think it is likely that the "classic" self-control conflict is especially difficult in part because of the way it is typically represented within the hierarchy. Donald experiences a conflict between a higher-order goal (his goal to be fit) and a lower-order temptation (the different gourmet desserts at the buffet). Traditionally, such a conflict has been represented as a particular type of vertical conflict: a long-term goal versus an incompatible means (temptation). Both the way in which people colloquially discuss self-control conflicts, as well as the way in which they are typically experimentally presented within the lab, highlight this seemingly inherent verticality. Indeed, this type of conflict may have become the "classic" self-control dilemma precisely because it is so difficult to resolve (in favor of the long-term interest) when represented in this vertical manner. However, we propose that if individuals were to transform this same objective conflict from a vertically represented conflict into a horizontally represented conflict, either at the level of goals or at the level of means, it would make the nature of the conflict more clear and thus make its resolution more likely.

Why might the "classic" representation be more difficult? Although speculative at this point, we think it is likely that a vertical conflict of this nature (between a goal and a temptation) is more difficult, at least in part, because the representation obscures the nature of the "real" conflict. In the classic vertical representation, all that is real is the temptation right in front of you (Higgins, 2009); trying to compare the temptation to some abstract, long-term goal is akin to a conflict between real apples and less real oranges. It is hard to know the meaning or priority of the short-term temptation without making it comparable to the long-term interest. Consequently, representing the conflict as a horizontal conflict either between apples or between oranges, i.e., either between goals or between means, may be one way to make the essence or reality of the conflict clear.

As an example of how this could play out, we return to Donald at the dessert buffet, confronting a commonplace self-control conflict. Donald's vertical conflict between his goal to be fit and the temptation of indulging in dessert as an incompatible means could also be represented as a horizontal conflict between two goals: "being healthy" versus "enjoying life's pleasures." Alternately, the conflict could be represented as a horizontal conflict between two possible means— "eating healthy" versus "eating dessert" —for attaining the goal of "having a happy, fulfilling life." Representing the conflict in these different ways may have significant implications for how individuals experience the conflict, how individuals are able to resolve the conflict, and how individuals feel about the resolution. For instance, transforming the conflict into a horizontal conflict between tactics or strategies may make more salient the multifinality of one means versus another (cf. Kruglanski et al., 2002), assisting in resolution. Transforming the conflict into a horizontal conflict between goals may allow for clearer prioritization of varied concerns. It is also possible that certain types of horizontal goal transformations (e.g., conflict between two promotion goals versus a conflict between a promotion and a prevention goal) may be easier to resolve than others. Thus, interventions could take advantage of the possibility that transforming the representation of conflicts in the hierarchy may help people effectively (and creatively) exert self-control.

We believe that there is much yet to be learned about how individuals can most effectively engage in control that is targeted at the level of goals or at the level of strategies (e.g., how to adopt more self-concordant goals or how to self-prime a motivational orientation that matches the strategic demands of a task). As we have touched upon briefly, some research suggests that there are important antecedents that moderate or predict the ability of individuals to target control at the level of goals or strategies that are most effective. For example, individuals high in neuroticism may be less effective at adopting approach goals (Elliot, Sheldon, & Church, 1997). Individuals with a high need for closure or a strong prevention focus may be more effective at shielding focal goals from competing alternatives (Shah et al., 2002). Exploring how underlying cognitive and motivational factors impact the execution of control may be fruitful for elucidating how these control processes operate.

Certainly, individuals who have the skills to flexibly shift goals, strategies, tactics or behaviors given different demands are at an advantage (cf. Cheng, 2003; Chiu et al., 1995). What is less clear is at what levels individuals have the most (and least) flexibility if things are not working, which itself could vary across situations and across individuals. Understanding the similarities and differences in the ways that control is enacted at different levels in the hierarchy in both conscious and nonconscious ways will be an exciting avenue for future research (cf. Hassin et al., 2009). Furthermore, as we noted earlier in the chapter, it will be important to investigate more precisely how tactical control is exerted—how control at the tactical level differs from control at the strategic versus behavioral levels, and how such differences are constrained or amplified by situational demands.

The approach we have taken in this chapter challenges the notion that self-regulation and self-control are fundamentally about stopping some lower-order behavioral impulse or response. Rather, self-control is fundamentally about the management and resolution of conflict at all levels of the hierarchy. It is just as much about activation, facilitation, or initiation at tactical, strategic, and goal levels. It is also about intervening at more than one level simultaneously to take advantage of the motivational relations between or among levels. We believe that thinking about control in this way can expand the possibilities for interventions that improve significantly the ways in which individuals triumph in regulating themselves.

NOTE

1 Although we are arguing that the "classic" self-control conflict is most typically represented vertically, we recognize that it could be viewed as a horizontal conflict within our hierarchy given a construal level theory (CLT) framework (c.f., Fujita, 2008; Fujita et al., 2006). A CLT analysis of self-control argues that the classic self-control conflict is a conflict between a higher-order concern and a lower-order concern. As defined within CLT, this may be vertical (goals versus incompatible means), but it could also be represented horizontally as a conflict between two goals—a higher-order goal (e.g., to learn about myself) versus a lower-order goal (e.g., to avoid hurting my arm muscles). The orthogonality of construal levels to the self-regulatory hierarchy (goals, strategies, tactics, behavior) makes this possible. However, given our belief that the classic self-control conflict is most often represented and experienced as a vertical conflict between a goal and an incompatible means (temptation) and given the constraints of what it is possible to cover in the current discussion, we are not able to discuss this complexity. Rather, we rely on the more general interpretation of the classic self-control conflict as vertically experienced as the jumping-off point for our exploration.

REFERENCES

Aldwin, C. *Stress, coping, and development.* New York, NY: Guilford Press, 1994.

Avnet, T., & Higgins, E. T. Locomotion, assessment, and regulatory fit: Value transfer from "how" to "what." J Exp Soc Psychol 2003; 39: 525–530.

Bargh, J.A., Gollwitzer, P.M., Lee-Chai, A., Barndollar, K., & Trotschel, R. The automated will: Nonconscious activation and pursuit of behavioral goals. J Pers Soc Psychol 2001; 81: 1014–1027.

Bandura, A. Self-efficacy: Toward a unifying theory of behavioral change. Psychol Rev 1977; 84: 191–215.

Bandura, A. *Self-efficacy: The exercise of control.* New York, NY: W. H. Freeman, 1997.

Baumeister, R. F., Bratslavsky, E., Muraven, M., & Tice, D. M. Ego depletion: Is the active self a limited resource? J Pers Soc Psychol 1998; 74: 1252–1265.

Baumeister, R. F., Schmeichel, B. J., & Vohs, K. D. Self-regulation and the executive function. The self as controlling agent. In: Kruglanski, A. W., & Higgins, E. T. (Eds.), *Social Psychology: Handbook of basic principles,* 2nd ed. New York, NY: Guilford, 2007: pp. 516–539.

Bianco, A. T., Higgins, E. T., & Klem, A. How "fun/importance" fit affects performance: Relating implicit theories to instructions. Pers Soc Psychol Bull 2003; 29: 1091–1103.

Bowlby, J. *Attachment* (Attachment and loss, Vol.1). New York, NY: Basic Books, 1969.

Bowlby, J. Separation: Anxiety and anger (Attachment and loss, Vol.2). New York, NY: Basic Books, 1973.

Cantor, N. Life task problem solving: Situational affordances and personal needs. Pers Soc Psychol Bull 1994; 20: 235–243.

Cantor, N., & Kihlstrom, J. F. *Personality and social intelligence.* Englewood Cliffs, NJ: Prentice Hall, 1987.

Carver, C. Some ways in which goals differ and some implications of those differences. In: Gollwitzer, P., & Bargh, J. (Eds.), *The psychology of action: Linking cognition to motivation and behavior.* New York, NY: Guilford, 1996; pp. 645–672.

Carver, C. S., & Scheier, M. F. *Attention and self-regulation: A control theory approach to human behavior.* New York, NY: Springer-Verlag, 1981.

Carver, C. S., & Scheier, M. F. Origins and functions of positive and negative effect: A control-process view. Psychol Rev 1990; 97: 19–35.

Carver, C. S., & Scheier, M. F. Situational coping and coping dispositions in a stressful transaction. J Pers Soc Psychol 1994; 66: 184–195.

Carver, C. S., & Scheier, M. F. *On the self-regulation of behavior*. Cambridge: Cambridge University Press, 1998.

Carver, C. S., Scheier, M. F., & Weintraub, J. K. Assessing coping strategies: A theoretically based approach. J Pers Soc Psychol 1989; 56: 267–283.

Chartrand, T. L., & Bargh, J. A. Nonconscious motivations: Their activation, operation, and consequences. In: Tesser, A., Stapel, D., & Wood, J. V. (Eds.), *Self and motivation: Emerging psychological perspectives*. Washington, DC: American Psychological Association, 2002: pp. 13–41.

Cheng, C. Cognitive and motivational processes underlying coping flexibility: A dual process model. J Pers Soc Psychol 2003; 84: 425–238.

Chui, C., Hong, Y., Mischel, W., & Shoda, Y. Discriminative facility in social competence: Conditional versus dispositional encoding and monitoring-blunting of information. Soc Cogn 1995; 13: 49–70.

Crowe, E., & Higgins, E. T. Regulatory focus and strategic inclinations: Promotion and prevention in decision-making. Organ Behav Hum Decis Process 1997; 69: 117–132.

Deci, E. L., & Ryan, R. M. *Intrinsic motivation and self-determination in human behavior*. New York: Plenum, 1985.

Dweck, C. *Self-theories: Their role in motivation, personality, and development*. Philadelphia: Psychology Press, 1999.

Elliot, A. J. The hierarchical model of approach-avoidance motivation. Motiv Emot 2006; 30: 111–116.

Elliot, A. J., & Church, M. A. A hierarchical model of approach and avoidance achievement motivation. J Pers Soc Psychol 1997; 72: 218–232.

Elliot, A. J., & Harackiewicz, J. M. Approach and avoidance achievement goals and intrinsic motivation: A mediational analysis. J Pers Soc Psychol 1996; 70: 461–475.

Elliot, A. J., & McGregor, H. A 2×2 achievement goal framework. J Pers Soc Psychol 2001; 80: 501–519.

Elliot, A. J., & Sheldon, K. M. Avoidance achievement motivation: A personal goals analysis. J Pers Soc Psychol 1997; 73: 171–175.

Elliot, A. J., & Sheldon, K. M. Avoidance personal goals and the personality-illness relationship. J Pers Soc Psychol 1998; 75: 1282–1299.

Elliot, A. J., Sheldon, K. M., & Church, M. A. Avoidance personal goals and subjective well-being. Pers Soc Psychol Bull 1997; 23: 915–927.

Elliot, A. J., & Thrash, T.M. Approach-avoidance motivation in personality: Approach and avoidance temperaments and goals. J Pers Soc Psychol 2002; 82: 804–818.

Emmons, R., & McAdams, D. Personal strivings and motive dispositions: Exploring the links. Pers Soc Psychol Bull 1991; 17: 648–654.

Fitzsimons, G. M., & Bargh, J. A. Thinking of you: Nonconscious pursuit of interpersonal goals associated with relationship partners. J Pers Soc Psychol 2003; 84: 148–163.

Förster, J., Higgins, E. T., & Bianco, A. T. Speed/accuracy decisions in task performance: Built in trade-of or separate strategic concerns? Organ Behav Hum Decis Process 2003; 90: 148–164.

Freitas, A. L., & Higgins, E. T. Enjoying goal-directed action: The role of regulatory fit. Psychol Sci 2002; 13: 1–6.

Freud, S. *A general introduction to psychoanalysis*. New York, NY: Washington Square Press, 1952. (Original work published 1920).

Freud, S. The ego and the id. In: Strachey, J. (Ed. and Trans.), *Standard edition of the complete psychological works of Sigmund Freud*, Vol. 19. London: Hogarth Press, 1961: pp. 3–66. (Original work published 1923).

Friedman, R. S., & Förster, J. The effects of promotion and prevention cues on creativity. J Pers Soc Psychol 2001; 81: 1001–1013.

Fujita, K. Seeing the forest beyond the trees: A construal-level approach to self-control. Soc Personal Psychol Compass 2008; 2/3: 1475–1496.

Fujita, K., Trope, Y., Liberman, N., & Levin-Sagi, M. Construal levels and self-control. J Pers Soc Psychol 2006; 90: 351–367.

Gable, S. L. Approach and avoidance social motives and goals. J Pers 2006; 74: 175–222.

Gaillot, M. T., Baumeister, R. F., DeWall, C. N., et al. Self-control relies on glucose as a limited energy source: Willpower is more than a metaphor. J Pers Soc Psychol 2007; 92: 325–336.

Gollwitzer, P. M., Fujita, K., & Oettingen, G. Planning and the implementation of goals. In Baumeister, R. F., & Vohs, K. D. (Eds.), *Handbook of self-regulation: Research, theory and applications*. New York, NY: Guilford Press, 2004: pp. 211–228.

Grant, H., Higgins, E. T., Baer, A., & Bolger, N. Coping style and regulatory fit: Emotional ups and downs in life. Unpublished manuscript, 2008.

Gray, J. A. *The neuropsychology of anxiety: An enquiry into the functions of the septo-hippocampal system.* New York, NY: Oxford University Press, 1982.

Gross, J. J. The emerging field of emotion regulation: An integrative review. Rev Gen Psychol 1998; 2: 271–299.

Hassin, R. R., Aarts, H., Eitam, B., Custers, R., & Kleiman, T. Non-conscious goal pursuit and the effortful control of behavior. In: Morsella, E., Gollwitzer, P. M., & Bargh, J. A. (Eds.), *The psychology of action, vol II.* New York, NY: Oxford University Press, in press.

Hassin, R. R., Bargh, J. A., & Zimerman, S. Automatic and flexible: The case of nonconscious goal pursuit. Soc Cogn 2009; 27: 20–36.

Heckhausen, H., & Gollwitzer, P. M. Thought contents and cognitive functioning in motivational versus volitional states of mind. Motiv Emot 1987; 11: 101–120.

Hembree, R. Correlates, causes, effects, and treatment of test anxiety. Rev Educ Res 1988; 58: 47–77.

Higgins, E. T. Self-discrepancy: A theory relating self and affect. Psychol Rev 1987; 94: 319–340.

Higgins, E. T. Beyond pleasure and pain. Am Psychol 1997; 52: 1280–1300.

Higgins, E.T. Knowledge accessibility and activation: Subjectivity and suffering from unconscious sources. In: Uleman, J., & Bargh, J. (Eds.), *Unintended thought.* New York, NY: Guilford, 1998: pp. 75–123.

Higgins, E. T. Making a good decision: Value from fit. Am Psychol 2000; 55: 1217–1230.

Higgins, E. T. Value from hedonic experience and engagement. Psychol Rev 2006; 113: 439–460.

Higgins, E.T. Organization of motives: Value, truth, and control working together. Unpublished manuscript, Columbia University, 2009.

Higgins, E. T., Bond, R. N., Klein, R., & Strauman, T. Self-discrepancies and emotional vulnerability: How magnitude, accessibility, and type of discrepancy influence affect. J Pers Soc Psychol 1986; 51: 5–15.

Higgins, E.T., Friedman, R. S., Harlow, R. E., Idson, L. C., Ayduk, O. N., & Taylor, A. Achievement orientations from subjective histories of success: Promotion pride versus prevention pride. Eur J Soc Psychol 2001; 31: 3–23.

Higgins, E. T., Idson, L. C., Freitas, A. L., Spiegel, S., & Molden, D. C. Transfer of value from fit. J Pers Soc Psychol 2003; 84: 1140–1153.

Higgins, E. T., Roney, C. J. R., Crowe, E., & Hymes, C. Ideal versus ought predilections for approach and avoidance: Distinct self-regulatory systems. J Pers Soc Psychol 1994; 66: 276–286.

Higgins, E. T., Shah, J., & Friedman, R. Emotional responses to goal attainment: Strength of regulatory focus as moderator. J Pers Soc Psychol 1997; 72: 515–525.

Hong, J., & Lee, A. Y. Be fit and be strong: Mastering self-regulation with regulatory fit. J Consum Res 2008; 34: 682–695.

Howell, A. J., & Watson, D. C. Procrastination: Associations with achievement goal orientation and learning strategies. Pers Individ Dif 2007; 43: 167–178.

Janis, I. L., & Mann, L. *Decision-making: A psychological analysis of conflict, choice, and commitment.* New York, NY: Free Press, 1977.

Kahneman, D., & Tversky, A. Prospect theory: An analysis of decision under risk. Econometrica 1979; 47: 263–291.

Kasser, T., & Ryan, R. M. Further examining the American dream: Well-being correlates of intrinsic and extrinsic goals. Pers Soc Psychol Bull 1996; 22: 281–288.

Kruglanski, A. W. Goals as knowledge structures. In: Gollwitzer, P. M., & Bargh, J. A. (Eds.), *The psychology of action: Linking cognition and motivation to behavior.* New York, NY: Guilford Press, 1996: pp. 599–619.

Kruglanski, A. W., Shah, J. Y., Fishbach, A., Friedman, R., Chun, W., & Sleeth-Keppler, D. A theory of goal systems. In: Zanna, M. P. (Ed.), *Advances in experimental social psychology.* 2002; 34: 331–378.

Lang, P. J. The emotion probe: Studies of motivation and attention. Am Psychol 1995; 50: 372–385.

Lazarus, R. S., & Folkman, S. *Stress, appraisal, and coping.* New York, NY: Springer, 1984.

Lewin, K. *A dynamic theory of personality.* New York, NY: McGraw-Hill, 1935.

Linville, P. W., & Clark, L. F. Production systems and social problem-solving: Specificity, flexibility, and expertise. In: Wyer, J. R. S., & Srull, T. K. (Eds.), *Social intelligence and cognitive assessments of personality.* Hillsdale, NJ: Erlbaum, 1989: pp. 131–151.

Metcalfe, J., & Mischel, W. A hot/cool-system analysis of delay of gratification: Dynamics of willpower. Psychol Rev 1999; 106: 3–19.

Miller, G. A., Galanter, E., & Pribram, K. H. *Plans and the structure of behavior.* New York, NY: Henry Holt, 1960.

Miller, N. E. Experimental studies of conflict. In: McV. Hunt J. (Ed.), *Personality and the behavior disorders*, Vol. 1. New York, NY: Ronald Press, 1994: pp. 431–465.

Mischel, W., & Baker, N. Cognitive appraisals and transformations in delay behavior. J Pers Soc Psychol 1975; 31: 254–261.

Mischel, W., Cantor, N., & Feldman, S. Principles of self-regulation: The nature of willpower and self-control. In: Higgins, E. T., & Kruglanski, A. W. (Eds.), *Social psychology: Handbook of basic principles.* New York, NY: Guilford, 1996: pp. 329–360.

Mischel, W., & Ebbesen, E. B. Attention in delay of gratification. J Pers Soc Psychol 1970; 16: 329–337.

Moore, B., Mischel, W., & Zeiss, A. Comparative effects of the reward stimulus and its cognitive representation in voluntary delay. J Pers Soc Psychol 1976; 34: 419–424.

Moskowitz, G. B. Preconscious effects of temporary goals on attention. J Exp Soc Psychol 2002; 38: 397–404.

Mowrer, O. H. *Learning theory and behavior.* New York, NY: Wiley, 1960.

Norem, J. K., & Cantor, N. Anticipatory and post hoc cushioning strategies: Optimism and defensive pessimism in "risky" situations. Cognit Ther Res 1986a; 10: 347–362.

Norem, J. K., & Cantor, N. Defensive pessimism: Harnessing anxiety as motivation. J Pers Soc Psychol 1986b; 51: 1208–1217.

Norem, J. K., & Illingworth, K. S. S. Strategy-dependent effects of reflecting on self and tasks: Some implications of optimism and defensive pessimism. J Pers Soc Psychol 1993; 65: 822–835.

Oettingen, G., Grant, H., Smith, P. K., Skinner, M., & Gollwitzer, P. M. Nonconscious goal pursuit: Acting in an explanatory vacuum. J Exp Soc Psychol 2006; 42: 668–675.

Penley, J. A., Tomaka, J., & Wiebe, J. S. The association of coping to physical and psychological health outcomes: A meta-analytic review. J Behav Med 2002; 25: 551–603.

Pervin, L. A. Goal concepts in personality and social psychology: A historical introduction. In: Pervin, L. A. (Ed.), *Goal concepts in personality and social psychology.* Hillsdale, NJ: Erlbaum, 1989: pp. 1–17.

Pervin, L. A. A dynamic systems approach to personality. Eur Psychol 2001; 6: 172–176.

Scholer, A. A., & Higgins, E. T. Distinguishing levels of regulatory focus: An analysis using regulatory focus theory. In: Elliot, A. J. (Ed.), *Handbook of approach and avoidance motivation.* Hillsdale, NJ: Lawrence Erlbaum, 2008: pp. 489–504.

Scholer, A. A., Stroessner, S. J., & Higgins, E. T. Responding to negativity: How a risky tactic can serve a vigilant strategy. J Exp Soc Psychol 2008; 44: 767–774.

Shah, J. Y. The motivational looking glass: How significant others implicitly affect goal appraisals. J Pers Soc Psychol 2003; 85: 424–439.

Shah, J. Y., Friedman, R., & Kruglanski, A. W. Forgetting all else: On the antecedents and consequences of goal shielding. J Pers Soc Psychol 2002; 83: 1261–1280.

Shah, J., & Higgins, E. T. Regulatory concerns and appraisal efficiency: The general impact of promotion and prevention. J Pers Soc Psychol 2001; 80: 693–705.

Shah, J., Higgins, E. T., & Friedman, R. S. Performance incentives and means: How regulatory focus influences goal attainment. J Pers Soc Psychol 1998; 74: 285–293.

Shah, J. Y., & Kruglanski, A. W. Priming against your will: How accessible alternatives affect goal pursuit. J Exp Soc Psychol 2002; 83: 368–383.

Sheldon, K. M., & Elliot, A. J. Not all personal goals are personal: Comparing autonomous and controlled reasons as predictors of effort and attainment. Pers Soc Psychol Bull 1998; 24: 531–543.

Sheldon, K. M., & Elliot, A. J. Goal striving, need satisfaction, and longitudinal well-being: The self-concordance model. J Pers Soc Psychol 1999; 76: 482–497.

Sheldon, K. M., Ryan, R. M., Deci, E. L., & Kasser, T. The independent effects of goal contents and motives on well-being: It's both what you pursue and why you pursue it. Pers Soc Psychol Bull 2004; 30: 475–486.

Showers, C. The motivational and emotional consequences of considering positive or negative possibilities for an upcoming event. J Pers Soc Psychol 1992; 63: 474–484.

Showers, C., & Cantor, N. Social cognition: A look at motivated strategies. Ann Rev Psychol 1985; 36: 275–305.

Skinner, B. F. *Beyond freedom and dignity.* New York, NY: Knopf, 1971.

Speilberger, C., & Vagg, P. *Test anxiety: Theory, assessment, and treatment.* Washington, DC: Taylor & Francis, 1995.

Spiegel, S., Grant-Pillow, H., & Higgins, E. T. How regulatory fit enhances motivational strength during goal pursuit. Eur J Soc Psychol 2004; 34: 39–54.

Strauman, T. J., & Higgins, E. T. Automatic activation of self-discrepancies and emotional syndromes: When cognitive structures influence affect. J Pers Soc Psychol. Special Issue: Integrating personality and social psychology, 1987; 53: 1004–1014.

Strauman, T. J., & Higgins, E. T. Self-discrepancies as predictors of vulnerability to distinct syndromes of chronic emotional distress. J Pers 1988; 56: 685–707.

Swets, J. A. The relative operating characteristic in psychology. Science 1973; 182: 990–1000.

Tangney, J. P., Baumeister, R. F., & Boone, A. L. High self-control predicts good adjustment, less pathology, better grades, and interpersonal success. J Pers 2004; 72: 271–322.

Taylor, S. E., & Aspinwall, L. G. Mediating and moderating processes in psychosocial stress: Appraisal, coping, resistance, and vulnerability. In Kaplan, H. B. (Ed.), *Psychosocial stress: Perspectives on structure, theory, life-course, and methods.* San Diego, CA: Academic Press, 1996: pp. 71–110.

Trope, Y., & Fishbach, A. Counteractive self-control in overcoming temptation. J Pers Soc Psychol 2000; 79: 493–506.

Vallacher, R. R., & Wegner, D. M. *A theory of action identification.* Hillsdale, NJ: Lawrence Erlbaum Associates, 1985.

Vallacher, R. R., & Wegner, D. M. What do people think they're doing? Action identification and human behavior. Psychol Rev 1987; 94: 3–15.

Vieth, A. Z., Strauman, T. J., Kolden, G. G., Woods, T. E., Michels, J. L., & Klein, M. H. Self-system theory (SST): A theory-based psychotherapy for depression. Clin Psychol Sci Pract 2003; 10: 245–268.

Webster, D. M., & Kruglanski, A.W. Individual differences in need for cognitive closure. J Pers Soc Psychol 1994; 67: 1049–1062.

CHAPTER 18

Getting Our Act Together: Toward a General Model of Self-Control

Eran Magen and James J. Gross

ABSTRACT

Research on self-control has enjoyed tremendous growth over the past few decades, as researchers from a variety of disciplines have tested different self-control techniques in different domains of self-control. The result has been a proliferation of theories, models, and approaches, each offering important, but so far largely unrelated insights. The lack of a unifying framework has been an impediment to the development of an incremental science of self-control, and has left researchers struggling to relate their work to that of others. In this chapter, we present a general model of self-control, which we call the cybernetic process model of self-control. This model integrates two existing models—Cybernetic control theory (Carver & Scheier, 1982) and the process model of emotion-regulation (Gross, 1998b)—and describes the process through which tempting impulses arise and may be regulated. The cybernetic process model of self-control provides a conceptual framework for organizing disparate findings from research on self-control, and serves as a useful aid in selecting and designing appropriate self-control techniques.

Keywords: Self-control, self-regulation, emotion-regulation, delay of gratification, cybernetic process, emotion, temptation, reconstrual, suppression, reappraisal, response modulation, intervention design

GETTING OUR ACT TOGETHER: TOWARD A GENERAL MODEL OF SELF-CONTROL

Research on self-control has been growing at an incredible pace. Based on current publication trends, the number of peer-reviewed articles on self-control for 2001 to 2010 will be greater than that of the four preceding decades put together.[1] We now know that an individual's level of self-control predicts important life outcomes including school performance, health behaviors, and substance abuse (Duckworth & Seligman, 2005; Mischel & Ayduk, 2004; Mischel, Shoda, & Rodriguez, 1989). In the clinical domain, low self-control is a central feature of many clinical disorders (Heiby & Mearig, 2002; Strayhorn, 2002a; Tangney, Baumeister, & Boone, 2004), as well as a reliable predictor of psychopathology and problematic behaviors in children (Eisenberg et al., 2001; Krueger et al., 1996).

With so much research on self-control taking place, what do we know about ways in which people can bolster their own self-control? What options does an individual have when facing temptation? And which of these options, or what combination of them, is best? Despite their importance, the answers to these questions are not as clear as we would like. Part of the reason for this lies in the fact that our understanding of self-control is currently hampered by the lack of an integrative framework. Presently, researchers have no way of meaningfully relating their findings to those of others in the field, and the result is a hodgepodge of techniques with which to bolster self-control in specific domains or in specific situations, rather than a structured method of prioritizing and choosing the most appropriate techniques for the challenges at hand.

Our goal in this chapter is to present an integrative framework that will enable researchers and practitioners to use a shared language when communicating their insights and findings. This, in turn, will allow practitioners and researchers to systematically examine the etiology of the difficulty that their client (or research participant) is facing, determine the points that are most amenable to intervention, and then select or develop the most appropriate intervention. Our main interest is in the behavioral-experiential aspects of temptations and self-control. However, our integrative framework may also serve as a basis for neuropsychological study of self-control, and thus help build bridges between researchers studying basic self-regulatory processes, researchers studying naturalistic human behavior, and practitioners who develop interventions for real-life temptations that people face in everyday life.

The framework that we propose represents the integration of two prior models: Cybernetic control theory (Carver & Scheier, 1982) and the process model of emotion-regulation (Gross, 1998b, Gross & Thompson, 2007). The integrated model provides an overarching framework that clarifies the relations among different self-control techniques and establishes a way to prioritize diverse interventions. To provide a foundation for this model, we begin by defining temptations and self-control. We then review and compare the two existing models of self-control, integrate them, and present a general model of self-control. We explain the different types of self-control methods and provide examples and empirical evidence relating to each one. We conclude by proposing a system of prioritization for the different types of self-control methods, based on features of the methods themselves, as well as external and internal contextual factors.

TEMPTATIONS AND SELF-CONTROL

We define temptation as *the impulse to behave in a way that one fully expects to regret at a later time*. Although people frequently behave in ways that are potentially regrettable, our focus is on behaviors that people fully expect to regret, even before they perform them. Note that this definition revolves around the belief that the individual holds about future regret *prior* to emitting the behavior. If one fully expects to regret a certain behavior and yet desires to perform it, one is experiencing temptation. It is possible that one would take the action that one expects to regret and later discover that one does not regret it (e.g., John chooses to stay in bed a few extra minutes despite knowing that he would be late for school and expecting to regret his decision, but upon arriving late at the school, he learns that his first class was cancelled). For our purposes, what matters is whether in the moment of choice the individual believes that the more immediately appealing alternative will lead to regret.

Note also that this definition does not include an element of probability ("Maybe I'll regret it, and maybe I won't.") —instead, one is certain that acting in accordance with the desired behavior will lead to regret. More specifically, people experience temptations when the goal of experiencing a relatively small short-term gain[2] is competing with the goal of experiencing a relatively large long-term gain (see Table 18–1). Thus, the short-term goal of feeling less upset (achievable by drinking alcohol) may compete with the long-term goal of staying sober (achievable by avoiding alcohol). Similarly, the short-term goal of feeling comfortable (achievable by staying in bed) may compete with the long-term goal of improving physical fitness (achievable by getting up to exercise). Generally speaking, despite having a clear long-term goal in mind (staying sober, improving physical fitness), one

TABLE 18-1. EXAMPLES OF TEMPTATIONS AND NONTEMPTATIONS

Short-Term Goal	Long-Term Goal	Self-Control Task
Temptations (conflict between short-term and long-term goals)		
Consuming alcohol	Remaining sober	Inhibiting alcohol consumption
Staying in bed	Improving physical fitness	Initiating exercising
Nontemptations (no conflict between short-term and long-term goals)		
Drinking water	Avoiding dehydration	N/A
Taking a nap	Being well-rested	N/A

Note: Temptations involve a conflict between a short-term goal and a long-term goal. In tempting situations, the person realizes that acting in line with the short-term goal will result in failure to attain the long-term goal and consequently lead to regret.

may feel drawn to act in a way that prevents the attainment of this goal (drink alcohol, stay in bed and skip exercise), to experience a relatively small but more immediate short-term gain.

Having defined temptation, we are now able to define self-control. Researchers have previously defined self-control as the ability to override pre-potent responses (Vohs, Baumeister, & Ciarocco, 2005), to overcome threats that short-term goals pose to long-term goals (Fishbach & Trope, 2005), or to act in accordance with perceived self-interests (Loewenstein, 1996). In the present context, we will use the term to denote *the ability to resist temptations*. The form that self-control takes depends on the temptation at hand. Referring to the examples we listed in Table 18-1, one might need to inhibit a certain impulse (e.g., avoid drinking alcohol despite the urge to do so) or to initiate a behavior despite the impulse to avoid doing so (e.g., starting to exercise despite the urge to stay in bed). Self-control refers to acting in line with one's long-term goal, despite the allure of a contradictory short-term goal.

EXISTING MODELS OF SELF-CONTROL

People attempt to control various aspects of their lives. Broadly speaking, people may try to control *extra*-personal factors (such as the temperature of the room, the Web site that their computer is displaying, the behavior of people around them, etc.), as well as *intra*personal factors (such as their mood, the tone of their voice, their level of hunger, etc.). Such extrapersonal and intrapersonal factors are typically tightly linked, and people often attempt to influence one by manipulating the other. Thus, to reduce my hunger (intrapersonal), I may order a burrito at a local store (thereby manipulating the behavior of the sales person, which is extrapersonal for me). Similarly, to receive flight details from a finicky voice-activated phone system (extrapersonal), I may have to put effort into enunciating clearly rather than screaming with frustration (thereby controlling my own speech, which is intrapersonal).

Two influential models describe ways in which people exercise extra- and intrapersonal control, with each model focusing mainly on one of these. The *cybernetic control model* (Carver, 2004; Carver & Scheier, 1982) describes ways in which interactions with the environment give rise to behavioral impulses, and then details how such impulses are translated into behaviors that gradually change the environment. The cybernetic control model is largely concerned with the ways people manipulate and shape their environments—how people exert *extra*personal control. In addition, the cybernetic control model, although broadly applicable, does not specify how individuals may change the regulation process itself, thus rendering the regulation process blind to itself.

The *process model of emotion regulation* (Gross, 1998b; Gross & Thompson, 2007) complements the cybernetic control model by delineating ways in which impulses can be modulated, albeit in the more specific field of emotion research. The process model of emotion regulation is largely concerned with ways in which people attempt to alter their own emotional experience and control their expression of this experience—how people exert *intra*personal control. Although the process model of emotion-regulation applies to a comparatively narrow domain (dealing only with emotions), it offers important insights regarding ways in which individuals regulate their own experience. We now turn to a more detailed overview of the two models.

The Cybernetic Control Model

Behavior is commonly guided by goals. People may want to help a friend, drink a glass of juice, drive safely, avoid humiliation, stay healthy, relax, or achieve any other goal. Goals typically require multiple steps to be achieved. These steps can themselves be thought of as smaller, intermediate sub-goals (Carver & Scheier, 1982; Vallacher & Wegner, 1987). Thus, eating watermelon (top goal) requires that I get it out of the refrigerator (sub-goal), which in turn requires that I stand next to the refrigerator (sub-sub-goal), which in turn requires that I get up from the sofa (sub-sub-sub-goal), and so on.

People continuously monitor their progress toward (or away from) goals by attending to their environment, and adjust their behavior in response to stimuli that seem relevant to the achievement of their various goals and sub-goals. After getting up from the sofa, I start walking towards the refrigerator. I can see that I am still too far away to reach it (my sub-goal of reaching the refrigerator hasn't been achieved), and so I continue to walk. As I walk, I hear the sound of an airplane flying by. This information is irrelevant with respect to my current goal, and so I continue on my way to the refrigerator. After a few steps, I find myself close enough to the refrigerator and I stop—I have achieved my sub-goal. I now switch to a different sub-goal (taking out some watermelon). In this way, I continuously compare the environment that I perceive ("I am standing next to the refrigerator with no watermelon in my hands.") to my current goal ("having watermelon in my hands"), and act on my environment to achieve my goal. If there is a discrepancy between my current goal and the environment I perceive, the comparison process produces an impulse that calls for a certain behavior, which is aimed at reducing the discrepancy between my situation (my perceived environment) and my goal.

Carver and Scheier (1982) formally presented this notion to the psychological community in the form of cybernetic control theory,[3] as depicted in Figure 18–1a.[4] In this model, each stage of the process receives input from the preceding stage, processes it in some way, and feeds an output to the subsequent stage of the process. Thus, the environment is perceived, and this impression of the environment is fed to the comparator. The comparator compares the perceived environment to a goal (or standard), and outputs an impulse, aimed at generating behavior that would influence the environment, so that it would more closely match the goal on the next comparison.[5] This impulse influences behavior, which in turn impacts the environment. The acted-upon environment is perceived again, compared again with the goal, the comparator outputs another impulse, and so forth.

The role of the comparator bears elaboration, as it is subtler than that of the other elements of the process. In the cybernetic control model, the comparator's work is described as determining whether there is a discrepancy between the environment (which the perception element provides) and the criterion or standard (which the goal element provides). We propose a small but critical addition to the "job description" of the comparator. After receiving input from the perception element, the comparator determines the *relevance* of that input to the goal and then, if the input is deemed relevant, compares the perceived environment to the goal. Making this additional role explicit helps to connect the cybernetic control model with the extensive literatures on emotion and motivation, in which researchers have suggested that emotions arise as a response to events that are seen as relevant to one's goals (Frijda, 1988; Gross & Thompson, 2007; Lazarus, 1991), and that the intensity of emotions is related to the rate and direction at which this discrepancy is changing (Carver, 2004; Carver & Scheier, 1990; Hsee & Abelson, 1991; Lawrence, Carver, & Scheier, 2002). In the example above, distance from the refrigerator was a relevant input (and therefore used to determine the behavioral impulse), whereas the sound

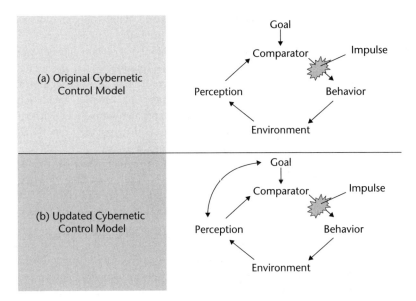

Figure 18–1. Cybernetic control model (Adapted from Carver & Scheier, 1982). The gray explosion (added by us) represents the generation of an impulse according to this model.

of a passing airplane was not (and therefore was not used to determine the behavioral impulse).

We have also updated the diagram of the cybernetic control model to represent research findings that are relevant to our discussion of self-control, by adding a bi-directional link between the *perception* and *goal* elements (*see* Figure 18–1b). A strong and consistent body of research has demonstrated how goals that people hold can bias their perception, often in a way that would preserve or bolster these goals (Jonas et al., 2001; Kunda, 1990; Lord, Ross, & Lepper, 1979; Nickerson, 1998). Conversely, a growing body of research has been exploring ways in which environmental cues can activate goals, even without conscious awareness (Bargh et al., 2001; Bonson et al., 2002; Fishbach, Friedman, & Kruglanski, 2003; Forster, 2007; Kavanagh, Andrade, & May, 2004; Lowe & Levine, 2005; Mauss, Cook, & Gross, 2007; Wansink, Painter, & Lee, 2006). As will become apparent in following sections, the powerful connection between goals and perception plays an important role when considering various methods of self-control.

The canonical example of a cybernetic control system is a thermostat, which compares the ambient temperature (the environment) to the target temperature (the goal). If the ambient temperature is higher than the goal, the mechanism works to lower the temperature, and if the ambient temperature is lower than the goal, the mechanism works to raise the temperature. But people differ from thermostats in a number of important ways—not the least of which is people's capacity to self-reflect, and their ability to hold conflicting goals. Thus, a person may experience an urge to heat a room (because he is uncomfortably cold), but at the same time realize that this would be a bad idea (because this would result in a large gas bill, which he cannot afford). This type of internal conflict, and the ways in which it may be resolved, is missing from the cybernetic control model. Although the cybernetic model provides a compelling account of the way in which people attempt to exert extra-personal control by regulating their *environment* (namely, reducing the discrepancy between it and their goal), the model is unable to adequately represent the way in which people attempt to exert intra-personal control by regulating their own *behavior*, in part because it focuses on ways in which impulses arise and are acted upon, rather than ways in which these impulses may be regulated.

The Process Model of Emotion-Regulation

Unlike thermostats, people are capable of simultaneously having multiple and even conflicting goals. As a result, people do not immediately act on all impulses that they experience; instead, they may try to postpone or change either the impulse itself or its expression, using a variety of methods. One domain in which researchers have intensively studied such efforts has been the field of emotion regulation. In subsequent sections, we will generalize findings from this field to the broader domain of self-control, to discuss ways in which people exert intra-personal control by regulating their own impulses and behavior.

Emotions are coordinated sets of responses (experiential, physiological, behavioral) that arise as a result of interacting with the environment and perceiving stimuli that are seen as relevant to one's goals, and prepare or propel individual to act in a specific manner (Frijda, 1988; Gross & Thompson, 2007; Lazarus, 1991). Thus, individuals experiencing anger become more likely to aggress, whereas individuals experiencing amusement become more likely to smile. In this manner, emotions function in a similar manner to "impulses" in the cybernetic control model.

People often try to control which emotions they experience, when they experience them, and how they express them (Gross & John, 2003). Giggling during a solemn religious ceremony can be awkward. Showing envy at a friend's good fortune is a good way to make everybody feel upset. Getting angry at the driver who just cut you off can be a very bad idea if you suffer from hypertension (or if you are driving in certain parts of L.A.). In general, people often attempt to up-regulate (i.e., have more of) or down-regulate (i.e., have less of) certain emotions, either to feel good, or because they believe that certain emotions are more beneficial in specific situations (Tamir, 2005; Tamir & Robinson, 2004).

People engage in emotion-regulation for a variety of reasons, including the motivation to avoid the unpleasant experience of negative emotions, to display more socially appropriate behavior, or to avoid dangerous physiological arousal. Sometimes the very experience (or behavioral expression) of a certain emotion can be considered a temptation, as an individual may expect to regret doing so. At other times, people may engage in emotion-regulation without experiencing temptation (i.e., without believing that they will regret their emotional experience in the future).

People attempt to regulate their own emotions in many ways. They may imagine their "happy place," breathe deeply and count to 10, smile at their conversation partner while planning exquisite revenge, or simply force themselves to display an emotion that they are not experiencing. To organize these diverse forms of emotion regulation, Gross (1998b) presented a process model of emotion regulation, which divided the various methods of regulating emotions into two broad categories according to the stage of the emotion-generative process during which they take place. *Antecedent-focused* emotion-regulation takes place before the emotion is generated, whereas *response-focused* emotion-regulation takes place after emotion is generated. These broad categories of emotion regulation methods can be broken down further into subcategories, each influencing a particular component of the emotion-generative process (*see* Fig. 18-2). The five general families of emotion regulation methods are: Situation selection, situation modification, attention deployment, cognitive change, and response modulation.

The process model of emotion regulation is inherently descriptive, rather than prescriptive. It specifies different types of emotion-regulation strategies and predicts differential effects of using different strategies, but it does not indicate when specific regulation strategies should be used. Research has demonstrated clear divergence in the consequences of using at least two of the strategies proposed by the model—namely, cognitive change and response modulation—on an experiential, physiological, cognitive, and social level (Gross, 2002; Gross & John, 2003; Richards & Gross, 2000; Butler et al., 2003; Gross, 1998a; Richards, Butler, & Gross, 2003). Nevertheless, the process model of emotion-regulation remains silent with respect

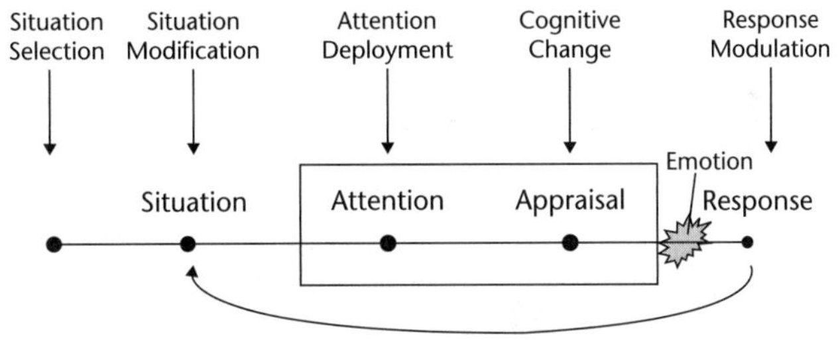

GETTING OUR ACT TOGETHER

Figure 18–2. Process model of emotion regulation (Adapted from Gross & Thompson, 2007). The gray explosion (added by us) represents the generation of emotion according to this model.

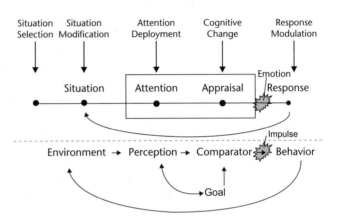

Figure 18–3. Juxtaposition of the process model of emotion regulation (top half: Gross, 1998b; Gross & Thompson, 2007) and the cybernetic control model (bottom half: Carver & Scheier, 1982). The gray explosion (added by us) represents the generation of an emotion (according to the process model of emotion regulation) or an impulse (according to cybernetic control theory).

to the levels of effort and benefit associated with each strategy of emotion regulation in particular situations, in part because it focuses on ways in which impulses are regulated, rather than ways in which these impulses arise in the first place.

The Cybernetic Process Model of Self-Control

Overall, these two models of self-control bear remarkable similarity to one another in terms of the process that they describe (*see* Fig. 18–3). At the heart of both models is the view that one's behavior is motivated by the difference between how things are (one's perceived environment) and how one would like things to be (one's goals). Although each model uses slightly different language to describe this process, both models trace the root of this impulse (emotion) to the comparison between (appraisal of) the real and the ideal.

Although the two models describe a similar underlying process by which impulses (emotions) are produced, they differ in their focus on possible targets for the regulatory process (*see* Fig. 18–4). The cybernetic control model focuses on ways in which people regulate their environments, whereas the process model of emotion regulation focuses on ways in which people regulate their responses to these environments—their impulses and behaviors. Moreover, whereas the cybernetic control model is broadly applicable, it does not explicitly address the problem at the heart of self-control challenges: The need to override prepotent responses that may undermine important long-term goals. In contrast, the process model of emotion-regulation offered a categorization system for ways in which individuals regulate a fairly specific type of impulse—namely, emotional reactions.

Considering both models together allows us to describe in detail the process by which impulses arise, as well as ways in which impulses

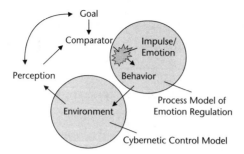

Figure 18-4. Different regulatory targets of the two models. The cybernetic control model (Carver & Scheier, 1982) focuses on regulation of the extrapersonal, whereas the process model of emotion regulation (Gross, 1998b; Gross & Thompson, 2007) focuses on regulation of the intrapersonal. The gray explosion represents the generation of an impulse (according to cybernetic control theory) or an emotion (according to the process model of emotion regulation).

can be self-regulated. In this sense, the cybernetic control model and the process model of emotion regulation complement each other well, since each describes a part of the phenomenon that self-control researchers are interested in. The former describes ways in which impulses are *externally regulated* and the *environment* is shaped, whereas the latter describes ways in which such impulses may be *self-regulated* and *behavior* may be shaped. The correspondence between the two becomes apparent when comparing the constituent elements of both models (*see* Fig. 18-3). Indeed, these general categories of emotion-regulation methods can be applied to the stages of the cybernetic process with respect to any domain that can be described using a cybernetic process, and can be used to analyze any process that requires or involves self-control. Applying the categories of regulatory acts that the process model of emotion-regulation offers with the broadly applicable cybernetic process model results in an integrated model that combines the strengths from both of its predecessors: Although maintaining an explicit focus on self-control processes, the new model can easily accommodate, describe, and analyze a broad array of temptation situations and self-control behaviors from a wide range of domains.

By integrating these two models, we have created the cybernetic process model of self-control (*see* Fig. 18-5). According to this general model of self-control, control of behavior may be achieved by applying one of the following methods, or any combination of them: *(1)* Situation selection, *(2)* Situation modification, *(3)* Attention deployment, *(4)* Cognitive change, or *(5)* Response modulation (*see* Table 18-2). Numerous common and effective responses to temptations involve using more than one type of method. A common example is engaging in an alternative activity to distract oneself (e.g., Feindler, Marriott, & Iwata, 1984; Patterson & Mischel, 1976), which combines elements of both attention deployment and response modulation. Successful application of any of these interventions will result in behaving in a manner that is better aligned with long-term goals (e.g., remaining sober, or improving physical fitness), despite competing short-term goals (e.g., drinking alcohol, staying in bed).

In the remainder of this section, we elaborate on each of the five families of self-control. We explain the general principal behind each method, and provide examples to illustrate when and how each may be used, based on examples from the research literature (*see* Table 18-2). To provide examples from a broad range of possible internal conflicts, we demonstrate the application of each method to two cases. In the first case, we present an individual who attempts to down-regulate (i.e., reduce or inhibit) the impulse to consume alcohol. In the second case, we present an individual who attempts to up-regulate (i.e., initiate or increase) the impulse to exercise. Following these examples, we provide a brief overview of research findings related to the methods of self-control that we presented.

Situation Selection

The most forward-looking approach to self-control is *situation selection*. This form of self-control refers to people's attempts to choose situations that make it more (or less) likely that they will experience impulses that lead to desirable (or undesirable) behaviors. In terms of the cybernetic control model, this technique operates on the *environment* element of the loop.

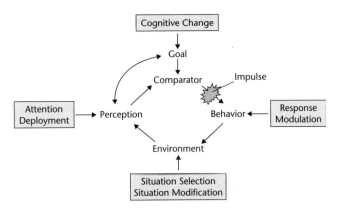

Figure 18–5. The cybernetic process model of self-control, based on cybernetic control theory (inner circle: Carver & Scheier, 1982) and the process model of emotion regulation (outer circle, gray boxes: Gross, 1998b; Gross & Thompson, 2007). The gray explosion represents the generation of an impulse (according to cybernetic control theory) or an emotion (according to the process model of emotion regulation).

TABLE 18–2. EXAMPLES OF EXISTING SELF-CONTROL TECHNIQUES AND THEIR CORRESPONDING STAGES AND INTERVENTION TYPES IN THE CYBERNETIC PROCESS MODEL OF SELF-CONTROL

Cybernetic Stage	Intervention Type	Specific Techniques
Environment	Situation selection	Breaking ties with drug-using associates[1]
		Food-related stimulus control[2]
		Hiding tempting object[3]
	Situation modification	Chemical pleasure blockers[4]
		Community reinforcement[5]
		Precommitment[6]
Perception	Attention deployment	Engaging in alternative activity[7]
		Goal verbalization[8]
		Cognitive load + self-control cues[9]
		Relaxation (e.g., deep breathing, imagery)[10]
Goal/Comparator	Cognitive change	Cognitive reconstrual[11]
		Covert modification[12]
		Soft commitment[13]
		Implementation intentions[14]
		Acceptance and defusion[15]
Behavior	Response modulation	Behavioral suppression[16]
		Engaging in alternative activity[7]
		Relaxation (e.g., deep breathing, imagery)[10]

Note: [1]Schroeder et al., 2001; [2]Foreyt & Goodrick, 1993; Poston 2nd & Foreyt, 2000; [3]Metcalfe & Mischel, 1999; Wansink, Painter, & Lee, 2006; [4]Drugs meant to reduce pleasure from alcohol consumption (e.g., Acamprosate/ Disulfiram/Naltrexone); Luty, 2006; [5]Sisson & Azrin, 1986; [6]Ariely & Wertenbroch, 2002; [7]Kavanagh, Andrade, & May, 2004; Patterson & Mischel, 1976; [8]Genshaft, 1983; Patterson & Mischel, 1976; [9]Mann & Ward, 2004; Parent, Ward, & Mann, 2007; Westling, Mann, & Ward, 2006; [10]Feindler, Marriott, & Iwata, 1984; Mann & Ward, 2004; [11]Fujita, Trope, Liberman, & Levin-Sagi, 2006; Magen & Gross, 2007; Mischel & Moore, 1980; [12]Feindler, Marriott, & Iwata, 1984; [13]Kirby & Guastello, 2001; Rachlin, this volume, but cf. Khan & Dhar, 2007; [14]Gollwitzer & Schaal, 1998; Milne, Orbell, & Sheeran, 2002; Sheeran & Orbell, 1999; [15]Forman et al., 2007; Gifford et al., 2004; Kavanagh, Andrade, & May, 2004; [16]Baumeister, Muraven, & Tice, 2000; Feindler, Marriott, & Iwata, 1984; Muraven & Baumeister, 2000.

For a number of days, Ken has been successfully abstaining from drinking alcohol. Before going to the bank, Ken realizes that his usual route will take him right past the bar in which he often used to drink. Ken decides to take a different, longer route through a local park, which does not have any bars or liquor stores next to it. This way, Ken reasons, he would not have to confront temptations that may be too powerful for him at this stage.

Before moving to a new city, Barbie is determined to begin exercising regularly. As she chooses between two potential new homes, Barbie is left with two appealing alternatives: one that is located very close to her workplace in a central part of the city, and another that is located farther away from her workplace, but close to attractive hiking and biking trails. Barbie chooses the latter, hoping that the accessibility of the trails will help her exercise more often.

In these examples, Ken and Barbie chose environments that would facilitate their goals by eliciting (or not eliciting) impulses that they consider more (or less) desirable. In the addiction literature, overwhelming evidence points to the power of environmental stimuli to evoke drug craving, if these stimuli were previously associated with drug consumption (Bonson et al., 2002; Weiss, 2005). In terms of the cybernetic process, if Ken spends time near certain stimuli (*environment*), he is more likely to notice them (*perception*). This may generate craving (a consumption *goal*), which may lead to alcohol consumption (*behavior*). Conversely, if Ken avoids such stimuli, he is less likely to perceive them, cravings are less likely to be activated, and drinking is less likely to be initiated.

A related line of classic research in social psychology discusses channel factors (Leventhal, Singer, & Jones, 1965), and demonstrates the strong influence that the accessibility of environmental facilitators and hindrances has on behavior. In Barbie's case, choosing to live in the second home will make her more likely to perceive the trails, exercise goals are more likely to become activated (hiking may seem like an attractive and available option), and she is more likely to go outside and enjoy the trails, thereby realizing her long-term goal of exercising more. For other examples of situation selection techniques, *see* Table 18–2.

Situation Modification

Being in a situation that may potentially elicit an undesirable impulse does not mean that this impulse is inevitable. *Situation modification* refers to strategically changing the situation to influence the impulses and subsequent behaviors that will result from it. Because modifying a situation beyond a certain extent can be said to produce a new situation, situation modification and situation selection are not easily separable. In terms of the cybernetic control model, situation modification operates on the *situation* element of the loop.

Ken, who recently decided to abstain from drinking alcohol, is going to his friend's house for dinner. After sitting down at the table, Ken realizes that his friend is wearing a T-shirt that features an advertisement for an alcoholic beverage. Continuing to face the advertisement poses a risk to Ken's determination to abstain from drinking, as it may trigger craving that will be hard to resist. Ken prefers not to cancel the friendly dinner (doing that would qualify as an application of the situation selection technique, and also as potentially rude). Nevertheless, Ken does not want to face the advertisement throughout the dinner. Ken decides to modify the situation by politely explaining this problem to his friend and asking the friend to wear a shirt that does not display such an advertisement.

Barbie just started biking, but would like to go at a more vigorous pace. She switches on her MP3-player, to listen to energizing music while biking, hoping that listening to the music will sustain a higher level of effort on her part.

In both of these examples, Ken and Barbie altered the situations they were in so as to prevent the elicitation of undesirable impulses, or facilitate the elicitation of desirable impulses. In this way, even without avoiding the situation or choosing a new environment, Ken and Barbie successfully prevented impulses which they wished to avoid. Making small changes to tempting situations, such as hiding treats behind a screen or moving them a short distance away, can significantly impact the power that such temptations exert over children (Mischel & Ebbesen, 1970) as well as adults (Wansink, Painter, & Lee, 2006). For other examples of situation modification techniques, *see* Table 18–2.

Attention Deployment

People often find themselves in situations that they cannot easily choose or change (thus ruling out situation selection or situation modification), and which may give rise to problematic impulses and behaviors. Yet even without changing the external situation, it is possible to selectively attend to certain aspects of the situation, to influence the impulses that arise. Situations have many aspects, and *attention deployment* refers to the way in which individuals direct their attention within a given situation to influence their reactions to it. Although this method of self-control does not change the external situation, it can be thought of as an internal version of situation selection, as it changes the *internal* situation that is experienced. In terms of the cybernetic control model, this technique operates on the *perception* element of the loop.

> The train that Ken is riding on his way home from work stops between two stations. The conductor's voice on the PA system apologizes for the delay, and explains that they will be stopped for a few minutes. Looking out the window, Ken realizes that he is stopped right next to a liquor store. Ken knows that continuing to pay attention to this store is likely to result in alcohol craving, which may be too great of a challenge at this stage of his abstinence. The train is packed, and Ken is unable to move elsewhere, cannot make the train start going (although, like many other passengers, he dearly wishes he could), and is unable to change the location or appearance of the store. Shutting his eyes, he finds that his mind wanders to the store, and tempts him to look at it more. Ken shuts his eyes again, this time keeping himself engaged by imagining the furniture in his apartment and trying to reposition the various pieces in his mind's eye. After a few minutes, the train jolts back into motion, and the liquor store disappears behind Ken's back.
>
> After biking for some time, the batteries in Barbie's MP3 player have run out—and Barbie is beginning to feel sore. To shift her attention away from her aching muscles, Barbie listens to the sounds that her bike is making on the dirt path, and improvises a song that incorporates the rhythmic squeaks and clanks.

In both of these examples, Ken and Barbie selectively turned their attention towards certain aspects of the situations that they were in, and away from others. By doing this, Ken and Barbie promoted desirable impulses and behaviors (and prevented the elicitation of undesirable impulses and behaviors), even without changing the external situations that they were in. Ken engaged in mental imagery (Feindler, Marriott, & Iwata, 1984), whereas Barbie kept herself busy by engaging in an alternative behavior (Mischel & Ayduk, 2004; Patterson & Mischel, 1976). Both of these diversions served to prevent them from attending to the stimuli that were eliciting the undesirable impulse. In general, attention deployment can be performed externally (e.g. by covering the eyes or ears) or internally (e.g. through distraction or concentration). This method of self-control is one of the first self-regulatory processes to appear in development (Rothbart, Ziaie, & O'Boyle, 1992), and appears to be used from infancy through adulthood, particularly when it is not possible to select or modify one's situation. For other examples of attention deployment techniques, see Table 18–2.

Cognitive Change

Selecting or modifying our environment is not always an option, and there are times when we need to attend to problematic situations or objects, which may give rise to counter-productive impulses and behaviors. Nevertheless, even in such difficult situations, one can still change the way in which one thinks about the situation, to alter the impulses that are generated in response to perceiving it. *Cognitive change* refers to the way in which people can either strategically transform the relevance of a stimulus to their goal, or change the goal against which they compare the stimulus. In terms of the cybernetic control model, this technique operates on the *goal/comparator* elements of the cybernetic loop.

> Ken, who is carrying on with his efforts to abstain from drinking alcohol, is watching a movie in which one of the actors plays an alcoholic who is trying to stop drinking, and exerts superhuman efforts to this end. At the very end of the

movie, the character surprises most of the audience members by exclaiming that "no human being can possibly resist such cravings" and promptly embarking on a drinking binge. When the lights turn back on, Ken feels shaken—if this impressive character could not do it, how could he? And if he is about to eventually fail, what is the point of going through this suffering in the meantime? As self-doubt gnaws at him, his craving for alcohol grows—if I'm going to start drinking again at some point, I may as well do it now... But Ken manages to calm himself down, by reminding himself that the movie is made for dramatic effect, and doesn't really reflect anything about his own experience. Comforted by this thought, Ken continues with his alcohol-free evening.

Before going to sleep, Barbie decided that she would exercise early on the following morning. When she wakes, exercising seems unappealing, and the bed is so warm and inviting... Exercising somehow seems less important at that moment, and Barbie realizes she is likely to fall back asleep. She then decides to think about the situation differently, and to view getting up for exercise as a test of willpower. Can she do it? Barbie now feels that getting up would be a sign of strength, while staying in bed would be a sign of weakness. Staying in bed suddenly seems less appealing...

In both of these examples, Ken and Barbie strategically changed how they thought about their situations to elicit more desirable reactions (and less undesirable reactions), even without changing the situation or shifting their attention away from the situation. In the example above, Ken used a method that Feindler and colleagues termed "covert modification" (Feindler, Marriott, & Iwata, 1984), and which proved helpful for aggressive school children who learned to control their aggression when responding to the words of others. By doubting the realism of the events in the movie, Ken was able to dramatically reduce the relevance of this information for his goal of remaining sober, and preserve his sense of self-efficacy (Bandura & Locke, 2003; Gwaltney et al., 2002; Zimmerman, Bandura, & Martinez-Pons, 1992). Barbie employed a different strategy—she *reconstrued* her experience and changed

the meaning of her choice. By thinking about her situation as a test of an internal quality that was important to her (willpower), Barbie changed the appeal of her possible choices. In an empirical test of this method, undergraduate students who were performing a math task were distracted by comedy video clips. Students who were instructed to think of the distracting comedy clips as a test of willpower were less distracted by them, and showed less enjoyment when they did attend to them—possibly as a result of perceiving themselves as failing on their own test of willpower (Magen & Gross, 2007). For other examples of cognitive change techniques, *see* Table 18-2.

Response Modulation

There are times when one experiences undesirable impulses which push one to act in a way that is clearly not in one's best interests, such as attacking one's friend, overeating, engaging in unsafe sex, or generally acting in a way which one expects to regret. Fortunately, experiencing powerful undesirable impulses does not necessarily result in undesirable behavior. *Response modulation* refers to the way in which people can attempt to directly control their own behavior despite the impulse that they experience to act in a certain way, by figuratively (or literally) clenching their teeth and willing themselves to behave in a manner that is more aligned with their own long-term goals. In terms of the cybernetic control model, this technique operates on the behavior element of the cybernetic loop.

Shopping at a supermarket, Ken is surprised when a salesman walks up to him and offers a free wine sample from a clear plastic cup. Having been abstinent from alcohol for nearly a week, the powerful impulse to accept the drink almost overwhelms Ken. With the salesman is holding the drink up to him, Ken is unable to turn his attention away from the drink and does not have the wherewithal to think about the situation in a new way. Instead, Ken swallows hard and forces his legs to take him away from the maddeningly aromatic wine. Two aisles later, sweating and breathing deeply, Ken is able to start thinking again.

Biking up an especially difficult part of the trail, Barbie is having a hard time. She is out of breath and can't stop thinking about getting off her bike for a rest. She can't change her situation without abandoning her exercise goal, and she is unable to shift her attention away from her difficulty, or to think about it differently. Nevertheless, Barbie wills herself to continue biking without stopping, and eventually completes her exercise routine. A few minutes later, after completing her ride, Barbie smiles to herself as she stretches.

In both of these examples, Ken and Barbie acted in a way that was different from the impulse they were experiencing. After perceiving the wine, Ken experienced a strong urge to drink the wine (perceiving the alcohol activated a consumption goal), but did not comply with this urge. Similarly, Barbie experienced a strong impulse to stop exercising (perceiving her own pain activated a resting and rejuvenating goal), but acted in opposition to this urge. Baumeister and colleagues (Chapter 20; Muraven, Collins, & Neinhaus, 2002) have been studying this form of self-control in a wide variety of domains, promoting the theory that the capacity for response modulation relies on a limited internal resource, which becomes depleted as a result of prior efforts, much like a muscle that becomes tired following exertion. According to this theory, such depletion results in a short-term reduction in the capacity to successfully apply this form of self-control (but cf. Martijn, Tenbèult, Merckelbach, Dreezens, & de Vries, 2002). For other examples of response modulation techniques, see Table 18–2.

IMPLICATIONS FOR SELECTING AND DESIGNING SELF-CONTROL INTERVENTIONS

With such an array of methods for self-control, how do we choose the method that will serve us best? Which method is most likely to succeed, and at what cost? Are some methods better suited for certain contexts or people? In the remainder of this chapter, we address these questions by considering the properties of the different intervention methods, as well as the role of external and internal contextual factors.

High-Leverage versus Low-Leverage Methods

The ideal place to intervene is usually at the beginning. Unfortunately, the cybernetic process is a recursive loop, and therefore does not have a clear starting point. This difficulty notwithstanding, we assert that intervention at some stages can be more efficient, and thus more likely to be successful than at others. In particular, we propose that the cybernetic stages can be divided into two types: high-leverage and low-leverage. We further propose that self-control methods targeting high-leverage stages (high-leverage methods) are more likely to be successful, and require less sustained effort, than the interventions that target low-leverage stages (low-leverage methods). Table 18–3 presents our proposed prioritization when selecting and designing self-control interventions.

TABLE 18–3. GENERIC PRIORITIZATION OF SELF-CONTROL METHODS, BASED ON DIVISION TO HIGH-LEVERAGE AND LOW-LEVERAGE METHODS

Cybernetic stage	Self-control method
High-leverage methods	
1. Environment	Situation selection/Situation modification
2. Goal/Comparator	Cognitive change
Low-leverage methods	
3. Perception	Attention deployment
4. Behavior	Response modulation

Note: The order of the stages as listed here is different from their chronological order (e.g., Figure 18–5). See text for the definitions of high-leverage and low-leverage methods.

In the cybernetic process, a *high-leverage* stage is one that exerts a strong influence on the subsequent stage, but is not necessarily influenced as strongly by the preceding stage. We propose that the environment exerts a strong influence on perception (the subsequent cybernetic stage), but can clearly exist and change independently of one's behavior (the preceding cybernetic stage). Similarly, we propose that one's goal/comparator exerts a strong influence on one's behavior (the subsequent cybernetic stage), and yet the goal itself can be determined irrespective of the stimuli that are being perceived (the preceding cybernetic stage). Therefore, we consider the environment and the goal/comparator stages to be high-leverage stages.[6]

In contrast, a *low-leverage* stage is one that is strongly influenced by the preceding cybernetic stage, and does not necessarily exert a strong influence over the subsequent stage. We propose that the environment exerts a strong influence on perception, whereas perception does not necessarily exert a strong influence on the goal/comparator. Similarly, we propose that the goal/comparator exerts a strong influence on behavior, whereas behavior does not necessarily exert a strong influence on the environment. Therefore, we consider the perception and the behavior stages to be low-leverage stages.[6]

Research has demonstrated that intervening at any stage of the cybernetic process can be effective, in terms of avoiding unwanted behavior and promoting desirable behavior (Strayhorn, 2002a). Nevertheless, there are important implications to choosing self-control methods that target high-leverage vs. low-leverage stages of the cybernetic process. Interventions that target low-leverage stages (perception, behavior) are likely to require constant energy expenditure and vigilance, since influential input from the preceding stages will continue to feed into them unaltered. Distracting oneself away from a tempting object (attention deployment, e.g. Ken's mental rearrangement of his furniture while sitting on the train next to a liquor store) requires sustained effort for as long as the tempting object is perceivable. Directly controlling one's own behavior to stop oneself from acting in a detrimental manner (response modulation, e.g. biting one's lip to avoid shouting insults at a friend) also requires sustained effort, since as long as the environment does not change, the environment → perception → comparator → impulse flow will continue to produce the same behavioral impulse, and this impulse will have to be continuously overridden via direct response modulation. This type of continuous effort may cause such interventions to backfire, as it exerts a psychological (Gross, 2002; Muraven, Collins, & Neinhaus, 2002; Muraven, Tice, & Baumeister, 1998; Richards & Gross, 2000) and physiological (Gross, 2002) toll on individuals who employ them, and can only be sustained for a limited duration before breaking down (Muraven & Baumeister, 2000; Muraven, Collins, & Neinhaus, 2002; Shiffman, 1984).

In contrast, interventions that target high-leverage stages may effectively alter the trajectory of the process and result in a self-sustaining change. Such an intervention, if performed successfully, does not require the expenditure of additional resources and does not tie up precious psychic resources. Avoiding exposure to a tempting situation (situation selection, e.g. asking a waiter not to bring the dessert menu) removes any subsequent need to resist this temptation, since the option is simply not available. Thinking about a tempting object as an opportunity to display a valued internal quality (cognitive change, e.g. Barbie thinking about getting out of bed as a test of willpower, rather than as simply an opportunity to exercise) changes the relevant goal against which the environment is compared, thereby producing different impulses in response to the same environment (Fujita et al., 2006; Magen & Gross, 2007). Thus, successful implementations of high-leverage interventions can result in lasting change that does not require sustained effort, even when the tempting object remains nearby and available.

The Role of External and Internal Contextual Factors

Both external and internal contextual factors may affect the effectiveness and suitability of particular self-control interventions. With respect to external factors, there are times when high-leverage methods are inappropriate, or when low-leverage interventions may be sufficient for the task at hand. If Barbie realizes that she has difficulty with controlling her anger at work during disagreements with customers, avoiding disagreements altogether (situation selection, a high-leverage self-control method) may not be a satisfactory, or even possible, solution. In this case, Barbie may be wise to explore possibilities for cognitive change (the next-best intervention method). Conversely, a low-leverage intervention may be a perfectly acceptable solution when the self-control effort does not need to be sustained for a long time, as in situations in which in an environment is likely to change very soon (e.g., swallowing nasty-tasting cough syrup).

Internal factors that influence the effectiveness of self-control interventions include transient internal states such as cognitive load or intoxication. Cognitive load occurs when a person's cognitive resources are taxed (e.g., being asked to hold a random digit string in memory while naming the capitols of different countries). The effect of cognitive load on self-control is not straightforward, but a number of studies suggest that the main impact of cognitive load is to make people more reliant upon salient environmental cues to guide their behavior—a reliance which can promote either low or high degrees of self-control, depending on the cues that are present (Mann & Ward, 2004; Parent, Ward, & Mann, 2007; Westling, Mann, & Ward, 2006). Similarly, research suggests that the main effect of intoxication is similar to that of cognitive load (Casbon et al., 2003; Ditto et al., 2006; MacDonald et al., 2003), by causing behavior to become more dependent on external cues.

People appear to underestimate the magnitude of the effect that such changes will have on them (Gilbert, Gill, & Wilson, 2002; Loewenstein, 1996; Nordgren, van der Pligt, & van Harreveld, 2006), a phenomenon which Loewenstein (2005) labeled "empathy gap." This empathy gap is potentially the most pernicious aspect of transient vulnerabilities of the sort we discussed here. The bulk of the evidence suggests that the best strategy may be to rely heavily on situation selection in preparation for times in which cognition may be impaired (e.g., intoxication, cognitive load, fatigue), and situation modification while in these states. Unfortunately, people are not likely to structure environments when they do not realize the extent of their future dependence on external cues. Thus, before drinking with a group of friends, one would be wise to avoid carrying car keys, credit cards, or large amounts of cash, all of which could lead to a variety of problems in the hands of an individual who (temporarily) determines how to act on the basis of the objects around him. Similarly, before going on a date with an attractive but unknown stranger, one would be wise to ensure the availability of contraceptives, rather than relying on their own sound judgment in the event that sex becomes a viable possibility.

Conclusion

In modern society, the role of self-control is perhaps more important than it has ever been before. Increasingly sophisticated marketing techniques have set up an environment that some researchers consider "toxic" (Wadden, Brownell, & Foster, 2002, p. 510), and which exerts its influence on people of an ever-younger age, as evidenced by a recent study demonstrating that children 3–5 years of age reported that food wrapped in McDonald's wrapper tastes better than food wrapped in plain wrapper (Robinson et al., 2007). Such an environment promises an abundance of short-term pleasure—and long-term suffering. Harming ourselves and others is easier than ever, as dangerous foods, drugs and weapons all continue to become increasingly more available, and as physical activity becomes a matter of choice for many members of society. Throughout our lives, the ability to successfully navigate this veritable sea of temptations is of the utmost importance.

Despite decades of research, systematic answers about how to manage temptations have remained elusive, in part because there has been no clear way to organize the multitude of domain-specific findings. In this chapter, we have presented the cybernetic process model of self-control, which provides a domain-general framework for analyzing both the arising and regulation of tempting impulses. This model delineates five general families of self-control methods, and prescribes a system for prioritizing these methods, while considering the idiosyncratic features of the person and the situation at hand. The five families of self-control methods are by no means mutually exclusive—indeed, successful treatment programs often combine several interventions that correspond to a number of the methods in our model (Feindler, Marriott, & Iwata, 1984; Forman et al., 2007; Strayhorn, 2002b). The selection of the specific techniques will depend on the nature of the temptation, as well as the person who will be facing it. We hope that the model we have presented here will prove beneficial for researchers and practitioners alike, by facilitating clear communication regarding the general and domain-specific features of self-control.

Author Note

Eran Magen is a Robert Wood Johnson Foundation Health & Society Scholar at the Leonard Davis Institute of Health Economics in the University of Pennsylvania, Email: eranm@wharton.upenn.edu. James. J. Gross is a professor of psychology at Stanford University. The first author thanks the Robert Wood Johnson Foundation Health & Society Scholars program for its financial support, Rachel Anderson for her patience and support, as well as, of course, his mother.

Notes

1. We used *PsycInfo* to identify peer-reviewed publications containing the phrase "self-control" or "self-regulation" in their title, abstract, or descriptor from 1960 to the present. To project the total number of publications for 2001–2010, we computed the average yearly publication rate for the years 2001–2006 and then multiplied this yearly average by 10. Results were: 1960–1970: 233; 1971–1980: 1,000; 1981–1990: 1,543; 1991–2000: 1,324; 2001–2010: 4,550.
2. We use the word "gain" to mean either the experiencing of a pleasant state, or the avoidance of experiencing an unpleasant state.
3. The word "cybernetic" derives from a Greek word meaning "pilot" or "governor," and relates to ways in which systems (both living and non-living) use feedback to operate more efficiently. The term in its present meaning was coined by Norbert Weiner (1948).
4. The original model included another source of influence on the environment (separate from the individual's behavior) labeled "disturbance." We chose not to display this component to maximize the clarity of the general model of self-control (which we present later).
5. The same authors have also postulated the existence of "anti-goals," which are standards that people wish to avoid, rather than approach (e.g. Carver, 2004). For the sake of simplicity, we limit our present discussion to regular goals, although we believe that our discussion applies equally to both types of goals.
6. Our delineation of high/low leverage stages is purely hypothetical, as we are not aware of existing research that addressed this issue.

References

Ariely, & Wertenbroch. Procrastination, deadlines, and performance: Self-control by precommitment. Psychol Sci 2002; 13: 219–224.

Bandura, & Locke. Negative self-efficacy and goal effects revisited. J Appl Psychol 2003; 88: 87–99.

Bargh, Gollwitzer, Lee-Chai, Barndollar, & Troetschel. The automated will: Nonconscious activation and pursuit of behavioral goals. J Pers Soc Psychol 2001; 81: 1014–1027.

Baumeister, Muraven, & Tice. Ego depletion: A resource model of volition, self-regulation, and controlled processing. Soc Cogn 2000; 18: 130–150.

Bonson, Grant, Contoreggi, et al. Neural systems and cue-induced cocaine craving. Neuropsychopharmacology 2002; 26: 376–386.

Butler, Egloff, Wlhelm, Smith, Erickson, & Gross. The social consequences of expressive suppression. Emotion 2003; 3: 48–67.

Carver. Self-regulation of action and affect. In Baumeister, R. F., & Vohs, K. D. (Eds.), Handbook of self-regulation. New York, NY, US: Guilford Press, 2004: pp. 13–39.

Carver, & Scheier. Control theory: A useful conceptual framework for personality-social, clinical, and health psychology. Psychol Bull 1982; 92: 111–135.

Carver, & Scheier. Origins and functions of positive and negative affect: A control-process view. Psychol Rev 1990; 97: 19–35.

Casbon, Curtin, Lang, & Patrick. Deleterious effects of alcohol intoxication: Diminished cognitive control and its behavioral consequences. J Abnorm Psychol 2003; 112: 476–487.

Ditto, Pizarro, Epstein, Jacobson, & MacDonald. Visceral influences on risk-taking behavior. J Behav Decis Mak. Special Issue: Role Affect Decis Mak 2006; 19: 99–113.

Feindler, Marriott, & Iwata. Group anger control training for junior high school delinquents. Cogn Ther Res, 1984; 8: 299–311.

Fishbach, Friedman, & Kruglanski. Leading us not unto temptation: Momentary allurements elicit overriding goal activation. J Pers Soc Psychol 2003; 84: 296–309.

Fishbach, & Trope. The substitutability of external control and self-control. J Exp Soc Psychol 2005; 41(3): 256–270.

Foreyt, & Goodrick. Evidence for success of behavior modification in weight loss and control. Ann Intern Med 1993; 119(7): 698–701.

Forman, E., Hoffman, K., McGrath, K., Herbert, J., Brandsma, L., & Lowe, M. A comparison of acceptance- and control-based strategies for coping with food cravings: An analog study. Behav Res Ther 2007; 45(10): 2372–2386.

Forster. Seven principles of goal activation: a systematic approach to distinguishing goal priming from priming of non-goal constructs. Pers Soc Psychol Rev 2007; 11: 211.

Frijda. The laws of emotion. Am Psychol 1988; 43: 349–358.

Fujita, Trope, Liberman, & Levin-Sagi. Construal levels and self-control. J Pers Soc Psychol 2006; 90: 351–367.

Genshaft. A comparison of techniques to increase children's resistance to temptation. Pers Individ Dif 1983; 4: 339–341.

Gifford, Kohlenberg, Hayes, et al. Acceptance-based treatment for smoking cessation. Behav Ther 2004; 35: 689–705.

Gilbert, Gill, & Wilson. The future is now: Temporal correction in affective forecasting. Organ Behav Hum Decis Process 2002; 88: 430–444.

Gollwitzer, & Schaal. Metacognition in action: The importance of implementation intentions. Pers Soc Psychol Rev 1998; 2: 124–136.

Gross. Antecedent- and response-focused emotion regulation: Divergent consequences for experience, expression, and physiology. J Pers Soc Psychol 1998; 74: 224–237.

Gross. The emerging field of emotion regulation: An integrative review. Rev Gen Psychol Special New directions in research on emotion 1998; 2: 271–299.

Gross. Emotion regulation: Affective, cognitive, and social consequences. Psychophysiology 2002; 39: 281–291.

Gross, & John. Individual differences in two emotion regulation processes: Implications for affect, relationships, and well-being. J Pers Soc Psychol 2003; 85: 348–362.

Gross, & Thompson. *Emotion regulation: Conceptual foundations.* New York, NY, US: Guilford Press, 2007.

Gwaltney, Shiffman, Paty, et al. Using self-efficacy judgments to predict characteristics of lapses to smoking. J Consult Clin Psychol 2002; 70: 1140–1149.

Hsee, & Abelson. Velocity relation: Satisfaction as a function of the first derivative of outcome over time. J Pers Soc Psychol 1991; 60: 341–347.

Jonas, Schulz-Hardt, Frey, & Thelen. Confirmation bias in sequential information search after preliminary decisions: An expansion of dissonance theoretical research on selective exposure to information. J Pers Soc Psychol 2001; 80: 557–571.

Kavanagh, Andrade, & May. Beating the urge: Implications of research into substance-related desires. Addict Behav Special Issue: Cross Bound Implicat Adv Basic Sci Manag Addict 2004; 29: 1359–1372.

Khan, & Dhar. Where there is a way, is there a will? The effect of future choices on self-control. J Exp Psychol Gen 2007; 136: 277–288.

Kirby, & Guastello. Making choices in anticipation of similar future choices can increase self-control. J Exp Psychol Appl 2001; 7: 154–164.

Kunda. The case for motivated reasoning. Psychol Bull 1990; 108: 480–498.

Lawrence, Carver, & Scheier. Velocity toward goal attainment in immediate experience as a determinant of affect. J Appl Soc Psychol 2002; 32: 788–802.

Lazarus. Progress on a cognitive-motivational-relational theory of emotion. Am Psychol 1991; 46: 819–834.

Leventhal, Singer, & Jones. Effects of fear and specificity of recommendation upon attitudes and behavior. J Pers Soc Psychol 1965; 2: 20–29.

Loewenstein. Out of control: Visceral influences on behavior. Organ Behav Hum Decis Process 1996; 65: 272–292.

Loewenstein. Hot-cold empathy gaps and medical decision making. Health Psychol Special Issue: Basic Appl Decis Mak Cancer Control 2005; 24: S49–S56.

Lord, Ross, & Lepper. Biased assimilation and attitude polarization: The effects of prior theories on subsequently considered evidence. J Pers Soc Psychol 1979; 37: 2098–2109.

Lowe, & Levine. Eating motives and the controversy over dieting: Eating less than needed versus less than wanted. Obes Res 2005; 13: 797–806.

Luty. What works in alcohol use disorders? Adv Psychiatr Treat 2006; 12: 13–22.

MacDonald, Fong, Zanna, & Martineau. *Alcohol myopia and condom use: Can alcohol intoxication be associated with more prudent behavior?* New York, NY, US: Psychology Press, 2003.

Magen, & Gross. Harnessing the need for immediate gratification: Cognitive reconstrual modulates the reward value of temptations. Emotion 2007; 7: 415–428.

Mann, & Ward. To eat or not to eat: Implications of the attentional myopia model for restrained eaters. J Abnorm Psychol 2004; 113: 90–98.

Martijn, Tenbèult, Merckelbach, Dreezens, & de Vries. Getting a grip on ourselves: Challenging expectancies about loss of energy after self-control. Soc Cogn 2002; 20: 441–460.

Mauss, Cook, & Gross. Automatic emotion regulation during an anger provocation. J Exp Soc Psychol 2007; 43: 698–711.

Metcalfe, & Mischel. A hot/cool-system analysis of delay of gratification: Dynamics of willpower. Psychol Rev 1999; 106: 3–19.

Milne, Orbell, & Sheeran. Combining motivational and volitional interventions to promote exercise participation: Protection motivation theory and implementation intentions. Br J Health Psychol 2002; 7: 163–184.

Mischel, & Ayduk. Willpower in a cognitive-affective processing system: The dynamics of delay of gratification. In Baumeister, R. F. & Vohs, K. D. (Eds.), *Handbook of self-regulation*. New York, NY, US: Guilford Press, 2004: pp. 99–129.

Mischel, & Ebbesen. Attention in delay of gratification. J Pers Soc Psychol 1970; 16: 329–337.

Mischel, & Moore. The role of ideation in voluntary delay for symbolically presented rewards. Cognit Ther Res 1980; 4: 211–221.

Muraven, & Baumeister. Self-regulation and depletion of limited resources: Does self-control resemble a muscle? Psychol Bull 2000; 126: 247–259.

Muraven, Collins, & Neinhaus. Self-control and alcohol restraint: An initial application of the Self-Control Strength Model. Psychol Addict Behav 2002; 16: 113–120.

Muraven, Tice, & Baumeister. Self-control as a limited resource: Regulatory depletion patterns. J Pers Soc Psychol 1998; 74: 774–789.

Nickerson. Confirmation bias: A ubiquitous phenomenon in many guises. Rev Gen Psychol 1998; 2: 175–220.

Nordgren, van der Pligt, & van Harreveld. Visceral drives in retrospect: Explanations about the inaccessible past. Psychol Sci 2006; 17: 635–640.

Parent, Ward, & Mann. Health information processed under limited attention: Is it better to be "hot" or "cool?" Health Psychol 2007; 26: 159–164.

Patterson, & Mischel. Effects of temptation-inhibiting and task-facilitating plans on self-control. J Pers Soc Psychol 1976; 33: 209–217.

Poston 2nd, & Foreyt. Successful management of the obese patient. Am Fam Physician 2000; 61: 3615–3622.

Richards, Butler, & Gross. Emotion regulation in romantic relationships: The cognitive consequences of concealing feelings. J Soc Pers Relat 2003; 20: 599–620.

Richards, & Gross. Emotion regulation and memory: The cognitive costs of keeping one's cool. J Pers Soc Psychol 2000; 79: 410–424.

Robinson, Borzekowski, Matheson, & Kraemer. Effects of fast food branding on young children's taste preferences. Arch Pediatr Adolesc Med 2007; 161: 792–797.

Rothbart, Ziaie, & O'Boyle. Self-regulation and emotion in infancy. In.San Francisco, CA, US: Jossey-Bass, 1992. Illicit drug use in one's social network and in one's neighborhood predicts individual heroin and cocaine use. Ann Epidemiol 2001; 11: 389–394.

Sheeran, & Orbell. Implementation intentions and repeated behaviour: Augmenting the predictive validity of the theory of planned behaviour. Eur J Soc Psychol 1999; 29: 349–369.

Shiffman. Coping with temptations to smoke. J Consult Clin Psychol 1984; 52: 261–267.

Sisson, & Azrin. Family-member involvement to initiate and promote treatment of problem drinkers. J Behav Ther Exp Psychiatr 1986; 17: 15–21.

Strayhorn. Self-control: Theory and research. J Am Acad Child Adolesc Psychiatr 2002; 41: 7–16.

Strayhorn. Self-control: Toward systematic training programs. J Am Acad Child Adolesc Psychiatr 2002; 41: 17–27.

Tamir. Don't worry, be happy? Neuroticism, trait-consistent affect regulation, and performance. J Pers Soc Psychol 2005; 89: 449–461.

Tamir, & Robinson. Knowing good from bad: The paradox of neuroticism, negative affect, and evaluative processing. J Pers Soc Psychol 2004; 87: 913–925.

Vallacher, & Wegner. What do people think they're doing? Action identification and human behavior. Psychol Rev 1987; 94: 3–15.

Vohs, Baumeister, & Ciarocco. Self-regulation and self-presentation: Regulatory resource depletion impairs impression management and effortful self-presentation depletes regulatory resources. J Pers Soc Psychol 2005; 88(4): 632–657.

Wadden, Brownell, & Foster. Obesity: Responding to the global epidemic. J Consult Clin Psychol Special Issue: Behav Med Clin Health Psychol 2002; 70: 510–525.

Wansink, Painter, & Lee. The office candy dish: Proximity's influence on estimated and actual consumption. Int J Obes 2006; 30: 871–875.

Weiner. Cybernetics. Sci Am 1948; 179: 14–18.

Weiss. Neurobiology of craving, conditioned reward and relapse. Curr Opin Pharmacol 2005; 5: 9–19.

Westling, Mann, & Ward. Self-control of smoking: When does narrowed attention help? J Appl Soc Psychol 2006; 36: 2115–2133.

Zimmerman, Bandura, & Martinez-Pons. Self-motivation for academic attainment: The role of self-efficacy beliefs and personal goal setting. Am Educ Res J 1992; 29: 663–676.

CHAPTER 19

Implicit Control of Stereotype Activation

Gordon B. Moskowitz and Peizhong Li

ABSTRACT

Control typically is conceived of as a set of conscious responses to an activated goal and the subsequent attempts at goal pursuit. Control includes the self-regulatory steps one initiates by conscious willing to address a selected goal and to allow one to pursue that goal. However, these attempts to compensate for a desired end state that is not yet achieved are not limited to conscious acts of deliberation, planning, and the resulting behavior. Implicit cognition plays an important role in preparing the individual to act, in allowing the individual both to detect goal-relevant stimuli in the environment and to process goal-relevant stimuli in a fashion that will facilitate goal achievement and shield one from distractions that could potentially derail attempts at self-control. Such automatic thought includes processes of spreading activation and inhibition as well as attentional selectivity. Evidence in support of such automatic processes of self-control is provided from the domain of stereotype control and the preconscious cognitive operations that result when one attempts to regulate the goal to be egalitarian and non-stereotypic in one's dealings with others.

Keywords: automaticity, egalitarian goals, goal activation (priming), goal inhibition, goal shielding, implicit goal, stereotype activation, stereotype control, stereotype inhibition, temporary goals

Stereotypes are pervasive in human judgment. It has long been assumed that their use, for good or bad, arises from a default goal that guides interpersonal perception. That goal is to easily and quickly arrive at meaning (understanding another person and the ability to predict what that person is likely to think, feel, and do) by using existing categories, schemas, and stereotypes that represent the summation of one's past experiences and learned knowledge about the people and actions being observed (e.g., Allport, 1954; Bruner, 1957; Heider, 1944). This goal is derived from a motive to gain knowledge/attain meaning—an epistemic motive–which Heider (1944) equated to a need. Speed and efficiency (fast closure) as a goal for pursuing this motive arises from limitations facing the processing system (e.g., Bruner, 1957). These limitations arise both from *(1)* risks presented by stimuli in the environment, the vast amount of stimuli that need to be processed in any moment, and the speed with which one needs to react to stimuli (all stimulus-driven factors) and, *(2)* capacity restrictions inherent in the brain's ability to process information. This goal is so pervasive, and the use of stereotypes that enable attaining the goal so routine, that both the goal and the processes used to attain it achieve a state of invisibility—they are implicit.

It is argued that despite the fact people often do not know they have stereotypes primed, or that they even have goals that cause the processing system to trigger those stereotypes, these processes are still intended. Because it is in the service of goals, stereotyping is purposive in nature and, despite its invisibility to the perceiver, controlled (at some level). The focus of this chapter is on the issue of a stereotype's controllability even when triggered and used outside the person's conscious awareness. Such control may seem difficult and, according to some models of mind, impossible. We argue that implicit goals, however nonconscious, still reside within a self-regulatory system that manages the pursuit of that goal in the context of one's other goals, and such management—especially when conflict among goals exists—may lead to implicit control.

What type of goal conflict can lead to stereotype control? Although there are many possibilities, we focus on conflict between implicit goals arising during person perception—the goal aforementioned that makes use of existing stereotypic knowledge, and the goal to be fair and egalitarian in how one evaluates other people. Thus, epistemic motives give rise to goals that are pursued through cognitive processes/operations that yield the activation, and often the use, stereotypes. But it is also argued that epistemic motives are not limited to being serviced by goals that produce a reliance on stereotypes, nor are they the only motives that shape the goals one forms during interpersonal perception and interaction. When goals to be fair and egalitarian exist, these goals initiate cognitive processes/operations that inhibit the activation of stereotypes.

A perceiver may interact with a person from a stereotyped group who could potentially cue dramatically different goals, such as the goal to be fair and the goal to stereotype. These goals need not be explicitly recognized by the person to be triggered or to be pursued. We propose that even when these goals are implicitly activated, conflict between them is resolved through goal shielding processes (e.g., Shah, Hall, & Leander, 2009) that promote the goal of being egalitarian. This includes the inhibition (rather than activation) of stereotypes, heightened accessibility of the goal and a facilitated ability to detect means to that goal, and one's limited processing resources being managed in a manner such that they are diverted to operations required for the goal pursuit (draining their availability for processing unrelated to the goal).

The Implicit Nature of Stereotype Activation

A stereotype is typically defined as a mental representation—a cognitive construct that may be activated from the presence of a particular cue in the environment (Allport, 1954; Banaji, Hardin, & Rothman, 1993; Bargh, 1999; Correll, Park, Judd, & Wittenbrink, 2002; Devine, 1989; Duncan, 1976; Hamilton & Gifford, 1976; Stangor & Lange, 1994; Von Hippel, Sekaquaptewa, & Vargas, 1995). Its specific contents (in the form of beliefs about the group, beliefs thought to be held by the group, exemplars, traits, social forces thought to affect the group, and knowledge about the group's behavior and norms) may vary from group to group, but the processes through which the representation is activated are proposed to stay constant. Separate from the activation or priming of the stereotype is its *use* in shaping judgment, recall, and behavior. The myriad forms of bias that result from using stereotypes once activated are beyond the scope of this review (for reviews, see Hamilton & Trolier, 1986; Stangor & Lange, 1994; Von Hippel, Sekaquaptewa, & Vargas, 1995). Dissociating these two processes—use and activation—raises important distinctions regarding stereotype control (e.g., Blair & Banaji, 1996; Devine, 1989; Kunda & Spencer, 2003; Moskowitz, Gollwitzer, Wasel & Schaal, 1999).

Devine (1989) argued that regardless of one's goals, even if one is a low-prejudice person, knowledge of a social stereotype of a group will require that stereotype to be activated automatically upon the mere presence of a group member. Activation was said to be inescapable:

"The present model assumes that...because the stereotype has been frequently activated in the past, it is a well-learned set of associations that is automatically activated in the presence of a member (or symbolic equivalent) of the target group. The model holds that this unintentional activation is equally strong and equally inescapable for high- and low-prejudice persons" (emphasis in original, Devine, 1989, p. 6).

The argument that stereotyping has a phenomenology that renders the process invisible to the person engaged in such a process is at least as old as Allport (1954). The ease and efficiency of stereotype activation had already been illustrated in experiments by Duncan (1976) and Dovidio et al. (1986). Devine illustrated that this process was not merely efficient, but automatic, and thus unstoppable. This was distinguished from the controlled nature of stereotype use, and as such, regulation of stereotyping was said to require a focus on stereotype use.

THE IMPLICATIONS OF THE IMPLICIT ACTIVATION OF STEREOTYPES FOR STEREOTYPE CONTROL

Although the processes that promote stereotype use may require little attention or effort to be engaged, they are not, as Devine (1989) suggested with stereotype activation, inevitably engaged. Devine (p. 15) stated: "non-prejudiced responses are, according to the dissociation model, a function of intentional, controlled processes, and require a conscious decision to behave in a nonprejudiced fashion." Although stereotype activation is not controllable, a distinctly different form of processing can override the outputs of such automatic processing. This separate mode of thinking is based in consciousness and the effortful exertion of specific types of cognition that require cognitive resources. Because such processing is effortful, it is said to be initiated only when the regulatory system is engaged—when one has a goal that conflicts with stereotype use.

Therefore, goals determine whether one's response toward a target is biased by (compatible with) an automatically activated stereotype or impacted by effortful (resource-dependent) processing. From a dual process perspective, self-regulation in stereotyping is successful when one's explicit goals engage one in processing that is incompatible with the use of stereotypes (e.g., Brewer, 1988; Devine, 1989; Fiske & Neuberg, 1990). Much like the Stroop effect reveals that people automatically have color names activated from semantic content, yet they can control the expression of those color names with conscious intent (evidenced by a delay in responding to the naming of incompatible font names), people can control stereotype use, despite the activation of the stereotype's content. It is difficult (requires effort), as this model argues, but it is possible.

Although not the focus of this chapter, even this type of stereotype control is an example of goal conflict. Whence the conflict? Livingston and Brewer (2002) assert that "automatic" stereotype activation is actually not automatic but dependent on one having a specific, higher-order, processing goal—the goal to categorize. Their experiments expose research participants to faces of African-American men and reveal stereotypes are only activated when the faces are seen in the context of one having the goal to categorize. The fact that goals to categorize may be the norm, present in most interpersonal interactions, would render stereotype activation the usual consequence of interpersonal perception. However, despite stereotype activation being functionally automatic (because it usually occurs), in theory it is not automatic—"usually" is not the same as "always." This difference between what typically may happen and what may happen in theory instead is not trivial. It suggests that even an implicit and pervasive cognitive process such as stereotype activation will not occur if the goal to categorize is not in place (we return to this important issue soon). However, it also suggests that in most interpersonal contact situations such a goal is in place, and it gives rise to stereotype activation. Therefore, conscious goals that promote control over the use of stereotypes (such as accountability, accuracy, outcome dependency, and fairness goals) would create goal conflict with the implicitly triggered goal.

Problems with Explicit Control

The possibility for explicit regulation of stereotype use is not to be confused with its plausibility. Reviewing the notion of conscious control more generally, Wilson and Brekke (1994) focused on what they called *mental contamination*, an array of implicit biases on judgment, emotion, and behavior. Their review of the possibilities for control over such contaminants, what they called *mental correction*, is essentially the same as the dual process model of stereotype control described above. Because the source of the contaminant is implicit, its cure is assumed to come after the fact by conscious control. To do so requires the person to be aware of the contaminated judgment, motivated to correct the judgment (to remove this contamination), in possession of a theory of how one is being influenced (the magnitude and direction of influence) and a theory of how precisely to accurately remove that influence, and to possess the cognitive resources to implement that strategy. We would add that it also requires that the implemented strategy should not introduce a new contaminant, one that unintentionally biases judgment (such as the ironic effects of thought suppression; e.g., Wegner, 1994).

Unlike the possibility for control over the contaminating influence of a stereotype, Wilson and Brekke (1994) discuss its plausibility, with rather bleak predictions for successful negotiation of this regulatory conflict: "we are rather pessimistic about people's ability to avoid or correct for mental contamination. We suggest that the nature of human cognition, as well as the nature of lay theories about the mind, makes it difficult to satisfy all the conditions necessary to avoid contaminated responses" (p. 120). Bruner (1957) long ago suggested that implicit processes evolve largely because the prerequisites for more effortful cognition are typically not in place. One lacks awareness of one's biases, and even if aware, one typically lacks an accurate theory of how to correct for the bias. Yet even if aware of bias and possessing the knowledge of how to correct it, the press of the situation often leaves one without the time or cognitive resources to implement that correction strategy (Moskowitz, Skurnik, & Galinsky, 1999).

Finally, even if one does have the awareness, knowledge, and ability to implement a regulatory strategy, the regulatory conflict between the goal to stereotype and the goal to not stereotype may still not be successfully resolved. Research has shown that even successful stereotype suppression can actually lead to subsequent levels of stereotype activation and use that are increased compared to control conditions (Wegner & Erber, 1992) and even increased compared to people who are explicitly using stereotypes (Galinsky & Moskowitz, 2000; 2007). Macrae et al. (1994) provided an example of such a failure at explicit stereotype control. Participants told not to use stereotypes when writing a story successfully suppressed the stereotype in a first story compared to a control group. However, the use of the stereotype was heightened when writing a second story (when the explicit goal to suppress stereotypes was no longer in place). An implicit measure of accessibility (response times on a lexical decision task) revealed a similar finding. People suppressing their stereotypes later had increased accessibility (faster reaction times) to stereotype-relevant words compared to a control group. Unfortunately, although people can control stereotypes through the conscious goals they adopt, goal conflict alone does not guarantee success. One must be able to implement the goal that conflicts with goals that promote implicit stereotyping. This is difficult and fraught with error.

The Implicit Nature of Stereotype Control: Stereotyping Conflicts with Egalitarian Goals

Given the many threats to successful pursuit of one's explicit goals to control stereotypes (including the large literature on ambivalent prejudice revealing that modern forms of prejudice are subtle and leave people lacking awareness they are biased and are in need of control; e.g., Dovidio & Gaertner, 1986), our research attention turned toward examining a more proactive form of control over stereotyping to complement the conscious control models existing in the literature. This research poses the question as to whether preconscious control is possible—is stereotype activation controllable, and if so, what is the role of implicit goals?

It is important to note at the outset that stereotype activation, although implicit, has been shown from a variety of different approaches to not be inevitable. The activation of a stereotype upon exposure to a cue that is associated with the stereotype can be disrupted by expectancies that lead one to associate nonstereotypical knowledge with the cue (Blair & Banaji, 1996). It can be disrupted when a nonstereotypical response is made habitual upon exposure to the cue (Kawakami, Dovidio, Moll, Hermsen, & Russin, 2000). It can be disrupted when cognitive load exists that prevents the person from having the resources that these processes, although implicit, require (Gilbert & Hixon, 1991). However, these many examples of the disruption of stereotype activation are not examples of goal conflict, nor are they examples of control being exerted in the name of self-regulation. Not all disruptions of the implicit operations that give rise to stereotype accessibility are the result of a goal to inhibit or prevent such processes from arising. In some cases, the triggering of associations that are counter-stereotypical occurs in the absence of a goal to initiate control, with the disruption instead resulting from associative learning. In other cases, disruption of a stereotype's activation is indeed a case of control. It is precisely such a case that we examine in this chapter by exploring the conflict between goals that promote stereotyping versus fairness.

In this sense, the control of stereotyping explored in this chapter falls within a general class of regulatory processes that contain the regulation of unwanted responding, emphasizing the implicit operations that help the individual to control an undesired thought or action. Classic examples include the regulation of one's weight (the goal to lose weight and avoid eating tempting foods; e.g., Fishbach, Friedman, & Kruglanski, 2003; Papies, Stroebe, & Aarts, 2007) and the regulation of one's thought (the goal to eliminate specific thoughts from entering consciousness; e.g., Wegner, 1994). Despite the fact that the unwanted response may share the same triggering stimulus as the desired response, and despite the fact that the triggering stimulus thus represents a temptation away from the desired response (and may have a history of being associated with the unwanted response/temptation), the regulatory system can learn new associations to this stimulus in the form of new goals and the implicit operations associated with these goals. This is also true of goals relating to stereotyping and stereotype control.

A person from a particular group (e.g., Women, Turks, Jews, frat boys, lawyers, etc.) may have consistently been paired with a stereotype throughout one's experience, yet it is possible for goals inconsistent with the stereotype to be selected, pursued, and associated with the group. In essence, the same stimulus person may represent both a trigger for the goal to be egalitarian, as well as a temptation away from that goal and toward the goal of attaining meaning quickly be relying on stereotypes. This struggle between competing goals is a regulatory battle we contend will get fought at a largely implicit level, resulting in the inhibition of stereotypes.

It is also important to note at the outset the distinction between the form of stereotype control being reviewed in this chapter and that suggested by Livingston and Brewer (2002). They argued that goals determine whether stereotypes are activated or not, illustrating that when a goal to categorize was not adopted by perceivers there was no processing initiated that resulted in stereotype activation—stereotypes need not be inhibited or suppressed because the category that might trigger the stereotype was never activated in the first place. This is not the type of control over stereotype activation of concern in this chapter. It is akin to saying people can control seeing something by closing their eyes. Such control is important, because it argues that people have the ability to look at others and not see group membership (e.g., they can be "color-blind"), even when the groups in question are categories such as race and gender and age. It suggests people have the ability to see others as individuals, not as category members.

However, our concern is with people whose eyes are wide open, who use the categories that are evidenced in the person's actions and features. Rather than examining the fact that people at times do not stereotype because they do not categorize others according to stereotypic groups, we instead examine the possibility that after having categorized a person as a member of a group, the stereotype that may be otherwise associated with, and triggered by, the category is instead inhibited and that this occurs as a result of goal conflict. When there is a conflict between two (or more) goals potentially associated with a category, the focal, or dominant, goal in a given context will be the one triggered by the category in that context. The movement from category activation to the activation of stereotypes involves several mediating steps. And the type of processing engaged in after category activation depends upon which goals associated with the category are directing the executive functions. Thus, people who categorize a person with a group label that may typically activate an associated stereotype will be able to control the stereotype activation part of this associative chain. This is possible if they have goals that compete with stereotype activation also associated to the category, such as the goal to be egalitarian. Egalitarian goals alter the route from category to stereotype, and we propose inhibit the competing goal.

Implicit Goals

The logic of this type of implicit goal conflict—where goals one does not know one has direct cognition one does not know one is engaged in—only became plausible as an argument following developments in the past two decades on the nonconscious nature of the self-regulatory system. Bargh (1990) introduced the concept of automotives—goals that are automatically triggered. It had long been known that when a goal had been selected, implicit cognitive operations that service the attainment of that goal may be triggered (for a review, see Srull & Wyer, 1986). Indeed, the history of cognitive psychology (and social cognition) is replete with such examples.[1] As Bargh noted, such illustrations begin with an explicit instruction to pursue a goal and are thus examples of the implicit operations that accompany explicit goals. What Bargh introduced was the possibility of wholly implicit goal pursuit, where not merely are the steps taken toward attaining the goal outside awareness but the "selection" of the goal itself is implicit.

Goal Priming

Bargh (1990) described goals as cognitive structures, similar to (but with important differences from) other social constructs such as schemas and stereotypes. Because they are stored as mental representations, goals can be triggered by the environment (both the external environment and the internal, mental life of the individual). Moskowitz and Gesundheit (2009) review a variety of ways that such goal priming can be attained without the individual's awareness. These include implicitly inferring goals from the action of others and having them "caught" by the perceiver (Aarts, Gollwitzer, & Hassin, 2004), subliminal exposure to goal-relevant words (Chartrand & Bargh, 1996; Shah & Kruglanski, 2003; Strahan, Spencer, & Zanna, 2002), subliminal exposure to relevant others who hold specific goals for the individual (Fitzsimons & Bargh, 2003; Shah, 2003), implicitly inferring the meaning of the difficulty one experiences when attempting to suppress a goal (Förster & Liberman, 2001; Liberman & Förster, 2000), and enacting behaviors that are associated with a specific goal (Li & Moskowitz, 2008). Finally, goals can have chronic states of heightened accessibility such that the individual need not have the goal triggered by any environmental cue for the regulatory system to be monitoring pursuit of the goal (e.g., Custers & Aarts, 2007; Moskowitz, 1993; Moskowitz, Salomon, & Taylor, 2000).

Moskowitz and Gesundheit (2009) also note that goals can have heightened accessibility without one's awareness even when the goal had been consciously selected at some prior point. This can occur when the person does not realize that the information they are consciously attending to is triggering a goal (e.g., Bargh et al., 2001; Chartrand & Bargh, 1996; Sassenberg & Moskowitz, 2005). It may also occur when the individual does not recognize that a subsequent task is relevant to, and their responses are being influenced by, the goal that had previously been consciously triggered (e.g., Aarts, Dijksterhuis, & Midden, 1999; Dijksterhuis, Bos, Nordgren, and Van Barren, 2005; Gollwitzer & Brandstätter, 1997; Koole, Smeets, van Knippenberg, & Dijksterhuis, 1999; Moskowitz, 2002; Woike, Lavezzary, & Barsky, 2001).

Goal Systems and Goal Shielding

A generation of research following the proposition of the automotive concept has empirically demonstrated that implicitly held goals may also be implicitly regulated through operations associated with the goal (e.g., Chartrand & Bargh, 1996; Moskowitz, 1993; Shah, 2003). Such preconscious control can impact how one thinks and acts without conceptualizing control as the "overturning" of a previously produced response. In the case of stereotype activation, we propose that the implicit operations that often give rise to stereotype activation are not inevitable, even when the goal to categorize is in place. Rather, other goals of the individual may be compatible (e.g., to restore self-esteem) or incompatible (e.g., to be creative; to be egalitarian) with stereotype activation, triggering cognitive operations that inhibit the operations that might otherwise produce the heightened accessibility of a stereotype. This logic has been supported by research outside of the domain of stereotyping.

Kruglanski et al. (2002) described a given goal as existing within a system of related representations. Such goal systems require coordination among goals such that movement toward one goal can impact movement toward another. This impact can be facilitative or inhibitory in nature—goals can be compatible or can compete. Goal compatibility arises in instances such as when one lower order goal is a step needed along the way to attaining a higher order goal. It arises when the same behavior (or means) is useful for accomplishing two distinct goals (what Kruglanski et al. called "multifinality"). It arises when the same cognitive operations are useful for preparing one to pursue two distinct goals (e.g., Hassin, Aarts, Eitam, Kleinman & Custers, 2009; Shah, 2003). The goal systems approach suggests that an incompatible goal can have an inhibitory effect on the accessibility, commitment, and pursuit of a target goal.

For example, Shah (2003) illustrated the inhibitory effect of an incompatible goal on the accessibility and pursuit of a target goal. Participants were primed either with a goal compatible or incompatible with the goal of verbal fluency. They later completed an anagram task that was described as a measure of verbal fluency. The primed goal representations increased and decreased goal accessibility as predicted—if a goal incompatible with verbal fluency was primed then the goal of verbal fluency was inhibited. This was evidenced by poorer performance on the anagram task. Aarts, Custers, and Holland (2007) also illustrated intergoal inhibition. People were given either the goal of socializing or a control goal. Next a sequential priming task was used in which participants were primed with a goal incompatible with socializing—studying. The results were that the goal to socialize was inhibited when the goal to study was primed, whereas no inhibition was found to content related to socializing for people in the control group.

Counteractive Control

The incompatibility between goals, and the inhibitory processes that arise as a result, is also illustrated by research on counteractive control (Trope & Fishbach, 2000; Fishbach et al., 2003). The name given to this phenomenon makes goal conflict explicit—the operations performed in the service of one goal need to be counteracted, or altered, because of their incompatibility with another goal of the individual. This work takes as a starting point the assumption that the goal system is hierarchically organized, with higher order goals specifying lower order goals, which in turn specify means to the goal and temptations away from the lower order goal. A cue associated with one goal, or a means to one goal, might be a temptation away from another goal to which it is also associated. If one encounters the cue, does it trigger the goal it promotes, or does it trigger the goal it detracts from (as a temptation away from that goal), thus allowing the regulatory system to counteract the pull of the temptation and shield/bolster the competing goal? Can a means to one goal actually trigger a competing goal that inhibits the desired end-state to which that cue is typically associated?

Let us consider a concrete example. The higher order goal to "achieve," a lower order goal "to run 3 miles every other day," a lower order goal "to consume comfort foods," and a particular cue that could serve as a means to some of these goals, "chocolate cake with cream cheese icing," might all be organized hierarchically within an individual's goal system, with specific intergoal relationships. In one person the goal of running 3 miles could actually facilitate the goal of eating comfort food, because the exercise could justify the consumption of the extra calories. For this person, the opportunity to devour the chocolate cake would be embraced because it is compatible with two lower order goals (eating comfort food and running). Further, the exercise goal, thus facilitated, might allow the person to address the higher order goal to achieve, because exercise is one route to better achievement (a healthy body to this individual facilitates a healthy and productive mind). Additionally, eating comfort food might have a similar effect, with a satisfied stomach promoting better work (or at least a reduction in stress).

However, this example provides many possibilities for goal conflict among these very same goals and means in the goal systems of different individuals. For one person the "running goal" would inhibit the goal of "eating comfort food," because such goals are incompatible with the higher order goal that produced the goal to run (such as the goal of maintaining good health). To such a person the chocolate cake could actually trigger the goal to run and inhibit the goal to eat, as a way to counteract the temptation from the cake to one's goals of eating healthfully and being in good shape. To another person the goal to "run" might be incompatible with achievement because such an activity, and the showering and changing of clothes associated with it, takes time away from the overwhelming number of activities one may perceive achievement to require. Thus the opportunity to run might actually make one more likely to work and to inhibit thoughts of running. For this same person, the goal to eat comfort food might be compatible with achievement, although the goal to exercise is not—the person works better when indulging in the food domain. For this person the cake could trigger the eating goal as well as the achievement goal and thus result in a goal system that counteracts the desire to exercise that eating cake may normally trigger. Exercise, being incompatible with achievement, gets inhibited when a cue (cake) that normally triggers the exercise is encountered.

In summary, whether exercise is organized in the system as an interference to or facilitator of achievement will dictate whether inhibition or facilitation of exercise is triggered when one encounters chocolate cake. Thus, in this model, because goals exist in a connected and hierarchically organized system, any goal can constitute an interfering temptation with respect to another higher order goal, whereas that same goal could be an overriding goal with respect to another interfering temptation. The system counteracts for the pull of one goal when that "pull" is incompatible with other goals. In the domain of stereotyping, think of people as the chocolate cake, egalitarian goals as the goal to run, and goals to stereotype as the goal to eat comfort food (with epistemic goals and categorization goals playing the part of achievement goals).

Before examining this sort of goal conflict and goal-shielding in the domain of stereotyping, experimental evidence for counteractive control is summarized to bolster the illustrative example described above. Trope and Fishbach (2000) investigated how needing to overcome short-term costs influenced the likelihood that individuals engaged in behaviors with implications for long-term goals (counteracting detrimental action in the short term with behavior beneficial for the long term). Participants were offered money to take a blood glucose test as part of a study of its influence on cognitive functioning. The test required them to refrain from eating foods containing glucose for either 6 hours or 3 days, and failure to do so (as the blood test would reveal) would mean not being paid and being penalized. Then they were asked to report how much money they would be willing to pay (ranging from $0 to $18) for failing to follow the study requirements (to cover costs of the canceled appointment). Participants penalized themselves for failure to comply during a long period of abstinence with higher penalty fees than for failure during a short period of abstinence. These results indicate that participants used the fee as a motivational tool, setting higher fees for failure to comply during a longer period of abstinence, which they perceived as more difficult. They counteract for the costs associated with the test by the use of self-control strategies.

Another experiment examined how the value placed on attaining a long-term goal influences the use of counteractive control in the face of short-term costs. Participants rated how important assessing and improving one's health was to them. Then they were told about a test that could assess one's risk for heart disease. The experimenters varied the supposed discomfort of the test from extremely high to low. Finally, participants were also told they would receive extra credit and had to decide if they wanted to receive the extra credit before or after the test. If before, then they did not actually have to complete the test; however, if after, then they could lose the extra credit if they did not finish the test. If health was important to the participants and they were in the high discomfort group, they tended to request getting the extra credit after completion of the test. People evidenced counteractive control because self-control was only elicited when participants were experiencing high discomfort in the service of long-term goals that were important to them.

Fishbach et al. (2003) extended these behavioral control strategies to the domain of implicit cognition. The interesting prediction here is that temptations away from a goal will actually trigger that incompatible higher order goal, setting in motion processes that override the temptation or counteract the value of the temptation. This was examined using dieting goals and tempting/fattening foods. A first experiment once again focused on counteractive behavior arising from goal conflict. Participants were primed with a temptation to the goal of dieting by having them arrive to a room containing images everywhere (cues) relating to delicious but fattening food. Others arrived to a room that, through the images it portrayed, primed participants with the goal of dieting. Thus, people with dieting goals were split into two groups—one whose members experienced goal conflict and one whose members experienced goal compatibility as a result of the goal-priming manipulation. After working on some tasks in one of these two types of rooms, participants were asked to choose between a chocolate bar and an apple as a departure gift for having participated in the experiment. Both the food-primed and diet-primed participants were more likely than control participants to choose the apple rather than chocolate.

A second experiment extended this finding of counteractive behavior to counteractive implicit cognition. Participants performed a serial priming procedure with a lexical decision task (where letter strings are presented, sometimes forming words, and one must indicate as quickly as possible whether the string forms a word). On the critical trials words were presented that related to one of the goals. A prime preceded each lexical decision with the key manipulation being whether the prime word was a known temptation away from, something incompatible to, the target goal. They found that when primed with a temptation, response times were faster on words associated with an incompatible goal relative to an irrelevant goal. Thus, temptations seemed to prime goals with which the temptation was incompatible. The goal system counteracts the presence of the undesired cue, what should be a temptation away from a goal, with the triggering of that focal goal that the cue would otherwise undermine.

Goal Shielding and (Counteractive) Control Over Stereotype Activation

The logic of the preceding discussion is that temptations away from a goal actually trigger that incompatible goal, initiating goal shielding—counteracting the pull away from the goal with processing that promotes pursuing the goal. This may include processes of inhibition that actually lower the accessibility of the competing goal (e.g., Shah, 2003). Thus, cake, rather than triggering "eating comfort food" goals, may inhibit those goals via the activation of health-conscious goals. This logic can be applied to the domain of stereotype control, using people as the temptations away from and toward one's goals. People, as social targets, are associated with both lower and higher order goals, and the particular goal that is activated (either implicitly or not) is capable of triggering implicit operations that inhibit incompatible lower order goals. As an example, an African American male represents a means toward the goal of being egalitarian and fair to members of stereotyped groups. The same person may also represent a temptation away from the goal of being egalitarian and toward the goal of stereotyping. If an individual is to successfully be egalitarian, the presence of an African American man should trigger the egalitarian goal and inhibit the competing, conflicting goal to stereotype. In this example, the temptation away from egalitarianism and toward stereotyping is also a means toward egalitarianism, a goal conflict that requires resolution.

It is posited not merely that the presence of a person, as a temptation away from the goal of being egalitarian and toward the goal of stereotyping, will trigger the egalitarian goal. It is additionally posited that the cognitive operations associated with this goal will include: inhibition of incompatible goals such as goals associated with stereotyping, heightened readiness to detect goal-relevant people because they represent opportunities to move toward their goal, and managing cognitive resources to facilitate goal pursuit. Therefore, rather than stereotype activation, African American men will trigger stereotype inhibition and increased accessibility of the egalitarian goal. The stronger one's egalitarian goal, the more likely stereotype activation will be inhibited. In summary, our experimental approach assumes that the cognitive system is already wired in a way that allows us to "not stereotype" as part of the implicit operations that accompany goal pursuit and that are used in resolving goal conflict. We do not only use goals consciously to overturn stereotypes that have already been activated. Implicit goals are serviced by implicit cognition that can prevent the activation of stereotypes as part of the operations used in resolving a goal conflict.

Chronic Egalitarian Goals

Support for this goal shielding approach to implicit control over stereotype activation has been provided across a range of experiments. Moskowitz et al. (1999) illustrated such control using individuals with chronic egalitarian goals toward women. For such individuals the more dominant association to the group would be their egalitarian goal; goals that promote stereotyping are incompatible with their egalitarian goals relating to women. Participants performed a word-pronunciation task. On critical trials the words to be pronounced were either stereotype-relevant or control words. On all trials these words were preceded by photographs of women. The results revealed that nonchronics had facilitated response times when stereotype-relevant words followed female primes (relative to control primes and to stereotype-irrelevant words following female primes). People with chronic egalitarian goals did not differ in their responses to target words as a function of the type of prime or the type of word. This pattern reveals stereotype activation for nonchronics and control for chronics.

Does the triggering of an egalitarian goal, resulting from its incompatibility with stereotyping, inhibit stereotype activation? Such spreading inhibition from an activated goal was evidenced by Moskowitz, et al. (1999, Experiment 4). The procedure was a negative priming task (e.g., Tipper, 1985) in which stereotype-relevant primes were to be ignored. Participants once again were asked to perform a pronunciation task. Prior to this, each trial presented participants with two words in either red or blue font on opposing sides of a fixation cross. The task was to remember the red item for a memory test that would occur soon after, and to ignore the blue item. These blue "distractors" were manipulated so that they were female names on half of the critical trials and gender-neutral items on the remainder of the trials. The data supported the prediction of implicit inhibition of the stereotype for people with egalitarian goals if—and only if—a stereotype-relevant cue had been presented. Ignoring female names led people with chronic egalitarian goals to have slower responses to stereotype-relevant words on the pronunciation task relative to control words (and relative to when stereotypic words were preceded by gender-neutral primes). The accessible focal goal leads to inhibition of responses incompatible with that goal when cues relevant to that incompatible response are encountered. This illustrates goal shielding in the important domain of stereotype control. A stereotype has a decreased likelihood of being triggered when the appropriate higher order goals are in place.

People with chronic egalitarian goals should not only control stereotype activation when encountering a member of a stereotyped group, they should also have a boost to the accessibility of the egalitarian goal. In a first experiment, Moskowitz et al. (2000) replicated the control of stereotype activation found by Moskowitz et al. (1999) using stereotypes of African Americans instead of women. Participants with chronic egalitarian goals toward African Americans and non-chronic participants were first primed with faces under the guise of a task regarding memory for faces and then performed a pronunciation task to assess stereotype accessibility. Trials started with a fixation cross, followed by two faces presented side by side. One face was to be memorized and one face was to be ignored. After an interval of 200 milliseconds, an attribute appeared and was to be pronounced as fast as possible. The attributes were either relevant or irrelevant to the stereotype of African Americans. On critical trials, the primes were faces of either African American or Caucasian men that were to be memorized that were simultaneously presented along with (paired with) a picture of a Caucasian man that was to be ignored.

Replicating Moskowitz et al. (1999), they found that nonchronics responded faster (facilitated pronunciation times) when presented with stereotypic attributes following an African American face relative to the same words following Caucasian faces. They also responded faster to stereotype-relevant words following African American faces than to stereotype-irrelevant words following African American faces. No such effects were found for people with chronic egalitarian goals. There was no evidence of stereotype activation upon exposure to a face of an African American man.

Moskowitz et al. (2000) further illustrated that the implicit operations associated with the egalitarian goal included the heightened accessibility of the goal given the presence of a goal-relevant target. Chronic egalitarian and nonchronic individuals performed a lexical decision task preceded by a priming task. The primes were pictures of African American and Caucasian male faces. The words on critical trials assessed goal activation—they were either related to the goal of being egalitarian or positive value words. When relevant to egalitarianism, participants with chronic goals showed facilitated response times if the words had been preceded by faces of African American men. Facilitation did not occur for the nonchronic participants, nor was it found in response times to control words regardless of prime type or goals. The results showed that in addition to stereotypes being inhibited, egalitarian goals were triggered by African American faces, faces that for other individuals were associated more dominantly with a stereotype.

Implicit control resulting from conflict with a chronic egalitarian goal is also evidenced in work on implicit prejudice. Plant and Devine (1998) have suggested that implicit goals to be egalitarian can have as their source either an internal or external motivation. They distinguish the internal motivation to respond without prejudice (IMS) and the external motivation to respond without prejudice (EMS). These implicit goals can similarly direct implicit cognition and shield goal pursuit. Devine et al. (2002) examined implicit prejudice (as opposed to stereotype activation) as it is impacted by EMS and IMS. One experiment primed participants with faces and then had them respond to affectively laden stimuli by indicating if they were good or bad. Participants without chronic goals to be nonprejudiced revealed a race bias in the priming task. Their responses to negative words were facilitated when those words were preceded by African American faces. However, people who could be labeled as having chronic egalitarian goals (those both high in IMS and low in EMS) did not show this bias. Their responses to negative words were not facilitated after being primed with African American faces relative to nonchronics (high IMS/high EMS people, low IMS/high EMS people, low IMS/low EMS people).

Although the focus of this chapter is on the impact on stereotype-relevant, implicit cognition resulting from a specific type of goal conflict (conflict between implicit stereotyping goals and implicit goals opposing stereotyping), work also exists illustrating how this implicit goal conflict impacts explicit responding. Glaser and Knowles (2008) examined what they labeled the implicit motivation to control prejudice (IMCP) on the expression of prejudice and discrimination. Participants high and low in IMCP performed the Correll et al. (2002) "Shooter Task" where participants played a video game in which they were to shoot only "armed" people (control targets hold objects like a telephone as opposed to a gun). Correll et al. found participants were more likely to shoot African Americans who were armed and were simultaneously more likely to refrain from shooting unarmed Caucasians (as opposed to unarmed African Americans). Glaser and Knowles found that although for nonchronics the strength of the shooter bias was determined by whether people associated African Americans with weapons, for chronics (people high in ICMP) a strong association between African Americans and weapons did not relate to the degree of shooter bias. Implicit goals determined the manner in which this discriminatory behavior was expressed.

Peruche and Plant (2006) also illustrated the impact on explicit bias resulting from chronic (implicit) goals to be egalitarian. They used a task conceptually similar to the shooter task. Pictures of sports equipment or neutral objects were superimposed on photographs of African American or Caucasian male faces. Participants had to indicate whether the object was sports-related or not. Racial bias was exhibited in the early trials (e.g., more mistakes were made when an African American face was presented with a neutral object) than in the later trials. Over time the bias was controlled. However, chronics exhibited a greater reduction in bias, as they were better able to control their explicit responses on this race-relevant task.

Temporary Egalitarian Goals

Moskowitz and Li (2009) illustrated that the inhibition of stereotype activation is not limited to specific types of people—those who possess chronic egalitarian goals. The goal to be egalitarian can be primed in any individual and similar processes of stereotype inhibition emerge. Their participants were individuals who valued the goal of egalitarianism, assigning ratings of seven or higher (from not at all relevant to extremely relevant) to egalitarian goals when listed among a large set of goals. These people were not chronically in pursuit of egalitarianism. Chronicity in goal pursuit indicates the dominance of a goal in one's goal hierarchy, and the accessibility of the goal across time and contexts without the need for external cuing. These "nonchronic" individuals merely valued the goal of being egalitarian, a criterion implemented in these experiments simply because people were not expected to adopt goals in domains in which they had little interest. Thus, the criterion here was merely to find people who thought being egalitarian was a good thing as opposed to the "main" thing.

In a first experiment, goals were manipulated by having half of the participants contemplate a personal failure at being egalitarian toward African Americans at some point in their recent life history (priming the goal via inducing a discrepancy; e.g., Gollwitzer, Hilton, & Wicklund, 1982; Koole et al., 1999; Moskowitz, 2002). The remaining participants contemplated failure relating to a goal irrelevant to egalitarianism. Participants next performed a lexical decision task. Each critical trial presented attributes either relevant or irrelevant to the stereotype of African Americans. The words were preceded by primes—faces of either African American or Caucasian men. Participants in the control condition had facilitated reaction times to stereotypical words after African American faces relative to the same words following Caucasian faces (and relative to control words following African American faces). Participants who had contemplated a previous failure at pursuing egalitarian goals showed a slow down in their responses to stereotype-relevant words after faces of African American men relative to the same words following faces of Caucasian men (and relative to stereotype-irrelevant words that followed faces of African American men). The stereotype was inhibited rather than activated but only when an egalitarian goal was primed.

Moskowitz and Li (2008) performed a conceptually similar experiment that altered how stereotype inhibition was assessed. The task required participants to indicate whether two words appearing simultaneously were written in the same color. On critical trials the words were either related to the stereotype of African Americans or control words. Similarly to the Stroop (1935) task, a slow down in response time illustrated activation. Faster reaction times provided evidence of stereotype inhibition. All trials were preceded by faces of African American or Caucasian men. Goals were manipulated prior to this task, in a fashion identical to the previous experiment. People with egalitarian goals showed facilitated response times to stereotype-relevant words after faces of African American men relative to the same words following faces of Caucasian men (and relative to stereotype-irrelevant words that followed faces of African American men). Control participants showed facilitation in that their reaction times were slower to stereotype-relevant words following African American faces. Thus, across two experiments, there was a pattern of inhibition of stereotypes following the presence of a stereotype-relevant target. This finding is identical to that seen among participants with chronic goals; however, in these studies the goals were temporary.

Moskowitz, Li, and Ignarri (2008) focused on the question of whether goal shielding occurs via the heightened accessibility of the egalitarian goal upon exposure to cues in the environment that are associated with the goal. Following the same goal priming manipulation described above (to instill egalitarian goals relating to African American men in half the participants), participants completed an ostensibly different experiment in which faces were presented as primes and accessibility of egalitarian goals was assessed. As predicted, the egalitarian goal was more accessible when the primes were faces of African American versus Caucasian men but only when participants had previously experienced a goal discrepancy relating to being egalitarian toward African Americans. When participants were exposed to control words, or when participants were in the control group (no egalitarian goal), this pattern did not emerge.

Moskowitz et al. (2008) further explored the question of whether goal shielding occurs by conceptualizing the people being presented in these experiments not simply as primes that can cue their egalitarian goals but as means to the goal. Research has shown that an association exists between a goal and opportunities to pursue that goal (Kruglanski et al., 2002). If such an association exists, then participants should display an increased ability to detect goal-relevant means in their environment when the goal is primed. In one experiment, four faces of men were presented simultaneously. Participants were asked to detect the individual wearing a bow tie. The faces were presented so that in some instances an African American male was in the set of faces at the same time as the focal stimulus (a Caucasian man in a bow tie). They found a facilitated ability to detect these goal-relevant cues in the environment if, and only if, an egalitarian goal had previously been triggered. This was evidenced by a slower response to the focal task (identifying where on the screen the man in a bow tie appeared). This did not occur when only Caucasian men were in the set of four images, nor did it occur for participants primed with control goals irrelevant to egalitarianism. The face of an African American represents an opportunity to start pursuing one's egalitarian goals and disrupts focused attention.

Are what is being triggered goals (and thus the stereotype inhibition being exhibited the effects of goal conflict) or semantic constructs (and thus the stereotype inhibition being exhibited the effects of inhibitory processes in concept priming)? Further, are these goals implicit, given that the individual is explicitly asked to contemplate failure at being egalitarian? The latter question is addressed first. As noted above (in the section on goal priming) goals can have implicit accessibility without one's awareness even if a goal is consciously selected at some prior point. This can occur if the individual does not recognize that a subsequent task is relevant to, and their responses are being influenced by, the goal that was previously consciously triggered. Such is the case in the Zeigarnik effect (Zeigarnik, 1927) where a goal is disrupted, yet continues to operate implicitly while the individual is engaged in a separate, attention-absorbing, other task. This is also seen in the "deliberation-without-attention" effect (Dijksterhuis, Bos, Nordgren, & van Baaren, 2006), where once again a goal is explicitly adopted (the goal to deliberate among several choice options), and attention is absorbed by a goal-irrelevant activity, yet the goal remains active and pursued implicitly (nonconscious deliberation).

In our stereotyping experiments (see previous section), the individual did not recognize that the lexical decision tasks they performed had anything to do with (a) stereotyping or (b) the egalitarian goals that had been triggered in the ostensibly separate experiment performed earlier in the session. They were not explicitly pursuing an egalitarian goal, although the inhibition of stereotypes indicated they were doing so implicitly. This conclusion has been bolstered by other aspects of the experimental design—mainly the short interval between the presentation of the prime and the response (200 ms) and the speed with which the participant responses are made (around 500 ms). These response intervals are too fast for conscious control to be exerted (e.g., Fazio, Jackson, Dunton, & Williams, 1995; Bargh & Chartrand, 2000).

The former question regarded the nature of the inhibition response observed. Inhibition was posited to be caused by goal conflict, yet it might be argued that the concept "egalitarian" inhibits the concepts linked to the stereotype. Kunda and Thagard's (1996) parallel-constraint-satisfaction theory states that during any concept activation, competing constructs are inhibited. Concept activation is characterized by an interplay of excitatory and inhibitory mechanisms (Andersen, Moskowitz, Blair, & Nosek, 2007; Macrae, Bodenhausen, & Milne, 1995) such that priming a stereotype inhibits stereotype-inconsistent semantic information, such as traits that are inconsistent with the stereotypical qualities of a group (Dijksterhuis & van Knippenberg, 1996). The nature of the inhibition was addressed in experiments by Moskowitz and Ignarri (2009).

It is well-established that goals have specific properties, or markers, not exhibited by semantic constructs (e.g., Martin & Tesser, 2009). One such property is that the accessibility of the goal does not decrease as time passes, as is the case with semantic activation; another is that attaining the goal, which may involve increased exposure to semantic constructs relating to the goal, does not increase goal accessibility but decreases it (e.g., Bargh et al., 2001; Förster, Liberman, & Friedman, 2007; Koole et al., 1999; Martin & Tesser, 2009; Spencer, Fein, Wolfe, Fong, & Dunn, 1998). Moskowitz and Ignarri (2009) used this marker of a goal to illustrate that the inhibitory processes observed in stereotype control result from goals in conflict, as opposed to interconstruct inhibition arising from semantic activation.

In one experiment, participants contemplated either success or failure at being egalitarian. This created two groups of participants with equal amounts of activation for the semantic concept "egalitarian." But one of the groups was posited to not merely have semantic activation but goal activation (arising from the failure and its associated state of goal "incompleteness"). The other group, having attained/affirmed the goal, should not have exhibited increased accessibility of the goal. Thus, if mere inhibition among semantic concepts causes inhibition, each group should have exhibited stereotype control. However, if control merges due to operations associated with a goal conflicted, only the people who had contemplated failure at being egalitarian would exhibit control. After the goal manipulation, the participants next completed an ostensibly separate experiment in cognitive psychology—a serial priming procedure with a lexical decision task. Faces of either African American or Caucasian men were presented on each trial as part of a supposed memory test and were followed immediately on critical trials by stereotypical or control attributes presented as part of a lexical decision task. Response times to the words yielded the measure of activation and inhibition.

They found that those people who contemplated success at being egalitarian had faster reaction times to stereotypical attributes after African American faces relative to their reaction times to the same words following Caucasian faces. Additionally, they had facilitated reaction times to stereotype-relevant words following African American faces relative to control words following African American faces. Further, reaction times to stereotypical words following stereotypical primes were faster for people who contemplated success at being egalitarian relative to people who had contemplated failure. Most importantly, people who had contemplated a failure relating to their egalitarian goals showed a slow down in their responses to stereotype-relevant words after faces of African American men relative to the same words following faces of Caucasian men—an inhibition of the stereotype.

Moskowitz et al. (2008) used the same marker of goal pursuit to illustrate goal shielding processes at work. Once again, participants who contemplated success at being egalitarian toward African Americans were used as a control group to be compared against people who contemplated failure at egalitarianism. In this experiment, the measure of goal shielding was not stereotype inhibition but displaced attention to means associated with the goal. They used the same paradigm described above, where the task was to identify which individual was wearing a bow tie in an array of four men (and indicate where on the screen that person was located). They found a facilitated ability to detect goal-relevant stimuli (means) in the environment only when failure—not success—at being egalitarian (a goal had been triggered)

had been contemplated. This was manifested by slowed responding to the Caucasian a man in a bow tie if—and only if—an African American man was also in the stimulus array. The effect resulted from operations specific to goals and did not occur following mere semantic activation of the concept "egalitarian."

Behavioral Inhibition and Egalitarian Goals

While illustrating that temporary goals to be egalitarian can be triggered through a "failure experience" and subsequently direct implicit cognition, similar illustrations of goal conflict impacting response are seen with explicit opportunities to address one's goal. Although the goal and the response may be explicit, the goal conflict described in such research is similar to the implicit conflict reviewed above. The research of Monteith et al. (2002) is a good example.

Monteith and Voils (2001) have argued that a set of processes aimed at controlling prejudice are triggered by cues in the environment associated with this goal. An association is said to develop among one's goal to be unbiased (triggered when one acts in a non-egalitarian fashion), the associated feelings and tension state linked to this goal, the cues in the environment that trigger these feelings and goals (e.g., the presence of an African American or homosexual person), and appropriate responses that promote the goal being attained (removing bias). Initially this process is said to be quite explicit to the person. Cues associated with the unwanted state of bias lead to conscious attempts to compensate for it. The cues (people who may potentially be evaluated in a biased way) serve as a warning to the discrepancy between how one should respond and how one does respond, triggering a response. They argue that the response triggered is the operation of the behavioral inhibition system (BIS), a motivational system that causes a pausing/interruption of current behavior. This allows one to increase attention to stimuli relevant to one's goal and to determine what might have caused the bias in one's reaction.

Monteith et al. (2002) led some participants to believe they responded with negative arousal to pictures with racial content, whereas control participants believed they had low negative arousal to the same stimuli. The task had participants exposed to these images while hooked to equipment that supposedly measured their arousal. Their task was to press the space bar on the keyboard after receiving feedback regarding their arousal levels to a given image. The time it took them to press the spacebar after receiving the feedback was the measure of behavioral inhibition. They found that people who believed they were negatively aroused by images of African Americans had their responses to pressing the spacebar slowed relative to the control group. The perception of explicit bias led to a conscious slowing down. Similar findings of behavioral inhibition were exhibited using a very different paradigm. Monteith (1993, Experiment 1) told some participants that their evaluation of an applicant to law school revealed anti-homosexual bias. Measures were then taken to see if explicit operations were performed to reduce the goal conflict this produced by assessing the extent to which participants ruminated on their bias in evaluating the applicant. This thought-listing task revealed that participants who were told they were biased focused their explicit thoughts on their discrepant feelings/behavior. They also dedicated more time to reading an essay that was provided for them about why such bias occurs and had better memory for the evidence provided in the essay.

Temporary Creativity Goals

The inhibition of stereotyping is not limited to egalitarian goals being in place. So long as the implicit operations associated with a goal are incompatible with stereotype activation, similar inhibition of stereotypes should be observed. For example, Sassenberg and Moskowitz (2005) argued that the operations linked to creativity goals include the inhibition of typical associates to a target and the triggering instead of more atypical associates. The typical associates to a target, those that might be triggered in the absence of processing goals, would be disrupted if the person's current goals initiated new, unique, innovative, and creative ways of thinking about the stimulus. Thus, the cognitive operations associated with these goals are incompatible with goals to stereotype; therefore, these operations include the inhibition of these incompatible stereotyping goals.

Sassenberg and Moskowitz (2005) illustrated this set of operations in the domain of stereotyping. The task involved a simple serial priming procedure preceded by the implicit priming of creativity goals. In the priming phase of the experiment, participants were exposed once again to pictures of faces of African American and Caucasian men. A lexical decision task followed the priming, and on critical trials the words presented were attributes relevant to the stereotype of African Americans and control attributes. The predicted pattern of inhibition and facilitation as a function of one's goals was revealed. Critically, participants primed with creativity goals had slower reaction times to attributes associated with African Americans when those attributes were preceded by African American faces relative to Caucasian faces. This pattern of responding did not emerge for control words. Conversely, the classic priming effect was found for people not primed with creativity. These individuals were faster to stereotypical attributes following African American (versus Caucasian) face primes, with no effect found for control attributes. When goal operations (inhibiting typical associates to a target) were incompatible with stereotype activation, the stereotype was inhibited rather than activated.

Temporary Perspective Taking Goals

In another illustration of the implicit operations that allow individuals to resolve goal conflict relative to stereotyping, Galinsky and Moskowitz (2000; 2007) focused on perspective-taking goals. Participants were asked to either take the perspective of a person in a photograph while writing an essay about a day in the life of the person, suppress thoughts about stereotypes during the task, to express stereotypes while performing the task, or were given no explicit goals. The person, a member of a stereotyped group, was described by participants with perspective-taking goals and thought-suppression goals in a far less stereotypical way. More important to the current discussion, an implicit measure of stereotype activation revealed that participants with perspective-taking goals did not have stereotypes accessible after the essay-writing task. This is in contrast to both control participants and to people with a suppression goal for whom the stereotype has heightened activation.

Galinsky and Moskowitz (2000) argued that different goals triggered different implicit operations. Perspective-taking goals were said to trigger operations that increased the perceived overlap between one's conception of the outgroup and the self as well as heightened attention to situational constraints that may have impacted behavior. Rather than activating stereotypes, the self-concept was being made accessible. Suppression goals, however, initiated a separate and unique set of implicit operations, such as triggering the concurrent operating and monitoring systems, operations that resulted in the stereotype being kept accessible.

However, Galinsky and Moskowitz (2007) additionally found that it was not merely the case that unwanted (stereotypical) thoughts had heightened accessibility following thought suppression. Goal-relevant thoughts—thoughts about concepts opposed to the stereotype—simultaneously had increased accessibility. The simultaneous activation of the to-be-suppressed thoughts as well as goal-relevant thoughts is a new finding that stands in contrast to what is typically found in the social cognition literature. As stated above, priming a stereotype (or any concept) usually facilitates access to stereotype-consistent traits while inhibiting stereotype-inconsistent traits (Dijksterhuis & van Knippenberg, 1996). Similarly, category label primes increase the ability to retrieve typical category members, while simultaneously limiting access to atypical members (Rothbart, Sriram, & Davis-Stitt, 1996). The findings of Galinsky and Moskowitz (2007) highlight the goal-specific nature of the implicit operations. A stereotype suppression goal may be one case in which processes of inhibition of counterstereotypical thoughts during stereotype activation might be surmounted.

Temporary Self-Esteem Goals

It is not meant to be suggested that goal conflicts always result in the inhibition of stereotypes. Depending on the nature of the goal, and the operations associated with the goal, stereotype control could be exhibited by increased stereotype activation. Fein and Spencer (1997) and Spencer et al. (1998) provided such an example. For example, Spencer et al. manipulated the goal to restore self-esteem by giving participants failure feedback on an intelligence test. As a means to restoring self-esteem, they heightened their implicit stereotyping. This was evidenced using images of African Americans as primes and word completions on a word fragment task as the measure of accessibility. Images of African American or Caucasian men were presented parafoveally for 17 milliseconds and masked, supposedly as a distraction to the word completion task. They found that participants used more stereotypical words following images of African Americans than Caucasians. However, this difference in stereotype activation only occurred for participants whose self-esteem had been undermined. More detailed data analyses showed that effects of self-image threat were detected only on stereotype-relevant words with derogative connotations, rather than all the stereotype words used. This result implies that the opportunity for stereotyping a target person had been seized upon by the participants for repairing their damaged esteem through derogating other people. Fein and Spencer found similar findings using an explicit measure of stereotyping – participants given the failure feedback rated a homosexual target as possessing more homosexual-stereotypical characteristics than a heterosexual target.

Conclusion

A particular person may be a woman, African American, a Kurd, a Jew, a mother, or all of the above. The point is that such a person is rarely encountered in a vacuum but, rather, encountered in contexts where perceivers have goals that the person could potentially address, and where one's resources for pursuing those goals are taxed to varying degrees. The cognitive operations triggered will be determined by the goals one has pertaining to that target in that context, as well as the resources available for goal pursuit. Stereotype activation is not inevitable.

The most thorough method of resolving goal conflict and shielding oneself from unwanted goals is to remove oneself from both the sensory stimulation and the cues that are associated with the unwanted goal. In intergroup contact, this escape is tantamount to racial segregation and avoidance of groups. Such escape is often not a desirable option in contact with outgroups (or in relation to unwanted goals irrelevant to stereotyping). One often has strong motives (both moral and practical) to engage in intergroup contact. Therefore, one is left with several options for control, some at the explicit level (correction attempts) and some at the implicit level. One option is to attempt to be "color-blind"—to not avoid the group in question but to avoid using the group's category label as a means to guiding one's response to the person. Another is to use categories but not the stereotypes that culture has taught one to associate with those categories. The focus in this chapter has been on the latter—a person may strategically manipulate the unwanted goal through implicit inhibition of its undesirable content and incompatible operations. When goal conflict exists between goals promoting stereotyping and goals opposing stereotyping, even if these goals are implicit, the cognitive operations used to resolve this conflict and shield one's goal to be unbiased can result in stereotype inhibition.

Note

1 Explicit goal priming, often in the form of task instructions, followed by implicit operations, is evidenced in experiments ranging from: dichotic listening tasks (Cherry, 1953), naming ink color (Logan, 1980), focusing on the letter "g" as hundreds of "g's" are used to construct a large "x" (Navon, 1977), responding with one hand to auditory stimuli and another to visual stimuli (Pashler, 1991), switching tasks when responding to a stimulus that can be categorized in multiple ways (Arrington & Logan, 2005), making a lexical decision (e.g., Neely, 1977), attending to a red triangle in a display of blue ones (e.g., Treisman & Gormican, 1988), forming impressions of people when presented with stimulus sentences and subsequently organizing that information in clusters according to person (Hamilton, Katz, & Leirer, 1980) or processing it more deeply if the information is inconsistent with the emerging impression (Hastie & Kumar, 1979), and memorizing information that is subsequently learned with the use of retrieved schemas (e.g., Bransford & Johnson, 1972; Cantor & Mischel, 1977).

References

Aarts, H., Custers, R., & Holland, R. W. The nonconscious cessation of goal pursuit: When goals and negative affect are coactivated. J Pers Soc Psychol 2007; 92: 165–178.

Aarts, H., Gollwitzer, P. M., & Hassin, R. R. Goal Contagion: Perceiving is for pursuing. Pers Soc Psychol 2004; 87: 23–37.

Aarts, H., Dijksterhuis, A. P., Midden, C. To plan or not to plan? Goal achievement of interrupting the performance of mundane behaviors. Eur J Soc Psychol 1999; 29: 971–979.

Allport, G. W. *The nature of prejudice*. Reading, MA: Addison-Wesley, 1954.

Andersen, S. A., Moskowitz, G. B., Blair, I. V., & Nosek, B. A. Automatic thought. In: Higgins, E. T., & Kruglanski, A. (Eds.), *Social psychology: Handbook of basic principles*, Vol 2. New York: Guilford; 2007: pp. 138–175.

Banaji, M. R., Hardin, C. D., & Rothman, A. J. Implicit stereotyping in person judgment. J Pers Soc Psychol 1993; 65: 272–281.

Bargh, J. A. Auto-motives: Preconscious determinants of thought and behavior. Multiple affects from multiple stages. In: Higgins, E. T., & Sorrentino, R. M. (Eds.), *Handbook of motivation and cognition: Foundations of social behavior*, Vol. 2. New York: Guilford; 1990: pp. 93–130.

Bargh, J. A. Automaticity in social psychology. In: Higgins, E. T., & Kruglanski, A. W. (Eds.), *Social psychology: Handbook of basic principles*. New York: Guilford; 1996: pp. 169–183.

Bargh, J. A. The cognitive monster: The case against the controllability of automatic stereotype effects. In: Chaiken, S., & Trope, Y. (Eds.), *Dual process theories in social psychology*. New York: Guilford; 1999: pp. 361–382.

Bargh, J. A., & Chartrand, T. L. The mind in the middle: A practical guide to priming and automaticity research. In: Reis, H. T., & Judd, C. M. (Eds.), *Handbook of research methods in social and personality psychology*. New York, NY: Cambridge University Press; 2000: pp. 253–285.

Bargh, J. A., Gollwitzer, P. M., Lee-Chai, A., Barndollar, K., & Trötschel, R. The automated will: Nonconscious activation and pursuit of behavioral goals. J Pers Soc Psychol 2001; 81: 1014–1027.

Bodenhausen, G. V. Stereotypes as judgmental heuristics: Evidence of circadian variations in discrimination. Psychol Sci 1990; 1: 319–322.

Blair, I., & Banaji, M. Automatic and controlled processes in stereotype priming. J Pers Soc Psychol 1996; 70: 1142–1163.

Brewer, M. B. A dual process model of impression formation. In: Srull, T. K., & Wyer, R.S. (Eds.), *Advances in social cognition*, Vol. 1. Hillsdale, NJ: Lawrence Erlbaum Associates; 1988: pp. 1–36.

Bruner, J. S. On perceptual readiness. Psychol Rev 1957; 64: 123–152.

Chartrand, T. L., & Bargh, J. A. Automatic activation of impression formation goals: Nonconscious goal priming reproduces effects of explicit task instructions. J Pers Soc Psychol 1996; 71: 464–478.

Correll, J., Park, B., Judd, C. M., & Wittenbrink, B. The police officer's dilemma: Using ethnicity to disambiguate potentially threatening individuals. J Pers Soc Psychol December 2002; 83(6): 1314–1329.

Custers, R., & Aarts, H. Goal-discrepant situations prime goal-directed actions if goals are temporarily or chronically accessible. Pers Soc Psychol Bull 2007; 33: 623–633.

Devine, P. G. Stereotypes and prejudice: Their automatic and controlled components. J Pers Soc Psychol 1989; 56: 5–18.

Devine, P. G., Plant, E. A., Amodio, D. M., Harmon-Jones, E., & Vance, S. L. The regulations of explicit and implicit race bias: The role of motivations to respond without prejudice. J Pers Soc Psychol 2002; 82(5): 835–848.

Dijksterhuis, A., Bos, M., Nordgren, L., & Van Baaren, R. B. On making the right choice: The deliberation-without-attention effect. Science 2006; 311: 1005–1007.

Dijksterhuis, A., & van Knippenberg, A. The knife that cuts both ways: Facilitated and inhibited access to traits as a result of stereotype activation. J Exp Soc Psychol 1996; 32: 271–288.

Dovidio, J. F., Evans, N., & Tyler, R. B. Racial stereotypes: The contents of their cognitive representations. J Exp Social Psychol 1986; 22: 22–37.

Dovidio, J. F., & Gaertner, S. L. Prejudice, discrimination, and racism: Historical trends and contemporary approaches. In: Dovidio, J. F., & Gaertner, S. L. (Eds.), *Prejudice, discrimination, and racism*. New York: Academic Press; 1986: pp. 1–34.

Fazio, R. H., Jackson, J. R., Dunton, B. C., & Williams, C. J. Variability in automatic activation as an unobtrusive measure of racial attitudes: A bona fide pipeline? J Pers Soc Psychol 1995; 69: 1013–1027.

Fein, S., & Spencer, S. J. Prejudice as self-image maintenance: Affirming the self through negative evaluations of others. J Pers Soc Psychol 1997; 73: 31–44.

Fishbach, A., Friedman, R. S., & Kruglanski, A. W. Leading us not unto temptation: Momentary allurements elicit overriding goal activation. J Pers Soc Psychol 2003; 84: 296–309.

Fiske, S. T., & Neuberg, S. L. A continuum of impression formation, from category-based to individuating processes: Influences of information and motivation on attention and interpretation. In: Zanna, M. P. (Ed.), *Advances in experimental social psychology*, Vol. 23. New York: Academic Press; 1990: pp. 1–74.

Fitzsimons, G. M., & Bargh, J. A. Thinking of you: Nonconscious pursuit of interpersonal goals associated with relationship partners. J Pers Soc Psychol 2003; 84: 148–164.

Förster, J., & Liberman, N. The role of attribution of motivation in producing postsuppressional rebound. J Pers Soc Psychol 2001; 81: 377–390.

Förster, J., Liberman, N., & Friedman, R. S. Seven principles of goal activation: A systematic approach to distinguishing goal priming from priming of non-goal constructs. Pers Soc Psychol Rev 2007; 11(3): 211–233.

Galinsky, A. D., & Moskowitz, G. B. Perspective taking: Decreasing stereotype expression, stereotype accessibility and in-group favoritism. J Pers Soc Psychol 2000; 78: 708–724.

Galinsky, A. D., & Moskowitz, G. B. Further ironies of suppression: Stereotype and counterstereotype accessibility following suppression. J Exp Soc Psychol 2007; 43: 833–841.

Gilbert, D. T., & Hixon, J. G. The trouble of thinking: Activation and application of stereotypic beliefs. J Pers Soc Psychol 1991; 60: 509–517.

Glaser, J., & Knowles, E. D. Implicit motivation to control prejudice. J Exp Soc Psychol 2008; 44(1): 164–172.

Gollwitzer, P. M., Wicklund, R. A., & Hilton, J. L. Admission of failure and symbolic self-completion: Extending Lewinian theory. J Pers Soc Psychol 1982; 43: 358–371.

Hamilton, D. L., & Trolier, T. K. Stereotypes and stereotyping: An overview of the cognitive approach. In: Dovidio, J. F., & Gaertner, S. L. (Eds.), *Prejudice, discrimination, and racism*. Orlando, FL: Academic Press; 1986: pp. 127–163.

Hamilton, D. L., & Gifford, R. K. Illusory correlation in interpersonal perception: A cognitive basis of stereotypic judgments. J Exp Soc Psychol 1976; 12: 392–407.

Hamilton, D. L., Katz, L. B., & Leirer, V. O. Cognitive representation of personality impressions: Organizational processes in first impression formation. J Pers Soc Psychol 1980; 39: 1050–1063.

Hassin, R. R., Aarts, H., Eitam, B., Custers, R., & Kleiman, T. Nonconscious goal pursuit and the effortful control of behavior. In: Morsella, E., Bargh, J. A., & Gollwitzer, P. M. (Eds.), *Oxford handbook of human action. Social cognition and social neuroscience*. New York: Oxford University Press; 2009: pp. 549–566.

Hastie, R., & Kumar, P. A. Person memory: Personality traits as organizing principles in memory for behaviors. J Pers Soc Psychol 1979; 37: 25–38.

Heider, F. Social perception and phenomenal causality. Psychol Rev 1944; 51: 358–374.

Kawakami, K., Dovidio, J., Moll, J., Hermsen, S., & Russin, A. Just say no (to stereotyping): Effects of training in the negation of stereotype associations on stereotype activation. J Pers Soc Psychol 2000; 78: 871–888.

Koole, S. L., Smeets, K., van Knippenberg, A., & Dijksterhuis, A. The cessation of rumination through self-affirmation. J Pers Soc Psychol 1999; 77: 111–125.

Kunda, Z., & Spencer, S. J. When do stereotypes come to mind and when do they color judgment? A goal-based theory of stereotype activation and application. Psychol Bull 2003; 129: 522–544.

Kunda, Z., & Thagard, P. Forming impressions from stereotypes, trait, and behaviors: A parallel constraint satisfaction theory. Psychol Rev 1996; 103: 284–308.

Livingston, R. W., & Brewer, M. B. What are we really priming? Cue-based versus category-based processing of facial stimuli. J Pers Soc Psychol 2002; 82(1): 5–18.

Macrae, C. N., Bodenhausen, G. V., & Milne, A. B. The dissection of selection in person perception: Inhibitory processes in social stereotyping. J Pers Soc Psychol 1995; 69: 397–407.

Macrae, C. N., Bodenhausen, G. V., Milne, A. B., & Jetten, J. Out of mind but back in sight: Stereotypes on the rebound. J Pers Soc Psychol 1994; 67: 808–817.

Macrae, C. N., Bodenhausen, G. V., Schloerscheidt, A. M., & Milne, A. B. Tales of the unexpected: Executive function and person perception. J Pers Soc Psychol 1999; 76: 200–213.

Macrae, C. N., Milne, A. B., & Bodenhausen, G. V. Stereotypes as energy-saving devices: A peek inside the cognitive toolbox. J Pers Soc Psychol 1994; 66: 37–47.

Martin, L. L., & Tesser, A. Five markers of motivated behavior. In: Moskowitz, G. B., & Grant, H. (Eds.), *The psychology of goals*. New York: The Guilford Press; 2009: pp. 257–276.

Monteith, M. J. Self-regulation of prejudiced responses: Implications for progress in prejudice reduction efforts. J Pers Soc Psychol 1993; 65: 469–485.

Monteith, M. J., Ashburn-Nardo, L., Voils, C. I., & Czopp, A. M. Putting the brakes on prejudice: On the development and operation of cues for control. J Pers Soc Psychol 2002; 83: 1029–1050.

Monteith, M. J., & Voils, C. I. Exerting control over prejudiced responses. In: Moskowitz, G. B. (Ed.), *Cognitive social psychology: The Princeton symposium on the legacy and future of social cognition*. NJ: Lawrence Erlbaum Associates, Inc.; 2001: pp. 375–388.

Moskowitz, G. B. Individual differences in social categorization: The effects of personal need for structure on spontaneous trait inferences. J Pers Soc Psychol 1993; 65: 132–142.

Moskowitz, G. B. Preconscious effects of temporary goals on attention. J Exp Soc Psychol 2002; 38: 397–404.

Moskowitz, G. B., Gollwitzer, P. M., Wasel, W., & Schaal, B. Preconscious control of stereotype activation through chronic egalitarian goals. J Pers Soc Psychol 1999; 77: 167–184.

Moskowitz, G. B., & Li, P. Stereotype inhibition from implicit goal shielding: A proactive strategy of stereotype control. Unpublished Manuscript, 2009.

Moskowitz, G. B., Li, P., & Kirk, E. The implicit volition model: On the preconscious regulation of temporarily adopted goals. In: Zanna, M. P. (Ed.), *Advances in experimental social psychology*, Vol. 36. San Diego, CA: Academic Press; 2004: pp. 317–413.

Moskowitz, G. B., & Ignarri, C. Implicit volition and stereotype control. Eur Rev Soc Psychol 2009; 20: 97–145.

Moskowitz, G. B., Salomon, A. R., & Taylor, C. M. Implicit control of stereotype activation through the preconscious operation of egalitarian goals. Soc Cogn 2000; 18: 151–177.

Moskowitz, G. B., Skurnik, I., & Galinsky, A. D. The history of dual-process notions, and the future of preconscious control. In: Chaiken, S., & Trope, Y. (Eds.), *Dual-process theories in social psychology*. New York: Guilford; 1999: pp. 12–36.

Neuberg, S. L. The goal of forming accurate impressions during social interactions: Attenuating impact of negative expectancies. J Pers Soc Psychol 1989; 56: 374–386.

Papies, E. K., Stroebe, W., & Aarts, H. Pleasure in the mind: restrained eating and spontaneous hedonic thoughts about food. J Exp Soc Psychol 2007; 43: 810–817.

Peruche, B. M., & Plant, E. A. Racial bias in perceptions of athleticism: The role of motivation in the elimination of bias. Soc Cogn 2006; 24(4): 438–452.

Plant, E. A., & Devine, P. G. Internal and external motivation to respond without prejudice. J Pers Soc Psychol 1998; 75(3): 811–832.

Rothbart, M., Sriram, N., & Davis-Stitt, C. The retrieval of typical and atypical category members. J Exp Soc Psychol 1996; 32: 1–29.

Sassenberg, K., & Moskowitz, G. B. Don't stereotype, think different! Overcoming automatic stereotype activation by mindset priming. J Exp Soc Psychol 2005; 41: 506–514.

Shah, J. Automatic for the people: How representations of significant others implicitly affect goal pursuit. J Pers Soc Psychol 2003; 84: 661–681

Shah, J. Y., Hall, D., & Leander, N. P. Moments of motivation: Towards a model of regulatory rotation. In: Moskowitz, G. B., & Grant, H. (Eds.), *The psychology of goals*. New York, NY: The Guilford Press; 2009: pp. 234–254.

Shah, J. Y., & Kruglanski, A. W. When opportunity knocks: Bottom-up priming of goals by means and its effects on self regulation. J Pers Soc Psychol 2003; 84: 1109–1122.

Spencer, S. J., Fein, S., Wolfe, C. T., Fong, C., & Dunn, M. A. Automatic activation of stereotypes: The role of self-image threat. Pers Soc Psychol Bull 1998; 24: 1139–1152.

Srull, T. K., & Wyer, R. S. The role of chronic and temporary goals in social information processing. In: Sorrentino, R. M., & Higgins, E. T. (Eds.), *Handbook of motivation and cognition*. New York: Guilford; 1986: pp. 503–547.

Stangor, C., & Lange, J. E. Mental representations of social groups: Advances in understanding stereotypes and stereotyping. In: Zanna, M. P. (Ed.), *Advances in experimental social psychology*, Vol. 26. San Diego, CA: Academic Press; 1994: pp. 357–416.

Strahan, E. J., Spencer, S. J., & Zanna, M. P. Subliminal priming and persuasion: Striking while the iron is hot. J Exp Soc Psychol 2002; 38: 556–568.

Stroop, J. R. Studies of interference in serial verbal reactions. J Exp Psychol 1935; 18: 643–662.

Tipper, S. P. The negative priming effect: Inhibitory priming by ignored objects. Q J Exp Psychol 1985; 37A: 571–590.

Trope, Y., & Fishbach, A. Counteractive self-control in overcoming temptation. J Pers Soc Psychol 2000; 79(4): 493–506.

von Hippel, W., Sekaquaptewa, D., & Vargas, P. On the role of encoding processes in stereotype maintenance. In: Zanna, M. P. (Ed.), *Advances in experimental social psychology*, Vol. 27. San Diego, CA: Academic Press; 1995: pp. 177–254.

Wegner, D. M. Ironic processes of mental control. Psychol Rev 1994; 101: 34–52.

Wegner, D. M., & Erber, R. The hyperaccessibility of suppressed thoughts. J Pers Soc Psychol 1992; 63: 903–912.

Wilson, T. D., & Brekke, N. Mental contamination and mental correction: Unwanted influences on judgments and evaluations. Psychol Bull 1994; 116: 117–142.

Woike, B. A., Lavezzary, E., & Barsky, J. The influence of implicit motives on memory processes. J Pers Soc Psychol 2001; 81: 935–945.

Zeigarnik, B. Das Behalten erledigter und unerledigter Handlungen [The retention of completed and uncompleted actions]. Psychologische Forschung 1927; 9: 1–85.

CHAPTER 20

Ego Depletion and the Limited Resource Model of Self-Control

Nicole L. Mead, Jessica L. Alquist, and Roy F. Baumeister

ABSTRACT

People break diets, procrastinate in the face of looming deadlines, imbibe too much alcohol the night before a midterm, struggle to save money, and lash out at loved ones and family members. They do all these things despite their best intentions not to. Why do people engage in such personally, interpersonally, and socially destructive behaviors? This chapter suggests that a major reason why people fail at self-control is because it relies on a limited resource. We define self-control as the capacity to alter one's responses; it is what enables people to forego the allure of short-term pleasures to institute responses that bring long-term rewards. One of the core functions of self-control may be to facilitate culture, which often requires that people curtail selfishness for the sake of effective group functioning. The first part of the chapter gives an overview of how self-control operates, including the possible biological basis of self-control. It covers a substantial body of literature suggesting that self-control operates on a limited resource, which becomes depleted with use. The second part of the chapter reviews the benefits of good self-control and the costs of bad self-control across a large variety of domains, such as consumption, self-presentation, decision making, rejection, aggression, and interpersonal relationships.

Keywords: Self-control, ego depletion, willpower, impulse, motivation

EGO DEPLETION AND THE LIMITED RESOURCE MODEL OF SELF-CONTROL

People break diets, make impulsive purchases, procrastinate in the face of looming deadlines, engage in unsafe sex, and yell at their spouses, children, and friends. They do all these things despite their best intentions to do otherwise. Why are people not very good at controlling their consumption, emotions, and impulses? Or, more specifically, why do people fail at self-control? This chapter features one answer to these questions, based on the understanding that self-control consumes a limited resource.

We define self-control as the self's ability to override unwanted thoughts, emotions, impulses, and automatic or habitual behaviors. When people are faced with impulses that conflict with longer-term goals, self-control is the capacity that allows the self to stop those unwanted impulses from developing into full-blown behaviors. In this way, self-control allows people to achieve goals that they have set for themselves and to conform to rules and standards imposed by their social environment.

Accumulating research findings suggest that self-control works much like a muscle: It can become fatigued when overused, making subsequent attempts at self-control difficult and making self-control failure probable. Indeed, the limited resource model of self-control states that all acts of self-control draw on the same limited resource and this resource becomes depleted with use (Baumeister & Heatherton, 1996; Baumeister, Heatherton, & Tice, 1994). Resource depletion increases the chance that people give in to unwanted urges, impulses, and desires.

The purpose of this chapter is to provide an overview of how self-control is needed to help people resolve inner motivational conflicts. Whenever a person is pulled between two impulses, motivations, or standards, self-control can be helpful, especially to restrain the self from acting on the impulse that would bring short-term benefits but long-term costs.

The chapter is comprised of two main sections. The first focuses on how self-control helps people resolve motivational conflicts that pit short-term gains against long-term costs or personal pleasures against interpersonal costs. It also describes how self-control operates, including the possible biological basis for self-control. The second part of the chapter presents evidence demonstrating the benefits of good self-control and the costs of bad self-control across a variety of domains. For example, it covers how self-control influences interpersonal relationships, addictive behaviors, and reactions to social rejection.

Self-Control Resolves Inner Motivational Conflicts

Self-control is vital to success in life and is regarded as one of the most important parts of the self (e.g., Baumeister, 1998; Higgins, 1996). Research indicates that people with good self-control do better in work and in social life, and they have fewer psychopathological problems than other people with relatively poor self-control (Duckworth & Seligman, 2005; Mischel, Shoda, & Peake, 1988; Shoda, Mischel, & Peake, 1990; Tangney, Baumeister, & Boone, 2004). In contrast, poor self-control is regarded as one of the most important causes of crime (Gottfredson & Hirschi, 1990; Pratt & Cullen, 2000) and has also been implicated as a major factor in other problematic behaviors such as impulsive spending and alcohol abuse (e.g., Muraven et al., 2005; Vohs & Faber, 2007).

Whenever two motivations or standards compete against one another, self-control is needed to restrain the unwanted impulse from manifesting itself in to behavior. Such inner motivational conflicts can be characterized as ongoing cost–benefit analyses. Often there are benefits of acting on the emerging impulse, which contributes to why such impulses can be difficult to override. However, the benefits are usually short-term and acting on the impulse can be associated with longer-term costs. A classic example of a motivational conflict is that of a dieter who is offered a piece of cake. The dieter may have the urge to say "yes" because eating cake would bring him or her short-term pleasure. However, the long-term goal of maintaining a fit physique, which is highly valued by Western culture, is also salient, and so the dieter may also be motivated to answer "no." These two motivations (pleasure of eating vs. goal to maintain physique) are in conflict, and self-control is needed to refrain from saying "yes" and to institute the answer "no." Motivation to stick to the diet, the social norm of the situation, and the amount of self-regulatory resources available all influence the cost–benefit analysis of motivational conflicts, but it is self-control that is the vital capacity needed to suppress the emerging undesired impulse.

The Importance of Self-Control

Good self-control benefits people at the individual level, the interpersonal level, and the societal level. People must override their own selfish impulses to cooperate with others, and adhere to morals, social norms, and other rules more generally. At the individual level, a person with good self-control is able to succeed at school and work, save money for a desired vacation, avoid self-destructive behaviors, and maintain an attractive physique. At the interpersonal level, good self-control facilitates relationship maintenance behaviors (e.g., Finkel & Campbell, 2001). At the societal level, good self-control helps people refrain from cheating, committing crimes, and engaging in unsafe sex practices. For example, people refrain from stealing because the short-term benefits associated with stealing often do not outweigh the long-term costs, which can include social disapproval and imprisonment. Indeed, poor self-control is often regarded as one of the most important causes of crime in that criminals often lack self-discipline (Gottfredson & Hirschi, 1990; Pratt & Cullen, 2000).

People who fail at self-control often claim that their behavior was not under their control (e.g., succumbing to addiction). Scholars, however, argue that most impulses are more or less under our control (e.g., Baumeister & Heatherton, 1996). If a gun were put to a person's head, he or she would probably be able to forego a tempting piece of chocolate cake, refrain from smoking, or refrain from arguing with a spouse. Uncontrollable factors, such as societal forces, environmental primes, and latent motivations all have the ability to arouse the impulse to perform such behaviors, but whether people engage in such behaviors is under their volitional control. Given that people often break diets, fail to quit smoking, and otherwise actively exert energy to engage in behaviors they do not want to, it is clear that people do not always regulate such impulses. That people could likely refrain from engaging in such behaviors if their life were on the line (i.e., they are given sufficient motivation to control the impulse) suggests that most impulses are not irresistible but are actually under people's control. Whether people do restrain impulses, however, depends on factors that reside in both the person (e.g., motivation or trait ability to self-regulate) as well as the situation (e.g., environmental stress factors or prospect of social approval from others). One of the main reasons people fail at self-control is insufficient resources to overcome the impulse (Baumeister & Heatherton, 1996), although people can still self-regulate if they are motivated to do so.

Hoch and Loewenstein (1991) suggested that whether people will consume at any given time depends on two things: the strength of the urge and the ability (strength) of the self to resist the urge. As we have already mentioned, self-control strength fluctuates. However, the strength of the impulse also varies, and this likely influences the chance of successful self-control.

Recent research suggests that people may feel stronger emotions and urges when they are depleted of their self-control resources as compared to when their self-control resources are intact (Vohs, Mead, & Schmeichel, 2008). For example, participants who were required to exert self-control on an initial task felt more negative emotions and less positive emotions toward a sad film clip than participants who were not required to exert self-control on an initial task.

Additional studies conducted by Vohs et al. (2008) found that increased urges after exerting self-control statistically mediated the negative effect of prior self-control exertion on subsequent self-control attempts (known as ego-depletion). In one study, participants who were required to engage in an initial act of self-control reported a stronger urge to eat cookies available to them (ostensibly as part of a taste test) as compared to participants who were not required to engage in an initial act of self-control. Replicating previous research on ego-depletion, participants who initially exerted self-control ate more cookies than participants who had not initially exerted self-control. Increased urge to eat the cookies after the initial self-control task was further found to mediate the effect of the first self-control task on the number of cookies eaten. Thus, at the end of a long day, not only do people have fewer resources to overcome the urge to consume a tempting snack than at the beginning of the day, but their desire to consume an unhealthy and tempting snack could be stronger as well. The two components that influence the outcome of a self-control conflict—the strength of the urge and the strength needed to overcome the urge—are separate, but they do appear to influence one another. In other words, when an urge becomes stronger, the self probably needs to muster up greater self-regulatory powers to overcome the urge.

How Self-Control Operates

There are three main ingredients of self-control: standards, monitoring, and strength. Each ingredient contributes to successful self-control, but each ingredient on its own does not guarantee successful self-control and represents a possible pathway for self-control failure (Baumeister et al., 1994; Carver & Scheier, 1981). We begin by providing a brief overview of the three main components and then describe a fourth ingredient that was recently proposed by Baumeister and Vohs (2007). After providing an overview, we give greater detail on the strength component, which is argued to be one of the main reasons for failed self-control (Baumeister & Heatherton, 1996).

Standards

The first component of self-control is standards. Standards are concepts of desired states, including expectations, values, and goals that a person wants to achieve. If a person has the goal to lose weight, he or she may set a desirable weight as the desired goal. For effective self-control, standards need to be clear and well-defined. If the standards are vague, ambiguous, inconsistent, or conflicting, problems achieving the goal can occur, and failed self-control becomes likely (Baumeister et al., 1994).

Monitoring

The second component of self-control is monitoring of the target behavior, especially in relation to the set standard or goal. Similar to standards, monitoring can influence the likelihood that one fails at self-control. Good monitoring of the self can facilitate self-control, but poor monitoring can result in self-control failure. For example, eating binges seem to occur when people stop keeping track of what they have eaten (for a review, see Heatherton & Baumeister, 1991). It has even been argued that one of the main functions of self-awareness is to increase monitoring so as to facilitate self-control (Carver & Scheier, 1981). Self-awareness draws attention to standards and goals and may therefore facilitate effective self-control. In contrast, alcohol has been argued to increase the chance of self-control failure (Baumiester et al., 1994) because alcohol impairs people's ability to monitor themselves (e.g., Hull, 1981).

Strength

In addition to appropriate goal setting and effective monitoring, successful self-control requires active exertion by the self toward achieving those goals. Even when people set appropriate standards and successfully monitor their progress towards their standard, they can still fail at bringing about the desired change because they do not expend the energy necessary to implement the desired behavior (Baumeister & Heatherton, 1996). Evidence indicates that the capacity to exert control over the self is limited in that it relies on a limited energy source and thus operates like a muscle (Baumeister & Heatherton, 1996; Baumeister et al., 1994; Gailliot et al., 2007). The limited resource model of self-control states that all acts of self-control draw from a common, but finite, pool of resources (Muraven & Baumeister, 2000). Thus, when people exert self-control, the resource becomes depleted and self-control failure becomes more likely than when this resource is intact (e.g., Muraven & Baumeister, 1998; Baumeister et al., 1998; Vohs & Schmeichel, 2003).

For example, in one study, participants were seated in front of a bowl of radishes and a bowl of chocolates (Baumeister et al., 1998). Participants randomly assigned to the depletion condition were told to avoid eating the chocolates and instead eat only the radishes. To comply with the researcher's request, the participants therefore had to suppress any desire to eat the chocolates and force themselves to eat the relatively unpleasant alternative instead. Participants randomly assigned to the no-depletion condition were allowed to eat freely and therefore exerted relatively less self-control than participants in the depletion condition. To assess participants' ability to exert self-control on a subsequent and unrelated task, the researchers measured how long participants persisted on an impossible figure-tracing task. To keep working on the task, even in the face of failure and discouragement, required self-control insofar as participants were required to override the urge to quit and instead kept working on the task. As predicted, participants who had previously forced themselves to eat the radishes instead of the chocolates quit much sooner on the figure-tracing task than participants who were allowed to eat whatever they wanted. Thus, participants who had to exert self-control to successfully complete the first task had less energy to exert self-control on the second task than participants who did not have to exert self-control on the first task.

Limited resources appear to be needed for the self's executive function, and thus they are used for behaviors that involve effortful control. Overriding impulses, interrupting behavioral sequences, and other acts of self-regulation that require conscious effort will deplete these resources, whereas automatic and straightforward behaviors will consume much less of them. At present, the most likely integration is that behaviors requiring active and effortful guidance by the self deplete the limited resource, whereas relatively habitual, overlearned, automatic, easy, and unconflicted actions do not consume it. The fact that controlled and executive processes are expensive in terms of this limited resource may be an important reason that so much of life follows routines and habits and that people readily automatize patterns of response that occur frequently, because such styles of action conserve resources.

Motivation

Baumeister and Vohs (2007) recently proposed that level of motivation to achieve a goal or meet a standard should be considered a fourth component of self-control. As mentioned previously, impulses such as sleeping and urinating are likely irresistible, given that no level of motivation could prevent them from occurring at some point. However, most impulses are probably not entirely irresistible and motivation conceivably influences whether a person does exert control over the self. That is, even when the self-control muscle tires, making exertion more difficult, it can still be flexed when given sufficient motivation.

Limited Resource Model of Self-Control

Self-regulatory strength fluctuates from person to person and from situation to situation. Early on, researchers recognized that because self-control seemed much like a muscle, the strength of one's self-control probably could gradually increase or decrease over time. Research supports the notion that self-regulating regularly may reduce susceptibility to self-regulatory depletion. In one study, some participants were asked to maintain good posture over the course of 2 weeks, whereas other participants were not asked to engage in a specific act of self-control over the 2 weeks (Muraven, Tice, & Baumeister, 1998). Results indicated that participants who had completed the regular exercises persisted longer on a laboratory persistence task (used to measure self-regulatory performance) than participants who had not completed the regular exercises.

A series of studies by Gailliot, Plant, Butz, and Baumeister (2007) found similar improvement in self-control from regular self-control exercise. Specifically, participants who completed 2 weeks of self-regulatory exercises performed better on (quite different) self-control tasks than participants who did not complete 2 weeks of self-regulatory exercises. Presumably, regularly monitoring and controlling the self improved participants' self-regulatory stamina, such that their self-regulatory strength was not drained as easily (for a review of studies that provide converging evidence, see Baumeister et al., 2006).

If self-regulatory strength is similar to a muscle, then there should be ways that the strength is replenished or ways to increase exertion despite exhaustion. Regarding replenishment of resources, one route suggested has been sleep and rest. People are much more likely to break their diets, commit impulsive crimes, or go on alcohol binges later in the evening than, say, in the morning (Baumeister et al., 1994). Presumably, people have exerted self-control throughout the day and have fewer resources remaining at the end of the day compared to the beginning of the day to control such impulsive acts. Perhaps more illuminating is research indicating that people were more effective at self-control at work if they had rested and recovered after work than if they did not get such rest (Sonnentag, 2003). For example, people who rested after they finished working for the day took more initiative and performed better at their job the next day than those who had not rested after they finished working for the day. Indeed, following a vacation, many people come back to work with renewed energy, saying that they feel refreshed and ready to delve back into work.

Biology of Self-Control

Recent research suggests that self-control may rely on glucose that is available in the bloodstream (Gailliot et al., 2007). Consistent with the theory that self-control depends on and depletes a limited resource, acts of self-control have been shown to reduce the amount of available glucose in the bloodstream and impair subsequent self-control attempts. For example, one study showed that blood glucose level dropped for participants who had to control their attention while watching a video, but it did not drop for participants who watched the video as they normally would (i.e., without trying to regulate their attention). The amount of glucose available in the bloodstream positively predicted performance on a subsequent self-control task: the lower the glucose, the poorer the self-control. The researchers further found that when they restored glucose to optimal levels, self-control decrements were eradicated. In other words, participants who engaged in initial acts of self-control but imbibed a drink that contained sugar performed better on a self-control task than participants who had engaged in an initial act of self-control but did not ingest glucose. Thus, the glucose drink seemed to replenish people's capacity for self-control.

EXECUTIVE FUNCTION OF THE SELF

As previously mentioned, self-control enables individuals to exert volitional control over their lives. The part of the self that allows for this volitional control is referred to as the executive function of the self (*see* Baumeister, 1998).

In addition to being responsible for enabling people to overcome urges and habits (as in self-control), the executive function of the self also seems to be vital to decision making. Indeed, research indicates that decision making may also draw on the same limited resource as that which self-control draws from (e.g., Baumeister et al., 1998; Vohs et al., 2008). In one study conducted by Baumeister et al. (1998), some participants were asked to give a counterattitudinal speech whereas other participants were asked to take the position on the speech but were led to believe it was ultimately their choice which position they took. Participants who believed they chose to give a counterattitudinal speech persisted for a shorter period of time on a subsequent self-regulatory task as compared to participants who had been told which position to take on the speech. In a different study, participants who made a series of choices subsequently drank less of a healthy but bad-tasting drink than participants who were simply asked to rate the same items without making choices (Vohs et al., 2008). Thus, the act of choosing seems to reduce people's self-regulatory powers temporarily.

Ego depletion also alters the way that people choose, which has implications for people's memory. Previous research on the self-choice effect indicates that individuals have superior memory for items that they chose than for items that have been chosen for them (e.g., Kuhl & Kazén, 1994). However, participants who engage in an act of self-control prior to choosing no longer show increased memory for items they chose (Schmeichel, Gailliot, & Baumeister, 2005). The lack of enhanced memory for items chosen by the self suggests that ego depletion may cause participants to choose in a relatively arbitrary manner, which makes them less likely to remember their choice than if it had been chosen in a more meaningful way.

The executive function also allows people to resist unwanted influence from others. Knowles, Brennan, and Lynn (2004) found that as participants went through a series of political advertisements they became progressively less skeptical of the information presented in the advertisements. This suggests that resisting persuasive messages depletes people's limited self-regulatory abilities, making it progressively more difficult to resist persuasive messages. More recently, research by Wheeler, Briñol, and Hermann (2007) has indicated that depletion increases the chance that people are easily persuaded. In one study, all participants first formed the habit of crossing out every instance of the letter "e" in a passage of text. Participants were then given a second piece of text. Half of the participants were asked to break their newly formed habit by crossing out the letter "e" according to a new set of rules, whereas the other half of participants continued following the initial instructions. Overriding habits requires active self-control and so participants who were asked to break the habit expended relatively more self-regulatory resources during the second task than participants who simply followed the initial rule. Results indicated that the participants who had to break the habit were more easily persuaded by a weak argument than participants who did not have to break the habit. Therefore, individuals' ability to evaluate and reject a poor argument may be another aspect of the executive function that depends on self-regulatory resources.

INTERPERSONAL INTERACTIONS

From a very young age, children are instructed to wait their turn, bite their tongue, and keep their hands to themselves. Doing these things will often help children get along socially, yet many children struggle to exert the necessary self-control these tasks require. In adult life, getting along socially also often requires individuals to exercise self-control. Adult interactions are facilitated by a wide range of stated and unstated social rules, many of which demand that individuals forego their immediate impulses in favor of long-term social gain. Self-control seems to be essential for establishing and maintaining these good relationships with others.

Self-Presentation

It is almost always beneficial for individuals to present their beliefs, goals, and personality in such a way that others will like and approve. However, considering and responding to the needs of a social audience can sometimes be a difficult process. Research has shown that effortful self-presentation taxes people's self-regulatory resources (Vohs, Baumeister, & Ciarocco, 2005). For example, participants who were asked to present themselves as competent and likeable to a skeptical audience were subsequently less able to control their emotions than participants who were told to "act naturally." Additional studies have demonstrated that individuals may present themselves in a less than optimal fashion when they are low on self-regulatory resources. For example, participants who engaged in an initial act of self-control scored higher in narcissism than participants who had not engaged in an initial act of self-control. People may automatically think very highly of themselves (e.g., Greenwald & Banaji, 1995; Pelham, Mirenberg, & Jones, 2002), but they typically restrain such egotistical thoughts and present themselves in a more modest light to others. Thus, ensuring that one's behavior is contextually appropriate can be demanding on the self.

Rejection and Ostracism

Unfortunately not all social interactions run smoothly and from time to time people reject offers of friendship. The act of ostracizing other people, however, appears to exact a cost upon the self. Ciarocco, Sommer, and Baumeister (2001) found that participants who were instructed to refrain from speaking to a confederate later spent less time persisting on unsolvable anagrams than participants who were not required to ostracize the confederate. Participants apparently needed to exert self-control to give other individuals the silent treatment.

Being socially rejected by others also impairs self-control. A series of studies by Baumeister et al. (2005) demonstrated that people who experienced an instance of social rejection or exclusion showed poorer performance on a variety of tasks (dichotic listening, drinking a healthy but bad-tasting beverage, and unsolvable puzzles) than participants who were accepted or who were given bad news that was not social. In one study, participants who were told that no one in their group wanted to work with them later ate more cookies than participants who were told that every member of their group wanted to work with them. The authors argued that rejected people become less motivated to restrain selfish impulses and behaviors given that we typically do so in return for social acceptance. Indeed, recent research indicates that when rejected individuals are given a nonsocial incentive to control themselves, self-control decrements after social rejection disappear (DeWall, Baumeister, & Vohs, 2008). Positive social relationships are thus one of the primary motivations for exercising self-control. If positive relationships seem unattainable, individuals will be less inclined to self-regulate unless given sufficient motivation to exert self-control.

Esteem Threat

In the course of daily life, individuals are sometimes faced with information that does not reflect well on themselves. People get bad reviews from bosses, endure complaints from spouses, and get poor marks on tests and papers. Research on self-control offers insight into why some people seem to be better at handling criticism and failure than others. For example, Baumeister, Heatherton, and Tice (1993) found that people with high self-esteem who were first given negative feedback performed worse on a subsequent complex self-regulatory task than people with low self-esteem. Research by Hoffman and Mann (2006) further demonstrated that this effect is specific to individuals with defensive high self-esteem. Specifically, successful self-regulatory outcomes require individuals to be able to predict accurately and set goals that are attainable. If a person sets a goal that is too high for himself or herself, the chance of self-regulatory failure increases because he or she may not be able to perform at the level required to achieve the goal. For individuals with defensive high self-esteem, having their ego threatened may cause them to have difficulty setting appropriate goals for themselves, thereby paving the way for failed self-control.

Aggression

As cultural beings, humans reap countless benefits from the knowledge and efforts of other people (*see* Baumeister, 2005). Aggression, which involves harming or even killing other individuals, poses a huge threat to this interdependent system. Individuals are therefore expected to avoid acting on harmful urges in order to live harmoniously with others. Recent research indicates that controlling aggressive impulses may require self-control. In one study, participants were first asked to write an essay that ostensibly would be evaluated by another participant (DeWall et al., 2007). Participants randomly assigned to the depletion condition were asked to avoid eating a donut that was placed in front of them, whereas participants in the no-depletion condition were asked to avoid eating much less tempting radishes. After this, participants were given very negative feedback from their partner on their essay ("This is one of the worst essays I've ever read") and were then given the opportunity to respond aggressively toward him or her. Participants were given a variety of ingredients, including hot sauce, to make a snack for their partner who purportedly did not like spicy foods. As predicted, participants who were low in self-regulatory resources added significantly more hot sauce to their critical partner's snack than participants whose self-regulatory resources were intact. Self-control thus helps people restrain aggressive impulses, but when people's self-regulatory powers are low they have a more difficult time controlling their impulses, so that they respond more aggressively upon provocation (Stucke & Baumeister, 2006; DeWall et al., 2007).

Relationships

Self-control has been implicated in relationship maintenance strategies. People often attribute success to themselves and failure to others (*see* Campbell & Sedikides, 1999), but a harmoniously close relationship often requires people to go against this tendency. Sometimes close relationships require individuals to take a little less of their share of the credit and a little more of their share of the blame than they normally would. Preliminary research has shown that self-control may influence whether individuals in close relationships share both the credit and the blame (Vohs, Finkenauer, & Baumeister, 2008). Specifically, when people are depleted, they are more likely to blame their partner for failure and to hog the credit for success.

Remaining in one's current relationship often requires an individual to avoid being tempted by other potential mates. Miller (1997) has shown that the longer participants spent looking at pictures of attractive individuals, the more likely their relationship was to break up during the subsequent months. Vohs et al. (2008) predicted that self-control governs whether individuals in relationships prevent themselves from being distracted and tempted by alternative partners. In a laboratory study, participants who had engaged in an initial act of self-control, and were therefore low in self-regulatory resources, spent more time looking at pictures of scantily clad opposite-sex individuals than participants who had not engaged in an initial act of self-control. This effect was only found among participants who were currently involved in romantic relationships (and who therefore presumably had reason to resist exposing themselves to such tempting images). These findings suggest that self-control helps individuals turn away from potential alternative partners to maintain their current relationship.

Although all relationships seem to require a certain amount of self-regulatory resources, certain kinds of interactions require more resources than others. High-maintenance interactions, which are demanding and inefficient, use more self-regulatory resources than low-maintenance interactions, which are effortless and efficient (Finkel et al., 2006). In one study by Finkel et al. (2006), participants were put in a situation with an unhappy confederate who was either receptive or not receptive to the suggestions of the participant. Participants who were forced to interact with the confederate who was not receptive later were less able to overcome physical fatigue when squeezing a handgrip than participants who interacted with the receptive confederate. The extra effort required to interact with a high-maintenance confederate therefore depleted resources that were later needed for self-control.

Interpersonal relationships are a fundamental part of human existence, and self-control helps us fulfill our need to connect with others. Research suggests that establishing positive first impressions, responding constructively (and not aggressively) to criticism, and maintaining close relationships all require a certain amount of self-control. When self-control is low, interpersonal relationships suffer, and when interpersonal relationships fail (as in rejection), self-control suffers.

Emotion Regulation

Emotions play a large role in informing people about their environment and their choices (for a review, *see* Baumeister et al., 2007). The fact that emotions may serve such a crucial role in human social life may be part of the reason that humans often have such difficulty changing them (*see* Baumeister, 2005). In a variety of studies, researchers have shown that both suppressing emotions (e.g., Baumeister et al., 1998) and exaggerating them (e.g., Schmeichel et al., 2006) consume self-control resources. For example, people who are required to read passages of boring text in an exaggerated manner show decrements in self-control performance on a subsequent task, indicating that instituting emotional responses can consume resources. Conversely, people's ability to control their emotions is impaired if they had previously expended self-control resources (Muraven et al., 1998).

Research by Vohs and Schmeichel (2003) suggests that reappraising one's emotions may not consume as many self-control resources as controlling one's emotions. Participants who were instructed to watch a film "with the detached interest of a medical professional" were as accurate in their time perception as participants who were not asked to control their emotions. However, participants who were asked to suppress their emotional reactions overestimated the amount of time they spent watching an emotional video. Time perception was taken as an indicator of depletion, as it was established that depletion caused people to perceive that time was moving slower than it actually was. Thus, the finding that participants who suppressed their emotions felt that time was moving slower than participants who reappraised their emotions suggests that suppressing emotions is more taxing on self-control resources than reappraising emotions.

Indeed, recent research suggests that the way in which people reappraise their emotional experience has implications for their subsequent processing of the emotions (see Chapter 23). Evaluating an emotional experience in an unattached fashion, which is different from suppressing emotions, leads to abstract and rational thinking (Kross, Ayduck, & Mischel, 2005). In contrast, reappraising a negative event from an immersed perspective tends to enhance negative feelings associated with the event. The way people evaluate emotional experiences may therefore have implications for subsequent attempts at self-control given that abstract construal of events can facilitate people's self-control (Fujita et al., 2006).

Stress

The popular media often caution about the negative effects of stress on individuals' physical health, and research shows that stress can also cause decrements in individuals' ability to exercise self-control. In an early study of self-control, Glass, Singer, and Friedman (1969) found that exposing people to unpredictable noise led to poorer performance on a proofreading task than exposing them to predictable noise. Further, if participants were lead to believe that they could stop the noise, they no longer showed subsequent decrements in performance (even though none of them actually did stop the noise). Bad odors (Rotton, 1983), crowding (Evans, 1979; Sherrod, 1974), and electric shock (Glass & Singer, 1972) are all stressors that have also been shown to decrease individuals' self-control abilities. A review of studies that investigated the relationships between stress and addictive behaviors found that coping with stress lead to relapse for a variety of addictive behaviors (for a review, see Muraven & Baumeister, 2000).

Addictive Behavior

As discussed earlier, many individuals view certain behaviors as the result of irresistible urges, termed *addictions*. Although these urges are not, in fact, irresistible, many individuals seem to have extensive difficulty using self-control to overcome overeating, compulsive shopping, and drug and alcohol abuse.

Overeating

Many individuals struggle with overeating. Indeed, weight loss centers across the country are designed to help individuals overcome bad eating habits. Resisting delicious but unhealthy food has been shown to require self-regulatory resources (Baumeister et al., 1998). As self-control is generally used to bring individuals' behavior in line with their values, attitudes toward food play a large role in how ego-depletion affects their eating habits. Research by Hofmann, Rauch, and Gawronski (2007) found that for participants whose self-regulator resources were intact, the amount of candies they ate reflected their self-reported views toward food. However, for participants who were low in self-regulatory resources, the amount of candies they ate reflected their automatic responses to food, rather than their stated attitude.

Vohs and Heatherton (2000) tested how resisting food would affect chronic dieters. They found that dieters who were invited to "help themselves" to a nearby bowl of unhealthy snacks later ate more ice cream than dieters who had external constraints not to eat the snacks (the experimenter told them not to touch or the snacks were 10 feet away). For nondieters, the distance and availability of the snacks did not affect participants' subsequent self-control. These findings suggest that, for dieters, the internal struggle to resist food consumed self-control resources, thereby impairing subsequent attempts at self-control. Research by Kahan, Polivy, and Herman (2003) also showed that dieters who had to engage in an initial act of self-control ate significantly more than dieters who did not engage in an earlier act of self-control. However, among nondieters, there was no influence of the self-control manipulation on eating. Findings suggest that whereas nondieters do not normally restrict their caloric intake, dieters use self-control to limit their caloric intake, leading them to eat more when resources to exert self-control are low.

Drugs and Alcohol

Restraining from alcohol and drug abuse can also require self-control. Muraven et al. (2005) found that underage drinkers were more likely to break their self-imposed drinking limit on days when their self-control resources had been heavily taxed than on days when their resources had not been heavily taxed. Testing the relationship between self-control and alcohol experimentally, researchers found that participants who had previously engaged in an act of self-control subsequently drank more alcohol preceding a driving test than participants who had not previously engaged in an act of self-regualtion (Muraven, Collins, & Nienhaus, 2002). Further, this effect was particularly pronounced for individuals who reported a generally high temptation to drink. This suggests that depletion may have the greatest effects for individuals who are particularly preoccupied with restricting a particular behavior.

Consumer Behavior

Continually resisting temptation has also been shown to cause self-control decrements in individuals' buying behaviors. Hoch and Loewenstein (1991) suggested that resistance could lead to greater temptation. They found that consumers' desire for a product increased after they resisted buying a product. Research has also shown that lowered self-regulatory resources can affect individuals' buying behavior. For example, individuals who were asked to make many decisions (and were therefore depleted) were more likely to splurge on attractive but expensive items than participants who had not been depleted (Bruyneel et al., 2006). Other research has indicated that self-control allows people to refrain from impulsive spending (Vohs & Faber, 2007). When self-control resources are low, people have a difficult time controlling their spending. In one study, participants were asked to read a long passage of boring text aloud. Half of the participants were asked to read the text in a very animated fashion, whereas the other half of participants read the passage in a natural way. As mentioned earlier, emotional exaggeration has been shown to tax self-regulatory resources (Schmeichel et al., 2006), and so the requirement to inject emotion into reading the passage was likely to consume self-regulatory resources. After the manipulation, participants were given $10 that they could keep or spend on a variety of snacks. Participants who had to read the passage in an animated fashion later spent more of the money they were given on snacks than participants who were not given any special instructions for reading the passage. This research suggests that self-control plays a large role in allowing individuals to control their spending behaviors, because when individuals are low in self-control, they are likely to spend imprudently or impulsively.

Conclusion

Throughout the course of everyday life people inevitably experience myriad conflicting motivations. People often feel the temptation, urge, or impulse to engage in behaviors that bring short-term benefits but long-term costs. Self-control, the capacity to restrain unwanted thoughts, impulses, and desires, is the vital part of the self that allows people to forego such immediate pleasures in the service of their long-term goals. A growing body of evidence suggests that people's ability to exert self-control depends on a limited resource. If the resource is taxed, especially by previous acts of self-control, it becomes more likely that people will give in to their impulses and temptations than if the resource had not been taxed. Although people's self-control strength is a powerful determinant of whether they successfully resolve self-control conflicts, several additional factors, such as standards set by the self and society, motivation to perform the behavior, and personal monitoring, all seem to work in concert with self-control strength to influence behavioral outcomes. The benefits of self-control are reaped at the personal level, interpersonal level, and societal level. It therefore behooves researchers to continue investigating the ways in which people can increase their chance of self-control success.

REFERENCES

Bargh, J. A., & Ferguson, M. J. Beyond behaviorism: On the automaticity of higher mental processes. Psychol Bull 2000; 126: 925–945.

Baumeister, R. F. The self. In: Gilbert, D. T., Fiske, S. T., & Lindzey, G. (Eds.), *Handbook of social psychology*, 4th ed. New York: McGraw-Hill, 1998: pp. 680–740.

Baumeister, R. F. *The cultural animal: Human nature, meaning, and social life.* New York: Oxford University Press, 2005.

Baumeister, R. F., Bratslavsky, E., Muraven, M., & Tice, D. M. Self-control depletion: Is the active self a limited resource? J Pers Soc Psychol 1998; 74: 1252–1265.

Baumeister, R. F., DeWall, C. N., Ciarocco, N. J., & Twenge, J. M. Social exclusion impairs self-control. J Pers Soc Psychol 2005; 88: 589–604.

Baumeister, R. F., Gailliot, M., DeWall, C. N., & Oaten, M. Self-control and personality: How interventions increase regulatory success, and how depletion moderates the effects of traits on behavior. J Pers 2006; 74: 1773–1801.

Baumeister, R. F., & Heatherton, T. F. Self-control failure: An overview. Psychol Inq 1996; 7: 1–15.

Baumeister, R. F., Heatherton, T. F., & Tice, D. M. When ego threats lead to self-control failure: Negative consequences of high self-esteem. J Pers Soc Psychol 1993; 64: 141–156.

Baumeister, R. F., Heatherton, T. F., & Tice, D. M. *Losing control: How and why people fail at self-control.* San Diego, CA: Academic Press, 1994.

Baumeister, R. F., & Vohs, K. D. Self-control, ego depletion, and motivation. Soc Personal Psychol Compass 2007; 1: 115–128.

Baumeister, R. F., Vohs, K. D., DeWall, C. N., & Zhang, L. How emotion shapes behavior: Feedback, anticipation, and reflection, rather than direct causation. Pers Soc Psychol Rev 2007; 11: 167–203.

Bruyneel, S., Dewitte, S., Vohs, K. D., & Warlop, L. Repeated choosing increases susceptibility to affective product features. Int J Res Market 2006; 23: 215–225.

Campbell, W. K., & Sedikides, C. Self-threat magnifies the self-serving bias: A meta-analytic integration. Rev Gen Psychol 1999; 3: 23–43.

Carver, C. S., & Scheier, M. E. *Attention and self-control: A control theory approach to human behavior.* New York: Springer-Verlag, 1981.

Ciarocco, N. J., Sommer, K. L., & Baumeister, R. F. Ostracism and ego depletion: The strains of silence. Pers Soc Psychol Bull 2001; 27: 1156–1163.

DeWall, C. N., Baumeister, R. F., Stillman, T. F., & Gailliot, M. T. Violence restrained: Effects of self-control and its depletion on aggression. J Exp Soc Psychol 2007; 43: 62–76.

DeWall, C. N., Baumeister, R. F., & Vohs, K. D. Satiated with belongingness? Effects of acceptance, rejection, and task framing on self-regulatory performance. J Pers Soc Psychol 2008; 95: 1367–1382.

Duckworth, A. L., & Seligman, M. E. P. Self-discipline outdoes IQ in predicting academic performance of adolescents. Psychol Sci 2005; 16: 939–944.

Evans, G. W. Behavioral and physiological consequences of crowding in humans. J App Soc Psychol 1979; 9: 27–46.

Finkel, E. J., & Campbell, W. K. Self-control and accommodation in close relationships: An interdependence analysis. J Pers Soc Psychol 2001; 81: 263–277.

Finkel, E. J., Campbell, W. K., Brunell, A. B., Dalton, A. N., Scarbeck, S. J., & Chartrand, T. L. High-maintenance interaction: Inefficient social coordination impairs self-control. J Pers Soc Psychol 2006; 91: 456–475.

Fujita, K., Trope, Y., Liberman, N., & Levin-Sagi, M. Construal levels and self-control. J Pers Soc Psychol 2006; 90: 351–367.

Gailliot, M. T., Baumeister, R. F., DeWall, C. N., et al. Self-control relies on glucose as a limited energy source: Willpower is more than a metaphor. J Pers Soc Psychol 2007; 92: 325–336.

Gailliot, M. T., Plant, E. A., Butz, D. A., & Baumeister, R. F. Increasing self-regulatory strength can reduce the depleting effect of suppressing stereotypes. Pers Soc Psychol Bull 2007; 33: 281–294.

Glass, D. C., & Singer, J. E. *Urban stress: Experiments on noise and social stressors.* New York: Academic Press, 1972.

Glass, D. C., Singer, J. E., & Friedman, L. N. Psychic cost of adaptation to an environmental stressor. J Pers Soc Psychol 1969; 12: 200–210.

Gottfredson, M. R., & Hirschi, T. *A general theory of crime.* Stanford, CA: Stanford University Press, 1990.

Gordijn, E. H., Hindriks, I., Koomen, W., Dijksterhuis, A., & Van Knippenberg, A. Consequences of stereotype suppression and internal suppression motivation: A self-control approach. Pers Soc Psychol Bull 2004; 30: 212–224.

Greenwald, A. G., & Banaji, M. R. Implicit social cognition: Attitudes, self-esteem and stereotypes. Psychol Rev 1995; 102: 4–27.

Higgins, E. T. The "self digest": Self-knowledge serving self-regulatory functions. J Pers Soc Psychol 1996; 71: 1062–1083.

Hoch, S. J., & Loewenstein, G. F. Time-inconsistent preferences and consumer self-control. J Consum Res 1991; 17: 492–507.

Hoffman, K. L., & Mann, T. When do ego threats lead to self-control failure? Negative consequences of defensive high self esteem. Pers Soc Psychol Bull 2006; 32: 1177–1187.

Hofmann, W., Rauch, W., & Gawronski, B. And deplete us not into temptation: Automatic attitudes, dietary restraint, and self-regulatory resources as determinants of eating behavior. J Exp Soc Psychol 2006; 43: 497–504.

Hull, J. G. A self-awareness model of the causes and effects of alcohol consumption. J Abnorm Psychol 1981; 90: 586–600.

Kahan, D., Polivy, J., & Herman, C. P. Conformity and dietary disinhibition: A test of the ego-strength model of self-control. Int J Eat Disord 2003; 32: 165–171.

Knowles, E. S., Brennan, M., & Linn, J. A. *Consuming resistance to political ads*. Manuscript in preparation, 2004.

Kross, E., Ayduk, O., & Mischel, W. When asking "why" does not hurt. Psychol Sci 2005; 16: 709–715.

Kuhl, J., & Kazén, M. Self-discrimination and memory: State orientation and false self-ascription of assigned activities. J Pers Soc Psychol 1994; 66: 1103–1115.

Miller, R. S. Inattentive and contented: Relationship commitment and attention to alternatives. J Pers Soc Psychol 1997; 73: 758–766.

Mischel, W., Shoda, Y., & Peake, P. The nature of adolescent competencies predicted by preschool delay of gratification. J Pers Soc Psychol 1988; 54: 687–696.

Muraven, M. R., & Baumeister, R. F. Self-control and depletion of limited resources: Does self-control resemble a muscle? Psychol Bull 2000; 126: 247–259.

Muraven, M., Collins, R. L., & Nienhaus, K. Self-control and alcohol restraint. An initial application of the self-control strength model. Psychol Addict Behav 2002; 16: 113–120.

Muraven, M., Collins, L. R., Shiffman, S., & Paty, J. A. Daily Fluctuations in self-control demands and alcohol intake. Psychol Addict Behav 2005; 19: 140–147.

Muraven, M., Tice, D. M., & Baumeister, R. F. Self-control as limited resource: Regulatory depletion patterns. J Pers Soc Psychol 1998; 74: 774–789.

Pelham, B. W., Mirenberg, M. C., & Jones, J. T. Why Susie sells seashells by the seashore: Implicit egotism and major life decisions. J Pers Soc Psychol 2002; 82: 496–487.

Pratt, T. C., & Cullen, F. T. The empirical status of Gottfredson and Hirschi's general theory of crime: A meta-analysis. Criminology 2000; 38: 931–964.

Richards, J. M., & Gross, J. J. Composure at any cost? The cognitive consequences of emotion suppression. Pers Soc Psychol Bull 1999; 25: 1033–1044.

Richards, J. M., & Gross, J. J. Emotion regulation and memory: The cognitive costs of keeping one's cool. J Pers Soc Psychol 2000; 79: 410–424.

Rotton, J. Affective and cognitive consequences of malodorous pollution. Basic App Soc Psychol 1983; 4: 171–191.

Schmeichel, B. J., Gailliot, M. T., & Baumeister, R. F. *Ego depletion undermines the benefits of the active self to memory*. Manuscript submitted for publication, 2005.

Schmeichel, B. J., Demaree, H. A., Robinson, J. L., & Pu, J. Ego depletion by response exaggeration. J Exp Soc Psychol 2006; 42: 95–102.

Schmeichel, B. J., Vohs, K. D., & Baumeister, R. F. Intellectual performance and ego depletion: Role of the self in logical reasoning and other information processing. J Pers Soc Psychol 2003; 85: 33–46.

Sherrod, D. R. Crowding, perceived control, and behavioral aftereffects. J App Soc Psychol 1974; 4: 171–186.

Sonnentag, S. Recovery, work engagement, and proactive behavior: A new look at the interface between nonwork and work. J Appl Psychol 2003; 88: 518–528.

Stucke, T. S., & Baumeister, R. F. Ego depletion and aggressive behavior: Is the inhibition of aggression a limited resource? Eur J Soc Psychol 2006; 36: 1–12.

Tangney, J. P., Baumeister, R. F., & Boone, A. L. High self-control predicts good adjustment, less pathology, better grades, and interpersonal success. J Pers 2004; 72: 271–322.

Vohs, K. D., Baumeister, R. F., & Ciarocco, N. J. Self-control and self-presentation: Regulatory resource depletion impairs impression management and effortful self-presentation depletes regulatory resources. J Pers Soc Psychol 2005; 88: 632–657.

Vohs, K. D., Baumeister, R. F., Schmeichel, B. J., Twenge, J. M., Nelson, N. M., & Tice, D. M. Making choices impairs subsequent self-control: A limited resource account of decision making, self-regulation, and active initiative. J Pers Soc Psychol 2008; 94: 883–898.

Vohs, K. D., & Faber, R. J. Spent resources: Self-regulatory resource availability affects impulse buying. J Consum Res 2007; 33: 537–547.

Vohs, K. D., Finkenauer, C., & Baumeister, R. F. *Depletion of self-regulatory resources makes people selfish*. Unpublished manuscript, University of Minnesota, 2007.

Vohs, K. D., & Heatherton, T. F. Self-regulatory failure: A resource-depletion approach. Psychol Sci 2000; 11: 249–254.

Vohs, K. D., Mead, N. L., Schmeichel, B. J. *Self-regulatory resource depletion makes people more extreme in their urges: A possible mechanism for ego depletion*. Unpublished manuscript, University of Minnesota, 2008.

Vohs, K. D., & Schmeichel, B. J. Self-control and the extended now: Controlling the self alters the subjective experience of time. J Pers Soc Psychol 2003; 85: 217–230.

Wheeler, S. C., Briñol, P., & Hermann, A. D. Resistance to persuasion as self-control: Ego-depletion and its effects on attitude changes processes. J Exp Soc Psychol 2007; 43: 150–156.

CHAPTER 21

Walking the Line between Goals and Temptations: Asymmetric Effects of Counteractive Control

Ayelet Fishbach and Benjamin A. Converse

ABSTRACT

Research on counteractive control explores the processes that individuals employ to increase the motivational strength of their high-order goals and decrease the motivational strength of their low-order temptations. In this chapter, we first describe the basic assumption of counteractive control theory: that self-control is an instrumental response to motivational conflicts. People only exercise self-control when a significant conflict is expected between high- and low-order motives. We then describe how self-control operations bolster the motivational strength of goal pursuit via one path and asymmetrically undermine the motivational strength of temptation pursuit via a second path. Next, we discuss the specific self-control strategies, including those that modify the choice situation, shift attainment expectations, and change the psychological representation of choice alternatives. We further distinguish between explicit self-control operations that rely on conscious processing, and implicit operations that do not require explicit consideration. We end with a broader discussion of the conditions under which goals and temptations appear to be in conflict or not.

Keywords: counteractive control, self-control dilemma, goal, temptation, goal conflict identification

People rarely desire one thing at a time. Rather, the process of goal pursuit involves constantly prioritizing the many goals that a person wishes to pursue and resolving goal conflicts (e.g., navigating career, leisure, and family activities). This chapter focuses on a specific type of goal conflict: *the self-control dilemma*. People face a self-control dilemma whenever the attainment of a high-order goal would come at the expense of a low-order, yet alluring temptation (Ainslie, 1992; Baumeister, Heatherton, & Tice, 1994; Loewenstein, 1996; Metcalfe & Mischel, 1999; Rachlin, 2000; Thaler & Shefrin, 1981). For example, a dieter's desire to finish a meal with a sweet dessert may not coincide with his desire to maintain a low-fat diet; a saver's wish to get the new gadget in stores may interfere with her saving plans; and a student's urge to procrastinate may not allow him to complete his assignments on time. Whenever these individuals contemplate the conflicting motives (i.e., goal vs. temptation), they experience a self-control dilemma. In response to self-control dilemmas, people exercise self-control to facilitate the attainment of the more important goal (Ainslie, 2001; Baumeister & Vohs, 2004; Gollwitzer &

Moskowitz, 1996; Kuhl & Beckmann, 1985; Mischel, Shoda, & Rodriguez, 1989; Rachlin, 2000).

Goals and temptations are both motivational states but they have different status. High-order motives (or "goals") serve central interests that are more important in the long run, and low-order motives (or "temptations") serve peripheral interests that are beneficial in the short-term. In this view, self-control is a tool for pursuing goals that are given high priority in a person's subjective goal hierarchy. Thus, the processes of goal setting (determining the goal hierarchy) do not require self-control. In addition, self-control is rarely a goal in itself, in the sense that people do not have a goal to exercise self-control. Instead, self-control is an instrumental response that improves goal striving when high-order goals are in conflict with low-order temptations.

Self-control problems are common and have a long history in human conduct and thought. Self-control problems have been documented and studied in philosophy, religion, and more recently in the social sciences by researchers in economics (e.g., Becker, 1960; O'Donoghue & Rabin, 2000; Thaler & Shefrin, 1981), political science (e.g., Elster, 1977; Schelling, 1984), and psychology (e.g., Ainslie, 2001; Baumeister & Vohs, 2004; Kuhl & Beckmann, 1985; Rachlin, 2000). In light of this rich history and the various perspectives presented in this volume, the purpose of the current chapter is to shed light on the processes of self-control from a structural point of view, proposed by counteractive control theory (Fishbach & Trope, 2005; Fishbach & Trope, 2007; Trope & Fishbach, 2000, 2005).

Counteractive control theory addresses the process by which people proactively counteract the threat that temptations pose to the attainment of goals. This threat arises only when important goals are in competition with similarly strong temptations and when external means are not in place to protect the goals. When people anticipate a self-control problem, they increase the motivational strength of pursuing a goal over giving in to temptations. Thus, the presence of tempting alternatives influences behavior in two opposing directions. Directly, it decreases the likelihood of adhering to goals. Indirectly, the perception of tempting alternatives triggers the operation of self-control, which acts to increase the likelihood of adhering to goals. By influencing the motivational strength of choice alternatives, counteractive control offsets the effect that the temptations may have on one's behavior.

In what follows, we define the structure of the counteractive process according to counteractive control theory and draw similarities between different self-control operations that are documented in the literature. Our main proposition is that counteractive control works to resolve the tension between high- and low-order motives by asymmetrically shifting their motivational strengths. High-order goals are strengthened so they may override low-order temptations. Low-order temptations are weakened so they may be overridden by high-order goals. These asymmetric shifts in motivational strength may involve behavioral strategies (imposing penalties, rewards) or mental operations (devaluing or bolstering the value of activities). These shifts may further involve explicit operations that require conscious awareness and planning or implicit processes that operate with minimal awareness and conscious planning. Regardless of the specific type of self-control operation, its function is similar: It either increases the tendency to operate on a personal motive or decreases the tendency to operate on it, depending on the status of the motive as a goal or temptation.

We divide our review into four major parts. We first lay the basic assumptions of counteractive control theory (Instrumental Self-Control)—namely, that self-control is an instrumental response to motivational conflicts. In the next section (Asymmetric Self-Control), we introduce the basic idea of counteractive control theory; that self-control results in asymmetric shifts in motivational strength. We then address the specific self-control strategies that are documented in the literature (The Strategies of Self-Control) and that follow the counteractive control process of increasing the motivational strength of goals relative to temptations. We end this review with a broader discussion of the conditions under which goals and temptations appear to be in conflict or not. In this section (Self-Control Failure), we propose that when people view these motivations as complementing each other, rather than conflicting with each other, they do not exercise self-control.

INSTRUMENTAL SELF-CONTROL

According to counteractive control theory, people exercise self-control only when a significant conflict is expected between their high-order goals and low-order temptations. That is, self-control is only needed when important goals are threatened by similarly strong temptations. For example, a trip down the buffet line will only engage self-control operations for someone who desires to enjoy delicious, but unhealthy tiramisu and to maintain a healthy diet. If the diner saw someone sneeze on the tiramisu, its value would be extremely weak compared to the health goal, and the threat to the goal would be minimal. No counteractive strategies to devalue the tiramisu or to bolster the value of healthier alternatives would be necessary in this case. In addition, if the diner was not concerned with weight-watching or healthy eating, the value of the delicious tiramisu would be extremely strong compared to the health goal. In this case, too, the conflict experience would be minimal, so counteractive control would not be used to boost the relative value of healthy options. More generally, if the relative value of goal compared to temptation is extreme, then there is no need for self-control.

To test this assumption, research on counteractive control has varied the strength of the tempting alternative, the goal alternative, or both. Results indicate that when the cost of pursuing the goal is minimal (i.e., when temptations are weak), self-control activation is minimal. People increase their self-control efforts as the temptation to disengage from the goal increases. For example, participants exercised self-control by imposing potential fines on themselves to complete an academic task at an inconvenient, as opposed to a convenient, time (Fishbach & Trope, 2005). In addition, when the perceived benefit of pursuing a goal is minimal (i.e., when goals are weak), self-control activation is also minimal. For example, only committed dieters (vs. nondieters) exercised self-control when they were exposed to fatty food (Fishbach, Friedman, & Kruglanski, 2003).

Even in the presence of similarly valuable goals and temptations, self-control only operates when it is needed to overcome a potential threat to important goals. If other means are in place, it is not needed. Thus, self-control and external means of control are often substitutable (Fishbach & Trope, 2005), and possibly, self-control operations also substitute for each other. Back at the buffet, for example, if the weight-watching tiramisu lover had foreseen her potential dilemma and ordered the "$6.99 soup-and-salad package" rather than the "$10.99 all-you-can-eat package," then there would be no need to enact other strategies of self-control at the end of the meal.

In a series of studies that tested this substitutability assumption, Fishbach and Trope (2005) found a significant decline in self-control efforts such as bolstering the value of high-order goals when participants believed that an experimenter was monitoring their behavior, when they were primed with controlling individuals (e.g., parents, professors), or when they were offered a large amount of money in return for pursuing their personal goals. Each of these externally imposed control methods—social monitoring, primed controlling figures, and monetary compensation—was sufficient to secure goal pursuit in the absence of self-control.

Finally, as discussed later, the goal and the temptation must be perceived as competing options if self-control is to be exercised. To the extent that people perceive these motivations as complementing one another, there is no need to exercise self-control (Fishbach & Zhang, 2008). For example, if the weight-watcher perceives unhealthy and healthy foods as complementing each other in a balanced meal, she would not exercise self-control when offered the unhealthy tiramisu, because presumably, she could compensate by choosing a healthier entrée.

ASYMMETRIC SELF-CONTROL

According to counteractive control theory, the route to self-control success (i.e., goal-congruent choice) can be divided into two distinct

TABLE 21–1. ASYMMETRIC SELF-CONTROL STRATEGIES

	Temptations	Goals
Changing the choice situation	• Self-imposed penalty • Pre-commitment to forgo • Avoidance	• Self-imposed rewards • Pre-commitment to pursue • Approach
Changing expectations and standards	• Low performance	• High performance
Changing the psychological meaning	• Devalue • Inhibit • Cool and abstract construal	• Bolster • Activate • Hot and concrete construal

paths: Via one path, self-control operations bolster the motivational strength of goal pursuit, and via another path, self-control operations undermine the motivational strength of temptation pursuit. By asymmetrically responding to goals versus temptations, self-regulators can thus increase the relative strength of the goal and secure goal pursuit.

It is important to clarify that temptations and goals are defined within a given conflict situation and with respect to each other. This structural (as opposed to content-based) definition suggests that any personal motivation can potentially constitute an interfering temptation with respect to a higher level goal, or it can constitute an overriding goal with respect to a lower order temptation. For example, "working out" may be perceived as interfering with the pursuit of higher order academic objectives, and it may be perceived as a goal that is interfered with by the pursuit of lower order relaxation objectives. An asymmetric counteractive process means that when working out is a tempting alternative relative to the goal of studying in the library, self-control acts to decrease its motivational strength. On the other hand, when working out is a high-order goal relative to the tempting alternative of lounging on the couch with a box of donuts, self-control acts to increase its motivational strength.

Therefore, cues for a competing motivation can increase or decrease the strength of a focal motivation, depending on their relative status in the motivational conflict. Specifically, people may be working on a focal goal (e.g., exercising) when cues for a higher-order goal (e.g., "studying") or lower-order temptation (e.g., "couch") manifest. In these situations, an accessible goal cue would undermine the motivational strength of a low-order temptation ("studying" undermines exercising), whereas an accessible temptation cue would augment the motivational strength of a high-order goal ("couch" augments exercising).

Self-control research has documented many of the specific operations that people employ to asymmetrically shift motivational strength. In what follows, we review the basic categories of self-control operations. We demonstrate how each of these operations creates asymmetric motivational shifts, working for goal pursuit and against temptation indulgence, as predicted by counteractive control theory.

THE STRATEGIES OF SELF-CONTROL

Asymmetric responses to goals and temptations have been observed in a number of forms. We discuss asymmetric changes to the choice situation, asymmetric expectations of future goal pursuit, and asymmetric construals of the psychological meaning of choice alternatives. We summarize these categories and the related self-control strategies in Table 21–1.

Changing the Choice Situation

People may increase the likelihood of goal-congruent choice by changing the situations in which choices occurs. For example, the objective value of an option can be changed by attaching a penalty or a reward to it, the available choice set can be altered, or an implicit disposition to approach or avoid certain options can develop. Further, each situational change increases goal-congruent choice by means of increasing the motivational strength of goals or, asymmetrically, undermining the motivational strength of temptations. For example, the value of a goal-consistent option can be increased by associating it with a contingent bonus, or the value of a temptation-related option can be decreased by associating it with a consumption penalty. Similarly, the choice set can be altered to favor goal pursuit by increasing the number or availability of goal-consistent options or by decreasing the number or availability of temptation-consistent options. On an implicit level, people can develop implicit dispositions toward goals or away from temptations.

Rewarding goal pursuit and penalizing temptation pursuit

In a study that demonstrated self-imposed rewards, Trope and Fishbach (2000) examined participants' willingness to make a bonus contingent on goal-consistent behavior. Participants who indicated that health goals were important to them were given the opportunity to receive reliable and accurate feedback about their future heart disease risks. The instrumentality assumption of counteractive control theory specifies that self-control will be activated only when there is conflict between similarly valued goals and temptations. Thus, the temptation of avoiding this test was manipulated across conditions. Some participants were told that the medical test would be highly uncomfortable, requiring them to engage in an hour of strenuous exercise and a painful hormone sampling procedure. Other participants were told that the test would be comfortable, requiring them to engage in an hour of relaxing reading and an easy hormone sampling procedure. To assess self-imposed rewards, participants were given a choice between receiving their compensation before or after the test, thus making it noncontingent or contingent on test completion. As predicted, participants were more likely to make their compensation contingent on completing the uncomfortable medical procedure (strong temptation) than the easy medical procedure (weak temptation). By self-imposing this contingency, participants risked losing the bonus, but they also increased the value, and hence the likelihood of completing the more painful cardiovascular test.

According to the asymmetry claim, people can achieve similar ends by imposing penalties that would be associated with temptation indulgence. Indeed, another study (Trope & Fishbach, 2000) investigated whether people imposed penalties on themselves for giving in to temptations. Participants were offered an opportunity to take a test that required abstinence from food that contained glucose. For some participants, the period of required abstinence was only 6 hours, and thus the temptation to skip was relatively weak; for others, the period of required abstinence was 3 days, and thus the temptation to skip was relatively strong. Participants were then asked to indicate the amount of money they would be willing to pay as a penalty for failing to complete the test. Consistent with our argument, they chose to set higher penalties for failure to complete a long period of abstinence than for failure to complete a short period of abstinence. Therefore, it seems that the differing responses to goals and temptations can achieve comparable ends in terms of increasing goal-congruent choice. When there is some tension between the value of goals and the value of competing temptations, contingent bonuses change the choice set to favor goals, whereas contingent penalties change the choice set to disfavor temptations; both lead to increased pursuit of higher order goals.

Pre-committing to pursue goals and avoid temptations

Individuals may precommit themselves to act according to their high-order goals by eliminating tempting alternatives or by adding goal alternatives to the choice set, thus biasing the choice set in favor of the goal (Ainslie, 1992; Green & Rachlin, 1996; Rachlin & Green, 1972; Schelling, 1984; Strotz, 1956; Thaler, 1991; Thaler & Shefrin, 1981). For example, students may choose to refrain from signing up for extracurricular activities to help ensure their study habits, and dieters may choose to buy only healthy options at the grocery store to increase healthy snacking later.

As an example of these asymmetric shifts of choice availability, Wertenbroch (1998) found that smokers prefer to buy their cigarettes by the pack, rather than in 10-pack saving cartons, to limit consumption. In another study (Ariely & Wertenbroch, 2002), students who were given the opportunity to set their own deadlines for class assignments committed themselves to due dates that were earlier than necessary. They did so at great potential cost (a grade penalty for being late) and no obvious benefit (e.g., extra feedback from the instructor), other than the increased motivation to complete their studies. By adopting this strategy, the smokers and the students eliminated their future freedom of choice, which people ordinarily seek to maintain (Brehm, 1966), for the sake of decreasing their exposure to temptations and securing the attainment of their higher order goals. The students' behavior provides an example of a single act that simultaneously, and asymmetrically, affects goal and temptation pursuits. On any given day, their early deadlines both increased the probability of working on assignments and decreased the probability of pursuing tempting alternatives.

Approaching goals and avoiding temptations

Another way people increase the motivational strength of high-order goals is by keeping their distance from tempting objects and maintaining close physical proximity to objects that are associated with their goals (Ainslie, 1992; Schelling, 1984; Thaler & Shefrin, 1981). For example, foreseeing the problem that a previous romantic partner may impose, people sometimes move to a different city or job. The process of avoiding temptations may be deliberative and effortful, such as relocating, but at other times, it may need to be quick and efficient. For example, the conflict between avoiding a long night of heartburn and finishing the rest of a delicious dessert may be resolved merely by pushing the half-eaten plate toward the center of the table without paying much attention to one's actions. This act may be asymmetrically complimented by reaching for the check and pulling it closer to bring the meal to a close.

As these examples demonstrate, explicit self-control operations are often accompanied by similar implicit responses. Recent self-control research has investigated a set of implicit self-control strategies (e.g., Fishbach et al., 2003; Fishbach & Shah, 2006; Fishbach & Trope, 2007; Gollwitzer, Bayer, & McCulloch, 2004; Moskowitz, Gollwitzer, Wasel, & Schaal, 1999). It shows that the intentional self-control strategies described thus far are often accompanied by, or replaced with, some unconscious strategies that create implicit changes in the motivational strength of goals and temptations.

In many respects, implicit self-control operates by principles similar to those described by research on unconscious goal pursuit (Aarts & Dijksterhuis, 2000; Bargh & Chartrand, 1999; Bargh & Ferguson, 2000; Bargh, Gollwitzer, Lee-Chai, Barndollar, & Troetschel, 2001; Shah & Kruglanski, 2003). Although, whereas unconscious goal pursuit promotes behavior that corresponds to situationally primed goals, unconscious self-control cancels out the influence of these primes if they are perceived as potential conflicts for one's higher order goals. For example, according to the basic principles of unconscious goal pursuit, priming an upstanding citizen with "sex," "drugs," and "rock 'n roll" should increase the likelihood that the citizen would engage in these iniquitous behaviors. Research on implicit self-control, however, predicts that these primes would increase self-control and more conservative behavior when they conflict with preexisting long-term goals. Thus, implicit self-control is elicited by situational cues and, at the very same time, acts to offset the influence of these cues on behavior and regain personal control (for a review, see Fishbach & Trope, 2007).

In terms of implicit counteractive changes to the choice situation, individuals who resist temptations explicitly and repeatedly may develop an implicit disposition toward approaching goals and avoiding temptations. A series of studies by Fishbach and Shah (2006) examined these implicit self-control tactics. Building

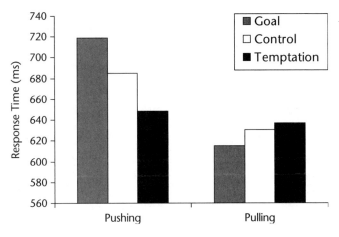

Figure 21–1. Response times for pushing versus pulling self-generated goal, control, and temptation targets. (Note: Lower values indicate stronger predisposition.)

from the finding that people are faster to pull a lever toward them to indicate an approach orientation and to push a lever away from them to indicate an avoidance orientation (Chen & Bargh, 1999; Markman & Brendl, 2005; Solarz, 1960), Fishbach and Shah found that participants automatically approached goal-related stimuli (through faster pulling responses) and avoided temptation-related stimuli (through faster pushing responses). For example, in one study, participants completed a lexical decision task with their own goals and temptations embedded in it, as well as control activities that they did not pursue (e.g., one participant listed exercising vs. alcohol vs. internship). In this task, participants had to identify a number of letter-strings as words or nonwords by either pulling or pushing a joystick. The critical trials included response times for self-generated goals, temptations, and control activities. It was demonstrated that participants are faster to pull (approach) goals and push (avoid) temptations (see Fig. 21-1).

Moreover, the tendency to automatically approach goals and avoid temptations facilitated goal attainment. For example, in one study, reaction times for pulling academic goals (e.g., library, homework) and pushing nonacademic temptations (e.g., travel, party) predicted student participants' GPA scores. Similarly, participants who were asked to pull (approach) academic-related concepts and push (avoid) nonacademic concepts planned to invest more time on their homework, compared with students who completed the opposite categorization task. Overall, the asymmetric response to goals and temptations seems to manifest even in these more implicit situational changes, and it plays an important role in adherence to high-order goals.

Changing Attainment Expectations

The aforementioned lines of research demonstrate how people secure the attainment of high-order goals by changing the choice situation. As a result of self-control, people's expectation to achieve the goal increases. For example, a person may correctly assume that the objective likelihood of undergoing a medical checkup is higher if there is a penalty imposed on failing to complete the checkup or a reward contingent on completing the checkup. Whereas the objective likelihood of pursuing a goal may increase as a result of self-control operations, the subjective likelihood of goal achievement may serve a self-control function in itself.

Recent research on counteractive optimism explores this possibility. It documents asymmetric effects on people's anticipated goal- and temptation-pursuit, which in turn influence their actual motivation to pursue goals or temptations (Zhang & Fishbach, in press). Specifically, counteractive optimism refers to a tendency to provide optimistic predictions of future engagement with goals and disengagement from temptations. For example, a student who has just received an invitation to next weekend's big party may make optimistic predictions about the amount and quality of studying that she will do before an upcoming exam. Asymmetrically, she may strategically underestimate how much time she will spend at the party. These estimates are instrumental when they act as performance standards to motivate studying over partying.

Previous research on goal-setting theory (Locke & Latham, 1990) and energization theory (Wright & Brehm, 1989) demonstrates that people adjust their effort to match their anticipated level of performance. High (i.e., more difficult) performance standards elicit greater motivation than low standards (Atkinson & Feather, 1966; Brehm & Self, 1989; Heath, Larrick, & Wu, 1999; Oettingen & Mayer, 2002; Taylor & Brown, 1988). Therefore, whenever optimistic predictions act as performance standards, they motivate actions that help to reach these standards, whereas pessimistic predictions undermine one's motivation. Specifically, expectations of greater goal pursuit and lesser temptation indulgence can be two sides of the same adaptive strategy—motivating actions via expectations.

Notably, the tendency for people to be optimistic in their predictions is well-documented. Research on the "planning fallacy" describes a common underestimation of how long it will take one to complete goal-related tasks (Buehler, Griffin, & Ross, 1994) and people routinely overestimate their performance on such tasks (Allison, Messick, & Goethals, 1989; Brown, 1986; Chambers & Windschitl, 2004; Kruger & Dunning, 1999; Kunda, 1987; Svenson, 1981; Windschitl, 2004). Although such optimistic tendencies do not always have a motivational origin or function, we next discuss the evidence that these biases do often provide strategic motivational benefits.

Increasing estimates of goal engagement and decreasing estimates of temptation engagement

Counteractive optimism predicts that when performance goals are important to people and they anticipate obstacles in goal pursuit, they become more optimistic about future goal attainment and temptation avoidance. Notably, however, predicting better performance in the face of great obstacles might mean that one's prediction is potentially less accurate. Hence, there is a trade-off between counteractive and accurate predictions. The pattern predicted by counteractive optimism holds as long as people are less concerned about the accuracy of their predictions. If people try to provide accurate (rather than motivating) predictions, anticipated obstacles should lead to more conservative predictions of goal attainment and temptation indulgence. In this way, accurate (vs. motivating) predictions can undermine the motivation to adhere to the goal.

To test these predictions, Zhang and Fishbach (in press) explored how students' anticipation of tempting alternatives affected their predictions about future coursework and leisure activities. Participants were recruited for a study about college students' "time allocation," in which they indicated the average number of hours they expected to spend each day in the upcoming week on goal-related activities (homework and class reading) and tempting alternatives (surfing the internet and hanging out with friends).

Consistent with a structural definition, the motivation to study does not represent a high-order goal on its own. It is only a high-order goal in relation to tempting alternatives such as leisure activities. Similarly, leisure activities on their own do not represent a low-order temptation—they only become so in relation to high-order goals such as studying. Thus, a subtle order manipulation determined the presence or absence of conflict for predictions of goal pursuit and temptation pursuit. In terms of predicting goal pursuit, participants were aware of potential conflict if they predicted how much time they would spend on leisure activities before they predicted how much time they would spend on academic activities. As shown in Figure 21–2, when accuracy motivation was low, awareness of the leisure temptation led to more optimistic predictions about academic pursuits. Participants who were aware of potential conflict, compared with those who were unaware, predicted that they would spend more time studying.

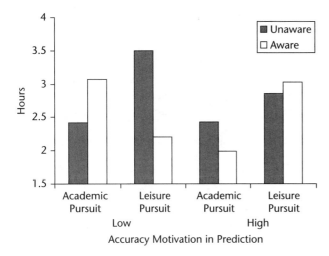

Figure 21-2. Predicted academic pursuit and leisure pursuit as a function of awareness of the self-control conflict and accuracy motivation.

According to the asymmetry claim, counteractive optimism would also be expected to operate by decreasing estimates of time spent pursuing temptations. In terms of predicting temptation pursuits, participants were aware of potential conflict if they predicted how much time they would spend on academic activities before they predicted how much time they would spend on leisure activities. As expected, when accuracy motivation was low, awareness of the academic goal led to more optimistic predictions about leisure pursuits. Participants who were aware of the potential conflict, compared with those who were unaware, predicted that they would spend fewer hours over the course of the week surfing the internet and hanging out with friends. Thus, yet another asymmetry of counteractive control is observed. The perceived self-control dilemma enacts counteractive optimism via inflated expectations about future goal pursuit and deflated expectations about future temptation indulgence.

Counteractive optimism does not influence predictions when people wish to be accurate. Accordingly, the aforementioned effects were not observed when participants were explicitly instructed to make accurate estimates. When accuracy motivation is high, awareness of the self-control conflict leads to more conservative predictions, expressed by less time anticipated for academic pursuits. There is a clear trade-off between people's desire to generate motivating predictions that secure a high level of performance when confronting obstacles and their desire to generate accurate predictions that account for these obstacles by decreasing performance expectations.

Optimism in action: Increasing performance as a result of optimistic predictions

Of course, optimistic predictions should not be considered adaptive self-control strategies unless they translate to goal-congruent choice and action. Several studies (Zhang & Fishbach, in press) have documented counteractive optimism in action. In one study, participants provided a rough prediction (i.e., they had low accuracy motivation) for the completion time of a take-home exam. Those who believed that it was going to be difficult, relative to those who believed it was going to be easy, expected to finish sooner and, importantly, ended up completing it sooner. As before, these effects were reversed when accuracy was emphasized and participants were not engaging in counteractive optimism. Participants who tried to be accurate and who believed the exam was going to be difficult, relative to those who believed it was going to be easy, provided more conservative predictions of completion time and ended up completing the exam later.

In another study, participants predicted how long they would persist at and how well they would perform in a novel anagram task. Participants who wanted to motivate high levels of performance stated more optimistic predictions and persisted longer on a task when they expected distracting music to play than when they did not expect distractions. Taken together, these studies demonstrate that people make asymmetric predictions about future choice dilemmas and that these predictions motivate goal-congruent action in the face of tempting alternatives. Specifically, people strategically overestimate their engagement with goal-related activities and underestimate their engagement with temptation-related activities. These optimistic predictions translate to increased goal-congruent choice and improved performance.

Changing the Psychological Meaning of Choice

Another strategy people employ to help protect their high-order goals from tempting alternatives involves changing the psychological meaning of future choices. People may selectively attend to, encode, and interpret information about the choice alternatives to bolster the value of high-order goals and, asymmetrically, discount the value of low-order temptations (Kuhl, 1986; Mischel, 1984). These changes in the psychological meaning of the choice situation can take several forms. For example, people can attend to what makes their goal valuable while devaluing temptations (Fishbach, Zhang, & Trope, 2010; Myrseth, Fishbach, & Trope, 2009a). Alternatively, they can employ a cool and abstract construal of tempting alternatives, which decreases their motivational strength (Fujita et al., 2006; Kross, Ayduk, & Mischel, 2005; Metcalfe & Mischel, 1999). Presumably, they can also employ a hot and concrete construal of goals to increase their motivational strength. In addition, as an implicit and rudimentary form of self-control, people can activate the representation of high-order goals in the face of low-order temptations and, asymmetrically, inhibit low-order temptations in the face of high-order goals (Fishbach et al., 2003).

Bolstering goals and devaluing temptations

The likelihood of goal-congruent choice can be increased by bolstering the value of goals or by devaluing temptations. People may bolster the value of high-order goals by linking the attainment of these goals to their self-standards (Bandura, 1989) or by elaborating on what makes them important (Beckmann & Kuhl, 1985; Fishbach, Shah, & Kruglanski, 2004; Kuhl, 1984). They may further devalue temptations by disassociating these motives from the self or ignoring aspects that make temptations important. This asymmetric bolstering and devaluation may take an explicit or implicit form.

In a study that tested for explicit counteractive evaluations (Myrseth et al., 2009b), participants in a romantic relationship rated the sex appeal of several attractive members of the opposite sex, based on ostensible profiles from "facebook.com." These target people were either listed as "single" or "in a relationship;" hence, they represented a potential threat to one's relationship or not. As shown in Figure 21–3, participants evaluated people as less

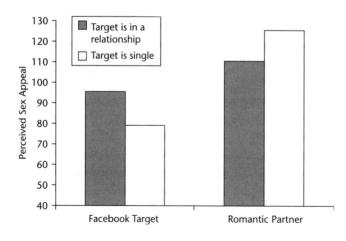

Figure 21–3. Perceived sex appeal of Facebook targets and participants' romantic partners as a function of Facebook targets' availability.

attractive when they were listed as single than when these same people were said to be in a relationship. The romantically involved participants devalued available "temptations." In addition, when asked to rate the sex appeal of their own romantic partners, participants rated their partners as more attractive after evaluating single people than after evaluating people said to be in a relationship. The romantically involved participants bolstered the value of the "goal related object" (i.e., one's own partner). Other studies have documented a congruent pattern of bolstering the value of academic tasks that were scheduled at an inconvenient (1 A.M.) versus convenient (9 P.M.) time or of bolstering the value of studying for a test when competing social motives were primed versus not primed (Trope & Fishbach, 2000).

The evidence for implicit counteractive evaluations comes from a set of recent studies conducted by Fishbach, Zhang, and Trope (2010). Using an evaluative priming procedure (Bargh et al., 1992; Fazio et al., 1986, 1995; Neely, 1977), Fishbach and colleagues found that making achievement goals highly accessible (vs. not making them accessible) led to more negative, implicit evaluations of tempting constructs. For example, students to whom achievement goals had been made highly accessible implicitly expressed more negative evaluations of nonacademic tempting concepts, such as television and beach, than students to whom achievement was not made highly accessible. It appears that goal accessibility is associated with spontaneous negative implicit evaluation of temptations—that is, accessible academic goals lead to implicit devaluation of nonacademic temptation concepts.

Consistent with the asymmetry claim, it was also expected that the perceived presence of nonacademic temptations would enhance the implicit evaluation of academic goals. In another study, some participants were asked to elaborate on how much they would enjoy engaging in numerous, tempting activities that commonly threaten university students' studying (e.g., watching movies, drinking alcohol). These participants, relative to participants who did not elaborate on tempting alternatives, expressed more positive implicit evaluations of academic concepts such as study and library. Consistent with the asymmetry prediction, these results suggest that temptation accessibility is associated with positive evaluation of goals.

Together, these two studies demonstrate the asymmetry in counteractive evaluation. When goal-related activities such as studying and reading are made accessible to students who are tempted by various nonacademic alternatives, they respond by devaluing the tempting alternatives. Asymmetrically, when tempting activities such as partying and vacation are made accessible to students with active academic goals, they respond by augmenting the value of their academic goals.

Heating up goals and cooling down temptations

Processing level is another variable that creates asymmetric shifts in motivational strength. Individuals may form a "cool," abstract, or psychologically distanced representation of their motivations, or they may form a "hot," concrete, or psychologically proximal representation of these goals. When confronting low-order temptations, one can attenuate the impact of these alternatives and secure overriding goal pursuit by employing a cool and abstract representation of the temptations (Fujita et al., 2006; Kross et al., 2005; Metcalfe & Mischel, 1999; Mischel & Ayduk, 2004). For example, congruent with the hot/cool systems analysis of self-control (e.g., Metcalfe & Mischel, 1999), children are more likely to avoid eating tempting marshmallows by thinking of them as "white clouds" (Mischel & Baker, 1975). This cool and nonappetitive representation improves their ability to delay gratification. In another study (Fujita et al., 2006), adults were willing to pay a smaller premium for immediate, rather than delayed, attractive gifts when they were construing temptation in a high-level, abstract fashion, as opposed to a low-level, concrete fashion.

According to a counteractive control analysis, one could also increase goal-congruent choice by forming a "hot," concrete, or psychologically proximal representation of the benefits from pursuing goals. This hypothesis is consistent with research on the self-regulatory benefits of setting concrete implementation intentions (e.g., Gollwitzer, 1999; Gollwitzer, & Brandstaetter, 1997), and it follows from the assumption of asymmetry. Thus we predict that a goal stimulus would gain motivational strength if people thought of it in "hot" and concrete terms. For example, a student would be more likely to study if she considered the "hot" (e.g., pride) versus "cool" (e.g., good grades) features of academic success. Notably, this should be the case only if the goal and the temptation represent distinct motivations (e.g., party vs. study). At other times—for example, in the delay of gratification paradigm—the conflict is between small-and-immediate versus large-and-delayed goal fulfillment (e.g., small candy now vs. large candy later). In these situations, considering the appetitive, "hot" characteristics of the goal simultaneously activates the appetitive, "hot" characteristics of the temptation and may induce impatience and a stronger desire to fulfill this motive by choosing the immediate, small reward (Metcalfe & Mischel, 1999).

Activating goals and inhibiting temptations

Counteractive control also entails changes in the activation level of constructs related to goals and temptations. In particular, people can secure their high-order goals by activating related constructs in response to interfering temptations and by inhibiting tempting constructs in response to cues for the overriding goal.

Research by Fishbach, Friedman, and Kruglanski (2003) found that subliminal presentation of a temptation-related construct facilitated the activation level of a construct representing a potentially obstructed goal. For example, one of these studies used participants' self-reported goals and temptations to obtain goal-temptation pairs (e.g., study–basketball). Using a sequential priming procedure, these researchers found that goal-related keywords (e.g., study) were more quickly recognized following subliminal presentation of relevant temptation-related keywords (e.g., basketball) than following subliminal presentation of irrelevant temptation-related keywords (e.g.,

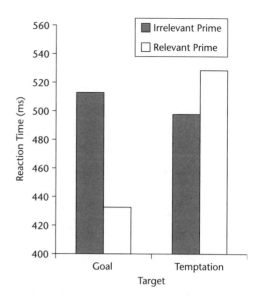

Figure 21–4. Reaction time for recognizing self-entered temptations and goals, following relevant versus irrelevant primes.

chocolate). As shown in Figure 21–4, this effect was asymmetric such that goal recognition was facilitated by temptation primes (e.g., academic targets were facilitated by nonacademic primes) and temptation recognition was inhibited by goal primes (e.g., nonacademic targets were inhibited by academic primes). Another subsequent study demonstrated that temptation–goal activation was further obtained under cognitive load; hence, it was independent of available cognitive resources. This result further supports the implicit and resource-free nature of this self-control response.

In another study, temptation-related cues were primed supraliminally as incidental aspects of the situation (Fishbach at al., 2003). In this study, dieters spent time in one of three situations: a room with popular fatty foods (temptation prime), a room with weight-watching magazines (dieting prime), and a room with general interest magazines (neutral prime). These priming stimuli were allegedly irrelevant to the purpose of the experiment. Priming the temptation to consume fattening food, like priming the goal of dieting directly, facilitated recognition of the word "diet" in a subsequent lexical decision task. Moreover, when offered a gift, participants in the temptation prime and in the diet prime conditions preferred to get an apple rather than a chocolate bar, whereas participants in the neutral condition preferred to get chocolate. Thus, asymmetric activation patterns seem to affect motivational strength to the extent of impacting choice outcomes. Temptations do so by facilitating the activation of high-order goals, whereas goals do so by inhibiting the activation of low-order temptations.

In sum, research reviewed in this section addresses the main self-control strategies that enable the counteractive control process. Counteractive control theory predicts that in response to a self-control dilemma, people employ a variety of strategies meant to increase the motivational strength of their high-order goals relative to their low-order temptations. In line with this premise, we reviewed several lines of research demonstrating asymmetric shifts in motivational strength.

SELF-CONTROL FAILURE

On the basis of our review, we propose that people do not confront temptations unguarded. Rather, they hold an array of strategies that help them secure the attainment of a more important goal in the face of the obstacles and hindrances posed by temptations. The processes of self-control allow people to pursue their long-term interests, such as achieving career goals, controlling body weight, suppressing the escalation of personal conflicts, and more. In general, people are rather effective at adhering to these goals. However, everyday observations suggest that self-control operations can also fail. Thus, even effective self-regulators sometimes fail to adhere to their self-selected goals.

A tacit assumption in the self-control literature is that the mutual presence of goals and temptations invariably activates a self-control conflict. But that assumption might not be warranted. As indicated before, people exercise self-control when there are no substitutable, external means of control. In addition, as we discuss next, people exercise self-control when they perceive goals and temptations as competing with each other, rather than as complementing each other, and when the goals clearly override the temptations in their goal hierarchy.

Failures to Recognize versus Failures to Exercise

Self-control research has traditionally referred to situations in which goals and temptations appear to be in competition. In this situation, researchers have assumed that people either exercise self-control, and resolve the conflict in favor of the more valuable goal, or that they fail to exercise self-control, and thus succumb to temptation (Baumeister et al., 1994). Relaxing this assumption, however, allows the possibility that some temptation-consistent choices do not result from a failure to exercise self-control but instead from a failure to identify a particular self-control dilemma as such (Myrseth & Fishbach, 2009). When people do not experience a conflict between goals and temptations, they may plan to balance between these motivations in the same way they balance between any equally central goals (e.g., career and family). Moreover, whereas cues for goals often remind people of the self-control problem and help them to guard against competing desires, the presence of goal cues can have the opposite effect, liberating one to pursue desires or temptations in the present (*see also*, Monin & Miller, 2001; Steele, 1988).

Research on the dynamics of self-regulation addresses the simultaneous pursuit of multiple goals, and it distinguishes between a dynamic of highlighting the more important goal by consistently choosing alternatives that are congruent with this motivation and a dynamic of balancing between the goal and the temptation (or two equally important goals) by making successive choices that alternate between the two (Fishbach & Dhar, 2005; Fishbach, Dhar, & Zhang, 2006; Fishbach, Zhang, & Koo, 2009; Koo & Fishbach, 2008; Zhang, Fishbach, & Dhar, 2007). The need for self-control emerges in a dynamic of highlighting, when people desire to focus on the goal and resist the temptation. But this pattern of highlighting is not the only way in which people approach their choices. In addition to highlighting the more important goal through consistent goal-congruent choice, people may also adopt a dynamic of balancing between goals and temptations by sampling from both goals and temptations over multiple choice opportunities. For example, most students do not choose the enlightened path at every turn—even the most studious will sometimes decide to relax with a movie or have some fun with friends, postponing their reading for later. In another domain, most health-conscious people do not choose the nutritious path at every turn—even the most fit will sometimes choose the chocolate shake now, promising to opt for the wheatgrass smoothie later.

Highlighting and balancing produce different patterns of choice that unfold over time. These dynamics are further associated with a framing of any particular choice as signaling goal commitment versus goal progress. In a highlighting dynamic, a particular goal-congruent choice is interpreted as commitment to that goal and subsequently increases its priority over competing motivations (Aronson, 1997; Atkinson & Raynor, 1978; Bem, 1972; Feather, 1990; Festinger, 1957; Locke & Latham, 1990). In this framing of self-regulation, goals and temptations compete with each other. On the other hand, in a balancing dynamic, a particular choice is interpreted as progress toward that particular pursuit (e.g. Byrne & Bovair, 1997; Carver & Scheier, 1998; Higgins, 1989), which serves as justification for moving to other, neglected pursuits. In this framing of self-regulation, goals and temptations complement each other.

People only engage the self-control process when it is needed to protect high-order interests from low-order temptations. Therefore, the usual asymmetry should be observed under the dynamic of highlighting. It is precisely when a goal seems to be in competition with a temptation that counteractive control should act to increase the likelihood of goal-congruent choice by increasing the relative attractiveness of goal pursuit. The motivational strength of goals should be increased and the motivational strength of temptations should be decreased. On the other hand, the dynamic of balancing should produce an opposite kind of asymmetry. When goals and temptations seem to complement each other, the immediate motivational strength of the temptation relative to the goal should be increased.

Specifically, when goals and temptations complement each other and a person wishes to balance between them, there are two possible patterns of choice: A person can first choose a goal option and later choose a temptation option, or the person can first choose a temptation option and later choose a goal option. Because goals offer delayed benefits and temptations offer immediate benefits (Ainslie, 1975; Loewenstein, 1996; Rachlin, 1997), under a balancing dynamic the latter sequence (temptation, then goal) is more advantageous. That is, people can expect to maximize the attainment from both by expressing an immediate preference for a tempting option with an intention to choose a goal option at the next opportunity. As such, they acquire the value of the temptation in the present and expect to obtain the value of the goal in the future. It follows that when goals and temptations seem to complement each other in a dynamic of balancing, people should express an immediate preference for tempting items and expect to balance on the subsequent choice. In this case, the motivational strength to pursue goals will likely be less than the motivational strength to pursue temptations, yielding the reverse asymmetry.

In one study that assessed evaluation of goals and temptations under each of the self-regulatory dynamics, Fishbach and Zhang (2008) presented healthy and fatty food images in two separate images, to induce a sense of competition and a dynamic of highlighting; together in one image, to induce a sense of complementarity and a dynamic of balancing; or in two separate experimental sessions, as a control condition. For example, in one stimuli set, some participants evaluated separate images of healthy berries and unhealthy soda; others evaluated berries and soda in a single image that featured both; and the rest evaluated berries in one session and soda in another session. As shown in Figure 21-5, across all stimuli sets, when healthy and fatty items were presented apart, in two separate images, the value of healthy items was higher

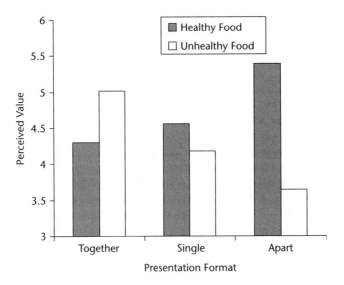

Figure 21–5. Perceived value of healthy and unhealthy food as a function of presentation format.

than unhealthy items—That is, consistent with the counteractive control process, when self-control was needed to protect healthy options from the allure of unhealthy options, participants counteractively bolstered the value of the goal-consistent choices relative to the temptations. However, when healthy and unhealthy items were presented together and appeared complementing, the value of the fatty items was higher than the value of the healthy items. When the balancing dynamic was active and there was no need to protect healthy choices against unhealthy choices (because participants acted as if they would balance these competing goals over time), participants bolstered the value of the temptations relative to goal-consistent choices. Finally, when the healthy and unhealthy items were evaluated in two separate experimental sessions ("single," control condition), their value was similar.

A follow-up study demonstrated that these evaluation effects have implications for the choices that people make in potential self-control dilemmas. In this study, Fishbach and Zhang assessed choices between healthy snacks (carrots) and unhealthy snacks (chocolate bars) that were either presented together in one pile, and appeared to be complementing, or apart in two separate piles, and appeared to be competing. Participants were more likely to select unhealthy, tempting snacks when they were included in the same (vs. different) pile with healthy snacks and appeared to complement rather than compete with each other.

Additional studies illustrated the full course of the different dynamics more explicitly by comparing immediate choices with future choices. As expected, when alternatives were presented together as complements, people tended to act according to a balancing dynamic, preferring temptations immediately and planning for goal-related choices in the future. On the other hand, when alternatives were presented apart as competitors, people tended to act according to a highlighting dynamic, preferring goal-related choices both immediately and for the future. For example, participants who chose two magazines from a mixed set of lowbrow (i.e., entertainment) magazines (e.g., *Maxim*, *Cosmopolitan*) and highbrow (i.e., news) magazines (e.g., *Time*, *The Economist*), preferred a lowbrow magazine immediately and a highbrow magazine for later reading; hence, they were balancing. In contrast, those who encountered the lowbrow and highbrow magazines in separate sets preferred highbrow magazines for both immediate and later readings; hence, they were highlighting and exercising self-control.

By expressing an immediate preference for tempting items, people may end up repeatedly giving in to temptations. For example, a desire to balance between the chocolate shake and the wheatgrass smoothie could result in a consistent preference for the chocolate. However, this choice would not necessarily reflect a failure to exercise self-control, because the person did not attempt to do so. When the goal seems to complement the temptation—that is, when one fails to recognize a potential self-control dilemma—self-control processes would not be instrumental and would not be evoked. One would feel justified to pursue attractive temptations.

Effects of Temptation Indulgence

To explain self-control failures, researchers have generally referred to an energy model of self-control. Baumeister, Vohs, and colleagues documented a phenomenon they refer to as "ego depletion"—a failure at self-control resulting from lack of resources. According to their theorizing, an act of prior self-control results in subsequent depletion and increased likelihood of self-control failure (Baumeister et al., 1994; Muraven & Baumeister, 2000; Vohs & Heatherton, 2000; Vohs & Schmeichel, 2003). In our terms, when high-order goals are in conflict with low-order temptations, any action toward the high-order goal would result in a weakened ability to pursue other congruent actions. For example, after studying for an upcoming exam a student may fail to adhere to her other goals, including exercising, eating healthy, or studying for other classes, because she lacks the resources to do so.

If the pursuit of goals depletes self-control resources, does this imply that the pursuit of temptations replenishes them? Research by Converse and Fishbach (2009) addressed this possibility, assessing whether fulfilling temptation experiences resulted in increased adherence to goal pursuit. This research found that people's lay theories were consistent with this expectation but that their actual goal pursuit was not. Specifically, Converse and Fishbach manipulated the quality of the temptation experience (fulfilling vs. unfulfilling) and then measured participants' expectations to revert to the goal following this experience and their actual goal pursuit. For example, in one study, participants indicated their expected and actual motivation to return to academic pursuits after a good (vs. mediocre) weekend break during the term. Although participants expected a good (vs. mediocre) weekend to leave them more motivated for their studies, they were actually no more motivated following a good break than following a mediocre break. Thus, whereas goal actions are depleting, giving in to temptation is not replenishing.

Giving in to temptations may alter people's priorities in favor of tempting activities at the expense of goal-congruent activities. As a result, a person's goal hierarchy may change such that the person values "temptations" more than the competing "goals." When the conventionally tempting activity acquires the status of a high-order goal, one could fail to adhere to a previously defined goal because the person does not recognize it as such. For example, students who have a pleasant vacation may change their priorities. Whereas academic success once trumped vacation vices for these students, world travel could become a high-order goal relative to their school work. Just as goal-temptation complementarity can alleviate the instrumentality of self-control, so too can changes in the goal hierarchy. When conventional temptations are no longer recognized as such, counteractive control will not be enacted to combat their influence.

CONCLUSIONS

Research on counteractive control explores the processes that individuals employ to increase the motivational strength of their high-order goals and decrease the motivational strength of their low-order temptations. It identifies various strategies that lead to this asymmetric motivational shift by operating on the choice situation, one's performance expectations, and the psychological representation of choice alternatives. It further distinguishes between explicit self-control operations that rely on conscious processing and implicit operations that do not require explicit consideration.

Whereas research on counteractive control focuses on adaptive self-control operations, there are several reasons why people may nonetheless fail at self-control. First, there is motivational depletion following goal pursuit and, in addition, temptation pursuit can change one's goal hierarchy in favor of the temptation. Second, a potential self-control conflict can go unrecognized, resulting in lesser attempts to exercise self-control. A failure to recognize a self-control conflict is expected if goals and temptations seem to complement (rather than compete with) each other or if people do not relate a single tempting episode (e.g., smoking one cigarette) with a pattern of giving in to temptations (e.g., being a smoker; Rachlin, 2000). Possibly, there are also other factors that prevent people from recognizing a self-control conflict or exercising self-control. One of the challenges of future self-control research will be to account for people's failures to exercise self-control as opposed to their failures to recognize particular situations as requiring self-control. A better understanding of the different causes of self-control failures can further advance the research on successful and adaptive self-control operations.

References

Aarts, H., & Dijksterhuis, A. Habits as knowledge structures: Automaticity in goal-directed behavior. J Pers Soc Psychol 2000; 78(1): 53–63.

Ainslie, G. *Picoeconomics: The strategic interaction of successive motivational states within the person.* Cambridge, England: Cambridge University Press, 1992.

Ainslie, G. *Breakdown of will.* New York: Cambridge University Press, 2001.

Ariely, D., & Wertenbroch, K. Procrastination, deadlines, and performance: Self control by precommitment. Psychol Sci 2002; 13(3): 219–224.

Atkinson, J. W., & Feather, N. T. *A theory of achievement motivation.* New York: Wiley, 1966.

Bandura, A. Self-regulation of motivation and action through internal standards and goal systems. In: Pervin, L. A. (Ed.), *Goal concepts in personality and social psychology.* Hillsdale, NJ: Erlbaum, 1989: pp. 19–85.

Bargh, J. A., Chaiken, S., Govender, R., & Pratto, F. The generality of the automatic attitude activation effect. J Pers Soc Psychol 1992; 62(6): 893–912.

Bargh, J. A., & Chartrand, T. L. The unbearable automaticity of being. Am Psychol 1999; 54(7): 462–479.

Bargh, J. A., & Ferguson, M. J. Beyond behaviorism: On the automaticity of higher mental processes. Psychol Bull 2000; 126(6): 925–945.

Bargh, J. A., Gollwitzer, P. M., Lee-Chai, A., Barndollar, K., & Troetschel, R. The automated will: Unconscious activation and pursuit of behavioral goals. J Pers Soc Psychol 2001; 81(6): 1014–1027.

Baumeister, R. F., Heatherton, T. F., & Tice, D. M. *Losing control: How and why people fail at self-regulation.* San Diego: Academic, 1994.

Baumeister, R. F., & Vohs, K. D. *Handbook of self-regulation: Research, theory, and applications.* New York: Guilford Press, 2004.

Becker, H. S. Notes on the concept of commitment. Psychol Bull 1960; 66: 32–40.

Beckmann, J., & Kuhl, J. *Action control: From cognition to behavior.* Berlin, Germany: Springer-Verlag, 1985.

Brehm, J. W. *A theory of psychological reactance.* New York: Academic Press, 1966.

Brehm, J. W., & Self, E. A. The intensity of motivation. Ann Rev Psychol 1989; 40: 109–131, Annual Reviews, US. 1989.

Buehler, R., Griffin, D., & Ross, M. Exploring the "planning fallacy": Why people underestimate their task completion times. J Pers Soc Psychol 1994; 67(3): 366–381.

Chen, M., & Bargh, J. A. Consequences of automatic evaluation: Immediate behavioral predispositions to approach or avoid the stimulus. Pers Soc Psychol Bull 1999; 25(2): 215–224.

Converse, B. A., & Fishbach, A. Effects of fulfilling and unfulfilling temptation experiences on subsequent goal pursuit. Unpublished manuscript, University of Chicago. 2009.

Elster, J. *Ulysses and the sirens.* Cambridge, England: Cambridge University Press, 1977.

Fazio, R. H., Jackson, J. R., Dunton, B. C., & Williams, C. J. Variability in automatic activation as an unobstrusive measure of racial attitudes: A bona fide pipeline? J Pers Soc Psychol 1995; 69(6): 1013–1027.

Fazio, R. H., Sanbonmatsu, D. M., Powell, M. C., & Kardes, F. R. On the automatic activation of attitudes. J Pers Soc Psychol 1986; 50(2): 229–238.

Fishbach, A., & Dhar, R. Goals as excuses or guides: The liberating effect of perceived goal progress on choice. J Consum Res 2005; 32: 370–377.

Fishbach, A., Dhar, R., & Zhang, Y. Subgoals as substitutes or complements: The role of goal accessibility. J Pers Soc Psychol 2006; 91(2): 232–242.

Fishbach, A., Friedman, R. S., & Kruglanski, A. W. Leading us not unto temptation: Momentary allurements elicit overriding goal activation. J Pers Soc Psychol 2003; 84(2): 296–309.

Fishbach, A., & Shah, J. Y. Self control in action: Implicit dispositions toward goals and away from temptations. J Pers Soc Psychol 2006; 90(5): 820–832.

Fishbach, A., Shah, J. Y., & Kruglanski, A. W. Emotional transfer in goal systems. J Exp Soc Psychol 2004; 40, 723–738.

Fishbach, A., & Trope, Y. The substitutability of external control and self control. J Exp Soc Psychol 2005; 41(3): 256–270.

Fishbach, A., & Trope, Y. Implicit and explicit mechanisms of counteractive self control. In: Shah, J. Y., & Gardner, W. (Eds.), *Handbook of motivation science*. New York: Guilford, 2007: pp. 281–294.

Fishbach, A., & Zhang, Y. Together or apart: When goals and temptations complement versus compete. J Pers Soc Psychol 2008; 94: 547–559.

Fishbach, A., Zhang, Y., & Koo, M. The dynamics of self-regulation. Eur Rev Soc Psychol 2009; 20: 15–344.

Fishbach, A., Zhang, Y., & Trope, Y. Counteractive evaluation: Asymmetric shifts in the implicit value of conflicting motivations. J Exp Soc Psychol 2009; 46: 29–38.

Fujita, K., Trope, Y., Liberman, N., & Levin-Sagi, M. Construal levels and self control. J Pers Soc Psychol 2006; 90(3): 351–367.

Gollwitzer, P. M. Implementation intentions: Strong effects of simple plans. Am Psychol 1999; 54: 493–503.

Gollwitzer, P. M., Bayer, U. C., & McCulloch, K. C. The control of the unwanted. In: Bargh, J. A., Uleman, J., & Hassin, R. (Eds.), *The new unconscious*. New York: Oxford University Press, 2004.

Gollwitzer, P. M., & Moskowitz, G. B. Goal effect on thought and behavior. In: Higgins, E. T., & Kruglanski, A. W. (Eds.), *Social psychology: Handbook of basic principles*. New York: Guilford Press, 1996: pp. 361–399.

Gollwitzer, P. M., & Brandstaetter, V. Implementation intentions and effective goal pursuit. J Pers Soc Psychol 1997; 73: 186–199.

Green, L., & Rachlin, H. Commitment using punishment. J Exp Anal Behav 1996; 65(3): 593–601.

Heath, C., Larrick, R. P., & Wu, G. Goals as reference points. Cogn Psychol 1999; 38(1): 79–109.

Koo, M., & Fishbach, A., Dynamics of self-regulation: How (un)accomplished goal actions affect motivation. J Pers Soc Psychol 2008; 94: 183–195.

Kross, E., Ayduk, O., & Mischel, W. When asking "why" does not hurt: Distinguishing rumination from reflective processing of negative emotions. Psychol Sci 2005; 16(9): 709–715.

Kuhl, J. Volitional aspects of achievement motivation and learned helplessness: Toward a comprehensive theory of action control. In: Maher, B. A. (Ed.), *Progress in experimental personality research* (Vol. 13). New York: Academic Press, 1984: pp. 99–171.

Kuhl, J., & Beckmann, J. *Action control from cognition to behavior*. New York: Springer-Verlag, 1985.

Locke, E. A., & Latham, G. P. *A theory of goal setting & task performance*. Upper Saddle River, NJ: Prentice-Hall, 1990.

Loewenstein, G. Out of control: Visceral influences on behavior. Organ Behav Hum Decis Process 1996; 65(3): 272–292.

Markman, A. B., & Brendl, C. Constraining theories of embodied cognition. Psychol Sci 2005; 16(1): 6–10.

Metcalfe, J., & Mischel, W. A hot/cool-system analysis of delay of gratification: Dynamics of willpower. Psychol Rev 1999; 106(1): 3–19.

Mischel, W., & Ayduk, O. Willpower in a cognitive-affective processing system: The dynamics of delay of gratification. In: Baumeister, R. F., & Vohs, K. D. (Eds.), *Handbook of self-regulation: Research, theory, and applications*. New York, NY: Guilford Press, 2004: pp. 99–129.

Mischel, W., Shoda, Y., & Rodriguez, M. L. Delay of gratification in children. Science 1989; 244(4907): 933–938.

Moskowitz, G. B., Gollwitzer, P. M., Wasel, W., & Schaal, B. Preconscious control of stereotype activation through chronic egalitarian goals. J Pers Soc Psychol 1999; 77(1): 167–184.

Muraven, M., & Baumeister, R. F. Self-regulation and depletion of limited resources: Does self control resemble a muscle? Psychol Bull 2000; 126(2): 247–259.

Myrseth, K. O. R., & Fishbach, A. Self-control: A function of knowing when and how to exercise restraint. Curr Dir Psychol Sci 2009; 18: 247–252.

Myrseth, K., Fishbach, A., & Trope, Y. Counteractive self-control theory: When making temptation available makes temptation less tempting. Psychol Sci 2009a; 20: 159–163.

Myrseth, K., Fishbach, A., & Trope, Y. Counteractive control effects on perceived attraction. Unpublished manuscript, University of Chicago, 2009b.

Neely, J. H. Semantic priming and retrieval from lexical memory: Roles of inhibitionless spreading activation and limited-capacity attention. J Exp Psychol Gen 1977; 106(3): 226–254.

O'Donoghue, T., & Rabin, M. The economics of immediate gratification. J Behav Decis Mak 2000; 13: 233–250.

Oettingen, G., & Mayer, D. The motivating function of thinking about the future: Expectations versus fantasies. J Pers Soc Psychol 2002; 83(5): 1198–1212.

Rachlin, H. The science of self control. Cambridge, MA: Harvard University Press, 2000.

Rachlin, H., & Green, L. Commitment, choice and self control. J Exp Anal Behav 1972; 17: 15–22.

Schelling, T. C. Self-command in practice, in policy, and in a theory of rational choice. Am Econ Rev 1984; 74(2): 1–11.

Shah, J. Y., Friedman, R., & Kruglanski, A. W. Forgetting all else: On the antecedents and consequences of goal shielding. J Pers Soc Psychol 2002; 83(6): 1261–1280.

Shah, J. Y., & Kruglanski, A. W. When opportunity knocks: Bottom-up priming of goals by means and its effects on self-regulation. J Pers Soc Psychol 2003; 84(6): 1109–1122.

Solarz, A. K. Latency of instrumental responses as a function of compatibility with the meaning of eliciting verbal signs. J Exp Psychol 1960; 59: 239–245.

Strotz, R. H. Myopia and inconsistency in dynamic utility maximization. Rev Econ Stud 1956; 23: 166–180.

Taylor, S. E., & Brown, J. D. Illusion and well-being: A social psychological perspective on mental health. Psychol Bull 1988; 103(2): 193–210.

Thaler, R. H. Quasi rational economics. New York: Russel Sage Foundation, 1991.

Thaler, R. H., & Shefrin, H. M. An economic theory of self control. J Polit Econ 1981; 89: 392–406.

Trope, Y., & Fishbach, A. Counteractive self control in overcoming temptation. J Pers Soc Psychol 2000; 79(4): 493–506.

Trope, Y., & Fishbach, A. Going beyond the motivation given: Self control and situational control over behavior. In: Hassin, R. R., Uleman, J., & Bargh, J. A. (Eds.), *The new unconscious*. New York: Oxford University Press, 2005: pp. 537–565.

Vohs, K. D., & Heatherton, T. F. Self-regulatory failure: A resource-depletion approach. Psychol Sci 2000; 11(3): 249–254.

Vohs, K. D., & Schmeichel, B. J. Self-regulation and extended now: Controlling the self alters the subjective experience of time. J Pers Soc Psychol 2003; 85(2): 217–230.

Wright, R. A., & Brehm, J. W. Energization and goal attractiveness. In: Pervin, L. A. (Ed.), *Goal concepts in personality and social psychology*. Hillsdale, NJ, England: Lawrence Erlbaum, 1989: pp. 169–210.

Zhang, Y., & Fishbach, A. Counteracting obstacles with optimistic predictions. J Exp Psychol Gen, in press.

Zhang, Y., Fishbach, A., & Dhar, R. W. When thinking beats doing: The role of optimistic expectations in goal-based choice. J Consum Res 2007; 34: 567–578.

CHAPTER 22

Seeing the Big Picture: A Construal Level Analysis of Self-Control

Kentaro Fujita, Yaacov Trope, and Nira Liberman

ABSTRACT

People frequently make decisions and act in a manner contrary to their goals and values. These self-control failures are widely prevalent, troubling, and implicated in some of the most pressing social issues, ranging from obesity and addiction to environmentalism and poor financial planning. Given humans' remarkable intellectual and reasoning capacities, why do people fail to express their admirable aims in their everyday lives? This chapter briefly reviews several prominent theoretical accounts that have been proposed to explain self-control failures. It then describes and reviews supporting evidence for an emerging new perspective inspired by construal level theory (Liberman, Trope, & Stephan, 2007; Trope & Liberman, 2003).

Drawing from decades of research indicating the central role of people's subjective construals in judgment and decision making (e.g., Griffin & Ross, 1991), we propose that self-control is a construal-dependent phenomenon. That is, whether one chooses to act in a manner consistent with one's global goals and values depends on how one has subjectively interpreted and constructed the event in his or her mind. Construal level theory suggests that people's subjective mental construals can differ in abstractness. Whereas low-level construals highlight the idiosyncratic, incidental, concrete, local features of an event, high-level construals extract the core, central, abstract, and global features. Self-control conflicts occur when the behavioral connotations of these two construals are mutually exclusive. The preferred action depends on which construal people adopt at the time of decision making. This chapter discusses how this proposed construal level perspective relates to extant theoretical perspectives, reviews supporting empirical evidence, and discusses the implications for our understanding of self-control.

Keywords: construal level theory, self-control, subjective construals, delay of gratification, temporal discounting, self-regulation

"Think globally, act locally" is an adage frequently used to highlight the discrepancy between people's goals and their actions. For example, environmentalists use the phrase to call attention to the fact that despite endorsing environmentalism, many people drive gas-guzzling cars and fail to recycle. Self-help gurus use the slogan to exhort smokers and dieters who express a desire to change their lives and yet act in a contrary manner. Given humans' remarkable intellectual capacities, why is it that people who have such admirable aims so often fail to act on them in their day-to-day lives? Researchers refer to this inability to act in accordance with one's global concerns and interests as self-control failure. When faced with more salient incentives and rewards in their immediate local context, people seem unable to resist them and forfeit their more global objectives. These failures have immense costs both to the individual and to society and are implicated in a number of the nation's biggest problems, such as obesity, substance abuse, aggression, unsafe sexual practices, and poor financial savings.

Not surprisingly, as exemplified by the contributors to this volume, self-control failures have become the focus of a multidisciplinary research effort among social, clinical, and developmental psychologists, as well as behavioral economists, neuroscientists, and political scientists. This research has sought to determine which factors lead people to make choices on the basis of local rather than global concerns. After broadly reviewing this literature, we describe a new approach (*see also* Fujita, Trope, & Liberman, 2006; Fujita, Trope, Liberman, & Levin-Sagi, 2006) that highlights the role that subjective mental construals, or representations of events, have on self-control decisions and action.

Extant Approaches to Self-Control

Temporal Discounting

The rewards to be reaped by our global goals are often not experienced in the present but instead will be enjoyed in the distant future. For example, one might not benefit immediately from avoiding chocolate cake but will reap the benefits of dieting in the future. Noting this, many researchers have suggested that self-control is a problem of intertemporal choice—making decisions between choices that differ in the timing of their rewards (e.g., Ainslie, 1975; Frederick, Loewenstein, & O'Donoghue, 2002; Thaler, 1991; Wertenbroch, 1998). Researchers studying intertemporal choice have focused on a psychophysical phenomenon known as temporal discounting—that is, people discount the value of distant future rewards. For example, people prefer to receive $10 today versus $10 a year from now. Not only do people discount value over time, they do so in a hyperbolic fashion (e.g., Ainslie, 1975; Frederick et al., 2002). In other words, an increase in delay of one unit of time causes a greater perceived drop in the value of a reward in the near versus distant future. As a result of this hyperbolic temporal discounting, people often prefer smaller, more immediate rewards rather than larger, delayed rewards. The value of the larger reward is discounted because of the delay, causing the smaller reward to be preferred. To enhance self-control, these models suggest changing the time frame of decisions—making decisions when both rewards are in the distant future and are equally discounted. Research has shown, for example, that although people overwhelmingly prefer to receive $20 now versus $50 in a year from now, the addition of a constant delay to both choice options leads them to reverse their preferences (e.g., Green, Fristoe, & Myerson, 1994; Kirby & Herrnstein, 1995). That is, adding a year's delay to both options leads people to prefer $50 two years from now rather than $20 one year from now. Results like this thus suggest that it is the temporal immediacy of the local reward that causes people to fail to act in line with their global concerns.

However, the psychological mechanisms of self-control are left unspecified from a temporal discounting perspective. That is, time demonstrably impacts the emphasis placed on local versus global concerns but it is less clear by what cognitive and motivational mechanisms. Moreover, the scope of time-discounting approaches is limited to situations involving temporal differences in the rewards. Self-control situations, however, need not always involve rewards with differences in time (*see also* Ainslie, 1975; Chapter 27). For example, an official offered a bribe experiences a local–global value conflict between taking money versus being honest. If one takes a bribe and benefits financially, one is simultaneously also violating values of honesty and integrity. If one refuses, one may not have benefited financially but is honest. We argue that such conflicts reveal differences not in the timing of rewards but the globality of their implications.

Automatic vs. Effortful Control

Self-control models that posit a distinction between automatic, nonconscious versus more effortful, consciously controlled responses are more explicit with respect to psychological mechanisms (e.g., Baumeister & Heatherton, 1996; Devine, 1989; Strack & Deutsch, 2004; Wegner, 1994; Wiers & Stacy, 2006). These models suggest that automatic processes, which promote thought and action tendencies that undermine one's more global goals and concerns, are engaged in the presence of salient local rewards or temptations in one's environment. For example, a succulent dessert might automatically initiate thoughts and actions related to eating, in spite of one's dieting goals (e.g., Papies, Stroebe, & Aarts, 2007). To prevent these automatic responses from undermining global considerations, these models suggest that more effortful conscious processes are required to override them. These effortful conscious processes, however, require sufficient cognitive resources. When these are taxed, impulsive automatic processes can run unchecked and lead to self-control failures. For example, research has shown that cognitive load or distraction leads to poorer ability to suppress undesired thoughts and actions (e.g., Macrae, Bodenhausen, Milne, & Jetten, 1994; Shiv & Fedorikhin, 1999; Wegner & Erber, 1992). These models thus suggest that automatic processes activated in the presence of local rewards in one's immediate context are responsible for breakdowns of self-control, particularly when one's cognitive resources are burdened.

Baumeister and colleagues (e.g., Baumeister & Heatherton, 1996; Muraven & Baumeister, 2000) have recently proposed a variant of the automatic versus effortful control model that stresses the role of motivational rather than cognitive resources in the resolution of self-control conflicts. This ego-depletion model suggests that effortful control draws from a limited motivational resource. Once this limited resource is depleted, it takes time to replenish. During this depleted state, effortful control is not possible and people fall prey to their automatic tendencies. Research supporting this model has demonstrated, for example, that exerting self-control on one task leads to poorer self-control on immediately subsequent tasks (for review, *see* Muraven & Baumeister, 2000). Like other automatic versus effortful control models, however, the ego-depletion model suggests that it is the automatic processes initiated by exposure to a local reward that undermine self-control, particularly when the ability to exert effortful control is hampered.

Although the automatic versus effortful control model is intuitively appealing, there are still enduring conceptual issues that require addressing. Perhaps the most critical conceptual issue is that any action tendency, in principle, can be automatized. This includes psychological processes and responses that promote, rather than undermine, one's valued goals and objectives. Empirically supporting this assertion, research has demonstrated that one's global goals and concerns can also be automatically activated by situational cues and dictate behavior (e.g., Chartrand & Bargh, 1996; Bargh, Gollwitzer, Chai, Barndollar, & Troetschel, 2001). Critically, the presence of local rewards can automatically activate processes and behaviors that promote rather than impair self-control (e.g., Fishbach, Friedman, & Kruglanski, 2003; Fishbach & Shah, 2006; Moskowitz, Gollwitzer, Wasel, & Schaal, 1999). Fishbach and colleagues (2003) have demonstrated that exposure to temptations can paradoxically automatically activate overriding goals and enhance self-control. For example, exposure to cues of indulgent eating (e.g., "chocolate" and "cake") facilitates activation of diet-related thoughts (e.g., "thin" and "slim"), which in turn inhibit thoughts (and actions) associated with consuming those temptations. Given that the same situation can automatically activate processes that both undermine and promote one's global interests, automatic versus effortful control models will need to specify the antecedent conditions that determine when automatic processes impair rather than promote self-control.

Hot vs. Cool

As an alternative to automatic versus effortful control accounts of self-control are those models that posit a distinction between affect and cognition. The most extensively studied of these latter approaches is the hot versus cool systems model (Metcalfe & Mischel, 1999; Mischel, Shoda, & Rodriguez, 1989; *see also* Loewenstein, 1996; Chapter 23). Mischel and his colleagues suggest that there are two systems of mental representation. The hot system is composed of affective representations that respond to the appetitive, emotional features of objects and events. The cool system, in contrast, is comprised of emotionally neutral cognitive representations that initiate deliberative, contemplative actions. Self-control failures occur when the hot system is differentially activated and co-opts the cool system. For example, children who were instructed to think about the hot, appetitive qualities of a temptation ("think about how crunchy and salty the pretzels are") were unable to delay gratification as well as those instructed to think about the cool, nonappetitive qualities ("think about how the pretzels are thin and long like logs;" Mischel & Baker, 1975). That is, the activation of affective mental representations caused people to focus on local rewards, thus causing them to fail to act in line with more global concerns.

Like automatic versus effortful control models, affect versus cognition accounts of self-control are intuitively appealing yet still need to address enduring conceptual issues. One conceptual problem lies with a lack of specificity in what is meant by affective, or "hot," representations. For example, it is not clear whether one should consider pride "warmer" than disgust or love "colder" than lust. Similarly, there are types of affective experiences and representations that, in principle, should promote rather than impair self-control. Supporting this assertion are empirical findings suggesting that affective representations strongly associated with guilt or shame enhance self-control (e.g., Amodio, Devine, & Harmon-Jones, 2007; Giner-Sorolla, 2001; Kivetz & Zheng, 2006). Research by Fishbach and Shah (2006) also suggests that appetitive, approach responses not only impair self-control as affect versus cognition models have suggested, but can also promote self-control to the extent that they are associated with more global goals. Findings such as these suggest that people appear capable of strategically managing their affective representations and reactions not simply to avoid falling prey to local temptations but also to promote more global interests. Affect versus cognition models thus still require theoretical and empirical refinement to determine under what conditions affective mental representations promote decisions with those that confer with local, rather than global, concerns.

CONSTRUAL LEVEL ANALYSIS

We have recently proposed an approach to self-control that, similarly to the hot versus cool model, focuses on the mental representations, or construals, that people form of objects and events (Fujita et al., 2006a, 2006b). By construals, we refer to people's subjective understanding or interpretation of an event. Extensive research in psychology has demonstrated that despite the same objective information, people can generate very different inferences and judgments about an event based on their own subjective construal of the event (e.g., *see* Fiske & Taylor, 2007; Griffin & Ross, 1991; Kunda, 1990). We model self-control as such a construal-dependent decision—that is, a critical determinant of whether people exert self-control is how they mentally represent or understand the situation. However, rather than distinguish affective versus cognitive representations as in the hot versus cool model, our approach focuses on the level of abstractness of one's construal. In other words, whether one views the proverbial trees or the forest beyond determines whether one makes decisions in line with one's local versus global concerns.

Construal Level Theory

Construal level theory (Liberman, Trope, & Stephan, 2007; Trope & Liberman, 2003; Trope, Liberman, & Wakslak, 2007) posits that the same event or object can be mentally represented at different levels of abstraction. High-level construals are mental representations generally used to construe psychologically distant events (e.g., those distant in time, space, social distance, or hypotheticality; Liberman et al., 2007). As information about distant entities is usually unreliable or unknown, high-level construals are more schematic and prototypical, extracting the central, defining features that are common to many instances of an event. Constructed through a process of abstraction, high-level construals are structured and coherent, selectively including relevant features and excluding irrelevant features. Low-level construals, on the other hand, are often used to construe psychologically near events and specify the concrete, incidental, secondary features that distinguish one event from another. For example, whereas the lighting and sound quality of a lecture hall might differentiate one academic course from another at a low-level construal, conceptualizing the same event at a high-level construal (fulfilling class requirements for a major) might render such features irrelevant and highlight other more central features, such as the course material.

The distinction between between high- versus low-level construals is supported by extensive research demonstrating changes in mental construal as a function of psychological distance (see Liberman et al., 2007; Trope & Liberman, 2003; Trope, Liberman, & Wakslak, 2007). For example, when considering psychologically distant versus near events, people categorize objects into fewer, broader groups, suggesting more prototypical, abstract processing (e.g., Liberman, Sagristano, & Trope, 2002; Smith & Trope, 2006; Wakslak, Trope, Liberman, & Alony, 2006). Similarly, individuals experiencing psychological distance (versus proximity) tend to make more gist-based intrusion errors in memory tests (Smith & Trope, 2006). Those who construe psychologically distant versus near events are also faster at detecting the Gestalt and slower at detecting differences in fine-grain details when performing visual perception tasks (Wakslak et al., 2006). Perceivers encode behaviors of psychologically distant others in terms of general, decontextualized characteristics (global traits) rather than context-specific actions (e.g., Fujita, Henderson, Eng, Trope, & Liberman, 2006; Henderson, Fujita, Trope, & Liberman, 2006; Nussbaum, Trope, & Liberman, 2003). Psychological distance, moreover, has been shown to promote the identification of actions in terms of the superordinate ends they achieve rather than the subordinate means that serve them (Fujita, Henderson, et al., 2006; Liberman & Trope, 1998; Smith and Trope, 2006; Wakslak et al., 2006). There is thus a substantial body of evidence that psychological near versus distant events are systematically associated with mental representations that differ in abstractness and globality and that these mental construals (high- versus low-level) are distinct.

Of critical importance is that research has demonstrated that changes in construal as a function of psychological distance systematically affect judgment, decision, and behavior. When events are psychologically distant, and therefore construed at high- versus low-levels, more weight is given to high-level features of the events. For example, studies have shown when considering psychologically distant events, people predict that typical events are more likely and atypical events are less likely, demonstrating judgments that are increasingly based on the prototypical likelihood of events (Henderson et al., 2006). People also base decisions to a greater extent on desirability (which reflect concerns about the superordinate end of an action) versus feasibility features (which reflect concerns about the subordinate means) when construing distant versus near events (Fujita, Eyal, Chaiken, Trope, & Liberman, 2008; Liberman & Trope, 1998; Sagristano, Trope, & Liberman, 2002). Moreover, people are better able to distinguish and make decisions that reflect primary, essential features of a task over secondary, incidental features when events are distant versus near. For example, when considering whether to buy a new radio to listen to music in the distant versus near future, people increasingly base decisions on central features of the radio, such as its sound quality, as opposed to incidental features, such as how eye-catching the display of a built-in clock is (Fujita et al., 2008; Trope & Liberman, 2000). Further demonstrating the impact of construals on judgment, people construing distant versus near events display greater sensitivity to persuasive messages that appeal to superordinate categories of objects (e.g., "whales") rather than specific exemplars (e.g., "Simoon, the whale;" Fujita et al., 2008). These findings further distinguish high- versus low-level construals and demonstrate their important role in social judgment and decision making.

Although high- and low-level construals specialize in distant and near events, respectively, they may also be activated in the absence of any differences in psychological distance. For example, the tendency to construe events at high- versus low-level construals can be primed as a "mindset" (e.g., Gollwitzer, 1990), activated by the mental construal of unrelated prior contexts. Imagining unrelated distant future events, for example, has been shown to enhance abstract, global information processing, whereas imagining unrelated near future events has been shown to promote more concrete, detailed processing (Förster, Friedman, & Liberman, 2004). Considering the superordinate ends versus subordinate means of an action in prior unrelated contexts has also been shown to influence judgment and decision making in a manner consistent with high- versus low-level construals, respectively (Freitas, Gollwitzer, & Trope, 2004; Fujita, et al., 2006b). Additionally, there is evidence of individual differences in the tendency to construe events at high- versus low-level construals (Freitas, Salovey, & Liberman, 2001; Levy, Freitas, & Salovey, 2002; Vallacher & Wegner, 1989). Thus, even in the absence of differences in distance (e.g., time, space, social distance, and hypotheticality), construal levels have been shown to impact judgment and decision making.

Applying Construal Level Theory to Self-Control

The central premise of the present construal level analysis is that self-control conflicts arise when high- and low-level construals of the same event suggest responses that are in direct opposition to each other. Consider, for example, a dieter who must choose between an apple and a candy bar as a snack. Lower level construals highlight concrete features such as taste and texture in choice and would promote the candy bar as the superior option. In contrast, higher level construals would lead the dieter to consider the more abstract, goal-relevant implications of both choice options. As dieters presumably value weight loss over hedonism, which represent the abstract implications of the two rewards respectively, higher level construals should promote preferences for the apple over the candy bar. Construing an event at one level or the other should thus tip decisions and behavioral responses in favor of the concerns highlighted by that respective construal—that is, the level at which an event is construed should impact self-control.

It should be noted that such theoretical predictions apply only to those for whom apples versus candy bars represents a meaningful self-control conflict. Among those unconcerned about weight loss, construing the choice at high- versus low-level construals should have little effect. Both the concrete features (taste and texture) and abstract implications (lack of concern for weight loss) of the choice suggest eating the candy bar. Therefore, the present model describes specifically those choices that represent a meaningful conflict between global goals and local rewards, rather than all types of decisional conflicts.

Before reviewing relevant supporting evidence for this hypothesis, however, it is important to distinguish the present approach from previous work and to highlight its novel and unique contributions. A construal level approach builds upon and extends time-discounting models of self-control by specifying more clearly the psychological mechanisms by which increasing temporal distance causes people to make decisions on the basis of more global versus local rewards. Time is but one dimension of psychological distance, and increases in psychological distance are associated with high-level construals (Liberman et al., 2007; Trope & Liberman, 2003; Trope, Liberman, & Wakslak, 2007). A construal level approach, however, also suggests that other dimensions of distance should also promote self-control, such as spatial or social distance, an observation supported by recent research (Jones & Rachlin, 2006; Kross, Ayduk, & Mischel, 2005; Libby, Shaeffer, Eibach, & Slemmer, 2007; Prencipe & Zelazo, 1995; Pronin, Olivola, & Kennedy, 2008; Vohs & Heatherton, 2000). It also suggests that even when the time frame of a self-control problem is kept constant, changes in the mental construal of an event should impact self-control, a prediction supported by studies that we describe shortly (Fujita et al., 2006b; see also Mischel & Baker, 1975; Moore, Mischel, & Zeiss, 1976).

As noted earlier, recent research has suggested that greater specification is required to understand when automatic versus effortful controlled processes will cause people to be tempted by local rewards or maintain pursuit of more global concerns (Fishbach et al., 2003; Fishbach & Shah, 2006). A construal level approach does not require that automatic versus effortful control necessarily be associated with local temptations versus global concerns and therefore circumvents that problem. Rather, it suggests that both types of processing might lead to self-control success or failure depending on construal. Indeed, people's construals of events, whether high- or low-level, are probably an amalgamation of both automatic and controlled processes. One can process both local and global features of events through effortless and effortful mechanisms (see also Fujita et al., 2008; Smith & Trope, 2006; Wakslak et al., 2006). Thus, what is critical to self-control may not be automatic versus effortful control of behavior but, rather, the manner in which the object or events in question are construed.

Finally, as stated earlier, a construal level perspective of self-control builds upon the hot versus cool systems model proposed by Mischel and colleagues (Mischel et al., 1989; Metcalfe & Mischel, 1999). Whereas the hot versus cool model stresses the role of affect in mental representation, a construal level approach emphasizes the role of superordinate abstraction. Although affective representations tend to be more concrete and contextually specific (thus reflecting low-level construals), it is important to recognize that affectivity and abstraction are independent. For example, the present construal level approach allows affect to be associated with both high- and low-level construals. This is noteworthy, as it suggests that affect can both promote and impair self-control as the literature suggests (e.g., Ferguson & Bargh, 2004; Fishbach & Shah, 2006; Giner-Sorolla, 2001; Kivetz & Zheng, 2006). One can also induce construals in a manner that does not require manipulating affect, a technique capitalized in the studies described in the following sections. Thus, the present approach can be considered a refinement of the hot versus cool systems model in that they both emphasize the central role mental representation has in self-control. However, the present approach emphasizes the level of abstraction of one's construal rather than affectivity.

Effect of Construals on Self-Control

The proposed construal level analysis predicts that mental construals should affect self-control. That is, the activation of high-level construals should promote global over local concerns in evaluation, decision, and action. In self-control conflicts, when global versus local concerns are pitted against one another, high-level construals should promote self-control. In what follows, we review empirical studies that support this prediction. In many of the studies we describe, the tendency to construe events at high- versus low-levels was induced as a mindset. That is, participants completed a task designed to induce high- versus low-level construals, and then were presented with a self-control dilemma in a subsequent ostensibly unrelated task. As noted earlier, the tendency of construal levels to "carry over" from one situation to another had been documented in several studies (e.g., Förster et al., 2004, Freitas et al., 2004; Fujita et al., 2006b). For example, generating superordinate ends versus subordinate means of an action induces a tendency to construe subsequent situations at high- versus low-level construals, respectively (Freitas et al.,

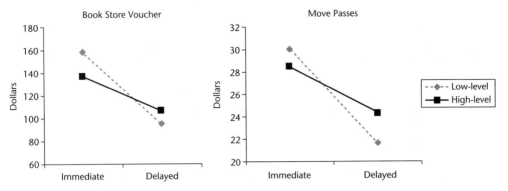

Figure 22–1. Preferences for immediate overdelayed outcomes as a function of construal level. Participants experimentally induced to represent situations at high-level construals evidenced a reduced tendency to prefer immediate over delayed outcomes (i.e., smaller difference in value between immediate vs. delayed outcomes), suggesting greater self-control. From Fujita, K., Trope, Y., Liberman, N., & Levin-Sagi, M. Construal levels and self-control, J Personal Soc Psychol, 2006; 90: 351–367. Copyright by the American Psychological Association. Adapted by permission.

2004; Fujita et al., 2006b). Generating superordinate category labels versus subordinate exemplars for a series of everyday objects has also been demonstrated to induce high- versus low-level construals in subsequent unrelated tasks (Fujita et al., 2006b). Notably, with these construal level inductions, high- versus low-level construals are activated without changing long-term (vs. short-term) temporal perspectives, boosting (vs. depleting) cognitive and motivational resources, nor activating cool (vs. hot) representations. In other words, they allow one to directly test the impact of mental construals on self-control.

In the first study to directly test the predictions of a construal level analysis of self-control (Fujita et al., 2006b, Experiment 1), participants were first asked to generate superordinate ends (high-level construals) versus subordinate means (low-level construals) of engaging in an action ("Promoting and maintaining good physical health;" see Freitas et al., 2004). They were then presented with temporal discounting items, which measured preferences for immediate over delayed outcomes. Each item presented participants with a positive outcome (e.g., $150 gift certificate to their favorite restaurant) and asked them to indicate in dollars how much they would pay to receive that outcome both immediately and delayed in time. The difference in these two dollar amounts provided an estimate of participants' preferences for immediate over delayed outcomes. As discussed earlier, measures of temporal discounting such as these represent good measures of self-control, as the tendency to prefer immediate over delayed outcomes is systematically associated with poorer self-control (e.g., Ainslie, 1975; Thaler, 1991; Frederick et al., 2002). Although in general, participants preferred immediate over delayed outcomes, this effect was moderated by construal level (see Figure 22–1). The activation of high- versus low-level construals led to decreased preferences for immediate over delayed outcomes, a pattern of preferences consistent with greater self-control. It bears noting that the effects of construal on temporal discounting preferences have been independently replicated by researchers using a variety of other construal level mindset inductions and temporal discounting items (Malkoc, Zauberman, & Bettman, 2008). Together, these studies indicate that even when the time frame of rewards is held constant, preferences for immediate over delayed rewards are moderated by construal level, demonstrating the unique and novel contributions to our understanding of temporal discounting that studying mental construals in the context of self-control can make.

A second study moved beyond preferences and tested the effects of construal on a behavioral measure of self-control (Fujita et al., 2006b, Experiment 2). Participants again were induced to construe subsequent events at high- versus low-level construals by generating superordinate ends versus subordinate means for engaging in a task ("Promoting and maintaining good relationships"). As a behavioral measure of self-control, participants were then asked to grip a handgrip. A handgrip is an exercise tool designed to strengthen forearm strength, and prolonged use causes physical discomfort. Previous research has demonstrated that the duration of time one is able to grip a handgrip is sensitive to changes in self-control (e.g., Muraven, Tice, & Baumeister, 1998). Participants were informed that gripping the handgrip was part of a new psychophysiological measure of personality. Dummy electrodes were taped to the participants' forearms, and they were told that by gripping the handgrip, their muscles would emit an electrical signal that the computer would use to diagnose an important aspect of their personality. Importantly, participants were told that they would be receiving this information at the end of the study, and that although it was uncomfortable, the longer they gripped the handgrip, the more accurate this information would be. Thus, participants experienced a self-control conflict between receiving increasingly accurate information about themselves and relieving the discomfort of their hand. Results indicated that as predicted, the induction of high- versus low-level construals led to greater self-control, as measured by surreptitious timing of how long participants gripped the handgrip.

Although the induction of high-level construals led to preferences and behavior consistent with greater self-control in these first two studies, it is not clear whether these effects resulted from true changes in self-control or from some more generalized psychological process. For example, it is possible that the induction of high-level construals made people more compliant or distractible, induced greater cognitive load, or in some way energized them motivationally. It may be these more general processes, rather than self-control specifically, that led to changes in preferences and behavior. If construal levels exert their effect uniquely on self-control by enhancing the weighing of high- versus low-level features in decisions and actions, they should have no effect in situations that do not involve self-control conflicts. That is, the effects of construal level should be apparent when an event evokes a meaningful self-control conflict.

A third study tested this boundary condition (Fujita et al., 2006, Experiment 4). Undergraduate student-participants were induced to construe events at high- versus low-level construals by having them generate superordinate (high-level) category labels versus subordinate (low-level) exemplars for a series of objects. For example, given "dog," participants were asked to generate "animal" versus "poodle." These student-participants were then asked to evaluate a series of words. Embedded within these words were words associated with temptations that undermine the goal to study (e.g., "television," "party," "phone"). To the extent that academic achievement represents a valued goal to student participants, negative evaluations of these words should reflect greater self-control. To ensure the temptations presented reflected a meaningful self-control conflict to our student-participants, participants were later asked to report how important it was for them to do well in school, how important it was to find time to study, how committed they were to studying daily, and how helpful or instrumental studying was to getting good grades. High-level construals should have lead to more negative evaluations of temptations. There should have been no effect of construal on evaluations of nontemptation words, as they were not associated with self-control conflicts. Moreover, the effect of construal should have been evident only among students who valued studying—that is, those to whom the temptation words actually represented meaningful self-control temptations. Indeed, the results of this study confirmed these predictions (see Fig. 22–2). Note that the effect of construals was specific to temptation (versus nontemptation) words and apparent only among those for whom they represented meaningful self-control conflicts (i.e., those who value studying). These data

As noted earlier, a major determinant of how people construe events is the psychological distance of those events (Liberman et al., 2007; Trope & Liberman, 2003; Trope et al., 2006). Construal level theory posits that social stimuli distal on any psychological dimension (temporal distance, spatial distance, social distance, hypotheticality) activate high-level construals, whereas those that are proximal activate low-level construals. This suggests that psychological distance from a self-control conflict should promote self-control. Indeed, as reviewed earlier, much of the work in self-control done by behavioral decision-making researchers has focused on temporal distance from a self-control conflict as a critical factor in self-control (e.g., Ainslie, 1975; Frederick et al., 2002; Freitas et al., 2001; Green et al., 1994; Thaler, 1991). However, a construal level approach also suggests that distancing along other dimensions should have analogous effects. Indeed, there is some suggestive evidence that increasing spatial distance from a temptation leads to greater self-control (e.g., Vohs & Heatherton, 2000). Moreover, research on visual perspective suggests that distancing one's self from an event by adopting a third- versus first-person visual perspective can enhance self-control (Kross et al., 2005; Libby et al., 2007; Principe & Zelazo, 1995). For example, Kross and colleagues (2005; see also Chapter 23) examined the control of negative ruminative emotions while working through and understanding the event that provoked those emotions. Their research suggests that taking a "distanced" third-person versus "immersed" first-person perspective enhanced people's ability to control ruminative emotions and successfully process negative events. Moreover, Kross and colleagues have demonstrated that the successful control of ruminative emotions by third- versus first-person visual perspectives appears to be mediated by abstract, high-level versus concrete, low-level mental construals of the event. Research has also indicated that social distancing has parallel effects on self-control to temporal distancing (e.g., Jones & Rachlin, 2006; Pronin et al., 2008), again suggesting construal levels as a common mechanism.

An intriguing possibility that we are currently testing in our labs is that people may strategically manage the psychological distance of objects and events associated with self-control conflicts as a prospective self-control strategy. That is, people may distance themselves from temptations and vices in an effort to maintain high-level construals through increased psychological distance. There is already evidence that people do indeed strategically perceive temporal and social distance in a way that serves self-esteem and other self-protective concerns (e.g., Libby & Eibach, 2002; Ross & Wilson, 2002). It remains to be seen, however, whether people may do so in service of self-control as well. The systematic study of psychological distance thus promises to be a fruitful and exciting area of self-control research.

Relation to Extant Models

As noted earlier, a construal level perspective to self-control provides several advancements over extant models. The present approach builds on temporal discounting by specifying the psychological mechanisms by which time impacts self-control. In other words, temporal distance is systematically associated with changes in mental construal (Trope & Liberman, 2003). However, as noted already, the present approach goes beyond temporal discounting by suggesting that other dimensions of distance beyond time should have analogous effects (Liberman et al., 2007; Trope et al., 2007), a claim supported by current research (e.g., Kross et al., 2005, Libby et al., 2007; Prencipe & Zelazo, 1995; Vohs & Heatherton, 2000). Moreover, as already described, construals moderate temporal discounting, indicating the central role construals have in self-control even in the absence of any differences in time frame (Fujita et al., 2006b; Malkoc et al., 2007; see also Mischel & Baker, 1975; Moore, Mischel, & Zeiss, 1976). Thus, a construal level approach integrates and extends extant work on temporal discounting.

As suggested earlier, recent research has suggested that automatic versus effortful control models of self-control require greater specification as to when each process leads to self-control failure versus success (e.g., Fishbach et al., 2003; Fishbach & Shah, 2006). The present approach suggests that a key factor determining self-control failure versus success is the level of construal of the event. Notably, low- versus high-level construals are independent of automatic versus effortful control. Construals recruit both automatic and effortful controlled processes, and one can process low- and high-level information through effortless or effortful mechanisms (*see also* Fujita et al., 2008; Smith & Trope, 2006; Wakslak et al., 2006). One intriguing possibility, however, is that whether automatic processes promote or undermine self-control may be determined by the person's construal level. Preliminary research in our lab indicates that higher level construals can prompt automatic processes that counter temptation impulses. For example, rather than automatically prompting thoughts about eating cake (e.g., Papies et al., 2007), construing a chocolate cake at higher level construals engage processes that promote thoughts about dieting (Fujita & Sasota, 2009; *see also* Fujita & Han, 2009). Therefore, whether automatic processes undermine or promote self-control may be determined by people's level of construal, suggesting that a construal level approach may provide theoretical integration to the current literature on automatic versus effortful control processes in self-control.

The present approach also contributes to the literature on affect versus cognition models of self-control. Although hot, affective mental representations have been shown to undermine global goals (Mischel et al., 1989; Metcalfe & Mischel, 1999), they can promote self-control as well (e.g., Amodio et al., 2007; Ferguson & Bargh, 2004; Fishbach & Shah, 2006; Giner-Sorolla, 2001). This suggests that the affect versus cognition models require greater specification as to when affective representations promote, rather than impair, self-control. The present approach suggests that the level of construal of affective representations may be the critical factor that determines self-control success versus failure. Affective representations tend to be more concrete and context-specific (i.e., associated with low-level construals) and hence frequently (but not always) lead to poorer self-control. Abstraction and affectivity, however, may be manipulated independently, and when affect is associated with higher level construals, affective representations may promote self-control. It is also important to note that the abstractness of mental construals can be manipulated and impact self-control without necessarily manipulating affectivity, as was the case in the studies described earlier (Fujita et al., 2006b). Together, these suggest that the present approach is distinct from the hot versus cool systems model but integrate conflicting findings within the hot versus cool systems literature and provide theoretical refinement by highlighting the importance of superordinate abstraction as a critical factor in self-control conflicts.

It would be amiss not to relate the current construal level approach to recent work on ego-depletion (e.g., Muraven & Baumeister, 2000). As noted earlier, the ego-depletion model is largely predicated on automatic versus effortful control models of self-control in that it suggests that self-control success or failure depends on effortfully controlling automatic processes. Where the ego-depletion model differs from traditional automatic versus effortful control models is that it suggests that effortful control draws from a limited motivational resource rather than limited cognitive capabilities. Once depleted, effortful control is no longer possible and individuals fall prey to automatic tendencies. Although a great deal of evidence suggests that exertion of self-control taxes this limited resource and leaves people vulnerable to self-control lapses in immediately subsequent tasks (for review, *see* Muraven & Baumeister, 2000), the specific mechanism for this phenomenon is not well-understood. Vohs & Schmeichel (2003) have recently suggested that ego-depletion effects are driven by changes in perceptions of time—that is, when regulatory resources are depleted, people experience time moving more slowly and are more focused on the immediate present. This preferential attention to the present may activate a more local, low-level construal, leading to the frequently documented self-control failures in subsequent tasks, as demonstrated in a number of ego-depletion studies. However, it may also be the case that self-regulatory resources and construals are completely independent factors. Whereas ego-depletion models are concerned with the energy required for self-control, a construal level analysis focuses on the mental construal of self-control conflicts. It may be the case that although high-level construals might increase the tendency for individuals to make decisions that are consistent with self-control concerns, if they are already ego-depleted, then they may not actually have the energy necessary to carry out or implement those decisions. Similarly, having self-regulatory energy may not lead to greater self-control unless individuals have decided to act on the basis of high-level construals of a situation. Future work, both theoretically and empirically, on integrating the two theoretical approaches is clearly warranted.

It is also important to distinguish the present construal level analysis from action identification theory (Vallacher & Wegner, 1987). Although action identification theory has not been directly applied to the problem of self-control, it has implications for self-control that are consistent with a construal level approach. For example, action identification theory posits that there are high and low levels of mental representation. Action identification, however, is a theory of mental representation specific to the representation of actions in hierarchical means—end relationships. Construal levels encompass more than just actions and their means—end relationships. They can represent features of any target: objects, situations, or events. Construal levels may also differ on features beyond means—end relationships, such as whether they refer to primary versus secondary goals or to goal-relevant versus goal-irrelevant aspects (Fujita et al., 2008; Trope & Liberman, 2000). For example, the conflict for students studying for finals regarding whether to review their notes versus watch television does not involve a conflict between means versus ends but rather a conflict between primary, global versus secondary, local goals. Thus, the effect of levels of construal on self-control is not reducible to levels of action identification.

Similarly, it is important to distinguish a construal level analysis of self-control from research on implementation intentions (for review, see Gollwitzer, 1999; Gollwitzer, Fujita, & Oettingen, 2004). Implementation intentions refer to action plans that people generate to promote goal success by linking specific goal-striving behaviors to specific environmental cues. For example, Gollwitzer and Brandstätter (1997) found that participants who generated implementation intentions to complete a school assignment over winter break ("If I see my computer, I will work on my assignment") were more likely to do so than those who merely were committed to the goal of completing the assignment ("I will work on my assignment"). Findings such as this may, at first glance, appear discrepant with a construal level analysis of self-control. Superficially, the success of implementation intentions in self-control conflicts appears to demonstrate that focusing on low-level behaviors promotes self-control. It is possible, however, that it is not the specification of low-level behaviors in implementation intentions that promotes self-control but rather the cognitive linking of local cues and temptations to more global goals. For example, whereas the smell of warm cookies (local feature) among dieters may normally initiate an action tendency to eat cookies, an implementation intention ("If I smell cookies, I will ignore them") links this low-level feature with a course of action consistent with a global concern (i.e., dieting). The activation of high-level construals in situations prior to the self-control conflict may even promote planning and the generation of implementation intentions as a prospective self-control strategy. Therefore, research suggesting that implementation intentions promote self-control may be entirely consistent with the present construal level analysis.

Conclusion

This chapter describes a model of self-control that seeks to understand why people so often fail to act in line with their global interests and thus fail at self-control. This model suggests that self-control decisions are construal-dependent decisions—that is, they are influenced by how people subjectively understand and interpret events. This model proposes that people differ in how abstractly they understand the event, and that more abstract, high-level construals of an event promote self-control. The studies reviewed indicate that mental construals do play an important role in the successful resolution of self-control dilemmas. As self-control failures are hypothesized to be at the heart of a number of personal and societal issues (Baumeister & Heatherton, 1996), understanding better what self-control is, its underlying processes, and the factors that enhance or impair self-control may provide insight into how to address these important issues. Knowledge of how mental construals impact self-control might be used to improve interventions currently used to promote self-control in a number of domains and spur the development of even more effective personal and public policy programs.

Authors' Note

Kentaro Fujita, Department of Psychology, The Ohio State University; Yaacov Trope, Department of Psychology and Stern Business School Marketing Department, New York University; and Nira Liberman, Department of Psychology, Tel Aviv University.

This research was supported by in part by National Science Foundation Grant #0817360 to Kentaro Fujita, National Institute of Mental Health Grant #R01-MH59030-06A1 to Yaacov Trope, and United States-Israel Binational Science Foundation Grant #2001057 to Nira Liberman and Yaacov Trope.

Correspondence concerning this article should be addressed to Kentaro Fujita, Department of Psychology, The Ohio State University, 1827 Neil Avenue Mall, Columbus OH 43210, or by email at fujita.5@osu.edu.

References

Ainslie, G. Specious reward: A behavioral theory of impulsiveness and impulse control. Psychol Bull 1975; 82: 463–496.

Amodio, D. M., Devine, P. G., & Harmon-Jones, E. A dynamic model of guilt: Implications for motivation and self-regulation in the context of prejudice. Psychol Sci 2007; 18: 524–530.

Ariely, D., & Wertenbroch, K. Procrastination, deadlines, and performance: Self-control by precommitment. Psychol Sci 2002; 13: 219–224.

Bargh, J. A., Gollwitzer, P. M., Chai, A. L., Barndollar, K., & Troetschel, R. Automated will: Nonconscious activation and pursuit of behavioral goals. J Pers Soc Psychol 2001; 81: 1014–1027.

Baumeister, R. F., & Heatherton, T. F. Self regulation failure: An overview. Psychol Inq 1996; 7: 1–15.

Beukeboom, C. J., & Semin, G. R. Mood and representations of behavior: The how and why. Emotion 2005; 19: 1242–1251.

Beukeboom, C. J., & Semin, G. R. How mood turns on language. J Exp Soc Psychol 2006; 42: 553–566.

Butler, R. Effects of task and ego-achievement goals on information seeking during task engagement. J Pers Soc Psychol 1993; 65: 13–31.

Chartrand, T. L., & Bargh, J. A. Automatic activation of impression formation and memorization goals: Nonconscious goal priming reproduces effects of explicit task instructions. J Pers Soc Psychol 1996; 71: 464–478.

Devine, P. G. Stereotypes and prejudice: Their automatic and controlled components. J Pers Soc Psychol 1989; 56: 5–18.

Dweck, C. S., & Leggett, L. A social-cognitive approach to motivation and personality. Psychol Rev 1988; 95: 256–273.

Eyal, T., Sagristano, M. D., Trope, Y., Liberman, N., & Chaiken, S. When values matter: Expressing values in behavioral intentions for the near vs. distant future. J Exp Soc Psychol 2009; 45: 35–43.

Ferguson, M. J., & Bargh, J. A. Liking is for doing: The effects of goal pursuit on automatic evaluation. J Pers Soc Psychol 2004; 87: 557–572.

Fishbach, A., Friedman, R. S., & Kruglanski, A. W. Leading us not into temptation: Momentary allurements elicit overriding goals activation. J Pers Soc Psychol 2003; 84: 296–309.

Fishbach, A., & Shah, J. Y. Self-control in action: Implicit dispositions toward goals and away from temptations. J Pers Soc Psychol 2006; 90: 820–832.

Fiske, S. T., & Taylor, S. E. *Social cognition: From brains to culture.* New York: McGraw-Hill. 2007.

Förster, J., Friedman, R. S., & Liberman, N. Temporal construal effects on abstract and concrete thinking: Consequences for insight and creative cognition. J Pers Soc Psychol 2004; 87: 177–189.

Frederick, S., Loewenstein, G., & O'Donoghue, T. Time discounting and time preference: A critical review. J Econ Lit 2002; 40: 351–401

Freitas, A. L., Gollwitzer, P. M., & Trope, Y. The influence of abstract and concrete mindsets on anticipating and guiding others' self-regulatory efforts. J Exp Soc Psychol 2004; 40: 739–752.

Freitas, A. L., Salovey, P., & Liberman, N. Abstract and concrete self-evaluative goals. J Pers Soc Psychol 2001; 80: 410–412.

Fujita, K., Eyal, T., Chaiken, S., Trope, Y., & Liberman, N. Influencing attitudes towards near and distant events. J Exp Soc Psychol In press.

Fujita, K., & Han, H. A. Moving beyond deliberative control of impulses: Effects of construal levels on evaluative associations in self-control. Psychol Sci 2009; 20: 799–804.

Fujita, K., Henderson, M. D., Trope, Y., & Liberman, N. Spatial distance and the mental construal of social events. Psychol Sci 2006; 17: 278–282.

Fujita, K., & Roberts, J. C. The role of mental construals in the adoption of prospective self-control strategies. Unpublished data, The Ohio State University, 2009.

Fujita, K., & Sasota, J. A. Construal moderated associations between temptations and goals. Unpublished data, The Ohio State University, 2009.

Fujita, K., Trope, Y., & Liberman, N. The role of mental construal in self-control. In: DeCremer, D., Zeelenberg, M., & Murnighan, J. K. (Eds.), Social psychology and economics. New York: Sage Publications, 2006: pp. 193–211.

Fujita, K., Trope, Y., Liberman, N., & Levin-Sagi, M. Construal levels and self-control. J Pers Soc Psychol 2006; 90: 351–367.

Gasper, K., & Clore, G. L. Attending to the big picture: Mood and global versus local processing of visual information. Psychol Sci 2002; 13: 34–40.

Giner-Sorolla, R. Guilty pleasures and grim necessities: Affective attitudes in dilemmas of self-control. J Pers Soc Psychol 2001; 80: 507–526.

Gollwitzer, P. M. Action phases and mind-sets. In: Higgins, E. T., & Sorrentino, R. M. (Eds.), Handbook of motivation and cognition: Foundation of social behavior, Vol. 2. New York: Guilford Press, 1990: pp. 53–92.

Gollwitzer, P. M. Implementation intentions: Strong effects of simple plans. Am Psychol 1999; 54: 493–503.

Gollwitzer, P. M., & Brandstätter, V. Implementation intentions and effective goal pursuit. J Pers Soc Psychol 1997; 73: 186–199.

Gollwitzer, P. M., Fujita, K., & Oettingen, G. Planning and the implementation of goals. In: Baumeister, R. F., & Vohs, K. D. (Eds.), Handbook of self-regulation: Research, theory and applications. New York: Guilford Press, 2004: pp. 211–228.

Green, L., Fristoe, N., & Myerson, J. Temporal discounting and preference reversals in choice between delayed outcomes. Psychon Bull Rev 1994; 1: 383–389.

Griffin, D. W., & Ross, L. Subjective construal, social inference, and human misunderstanding. In: Zanna, M. P. (Ed.), Advances in experimental social psychology, Vol. 24. New York: Academic Press, 1991: pp. 319–359.

Henderson, M. D., Fujita, K., Trope, Y., & Liberman, N. Transcending the "Here": The effects of spatial distance on social judgment. J Pers Soc Psychol 2006; 91: 845–856.

James, W. Principles of psychology. New York: Holt, 1890.

Jones, B., & Rachlin, H. Social discounting. Psychol Sci 2006; 17: 283–286.

Kahnemann, D., & Lovallo, D. Timid choices and bold forecasts: A cognitive perspective on risk taking. Manage Sci 1993; 39: 17–31.

Kirby, K. N., & Herrnstein, R. J. Preference reversals due to myopic discounting of delayed reward. Psychol Sci 1995; 6: 83–89.

Kivetz, R., & Zheng, Y. Determinants of justification and self-control. J Exp Psychol Gen 2006; 135: 572–587.

Kross, E., Ayduk, O., & Mischel, W. When asking "why" does not hurt: Distinguishing rumination from reflective processing of negative emotions. Psychol Sci 2005; 16: 709–715.

Kudadjie-Gyambi, E. & Rachlin, H. Temporal patterning of choice among delayed outcomes. Organ Behav Hum Decis Process 1996; 65: 61–67.

Kunda, Z. The case for motivated reasoning. Psychol Bull 1990; 108: 480–498.

Levy, S. R., Freitas, A. L., & Salovey, P. Construing action abstractly and blurring social distinctions: Implications for perceived homogeneity among, but also empathizing with and helping, others. J Pers Soc Psychol 2002; 83: 1224–1238.

Libby, L. K., & Eibach, R. P. Looking back in time: Self-concept change affects visual perspective in autobiographical memory. J Pers Soc Psychol 2002; 82: 167–179.

Libby, L. K., Shaeffer, E. M., Eibach, R. P., & Slemmer, J. A. Picture yourself at the polls: Visual perspective in mental imagery affects self-perception and behavior. Psychol Sci 2007; 18: 199–203.

Liberman, N., Sagristano, M., & Trope, Y. The effect of temporal distance on level of mental construal. J Exp Soc Psychol 2002; 38: 523–534.

Liberman, N., & Trope, Y. The role of feasibility and desirability considerations in near and distant future decisions: A test of temporal construal theory. J Pers Soc Psychol 1998; 75: 5–18.

Liberman, N., Trope, Y., & Stephan, E. Psychological distance. In: Higgins, E. T., & Kruglanski, A. W. (Eds.), *Social psychology: Handbook of basic principles*, Vol. 2. New York: Guilford Press, 2007: pp. 353–381.

Lin, E. L., Murphy, G. L., & Shoben, E. J. The effects of prior processing episodes on basic-level superiority. Q J Exp Psychol 1997; 50: 25–48.

Loewenstein, G. F. Out of control: Visceral influences on behavior. Organ Behav Hum Decis Process 1996; 65: 272–292.

Macrae, C. N., Bodenhausen, G. V., Milne, A. B., & Jetten, J. Out of mind but back in sight: Stereotypes on the rebound. J Pers Soc Psychol 1994; 67: 808–817.

Magen, E., & Gross. J. J. Harnessing the need for immediate gratification: Cognitive reconstrual moderates the reward value of temptations. Emotion 2007; 7: 415–428.

Malkoc, S. A., Zauberman, G., & Bettman, J. R. Impatience is in the mindset: Carryover effects of processing abstractness in sequential tasks. Unpublished manuscript, University of Minnesota, 2008.

Metcalfe, J., & Mischel, W. A hot/cool system analysis of delay of gratification: Dynamics of willpower. Psychol Rev 1999; 106: 3–19.

Mikulincer, M., Kedem, P., & Paz, D. Anxiety and categorization—1. The structure and boundaries of mental categories. Pers Individ Diff 1990; 11: 805–814.

Mikulincer, M., Paz, D., & Kedem, P. Anxiety and categorization—2. Hierarchical levels of mental categories. Pers Individ Diff 1990; 11: 815–821.

Mischel, W., & Baker, N. Cognitive appraisals and transformations in delay behavior. J Pers Soc Psychol 1975; 31: 254–261.

Mischel, W., Shoda, Y., & Rodriguez, M. L. Delay of gratification in children. Science 1989; 244: 933–938.

Moore, B., Mischel, W., & Zeiss, A. Comparative effects of the reward stimulus and its cognitive representation in voluntary delay. J Pers Soc Psychol 1976; 34: 419–424.

Moskowitz, G., B., Gollwitzer, P. M., Wasel, W., & Schaal, B. Preconscious control of stereotype activation through chronic egalitarian goals. J Pers Soc Psychol 1999; 77: 167–184.

Muraven, M., & Baumeister, R. F. Self-regulation and depletion of limited resources: Does self-control resemble a muscle? Psychol Bull 2000; 126: 247–259.

Muraven, M., Tice, D. M., & Baumeister, R. F. Self-control as a limited resource: Regulatory depletion patterns. J Pers Soc Psychol 1998; 74: 774–789.

Nussbaum, S., Trope, Y., & Liberman, N. Creeping dispositionism: The temporal dynamics of behavior prediction. J Pers Soc Pscyhol 2003; 84: 485–497.

Papies, E., Stroebe, W., & Aarts, H. Pleasure in the mind: Restrained eating and spontaneous hedonic thoughts about food. J Exp Soc Psychol 2007; 43: 810–817.

Principe, A., & Zelazo, P. Development of affective decision making for self and other: Evidence for the integration of first- and third-person perspectives. Psychol Sci 1995; 16: 501–505.

Pronin, E., Olivola, C. Y., & Kennedy, K. A. Doing unto future selves as you would do unto others: Psychological distance and decision-making. Pers Soc Psychol Bull 2008; 34: 224–269.

Rachlin, H. Self control: Beyond commitment. Behav Brain Sci 1995; 18: 109–159.

Read, D., Loewenstein, G., & Kalyanaraman, S. Mixing virtue and vice: Combining the immediacy effect and the diversification heuristic. J Behav Decis Mak 1999; 12: 257–273.

Read, D., Loewenstein, G., & Rabin, M. Choice bracketing. J Risk Uncertain 1999; 19: 171–197.

Ross, M., & Wilson, A. E. It feels like yesterday: Self-esteem, valence of personal past experiences, and judgments of subjective distance. J Pers Soc Psychol 2002; 82: 792–803.

Sagristano, M., Trope, Y., & Liberman, N. Time-dependent gambling: Odds now, money later. J Exp Psychol Gen 2002; 131: 364–376.

Shiv, B., & Fedorikhin, A. Heart and mind in conflict: The interplay of affect and cognition in consumer decision making. J Consum Res 1999; 26: 278–292.

Smith, P. K., & Trope, Y. You focus on the forest when you're in charge of the trees: Power priming and abstract information processing. J Pers Soc Psychol 2006; 90: 578–596.

Strack, F., & Deutsch, R. Reflective and impulsive determinants of social behavior. Pers Soc Psychol Rev 2004; 8: 220–247.

Tangney, J. P., Baumeister, R. F., & Boone, A. L. High self-control predicts good adjustment, less pathology, better grades, and interpersonal success. J Pers 2004; 72: 271–322.

Thaler, R. H. *Quasi rational economics*. New York: Russel Sage Foundation, 1991.

Tice, D. M., Bratslavsky, E., & Baumeister, R. F. Emotional distress regulation takes precedence over impulse control: If you feel bad, do it! J Pers Soc Psychol 2001; 80: 53–67.

Trope, Y. Self-enhancement and self-assessment in achievement motivation. In: Sorrentino, R. M., & Higgins, E. T. (Eds.), *Handbook of motivation and cognition: Foundations of social behavior*, Vol. 1. New York: Guilford Press, 1986: pp. 350–378.

Trope, Y., & Fishbach, A. Counteractive self-control in overcoming temptation. J Pers Soc Psychol 2000; 79: 493–506.

Trope, Y., Igou, E. R., & Burke, C. Mood as recourse in structuring goal persuit. In: Forgas, J. (Ed.), *Affect in social thinking and behavior*. New York: Psychology Press, 2006: pp. 217–234.

Trope, Y., Ferguson, M., & Raghunathan, R. Mood as a resource in processing self-relevant information. In: Forgas, J. P. (Ed.), *Handbook of affect and social cognition*. Mahwah, NJ: Erlbaum, 2001: pp. 256–274.

Trope, Y., & Liberman, N. Time-dependent changes in preferences. J Pers Soc Psychol 2000; 79: 876–889.

Trope, Y., & Liberman, N. Temporal construal. Psychol Rev 2003; 110: 403–421.

Trope, Y., Liberman, N., & Wakslak, C. J. Construal levels and psychological distance: Effects on representation, prediction, evaluation, and behavior. J Consum Psychol 2007; 17: 83–95.

Trope, Y., & Neter, E. Reconciling competing motives in self-evaluation: The role of self-control in feedback seeking. J Pers Soc Psychol 1994; 66: 646–657.

Vallacher, R. R., & Wegner, D. M. What do people think they're doing? Action identification and human behavior. Psychol Rev 1987; 94: 3–15.

Vallacher, R. R., & Wegner, D. M. Levels of personal agency: Individual variation in action identification. J Pers Soc Psychol 1989; 57: 660–671.

Vohs, K., & Heatherton, T. F. Self-regulatory failure: A resource-depletion approach. Psychol Sci 2000; 11: 249–254.

Vohs, K. D., & Schmeichel, B. J. Self-regulation and the extended now: Controlling the self alters the subjective experience of time. J Pers Soc Psychol 2003; 85: 217–230.

Wakslak, C. J., Trope, Y., Liberman, N., & Alony, R. Seeing the forest when entry is unlikely: Probability and the mental representation of events. J Exp Psychol Gen 2006; 135: 641–653.

Wegner, D. M. Ironic processes of mental control. Psychol Rev 1994; 101: 34–52.

Wegner, D. M., & Erber, R. The hyperaccessibility of suppressed thoughts. J Pers Soc Psychol 1992; 63: 903–912.

Wertenbroch, K. Consumption self control by rationing purchase quantities of virtue and vice, Market Sci 1998; 17: 317–337.

Wiers, R. W., & Stacy, A. W. Implicit cognition and addiction. Curr Dir Psychol Sci 2006; 15: 292–296.

CHAPTER 23

From Stimulus Control to Self-Control: Toward an Integrative Understanding of the Processes Underlying Willpower

Ethan Kross and Walter Mischel

ABSTRACT

Self-control is fundamental to human survival and success in the modern world. Consequently, a critical challenge is to understand the processes that underlie it. The main goal of this chapter is to address this issue to demystify the self-control construct. The chapter begins with the assumption that to make sense of the psychological processes that enable self-control it is first necessary to understand how they operate within the larger cognitive affective processing system guiding people's thoughts, feelings, and behaviors (Mischel & Shoda, 1995, 1998; Shoda & Mischel, 1998). We begin by briefly describing this system and discussing its value for conceptualizing self-control dynamics. Drawing primarily from research on delay of gratification in children (Mischel, Shoda, & Rodriguez, 1989) and recent work that has begun to link this research with findings at the cognitive and neural levels of analysis the chapter then describes the psychological processes that enable people to effectively exert self-control. The second half of the chapter transitions from reviewing prior research on delay of gratification to current work examining how findings revealed from this paradigm generalize to different kinds of situations that require self-control. The chapter concludes by discussing how future research on self-control may benefit from work that bridges different levels of analysis.

Keywords: Self-control, dual processes, reconstrual, personality, psychological distance

INTRODUCTION

A key feature that distinguishes human beings from all other animal species is their capacity for "willpower"—the ability to deliberately exert control over one's thoughts, feelings, and actions. Like so many of the remarkable capacities people possess, however, they often fail to be actualized when they are most needed. Consider, for example, the dilemma of the desperate alcoholic who is trying to quit but is tempted to enter the bar, or the starving dieter faced with the decadent crème brule, or the test-anxious student before an important examination. These kinds of "hot," emotionally arousing situations rapidly generate intense feelings, bypassing self-regulatory controls when they are most needed and making the person the victim of what Skinner and others called "stimulus control." Without the essential "preliminaries" that enable self-control—as William James (1890) named them over a century ago—willpower becomes a myth rather than a protective factor. The main goal in this chapter is to identify these preliminaries to demystify the willpower construct. The chapter begins with the assumption that to make sense of the psychological processes that enable self-control, it is first necessary to understand how they operate within the larger cognitive affective processing system guiding people's thoughts, feelings, and behaviors (Mischel & Shoda, 1995, 1998; Shoda & Mischel, 1998). We thus begin by briefly describing this system and discussing its value for conceptualizing willpower dynamics. Drawing primarily from research on delay of gratification in children (Mischel, Shoda, & Rodriguez, 1989) as well as recent work that has begun to link this research with findings at the cognitive and neural levels of analysis, the chapter then addresses the main question motivating this review: What psychological processes enable people to transition from stimulus-control to self-control? The second half of the chapter transitions from reviewing prior research on delay of gratification to current work examining how basic findings revealed from this paradigm generalize to diverse situations and domains that require self-regulation—for example, situations in which adolescents and adults experience strong negative emotions that require regulation. The chapter concludes by discussing how current and future research on self-control may benefit from interdisciplinary work that bridges different levels of analysis.

Keywords: Stimulus control, self-control, emotion regulation, Cognitive-Affective Processing System.

A Framework for Understanding Willpower Dynamics: A Self-Regulatory Processing System

The model guiding our conceptualization of willpower is the Cognitive–Affective Processing System (CAPS), a connectionist framework that was designed for analyzing individual differences and basic psychological processes, such as self-control (Mischel & Shoda, 1995, 1998; Shoda & Mischel, 1998). The CAPS model conceptualizes the human mind as a network of mental representations whose distinctive pattern of

TABLE 23–1. TYPES OF COGNITIVE-AFFECTIVE UNITS IN THE COGNITIVE–AFFECTIVE PROCESSING SYSTEM

1. ENCODINGS: Categories (constructs) for the self, people, events, and situations (external and internal).
2. EXPECTATIONS AND BELIEFS: About the social world, about outcomes for behavior in particular situations, about one's self-efficacy.
3. AFFECTS: Feelings, emotions, and affective responses (including physiological reactions).
4. GOALS: Desirable outcomes and affective states, aversive outcomes, and affective states; goals and life projects.
5. COMPETENCIES AND SELF-REGULATORY PLANS: Potential behaviors and scripts that one can do and plans and strategies for organizing action and for affecting outcomes and one's own behavior and internal states.

Note: From "A Cognitive-Affective System Theory of Personality: Reconceptualizing Situations, Dispositions, Dynamics, and Invariances in Personality Structures," by W. Mischel & Y. Shoda, *Psychological Review*, 1995; 102: 253. Copyright by the American Psychological Association.

activation determines the thoughts and feelings they experience and the behaviors they display (e.g., Higgins, 1990; *see also* Shoda & Smith, 2004). The mental representations that are most relevant to situations that require willpower are captured by a set of *cognitive affective units* (CAUs) that encompass how people appraise and interpret situations, the feeling states they generate, their goals and values, expectancies and beliefs, and their self-regulatory competencies (for description of CAUs, *see* Table 23–1). Although each of these CAUs are themselves comprised of lower level processes (a point we return to later), we cast our discussion here at this relatively molar level of analysis because much of the work we review in this chapter has been performed at this level.

Within this model, different CAUs are interconnected within a stable associative network: this organization guides and constrains their activation with pathways of activation and deactivation. The relatively stable patterns of activation constitute the *processing dynamics* of the self-regulatory system. Situational features,

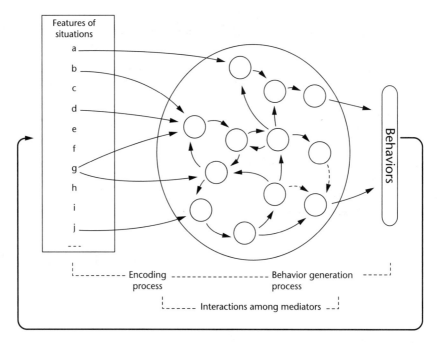

Figure 23-1. Illustrative Self-regulatory Dynamics in a Cognitive-Affective Processing System. Self-regulation in a cognitive–affective processing (CAPS) network is illustrated by the large circle, and the smaller circles within it represent the cognitive–affective units (CAUs). The CAUs are interconnected either through excitatory (solid lines) or inhibitory (broken lines); the darkness of a line indicates the strength of the association between any two CAUs. As illustrated, situational features are encoded by CAUs, which in turn, activates a subset of mediating units that are interconnected through a stable activation network. The dynamics of this network guide and constrain the individual's behavior in relation to particular situation features.

which may be events that are either encountered in the external world or created internally (i.e., through daydreaming, fantasies, rumination, etc.), are encoded by CAUs, which then activate additional CAUs through the activation network. In turn, the unique patterns of CAUs that become activated elicit a behavioral response pattern.

This kind of model offers a framework for understanding how different psychological processes interact to influence self-regulatory behavior. It thus allows researchers to move beyond broad trait level descriptions of self-control and focus instead on the interacting set of processes that underlie individual differences in self-control patterns. The next section discusses some of these processes, using the CAPS model as a framework to understand their operation.

Processes in Self-Control

When considering the processes that underlie self-control, it is important to distinguish between the individual's *motivation* to exert control and his or her *ability* to do so effectively (e.g., Mischel, 1974, 2004). This section describes the psychological mechanisms underlying each of these components of self-control, drawing primarily from research on children's delay of gratification ability (Mischel, 1974; Mischel, Shoda, & Rodriguez, 1989). The early phases of this research, which we focus on next, examined differences among children in their willingness to choose more valuable but delayed rewards rather than less valuable but immediately available ones. The emphasis was thus on the *motivational* factors influencing people's initial choice to delay gratification.

Processes Involved in the Decision to Delay Gratification

According to a CAPS analysis, whether people initially choose to exercise self-control in a given situation begins with their encodings. Do they judge a situation to be personally relevant? Is it meaningful and does it warrant further attention? To the extent that the answers to these questions are affirmative, additional CAUs become activated that directly influence goal commitment. Starting in the late 1950s, Mischel and colleagues began to investigate the specific patterns of CAU activation that impact people's decision to exert self-control by examining differences in their preferences for valuable but delayed rewards versus less valuable but immediately available ones. For example, a person's decision to take $1 today rather than get $1.50 tomorrow.

In one early set of studies conducted with children on the island of Trinidad, large differences were first observed in choice behavior between East Indian and Black participants, with the former group often preferring the delayed reward and the latter group the immediately available reward (Mischel, 1961). These differences disappeared, however, when the effect of father absence was statistically controlled. Those children who came from homes with absent fathers were likely to have fewer experiences with male social agents who kept their promises. Consequently, the same children showed less trust (an "expectancy" in CAPS language) that the male experimenter would provide the promised delayed rewards. These findings highlight the role that outcome expectancies and beliefs play in goal commitment. To the extent that participants encoded a situation as requiring self-control, they were likely to attempt to exercise it only if they trusted that the rewards would materialize (and were also valued to them, a point addressed below; Mischel, 1974).

Subsequent studies have indicated that an additional type of expectancy influencing people's motivation to delay gratification are their self-efficacy beliefs—whether they believe they can successfully exert self-control (Bandura, 1986; Mischel, Cantor, & Feldman, 1996). For example, in one experiment, participants were given bogus success or failure feedback for their performance on a series of verbal reasoning problems (Mischel & Staub, 1965). They then had to make a choice between a highly valued reward, the receipt of which depended on their successful performance on a similar reasoning task, and a less preferred but noncontingent reward. Participants who were given false-positive feedback chose to work for the more preferred delayed reward significantly more often than individuals who were given false-negative feedback. Moreover, in a control group in which participants were given no feedback, pre-experimental success expectancies were a significant determinant of people's choices to work for contingent rewards. Thus, how well the participants felt that they could perform the task determined whether or not they chose to try for the more difficult but preferred reward. These findings are consistent with research indicating that individuals who perceive themselves as having little control over the situations they find themselves in often feel powerless and choose to not engage in self-regulation (Dweck, 1986; Seligman, 1975).

Of course, goal commitment does not depend solely on peoples' trust and self-efficacy expectations. It is also influenced by the subjective value of the rewards in the situation. Through temporal discounting mechanisms, rewards that are delayed have less value than equivalent rewards that are immediately available (Ainslie, 2001; Loewenstein, Reid, & Baumeister, 2002; Rachlin, 2000). Therefore, the longer the future rewards are delayed, the less likely it is that children will choose to wait for them (Mischel & Metzner, 1962). Thus goal commitment in delay of gratification is enhanced with the relative magnitude of the delayed reward and decreases as the required time it takes to attain the reward increases (Mischel, 1966, 1974).

Summary

The findings reviewed in this section highlight a specific pattern of CAU activation that influence people's decisions to commit to attempting to exert self-control. Specifically, consistent with utility theories, they indicate that the choice to wait for a larger but delayed reward is determined largely by an expectancy-value mechanism (Mischel & Ayduk, 2004). In short, a person must value the delayed reward enough to commit to pursuing it, must believe that they possess the ability to successfully exert self-control should they choose to do so, and must trust that they will receive the valued reward upon successfully fulfilling their goal.

Processes Involved in the Ability to Delay Gratification

Whereas early research on delay of gratification focused on determinants underlying the decision to exert self-control, later research focused on the psychological processes that enable children to successfully delay their immediate gratification once they have committed to the goal of doing so. This section discusses the attentional and cognitive mechanisms that underlie this ability, using findings from the classic preschool delay of gratification paradigm (Mischel et al., 1989).

The delay of gratification task is an experimental method that has become a prototype for studying issues related to self-regulatory competency in the laboratory. In this method, a young child is presented with a desired treat—for example, tiny pretzel sticks, or little marshmallows, or shiny poker chips. A dilemma is then posed: the child can wait until the experimenter returns and get two of the desired treats, or ring a bell and the experimenter returns immediately but the child gets only one treat. The child prefers the larger outcome and commits herself to wait for it. As waiting for the chosen goal drags on, the child becomes increasingly tempted to ring the bell and take the immediately available treat.

A choice conflict between waiting for two treats or settling for one immediately may seem far removed from the choices adults confront in their worlds, but for the young child this type of problem creates a genuine and powerful conflict. Empirically, performance on this task has been shown to predict a number of consequential life outcomes, such as self-control and self-regulation in goal pursuit decades later, suggesting that this paradigm is capable of tapping into the processes that are needed to exert self-control in a variety of domains. For example, the number of seconds children can wait in certain diagnostic situations (i.e., when no regulatory strategies are provided by the experimenter and children have to access their own competencies) predicts higher SAT scores, better personal and interpersonal competencies years later and higher cognitive control ability (e.g., Ayduk et al., 2000; Eigsti et al., 2006; Mischel, Shoda, & Peake, 1988; Shoda, Mischel, & Peake, 1990). Further, it seems to be a protective buffer against the negative consequences of such dispositional vulnerabilities as anxious rejection sensitivity (e.g., Ayduk et al., 2000; Ayduk et al., 2007). The delay task thus provides a method of studying the psychological processes underlying willpower systematically, a methodology that taps the types of self-regulatory skills that enable people to effectively control hot impulsive reactions, which are necessary for successful pursuit of life goals.

Attentional Processes

Early work using the delay of gratification paradigm tested alternative predictions concerning the role that attention plays in people's ability to delay immediate gratification. On the one hand, Freud (1911) argued that the transition from infantile id-driven impulsivity to ego control and delay of gratification begins when the young child creates a "hallucinatory wish-fulfilling" image of the delayed object. The assumption is that by imagining the desired object and "binding time," delaying immediate gratification becomes possible. Behaviorists of the time made similar predictions, but for very different reasons (*see* Mischel, 1974). Their research suggested that when animals learn, behavior toward a goal is maintained by "fractional anticipatory goal responses" that cognitively represent the desired rewards and sustain goal pursuit—for example, as the animal tries to find its way back to the food at the end of a maze in a learning task (Hull, 1931). Again, the prediction was that focusing attention on the delayed rewards—thinking about them— should reinforce one's ability to sustain delay gratification to fulfill goal pursuit.

To examine the role that attention on rewards plays in self-control, Mischel and Ebbesen (1970) conducted a series of experimental studies in which they varied whether reward items were available for attention while children were waiting in the delay of gratification paradigm. In one condition, children waited with both the immediately available and the delayed reward exposed in full view. In a second condition, both options faced the child but were concealed from attention by an opaque cover positioned over them. In two other conditions, either the delayed reward alone or the immediately available reward alone was exposed during the delay period. On average, children waited more than 11 minutes when none of the rewards were exposed but waited only a few minutes when any of the rewards—either both rewards, just the delayed reward, or just the immediately available reward—were available to attention. Directly contradicting the predictions coming from both the psychodynamic and animal learning traditions, the results showed that focusing attention on a desired stimulus decreased the ability to delay gratification.

If attention is what matters, Mischel and colleagues next reasoned then distracting children from focusing on the rewards should have the same effect as removing the rewards from attention. This is precisely what has been found. In one experiment, for example, children were provided with a distracting toy (e.g., a Slinky toy) while the rewards were exposed on the table in front of them (Mischel, Ebbesen, & Zeiss, 1972). In this condition, more than half of the children waited the full amount of time until the experimenter returned, indicating that the experiment was over (15 minutes). In contrast, none of the children who were left waiting for the exposed rewards without the distracter toy were able to do so. In another experiment, the same effect of distraction on delay times was found when children were instructed to think about fun thoughts while they waited by receiving the following instruction: "While you're waiting, if you want to, you can think of mommy pushing you on a swing at a birthday party." Similar to the Slinky condition, more than half of the children who were cued to distract themselves with fun thoughts waited until the experimenter returned and indicated that the experiment was over (Mischel et al., 1972).

Collectively, these findings indicate that diverting attention away from the appetitive features of a stimulus helps children delay gratification. Note, however, that not all distracters were equally effective. Unsurprisingly, when the distracting object was not appealing (e.g., instructing individuals to think about sad thoughts), then attention was diverted back to the stimulus and delay of gratification was undermined. To be effective in keeping attention away from the temptations in the situation, attention to the distracter must itself be reinforcing.

Reconstrual Processes

Although strategically focusing attention away from a desired stimulus is an effective way of facilitating self-control in the face of temptation, that option often is not available or not sustainable. Consider, for example, the dieting pastry chef who has sworn off eating chocolate yet has to make delectable flourless chocolate cakes for dessert each night (creating one potential temptation after another) and still attend carefully to each or risk his reputation. However, as the cognitive revolution more than 40 years ago has made clear, even when the realities of life prevent sustained distraction, stimulus control does not necessarily have to trump self-control.

As one early step in that revolution, in the late 1960s, Mischel and colleagues began to investigate the role that cognition—specifically the mental representation of a stimulus—plays in self-control, examining how alternative ways of mentally representing the stimulus effect its emotional and behavioral impact. They drew on a distinction that had been made in the research literature between two different aspects of a stimulus: its motivational, consummatory, arousing, action-oriented, or motivating "go" features and its informational, cognitive cue, or discriminative stimulus functions (Berlyne, 1960; Estes, 1972). Given this distinction, Mischel and Moore (1973) reasoned that when a child thinks about the rewards in front of them as "real," attention is placed on their hot, arousing, consummatory features, which should, in turn, elicit the motivating effects of the stimulus, making delay of gratification more difficult, and leading quickly to the "go" response (e.g., ring the bell, get the treat now). In contrast, they predicted that thinking about the rewards in terms of their cooler, more abstract features should allow the child to focus on the reward without activating consummatory trigger reactions. For example, mentally representing the rewards as pictures emphasizes their cognitive, informational features rather than their consummatory features. Therefore, Mischel and Moore speculated that this kind of "cool" mental transformation would reduce the conflict between wanting to wait and wanting to ring the bell by shifting attention away from the arousing features of the stimulus and onto their informative meaning (*see also* Trope & Liberman, 2003).

To test this prediction, Mischel and Moore (1973) presented one group of children in the delay of gratification task with slide-presented life-size pictures of the rewards, formally called "iconic representations." The hypothesis was that the pictures of the rewards would be relatively more abstract than the actual rewards, and thus the temptation to reach for them should be attenuated. These iconic representations were pitted against the presence of the real rewards themselves during the delay period. As predicted, exposure to the pictures of the images of the rewards significantly increased children's waiting time, whereas exposure to the actual rewards decreased delay time.

As with distraction, however, individuals need not rely exclusively on changing the external world in efforts at self-control. Moore, Mischel, and Zeiss (1976) faced children with actual rewards, but this time the participants were cued in advance by the experimenters to pretend that they were pictures by "putting a frame around them in your head" (Moore, Mischel, & Zeiss, 1976). In a second condition, the children were shown pictures of the rewards (no "real" rewards were present) but this time asked to think about them as if they were real. The findings indicated that children were able to delay almost 18 minutes when they pretended that the rewards facing them were pictures. In contrast, they were able to wait less than 6 minutes if they pretended that pictures in front of them were real rewards. As one child put in the postexperimental inquiry when asked how she was able to wait so long : "You can't eat a picture."

The role that cognitive reconstrual processes play in facilitating delay of gratification was further demonstrated clearly in an early study by Mischel and Baker (1975) using a different type of cognitive manipulation to transform the mental appraisal or representation of the rewards. In this study, children were cued to represent the rewards available in front of them in either cool, informational or hot, consummatory features. For example, children in the cool focus condition who were waiting for marshmallows were cued (or "primed" in current terminology) to think of them as "white, puffy clouds." Those waiting for pretzels were cued to think of them as "little, brown logs." In the hot ideation condition, the instructions cued children to think about the marshmallows as "yummy and chewy" and the pretzels as "salty and crunchy." As expected, when children thought about the rewards in hot terms, they were able to wait only 5 minutes, whereas when they thought about them in cool terms, delay time increased to 13 minutes.

Summary

Collectively, the findings reviewed in this section highlight the critical role that mental representations plays in determining the outcomes of self-regulatory efforts. Specifically, although different types of cognitive and attentional deployment strategies were used to facilitate self-control in the delay of gratification research, they shared a key characteristic. Namely, they replaced mental representations of rewards that were emotionally "hot" and difficult to resist with alternative representations that were "cool" and did not elicit impulsive trigger reactions, thus enabling persistence for delayed but larger outcomes.

Hot/Cool Systems Within Cognitive–Affective Processing System

To integrate findings from the delay of gratification research with recent advances in cognitive psychology and neuroscience, two closely interacting systems—a cognitive "cool" system and an emotional "hot" system—have been proposed as components of the broader CAPS framework

(Mischel & Ayduk, 2004; Metcalfe & Mischel, 1999). The interactions between these systems underlie people's ability—or inability—to exert self-control in general and sustain effortful control in pursuit of delayed goals in particular.

In the hot/cool framework, the cool system is a "know" system. Attuned to the informational, cognitive, and spatial aspects of stimuli, the cool system consists of a network of informational, *cool nodes* that are elaborately interconnected to each other and generate rational, reflective, and strategic behavior. The cool system is the basis of self-regulation and self-control. In contrast, the hot system is conceptualized as a "go" system: emotional, simple, reflexive, and fast. It is tuned biologically to be responsive to innate releasing stimuli—both negative and positive—that elicit automatic, aversive, fear-and-flight reactions or appetitive and sexual approach reactions. It consists of relatively few representations or *hot spots*, which, when activated by trigger stimuli, elicit virtually reflexive avoidance and approach reactions. Impulsive and automatic, the hot system is the basis of emotionality. It undermines rational attempts at self-control. The two systems continuously interact, such that as one becomes accentuated, the other becomes attenuated.

Given the relative paucity of research available on the neural correlates of self-control at the time that the hot/cool model was first suggested, Metcalfe and Mischel (1999) deliberately eschewed making strong claims about the differences in brain structures underlying the operation of each system. Since then there has been an explosion of research in this area (for review, *see* Kross & Ochsner, this handbook; *see also* Lieberman, 2007; Ochsner & Gross, 2007). Although it is outside the scope of this paper to review this emerging field, collectively the findings suggest that the amygdala—a small, almond-shaped region in the forebrain thought to enable fight or flight responses—is critically involved in hot system processing (Gray, 1982, 1987; LeDoux, 1996; Metcalfe & Jacobs, 1996, 1998). This brain structure reacts almost instantly to stimuli that individuals perceive as arousing (Adolphs et al., 1999; LeDoux, 1996, 2000; Phelps et al., 2001; Winston et al., 2002), immediately cueing behavioral, physiological (autonomic), and endocrine responses. It mobilizes the body for action, readying it to fight or flight in response to a perceived threat. The cool system, in contrast, seems to be associated with prefrontal and cingulate systems involved in cognitive control and executive function (e.g., Jackson et al., 2003; Ochsner & Gross, 2007).

Interactions between Systems

According to the hot/cool model, cognition and affect operate in continuous interaction with one another in producing phenomenological experiences and behavioral responses (for closely related opponent process models see Epstein, 1994; Lieberman, Gaunt, Gilbert, & Trope, 2002). In this model, hot spots and cool nodes that have the same external referent are directly connected to one another, and thus link the two systems (Metcalfe & Mischel, 1999; *see also* Metcalfe & Jacobs, 1996, 1998). Hot spots can be evoked by activation of corresponding cool nodes; alternately, hot representations can be cooled through intersystem connections to the corresponding cool nodes. For example, instructing a child to think vividly about the taste of a marshmallow can heighten a craving response. Alternatively, hot representations can be cooled through cool system cognitive processes (e.g., attention switching, reconstrual). Thus children can also attenuate their craving response as they look at an appetizing cookie by focusing their attention on its shape, rather than its appetitive features, or by reconstruing it in ways that lead it to be less appealing—for example, imagining that the chocolate chips aren't chocolate but, rather, specks of dirt. Willpower becomes possible to the extent that cooling strategies generated by the cognitive cool system circumvent hot system activation.

Evidence supporting the idea that the two systems directly interact with each other in the manner proposed by Metcalfe and Mischel (1999) has come from a recent neuro-imaging study that used functional magnetic resonance imaging (fMRI) to examine the regions of neural activity that become active when people consciously regulate their emotional responses to viewing aversive images (e.g., a picture of a bloody corpse). In this study, Ochsner and colleagues (2004) found that when participants were instructed to cognitively reappraise potent negative stimuli in ways that decreased their negative response (i.e., imagine that the image is fake), activity in cingulate and prefrontal systems (regions thought to be involved in cool system cognitive processing) increased, whereas activity in the amygdala (a hot system structure according to the current analysis) decreased. In contrast, when participants were told to reinterpret the aversive images in ways that made them feel worse, activity in both the amygdala and prefrontal/cingulate systems increased, suggesting that individuals were using cognitive reappraisal processes to enhance their negative feelings.

Factors Influencing the Balance of Hot/Cool System Activation

Several factors influence the balance of hot/cool system processing, the first of which is the developmental level of the individual. The hot system develops and dominates early in life, whereas the cool system develops later (by age 4) and becomes increasingly dominant over the course of development. These developmental differences are consistent with evidence on the differential rates of development of the relevant brain areas for these two systems (for reviews, see Eisenberger et al., 2004; Rothbart, Ellis, & Posner, 2004). Consequently, early in development, young children are primarily under stimulus control, as they have not yet developed the cool system structures needed to regulate hot system processing. As the cool system develops over time, it becomes increasingly possible for children to generate cooling strategies to regulate impulses (Mischel et al., 1989). Empirical evidence from the delay of gratification research is consistent with these expectations. For example, whereas delay of gratification is virtually impossible for children younger than age 4 years (Mischel, 1974), by age 12 years almost 60% of children in some studies were able to wait the duration of the period to receive the awaited reward (25 minutes maximum; Ayduk et al., 2000, Study 2).

In the context of the impulsive responses and emotional reactions that fully developed adults commonly face, perhaps the most important determinant of hot/cool system balance is stress. At high levels, stress deactivates the cool system and creates hot-system dominance. At lower levels of stress, complex thinking, planning, and remembering are possible. When stress levels jump from low to very high, responding tends to be reflexive and automatic. Under conditions in which an animal's life is threatened, quick responses driven by innately determined stimuli may be essential. At the same time, such automatic reactions undo rational efforts at constructive self-regulation. Of course, whether a situation elicits stress in the first place depends critically on the individual's *encodings*. For example, the aspiring graduate student may interpret a rejection letter of her most recent submitted manuscript as a highly stressful event, whereas her tenured advisor may remain calm when exposed to the same rejection feedback. Therefore, to recognize when stress is likely to elicit hot system dominance and problematic responses, it is necessary to identify the specific types of psychological situations that individuals encode as stressful. As numerous recent studies have demonstrated, individuals are likely to differ both predictably and meaningfully in the way they encode such situations (Mischel & Shoda, 1995).

Chronic stress may serve not only to bias processing in terms of hot system functioning but also lead to physical changes that have direct implications for self-control ability. For example, recent studies have indicated that rats exposed to repeated stress demonstrate dendritic spine loss in medial prefrontal cortex (Brown, Henning, & Wellman, 2005; Radley et al., 2006, 2005)—a cellular feature of stress-related psychiatric disorders in which the prefrontal cortex is impaired—and dendritic spine growth in the amygdala (Mitra et al., 2005; Vyas, Bernal, & Chattarji, 2003)— a neuronal event that is thought to facilitate increased emotionality. In humans, severe and chronic stress (as in war and terror conditions) may result in dominant activation of the hot system as opposed to the cool system in ways that become relatively stable and difficult to reverse.

Hot/Cool System Interactions in Everyday Life

When considering how people can be helped to self-regulate adaptively in everyday life, there is an important caveat. In the real world, situations that require individuals to exert self-control often involve both strategic cooling processes that enable people to remain calm and reflective in the face of impulsive responses. But they also need strategic heating processes to maintain the motivation and commitment for pursuing goals rather than quitting. To illustrate, a study by Peake and colleagues (2002) examined the role of attention while delaying gratification when children were required to work to complete a task (rather than passively wait) to get the larger, delayed reward. They found that when the task was interesting (i.e., feeding a toy bird with marbles), deploying attention to the rewards was detrimental. Under such conditions, focusing attention on rewards disrupted a fun and engaging distracter, thus undermining delay of gratification ability. However, when the task was not engaging (i.e., sorting marbles into cups according to color), flexibly focusing attention on the rewards by glancing back and forth at them facilitated delay of gratification. Such flexible strategic attention deployment presumably served to remind children of why they were engaging in the nonengaging task (to get the desired treats), thus motivating instrumental work. Such flexibility in attention deployment is consistent with the idea that it is the balanced interactions between the hot and cool systems that sustain delay of gratification, as they exert their motivating and cooling effects in tandem (*see also* Mischel et al., 1989).

The critical importance of achieving a balance between hot/cool system processing in everyday life was further supported in a study by Bonnano and colleagues (2004) that examined how individual differences in the ability to flexibly suppress and enhance emotional expression were prospectively linked to adjustment across the first 2 years of college. In this study, people who possessed both of these abilities, and were thus able to both heat up and cool down their emotions, as the situations they found themselves in demanded it, demonstrated the best long-term adjustment (lower levels of distress 2 years after the 9/11 attack). In contrast, participants who were low in both abilities displayed the poorest long-term adjustment.

HOT/COOL PROCESSES IN NEGATIVE EMOTION REGULATION

Although the hot/cool framework is based largely on findings from research using the delay of gratification paradigm, new research is extending the predictions that derive from it to a wide variety of contexts in which people must cope with automatically triggered negative emotional reactions (for reviews, *see* Kross, 2009; Mischel, DeSmet, & Kross, 2006). For example, situations in which a person experiences a transgressor and quickly responds with anger and aggression or situations in which people become anxious and respond with fear related avoidance responses. This research was driven by the assumption that the dilemma activated in the delay of gratification studies, in which a child was required to wait for a delayed treat while facing an immediate temptation, had a basic similarity to many of the situations people regularly experience when trying to go from good intentions to actual self-control efforts in everyday life. Consider, for example, the resolution to maintain relationship harmony that becomes easily sabotaged by the explosion of anger, hostility, and jealousy that erupts virtually automatically. It is in the heat of the moment that the need to inhibit hot, automatic but potentially destructive reactions becomes both most important and most difficult.

Hot/Cool Processes Involved in Coping with Interpersonal Conflict

Preliminary evidence suggesting that the processes involved in delaying gratification also help people regulate automatically triggered defensive emotional reactions comes from work examining the role of delay ability in the context of rejection sensitivity (RS; Downey & Feldman, 1996). Rejection sensitivity is a chronic processing disposition characterized by anxious expectations of rejection. These expectations stem from prior rejection experiences and get activated when people encounter interpersonal situations in which rejection is possible. In such situations, people who are high in RS (HRS) feel threatened, leading to the activation of their defensive, fight-or-flight systems (Downey et al., 2004; Kross et al., 2007). Attention narrows on detection of threat-related cues, which in turn makes the HRS person ready to see the actualization of the threatening outcome. Anticipation of threat also creates action readiness so that people high in RS are likely to react automatically, defensively, and intensely when the threat is experienced. Unsurprisingly, when HRS people perceive rejection, they respond to it with hostility and aggression as well as depression and withdrawal symptoms (Ayduk et al., 1999; Ayduk, Downey, & Kim, 2001; Downey et al., 1998). These negative behaviors, in turn, elicit actual rejection from their partners, leading to a self-fulfilling prophecy, and their romantic relationships are likely to end sooner than people who are low in RS (Downey et al., 1998).

According to the hot/cool model, effective coping in threatening interpersonal contexts among HRS individuals should involve cooling the "hot," emotional features associated with the situation (Arriaga & Rusbult, 1998; Mischel et al., 1989). One study exploring these links was an adult follow-up of the participants who had participated in one of the original Bing delay of gratification studies (Ayduk et al., 2000). This study showed that among HRS individuals, the number of seconds that participants were able to wait as preschoolers in the delay situation predicted their adult resiliency against the potentially destructive effects of RS. That is, HRS adults who had high delay ability in preschool had more positive functioning (high self-esteem, self-worth, and coping ability) compared to similarly high RS adults who were not able to delay in preschool. High RS participants showed higher levels of cocaine/crack use and lower levels of education than those low in RS unless they were good delayers in preschool. In contrast, high RS people who had high delay ability in preschool had relatively lower levels of drug use and higher educational levels and in these respects were similar to low RS participants (*see also* Ayduk et al., 2007).

A similar pattern of results was found in a second study with middle school children. Specifically, whereas high RS children with low delay ability were more aggressive toward their peers and thus had less positive peer relationships than children low in RS, high RS children who were able to delay longer were even less aggressive and more liked than low RS children. Similarly, a cross-sectional study of pre-adolescent boys with behavioral problems characterized by heightened hostile reactivity to potential interpersonal threats showed that the spontaneous use of cooling strategies in the delay task (i.e., looking away from the rewards and self-distraction) predicted reduced verbal and physical aggression (Rodriguez et al., 1989).

Hot/Cool Process Involved in Coping with Intrapersonal Conflict

Many of the most troubling self-regulatory challenges do not involve struggling to get along with other people but, rather, battling to resolve intrapsychic conflicts occurring in one's own mind. The rejected lover in a romantic relationship may desperately want to "work through" his feelings to move on but instead finds himself continually brooding and ruminating. The mere thought of the recalled experience easily triggers a cascade of negative responses that make it difficult to think calmly without losing control (Nolen-Hoeksema, 1991). The question here is: How can a person focus on a painful experience in a cool way, so that it can be worked through without becoming overwhelmed with hot system activation and negative feelings?

To address this question, guided by the Hot/Cool framework, Kross, Ayduk, and Mischel (2005) proposed that a critical factor determining whether people's attempts to adaptively work through negative experiences succeed or fail is the *type of self-perspective* they adopt. Prior research indicates that when people focus on negative past experiences, they typically do so from a *self-immersed* perspective in which self-relevant events and emotions are experienced in the first-person, through one's own eyes (Nigro & Neisser, 1983). Drawing from this literature,

Kross et al. (2005) hypothesized that when individuals focus on negative feelings from a self-immersed perspective, episodic information concerning the specific chain of events (i.e., what happened?) and emotions experienced (i.e., what did I feel?) would become accessible (cf., McIsaac & Eich, 2004), serving to increase negative affect.

Focusing on negative experiences from a self-immersed perspective is not, however, the only vantage point people can adopt while thinking about past events. As James (1890) suggested long ago, and many others have since examined (e.g., Leary, 2002; Libby & Eibach, 2002; McIsaac & Eich, 2004; Nigro & Neisser, 1983; Pronin & Ross, 2006; Robinson & Swanson, 1993), experiences can also be focused on from a self-distanced perspective, in which the individual becomes an observer of the self. Kross et al. (2005) predicted that adopting a self-distanced perspective to analyze negative feelings would reduce people's tendency to reflexively *recount* what happened to them and instead allow them to *reconstrue* their experience in cool ways that reduce its aversiveness (Metcalfe & Mischel, 1999; Mischel, 1974; *see also* Gross, 2001, and Lazarus, 1991).

These hypotheses were supported in a set of studies that manipulated the type of self-perspective (self-immersed vs. self-distanced) participants adopted while focusing on the reasons underlying feelings associated with anger-related interpersonal experiences (Kross et al., 2005). Specifically, when participants analyzed their feelings from a *self-immersed* perspective (immersed-analysis from hereon), episodic information concerning the specific chain of events (e.g., "*He told me to back off;*" "*I remember watching her cheat on me...*") and emotions experienced (e.g., "*I was so angry...*") became more accessible. In contrast, participants who analyzed their feelings from a *self-distanced* perspective (distanced-analysis from hereon) focused relatively less on what happened to them (i.e., recounting) and relatively more on reconstruing the event (e.g., "*I understand why the fight happened;*" "*It might have been irrational but I understand his motivation now.*"). Moreover, this shift in the content of peoples' thoughts about their past experiences (less recounting, more reconstruing) mediated the effect of the perspective manipulations on negative affect (*see also* Strack, Schwarz, & Gschneidinger, 1985). Thus, the more reconstruing and less recounting participants engaged in, the less negative affect they displayed.

These findings provided initial clues about the processes that enable people to analyze negative experiences without becoming overwhelmed but also raised a number of additional questions. For example, do these different ways of analyzing negative experiences impact people on the physiological level? To the extent that distanced-analysis reduces emotional reactivity, we predicted that these manipulations would influence autonomic nervous system reactivity. To test this prediction, Ayduk and Kross (2008) randomly assigned participants to analyze a recent anger experience from either a self-immersed or self-distanced perspective while their blood pressure levels were continuously monitored. Consistent with predictions, participants in the distanced-analysis group displayed significantly lower levels of blood pressure reactivity (relative to baseline). This was found both during the experiment, when they were explicitly instructed to analyze their feelings, and 20 minutes after the experiment was over, during a recovery period (Ayduk & Kross, 2008; also see Ayduk & Kross, in press).

Another question raised by our initial studies concerned the incremental utility of distanced-analysis compared to distraction. Distracting individuals from thinking about negative feelings is an extremely effective means of cooling negative affect in the short-term (e.g., Nolen-Hoeksema 1991). It therefore provides a "gold standard" to compare distanced-analysis against. Motivated by the delay of gratification findings, which indicated that cognitive reconstrual strategies are at least as effective as distraction in facilitating impulse control (Mischel, Shoda, & Rodriguez, 1989; Mischel & Rodriguez, 1993), Kross and Ayduk (2008) hypothesized that distanced-analysis would be as effective in reducing negative affect as distraction. Findings were consistent with this prediction. Whereas both distraction and distanced-analysis led to significantly lower levels of negative affect relative to immersed-analysis, distraction and distanced-analysis participants displayed the same relatively low levels of negative affect (Kross & Ayduk, 2008).

Work in our lab also has begun to address the long-term effects of these perspective-taking manipulations. To the extent that a person's memory of a negative experience has been adaptively "processed" as a function of some psychological intervention (i.e., the individual's memory of a negative experiences is altered in ways that reduce its aversiveness), prior research suggests that the individual should display lower levels of emotional reactivity when that memory becomes reactivated in the future (Foa & Kozak, 1986; Rachman, 1980). To examine whether distanced-analysis facilitates such adaptive emotional processing, two short-term longitudinal experiments were conducted (Kross & Ayduk, 2007). During Session 1 of each study, participants recalled a negative experience and were then randomly assigned to an immersed-analysis, distanced-analysis, or distraction condition. Participants were then asked to return to the lab either 24 hours (Study 1) or 7 days (Study 2) later for additional testing. During Session 2, all participants were instructed to recall and think about the same experience they thought about during Session 1 without receiving any additional instructions. They then indicated their current level of negative affect and the amount of time they spent thinking about their past experience between the two sessions. Findings indicated that whereas both distanced-analysis and distraction participants displayed lower levels of negative affect than immersed-analysis participants during Session 1, during Session 2 only distanced-analysis participants were buffered against negative affect and even-related recurring negative thoughts (Kross & Ayduk, 2008).

Summary

Collectively, the findings reviewed in this section highlight the role that hot/cool processes play in self-control dilemmas. The dilemmas span situations ranging from those dealing with conflict and emotional turmoil in interpersonal situations to the intrapsychic conflicts commonly experienced in everyday life. Therefore, whether it is waiting for bigger cookies or marshmallows, dealing with the news that your partner no longer loves you, or the remembrance of a distressing, enraging event, the basic underlying processes involved in regulating impulses and emotional responses in a wide range of situations may be the same.

TOWARD AN INTEGRATED UNDERSTANDING OF SELF-CONTROL

Our primary goal in this chapter has been to help demystify the concept of willpower by describing some of the basic processes underlying the motivation to exert self-control and the ability to do so effectively. This section turns our discussion toward the future, and some of the research directions that, in our view, may be especially important for constructing a cumulative science of self-control. In describing these future directions, our goal is neither to be exhaustive nor pre-emptive nor prescriptive. Many exciting paths exist, and we simply point to some of the specific ways that our understanding of self-control may be enhanced.

Biological and Environmental Influences on Self-Control Ability

It is now well-known that biological predispositions (e.g., temperament) bias the development of self-regulatory ability in particular directions. However, their influences are constantly modulated by the affordances presented by cultural, social, and interpersonal contexts (Grigorenko, 2002; Mischel & Shoda, 1999). For example, children's "difficult" temperament is related to increased cortisol levels—a physiological marker of dysregulation—in the face of stress but only in the context of poor and unresponsive adult caring (for review, see Gunnar & Donzella, 2002). As such, as researchers from different disciplines continue to work together, one important issue will be to further unpack the biological and environmental influences that contribute to self-control ability and to specify how these developmental influences interact.

One particularly exciting direction for future work in this area concerns the genetic factors that play a role in the development of self-regulatory ability. Although gene association studies attempting to link genes to complex phenomena (e.g., self-control) remain in their infancy, there have been a number of recent advances trying to examine such gene–behavior relationships by focusing on subprocesses involved in self-control. Some studies have focused on the dopamine system—a neurotransmitter system that is believed to play an important role in interference resolution, which is an important feature of cognitive control (e.g., Grannon et al., 2000; Mehta et al., 2000; Brozoksi et al., 1979)—and have revealed associations between dopamine transporter and receptor genes and several clinical disorders related to self-control, such as attention deficit hyperactivity disorder (Cook et al., 1996; Gill et al., 1997; Swanson et al., 1999; 2000; LaHoste, 1996). The gains from these early gene association studies highlight the importance of interdisciplinary work. Without knowledge of a basic process involved in self-control, it would not have been possible to establish a gene–behavior link. The success of future work in this area thus hinges on the continued interaction of researchers operating at multiple levels.

Identifying the Cognitive and Neural Substrates of Cooling Strategies

There now exists much research identifying the different types of cooling strategies that help people regulate automatically triggered impulses and emotions (for reviews, see Gross, 1998; Mischel & Ayduk, 2004). It remains unclear precisely how these strategies compare and contrast in terms of the cognitive processes that underlie them. To illustrate, consider the case of distraction and reconstrual, as operationalized in the delay of gratification paradigm. Similar low-level cognitive processes (e.g., working memory, mental imagery, interference resolution, language) likely underlie the operation of both of these strategies, each of which has been shown to facilitate delay of gratification ability. For example, both strategies involve switching attention away from the appetitive features of a desired stimulus and then keeping that information out of mind. However these strategies also likely differ in a number of ways. Distraction, for example, which involves redirecting attention away from one stimulus and onto another one, is likely to recruit semantic processes to a lesser degree than reconstrual strategies that require people to actively re-represent how they think about a stimulus (i.e., imagine a marshmallow as a puffy white cloud rather than a gooey sweet treat).

Beyond enhancing our understanding of the processes involved in self-control, work examining this issue promises to have important clinical applications. To the extent that research can reliably identify the specific types of cognitive processes that underlie different types of cooling strategies and CAUs, it may be possible to improve the way individuals who experience self-control difficulties are treated and assessed. For example, self-control interventions could be tailored around assessing what specific types of executive functions people experience difficulty engaging in during situations that require self-control. Treatment then can focus on helping people to improve their ability to engage in those processes when they are needed. In this vein, fMRI and related brain-imaging techniques promise to play a valuable role. A good deal of research already links activity in specific networks of brain regions to cognitive and emotional processes involved in self-control (for reviews, see Lieberman, 2007; Ochsner & Gross, 2007). These techniques could provide researchers with tools for assessing specific deficits in self-control ability and monitoring the effectiveness of training interventions designed to improve people's skills.

From Basic to Applied: Self-Regulatory Training Interventions

The research reviewed in this chapter leaves us with large questions: Can young children be taught effectively and enduringly the skills needed to delay gratification? Could such interventions, particularly early in life, lead to the kinds of adaptive and protective longitudinal outcomes that have been associated with delay of gratification ability when it is assessed spontaneously? We already know that cooling strategies are experimentally modifiable both in children and adults (Ayduk et al., 2001; Kross et al., 2005; Mischel et al., 1989). Also, modeling effective control strategies can have positive consequences, generalizing to behavior outside of the lab in the short run for at least a month (Bandura & Mischel, 1965). We do not know whether and how socialization, education, and therapy can effectively be utilized to help individuals gain the necessary cooling competencies to make willpower more accessible over the life course. Given the dramatic long-term correlates of delay of gratification assessed in early childhood in the classic paradigm, the need for such research seems self-evident.

Relevant examples already exist for such intervention work, for example, by using implementation strategies and techniques (Gollwitzer, 1996; Mischel & Patterson, 1976). Implementation strategies connect general goals ("Don't eat the cookie") to a specific implementation intention ("If mommy says dinner is about to be served, don't eat any sweets"). In this way, a specific contingency (IF_____) is established that becomes connected to a specific planned response (THEN_____). Gollwitzer and colleagues (1996) have shown that creating implementation intentions helps ensure implementation of the plan by tying a hot trigger event to the intended response rather than the habitual response. Designing interventions of this sort can help automatize the way an individual responds to an impulsive response, altering the unique set of CAUs that become activated when people find themselves in situations requiring self-control, thus enabling them to quickly engage in the appropriate behavior. They also allow interventions to be focused around the specific types of psychological situations in which individuals experience difficulty in self-control, offering a more contextual approach to the treatment of self-regulatory deficits (Mischel & Shoda, 1995). With regard to delay of gratification, effective interventions are likely to require going much beyond teaching a child how to delay gratification in a few particular tasks. Rather, such programs would entail extensive rehearsal, planning, and generalization strategies for implementing the necessary self-regulatory action when it is needed in a variety of everyday life contexts.

Conclusion

In the opening of his chapter on the will over 100 years ago, William James (1890/1981) distinguished between wishing and willing thought into action. According to James, "Desire, wish, will, are states of mind which everyone knows and which no definition can make plainer...If with the desire there goes a sense that attainment is not possible we simply wish; but if we believe the end is in our power, we will that desired feeling, having, or doing shall be real...and real it presently becomes, either immediately upon the willing or after certain preliminaries have been fulfilled (p. 486)". As James noted, to transition from wishing to willing, certain preliminary conditions must be met. This chapter described some of these preliminary conditions. It focused on a set of cognitive and attentional processes that substantially enhance the ability to achieve self-control and showed the impressive long-term correlates of this ability. The chapter also outlined the broader CAPS framework in which these processes have been conceptualized, specifying some of the conditions under which the necessary psychological operations may be effectively implemented.

The demystification of "willpower" and human agency, and the development of an increasingly powerful analysis of the processes that underlie and undermine the human capacity to exert self-control, will long remain one of the great challenges within psychology and related disciplines. We hope this chapter has shown it is a challenge worth pursuing, with useful methods and models already in reach and with results that suggest some of the demystification is at least on the way.

Authors' Note

Preparation of this chapter was supported by a grant from the National Institute of Mental Health (MH39349) and National Science Foundation (BCS-0624262). Correspondence concerning this article should be addressed to Ethan Kross, University of Michigan, Department of Psychology and Research Center for Group Dynamics, 530 Church Street, Ann Arbor, MI 48109-1109 or Walter Mischel,

Columbia University, Department of Psychology, 1190 Amsterdam Avenue, Mail Code 5501, New York, NY 10025. Electronic mail may be sent to ekross@psych.columbia.edu or to wm@psych.columbia.edu

References

Adolphs, R., Tranel, D., Hamann, S., et al. Recognition of facial emotion in nine individuals with bilateral amygdala damage. Neuropsychologia 1999; 37: 1111–1117.

Ainslie, G. *Breakdown of will.* Cambridge: Cambridge University Press, 2001.

Arriaga, X. B., & Rusbult, C. E. Standing in my partner's shoes: Partner perspective taking and reactions to accommodative dilemmas. Pers Soc Psychol Bull 1998; 24(9): 927–948.

Ayduk, O., Zayas, V., Downey, G., Cole, A. B., Shoda, Y., & Mischel, W. Rejection sensitivity and executive control: Joint predictors of borderline personality features. 2007.

Ayduk, O., Downey, G., & Kim, M. Rejection sensitivity and depressive symptoms in women. Pers Soc Psychol Bull 2001; 27: 868–877.

Ayduk, O., Downey, G., Testa, A., Yen, Y., & Shoda, Y. Does rejection sensitivity elicit hostility in rejection sensitive women? Soc Cogn 1999; 17: 245–271.

Ayduk, O., & Kross, E. Enhancing the pace of recovery: Self-distanced analysis of negative experiences reduces blood pressure reactivity. Psychol Sci 2008; 9(3): 229–231.

Ayduk, O., & Kross, E. From a distance: Implications of spontaneous self-distancing for adaptive self-reflection. J Pers Soc Psychol in press.

Ayduk, O., Mendoza-Denton, R., Mischel, W., Downey, G., Peake, P., & Rodriguez, M. L. Regulating the interpersonal self: Strategic self-regulation for coping with rejection sensitivity. J Pers Soc Psychol 2000; 79(5): 776–792.

Bandura, A. *Social foundations of thought and action: A social cognitive theory.* Englewood Cliffs, NJ: Prentice Hall, 1986.

Bandura, A., & Mischel, W. Modification of self-imposed delay of reward through exposure to live and symbolic models. J Pers Soc Psychol 1965; 2(5): 698–705.

Berlyne, D. *Conflict, arousal, and curiosity.* New York: McGraw Hill, 1960.

Bonanno, G. A., Papa, A., Lalande, K., Westphal, M., & Coifman, K. The importance of being flexible: The ability to enhance and suppress emotional expression predicts long-term adjustment. Psychol Sci 2004; 157: 482–487.

Brown, S. M., Henning, S., & Wellman, C. L. Mild, short-term stress alters dendritic morphology in rat medial prefrontal cortex. Cereb Cortex 2005; 15: 1714–1722.

Brozoski, T. J., Brown, R. M., Rosvold, H. E., & Goldman, P. S. Cognitive deficit caused by regional depletion of dopamine in prefrontal cortex of rhesus monkey. Science 1979; 205(4409): 929–932.

Cook, E. H., Jr., Stein, M. A., Krasowski, M. D., et al. Association of attention-deficit disorder and the dopamine transporter gene. Am J Hum Genet 1995; 56(4): 993–998.

Downey, G., & Feldman, S. Implications of rejection sensitivity for intimate relationships. J Pers Soc Psychol 1996; 70(6): 1327–1343.

Downey, G., Freitas, A., Michealis, B., & Khouri, H. The self-fulfilling prophecy in close relationships: Do rejection sensitive women get rejected by romantic partners? J Pers Soc Psychol 1998; 75(2): 545–560.

Downey, G., Mougios, V., Ayduk, O., London, B. E., & Shoda, Y. Rejection sensitivity and the defensive motivational system: Insights from the startle response to rejection cues. Psychol Sci 2004; 15: 668–673.

Dweck, C. S. Motivational processes affecting learning. Am Psychol 1986; 41: 1040–1048.

Eigsti, I., Zayas, V., Mischel, W., et al. Predictive cognitive control from preschool to late adolescence and young adulthood. Psychol Sci 2006; 17: 478–484.

Eisenberger, N., Smith, C. L., Sadovsky, A., & Spinrad, T. L. Effortful control: Reactions with emotion regulation, adjustment, and socialization in childhood. In: Baumeister, R. F., & Vohs, K. D. (Eds.), *Handbook of self regulation.* New York: Guilford Press; 2004.

Epstein, S. Integration of the cognitive and psychodynamic unconscious. Am Psychol 1994; 49: 709–724.

Estes, W. K. Reinforcement in human behavior. Am Sci 1972; 60(6): 723–729.

Foa, E. B., & Kozak, M. J. Emotional processing of fear: Exposure to corrective information. Psychol Bull 1986; 99: 20–35.

Freud, S. Formulations regarding the two principles of mental functioning. In: *Collected Papers, Vol. IV.* New York: Basic Books; 1911/1959.

Gill, M., Daly, G., Heron, S., Hawi, Z., & Fitzgerald, M. Confirmation of association between attention deficit hyperactivity disorder and a dopamine transporter polymorphism. Mol Psychiatry 1997; 2(4): 311–313.

Gollwitzer, P. M. The volitional benefits of planning. In: Gollwitzer, P. M., & Bargh, J. A. (Eds.), *The psychology of action: Linking cognition and motivation to behavior.* New York, NY: Guilford Press; 1996.

Granon, S., Passetti, F., Thomas, K. L., Dalley, J. W., Everitt, B. J., & Robbins, T. W. Enhanced and impaired attentional performance after infusion of D1 dopaminergic receptor agents into rat prefrontal cortex. J Neurosci 2000; 20(3): 1208–1215.

Gray, J. A. *The neuropsychology of anxiety.* London: Oxford, 1982.

Gray, J. A. *The psychology of fear and stress*, 2nd ed. New York, NY: McGraw-Hill, 1987.

Grigorenko, E. L. In search of the genetic engram of personality. In: Mischel, W. (Ed.), *Advances in personality science.* New York, NY: Guilford Press; 2002: pp. 29–82.

Gross, J. J. The emerging field of emotion regulation: An integrative review. Rev Gen Psychol 1998; 2: 271–299.

Gross, J. J. Emotion regulation in adulthood: Timing is everything. Curr Dir Psychol Sci 2001; 10: 214–219.

Gunnar, M. R., & Donzella, B. Social regulation of the cortisol levels in early human development. Psychoneuroendocrinology 2002; 27: 199–220.

Higgins, E. T. Personality, social psychology, and person-situation relations: Standards and knowledge activation as a common language. In: Pervin, L. A. (Ed.), *Handbook of personality: Theory and research.* New York, NY: The Guilford Press; 1990: pp. 301–338.

Hull, C. L. Goal attraction and directing ideas conceived as habit phenomena. Psychol Rev 1931; 38: 487–506.

Jackson, D. C., Muller, C. J., Dolski, I., et al. Now you feel it, now you don't: Frontal brain electrical asymmetry and individual differences in emotion regulation. Psychol Sci 2003; 14: 612–617.

James, W. *The principles of psychology*, Vol. 2. Cambridge, MA: Harvard University Press, 1890/1981.

Kross, E. When the self becomes other: Towards an integrative understanding of the processes distinguishing adaptive self-reflection from rumination. Ann N Y Acad Sci 2009; 1167: 39–40.

Kross, E., & Ayduk, O. Facilitating adaptive emotional analysis: Distinguishing distanced-analysis of depressive experiences from immersed-analysis and distraction. Pers Soc Psychol Bull 2008.

Kross, E., Ayduk, O., & Mischel, W. When asking "why" does not hurt: Distinguishing rumination from reflective processing of negative emotions. Psychol Sci 2005; 16: 709–715.

Kross, E., Egner, T., Downey, G., Ochsner, K., & Hirsch, J. Neural dynamics of rejection sensitivity. J Cogn Neurosci 2007; 19: 945–956.

LaHoste, G. J., Swanson, J. M., Wigal, S. B., et al. Dopamine D4 receptor gene polymorphism is associated with attention deficit hyperactivity disorder. Mol Psychiatry 1996; 1(2): 121–124.

Lazarus, R. S. *Emotion and adaptation.* New York: Oxford University Press, 1991.

Leary, M. The self and emotion: The role of self-reflection in the generation and regulation of affective experience. In: Davidson, R., Scherer, K., & Goldsmith, H. (Eds.), *Handbook of affective sciences.* New York, NY: Oxford University Press, 2002: pp. 773–786.

LeDoux, J. *The Emotional Brain.* New York, NY: Simon & Schuster, 1996.

LeDoux, J. Emotion circuits in the brain. Ann Rev Neurosci 2000; 23: 155–184.

Libby, L., & Eibach, R. Looking back in time: Self-concept change affects visual perspective in autobiographical memory. J Pers Soc Psychol 2002; 82: 167–179.

Lieberman, M. D., Gaunt, R., Gilbert, D. T., & Trope, Y. Reflection and reflexion: A social cognitive neuroscience approach to attributional inference. In: Zanna, M. (Ed.), *Advances in experimental social psychology.* New York, NY: Academic Press; 2002.

Lieberman, M. D. The X- and C-systems: The neural basis of automatic and controlled social cognition. In: Harmon-Jones, E., & Winkelman, P. (Eds.), *Fundamentals of social neuroscience.* New York, NY: Guilford; 2007: pp. 290–315.

Loewenstein, G., Read, D., & Baumeister, R. *Time and decision.* New York, NY: Russel Sage Foundation, 2002.

McIsaac, H. K., & Eich, E. Vantage point in traumatic memory. Psychol Sci 2004; 15: 248–253.

Mehta, A. D., Ulbert, I., & Schroeder, C. E. Intermodal selective attention in monkeys.

Metcalfe, J., & Jacobs, J. W. A "hot/cool-system" view of memory under stress. PTSD Res Quarter 1996; 7: 1–6.

Metcalfe, J., & Jacobs, J. W. Emotional memory: The effects of stress on "cool" and "hot" memory systems. In: Medin, D. L. (Ed.), *Psychology of learning and motivations*. San Diego, CA: Academic; 1998.

Metcalfe, J., & Mischel, W. A hot/cool system analysis of delay of gratification: Dynamics of willpower. Psychol Rev 1999; 106: 3–19.

Mischel, W. Preference for delayed reinforcement and social responsibility. J Abnorm Psychol 1961; 62: 1–7.

Mischel, W. Theory and research on the antecedents of self-imposed delay of reward. In: Maher, B. A. (Ed.), *Progress in experimental personality research (Volume 3)*. New York, NY: Academic Press; 1966.

Mischel, W. Cognitive appraisals and transformations in self-control. In: Weiner, B. (Ed.), *Cognitive views of human motivation*. New York, NY: Academic Press, 1974.

Mischel, W. Toward an integrative science of the person. Ann Rev Psychol 2004; 55: 1–22.

Mischel, W., & Ayduk, O. Willpower in a cognitive-affective processing system: The dynamics of delay of gratification. In: Baumeister, R. F., & Vohs, K. D. (Eds.), *Handbook of self-regulation*. New York, NY: Guilford Press, 2004.

Mischel, W., & Baker, N. Cognitive appraisals and transformations in delay behavior. J Pers Soc Psychol 1975; 31(2): 254–261.

Mischel, W., Cantor, N., & Feldman, S. Principles of self-regulation: The nature of willpower and self-control. In: Higgins, E. T., & Kruglanski, A. W. (Eds.), *Social psychology: Handbook of basic principles*. New York, NY: Guilford, 1996.

Mischel, W., DeSmet, A., Kross, E. Self regulation in the service of conflict resolution. In: Deutsch, M., Coleman, P. T., & Marcus, E. C. (Eds.), *Handbook of conflict resolution: Theory and practice*, 2nd ed. San Francisco, CA: Josey-Basss; 2006.

Mischel, W., & Ebbesen, E. B. Attention in delay of gratification. J Pers Soc Psychol 1970; 16(2): 239–337.

Mischel, W., Ebbesen, E. B., & Zeiss, A. R. Cognitive and attentional mechanisms in delay of gratification. J Pers Soc Psychol 1972; 21(2): 204–218.

Mischel, W., & Metzner, R. Preference for delayed reward as a function of age, intelligence, and length of delay interval. J Abnorm Soc Psychol 1962; 64(6): 425–431.

Mischel, W., & Moore, B. Effects of attention to symbolically-presented rewards on self-control. J Pers Soc Psychol 1973; 28(2): 172–179.

Mischel, W., & Patterson, C. J. Substantive and structural elements of effective plans for self-control. J Pers Soc Psychol 1976; 34: 942–950.

Mischel, W., & Shoda, Y., A cognitive-affective system theory of personality: Reconceptualizing situations, dispositions, dynamics and invariance in personality structure. Psychol Rev 1995; 102(2): 246–268.

Mischel, W., & Shoda, Y. Reconciling processing dynamics and personality dispositions. Ann Rev Psychol 1998; 49: 229–258.

Mischel, W., & Shoda, Y. Integrating dispositions and processing dynamics within a unified theory of personality: The cognitive-affective personality system (CAPS). In: John, O. P. (Ed.), *Handbook of personality: Theory and research*. New York, NY: Guilford Press; 1999: pp. 197–218.

Mischel, W., Shoda, Y., & Peake, P. The nature of adolescent competencies predicted by preschool delay of gratification. J Pers Soc Psychol 1988; 54(4): 687–696.

Mischel, W., Shoda, Y., & Rodriguez, M. L. Delay of gratification in children. Science 1989; 244: 933–938.

Mischel, W., & Staub, E. Effects of expectancy on working and waiting for larger rewards. J Pers Soc Psychol 1965; 2(5): 625–633.

Mitra, R., Jadhav, S., McEwen, B. S., Vyas, A., & Chattarji, S. Stress duration modulates the spatiotemporal patterns of spine formation in the basolateral amygdala. Proc Natl Acad Sci 2005; 102: 9371–9376.

Moore, B., Mischel, W., & Zeiss, A. R. Comparative effects of the reward stimulus and its cognitive representation in voluntary delay. J Pers Soc Psychol 1976; 34: 419–424.

Morf, C. C., & Mischel, W. Special issue: Self-concept, self-regulation, and psychological vulnerability. Self Ident 2002; 1: 103–199.

Nigro, G., & Neisser, U. Point of view in personal memories. Cogn Psychol 1983; 15: 467–482.

Nolen-Hoeksema, S. Responses to depression and their effects on the duration of depressive episodes. J Abnorm Psychol 1991; 100: 569–582.

Ochsner, K. N., & Gross, J. J. The neural architecture of emotion regulation. In: Gross, J. J., & Buck, R. (Eds.), *The handbook of emotion regulation*. New York, NY: Guilford Press; 2007: pp. 87–109.

Ochsner, K. N., Ray, R. D., Robertson, E. R., et al. For better or for worse: Neural systems supporting the cognitive down- and up-regulation of negative emotion. Neuroimage 2004; 23(2): 483–499.

Peake, P., Hebl, M., & Mischel, W., Strategic attention deployment for delay of gratification in waiting and working situations. Dev Psychol 2002; 38(2): 313–326.

Phelps, E. A., O'Connor, K. J., Gatenby, J. C., Gore, J. C., Grillon, C., & Davis, M. Activation of the left amygdala to a cognitive representation of fear. Nat Neurosci 2001; 4: 437–441.

Pronin, E., & Ross, L. Temporal differences in trait self ascription: When the self is seen as an other. J Pers Soc Psychol 2006; 90: 197–209.

Rachlin, H. *The science of self-control*. Cambridge, MA: Harvard University Press, 2000.

Rachman, S. Emotional processing. Behav Res Ther 1980; 18: 51–60.

Radley, J. J., Sisti, H. M., Hao, J., et al. Chronic behavioral stress induces apical dendritic reorganization in pyramidal neurons of the medial prefrontal cortex. Neuroscience 2004; 125: 1–6.

Radley, J. J., Rocher, A. B., Miller, M., et al. Repeated stress induces dendritic spine loss in the rat medial prefrontal cortex. Cereb Cortex 2006; 16(3): 313–320.

Robinson, J. A., & Swanson, K. K. Field and observer modes of remembering. Memory 1993; 1: 169–184.

Rodriguez, M. L., Mischel, W., & Shoda, Y. Cognitive person variables in the delay of gratification of older children at risk. J Pers Soc Psychol 1989; 57(2): 358–367.

Rothbart, M. K., Ellis, L. K., & Posner, M. I. Temperament and self-regulation. In: Baumeister, R. F., & Vohs, K. D. (Eds.), *Handbook of self-regulation*. New York, NY: Guilford Press; 2004.

Seligman, M. E. P. *Helplessness: On depression, development, and death*. San Francisco, CA: Freeman, 1975.

Shoda, Y., & Mischel, W. Personality as a stable cognitive-affective activation network: Characteristic patterns of behavior variation emerge from a stable personality structure. In: Miller, L. C. (Ed.), *Connectionist models of social reasoning and social behavior*. Mahwah, NJ: Erlbaum; 1998.

Shoda, Y., Mischel, W., & Peake, P. Predicting adolescent cognitive and self-regulatory competencies from preschool delay of gratification: Identifying diagnostic conditions. Dev Psychol 1990; 26(6): 978–986.

Shoda, Y., & Smith, R. Conceptualizing personality as a cognitive affective processing system: A framework for maladaptive behavior patterns and change. Behav Ther 2004; 35: 147–165.

Strack, F., Schwarz, N., & Gschneidinger, E. Happiness and reminiscing: The role of time perspective, affect, and mode of thinking. J Pers Soc Psychol 1985;49: 1460–1469.

Swanson, J. M., Sunohara, G. A., Kennedy, J. L., et al. Association of the dopamine receptor D4 (DRD4) gene with a refined phenotype of attention deficit hyperactivity disorder (ADHD): A family-based approach. Mol Psychiatr 1999; 3(1): 38–41.

Swanson, J., Oosterlaan, J., Murias, M., et al. Attention deficit/hyperactivity disorder children with a 7-repeat allele of the dopamine receptor D4 gene have extreme behavior but normal performance on critical neuropsychological tests of attention. Proc Natl Acad Sci USA 2000; 97(9): 4754–4759.

Trope, Y., & Liberman, N. Temporal construal. Psychol Rev 2003; 110: 403–421.

Vyas, A., Bernal, S., & Chattarji, S. Effects of chronic stress on dendritic arborization in the central and extended amygdala. Brain Res 2003; 965: 290–294.

Winston, J. S., Strange, B. A., O'Doherty, J., & Dolan, R. J. Automatic and intentional brain responses during evaluation of trustworthiness of faces. Nat Neurosci 2002; 5: 277–283.

PART III

Social

CHAPTER 24

Self-Control in Groups

John M. Levine, Kira M. Alexander, and Thomas Hansen

ABSTRACT

This chapter reviews theoretical and empirical work on self-regulation at the group level of analysis. We examine how groups exert control over their members and how members respond to these control efforts. The first section of the chapter focuses on groups as agents of control. Following a discussion of the functions that groups serve, we examine how groups use norms and roles to control their members. The second section of the chapter focuses on individuals as targets of control. Here we examine two opposing ways in which individuals respond to perceived group pressure: capitulation and resistance. We conclude by examining two implicit assumptions underlying our analysis of group control—that groups initiate control for their own ends and that members view such control as an unwelcome constraint. Using the example of social support groups, we discuss the relationship between self- and group-control when individuals seek group help in regulating their behavior.

Keywords: Norms, reaction to deviance, roles, group socialization, conformity, group decision making, group performance, social dilemmas, nonconformity, minority influence, collective action, social support groups

SELF-CONTROL IN GROUPS

As this volume indicates, self-control has emerged as a major research topic in social psychology as well as other disciplines (*see* Baumeister & Vohs, 2004). It is noteworthy that in contrast to the substantial theoretical and empirical work on the cognitive and affective underpinnings of self-control, relatively little attention has been paid to its social nature—the fact that self-control occurs in social contexts and has social causes and consequences (for exceptions, *see* Abrams, 1994; Leary, 2004; Postmes & Spears, 1998; Vohs & Ciarocco, 2004).[1] In particular, the issue of self-regulation at the group level has been neglected. To address this latter issue, it is necessary to shift one's analytic perspective in two related ways—by using the group rather than the individual as the unit of analysis and by focusing on interpersonal as well as intrapersonal processes.

This chapter reviews theoretical and empirical work on how groups exert control over their members and how members respond to these control efforts. First, we discuss groups as agents of control, examining the impact of norms and roles. Then, we consider individuals as targets of control, examining cases in which they capitulate to versus resist group pressure. Although our discussion is applicable to groups of varying sizes, we focus primarily on small groups whose members interact regularly, are behaviorally interdependent, have emotional ties, and share a common frame of reference.

Groups as Agents of Control

To understand *how* groups exert control over their members, it is first necessary to understand *why* they desire to exert this control in the first place. To answer this latter question, we need to know the functions that groups serve. Three generic group functions have been identified (Arrow, McGrath, & Berdahl, 2000). The first is completing group projects, which include generating ideas or plans; solving problems with correct answers or deciding issues involving preferences; resolving conflicts of viewpoint or interest; and performing tasks against objective standards and resolving conflicts of power (McGrath, 1984). The second generic function is fulfilling members' needs, which include survival needs (e.g., obtaining food and shelter, fending off enemies); psychological needs (e.g., developing intimate relations, avoiding loneliness); informational needs (e.g., accurately assessing the environment and one's ability to cope with it); and identity needs (e.g., positively differentiating one's group from other groups) (Baumeister & Leary, 1995; Forsyth, 1999; Mackie & Goethals, 1987). Finally, the third function, which underlies a group's ability to fulfill both of the previous functions, is remaining a viable, intact social unit (i.e., maintaining system integrity).

Arrow et al. (2000) argue that although groups vary in the relative priorities they assign to these three functions and these priorities change over time, all the functions must be fulfilled if a group is to function effectively. Moreover, the functions are interdependent, such that changes in one affect the others. For these reasons, "all groups exercise some degree of *self-regulation* (italics ours) over how they pursue their multiple functions over time" (Arrow et al., 2000, p. 48). This emphasis on the importance of self-regulation at the group level is compatible with Arrow et al.'s definition of groups as complex, adaptive, dynamic, and open systems. It suggests that self-regulation in groups can be understood in terms of the basic characteristics of systems in general, including efforts to achieve desired end states (or goals), ways of monitoring movement toward these end states, and mechanisms for taking corrective action if necessary (cf. Carver & Scheier, 1998). Taken as a whole, this analysis provides a persuasive case for viewing self-control as a group (as well as an individual) phenomenon. The basic assumption underlying our analysis is that groups engage in self-control to the extent that they perceive threats to their ability to carry out the three generic functions discussed above.

How are the self-control efforts of groups manifested in behavior? Following Arrow et al. (2000), we suggest that groups engage in two distinct (although often interrelated) kinds of control—one involving their internal dynamics and the other involving their external dynamics (i.e., their relationships with the environments in which they are embedded). The former involves controlling group composition (the number and type of people who belong to the group), group conflict (competition between members to obtain scarce resources, both tangible and intangible), and group performance (cooperation among members to create joint products and achieve common goals) (cf. Levine & Moreland, 1998). The latter involves controlling group ecology—the physical, social, and temporal environments in which the group operates. In this chapter, we focus on groups' efforts to control their internal dynamics (for an analysis of how groups deal with their environments, *see* Arrow et al., 2000).

In exerting control, groups can utilize various tools and resources that comprise "group structure." These include norms, roles (including leadership), status systems, friendship and communication networks, and culture (Levine & Moreland, 2006). Group structure can be conceptualized as a relatively enduring set of forces that constrain and regularize interactions among members and thus produce patterned relations among them (cf. McGrath, 1984). A comprehensive discussion of the origins and consequences of group structure is beyond the scope of this chapter. Instead, we focus on two aspects of structure that are particularly relevant to group control: norms and roles. In so doing, we examine how norms and roles function at a particular point in time (for discussions of norm and role change, see Arrow et al., 2000 and Cialdini & Trost, 1998).

Norms

According to Cialdini and Trost (1998), "Social norms are rules and standards that are understood by members of a group, and that guide and/or constrain social behavior without the force of laws" (p. 152). These authors distinguish between descriptive norms, which reflect how group members *do* behave, and injunctive norms, which reflect how group members *should* behave. Both kinds of norms influence members' behavior, but they do so in different ways. Descriptive norms, which signal behavior that is typical or routine, produce informational influence. Injunctive norms, which signal behavior that is correct or proper, produce normative influence (Deutsch & Gerard, 1955). Because injunctive norms reflect the group's desire to exert influence, they are of particular relevance to our present concerns. In analyzing the operation of injunctive norms, two fundamental questions need to be answered. First, what is the content of these norms—that is, what domains of member behavior do they cover? And second, how do groups deal with people who violate these norms?

Content of Injunctive Norms

Recent evolutionary analyses of group responses to deviance are relevant to the first question (e.g., Kurzban & Leary, 2001; Neuberg, Smith, & Asher, 2000). These analyses assume that because exploitative (i.e., selfish) behavior threatens group welfare, humans have evolved "mechanisms to identify individuals who threaten or hinder successful group functioning, to label them as such, to motivate group members to withhold group benefits from them, and to separate such individuals from the group if necessary" (Neuberg et al., 2000, p. 36). Certain behaviors pose "generic" threats that elicit punishment or exclusion from virtually all groups (cf. Neuberg et al., 2000; Stangor & Crandall, 2000). Chief among these are behaviors that directly threaten group effectiveness by revealing low ability and/or motivation. Other threatening behaviors include failing to honor basic social norms, such as reciprocity, commitment keeping, property rights, and fair exchange (Kerr, 1995), and being an "imposter" (i.e., claiming a group identity but violating the expectations of that identity) (Hornsey & Jetten, 2003). Finally, group effectiveness is threatened by failure to sacrifice personal welfare in favor of group welfare and to show ingroup loyalty, especially in the context of intergroup conflict. Disloyalty (and its most extreme form, treason) is generally viewed as the worst form of deviance and elicits the strongest punishment (Hogg, Fielding, & Darley, 2005; Levine & Moreland, 2002).

Evolutionary analyses suggest that normative violations need not pose direct threats to group welfare to elicit punishment. Even indirect threats are sufficient, as indicated by the hostility directed toward people who act in an unpredictable fashion (Kurzban & Leary, 2001) or who violate implicit "residual rules" regarding appropriate dress, cleanliness, interaction distance, and so on (Scheff, 1966). Moreover, similar reactions are triggered by "abnormal" bodily conditions that imply inability to reciprocate benefits or produce viable offspring (e.g., physical disabilities, deformities) or that signal danger of infection (e.g., skin disorders) (Neuberg et al., 2000; Thornhill & Gangestad, 1993).

One kind of indirect threat to group effectiveness—deviations from group consensus on opinion issues—has been of special interest to social psychologists. According to Festinger (1950), groups desire consensus because it validates opinions that are not based on physical reality (social reality motive) and facilitates the attainment of collective goals (group locomotion motive). As discussed below, these motives have a normative component in that individuals who deviate from group consensus on opinion issues elicit negative reactions.

The foregoing analysis should not be interpreted to mean that a given behavior is equally threatening to all groups under all circumstances. Instead, the threat value of a behavior depends on the perceived likelihood that it will seriously undermine the group's ability to achieve a valued goal. This likelihood judgment, in turn, can be influenced by aspects of the behavior, the individual who emits it, and the group context in which it occurs (Levine, 1989; Levine & Moreland, 2002). For example, holding constant the value of the group goal, a norm violation is likely to be perceived as more threatening if it clearly undermines the group's ability to achieve its goal, the violator seems unconcerned about the implications of his/her behavior, and the group is in a competitive relationship with another group. This line of argument suggests that norm violations are defined through an inferential process (Levine, 1989; see also Hogg et al., 2005; Stangor & Crandall, 2000). This process can vary in complexity, depending on the amount and clarity of available information and the group's motivation and ability to consider it. Further complications are introduced because the inferential process occurs at the group level and hence may be influenced by a variety of social processes, such as majority and minority influence, group polarization, and coalition formation (Levine, 1989).

Reaction to Norm Violations

Groups have a strong inclination to punish people who deviate from norms. A recent meta-analysis of the strength of a wide array of social psychological phenomena revealed a mean effect size of 0.60 for the finding that "people who deviate from a group are rejected by that group." This compares to mean effect sizes of 0.32 for 27 group phenomena and 0.21 for 474 social psychological phenomena (Richard, Bond, & Stokes-Zoota, 2003).

Much of the social psychological work on reaction to norm violations has involved opinion deviance. Festinger (1950) argued that the presence of opinion deviates in a group produces pressure toward uniformity, which stimulates communication designed to convince the deviates to adopt the group's modal position. If this communication fails, the group redefines its boundaries by rejecting the deviates. Festinger hypothesized, for example, that communication to deviates increases with their level of disagreement, their desirability as members, and the likelihood that they will succumb to influence. In addition, he predicted that group rejection of deviates increases with their level of disagreement, the relevance of the issue to group function, and group cohesiveness. These and related hypotheses were partially confirmed in studies conducted by Festinger and his colleagues (see reviews by Levine, 1989; Turner, 1991).

Group reaction to opinion deviance was also investigated in a number of later experiments (see Levine, 1989). For example, these studies showed that extreme deviates receive more communication and less favorable evaluation than moderate deviates (e.g., Hensley & Duval, 1976; Levine & Ranelli, 1978) and the reasons deviates offer for their position influence their evaluations (e.g., Levine & Ruback, 1980). Moreover, deviates shifting toward the majority position are generally better liked than those shifting away (e.g., Levine, Saxe, & Harris, 1976), and evaluations of shifting deviates are affected by the social pressures ostensibly acting on them (e.g., Levine, Sroka, & Snyder, 1977).

Other studies focused on deviates who interfered with the group's locomotion toward valued goals, rather than simply cast doubt on the validity of the group's modal opinion. These studies found that evaluations of such deviates varied inversely with their level of interference and perceived responsibility for it (e.g., Jones & deCharms, 1957; Miller & Anderson, 1979) and positively with their perceived desire to help the group (Hornsey, 2006). Deviates' status within the group also matters. People who conform to group norms and demonstrate task competence gain status (idiosyncrasy credit) that allows them to deviate later (Hollander, 1958; however, see Bray, Johnson, & Chilstrom, 1982). And, in many cases, groups expect their leaders to produce innovation and reward them when they deviate in ways that further group goals (e.g., Suchner & Jackson, 1976).

The group context in which deviance occurs also is important. For example, although most groups have norms that prohibit deviance (e.g., Janis, 1982), some groups tolerate or even encourage it (e.g., McAuliffe et al., 2003). Not surprisingly, deviates are liked better in the latter groups (e.g., Hornsey et al., 2006). In addition, some groups view certain kinds of deviance (e.g., performing exceptionally well; taking a position more extreme than the modal group position) in a positive light and hence reward it (e.g., Abrams et al., 2002; Blanton & Christie, 2003; Hogg et al., 2005). Finally, group members' desire for consensus, or collective cognitive closure, can affect their reaction to deviance (Kruglanski & Webster, 1991), as can their perception of other members' feelings toward deviates (Wheeler & Caggiula, 1966).

Social identity theory provides a useful perspective on reaction to opinion and other forms of deviance. Work in this tradition downplays the importance of face-to-face interaction and behavioral interdependence and instead emphasizes collective self-definition (Marques, Abrams, Paez, & Hogg, 2001a). Initial studies focused on the "black sheep effect"—the fact that unlikable ingroup members are evaluated more negatively than similar outgroup members (e.g., Marques, Yzerbyt, & Leyens, 1988) because doing so maintains a positive group identity (Marques & Paez, 1994; however, see Eidelman & Biernat, 2003). More recent studies are based on the Subjective Group Dynamics (SGD) model (see Abrams et al., 2005; Marques et al., 2001a). These studies indicate, for example, that ingroup deviates are rejected more by people who identify strongly with the ingroup (e.g., Branscombe et al., 1993; Hutchison & Abrams, 2003), particularly when their own prototypicality is threatened (Schmitt & Branscombe, 2001). Such rejection is also more likely when deviance occurs in an intergroup context (Matheson, Cole, & Majka, 2003; but see Marques & Yzerbyt, 1988), the status of the ingroup is insecure (Marques, Abrams, & Serodio, 2001b), and perceived consensus among ingroup members is low (Marques et al., 2001b).

Group reaction to norm violations can vary in form and intensity (Levine & Kerr, 2007). Recall that Festinger (1950) distinguished between communication directed toward changing the deviate's opinion and rejection of that person (see also Schachter, 1951; Marques et al., 2001b). Rejection, in turn, can take several forms, including isolation within the group, deprivation of normal group privileges, and expulsion from the group (see Israel, 1956; Eidelman, Silvia, & Biernat, 2006; Williams, 2007). In addition, group members can employ a wide range of specific behaviors to produce change or signal rejection (e.g., persuasive messages or promises/threats in the former case; failure to acknowledge the deviate's presence or physical punishment in the latter case). When confronting deviates, then, the group must make a decision about which behavior(s) to use. Such decisions are based on the salience of particular rules for dealing with deviates (e.g., start with mild tactics and escalate if necessary), as well as the group's resources and goals. Some resources for dealing with deviates may be available and easy to use (e.g., emitting derisive laughter whenever deviates speak), whereas others may be unavailable or available but difficult to use (e.g., reducing salaries of tenured faculty members). And different group goals (e.g., modifying deviates' behavior, intimidating would-be dissenters, signaling to outsiders that the group has high standards) may lead to the use of different behaviors.

Our discussion of injunctive norms has emphasized broad norms that exist in virtually all groups and apply to all members. Other norms, however, are narrower in that they apply to only some members of virtually all groups (e.g., new members, leaders) or to only some members of some groups (e.g., anesthesiologists on surgical teams). Such specialized normative expectations are often labeled role expectations.

Roles

Roles can be defined as differentiated patterns of behavior exhibited by people occupying particular positions in a group (McGrath, 1984). Role expectations are the duties and obligations of occupants of one role vis-à-vis occupants of complementary roles (Sarbin & Allen, 1968). Thus, role expectations serve as normative standards for how subsets of group members should behave toward one another (e.g., leaders and followers, anesthesiologists and nurses). In the following sections, we discuss three roles that are of special relevance to group control. These are the roles of prospective member, new member, and leader (a particular kind of full member). Except in rare cases in which groups are "closed" to new members (Ziller, 1965), membership change is an unavoidable fact of group life. Because this change can have profound consequences for groups (Arrow & McGrath, 1995; Levine & Choi, 2004), it is not surprising that groups devote substantial time and energy to controlling when and how it occurs. Two important mechanisms for controlling this change are *(1)* recruiting prospective members who have the potential to enhance group effectiveness (selection) and *(2)* shaping new members' behavior to increase the likelihood that they will realize this potential (socialization). In addition, after new members have undergone socialization and been promoted to full membership, the group seeks to place them in specialized roles that will serve its interests. The most important such role is that of leader.

To fulfill the generic functions identified above, groups must move people through these three roles in a timely and efficient fashion. Group socialization theory describes and explains the process by which this occurs (*see* Levine & Moreland, 1994; Moreland & Levine, 1982). The theory postulates that groups and individuals evaluate the past, present, and anticipated future rewardingness of their own and alternative relationships. These evaluations produce feelings of commitment, which can rise and fall over time. When the commitment levels of both parties reach specific levels (decision criteria), indicating that a role transition should occur, the person moves from one phase of group membership to another. Following this role transition, the two parties engage in new evaluations, which produce further changes in commitment and, when new decision criteria are reached, additional role transitions. In this way, the individual can move through five phases of group membership (investigation, socialization, maintenance, resocialization, and remembrance), separated by four role transitions (entry, acceptance, divergence, and exit).

We focus here on the first three phases of group membership (investigation, socialization, and maintenance) and the role transitions that separate them (entry and acceptance). During *investigation*, when the individual is a prospective member, the group looks for people who can contribute to the attainment of collective goals (recruitment), and the individual looks for groups that can contribute to the satisfaction of personal needs (reconnaissance). If the evaluation process causes both parties' commitment to rise to their respective entry criteria, the role transition of *entry* occurs, and the person becomes a new member. During *socialization*, the group attempts to change the individual so that he/she contributes more to collective goal attainment, and the individual attempts to change the group so that it contributes more to personal need satisfaction. To the extent each party is successful, the individual undergoes assimilation and the group undergoes accommodation. If the evaluation process causes both parties' commitment to rise to their respective acceptance criteria, the role transition of *acceptance* occurs, and the person becomes a full member. During *maintenance*, the group and the individual engage in role negotiation to find a specialized role for the individual that maximizes the group's ability to achieve collective goals and the individual's ability to satisfy personal needs. Both entry and acceptance signal the individual's movement toward full membership and hence reflect the group's increasing inclusion of the individual (Levine, Moreland, & Hausmann, 2005).[2]

In the following discussion, we adopt the perspective of the group rather than the individual. This does not mean, however, that individuals are helpless pawns of the groups to which they belong. Individuals can influence groups by affecting the occurrence and timing of role transitions (Levine et al., 2005; Moreland & Levine, 1984) and by producing other kinds of group accommodation (Levine, Choi, & Moreland, 2003; Levine & Moreland, 1985).

Prospective Members

The first step in recruitment is identifying prospective members (Moreland & Levine, 1982). This can be more or less difficult, depending on how many people know the group is recruiting and how eager they are to join. Often the task of identifying prospective members is assigned to recruitment specialists, who are also charged with the managing the second step in the process—namely, evaluating candidates' attractiveness. This process also varies in difficulty, depending on how much members agree about important newcomer characteristics, how hard it is to assess these characteristics, and how costly evaluative errors are. Groups use several sources of information in evaluating candidates, including formal applications, interviews, and contacts with people who know the candidates.

To the extent a candidate is evaluated positively, the group's commitment to that person will increase. If this commitment meets or exceeds the group's entry criterion, the group will move to the third step in the recruitment process by trying to convince the individual to make the role transition of entry and become a new member. This will be easy if the individual's commitment to the group meets or exceeds his/her entry criterion, in which case the group need only extend an offer of membership. The group's eagerness to extend an offer and the individual's eagerness to accept it vary positively with how much each party's commitment exceeds its entry criterion. However, if the group is ready for entry but the individual is not, the group must try to convince the individual that he/she would benefit from joining. The group's effort to produce the transition varies positively with how much its commitment exceeds its entry criterion, whereas the individual's resistance to this effort varies positively with how much the person's commitment falls short of his/her criterion.

Groups can use two general strategies for convincing reluctant prospective members to join: persuading them to raise their commitment to the group and/or to lower their decision criterion for entry (Levine et al., 2005). Commitment-oriented tactics include increasing prospective members' participation in group activities, offering inducements (e.g., money, status) to join, and promising to alter the group to meet their needs. Criterion-oriented tactics include convincing prospective members that others in their position, who were later successful in the group, had relatively low entry criteria when they joined. Alternatively, the group can signal that, if prospective members refuse to lower their entry criterion, the group will lose commitment to them and perhaps withdraw its offer of membership.[3]

New Members

Entry into a group, like all role transitions, is typically signaled by a ceremony of some kind. Entry ceremonies are designed to test and increase new members' commitment to the group. Such commitment is important because oldtimers often do not trust new members' skills or allegiance to group goals. Entry ceremonies vary in pleasantness. Those involving positive treatment of newcomers include parties, gifts, and offers of future aid. Such ceremonies can motivate newcomers to work hard to merit the welcome they received. Those involving negative treatment are sometimes quite severe. An example is the entry ceremony suffered by a new member of a high school wrestling team: "Over the span of a month…the sophomore had been spat on, hogtied, imprisoned inside a gymnasium locker, slammed into walls and held down while other players forced the handle of a plastic knife into his rectum" (Jacobs, 2000, p. 30). Negative entry ceremonies can elicit cognitive dissonance in new members and thereby increase their attraction and commitment to the group (Aronson & Mills, 1959; Gerard & Matthewson, 1966; but *see* Lodewijkx & Syroit, 1997). In addition, negative ceremonies can reinforce new members' perceptions of their low status in the group and thereby increase their motivation to gain oldtimers' respect.

During the socialization phase of group membership, the group attempts to provide new members with the knowledge, ability, and motivation they will need to play the role of full member. The tactics that groups use to socialize newcomers vary greatly but can be classified along six dimensions (Van Maanen & Schein, 1979). Collective tactics involve processing newcomers together, whereas individual tactics involve processing them in relative isolation from one another. Formal tactics involve structured training outside the workplace, whereas informal tactics involve trial-and-error learning inside the workplace. Sequential tactics have discrete, identifiable steps, whereas random tactics have unknown, ambiguous, or continually changing steps. Fixed tactics specify the length of the socialization process, whereas variable tactics do not. Serial tactics allow newcomers to learn their roles from oldtimers, whereas disjunctive tactics force newcomers to learn these roles on their own. And investiture tactics affirm newcomers' personal characteristics, whereas divestiture tactics denigrate these characteristics.

In addition to these general tactics, a number of other factors can affect a group's ability to socialize newcomers (Levine & Moreland, 1999; Moreland & Levine, 1989). For example, newcomers who are familiar with the group, committed to it, and possess valuable skills are easier to socialize than are those who do not possess these qualities. In addition, newcomer socialization is facilitated when oldtimers have had experience with newcomers in the past, are committed to current newcomers (because the group needs their skills), and are knowledgeable about the group's values and practices.

If the group's commitment to the newcomer during socialization meets or exceeds its acceptance criterion, the group will try to convince the individual to make the role transition of acceptance and become a full member. This will be easy if the individual's commitment to the group also meets or exceeds his/her acceptance criterion. In this case, both parties' eagerness for the transition varies positively with how much their commitment exceeds their acceptance criteria. However, if the group is ready for the transition but the individual is not, then the group must work harder to convince the individual to become a full member. In this case, the group's effort to produce the transition varies positively with how much its commitment exceeds its acceptance criterion, whereas the individual's resistance to this effort varies positively with how much the person's commitment falls short of his/her criterion. Paralleling the situation for entry, groups can use a variety of strategies to persuade reluctant new members to raise their commitment and/or lower their acceptance criterion (Levine et al., 2005).

Leaders

As with entry ceremonies, acceptance ceremonies are designed to test and increase members' commitment to the group. Although people finishing socialization generally pose less threat to the group than those beginning this process, their commitment remains important because full members have the major responsibility for group welfare. Like entry ceremonies, acceptance ceremonies vary in pleasantness. An example of a positive ceremony was the phone call received by Michael Lewis, a young bond trader at Salomon Brothers. Following a big sale, Lewis heard from another Salomon trader nicknamed the Human Piranha, who said: "'I heard you sold a few bonds'....That is...awesome....You are one Big Swinging Dick, and don't ever let anybody tell you different.' It brought tears to my eyes....to be called a Big Swinging Dick by the man who, years ago, had given birth to the distinction and in my mind had the greatest right to confer it upon me" (Lewis, 1989, p. 184).

During the maintenance phase of group membership, when the commitment levels of the group and the individual are at their highest levels, the two parties engage in role negotiation to find a specialized role for the person that maximizes the group's ability to achieve collective goals and the person's ability to satisfy personal needs. This in turn leads to substantial role differentiation in the group, more than in any other phase. The most important role is that of leader. A great deal of theoretical and empirical work has been done on leadership, much of it involving leader effectiveness (*see* Chemers, 2001; Hogg, 2007). Here we focus on a different question that is particularly relevant to group control—namely, how groups select particular people to play leadership roles.

There is substantial evidence that groups give leadership roles to people who possess certain personal qualities, such as charisma, intelligence, and task expertise (Hogg, 2007). For example, group members who exhibit task competence and conformity to group norms gain legitimacy as leaders (e.g., Hollander, 1958), and people whose characteristics and behaviors fit leadership prototypes are more likely to be seen as leaders (e.g., Lord, Foti, & DeVader, 1984). In addition, people who exhibit "transformational" leadership behaviors (e.g., intellectual stimulation, individualized consideration of followers) inspire high trust and commitment (e.g., Bass, 1985; Hogg, 2007). It is perhaps not surprising, then, that people who want to be selected as leaders engage in what Chemers (2001) calls "image management" to convince others that they deserve power and responsibility.

The image management perspective is consistent with a social identity interpretation of leadership emergence (e.g., Hogg & van Knippenberg, 2003). This approach assumes that people who are highly identified with a group pay attention to the group prototype (i.e., characteristics that distinguish the group from outgroups) and to those members who exemplify this prototype. These prototypical members are liked as *group members* (rather than as individuals), which enhances their status. In addition, because their behavior is attributed to enduring personality characteristics (the fundamental attribution error), they are seen as having a charismatic leader personality. As a result, they are given power to exert influence in the group, and their status advantage over other members is strengthened. An interesting implication of this analysis is that, in order to maintain or enhance their power, leaders seek to *manage* prototypes—for example by emphasizing their own prototypicality or questioning that of rivals.

Individuals as Targets of Control

Our discussion so far has focused on groups as agents of control, with special attention to how they use norms and roles to influence their members' behavior. A full understanding of group control, however, necessitates consideration of the other side of the coin—how members respond to these influence attempts. These responses can take one of two basic forms: members can capitulate to group efforts to control their behavior or they can resist these efforts.

Members' Capitulation to Group Pressure

Stimulated by Baumeister and Leary's (1995) classic paper, there has been an explosion of recent work on the origins and consequences of the need to belong (for example, see Abrams, Hogg, & Marques, 2005; Levine & Kerr, 2007; Williams, Forgas, & von Hippel, 2005). According to Leary and Baumeister (2000), this need has produced an internal monitoring system—the sociometer—for assessing one's level of acceptance/rejection by others. The sociometer is highly sensitive to how one is treated by others, operates automatically (i.e., continuously, involuntarily, and unconsciously), and is more responsive to cues of rejection than acceptance. The operation of this system is illustrated by work on reactions to ostracism (Williams, 2007). In the remainder of this section, we discuss how desire for social inclusion can affect three important kinds of behavior in group settings: conformity, group decision making and problem solving, and behavior in mixed-motive situations (for a more detailed discussion, see Levine & Kerr, 2007).

Conformity

Given our interest in the operation of injunctive norms, we focus here on normative influence, which is defined as "influence to conform to the positive expectations of another" (Deutsch & Gerard, 1955, p. 629). Normative influence is based on group members' desire to gain or maintain social approval and their assumption that such approval is contingent on conformity to group norms. Several lines of work have suggested that people anticipate positive responses for conformity and negative responses for deviance, which in turn affects their behavior. For example, there is evidence that majority opposition is a stressful experience and that conformity can reduce this stress (e.g., Costell & Leiderman, 1968; Gerard, 1961). In addition, compared to people who conform to group consensus, those who dissent both expect to receive (e.g., Gerard & Rotter, 1961) and actually do receive negative reactions from others (see the earlier discussion of group reactions to norm violations). Finally, the dramatic effectiveness of social support (i.e., the presence of a single person agreeing with one's position) in reducing conformity rests, at least partly, on the partner's ability to reduce fear of group retaliation for deviance (Allen, 1975; Asch, 1952).

Research on other situational determinants of conformity also suggests that fear of group punishment is important. For example, group members conform more when working for a common goal than for individual goals (e.g., Deutsch & Gerard, 1955), unless their behavior reduces, rather than enhances, the group's likelihood of success (Sakurai, 1975). In addition, they conform more when responding publicly rather than privately (e.g., Asch, 1956; Deutsch & Gerard, 1955); when they anticipate future interaction with other group members (e.g., Lewis, Langan, & Hollander, 1972); and when the group is attractive, particularly if they feel insecure about group acceptance and believe conformity will increase it (Jetten, Hornsey, & Adarves-Yorno, 2006; Jones & Gerard, 1967). Finally, group identification increases conformity, although social identity researchers ascribe this effect to informational as well as normative influence (e.g., Abrams & Hogg, 1990; Turner, 1991).

Normative influence is often assumed to produce public compliance (overt behavioral change toward the majority's position) but not private acceptance (covert attitudinal change toward this position), whereas informational influence ("influence to accept information obtained from another as *evidence about reality*"—Deutsch & Gerard, 1955, p. 629) is often assumed to produce both compliance and acceptance. However, as Levine and Kerr (2007) discuss, the relationship between normative/informational influence and compliance/acceptance is not so simple. At the methodological level, it is often difficult to distinguish operationally between the two kinds of influence (Allen, 1965; Nail, MacDonald, & Levy, 2000). At the conceptual level, arguments can be made for questioning simple connections between processes underlying influence and outcomes of this influence. For example, Wood (1999) has suggested that, depending on circumstances, normative influence can produce either public or private change. Finally, Levine and Kerr (2007) have argued that the desire to hold accurate views (the foundation of informational influence) may derive, at least in part, from the desire to be accepted (the foundation of normative influence). This is because holding "accurate" (i.e., consensual) views on important issues is one way to gain group approval.

Group Decision Making and Problem Solving

In contrast to conformity situations, where there is typically little or no interaction between members of minority and majority factions and no explicit group goal, in decision-making and problem-solving situations group members work together to achieve common goals. If members of decision-making groups are concerned about others' evaluations, then small (minority) factions should be more susceptible to influence pressures than large (majority) factions. And this is indeed the case, as indicated by consistent evidence for majority influence ("strength-in-numbers") in such groups (Hastie & Kameda, 2005; Stasser, Kerr, & Davis, 1989). Although informational influence no doubt plays a role in majority influence, several lines of work suggest that normative influence is quite important. For example, people occupying minority positions in decision-making groups, such as political advisory committees and juries, often report fearing punishment and conforming as a result (e.g., Schlesinger, 1965). In addition, properties of group tasks moderate majority influence in ways consistent with the operation of normative influence. One such property is the basis for validating "correct" judgments. Festinger (1950) argued that some judgments can be validated using nonsocial criteria (physical reality), whereas others rely on social consensus (social reality). Because minorities should have more difficulty defending their position on issues of social than physical reality (i.e., on judgmental than on intellective issues), they should be more likely to fear punishment for dissent on judgmental issues. This in turn should cause them to conform more on such issues, and they do so (Laughlin, 1999). In addition, evidence suggests that normative factors underlie majority influence on judgmental issues. Kaplan and Miller (1987) found that whereas group discussions of intellective issues involved more informational than normative influence, discussions of judgmental issues involved more normative than informational influence.

As discussed below, majorities in decision-making groups do not always prevail, and the conditions under which this occurs shed light on the normative basis of social influence. On intellective tasks (e.g., simple arithmetic problems), one-person minorities holding deviant but clearly correct positions typically win over the majority to their position (Laughlin & Ellis, 1986). One interpretation of this effect is that the minority member's strong belief in the validity of his/her answer reduces fear of group punishment for deviance. In addition, on tasks with both intellective and judgmental features (e.g., vocabulary tests), two-person minorities giving deviant but correct answers typically prevail (Laughlin & Ellis, 1986), suggesting that a social supporter can reduce fear of punishment for deviance in decision-making, as well conformity, settings.

The group polarization literature also provides evidence for the operation of normative influence. For example, research on polarization indicates that group members' initial opinions become more extreme during discussion but shift back toward moderation afterward (e.g., Kerr et al., 1976), suggesting that people who initially disagreed with the majority position complied publicly to avoid exclusion. In addition, polarization can be produced by mere knowledge of others' positions, without discussion or exposure to supporting arguments (Isenberg, 1986). These findings are inconsistent with informational influence explanations of polarization (e.g., Burnstein & Sentis, 1981) and consistent with social comparison (i.e., normative) explanations, which emphasize group members' tendency to compare themselves with others and their desire to gain approval and see themselves in a positive light (Isenberg, 1986).

Behavior in Mixed-Motive Settings

In many situations, group members' personal interests are at least partially incompatible with one another or with the collective interests of the group. Such incompatibility occurs in most group performance settings, because the costs of working together produce or amplify conflicts of interest between members. In addition, group members' personal interests are incompatible in social dilemma settings, where members are motivated to compete rather than cooperate. In both settings, fear of group punishment can substantially influence members' behavior.

Group Performance

Groups have norms prescribing how hard members should work and how well they should perform on collective tasks. Therefore, if members are concerned about how they are evaluated, their motivation and performance should be influenced by whether others can monitor their behavior. Consistent with this hypothesis, work on social facilitation indicates that others' presence can energize performance (Zajonc, 1965). This effect is often attributed to evaluation apprehension and the attendant arousal it produces (Williams, Harkins, & Karau, 2003), which in turn facilitates performance on simple tasks and inhibits performance on difficult tasks. Evaluation apprehension is also important in social loafing, where the presence of others de-energizes performance. Here, people expend less effort when working collectively than coactively or individually (Latane, Williams, & Harkins, 1979; Williams et al., 2003). One explanation of loafing is that collective tasks reduce individuals' evaluative apprehension. Finally, as would be expected if evaluation apprehension influences members' conformity to performance norms, group productivity is greater when groups are cohesive (Mullen & Copper, 1994) and elicit strong identification (Haslam, 2004), especially when member interaction or interdependence is high (Gully, Devine, & Whitney, 1995).[4] Similarly, social loafing is lower in higher valence groups (Williams et al., 2003).

Concern about others' evaluations also influences how group members respond to basic social interaction norms, including the equity norm stipulating that members' outputs should be proportional to their inputs. There is evidence that members' task motivation is undermined if they believe their contributions are not essential to group success (i.e., are "dispensable") (Kerr & Bruun, 1983). This response is less likely, however, if members think that others can monitor their effort and are likely to interpret low effort as social loafing (Harkins & Petty, 1982). Moreover, when an ostensibly capable member is clearly slacking off, other members "punish" the person by reducing their own effort, even though such behavior is costly to them and the group as a whole (cf. Kerr, 1983). And the tendency to match others' effort is stronger when concerns about social relations are greater (e.g., Groenenboom, Wilke, & Wit, 2001).

Social Dilemmas

Concerns about group evaluation also influence cooperation in mixed-motive situations, such as social dilemmas (Kerr, 1983, 1986). In these situations, a personally beneficial (group-defecting) choice provides higher outcomes to an individual than does a cooperative choice, regardless of what other group members do, but defection by everyone causes worse outcomes for both the individual and the group than does universal cooperation (Orbell & Dawes, 1981). Just as attraction to and identification with the group increase effort in performance settings, these factors also increase cooperation in social dilemmas (e.g., Kramer & Brewer, 1984), as does the publicness (surveillance potential) of a person's behavior (e.g., DeCremer & Bakker, 2003). In addition, just as group members are expected to work hard on group tasks, they are expected to sacrifice their self-interest for the welfare of the group. Such sacrifice does indeed occur, both in natural settings (e.g., Fussel, 2003) and in the laboratory (e.g., Hertel & Kerr, 2001; Van Vugt & Hart, 2004; Zdaniuk & Levine, 2001). Moreover, uncooperative group members are punished (e.g., Fehr & Gächter, 2002), and eliminating the threat of such punishment increases the likelihood of uncooperative behavior (Kerr, 1999; Ouwerkerk et al., 2005). Also relevant is evidence that norms of commitment keeping developed during discussion (e.g. Kerr & Kaufman-Gilliland, 1994) and norms of fairness or reciprocity (e.g., Komorita, Parks, & Hulbert, 1992; van Dijk & Vermunt, 2000) influence behavior in social dilemmas.

Members' Resistance to Group Pressure

Having discussed how members are influenced by the groups to which they belong, we now consider the other side of the coin—namely, how members resist group efforts to control their behavior. Some examples of such resistance were briefly mentioned above. For example, in discussing conformity, we noted that the presence of a social supporter often frees individuals from group pressure, allowing them to remain independent when they would otherwise conform. And, in discussing group socialization theory, we noted that individuals can influence groups by affecting the occurrence and timing of role transitions and by causing groups to change in other ways. In this section, we consider in more detail how individuals resist group pressure.

Nonconformity

At the outset, it is important to note that, contrary to what is often portrayed in textbooks (Friend, Rafferty, & Bramel, 1990), conformity is not necessarily the dominant response to group pressure. For example, in Asch's (1956) classic "conformity" studies, approximately two-thirds of participants' total responses were independent (or correct), and almost five times as many participants were always independent as always yielded (24% vs. 5%). Moreover, Asch (1951) was able to substantially reduce the total number of yielding responses by having a single confederate (social supporter) dissent from the erroneous majority by giving correct responses (for later work on social support, *see* Allen, 1975).

Asch argued that independence is critical to group functioning because disagreement stimulates thinking, a notion reflected in later work on minority influence (e.g., Moscovici, 1976). However, Asch's view of why people exhibit independence differs from that of minority influence theorists (Levine, 1999). According to Moscovici, for example, people adopt independent positions because they are sure they are right and want their position to prevail. To accomplish this goal, they provoke conflict by consistently disagreeing with the majority and eschewing any compromise. From this perspective, independence is self-oriented, competitive, and closed-minded. In contrast, Asch suggested that people take independent positions because they want to help the group achieve a correct consensus and are willing to abandon their positions in the face of disconfirming evidence. From this perspective, independence is group-oriented, cooperative, and open-minded. Although both kinds of independence can benefit a group, they are likely to produce different interaction patterns (e.g., more or less interpersonal conflict) and different outcomes (e.g., less or more willingness to abide by group decisions).

An important determinant of whether people behave independently is their level of identification with the group. Although strongly identified group members often conform to group norms, the relationship between identification and conformity is not simple (Packer, 2008). For example, highly identified members of groups with individualist (rather than collectivist) norms may "conform" by "nonconforming" (cf. Jetten & Postmes, 2006). In addition, highly identified members may feel obligated to nonconform if they believe the group's norms are harmful, even if such behavior is personally costly (cf. Van Vugt & Hart, 2004; Zdaniuk & Levine, 2001). The costs of such behavior may be mitigated by the assumption that current (or future) members will eventually appreciate the dissenting behavior and view it as reflecting loyalty (cf. Hornsey, 2006; Levine & Moreland, 2002). Because high-status members (e.g., leaders) are likely to identify strongly with the group, feel responsible for its performance, and assume that they have (at least some) freedom to deviate, they may be particularly prone to challenge norms they perceive as maladaptive (cf. Abrams, Randsley de Moura, Marques, & Hutchison, 2008).

Although we have used the terms nonconformity and independence interchangeably, a case can be made for viewing independence as a subcategory of nonconformity. More specifically, independence, defined as giving zero weight to the group's response and therefore "standing pat" (i.e., showing no movement), can be differentiated from anticonformity, defined as giving substantial *negative* weight to the group's position and moving away from it (Willis, 1963).[5] In addition to desire to change the group or help it achieve a correct consensus, independence might be stimulated by desire to hold an accurate view of reality (e.g., Baron, Vandello, & Brunsman, 1996). In contrast, anticonformity might be produced by desire to regain threatened freedom (e.g., Brehm & Brehm, 1981) or to distance oneself from disliked others (e.g., Wood, Pool, Leck, & Purvis, 1996). And both kinds of nonconformity might be elicited by desire to signal one's uniqueness (e.g., Brewer, 1991; Hornsey & Jetten, 2004).

Most of the work on independence and anticonformity has focused on "passive" efforts by numerical minorities to resist group pressure. However, as suggested earlier, sometimes minorities exert "active" efforts to change the groups to which they belong. In the remainder of this section, we discuss three classes of active strategies: minority influence, factional conflict, and collective action.

Minority Influence

Interest in minority influence was stimulated by Moscovici (1976, 1980), who argued that minorities can be sources as well as targets of influence and that both majority influence (conformity) and minority influence (innovation) are based on conflict and behavioral style. According to Moscovici, minorities confronting disagreeing majorities focus on the social implications of the disagreement, which causes them to agree publicly, but not privately, with the majority. In contrast, majorities confronting disagreeing minorities focus on the content of the minority's position, which causes them to agree privately, but not publicly, with the minority. Since Moscovici's early work, a great deal of theoretical and empirical attention has been devoted to when and why minority influence occurs (*see* reviews by Martin & Hewstone, 2001; Wood et al., 1994).

In regard to the "when" question, several variables have been found to affect the likelihood of minority influence. For example, minorities tend to be more influential on indirect than on direct measures and are more effective if they are ingroup (as opposed to outgroup) members, consistently maintain their position (but not to point of "rigidity"), and espouse positions that are popular in the wider society. In addition to convincing majorities to adopt their position, minorities can also stimulate them to attend to new aspects of a problem, think divergently, and generate novel solutions (e.g., Gruenfeld, Thomas-Hunt, & Kim, 1998; Nemeth, 1995), which in turn can enhance group performance (e.g., Schulz-Hardt et al., 2006).

In regard to the "why" question, several explanations of minority (and majority) influence have been offered (*see* Martin & Hewstone, 2001). Although a comprehensive discussion of these models is beyond the scope of this chapter, two of their common characteristics are worth mentioning. First, the models generally assume that minority influence is based on informational rather than normative factors (but *see* Wood et al., 1994). Second, they focus on the cognitive dynamics of majority members rather than the social dynamics of groups containing majority and minority factions (for exceptions, *see* Choi & Levine, 2004; De Dreu & De Vries, 1997; Smith, Tindale, & Dugoni, 1996; Schulz-Hardt et al., 2006). As Levine and Kaarbo (2001) noted, this general neglect of group dynamics is surprising in light of Moscovici's initial interest in how small, powerless minorities produce revolutionary changes in art, science, religion, and politics.

A more "social" perspective on minority influence is provided by group socialization theory (Levine & Moreland, 1994; Moreland & Levine, 1982), which was outlined earlier. An important hallmark of this theory is its emphasis on reciprocal influence—the fact that individuals can influence groups as well as vice versa. As Levine and Moreland (1985) argued, individuals in all five phases of group membership (investigation, socialization, maintenance, resocialization, and remembrance) have the potential to change the group and thereby introduce innovation. For example, as suggested earlier, individuals can influence groups by making it easier or harder for them to produce desired role transitions. Thus, in trying to convince reluctant individuals to make particular transitions (by changing their commitment levels and/or decision criteria), groups may change *themselves*—for example by altering how they apportion responsibilities and privileges, which in turn may affect members' social relations and task performance.

Recently, Levine and his colleagues (Levine, Choi, & Moreland, 2003) analyzed the conditions under which newcomers (who are typically numerical minorities in the groups they join) can produce innovation. Such innovation may seem unlikely at first glance, because newcomers often have low power and status. However, under the right conditions, they may be able to produce substantial innovation (Choi & Levine, 2004; Levine & Choi, in press). According to Levine et al. (2003), the ability of newcomers to change the groups they enter depends on characteristics of the group and characteristics and behaviors of the newcomers. They argued, for example, that newcomers are more influential in groups that are understaffed, are in early stages of development, are low in cohesion, have a climate favoring innovation, and are performing poorly. In addition, they suggested that newcomers are more influential when they have high status, are relatively numerous, possess positive qualities (e.g., similarity to oldtimers, expertise), and employ assertive tactics in dealing with oldtimers (Levine & Kaarbo, 2001). These tactics include threatening to undermine group performance (e.g., by active sabotage or social loafing), inducing the group to recruit people who agree with the minority's views, and convincing the group to adopt a unanimity (rather than a majority) decision rule, which increases the minority's power.

Factional Conflict

As this discussion implies, in many groups majority-minority relations involve some form of *factional conflict*, in which the goal is gaining power to decide group policy and/or distribute group resources (e.g., money, jobs). In such cases, minorities can exert substantial influence. For example, in parliamentary political systems, minority parties are often members of ruling coalitions and exert considerable influence over these coalitions (Kaarbo, 1996). Viewing minority influence through the lens of factional conflict suggests several factors that have been overlooked in previous research. For example, Levine and Kaarbo (2001) noted that social psychologists tend to view minorities as heroic advocates of social change battling reactionary majorities that will do anything to maintain the status quo. Yet minorities can block as well as produce change, and this change can be either forward-oriented (i.e., reflect a new group policy) or backward-oriented (i.e., reflect a prior group policy). By crossing these two factors, Levine and Kaarbo differentiated four types of minorities—*progressive* minorities, who produce change to a new group policy; *reactionary* minorities, who produce change to a prior group policy; *conservative* minorities, who block the majority's efforts to produce change to a new group policy; and *modernist* minorities, who block the majority's efforts to produce change to a prior group policy. In addition, Levine and Kaarbo noted that social psychologists tend to study groups containing a majority and just one minority. However, groups often contain two or more minorities, which can cooperate or compete with one another. In such cases, the majority often goes to great lengths to prevent or disrupt minority coalitions that might threaten its dominance.

An implicit assumption of our analysis is that minorities not only want to prevail, but derive gratification from doing so. Although this is no doubt true when "winning" yields tangible benefits, it is not always the case. In a series of studies, Prislin and her colleagues found that members of minority factions that gain majority status as a result of (former) majority members defecting to their position remain alienated from the group and do not identify with it (Prislin & Christensen, 2005). This occurs because minorities that become majorities often doubt the sincerity of the people who converted to their position. Consistent with this argument, when converts signal their sincere agreement with the minority's position, members of a new majority are much more likely to identify with the group (Prislin, Levine, & Christensen, 2006).

Collective Action

As our discussion of factional conflict suggests, majority–minority relations often require analysis at the intergroup level (*see* Butera & Levine, 2009). Much of the early work on collective action by minority groups focused on the kinds of social comparison that produce intergroup conflict (*see* Levine & Moreland, 1987). For example, Davies (1969) offered a "J-curve" model of collective violence, and Gurr (1970) differentiated three types of expectation–outcome discrepancy (decremental, aspirational, progressive) that can lead to civil strife. In addition, Runcimann (1966) made an important distinction between egoistic relative deprivation (the perception that one's own outcomes are lower than those of other ingroup members) and fraternal relative deprivation (the perception that the ingroup's outcomes are lower than those of an outgroup). Evidence indicates that the latter kind of deprivation is more likely to produce collective action (e.g., Crosby, 1982; Martin & Murray, 1984).

In their classic presentation of social identity theory, Tajfel and Turner (1979) argued that members of low-status groups (who are often numerical minorities) can deal with their situation in three ways: individual mobility (moving to a higher status outgroup or psychologically disengaging from the ingroup); social creativity (cognitively altering intergroup comparisons to achieve positive distinctiveness for the ingroup); or social competition (engaging in direct competition with the outgroup). They suggested further that social competition (collective action) only occurs when members of a low-status group identify with it, view intergroup boundaries as impermeable, and view the status hierarchy as illegitimate and potentially modifiable.

Recent analyses of collective action have extended and elaborated Tajfel and Turner's formulation (see Wright, 2001). An interesting example is Sturmer and Simon's (2004) dual-pathway model of collective action, which combines social identity theory and Klandersmans's (1997) theory of participation in social movements. This model suggests that involvement in collective action is based on both identification with the ingroup and a cost–benefit analysis of how collective action will influence the attainment of group and personal goals.

Sturmer and Simon's (2004) research focuses on social movements designed to facilitate the integration of minority groups (e.g., older people, gays) into society. However, as they noted, social movements can also have other goals, such as improving the wider society (e.g., environmental activism). In addition, social movements can emerge when a minority faction perceives that the group it belongs to is abandoning its fundamental principles. In Levine and Kaarbo's (2001) terminology, such minorities have both reactionary and conservative tendencies, because they want to produce change to a prior group policy *and* block change to a new policy. Recent work on schisms sheds light on the conditions under which reactionary/conservative social movements emerge (*see* Hart & Van Vugt, 2006; Sani & Reicher, 1998; Sani & Todman, 2002. According to Sani (2005), schisms begin when a subset of group members perceives that those who speak for the group are endorsing values that subvert the core principles underlying the group's identity. This perception causes dissatisfied members to experience negative emotions, feel reduced identification with the group, and perceive the group as lower in entitativity. Weak identification and strong negative emotions, in turn, increase schismatic intentions, although these factors have less impact when dissatisfied members feel they have "voice" in the group.

Conclusion

In this chapter, we reviewed a wide range of theoretical and empirical work on how groups exert control over their members and how members respond to these control efforts. In regard to *groups as agents of control*, we discussed two aspects of group structure (norms and roles) that are relevant to groups' motivation and ability to control their members. In discussing norms, we examined the content of injunctive norms specifying how group members should behave and the ways groups react to violations of these norms, with special attention to opinion deviance. In discussing roles, we examined how group control is manifested during the recruitment of prospective members, the socialization of new members, and the selection of full members to play leadership roles. In regard to *individuals as targets of control*, we discussed cases in which members capitulate to group pressure and those in which they resist such pressure. In discussing capitulation, we examined conformity, group decision making and problem solving, and behavior in mixed-motive settings. In discussing resistance, we examined nonconformity, minority influence, factional conflict, and collective action. Although our review is not exhaustive, we believe it makes a strong case for conceptualizing self-regulation as a group, as well as an individual, phenomenon.

Two assumptions underlying our analysis of group control should be made explicit. The first is that this control is initiated by the group for its own ends. The second is that members view this control as an unwelcome constraint on their behavior. Although these assumptions are often true, exceptions can occur, and these are instructive for how we think about group control. Consider, for example, individuals who join and/or remain in a group because they *want* the group to help them control their own behavior. This kind of motivation may stimulate participation in "support groups," which have become extremely popular in face-to-face settings and on the Internet (*see* Alexander, Peterson, & Hollingshead, 2003; Forsyth, 2001; Helgeson & Gottlieb, 2000).

The array of support groups is staggering. Such groups exist for people with health problems (e.g., cancer, diabetes, HIV/AIDS, sickle cell disease), cognitive and physical disabilities (e.g., aphasia, visual impairment), psychological problems (e.g., depression, obsessive-compulsive disorder), problematic behaviors (e.g., alcohol abuse, smoking, eating disorders), and difficult life experiences (e.g., divorce, physical abuse, bereavement, and even abduction by unidentified flying objects). Still other groups exist for parents and caregivers of those with problems.

Analyses of support groups typically focus on their ability to provide emotional support and helpful information, as well as opportunities to express one's feelings, engage in social comparison, and provide assistance to others (Forsyth, 2001; Helgeson & Gottlieb, 2000). In addition, as suggested above, at least some groups help individuals regulate their behavior by decreasing the likelihood that they will engage in "bad" behavior and increasing the likelihood that they will engage in "good" behavior.

This self-regulation function can be either indirect or direct. Indirect self-regulation, in which other group members do not attempt to influence individual behavior, derives from the other functions that support groups serve. Thus, a person might learn about a strategy that others use to cope with a similar problem and then adopt this strategy. For example, a Weight Watchers' website advises members, "You don't have to participate in the discussion, but you're bound to hear helpful strategies and motivating advice that helps you through the week." In contrast, direct self-regulation, in which other group members do attempt to exert influence, only occurs if these members can monitor a person's behavior. In some cases, this may be easy, as when a person shows up drunk at an Alcoholics Anonymous meeting. In other cases, a person's behavior is not visible and hence can only be monitored on the basis of self-reports. In the latter cases (which include virtually all Internet groups), members who want the group to help them regulate their behavior must reveal personal information about themselves. For example, a smoking cessation group (Freedom from Tobacco Quit Smoking Now) urges members to engage in what it calls "crisis posting," admonishing them to "Post to the group should you ever feel that your recovery is threatened. We take offense to members turning to nicotine before us. Don't head for an ashtray, a pack, a smoker, or store. Head for Freedom and share a quick and simple post seeking help...."

The fact that some people participate in groups because they desire help in controlling their behavior raises broader questions about the relationship between self- and group-control. For example, which group processes promote self-control in general and which processes are more effective for particular kinds of control (e.g., promotion-based striving to attain positive outcomes vs. prevention-based striving to avoid negative outcomes)? Do people join groups to help them control their thoughts and feelings as well as (or instead of) their overt behaviors? Do support groups always have an "agenda," in the sense of trying to help members think, feel, or act in particular ways, or do some groups promote the absence of any constraints (e.g., "tune in, turn on, drop out")? Finally, what factors cause members of support groups to play an active role in helping others gain self-control within the confines of the group and outside it? These and related questions deserve systematic investigation by researchers interested in the social dimension of self-regulation.

NOTES

1. Because the focus of this chapter is adult behavior, we do not review research on social antecedents of self-control in children (e.g., Calkins, 2004; Eisenberg, Smith, Sadovsky, & Spinrad, 2004).
2. The last two role transitions specified by the theory divergence and exit signal the individual's movement away from full membership and hence reflect the group's increasing exclusion of the individual (Levine et al., 2005). The process of exclusion was [considered] in our earlier discussion of how groups deal with norm violations (deviance).
3. This analysis assumes that prospective members do not belong to another group that demands their time and energy. If they do, the group seeking to recruit them must also convince them to leave their current group by reducing their commitment to that group and/or raising their exit criterion (Levine, Moreland, & Ryan, 1998).
4. Exceptions occur when group norms prescribe low, rather than high, levels of performance (e.g., Roethlisberger & Dickson, 1939).
5. Although we focus here on public behavior, private beliefs can also be implicated in independence and anti-conformity. For a more complex typology of nonconformity based on an individual's public and private agreement/disagreement with a group before and after exposure to the group's position, see Nail et al. (2000).

REFERENCES

Abrams, D. Social self-regulation. Pers Soc Psychol Bull 1994; 20: 473–483.

Abrams, D., de Moura, G. R., Hutchison, P., & Viki, G. T. When bad becomes good (and vice versa): Why social exclusion is not based on difference. In: Abrams, D., Hogg, M. A., & Marques, J. M. (Eds.), *The social psychology of inclusion and exclusion*. New York: Psychology Press, 2005: pp. 161–190.

Abrams, D., & Hogg, M. A. Social identification, self-categorization and social influence. Eur Rev Soc Psychol 1990; 1: 195–228.

Abrams, D., Hogg, M. A., & Marques, J. M. (Eds.). *The social psychology of inclusion and exclusion*. New York: Psychology Press, 2005.

Abrams, D., Marques, J., Bown, N., & Dougill, M. Anti-norm and pro-norm deviance in the bank and on the campus: Two experiments on subjective group dynamics. Group Process Intergroup Relat 2002; 5: 163–182.

Abrams, D., Randsley de Moura, G., Marques, J. M., & Hutchison, P. Innovation credit: When can leaders oppose their group's norms? J Pers Soc Psychol 2008; 95: 662–678.

Alexander, S. C., Peterson, J. L., & Hollingshead, A. B. Help is at your keyboard: Support groups on the Internet. In: Frey, L. R. (Ed.), *Group communication in context: Studies of bona fide groups*, 2nd ed.. Mahwah, NJ: Erlbaum, 2003: pp. 309–334.

Allen, V. L. Situational factors in conformity. In: Berkowitz, L. (Ed.), *Advances in experimental social psychology*, Vol. 2. New York: Academic Press, 1965: pp. 133–175.

Allen, V. L. Social support for nonconformity. In: Berkowitz, L. (Ed.), *Advances in experimental social psychology*, Vol. 8. New York: Academic Press, 1975: pp. 1–43.

Aronson, E., & Mills, J. The effect of severity of initiation on liking for a group. J Abnorm Soc Psychol 1959; 59: 177–181.

Arrow, H., & McGrath, J. E. Membership dynamics in groups at work: A theoretical framework. In: Staw, B. M., & Cummings, L. L. (Eds.), *Research in organizational behavior*, Vol. 17. Greenwich, CT: JAI Press, 1995: pp. 373–411.

Arrow, H., McGrath, J. E., & Berdahl, J. L. *Small groups as complex systems: Formation, coordination, development, and adaptation.* Thousand Oaks, CA: Sage, 2000.

Asch, S. E. Effects of group pressure upon the modification and distortion of judgments. In: Guetzkow, H. (Ed.), *Groups, leadership, and men*. Pittsburgh, PA: Carnegie Press, 1951: pp. 177–190.

Asch, S. E. *Social psychology*. New York: Prentice Hall, 1952.

Asch, S. E. Studies of independence and submission to group pressure: I. A minority of one against a unanimous majority. Psychol Monogr 1956; 70 (Whole No. 417).

Baron, R. S., Vandello, J. A., & Brunsman, B. The forgotten variable in conformity research: Impact of task importance on social influence. J Pers Soc Psychol 1996; 71: 915–927.

Bass, B. M. *Leadership and performance beyond expectations*. New York: Free Press, 1985.

Baumeister, R. F., & Leary, M. R. The need to belong: Desire for interpersonal attachments as a fundamental human motivation. Psychol Bull 1995; 117: 497–529.

Baumeister, R. F., & Vohs, K. D. (Eds.). *Handbook of self-regulation: Research, theory, and applications*. New York: Guilford Press, 2004.

Blanton, H., & Christie, C. Deviance regulation: A theory of action and identity. Rev Gen Psychol 2003; 7: 115–149.

Branscombe, N. R., Wann, D. L., Noel, J. G., & Coleman, J. In-group or out-group extremity: Importance of the threatened social identity. Pers Soc Psychol Bull 1993; 19: 381–388.

Bray, R. M., Johnson, E., & Chistrom, J. T., Jr. Social influence by group members with minority opinions: A comparison of Hollander and Moscovici. J Pers Soc Psychol 1982; 43: 78–88.

Brehm, S. S., & Brehm, J. W. *Psychological reactance: A theory of freedom and control*. New York: Academic Press, 1981.

Brewer, M. B. The social self: On being the same and different at the same time. Pers Soc Psychol Bull 1991; 17: 475–482.

Burnstein, E., & Sentis, K. Attitude polarization in groups. In: Petty, R. E., Ostrom, T. M., & Brock, T. C. (Eds.), *Cognitive responses in persuasion*. Hillsdale, NJ: Erlbaum, 1981: pp. 197–216.

Butera, F., & Levine, J. M. (Eds.). *Coping with minority status: Responses to exclusion and inclusion*. New York: Cambridge University Press, 2009.

Calkins, S. D. Early attachment processes and the development of emotional self-regulation. In: Baumeister, R. F., & Vohs, K. D. (Eds.), *Handbook of self-regulation: Research, theory, and applications*. New York: Guilford Press, 2004: pp. 324–339.

Carver, C. S., & Scheier, M. F. *On the self-regulation of behavior*. Cambridge, UK: Cambridge University Press, 1998.

Chemers, M. M. Leadership effectiveness: An integrative review. In: Hogg, M. A., & Tindale, S. (Eds.), *Blackwell handbook of social psychology: Group processes*. Malden, MA: Blackwell, 2001: pp. 376–399.

Choi, H-S., & Levine, J. M. Minority influence in work teams: The impact of newcomers. J Exp Soc Psychol 2004; 40:, 273–280.

Cialdini, R. B., & Trost, M. R. Social influence: Social norms, conformity, and compliance. In: Gilbert, D., Fiske, S., & Lindzey, G. (Eds.), *The handbook of social psychology*, 4th ed., Vol. 2. Boston: McGraw-Hill, 1998: pp. 151–192.

Costell, R. M., & Leiderman. P. H. Psychophysiological concomitants of social stress: The effects of conformity pressure. Psychosom Med 1968; 30: 28–310.

Crosby, F. *Relative deprivation and working women*. New York: Oxford, 1982.

Davies, J. C. The J-curve theory of rising and declining satisfactions as a cause of some great revolutions and a contained rebellion. In: Graham, H. D., & Gurr, T. R. (Eds.), *Violence in American: Historical and comparative perspectives*. New York: New American Library, 1969: pp. 671–709.

De Cremer, D., & Bakker, M. Accountability and cooperation in social dilemmas: The influence of others' reputational concerns. Curr Psychol Dev Learn Pers Soc 2003; 22: 155–163.

De Dreu, C. K. W., & De Vries, N. K. Minority dissent in organizations. In: De Dreu, C. K. W., & Van De Vliert, E. (Eds.), *Using conflict in organizations*. London: Sage, 1997: pp. 72–86.

Deutsch, M., & Gerard, H. B. A study of normative and information social influences upon individual judgment. J Abnorm Soc Psychol 1955; 51: 629–636.

Eidelman, S., & Biernat, M. Derogating black sheep: Individual or group protection? J Exp Soc Psychol 2003; 39: 602–609.

Eidelman, S., Silvia, P. J., & Biernat, M. Responding to deviance: Target exclusion and differential devaluation. Pers Soc Psychol Bull 2006; 32: 1153–1164.

Eisenberg, J., Smith, C. L., Sadovsky, A., & Spinrad, T. L. Effortful control: Relations with emotion regulation, adjustment, and socialization in childhood. In: Baumeister, R. F., & Vohs, K. D. (Eds.), *Handbook of self-regulation: Research, theory, and applications*. New York: Guilford Press, 2004: pp. 259–282.

Fehr, E., & Gächter, S. Altruistic punishment in humans. Nature 2002; 415: 137–140.

Festinger, L. Informal social communication. Psychol Rev 1950; 57: 271–282.

Forsyth, D. R. *Group dynamics*, 3rd ed. Belmont, CA: Wadswort, 1999.

Forsyth, D. R. Therapeutic groups. In: Hogg, M. A., & Tindale, S. (Eds.), *Blackwell handbook of social psychology: Group processes*. Malden, MA: Blackwell, 2001: pp. 628–659.

Friend, R., Rafferty, Y., & Bramel, D. A puzzling misinterpretation of the Asch "conformity" study. Eur J Soc Psychol 1990; 20: 29–44.

Fussell, P. *The boys' crusade*. New York: Random House, 2003.

Gerard, H. B. Disagreement with others, their credibility, and experienced stress. J Abnorm Soc Psychol 1961; 62: 559–564.

Gerard, H. B., & Mathewson, G. C. The effect of severity of initiation on liking for a group: A replication. J Exp Soc Psychol 1966; 2: 278–287.

Gerard, H. B., & Rotter, G. S. Time perspective, consistency of attitude and social influence. J Abnorm Soc Psychol 1961; 62: 565–572.

Groenenboom, A., Wilke, H. A. M., & Wit, A. P. Will we be working together again? The impact of future interdependence on group members' task motivation. Eur J Soc Psychol 2001; 31: 369–378.

Gurr, T. R. *Why men rebel*. Princeton, NJ: Princeton University Press, 1970.

Gruenfeld, D. H., Thomas-Hunt, M. C., & Kim, P. H. Cognitive flexibility, communication strategy, and integrative complexity in groups: Public vs. private reactions to majority and minority status. J Exp Soc Psychol 1998; 34: 202–226.

Gully, S. M., Devine, D. J., &Whitney, D. J. A meta analysis of cohesion and performance: Effects of levels of analysis and task interdependence. Small Group Res 1995; 26: 497–520.

Harkins, S. G., & Petty, R. E. Effects of task difficulty and task uniqueness on social loafing. J Pers Soc Psychol 1982; 43: 1214–1229.

Hart, C. M., & Van Vugt, M. From fault line to group fission: Understanding membership changes in small groups. Pers Soc Psychol Bull 2006; 32: 392–404.

Haslam, S. A. *Psychology in organizations*, 2nd ed. London: Sage, 2004.

Hastie, R., & Kameda, T. The robust beauty of majority rules in group decisions. Psychol Rev 2005; 112: 494–508.

Helgeson, V. S., & Gottlieb, B. H. Support groups. In: Cohen, S., Underwood, L. G., & Gottlieb, B. H. (Eds.), *Social support measurement and intervention: A guide for health and social scientists*. New York: Oxford University Press, 2000: pp. 221–245.

Hensley, V., & Duval, S. Some perceptual determinants of perceived similarity, liking, and correctness. J Pers Soc Psychol 1976; 34: 159–168.

Hertel, G., & Kerr, N. L. Priming and in-group favoritism: The impact of normative scripts in the minimal group paradigm. J Exp Soc Psychol 2001; 37: 316–324.

Hogg, M. A. Social psychology of leadership. In: Kruglanski, A. W., & Higgins, E. T. (Eds.), *Social psychology: Handbook of basic principles*, 2nd ed.. New York: Guildford Press, 2007: pp. 716–733.

Hogg, M. A., Fielding, K. S., & Darley, J. Fringe dwellers: Processes of deviance and marginalization in groups. In: Abrams, D., Hogg, M. A., & Marques, J. M. (Eds.), *The social psychology of inclusion and exclusion*. New York: Psychology Press, 2005: pp. 191–210.

Hogg, M. A., & van Knippenberg, D. Social identity and leadership processes in groups. In: Zanna, M. P. (Ed.), *Advances in experimental social psychology*, Vol. 35. San Diego, CA: Academic Press, 2003: pp. 1–52.

Hollander, E. P. Conformity, status, and idiosyncracy credit. Psychol Rev 1958; 65: 117–127.

Hornsey, M. J. Ingroup critics and their influence on groups. In: Postmes, T., & Jetten, J. (Eds.), *Individuality and the group: Advances in social identity*. London: Sage, 2006: pp. 74–91.

Hornsey, M. J., & Jetten, J. Not being what you claim to be: Imposters as sources of group threat. Eur J Soc Psychol 2003; 33: 639–657.

Hornsey, M. J., & Jetten, J. The individual within a group: Balancing the need to belong with the need to be different. Pers Soc Psychol Rev 2004; 8: 248–264.

Hornsey, M. J., Jetten, J., McAulifffe, B. J., & Hogg, M. A. The impact of individualist and collectivist group norms on evaluations of dissenting group members. J Exp Soc Psychol 2006; 42: 57–68.

Hutchison, P., & Abrams, D. Ingroup identification moderates stereotype change in reaction to ingroup deviance. Eur J Soc Psychol 2003; 33: 497–506.

Isenberg, D. J. Group polarization: A critical review and meta-analysis. J Pers Soc Psychol 1986; 50: 1141–1151.

Israel, J. *Self-evaluation and rejection in groups: Three experimental studies and a conceptual outline*. Uppsala, Sweden: Almqvist & Wiksell, 1956.

Jacobs, A. The violent cast of hazing in schools mirrors society, experts say. New York Times, March 5, 2000, p. 30.

Janis, I. L. *Groupthink*, 2nd ed. Boston: Houghton Mifflin, 1982.

Jetten, J., Hornsey, M. J., & Adarves-Yorno, I. When group members admit to being conformist: The role of relative intragroup status in conformity self-reports. Pers Soc Psychol Bull 2006; 32: 162–173.

Jetten, J., & Postmes, T. "I did it my way": Collective expressions of individualism. In: Postmes, T., & Jetten, J. (Eds.), *Individuality and the group: Advances in social identity*. Thousand Oaks, CA: Sage, 2006: pp. 116–136.

Jones, E. E., & deCharms, R. Changes in social perception as a function of the personal relevance of behavior. Sociometry 1957; 20: 75–85.

Jones, E. E., & Gerard, H. B. *Foundations of social psychology*. New York: Wiley, 1967.

Kaarbo, J. Power and influence in foreign policy decision making: The role of junior coalition partners in German and Israeli foreign policy. Int Stud Q 1996; 40: 501–530.

Kaplan, M. F., & Miller, C. E. Group decision making and normative versus informational influence: Effects of type of issue and assigned decision rule. J Pers Soc Psychol 1987; 53: 306–313.

Kerr, N. L. Motivation losses in task-performing groups: A social dilemma analysis. J Pers Soc Psychol 1983; 45: 819–828.

Kerr, N. L. Motivational choices in task groups: A paradigm for social dilemma research. In: Wilke, H. A. M., Messick, D. M., & Rutte, C. G. (Eds.), *Experimental social dilemmas*. Frankfurt am Main: Lang GmbH, 1986: pp. 4–27.

Kerr, N. L. Norms in social dilemmas. In: Schroeder, D. (Ed.), *Social dilemmas: Perspectives on individuals and groups*. Westport, CT: Praeger, 1995: pp. 31–47.

Kerr, N. L. Anonymity and social control in social dilemmas. In: Foddy, M., Smithson, M., Schneider, S., & Hogg, M. A. (Eds.), *Resolving social dilemmas*. Philadelphia: Psychology Press, 1999: pp. 103–119.

Kerr, N. L., Atkin, R., Stasser, G., Meek, D., Holt, R., & Davis, J. H. Guilt beyond a reasonable doubt: Effects of concept definition and assigned decision rule on the judgments of mock jurors. J Pers Soc Psychol 1976; 34: 282–294.

Kerr, N. L., & Bruun, S. Dispensability of member effort and group motivation losses: Free-rider effects. J Pers Soc Psychol 1983; 44: 78–94.

Kerr, N. L., & Kaufman-Gilliland, C. M. Communication, commitment, and cooperation in social dilemmas. J Pers Soc Psychol 1994; 48: 349–363.

Klandermans, B. *The social psychology of protest*. Oxford, UK: Blackwell, 1997.

Komorita, S. S., Parks, C. D., & Hulbert, L. G. Reciprocity and the induction of cooperation in social dilemmas. J Pers Soc Psychol 1992; 62: 607–617.

Kramer, R. M., & Brewer, M. B. Effects of group identity on resource use in a simulated commons dilemma. J Pers Soc Psychol 1984; 46: 1044–1057.

Kruglanski, A. W., & Webster, D. M. Group members' reactions to opinion deviates and conformists at varying degrees of proximity to decision deadline and of environmental noise. J Pers Soc Psychol 1991; 61: 212–225.

Kurzban, R., & Leary, M. R. Evolutionary origins of stigmatization: The functions of social exclusion. Psychol Bull 2001; 127: 187–208.

Latané, B., Williams, K., & Harkins, S. Many hands make light the work: The causes and consequences of social loafing. J Pers Soc Psychol 1979; 37: 822–832.

Laughlin, P. R. Collective induction: Twelve postulates. Organ Behav Hum Decis Process 1999; 80: 50–69.

Laughlin, P. R., & Ellis, A. L. Demonstrability and social combination processes on mathematical intellective tasks. J Exp Soc Psychol 1986; 22: 177–189.

Leary, M. R. The sociometer, self-esteem, and the regulation of interpersonal behavior. In: Baumeister, R. F., & Vohs, K. D. (Eds.), *Handbook of self-regulation: Research, theory, and applications*. New York: Guilford Press, 2004: pp. 373–391.

Leary, M. R., & Baumeister, R. F. The nature and function of self-esteem: Sociometer theory. In: Zanna, M. P. (Ed.), *Advances in experimental social psychology*, Vol. 32. San Diego: Academic Press, 2000: pp. 1–62.

Levine, J. M. Reaction to opinion deviance in small groups. In: Paulus, P. B. (Ed.), *Psychology of group influence*, 2nd ed. Hillsdale, NJ: Erlbaum, 1989: pp. 187–231.

Levine, J. M. Solomon Asch's legacy for group research. Pers Soc Psychol Rev 1999; 3: 358–364.

Levine, J. M., & Choi, H-S. Impact of personnel turnover on team performance and cognition. In: Salas, E., & Fiore, S. M. (Eds.), *Team cognition: Understanding the factors that drive process and performance*. Washington, DC: American Psychological Association, 2004: pp. 153–176.

Levine, J. M., & Choi, H-S. Newcomers as change agents: Minority influence in task groups. In: Martin, R., & Hewstone, M. (Eds.), *Minority influence and innovation: Antecedents, processes, and consequences*. Psychology Press, in press.

Levine, J. M., Choi, H.-S., & Moreland, R. L. Newcomer innovation in work teams. In: Paulus, P. B., & Nijstad, B. (Eds.), *Group creativity: Innovation through collaboration*. Oxford, UK: Oxford University Press, 2003: pp. 202–224.

Levine, J. M., & Kaarbo, J. Minority influence in political decision-making groups. In: De Dreu, C. K. W., & De Vries, N. K. (Eds.), *Group consensus and minority influence: Implications for innovation*. Oxford, UK: Blackwell, 2001: pp. 229–257.

Levine, J. M., & Kerr, N. L. Inclusion and exclusion: Implications for group processes. In: Kruglanski, A. W., & Higgins, E. T. (Eds.), *Social psychology: Handbook of basic principles*, 2nd ed.. New York: Guildford Press, 2007: pp. 759–784.

Levine, J. M., & Moreland, R. L. Innovation and socialization in small groups. In: Moscovici, S., Mugny, G., & Van Avermaet, E. (Eds.), *Perspectives on minority influence*. Cambridge: Cambridge University Press, 1985: pp. 143–169.

Levine, J. M., & Moreland, R. L. Social comparison and outcome evaluation in group contexts. In: Masters, J. C., & Smith, W. P. (Eds.), *Social comparison, social justice, and relative deprivation: Theoretical, empirical, and policy perspectives*. Hillsdale, NJ: Erlbaum, 1987: pp. 105–127.

Levine, J. M., & Moreland, R. L. Group socialization: Theory and research. In: Stroebe, W., & Hewstone, M. (Eds.), *European review of social psychology*, Vol. 5. New York: John Wiley & Sons, 1994: pp. 305–336.

Levine, J. M., & Moreland, R. L. Small groups. In: Gilbert, D., Fiske, S., & Lindzey, G. (Eds.), *The handbook of social psychology*, 4th ed., Vol. 2. Boston: McGraw-Hill, 1998: pp. 415–469.

Levine, J. M., & Moreland, R. L. Knowledge transmission in work groups: Helping newcomers to succeed. In: Thompson, L. L., Levine, J. M., & Messick, D. M. (Eds.), *Shared cognition in organizations: The management of knowledge*. Mahwah, NJ: Erlbaum, 1999: pp. 267–296.

Levine, J. M., & Moreland, R. L. Group reactions to loyalty and disloyalty. In: Thye, S. R., & Lawler, E. J. (Eds.), *Group cohesion, trust and solidarity: Advances in group processes*, Vol. 19. Oxford, UK: Elsevier Science, 2002: pp. 203–228.

Levine, J. M., & Moreland, R. L. (Eds.). *Small groups: Key readings in social psychology*. New York: Psychology Press, 2006.

Levine, J. M., Moreland, R. M., & Hausmann, L. Managing group composition: Inclusive and exclusive role transitions. In: Abrams, D., Hogg, M. A.,

& Marques, J. M. (Eds.), *The social psychology of inclusion and exclusion*. New York: Psychology Press, 2005: pp. 137–160.

Levine, J. M., Moreland, R. L., & Ryan, C. S. Group socialization and intergroup relations. In: Sedikides, C., Schopler, J., & Insko, C. A. (Eds.), *Intergroup cognition and intergroup behavior*. Mahwah, NJ: Erlbaum, 1998: pp. 283–308.

Levine, J. M., & Ranelli, C. J. Majority reaction to shifting and stable attitudinal deviates. Eur J Soc Psychol 1978; 8: 55-70.

Levine, J. M., & Ruback, R. B. Reaction to opinion deviance: Impact of a fence straddler's rationale on majority evaluation. Soc Psychol Q 1980; 43: 73–81.

Levine, J. M., Saxe, L., & Harris, H. J. Reaction to attitudinal deviance: Impact of deviate's direction and distance of movement. Sociometry 1976; 39: 97–107.

Levine, J. M., Sroka, K. R., & Snyder, H. N. Group support and reaction to stable and shifting agreement/disagreement. Sociometry 1977; 40: 214–224.

Lewis, M. *Liar's poker*. New York: Penguin Books, 1989.

Lewis, S. A., Langan, C. J., & Hollander, E. P. Expectation of future interaction and the choice of less desirable alternatives in conformity. Sociometry 1972; 35: 440–447.

Lodewijkx, H. F. M., & Syroit, J. E. M. M. Severity of initiation revisited: Does severity of initiation increase attractiveness in real groups? Eur J Soc Psychol 1997; 27: 275–300.

Lord, R. G., Foti, R. J., & de Vader, C. L. A test of leadership categorization theory: Internal structure, information processing, and leadership perceptions. Organ Behav Hum Perform 1984; 34: 343–378.

Mackie, D. M., & Goethals, G. R. Individual and group goals. In: Hendrick, C. (Ed.), *Group processes*. Newbury Park, CA: Sage, 1987: pp. 144–166.

Marques, J. M., Abrams, D., Paez, D., & Hogg, M. A. Social categorization, social identification, and rejection of deviant group members. In: Hogg, M. A., & Tindale, R. S. (Eds.), *Blackwell handbook of social psychology: Group processes*. Malden, MA: Blackwell, 2001a: pp. 400–424.

Marques, J. M., Abrams, D., & Serodio, R. G. Being better by being right: Subjective group dynamics and derogation of in-group deviants when generic norms are undermined. J Pers Soc Psychol 2001b; 81: 436–447.

Marques, J. M., & Paez, D. The "Black Sheep Effect": Social categorization, rejection of ingroup deviates, and perception of group variability. Eur Rev Soc Psychol 1994; 5: 37–68.

Marques, J. M., & Yzerbyt, V. Y. The black sheep effect: Judgmental extremity towards ingroup members in inter- and intra-group situations. Eur J Soc Psychol 1988; 18: 287–292.

Marques, J. M., Yzerbyt, V. Y., & Leyens, J.-P. The "Black Sheep Effect": Extremity of judgments towards ingroup members as a function of group identification. Eur J Soc Psychol 1988; 18: 1–16.

Martin, J., & Murray, A. Catalysts for collective violence: The importance of a psychological approach. In: Folger, R. (Ed.), *The sense of injustice: Social psychological perspectives*. New York: Plenum, 1984: pp. 95–139.

Martin, R., & Hewstone, M. Conformity and independence in groups: Majorities and minorities. In: Hogg, M. A., & Tindale, S. (Eds.), *Blackwell handbook of social psychology: Group processes*. Malden, MA: Blackwell, 2001: pp. 209–234.

Matheson, K., Cole, B., & Majka, K. Dissidence from within: Examining the effects of intergroup context on group members' reactions to attitudinal opposition. J Exp Soc Psychol 2003; 39: 161–169.

McAuliffe, B. J., Jetten, J., Hornsey, M. J., & Hogg, M. A. Individualist and collectivist group norms: When its OK to go your own way. Eur J Soc Psychol 2003; 33: 57–70.

McGrath, J. E. *Groups: Interaction and performance*. Englewood Cliffs, NJ: Prentice-Hall, 1984.

Miller, C. E., & Anderson, P. D. Group decision rules and the rejection of deviates. Soc Psychol Q 1979; 42: 354–363.

Moreland, R. L., & Levine, J. M. Socialization in small groups: Temporal changes in individual-group relations. In: Berkowitz, L. (Ed.), *Advances in experimental social psychology*, Vol. 15. New York: Academic Press, 1982: pp. 137–192.

Moreland, R. L., & Levine, J. M. Role transitions in small groups. In: Allen, V. L., & van de Vliert, E. (Eds.), *Role transitions: Explorations and explanations*. New York: Plenum Press, 1984: pp. 181–195.

Moreland, R. L., & Levine, J. M. Newcomers and oldtimers in small groups. In: Paulus, P. B. (Ed.), *Psychology of group influence*, 2nd ed. Hillsdale, NJ: Erlbaum, 1989: pp. 143–186.

Moscovici, S. *Social influence and social change*. London: Academic Pres, 1976.

Moscovici, S. Toward a theory of conversion behavior. In: Berkowitz, L. (Ed.), *Advances in experimental social psychology*, Vol. 13. New York: Academic Press, 1980: pp. 209–239.

Mullen, B., & Copper, C. The relation between group cohesiveness and performance: An integration. Psychol Bull 1994; 115: 210–227.

Nail, P. R., MacDonald, G., & Levy, D. A. Proposal of a four-dimensional model of social response. Psychol Bull 2000; 126: 454–470.

Nemeth, C. J. Dissent as driving cognition, attitudes and judgments. Soc Cogn 1995; 13: 273–291.

Neuberg, S. L., Smith, D. M., & Asher, T. Why people stigmatize: Toward a biocultural framework. In: Heatherton, T. F., Kleck, R. E., Hebl, M. R., & Hull, J. G. (Eds.), *The social psychology of stigma*. New York: Guilford Press, 2000: pp. 31–61.

Orbell, J., & Dawes, R. Social dilemmas. In: Stephenson, G., & Davis, J. H. (Eds.), *Progress in applied social psychology*, Vol. 1. Chichester, UK: Wiley, 1981: pp. 37–66.

Ouwerkerk, J. W., Kerr, N. L., Gallucci, M., & van Lange, P. A. M. Avoiding the social death penalty: Ostracism and cooperation in social dilemmas. In: Williams, K. D., Forgas, J. P., & von Hippel, W. (Eds.), *The social outcast: Ostracism, social exclusion, rejection, and bullying*. New York: Psychology Press, 2005: *pp. 321–332.*

Packer, D. J. On being both with us and against us: A normative conflict model of dissent in social groups. Pers Soc Psychol Rev 2008; 12: 50–72.

Postmes, T., & Spears, R. Deinidividuation and antinormative behavior: A meta-analysis. Psychol Bull 1998; 123: 238–259.

Prislin, R., & Christensen, P. N. Social change in the aftermath of successful minority influence. In: Stroebe, W., & Hewstone, M. (Eds.), *European review of social psychology*, Vol. 16. New York: Taylor and Francis, 2005: pp. 43–73.

Prislin, R., Levine, J. M., & Christensen, P. N. When reasons matter: Quality of support affects reactions to increasing and consistent agreement. J Exp Soc Psychol 2006; 42: 593–601.

Richard, F. D., Bond, C. F. Jr., & Stokes-Zoota, J. J. One hundred years of social psychology quantitatively described. Rev Gen Psychol 2003; 7: 331–363.

Roethlisberger, F. J., & Dickson, W. J. *Management and the worker*. Cambridge, MA: Harvard University Press, 1939.

Runcimann, W. G. *Relative deprivation and social justice: A study of attitudes to social inequality in twentieth-century England*. London: Routledge and Kegan Paul, 1966.

Sakurai, M. M. Small group cohesiveness and detrimental conformity. Sociometry 1975; 38: 340–357.

Sani, F. When subgroups secede: Extending and refining the social psychological model of schism in groups. Pers Soc Psychol Bull 2005; 31: 1074–1086.

Sani, F., & Reicher, S. When consensus fails: An analysis of the schism within the Italian Communist Party (1991). Eur J Soc Psychol 1998; 28: 623–645.

Sani, F., & Todman, J. Should we stay or should we go? A social psychological model of schisms in groups. Pers Soc Psychol Bull 2002; 28: 1647–1655.

Sarbin, T. R., & Allen, V. L. Role theory. In: Lindzey, G., & Aronson, E. (Eds.), *The handbook of social psychology*, 2nd ed., Vol. 1. Reading, MA: Addison-Wesley, 1968: pp. 488–567.

Schachter, S. Deviation, rejection, and communication. J Abnorm Soc Psychol 1951; 46: 190–207.

Scheff, T. *Being mentally ill: A sociological theory*. Chicago: Aldine, 1966.

Schlesinger, A. M. Jr. *A thousand days*. Boston: Houghton-Mifflin, 1965.

Schmitt, M. T., & Branscombe, N. R. The good, the bad, and the manly: Threats to one's prototypicality and evaluations of fellow in-group members. J Exp Soc Psychol 2001; 37: 510–517.

Schulz-Hardt, S., Brodbeck, F. C., Mojzisch, A., Kerschreiter, R., & Frey, D. Group decision making in hidden profile situations: Dissent as a facilitator for decision quality. J Pers Soc Psychol 2006; 91: 1080–1093.

Smith, C. M., Tindale, R. S., & Dugoni, B. L. Minority and majority influence in freely interacting groups: Qualitative versus quantitative differences. Br J Soc Psychol 1996; 35: 137–149.

Stangor, C., & Crandall, C. S. Threat and social construction of stigma. In: Heatherton, T. F., Kleck, R. E., Hebl, M. R., & Hull, J. G. (Eds.), *The social psychology of stigma*. New York: Guilford Press, 2000: pp. 62–87.

Stasser, G., Kerr, N. L., & Davis, J. H. Influence processes and consensus models in decision-making groups. In: Paulus, P. B. (Ed.), *Psychology of group influence*, 2nd ed. Hillsdale, NJ: Erlbaum, 1989: pp. 279–326.

Sturmer, S., & Simon, B. Collective action; towards a dual-pathway model. In: Stroebe, W., & Hewstone, M. (Eds.), *European review of social psychology*, Vol. 15. New York: Psychology Press, 2004: pp. 59–99.

Suchner, R. W., & Jackson, D. Responsibility and status: A causal or only a spurious relationship? Sociometry 1976; 39: 243–256.

Tajfel, H., & Turner, J. An integrative theory of intergroup conflict. In: Austin, W. G., & Worchel, S. (Eds.), *The social psychology of intergroup relations*. Monterey, CA: Brooks/Cole, 1979: pp. 33–47.

Thornhill, R., & Gangestad, S. W. Human facial beauty: Averageness, symmetry, and parasite resistance. Hum Nat 1993; 4: 237–269.

Turner, J. C. *Social influence*. Pacific Grove, CA: Brooks/Cole, 1991.

van Dijk, E., & Vermunt, R. Strategy and fairness in social decision making: Sometimes it pays to be powerless. J Exp Soc Psychol 2000; 36: 1–25.

Van Maanen, J., & Schein, E. H. Toward a theory of organizational socialization. Res Organ Behav 1979; 1: 209–264.

Van Vugt, M., & Hart, C. M. Social identity as social glue: The origins of group loyalty. J Pers Soc Psychol 2004; 86: 585–598.

Vohs, K. D., & Ciarocco, N. J. Interpersonal functioning requires self-regulation. In: Baumeister, R. F., & Vohs, K. D. (Eds.), *Handbook of self-regulation: Research, theory, and applications*. New York: Guilford Press, 2004: pp. 392–407.

Wheeler, L., & Caggiula, A. R. The contagion of aggression. J Exp Soc Psychol 1966; 2: 1–10.

Willis, R. H. Two dimensions of conformity-nonconformity. Sociometry 1963; 26: 499–513.

Williams, K. D. Ostracism. Annu Rev Psychol 2007; 58: 15.1–15.28.

Williams, K. D., Forgas, J. P., & von Hippel, W. (Eds.). *The social outcast: Ostracism, social exclusion, rejection, and bullying*. New York: Psychology Press, 2005.

Williams, K. D., Harkins, S. G., & Karau, S. J. Social performance. In: Hogg, M. A., & Cooper, J. (Eds.), *The Sage handbook of social psychology*. Thousand Oaks, CA: Sage, 2003: pp. 327–346.

Wood, W. Motives and modes of processing in the social influence of groups. In: Chaiken, S., & Trope, Y. (Eds.), *Dual-process theories in social psychology*. New York: Guilford Press, 1999: pp. 547–570.

Wood, W., Lundgren, S., Ouellette, J. A., Busceme, S., & Blackstone, T. Minority influence: A meta-analytic review of social influence processes. Psychol Bull 1994; 115: 323–345.

Wood, W., Pool, G. J., Leck, K., & Purvis, D. Self-definition, defensive processing, and influence: The normative impact of majority and minority groups. J Pers Soc Psychol 1996; 71: 1181–1193.

Wright, S. C. Strategic collective action: Social psychology and social change. In: Brown, R., & Gaertner, S. (Eds.), *Blackwell handbook of social psychology: Intergroup processes*. Malden, MA: Blackwell, 2001: pp. 409–430.

Zajonc, R. B. Social facilitation. Science 1965; 149: 269–274.

Zdaniuk, B., & Levine, J. M. Group loyalty: Impact of members' identification and contributions. J Exp Soc Psychol 2001; 37: 502–509.

Ziller, R. C. Toward a theory of open and closed groups. Psychol Bull 1965; 64: 164–182.

CHAPTER 25

Justice as Social Self Control

Tom R. Tyler

ABSTRACT

This chapter examines the idea of justice from a self-control perspective. It argues that justice involves socially shared rules whose function is to facilitate people's efforts to manage social interactions. Because of the benefits of social interactions, people want to live in social groups and cooperate with others. However, doing so requires them to recognize what constitutes a reasonable balance between doing what benefits them and doing what benefits others. Rules of social justice define that reasonable balance and, in so doing, make social life more viable. This is directly true with principles of distributive justice, which indicate who should receive what. It is indirectly true of principles of procedural justice, which define how authorities should decide who should receive what. In both cases, reliance on justice makes the functioning of relationships and groups more efficient and effective.

Keywords: Distributive justice, procedural justice, social exchange, social identity, group engagement model

People live most of their lives within social contexts—that is, in interaction with other people in dyads, small groups, organizations, and societies. Those social contexts vary in many ways, but they all pose a similar challenge for human beings: to manage interactions that require the modification of one's own behavior based on the desires of others. Such modifications are inherently problematic when a core human motivation for interacting with others is the pursuit of personal self-interest—namely, the satisfaction of personal desires. To be able to participate in social settings, people have to be both willing and able to modify their actions to account for the interests of others—that is, they cannot simply do what they want. Successful social interaction involves these two key issues: motivation and ability.

People first have to be motivated to adjust their behavior to respond to the concerns of others; they have to be interested in cooperation. The traditional social exchange framework argues that their reason for so doing is to gain the personal advantages of joint action, advantages that are greater than those obtained by acting alone. In other words, people can gain more for themselves overall by cooperating with others, but they may not receive what they want in any particular situation.

It is obvious to most people that they are better off living in a social group than they would be if they lived alone even if they have to modify their behavior to some degree. Social groups allow for specialization of tasks and make it possible to engage in joint efforts that are beyond the capability of any single person, as anyone who has ever needed a lawyer, doctor, or accountant quickly realizes. They expand the resources that are available, leading to the possibility that everyone becomes better off in a group setting. Most of us cannot imagine a life outside of these dense social frameworks, nor would we want to have to live without the benefits of cooperation, despite any potential romantic allure of novels such as *Robinson Crusoe*. Although sometimes frustrated and denied, people are better off working with others, as evidenced by the course of social evolution over time into ever-denser social groupings.

Second, people who are motivated to cooperate with others need to have the ability to do so effectively. It is this *capacity* to cooperate that is typically associated with self-control. People have to be able to resist the motivation to act in their immediate self-interest in every setting. Many self-control problems are associated with actions that pit a person's own short- and long-term interests. For example, people who cannot resist the impulse to consume food in the short-term have the long-term problem of becoming overweight. People who spend in the moment do not save enough for their retirement. These problems of self-control are personal, not social, because they are not linked to the presence of others, but they illustrate the problem of capacity—people want to engage in actions that are beneficial in the long-term but are unable to do so.

There is also a similar set of self-control problems that are social in nature. These involve the need to defer one's personal desires to make cooperation with other people possible. If people want to interact successfully with others, then they have to be able to defer their own interests to those of others on some occasions. A classic example of such a social coordination problem is taking turns (Piaget, 1932). Children on a playground often ask their friends to push them when they are on a swing. And the pusher typically expects that there will be reciprocity, so that they will also be pushed. If a child lacks the ability to defer their own desires and insists on always being pushed, then they will rapidly run out of playmates. Hence, the ability of exercise self-control is central to cooperating with others. The self-control issue involves the ability to control one's own desires when interacting with others. Just as eating too much has negative long-terms costs, so does insisting on having one's way in all social settings.

Of course, there are other issues, for example social intelligence. This is the ability to take the role of the other, which involves having the capacity to effectively understand what others want. In addition to needing to have the ability to exercise self-control, it is important to have the capacity to understand social signals and to be able to effectively coordinate with others. People have to be able to read others so that when they are trying to cooperate they take actions that are effective. Effective actions are actions that produce reciprocal cooperation on the part of others. If I think that I will encourage cooperation by making crude jokes about someone's mother, then I have a poor understanding of what works in social settings. I may be highly motivated to coordinate with others, but do not understand the actions I should take to achieve that objective.

Problems of personal self-control are sometimes presented as asymmetrical—that is, they are framed as if the issue lies only with those who eat too much or save too little. Yet, we can also recognize that there are those who struggle with the opposite problems of eating too little or saving too much, patterns that may also have negative personal consequences. A person who never spends money and dies with millions in the bank may be viewed as also having a self-control problem, just as does someone who eats so little that they suffer health problems. The problem for a person is to establish an optimal level, neither too little nor too much. The effective person, in other words, neither starves nor gets fat.

Similar issues are involved in social aspects of self-control. People seek to establish an optimal level of self-interest when dealing with others. If people seek too much, others will not be willing to interact with them and they will lose the benefits of social interaction. If they seek too little, they will end up being exploited by others (a "sucker," something that Loewenstein, Thompson, and Bazerman [1989] show people generally consider a very undesirable state). As an example, the child who wants to be constantly pushed on a swing would love to find someone who is willing to constantly push them. However, the constant pusher is not benefiting from the social interaction, so their own motivation to leave ought to be strong, and in normal people it is. The key is to optimize, providing enough for others to keep them in the relationship, and enough for oneself to make the relationship desirable. And, of course, achieving this balance can be complex, because the value of the resources and services being exchanged must be estimated. How many hours of tax preparation does one owe the person who has just saved their life by performing brain surgery?

How do these comments about self-control relate to the topic of this chapter—justice? I argue that justice is a socially created shared judgment whose function is to help people to optimize their behavior in social settings by indicating to them what level of compromise is best within a given setting. Justice refers to the rules through which people evaluate whether something is ethical, appropriate, reasonable, or fair. For example, distributive justice involves the principles that tell you if outcomes are being allocated fairly to people in a group setting. If people say it is fair to divide a cake into equal slices for everyone, they are invoking a principle of distributive justice. Procedural justice involves assessments of the justice of the procedures used to make decisions. For example, a jury trial is a procedure for making a decision of guilt or innocence. Whether the trial itself is fairly conducted is distinct from whether the verdict is just. Finally, there is retributive justice, the judgments people make about what is a fair punishment when someone breaks a social rule. This chapter deals with issues of distributive and procedural justice.

I assume that people are motivated to cooperate but are more likely to do so when they can determine what reasonable forms of cooperation are. Hence, justice helps people because it allows them to be more able to see what a reasonable strategy for approaching social interactions looks like. If people follow principles of justice, they are more confident that they have solved the problem of optimization in a reasonable and appropriate way. Combined with their motivation to cooperate, justice-based optimizations create the conditions under which people can and do cooperate. Of course, these justice rules are social creations. They reflect the shared view of a group about which principles ought to dictate people's outcomes in social settings. Justice principles only facilitate social interaction when they are shared.

A core aspect of shared social rules is that they do not reflect the self-interest of any single individual in the group. Rather, they are a collective effort to define reasonable principles for social interaction. John Rawls captures this quality of justice rules when he talks about the idea of rules developed "behind the veil of ignorance." People try to make rules, he argues, without referencing their own situation. We would, for example, decide if we should reward people for having skills of a particular type without knowing if we would have those skills ourselves (Rawls, 1971). Of course, in reality people cannot separate their personal attributes and status from their thoughts, feelings, and preferences, so justice rules are a compromise between the interests of various people and groups, as are the rules and laws that guide them. However, the core argument is the same—justice principles are shared group assessments of what is reasonable as a compromise among competing self-interests within a given type of situation.

DISTRIBUTIVE JUSTICE AND SELF-CONTROL

What can be done to help people to establish and maintain optimal levels of seeking for the self versus deference to others in a society composed of self-interested people, each trying to optimize their gains and losses in social settings? One approach is to establish rules that communicate to people where the balance point lies between seeking too much and accepting too little. I suggest that these rules of reasonableness are framed as justice principles. They communicate what is appropriate and "fair." Such rules are beneficial to all, because they create a framework within which productive and profitable social interactions can occur. People can focus on whether they are getting what they deserve, and there are clear rules to indicate what that means. Adherence to these rules allows people to feel that their behavior will sustain mutually beneficial social exchanges. Consequently, they are more willing to make and adhere to agreements with others that follow these principles.

For justice rules to be effective as guides for social coordination, three things must be true. First, there must be consensus about what the principles of justice are. Second, people must be willing to follow those justice rules in the sense that they accept them as guides concerning appropriate compromises with others. Third, people must be willing to enforce these justice rules when others do not follow them. If others try to take "too much" or even "too little," people should be upset and take some form of action. We can first examine these ideas in the context of distributive justice. Rules of justice that pertain to the appropriate distribution of resources between people are referred to as *rules of distributive justice*. These rules tell us what a person is entitled to or "deserves" under particular social conditions.

Consensus is the first important idea underlying the social utility of justice rules. Justice is a social judgment. Unlike a table, a chair, or another person, there is no physical reality to justice. The concept of justice is created by people in a group to facilitate their interaction, and it only effectively fulfills that function when its meaning is shared among the members of a group. Justice is like a word in a language. A language spoken by one person has no communicative value—the meaning of words must be shared.

The principles of distributive justice are complex, and no single principle is used by people to reflect distributive justice in all types of settings (Deutsch, 1975). It is interesting, therefore, that evidence from studies conducted in the United States suggests that within a given context, there is a general social consensus about what principles govern. For example, in the United States people generally agree that work settings should be governed by equity, social settings by equality, and social welfare settings by need (Tyler, 1984). This consensus is crucial, because without it justice cannot effectively coordinate action. Of course, there are many other possible justice principles (Reis, 1987). However, research suggests that equity, equality, and need are the three core principles most widely used, at least in the United States (Deutsch, 1985). Of course, it is not necessary that people within all societies agree about principles of distributive justice. Just as everyone in the world need not speak the same language, the only issue is whether people who interact have a shared social framework within which to coordinate their actions. Hence, cross-cultural studies show that different societies have different rules of distributive fairness within particular arenas but that within those societies people tend to agree about a common rule (Tyler et al., 1997).

Evidence further suggests that people are willing to defer to the principles of distributive justice. In particular, they are most satisfied with outcomes if they think that the outcomes they are receiving are fair. So, a person is happier to receive $2 if they believe that $2 is the "fair" wage for a task. A typical example of this finding is the study by Pritchard, Dunnette, and Jorgenson (1972). In that study, students were paid a given amount of money to perform a task. However, they were then told that the wage was fair, unfairly high, or unfairly low. Those who thought they were being fairly paid were the most likely to express satisfaction, whereas those who were over- or underpaid were less satisfied.

Other studies have shown that people adjust their effort to reestablish justice. For example, if people are told they are overpaid, they work harder (Walster, Walster, & Berscheid, 1978). This can involve higher levels of output and/or working longer hours. Whatever actions are taken, however, they are motivated by the desire to reestablish justice. Conversely, the amount that employees steal from their workplace has been shown to be linked to their judgments about pay unfairness, with workers stealing more to restore justice if they feel more underpaid.

Finally, studies have suggested that people will leave situations in which they feel unfairly overpaid to go to situations in which they are paid less but feel that their pay is more fair (Schmitt & Marwell, 1972).

These findings all suggest that people accept principles of distributive justice. Of course, we need to recognize that in many of the experimental studies people are explicitly told that their allocations are fair or unfair. It is important, therefore, that there is also nonexperimental research to indicate that in field settings people's naturally occurring evaluations of the justice or injustice of their outcomes in groups shape their behavior within those groups, as well as their likelihood of leaving or staying (Tyler et al., 1997).

Finally, more recent research has also suggested that people will forgo personally beneficial outcomes in an effort to enforce adherence to fairness rules by others (Henrich et al., 2006). As an example, Gurerk, Irlenbusch, and Rockenbach (2006) gave people the opportunity to choose a setting in which they could enforce cooperative rules at a personal cost. They showed that many people chose this setting, and when they were in it, were willing to forgo personal gains to punish those people who deviated from cooperative choices. In other words, people were willing to pay costs to enforce fairness rules by punishing those who violated those rules. Recent research on such "altruistic" punishment suggests that personally rewarding neural processes are activated when people engage in behavior designed to protect group rules, suggesting that people's motivation to protect group rules may develop within fundamental human cognitive and emotional processes (Fehr & Gintis, 2007).

Another example of people's willingness to enforce rules of distributive justice is their willingness to support the redistribution of resources to make that distribution more fair. One example is Montada and Schneider (1989), who demonstrated that West Germans were willing to redistribute resources to East Germans when they viewed themselves as unfairly advantaged.

This is one of many examples showing that people who feel that they are unfairly advantaged are willing to voluntarily redistribute resources to benefit those whom they view as having too little. People also change their level of active effort on behalf of groups, working harder or slacking off when involved in collective efforts, again in response to justice judgments (Tyler & Blader, 2000). In either case, people incur personal losses to uphold principles of distributive justice.

Does justice work as intended? If it does, then relationships characterized by distributive fairness are more stable and longlasting. This is the finding of studies of long-term relationships. Those relationships in which the parties experience the interactions as consistent with distributive fairness have a better psychological quality and last longer (Sprecher, 2001; Ybema et al., 2001). Therefore, justice norms have the desired effect of facilitating satisfying and stable long-term social interactions.

These arguments suggest that one way people manage the self-control problem in social settings is by agreeing on the standards or rules that define the appropriate balance of self-versus-other concern that should govern interactions (Thibaut & Faucheux, 1965). Those rules then encourage people to exercise self-control by focusing upon adherence to these generally agreed-upon standards. Rather than seeking to maximize self-interest, people accept outcomes that reflect what they deserve. And, as we might expect, people are more likely to focus on fairness rules when they view the relationships involved as more potentially unstable (Barrett-Howard & Tyler, 1986). People recognize that it is especially important to act fairly when there are situational forces working against the stability of the relationship, because fairness strengthens those relationships and lowers the likelihood that they will collapse. Additionally, people care more about justice when they are dealing with people with whom they have valuable and productive exchange relationships (Deutsch, 1985).

Although research has supported the basic arguments of distributive justice, it also points to problems in the implementation of solutions to social coordination based on the use of distributive justice. One problem is that people exaggerate their contributions to cooperative efforts and, hence, have conceptions of entitlement that differ from those of objective observers. This makes it difficult to use principles besides equality. One consequence is that equality heuristics are often used to avoid conflict and to promote social harmony. When equity or need is involved, some mechanism must be found to rank people. This is hard for the involved parties, so it requires some form of third-party neutrality. As an example, *Getting to Yes,* a widely cited guide on how to negotiate effectively, advocates using neutral, fact-based standards when negotiating. To do so, someone might try to establish the fair price for a car by citing a table of used car values. Another approach, which is outlined below, is to use neutral authorities.

The problem of exaggerated perceptions of entitlement reflects the general tendency for people to define justice through the prism of self-interest. At the most extreme, "justice" simply reflects self-interest and has no ability to bridge interests. In that situation, justice cannot serve as a facilitator of self-control. Fortunately, people are typically found to distinguish between self-interest and justice, even when their views about justice are affected by self-interest. Hence, people are often found to compromise, shaping their sense of justice by their understanding of what is good for them but also recognizing and being influenced by appeals to justice that are not in their self-interest. The tendency to develop a self-serving view of justice is diminished when principles of justice are clearly articulated, either by an authority representing the group, by group norms, or in some other way. For example, studies have shown that clearly partitioned resources are divided more fairly, whereas ambiguously partitioned resources encourage more self-serving allocations (Allison et al., 1992; Herlocker et al., 1997). Hence, societies and social authorities facilitate effective social exchange by providing a clear definition of what is fair in a given setting.

As these comments suggest, self-control involves several problems. First, people have to be able to restrain their motivation to define justice in self-serving ways if they are to achieve standards of fairness that are mutually acceptable. Then, once rules are defined, people have to have the self-control to follow those rules rather than seeking to take "too much," which is behavior that is likely to provoke rejection and undermine relationships.

This tendency to take too much has been widely studied in recent years in the ultimatum game, a setting in which one person controls resources and makes an offer to another to divide them. If that offer is accepted, both can keep the resources allocated by the offer. Whereas self-interest might motivate the person controlling resources to offer the other person only a minor amount (e.g., $1 out of a total of $10), studies have suggested that such behavior is very likely to provoke a rejecting response, leading neither party to gain anything from the allocation. Instead, studies have suggested that those who control the resource seek to balance between self-interest and justice. They offer amounts that generally are less than an equal division but give people more than would be predicted by the argument that the receiver would accept anything more than zero. In making their offers, allocators balance between self-interest and justice (Fehr & Gintis, 2007).

The argument that distributive justice rules facilitate stable social interaction is consistent with the general suggestion that rules are helpful in maintaining productive relationships. This argument has been supported by the finding of other studies, noted above, which have indicated that groups both define justice (Barrett-Howard & Tyler, 1986) and create binding rules to enforce allocation by those justice norms when faced with vulnerable relationships that might easily collapse (Thibaut & Faucheux, 1965). Hence, it is important both to make formal rules and to have defined principles of justice as a basis for crafting them.

Procedural Justice and Self-Control

The problem with rules as a solution to social coordination problems is illustrated by the existence of different principles of distributive justice. No single rule can cover all situations. Hence, there are conflicts among rules. A worker with a sick child invokes the principle of need to justify leaving their job, whereas the manager invokes equity to argue that the lost productivity should be held against them. No single rule is sufficiently complex to handle the problems and issues arising in social interactions. Rules are not as flexible, for example, as appointing someone and giving them discretion to do what is right in whatever situation arises. Hence, the creation of rules is only one of the mechanisms that groups have for creating stable social interactions. A second approach is the creation of institutions and authorities with discretion to do what is right within the situation (Messick et al., 1983; Tyler & Degoey, 1995).

Authorities have the advantage of being able to exercise discretion—that is, they can adapt to unique or changing situations by deciding to apply different rules or even by changing what the rules are. Hence, it is not surprising that one commonly chosen approach that groups adopt to determine "fair" approaches to allocation is to choose an authority and then let that authority exercise discretion. That discretion provides a mechanism through which equity, need, or other similar principles can operate in complex situations (Tyler & Degoey, 1995).

An example of this view of leadership is contained within the work of Thibaut and Walker (1975). That research examines the use of authorities to resolve disputes. The authors argue that the key issue in the fair resolution of disputes is understanding how to apply the principle of equity. However, giving the parties what they deserve requires an understanding of their relative contributions to the relationship. To understand those contributions, they recommend that leaders use a procedure in which they allow the parties to provide evidence before they make a decision. By providing evidence of their contributions, the parties enable the authority to exercise their discretion and make a decision that gives each party to the dispute the outcome they "deserve." Additionally, the authority can better apply a justice rule (equity) that requires estimation of each person's contribution because they can apply that rule without self-serving exaggerations. The ability of authorities to apply principles such as equity and need, which require evaluations of people's unique situations, without self-serving biases is one of the reasons that authorities are important in resolving conflicts and are better able to do so effectively.

The use of authorities to make discretionary decisions helps make social life more flexible but also raises additional problems for groups. There have to be mechanisms for determining who will be the authority and for evaluating whether the decisions made by the authority are reasonable and fair. The first question, establishing an authority, lies at the heart of early work by Lewin on elected leaders (Gold, 1999), as well as being central to more recent studies of political leadership. Studies consistently find that people are more likely to defer to leaders elected to authority via fair procedures. Hence, people are sensitive to how authorities are empowered.

The second issue, the fairness of the manner in which authority is exercised is at the heart of most recent procedural justice research (Lind & Tyler, 1988; Tyler, 2000). Research on this question comes to a strikingly similar conclusion. Authorities are more likely to be deferred to when they make their decisions using fair procedures. Of course procedural justice involves many types of procedures, not just those dealing with conflicts. For example, when resources are allocated or burdens are imposed, then procedures are used. In fact studies show that even procedures that do not involve authorities, such as bargaining and market transactions, are evaluated in terms of procedural justice (Hollander-Blumoff & Tyler, 2008; Sondak & Tyler, 2007).

To distinguish procedural justice from distributive justice, we should consider core features of procedural justice. There are two central features. First, decisions should be made in fair ways. This includes allowing people to state their arguments; consistently applying neutral rules; using facts; and avoiding personal biases. Second, people should be treated fairly. They should be respected as people; treated with dignity and courtesy; shown respect for their rights; and dealt with by authorities who are sincerely trying to make decisions that are good for the people involved.

The existence of consensus about distributive justice rules has already been mentioned. There is also found to be widespread consensus about fair procedures for resolving conflicts. This does not mean that there is a universally fair procedure. As was the case with distributive justice, people view different procedures as fair ways to resolve different types of problems (Tyler, 1988). For example, they suggest that a procedure with opportunities for input is particularly important with disputes. When rule violations are involved, having a trustworthy authority to evaluate the wrong reflected in the violation is viewed as key. So, mediation is a good procedure for dispute resolution but adjudication is desirable for deciding how to punish criminals.

Although people's views about the fairest procedure for managing different types of problem differ, for any particular type of problem there is general consensus about what a fair procedure is. There are usually no differences linked to ethnicity, gender, class, income, education, ideology, or political orientation. In other words, the type of consensus is the same as that found with distributive justice. Additionally, it is the type of consensus required for justice to work for social coordination. The parties to a particular social situation will agree about whether the procedure being used to make decisions is a fair one, even when they differ in their social characteristics.

The finding that people will defer to fairly made decisions is strong (Lind & Tyler, 1988). Numerous studies have shown that people are more accepting of decisions made by third-party authorities when those authorities use fair procedures (Tyler, 2000). Further, in groups and organizations, people adhere to rules when those rules are fairly made and implemented. This is true of workplace rules (Tyler & Blader, 2000) and of laws (Tyler, 2006a). Additionally, as noted, people are also sensitive to procedures in relationships that do not involve authorities (Barrett-Howard & Tyler, 1986).

Will people incur costs to preserve procedural justice rules, as they have been shown to be willing to do with principles of distributive justice? There is no equivalent to the behavioral economics literature, in which people are asked to incur personal costs to preserve a rule. However, there is a general finding that people will more willingly defer to those leaders who uphold principles of procedural justice in their actions. Hence, people do show the willingness to support systems of authority that are procedurally fair, even when they do not personally gain from those systems (Tyler, 2006a, b). In fact, when people lose—for example, by having to pay a fine after being convicted of violating a rule—they do so more voluntarily when procedures are fair. In this sense, people will incur costs to support rules, because they pay even when they could realistically avoid payment or at least try to subvert it.

The link of procedural justice to authorities and institutions has been found to be especially strong. When people are evaluating authorities, their primary criterion for judging those authorities is their assessment of procedural justice (Tyler & Lind, 1992). This suggests that people support authorities who are exercising their authority fairly while rejecting those who do not. And, most importantly, these procedural justice judgments are more important than evaluations of either outcome favorability or outcome fairness. In other words, authorities and institutions are judged by their processes, not the outcomes of those processes. Of course, this distinction should not be exaggerated. People are found to believe that a fair procedure is more likely to produce a fair outcome, so by evaluating authorities in procedural terms they are reacting to a system that they believe generally supports distributive fairness.

THE TRANSFORMATION OF MOTIVES

The prior sections have focused the discussion of social interaction and the role of justice in such interaction in terms that are consistent with a social exchange perspective. People are argued to be involved in interaction to pursue long-term gains. They value justice rules because having those rules helps to regularize and stabilize interaction, so that those interactions are beneficial to themselves in the long run. All that has been said is based on the image of people as rational calculators, carefully assessing gains and losses in social settings. People are in exchange relationships. However, people can also become involved in long-term relationships in which they cease to focus on their immediate personal self-interest, and rather have a more general loyalty to the relationship. In such a situation people are still acting out of self-interest but define self-interest in terms of the long-term maintenance of the relationship. The transition from an exchange to a communal relationship is one example of moving from a focus on particular rewards to a general long-term loyalty to a relationship that, over time, will be valuable (Clark, 1984).

Perhaps the most striking aspect of social interaction is the argument that it has the possibility to change the conception of what constitutes self-interest. In the case of long-term relationships, for example, people are no longer calculating their own and others short-term inputs and outcomes in the relationship. Rather, they have a general commitment to do what is needed to make the relationship work effectively (DeCremer & Tyler, 2005). Their commitment shifts to a focus on the overall nature of the social connection, rather than the results of a particular interaction. This does not mean that people no longer care about themselves. Rather, they develop a way of thinking about their self-interest that extends over time. This has two consequences. First, it shifts people to a long-term focus. Second, it leads to a focus on the quality and nature of the relationship as a cue about one's long-term prospects, rather than on particular gains and losses.

The idea of a long-term focus can fit easily into the social exchange approach. People may want outcomes from others but be focused on outcomes over time. This idea is a core suggestion of the investment model proposed by Caryl Rusbult. This model suggests that people have a long-term focus on outcomes. They do not calculate within a particular situation but are acting on the desire to assure desirable outcomes over time (Rusbult & Van Lange, 1996). Here justice performs an important signaling function by communicating the likelihood of long-term gains and losses. Hence, justice helps people to calibrate the consequences of loyalty to a relationship. A just relationship is likely to neither lead one to be exploited nor to tempt one to take too much and undermine the long term viability of the relationship.

Beyond the argument that people shift from a short-term to a long-term focus, there is an additional idea that emerges from discussions of the socializing effects of groups on people. That idea is that the personal implications of the relationship for identity and self begin to substitute for a focus on gains or losses, short- or long-term. People increasingly focus on questions of their status in groups and/or the quality of their relationships with others, because high-quality relationships and high stature in groups are desirable due to their positive influence on identity (Tyler & Blader, 2000). So, the connection to the group becomes more than just an issue of long-term instrumental gains. People's connection to the group becomes an issue that is functionally autonomous and has an importance on its own. People want to have a positive identity and feelings of self-worth and define those in terms of their stature in groups. This fact is made very clear by minimal group studies in which loyalty to groups produces behavioral effects in situations designed to define groups in ways that have no implications for the distribution of resources (Hogg & Abrams, 1988). Once people define themselves in terms of a group membership and link their self-worth to the status of that group, enhancing group status becomes something that motivates action.

As noted, the possibility of social connections changing the concerns that people have in relationships is most clearly suggested by social identity theory, a theory that argues that status is central to self and identity and that status comes from groups. This theoretical framework argues that people merge their sense of self with groups, and self and group become one and the same. If people identify with others, either with a group, organization, or society (Hogg & Abrams, 1988) or with particular others with whom they have a personal relationship (Davis, Tyler, & Andersen, 2007), then they become interested in doing what benefits the collectivity and work on behalf of this larger entity (Tyler & Blader, 2000, 2003). People focus on identity, with outcomes only becoming important when they have identity-relevant implications. Fundamentally, people are interested in the ability of groups to help them create and maintain a favorable sense of self. Therefore, they want the group to succeed and they work on its behalf. This approach to groups alters the way we think about why people are in short- or long-term relationships.

The key point of this argument is that our traditional way of thinking about self-control loses its meaning when the psychology underlying people's relationship to others changes. If people cease to calculate their personal self-interest, and simply decide that their loyalty lies with the long-term well-being of the group, organization, or society of concern, then it is possible to see identification as a form of social coordination, and hence as an aspect of self-control. Identification, like justice, becomes a way for people to control their behavior to achieve long-term goals. Hence, identification with the group is itself a self-control mechanism, because it motivates people to act in ways that facilitate productive social interaction, which is what benefits everyone in the group in the long-term. Self-control involves having the ability to act in terms of group interest rather than pursuing personal gain. And, the merger of self and group enhances that ability because actions are increasingly motivated by group-based judgments. Further, a focus on identity facilitates the departure from a focus on short-term self interest by distancing people from outcome based concerns. Their loyalty becomes focused upon the rules, authorities and institutions that define the group.

Interestingly, this shift in focus toward identification and group level conceptualizations of self-interest also makes the personal problem of deciding if one is being exploited more difficult. In discussions of organizations and societies a critical perspective on social institutions emphasizes the possibility that this focus on status can lead people into social relationships that are fundamentally exploitative (Tyler, 2006b). For example, Marx coined the term "false consciousness" to suggest the possibility that people's loyalty could be maintained in situations in which, at least in outcome terms, they were being "suckers" to willingly continue within a particular relationship, organization or society.

And, of course, there is still the traditional question of self-control as a personal issue of being *able* to control one's impulses and desires. A person may want to do what is good for the group but lack the will to do so. So, even the most patriotic soldier may experience a self-interested survival-based panic and fail to fight in the heat of battle. However, people in general are not likely to experience conflicts between their own and their group's interests, because they will see them as the same. And, as noted earlier, people seem remarkably able to redirect themselves in ways that shape their behavior in response to group interests, which means that they are able to shift the framing of their personal desires. In addition to needing to be motivated to act for their group, people need to be able to do what is good for the group, just as they need to be able to do what is good for themselves when acting instrumentally. People may want to go to war and fight for their country, but they have to be able to actually do that to be helpful to their group.

When people have merged their sense of self with others, they become motivated to act voluntarily on behalf of the group. For example, the members of a community who are identified with their group are more willing to voluntarily follow its rules (Sunshine & Tyler, 2003; Tyler & Degoey, 1995) just as employees who identify with a company are more willing to voluntarily work on behalf of the company's welfare (Tyler & Blader, 2000). Society calls on its members to engage in various types of behavior that have a strongly voluntary component, ranging from paying taxes to fighting in wars. Such actions often have sanctions associated with rule-breaking, but for their effectiveness these sanctions depend on most people doing what is needed without having to be threatened or otherwise coerced (Tyler, in press).

In the group engagement model, Tyler and Blader (2000, 2003) argue that justice is important because it encourages identification with others, builds supportive attitudes and values, and motivates cooperative actions on behalf of collective entities. This is true of both distributive and procedural justice. Identification with others is a key focus because it defines the merger of self and other. Supportive attitudes and values matter because they are long-term dispositions and reflect a general commitment or loyalty to the group that separates people from immediate calculations of gain or loss within particular contexts.

The crucial question for people seeking to balance their own self-interest in relationship to the interests of others is to decide the degree to which they will merge their sense of self with a group, something that blurs the self–other distinction. Here, because they shape the degree to which people make such a merger procedural and distributive justice serve a key signaling function that guides people's decisions about how much to merge their sense of self with the group. Justice is the cue that people use to assess the quality of their connection to others, and those quality assessments shape the degree to which people then merge their self with the group. That merger, in turn, leads to a focus on the group and its concerns. Both justice and identification are mechanisms of social self-control, with justice acting as a cue that motivates identification.

In short-term instrumental relationships, justice acts to regulate the expression of self-interest and thereby facilitates cooperation. This short-term focus is often on distributive justice. In long-term instrumental relationships, there is a tendency for people to lengthen their time frame, leading to concerns about outcomes over time. This is reflected in the idea of long-term relationships and in the concept of investment in groups. People monitor justice over more issues and over time, a focus that leads to concerns about procedural justice. This shift in focus occurs for a variety of reasons. The need for flexibility in decision making has already been noted. The shift also occurs because a distributive justice focus becomes cumbersome when too many issues are involved. Think about all the benefits and burdens associated with membership in any group and imagine trying to compare them to those of others to decide whether one is receiving fair outcomes. In contrast, people can evaluate procedural fairness using simple principles. If they recognize that decisions are being made following these principles, then it is easy to conclude that the outcomes one is receiving from those procedures are fair. Finally, in long-term group interactions, people develop stronger levels of identification with the group. This stronger identification reflects a merger of self with the group, which leads personal self-interest to be less of an issue, minimizing problems of self-control. Justice cues encourage such an identification with the group because they signal to people that they are going to be advantaged if they lessen a self-interested vigilance toward group actions. If groups are fairly managed, people can be confident that their long-term well-being will be secure within them. This includes both their outcomes and their identity. Hence, justice is not just a set of principles for judging one's outcomes or of procedures for making decisions. It is also a signal that shows people that it is safe to merge their self into the group, forgoing a concern with immediate outcomes and focusing on a long-term connection to the group. People can then turn their attention to wholeheartedly working for group success, secure in the knowledge that they will fairly benefit from the results of that success.

Interestingly, identification also changes the way that people relate to the group. Consistent with the argument that people drop a focus on their outcomes and focus instead on their relationship to the group and its identity implications, empirical research shows that group identification shifts the basis of cooperation from instrumental judgments to judgments about the justice of group procedures. The more people merge their identity with a group the more they evaluate their connection to the group by focusing on how the institutions and authorities in the group exercise authority on behalf of the group. For example, Tyler and Degoey (1995) investigated community initiatives to cooperate by conserving water during a drought. People were asked to cooperate by complying with the policies instituted by the community-based water resources board. One reason for complying with these policies was instrumental in nature—the policies of the board benefited the resident in the long run (thus, the outcome was favorable). The second reason was procedural—that is, the judgment that compliance was due because the water board made its decision in a fair way. People who identified with their community based their decision to cooperate on procedural justice judgments, whereas people who did not identify with their community based their cooperation on instrumental judgments. These findings indicate that identification motivates a procedural justice perspective on the group—that is, a focus on the fairness of group procedures.

Other studies also support this perspective (for a review, see Tyler, 1999). Smith and Tyler (1996), for example, found that advantaged Americans were more willing to support policies to help the disadvantaged because those policies were fairly enacted *if* they identified with being Americans. In addition, Tyler and DeCremer (2005) showed that employees were more likely to cooperate with a new company following a merger because the merger was fairly conducted when they identified with the new company. In both cases, when people defined themselves in terms of group membership, their behavior was motivated by evidence of group fairness.

Although the group engagement model argues that identity is the key focus of concern that underlies people's attention to their relationship to a group, it can be more broadly suggested that justice matters because it provides the information that people need to assess the overall quality of their social connections, irrespective of whether people are concerned about identity or about the long-term resources the group will provide to them. This suggestion underlies the uncertainty management model (van den Bos & Lind, 2002). And it can be applied irrespective of whether it is identity or resources about which people are concerned. In other words, to the degree that people care about long-term outcomes, justice is important because it signals that those outcomes will be reasonable and appropriate. To the degree that people care about their identity, justice also provides identity-relevant cues.

In either case, the point is that people have a broader and more long-term connection to others but use justice as a cue to tell them about the status of this relationship. People focus on justice, rather than simply evaluating the short or long term quality of their outcomes when deciding how to act in relationship to a group . Hence, justice has a signaling function, and the importance of that signaling function increases over time as people become psychologically invested in groups. Justice communicates to people that it makes sense to merge their selves with the group. In other words, justice helps to eliminate issues of self-control in the short term by building a broader framework around them. It also builds identification, which further helps to eliminate the self-control problems that remain. Hence, both justice and identification work together in several ways to provide a social solution to problems associated with self-control.

The Two Functions of Justice

Why does justice play such an important role in shaping the degree of identification with the group? Justice plays two roles. Justice is first important because it communicates to people in groups that merging their identities with a group does not pose a threat to their sense of well-being (i.e., that their identities are secure)—justice shapes identity security. Second, justice suggests to people that if they merge their sense of self with the group, then there will be opportunities for their identities to be validated and enhanced and their well-being increased—so justice shapes opportunities for identity validation/enhancement.

As noted above, justice first provides evidence supporting feelings of *identity security*. A merger of the self with the group may provide people with support for positive feelings of self-worth and high self-esteem through their connection to the group. By being members of a group, people can first use the group as a source of identity-relevant categories through which they define themselves. In addition, they can use the status of the group as a source of self-affirmation—gaining confidence in their own identity through their association with the group. Thus, people have a great deal to gain by their association with groups, at least if that association has favorable identity implications.

Although using a group to determine one's identity can facilitate positive feelings of self-worth and self-esteem, it also contains risks. People can receive favorable identity-relevant information, but they can also have their identities damaged when they receive negative feedback from the group. For example, a person who identifies more strongly with a group is more psychologically damaged when they see that the group operates in negative ways (Brockner, Tyler, & Cooper-Schneider, 1992). Opening one's self to the group creates vulnerabilities and opens the possibility of receiving negative feedback that damages one's identity. People are sensitive to the potential pitfalls of identification with groups that provide them with negative status information.

The degree to which people identify with a group, an organization, or a society reflects their effort to balance the potential identity gains associated with merging their identities with a group against the potential risks of that same merger of the self and the group. The group engagement model argues that to the degree that people feel that the group acts via fair procedures, they are more likely to feel that their identity can be safely and securely merged with that of the group. And, of course, they can expect that those fair procedures will lead to fair outcomes.

Justice, in other words, alleviates people's concerns that group membership will result in negative consequences for the self. These elements of justice provide people with a sense of identity security. The existence of such security leads people to feel comfortable to engage psychologically and behaviorally in groups.

Each of the two functions of procedures—quality of decision making and quality of interpersonal treatment—contributes to people's assessment that it is safe for them to merge their identity with their group. Consider the quality of the decision making in the group. If a group makes decisions unfairly (i.e., inconsistently, based on biases or personal opinions instead of facts), then there is a risk that stereotypes or personal prejudices might potentially be applied to group members that belong to particular subgroups. In other words, neutrality encourages confidence in the security of including the group in one's sense of self, leading people to be more willing to engage in a group.

Fair procedures also involve a high quality of treatment, which is also protective of one's sense of self. Treatment with respect and dignity, for example, means that one will not be belittled or demeaned. Experiencing stereotyping and prejudice within the groups that people belong to is damaging to their sense of self, which may in turn lead them to maintain a psychological distance between their identity and group membership. Fair procedures reassure people that stereotypes are not and will not be applied. Fair procedures are also linked to the belief that authorities are benevolent and caring.

To summarize, there are clear benefits to shifting one's linkage to groups to one rooted in identity, rather than in exchange. However, there are also risks. In deciding how much to make a merger of self with the group people use cues provided by the manner in which the group functions. The fairness of procedures suggests the degree to which identities can be safely exposed to group dynamics by a merger of self with the group.

And, of course, although the group engagement model argues for the centrality of identity in people's connections with groups, the logic outlined above is the same if we view people are linked to others in long-term resource exchanges. People need some sign to tell them if throwing their lot in with the group makes sense. People need to see evidence that they will not be systematically disadvantaged in the outcomes that they will obtain by cooperation. For example, people do not want to feel that they are receiving unfairly low levels of outcomes either in relationship to others in the group or in comparison to what they might obtain on their own or in another group. However, as has been noted, it is hard to calculate outcome fairness in complex relationships. Hence, people use the existence of fair procedures, as well as evidence that authorities are trustworthy and motivated by benevolence, to reassure them that cooperating with the group makes sense.

In addition, procedural justice provides opportunities for identity validation/enhancement. People do not simply want to know that their identities will *not be diminished* by group membership, they also want to believe that their identities will be affirmed and strengthened through interactions with the group and group authorities. Of the elements of procedural justice, the element that is especially key to identity validation and enhancement is interpersonal respect. It is widely recognized throughout social psychology that people have a fundamental desire to have their sense of self affirmed via treatment with dignity and respect when dealing with others (Baumeister, Schmeichel, & Vohs, 2007). Treatment with dignity and respect both affirms that one is included within the group and entitled to its status and shows people that within the group, one is respected and valued by group members and authorities. In other words, it strengthens people's feelings of self-worth.

Inclusion in a group conveys entitlement to group status. As has been noted, group membership is relevant to identity. An American shares with others the high status of being a member of the group "Americans." Showing recognition for that status occurs by acknowledging a person's membership and, through that, their entitlement to the rights and standing of a group member. Therefore, demeaning treatment or harassment by the police, the courts, or any other societal authority is not just a material harm but also a denial of status. It is for this reason that it is most likely to be experienced by minority group members and as a reflection of a denial of their status as full-fledged members of society.

In addition to having the shared status of all group members, people occupy different status within a group and quality of treatment communicates what that status is (Tyler & Blader, 2000). Hence, people can also feel that it is important to have their high standing noted. People respond to having their unique positive status acknowledged by engaging in individualized efforts to help their group. As noted, when people are demeaned or disrespected, they respond to evidence of their low status by withdrawing from the group.

And, again, this enhancement argument is not dependent upon believing that people are linked to groups by a desire for status. It is equally true if the motivation to gain resources links people to groups. From that perspective people want an arena in which to seek resources that is governed by rules of fairness. For example, someone who has worked hard to achieve skills and become competent does not want to compete for promotion within a system that allows an authority to ignore their skills and deny them a promotion because of prejudice against women. A system of fair procedures for decision making ensures that decisions are made by the consistent application of neutral and fact based procedures, without the involvement of biases or prejudice. Hence, it encourages everyone to believe that there is a level playing field with clear rules of the game. The attainment of desirable rewards becomes something that occurs within a framework of understandable rules. In a fair procedure, for example, people have a chance to present evidence and the rules of decision making are transparent and neutral.

The main distinction between an identity based and a resource based view of the enhancing or facilitating role that procedures can play is the degree to which people are more concerned about fair decision making or fair interpersonal treatment, with treatment speaking more directly to identity and decision making to the ability to obtain fair outcomes. As an example, Thibaut and Walker (1975) talked about fair procedures from a social exchange perspective and focused upon evidence presentation as a mechanism for ensuring that rewards were equitably distributed. Hence, a person operating within a fair procedure should have confidence that merit will be rewarded and that the allocation of resources will not be capricious or arbitrary. So, a person interested in enhancing their resources will want to interact within a framework governed by principles of procedural fairness.

The Evolution of Institutions and Values

In an influential APA Presidential address, Donald Campbell provocatively argued that the development of moral and social values could be viewed as an example of social evolution (Campbell, 1975). He suggested that these values, and the institutions that maintain them, serve the useful function of helping people to control "human nature,"—that is, the biological tendency to act in one's immediate self-interest. Campbell suggested that behavior arising from self-interest was not adaptive in many situations and that those people who live in social groups that have developed effective mechanisms for minimizing such behavioral tendencies will be more likely to flourish. Those groups have taught their members to identify with their group and to hold group values and norms. Increasingly, such socially superior groups dominated over others, leading to our present highly socialized world.

The argument being made here fits well with Campbell's perspective. The argument in this chapter is that social mechanisms—principles of justice, identification with the group, the development of supportive values and attitudes—all facilitate people's efforts to control their motivation to pursue short-term self-interest, a motivation that may well have a biological basis but that nonetheless is an inferior way to approach interactions with others (Tyler, in press). Put simply, social institutions are designed to overcome biological tendencies.

Additionally, within this system, justice plays a key role because it allows people to monitor their relationship with complex social groups. In particular, because of their ability to evaluate how authority is exercised in groups—via procedural justice—people are able to enter into more complex social interactions than they would if they lacked the ability to make the evaluations of their situation needed to sustain long-term identity links to others.

Conclusion

In interactions shaped by people seeking to coordinate their efforts to pursue their self-interest, justice rules are important because they provide a set of shared rules about appropriate and reasonable ways to allocated resources. This encourages cooperation and discourages the type of destabilizing disputes that undermine ongoing exchange relationships. Although the image of the mysterious haggling at a Turkish bazaar brings drama to novels and movies, most interactions are facilitated when the principles of fair exchange are overt and, most importantly, are commonly understood by all parties. And, besides, in reality even haggling has justice principles behind it and is not as chaotic as it seems to people who are unfamiliar with its rules. This is important in both short- and long-term relationships. Rules indicating what is just facilitate social control.

However, in most ongoing social interactions, people move beyond the use of distributive justice rules. They create authorities and give those authorities discretion over how to make decisions and resolve conflicts. When reacting to these authorities, people focus on how they make decisions. This procedural justice focus shows that people understand the centrality of fair decision making to the attainment of both just outcomes and harmonious relationships. The people in groups recognize that when procedures are fair, group membership makes sense.

Finally, when in groups, people actually transform. They change in two ways. First, their conception of their interests changes from self-interest to group interest. People recognize that working on behalf of fairly run groups makes sense, and they do what is needed to help the group succeed. That success, in turn, assures personal benefit. This does not mean that people become insensitive to their own situation. But, procedural justice within a group tells people that group success will result in personal gains. It is interesting that although justice continues to be a central issue, the focus shifts from distributive to procedural justice. This occurs because a process is required to determine how to apply distributive justice norms in complex social settings, and people believe that a fair process leads to a just outcome.

People also transform when they are in groups in that they become more concerned about their sense of self and their personal identity and less directly concerned about their material outcomes. Issues of standing and status replace judgments of gain and loss as the core evaluations that guide people's thoughts, feelings, and actions in groups. Additionally, people develop attitudes and values that link them to groups and that become separate from material calculations. The desire to do what is right and to fulfill one's obligations to others become central factors shaping behavior and core elements of how people think of themselves as people. Also, as has been outlined, assessments of the procedural justice of groups guide both the merger of one's identity with a group and the development of these types of supportive values.

The development of identity and values reflects a highly evolved—and particularly human—approach to resolving the social control problem. People become motivated by the desire to do what is right and rewarded for such actions by having a favorable sense of self and a positive identity both of which are linked to the respect of others and high standing in groups. Self-control becomes a strategy of self-regulation that people learn through social institutions and via a set of socially established principles of justice that guide their reactions thoughts, feelings, and actions social settings.

REFERENCES

Allison, S. T., et al. Social decision making processes and the equal partitioning of shared resources. J Exp Soc Psychol 1992; 28: 23–42.

Barrett-Howard, E., & Tyler, T.R. Procedural justice as a criterion in allocation decisions. J Pers Soc Psychol 1986; 50: 296–304.

Baumeister, R. F., Schmeichel, B. J., & Vohs, K. D. Self-regulation and the executive function. In: Kruglanski, A., & Higgins, E. T. (Eds.), *Social psychology: Handbook of basic principles*, 2nd ed. New York: Guilford, 2007: pp. 516–539.

Brockner, J., Tyler, T. R., & Schneider, R. The higher they are, the harder they fall: The effects of prior commitment and procedural injustice on subsequent commitment to social institutions. Admin Sci Quarter 1992; 37: 241–261.

Campbell, D. T. On the conflicts between biological and social evolution and between psychology and moral tradition. Am Psychol 1975; 30: 1103–1126.

Clark, M. S. Record keeping in two types of relationship. J Pers Soc Psychol 1984, 47: 549–557.

Davis, A., Tyler, T. R., & Andersen, S. Building community one relationship at a time. Soc Justice Res 2007; 20(2): 181–206.

DeCremer, D., & Tyler, T. R. Managing group behavior: The interplay between procedural justice, sense of self, and cooperation (Volume 37). In: Zanna, M. (Ed.), *Advances in experimental social psychology*. New York: Academic Press, 2005: pp. 151–218.

Deutsch, M. Equity, equality, and need: What determines which value will be used as the basis of distributive justice? J Soc Issues 1975; 31: 137–149.

Deutsch, M. *Distributive justice*. New Haven: Yale, 1985.

Fehr, E., & Gintis, H. Human motivation and social cooperation. Annu Rev Sociol 2007; 33: 43–64.

Gold, M. *The complete social scientist: A Kurt Lewin reader*. Washington, D.C.: American Psychological Association, 1999.

Gurerk, O., Irlenbusch, B., & Rockenbach, B. The competitive advantage of sanctioning institutions. Science 2006; 312: 108–111.

Henrich, J., McElreath, R., Barr, A., et al. Costly punishment across human societies. Science 2006; 312: 1767–1770.

Herlocker, C. E., et al. Intended and unintended overconsumption of physical, spatial, and temporal resources. J Pers Soc Psychol 1997; 73: 992–1004.

Hogg, M., & Abrams, D. *Social identifications*. London: Routledge, 1988.

Hollander-Blumoff, R., & Tyler, T. R. Do nice guys finish last? Procedural justice and negotiation outcomes. Law Soc Inq 2008; 33: 473–500.

Lind, E. A., & Tyler, T. R. *The social psychology of procedural justice*. New York, NY: Plenum, 1988.

Loewenstein, G. F., Thompson, L., & Bazerman, M. H. Social utility and decision making in interpersonal contexts. J Pers Soc Psychol 1989; 57: 426–441.

Messick, D. M., Wilke, H., Brewer, M. B., Kramer, R. M., Zemke, P., & Lui, L. Individual adaptations and structural changes as solutions to social dilemmas. J Pers Soc Psychol 1983; 44: 294–309.

Montada, L., & Schneider, A. Justice and emotional reactions to the disadvantaged. Soc Justice Res 1989; 3: 313–334.

Piaget, J. *The moral judgment of the child*. New York, NY: Harcourt, 1932.

Pritchard, D., Dunnette, M. D., & Jorgenson, D. O. Effects of perceptions of equity and inequity on worker performance and satisfaction. J Appl Psychol 1972; 56: 75–94.

Rawls, J. *A theory of justice*. Cambridge: Harvard University Press, 1971.

Reis, H. The nature of the justice motive. In: Masters, J. C., & Smith, W. P. (Eds.), *Social comparison, social justice, and relative deprivation*. Hillsdale, NJ: Erlbaum, 1987: pp. 131–150.

Rusbult, C. E., & Van Lange, P. Interdependence processes. In: Higgins, E. T., & Kruglanski, A. W. (Eds.), *Social psychology*. New York, NY: Guilford, 1996: pp. 564–596.

Schmitt, D. R., & Marwell, G. Withdrawal and reward allocation as responses to inequity. J Exp Soc Psychol 1972; 8: 207–221.

Smith, H. J., & Tyler, T. R. Justice and power: When will justice concerns encourage the advantaged to support policies which redistribute economic resources and the disadvantaged to willingly obey the law? Eur J Soc Psychol 1996; 26: 171–200.

Sondak, H., & Tyler, T. R. How does procedural justice shape the desirability of markets. J Econ Psychol 2007; 28: 79–92.

Sprecher, S. Equity and social exchange in dating couples. J Marriage Fam 2001; 63: 599–613.

Sunshine, J., & Tyler, T. R. The role of procedural justice and legitimacy in shaping public support for policing. Law Soc Rev 2003; 37(3): 555–589.

Thibaut, J., & Faucheux, C. The development of contractual norms in a bargaining situation under two types of stress. J Exp Soc Psychol 1965; 1: 89–102.

Thibaut, J., & Walker, L. *The social psychology of procedural justice*. Hillsdale, N.J.: Erlbaum, 1975.

Tyler, T. R. Justice in the political arena. In: Folger, R. (Ed.), *Justice: Emerging psychological perspectives*. New York, NY: Plenum, 1984: pp. 189–225.

Tyler, T. R. What is procedural justice?: Criteria used by citizens to assess the fairness of legal procedures. Law Soc Rev 1988; 22: 103–135.

Tyler, T. R. Why people cooperate with organizations: An identity-based perspective. Res Organ Behav 1999; 21: 201–246.

Tyler, T. R. Social justice: Outcome and procedure. Int J Psychol 2000; 35: 117–125.

Tyler, T. R. *Why people obey the law: Procedural justice, legitimacy, and compliance*. Princeton: Princeton University Press, 2006a.

Tyler, T. R. Legitimacy and legitimation. Annu Rev Psychol 2006b; 57: 375–400.

Tyler, T. R. *Why people cooperate: The role of social motivations*. Princeton, N.J.: Princeton University Press, in press.

Tyler, T. R., & Blader, S. *Cooperation in groups: Procedural justice, social identity, and behavioral engagement*. Philadelphia, PA: Psychology Press, 2000.

Tyler, T. R., & Blader, S. Procedural justice, social identity, and cooperative behavior. Pers Soc Psychol Rev 2003; 7: 349–361.

Tyler, T. R., Boeckmann, R., Smith, H. J., & Huo, Y. J. *Social justice in a diverse society*. Denver, CO: Westview, 1997.

Tyler, T. R., & Degoey, P. Collective restraint in social dilemmas: Procedural justice and social identification effects on support for authorities. J Pers Soc Psychol 1995; 69: 482–497.

Van den Bos, K., & Lind, E. A. Uncertainty management by means of fairness judgments. Adv Exp Soc Psychol 2002; 34: 1–60.

Walster, E., Walster, G. W., & Berscheid, E. *Equity*. Boston: Allyn and Bacon, 1978.

Ybema, J. E., Kuijer, R. G., Buunk, B. P., DeJong, G. M., & Sanderman, R. Depression and perceptions of inequity among couples facing cancer. Pers Soc Psychol Bull 2001; 27: 8–13.

CHAPTER 26

System Justification and the Disruption of Environmental Goal-Setting: A Self-Regulatory Perspective

Irina Feygina, Rachel E. Goldsmith, and John T. Jost

ABSTRACT

Global warming and environmental destruction pose formidable social dilemmas. Although the contribution of each person to the problem through consumption, utilization, and waste is nearly invisible, the cumulative impact for the well-being of societies and individuals within those societies is potentially devastating. We propose that an important psychological factor contributing to the environmental commons dilemma is the motivation to justify and rationalize the status quo and the extant socioeconomic system. Rather than acknowledging and confronting environmental problems, we propose that people may engage in denial of environmental realities as a means of satisfying short-term needs associated with system justification. Denial, in turn, contributes to a failure to set pro-environmental goals and the tendency to perpetuate environmentally harmful behaviors that are detrimental in the long term. Four studies provided support for these predictions. People who exhibit chronically stronger (vs. weaker) tendencies to justify the system reported greater denial of environmental problems, less favorable attitudes toward the environment, and failure to set useful, ambitious goals or to engage in behaviors that would prevent further environmental deterioration. Differences in system justification tendencies helped to explain commonly observed differences in environmental attitudes and behaviors among liberals and conservatives, women and men, and those with more—compared to less—years of education. In addition, our experimental evidence suggested that engaging in denial of environmental realities serves a system-justifying function by re-establishing a view of the system as legitimate and just, but it interferes with setting goals to help the environment. Implications of a self-regulatory perspective on environmental attitudes and potential contributions of a social psychological analysis of commons dilemmas are discussed.

Keywords: System justification, commons dilemmas, self-regulation, intention formation, goal setting, global warming, environmental attitudes and behaviors

"The individual benefits as an individual from his ability to deny the truth even though society as a whole, of which he is a part, suffers."
—Garrett Hardin, 1968

Forty years ago, Garrett Hardin (1968) outlined the dynamics of the "Tragedy of the Commons," a social dilemma in which a group of people must share a limited resource in such a way as to prevent its overuse and disappearance. Hardin pointed out that giving people freedom of action with respect to common or shared resources often produces a tragic spiral: Each individual acts rationally by increasing his or her consumption of the limited resource to maximize personal gain, but the resource cannot be replenished at the high rate of consumption; as a result, the whole group (or community) pays a stiff price for the selfish, albeit rational, behavior of its individual members. Thus, what can be seen as rational behavior at the level of the individual decision-maker is irrational from the point of view of maintaining the well-being of the collective and therefore the well-being of each individual as well. There is ample reason to think that the situation posed by climate change and environmental degradation is a particularly worrisome instance of the commons dilemma and that its resolution will require interventions of a social psychological (as well as technological) nature.

Global Warming and Environmental Destruction

Hardin's (1968) analysis of social dilemmas was driven by a concern about the dangerous rate of population growth, which, if permitted to continue at an unregulated pace, would result in making the planet uninhabitable for all people. In addition, he pointed out that unregulated agricultural practices were bringing about destruction of land, uncontrolled fishing was causing extinction of many species of fish and whales, and the byproducts of contemporary production and life were polluting our commons. Not only has there been an abject failure to address these problems in the decades since Hardin's article appeared, but environmental threats have dramatically escalated in both magnitude and urgency.

The Nature and Severity of Environmental Concerns

Of chief concern are the realities of environmental destruction and global warming, the impact of which spans the entire planet (Keeling & Whorf, 2005; Oreskes, 2004; Watson, 2002; Webster et al., 2005). Current crises include depletion of natural resources, destruction of natural habitats, extinction of species, and loss of biodiversity that supports the intricate interdependence among humans, other animal species, and plant life; and accumulation of hazardous and nonbiodegradable waste that pollutes the air, water, and soil. These problems pale in comparison to global warming, however, which is partly caused by industrial greenhouse emissions that become trapped in the atmosphere and increase its temperature (Weart, 2004). These processes bring about the melting of polar ice, rising of sea levels, and loss of habitats; changes in weather patterns, including escalation in severe weather events, like hurricanes and cyclones; and a powerful effect on agriculture, the extent of which is not yet known but whose effects can already be felt across the globe as food crises and political conflict escalate in response to environmental challenges. There is no genuine doubt remaining in the scientific community that global warming is taking place and that industrial processes are contributing to its progression (e.g., Oreskes, 2004; Scheffer et al., 2001; Watson, 2002).

Global warming and environmental destruction pose a clear and formidable social dilemma. Each person who consumes goods and utilizes conveniences that result from the industrial process contributes to the emission of dangerous gases into the atmosphere; the chemical pollution of water, air, and soil; the degradation of our natural environment; and climate change at local and global levels. Whereas the contribution of each person is nearly invisible, the cumulative impact for the well-being of societies and individuals within those societies is potentially devastating. Moreover, the rate and progression of global warming is probabilistic (Heath & Gifford, 2006) and therefore experienced as uncertain. This uncertainty increases the likelihood that people will consume resources selfishly and make overly positive assessments about the state of those resources (Kortenkamp & Moore, 2006; Weber et al., 2004).

Global Warming and the Need for Collective Intervention

As argued by Hardin (1968) and many subsequent researchers of social dilemmas, solutions to a commons dilemma must be developed at the group or collective level of organization (e.g., Dawes & Messick, 2000; Komorita & Parks, 1997; Kramer, 1991; Tyler & Degoey, 1995). That is, a coordinated effort is needed to create a sustainable set of practices and distribution rules that allow individuals to access shared resources while maintaining the well-being of the common good and the collective. In this case and others, it seems necessary to transform social institutions and leadership practices to reverse potentially catastrophic outcomes and create a sustainable arrangement.

According to most experts, it is already impossible to stop global climate change, but there are opportunities to slow its progression and prevent its most dire consequences. It is clear that efforts to mitigate the worst effects of global climate change will require a large-scale shift in attitudes and actions. At the governmental level, it is imperative to institute policy changes that include incentives for conservation and the development and use of sustainable practices, as well as regulation of harmful processes entailed in production and consumption. That is, the use of exhaustible common resources must be kept in check by rules, procedures, and authorities that account for the aggregate effects of individual choices to prevent detrimental consequences for the community as a whole (Weber et al., 2004). At the level of the economic system, widespread changes are also necessary, and these entail extensive investments in the development of alternative energy sources and environmentally friendly industrial practices, as well as a sustained commitment and willingness to sacrifice on a global scale. An effective allocation of resources would involve the loss of hundreds of billions of dollars and 1% to 2% of the national capital of the world's richest countries (Nordhaus & Yang, 1996).

In sum, helping the environment will require significant changes and sacrifices at three levels of analysis: individual, group, and system (see Stangor & Jost, 1997). Most importantly, social structures and practices will not be transformed or effectively implemented without the initiative, support, and compliance of individuals who make up the collective (Tyler & Degoey, 1995). Therefore, the key to developing a solution is widespread awareness of the problem, its scope and progression, and one's own contribution and responsibility. Research indicates that this degree of awareness is still lacking in the general public. Despite vast scientific evidence that global climate change is rapidly occurring, caused by human activities, and poses numerous threats to the earth's ecosystems (Hansen, 2004; Keeling & Whorf, 2005; Oreskes, 2004; Scheffer et al., 2001; Weart, 2004; Webster et al., 2005), many people still deny the severity of the problem and resist efforts to address it.

According to public opinion surveys, a majority of U.S. respondents accord minimal importance to the problem of global warming and doubt that it will impact them or their way of life (Carroll, 2007; Gallup Poll, 2007). Many respondents believe that global warming claims are exaggerated and that human activity is not to blame for climate change, and as little as 6% of U.S. respondents believe that the environment should be given priority in policy (Gallup Poll, 2007; Pew, 2006). It is clear that addressing environmental problems will require questioning and altering the societal status quo, as is often the case with respect to commons dilemmas (Hardin, 1968). This raises a question that is of focal interest to us in this chapter—namely, What are the factors that facilitate or inhibit the successful development and implementation of a collective intervention aimed at changing the current state of affairs?

SOURCES OF SELF-REGULATORY FAILURE IN THE COMMONS DILEMMA

To address this broad question it is useful to examine the role of self-regulatory processes that are likely to operate in a commons dilemma situation, which requires an individual to exercise self-control in such a manner that is beneficial rather than detrimental to the collective (and, ultimately, to the individual). There are several interrelated conflicts or trade-offs that reflect the self-regulatory challenges posed by a commons dilemma. First, there is a conflict between the short- and long-term interests of the individual (e.g., Loewenstein, 1996; Trope & Fishbach, 2000). Whereas it may seem beneficial in the short run for an individual to fulfill his or her own immediate needs and desires, a long-range perspective reveals that depleting collective resources will eventually be very costly to the self and others. More generally, there is a trade-off between the concrete, immediately accessible needs of the individual and the more abstract—and therefore less accessible—needs of the community as a whole (Fujita et al., 2006; Trope & Fishbach, 2000).

Moreover, the commons dilemma suggests a potential conflict between a "hot," affectively compelling response to the gratification of those needs that drive consumption of the limited resource, and the results of a "cold," logical, removed analysis of the implications of the same behavior (Baumeister & Heatherton, 1996; Loewenstein, 1996; Metcalfe & Mischel, 1999). In other words, the satisfaction of a person's immediate need to consume can give rise to a positive affective response, but a cognitive analysis would suggest that forbearance and control of one's desires is needed, even if it elicits immediate negative affect. Finally, the pursuit of many goals occurs at an implicit or automatic level of awareness (e.g., Bargh, 1990; Bargh et al., 2001). This could mean that people are apt to satisfy their needs without necessarily engaging in a process of conscious deliberation and that a self-regulatory effort would be required to raise awareness of potentially problematic behaviors. Failure of self-regulation in one or more of these areas could easily lead to the perpetuation of environmentally destructive activities that will inflict lasting, if not permanent, harm at the level of the collective.

We propose that the self-regulatory challenge in confronting environmental problems is exacerbated by an additional, previously unexplored factor. Although resolving the environmental commons dilemma calls for collective (or aggregate-level) interventions that bring about significant changes to the status quo, extensive research has suggested that people become psychologically invested in current social, economic, and political arrangements and are motivated to defend, maintain, and justify the status quo and to resist major changes to it (e.g., Jost, Banaji, & Nosek, 2004). Questioning current practices and policies and transforming the status quo—both of which may be needed to overcome the commons dilemma with respect to environmental degradation—directly contradict the desire to see the status quo as just, valid, and legitimate. Therefore, a potent self-regulatory challenge that may hinder attitudinal and behavior change is to overcome the preference for the status quo in favor of abstract, long-term considerations that may require a substantial overhaul of what is presently a flawed system. This challenge can be better understood by considering findings from system justification theory.

MOTIVATION IN SERVICE OF THE STATUS QUO

According to system justification theory, individuals' evaluations of institutions and systems are influenced by epistemic needs to maintain a sense of certainty and stability, existential needs to feel safety and reassurance, and relational needs to affiliate with others who are part of the same social system (Jost et al., 2004; Jost & Hunyady, 2002, 2005). The motivation to justify the extant system may align with a person's objective social interests if the system works to one's personal or collective advantage (Jost, Burgess, & Mosso, 2001). However, system justification needs may lead people to support and justify the system even in situations in which they are harmed or placed at a disadvantage by the status quo (Henry & Saul, 2006; Jost et al., 2003c).

At an individual level of analysis, system justification can have short-term palliative effects. These include alleviating the anxiety, uncertainty, and fear that arise when the societal system is threatened (Jost & Hunyady, 2002) and providing a means of rationalizing problematic aspects of the status quo (Kay, Jimenez, & Jost, 2002). However, the long-term consequences of pursuing a system justification goal can sometimes be negative, especially for persons and groups who are disadvantaged by the socioeconomic system and its hierarchy (e.g., Jost & Thompson, 2000). Part of the problem is that system justification can interfere with forming intentions or taking action aimed at correcting problems or addressing injustices.

At a collective level of analysis, system justification can have salutary effects by contributing to the perceived legitimacy and stability of groups and institutions (Tyler, 2006) and increasing satisfaction and compliance with authorities (Feygina & Tyler, 2009). However, if there are defects or shortcomings in the system, or if social change is necessary to adapt to new realities or concerns, then system justification can lead to the perpetuation of ongoing problems and detrimental outcomes (e.g., Wakslak et al., 2007). Thus, engaging in system justification can serve to sustain a problematic status quo (Jost & Hunyady, 2005).

Motivational Properties of System Justification

The deleterious effects of system justification may be better understood by considering the self-regulatory properties of the goal to justify the system (Jost et al., 2007). The system justification goal is activated whenever there is a need to maintain or restore perceptions of the well-being, legitimacy, or stability of the social system in order to experience a sense of meaning and security. As such, it can be expected to shift focus toward short-term rewards rather than long-term outcomes (Loewenstein, 1996; Miller & Brown, 1991; Trope & Fishbach, 2000). Moreover, the desire to rationalize the status quo can acquire psychological urgency under some circumstances, such as when the social system is perceived as under attack (Jost & Hunyady, 2005; Ullrich & Cohrs, 2007). The need to alleviate emotional distress (Jost & Hunyady, 2002) can exacerbate one's focus on immediate goals, thereby deprioritizing a cognitive and integrative analysis of the situation (Fujita et al., 2006; Metcalfe & Mischel, 1999; Tice, Bratislavsky, & Baumeister, 2001). Because the system justification goal can operate at an implicit level of awareness (Liviatan & Jost, 2009; *see also* Jost et al., 2004), its effects can be especially difficult to recognize and overcome.

Individual and Situational Differences in System Justification Tendencies

The strength of the goal to justify the system can vary because of dispositional factors (Jost & Hunyady, 2005). Research has indicated that some people are more threatened than others by negative information about the social system and have a stronger need to defend it (e.g., Jost et al., 2003a). For example, political conservatives have stronger psychological needs to reduce uncertainty and threat (Jost et al., 2003b), and they are also more likely to engage in system justification and subscribe to fair market ideology (Jost et al., 2003a). Studies summarized by Jost, Nosek, and Gosling (2008) reveal a correlation of between 0.4 and 0.5 between participants' self-reported political conservatism and their scores on a general measure of system justification (see Kay & Jost, 2003). The strength of system justification goals can also vary as a result of situational factors (Jost & Hunyady, 2005). For example, situational threats directed at the legitimacy or stability of the status quo typically produce an increase in system justification, presumably because people are striving to relieve the negative implications of the threat (e.g., Jost & Hunyady, 2002; Kay, Jost, & Young, 2005; Lau, Kay, & Spencer, 2008; Ullrich & Cohrs, 2007).

It should be pointed out that information about environmental destruction is highly threatening, because it entails a personal threat to the health and well-being of oneself and one's family as well as a system-level threat to the societal status quo. Global climate change threatens both the legitimacy and stability of cultural practices and economic institutions, technological progress, and economic development, and it also challenges assumptions about capitalism and human efforts to master the natural world. Because climate change poses a powerful threat to many core aspects of the socioeconomic system, it is likely to elicit a strong psychological need to defend and justify the status quo (e.g., Jost & Hunyady, 2002). The desire to see national and international leaders, governments, corporations, and institutions as generally fair and legitimate may also prevent realistic assessments of environmental risks posed by common practices and continued environmental neglect (Jost et al., 2003a).

System Justification Has the Potential to Disrupt Environmental Self-Regulation

In this chapter, we propose that system justification motivation has the potential to interfere directly with setting and pursuing goals to develop and implement collective interventions designed to resolve commons dilemmas, such as the problem of global warming and environmental destruction. There are at least three specific ways in which we think that system justification goals could come into conflict with pro-environmental self-regulatory processes (*see* Fig. 26–1). First, system justification could prevent people from acknowledging the need for change and instead lead them to deny problems with the status quo. This may satisfy short-term hedonic needs, but it clearly inhibits pro-environmental action in the long term. Second, even if people are able to acknowledge environmental problems, the need to maintain and defend the social system may disrupt the process of setting firm pro-environmental goals. This is because people may be resistant—either consciously or unconsciously—to actually making changes in their personal lifestyle or their social and political priorities. Both denial and resistance to change are likely to contribute to the failure to set pro-environmental goals and the tendency to perpetuate environmentally harmful behavior. Third, even if a person does succeed in setting goals to help the environment, the pursuit of these goals may be hampered by goal interference, insofar as the desire to maintain favorable attitudes concerning existing institutions and practices may be incompatible with the execution of environmentally helpful behavior.

In summary, there are several reasons to expect system justification needs to exert a detrimental effect on long-term self-control processes in the environmental domain. The immediate, at times emotionally pressing, need to reaffirm the social system may interfere with acknowledging problematic aspects of the status quo, disrupt the setting of goals to change it, and inhibit taking action to change environmental practices that are ultimately detrimental to both the individual and the collective. Moreover, we suggest that persons who chronically experience a stronger need to justify the system will encounter a greater self-regulatory challenge when faced with information about environmental problems; therefore, they are expected to engage in greater denial and to form fewer intentions to help the environment.

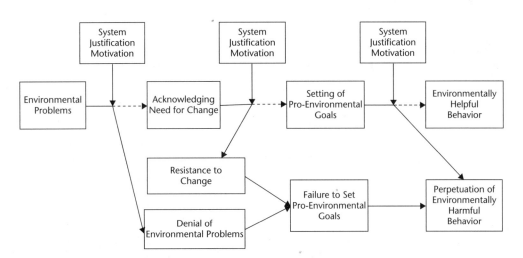

Figure 26–1. System justification motivation has the potential to disrupt pro-environmental self-regulatory processes at several different phases.

OVERVIEW OF THEORETICAL APPROACH

The foregoing discussion gives rise to several more specific hypotheses concerning the effects of system justification needs on environmental awareness and goal setting. First, we expect to find that chronic motivation to justify the system is one antecedent to the denial of environmental problems, as well as a hindrance to forming intentions to help the environment. We predict that people who exhibit stronger tendencies to justify the system will report *(1)* greater denial of environmental problems; *(2)* less favorable attitudes toward the environment; and *(3)* failure to set useful, ambitious goals or to engage in behaviors that would prevent further environmental deterioration. Second, we predict that engaging in denial of environmental realities will serve a system-justifying function by re-establishing a view of the system as legitimate and just, but it will interfere with setting goals to help the environment. The following sections present evidence from four studies that are consistent with these hypotheses (*see also* Feygina, Jost, & Goldsmith, 2010).

EMPIRICAL SUPPORT FOR THE THEORETICAL FRAMEWORK

System Justification and the Denial of Environmental Problems

Our first study was designed to establish the existence of a link between system justification and denial of environmental concerns (Feygina et al., 2010, Study 1). We hypothesized that individuals who report a greater tendency to engage in system justification would be more likely to deny environmental problems. As part of a larger survey, undergraduate participants at the University of Oregon filled out a measure of general system justification (Kay & Jost, 2003), which included items such as: "Most policies serve the greater good" and "Society is set up so that people usually get what they deserve." Environmental denial was assessed using items from the New Environmental Paradigm scale, including: "The so-called 'ecological crisis' facing humankind has been greatly exaggerated" and "If things continue on their present course, we will soon experience a major environmental catastrophe" (reverse-coded) (Dunlap et al., 2000).

Results indicated that engaging in system justification significantly predicted the denial of threatening environmental problems, as hypothesized. This relationship was observed for several different types of environmental attitudes. Specifically, people who scored higher on system justification were more likely to deny *(a)* the possibility of an ecological crisis; *(b)* limits to natural resources and the earth's sustainability; *(c)* the necessity to abide by the constraints of nature; and *(d)* the danger of disrupting balance in nature.

This study also helped to clarify the oft-reported (but seldom explained) finding that women report greater concern for the environment and are more willing than men to take action to help the environment (Cottrell, 2003; Dietz, Kalof, & Stern, 2002; Zelezny, Chua, & Aldrich, 2000). Considering prior findings that women report significantly lower levels of system justification than men (Jost & Kay, 2005; *see also* Sidanius & Pratto, 1999), we hypothesized that gender differences in system justification motivation would help to explain gender differences in pro-environmental attitudes. Indeed, we found that men reported greater denial of environmental problems, as well as stronger system justification tendencies. Moreover, a mediational analysis demonstrated that variability in system justification scores partially accounted for the observed relationship between gender and denial of environmental problems.

System Justification and Ideological Differences in Environmental Attitudes

Our next study was designed to further elucidate the relationship between system justification and denial of environmental problems, as well as to examine whether variability in system justification tendencies could help to account for several known group differences in environmental attitudes (Feygina et al., 2010, Study 2). Much prior evidence has suggested that liberal respondents report greater concern for and engagement with environmental issues, and are more willing to take action on behalf of the environment, compared to conservative respondents (Cottrell, 2003; Samdahl & Robertson, 1989; Thompson & Gasteiger, 1985; Van Liere & Dunlap, 1980). We hypothesized that this ideological difference in environmental attitudes could result from (at least in part) differences in the strength of the system justification motive as a function of political orientation (*see also* Jost, Nosek, & Gosling, 2008). Similarly, we expected that persons who identified more strongly with the national system, and therefore are more invested in its legitimacy and stability, would also be more motivated to deny environmental problems. We also predicted that greater denial of environmental problems resulting from system justification needs would hinder engagement in pro-environmental behaviors—that is, the negative relationship between system justification and behaviors helpful to the environment should have been mediated by denial of environmental problems.

This study made use of Kay and Jost's (2003) measure of general system justification, as well as a measure of economic system justification (Jost & Thompson, 2000), which included items such as: "Economic positions are legitimate reflections of people's achievements," and "There are many reasons to think that the economic system is unfair" (reverse-coded). Participants, who were students at New York University, also completed an ideological self-placement item by placing themselves on a scale ranging from "extremely liberal" to "extremely conservative." We also assessed strength of participants' national identification using items adapted from the Luhtanen and Crocker (1992) collective self-esteem scale, such as: "Being an American is an important reflection of who I am," and "Overall, being an American has little to do with how I feel about myself" (reverse-coded). Denial of environmental problems was again assessed using the NEP scale, and an additional measure assessed environmental behaviors with items such as: "How often do you recycle paper and bottles/cans?"; "How often do you give money to organizations that help the environment?"; and "How often do you encourage government representatives to adopt policies that are good for the environment?"

Consistent with our prior findings, a structural equation analysis indicated that engaging in general and economic system justification each predicted significantly greater denial of environmental problems. Moreover, respondents who endorsed a conservative political orientation and who had a stronger national identity engaged in greater system justification in general as well as greater economic system justification, and were more likely to deny environmental problems. Moreover, attitudinal differences in environmental denial were partially accounted for by differences in system justification tendencies associated with political orientation and national identification. As hypothesized, the degree to which respondents engaged in system justification, as well as denial of environmental problems, predicted significantly lower rates of engagement in behaviors that could improve environmental conditions. Moreover, denial completely mediated the relationship between system justification motivation and environmental behaviors.

Generalization to a Nationally Representative Sample

In a third study, we analyzed data from U.S. respondents to the 1999–2001 World Values Survey to determine whether endorsement of system-justifying ideologies such as the Protestant work ethic and opposition to equality (*see* Jost & Hunyady, 2005) would be associated with decreased support for environmental causes. We also sought to further investigate the effects of political orientation and education on environmental attitudes and behavioral intentions.

Prior research shows that more highly educated respondents tend to be more concerned about environmental problems and more willing to take action to address them (Cottrell, 2003; Ostman & Parker, 1987; Scott & Willits, 1994; Van Liere & Dunlap, 1980). We hypothesized that differences in system justification motivation could provide a partial explanation for the positive effect of education on environmental attitudes (cf. Jost et al., 2003c).

Endorsement of the Protestant work ethic was assessed with items such as: "Work is a duty towards society" and "It is humiliating to receive money without having to work for it." Regarding opposition to equality, participants were given two response options and asked to indicate how strongly they preferred one response option over another. The first pair of options read: "Incomes should be more equal" vs. "We need larger income differences as incentives for individual effort." The second pair of options was: "The government should take more responsibility to ensure that everyone is provided for" vs. "People should take more responsibility to provide for themselves." Political orientation was assessed using a single ideological self-placement item, ranging from "Left" to "Right."

In this study we were not able to assess denial of environmental problems *per se*. Rather, we focused on the holding of environmentally harmful attitudes by asking respondents to choose whether: "Economic growth and creating jobs should be the top priority, even if the environment suffers to some extent," or "Protecting the environment should be given priority, even if it causes slower economic growth and loss of jobs" and whether "Human beings should master nature" or "Humans should coexist with nature." In addition, participants reported on the extent to which they embraced the following behavioral intentions: "I would give part of my income if I were certain that the money would be used to prevent environmental pollution" and "I would agree to pay higher taxes if the extra money was used to prevent environmental pollution."

Results from a structural equation analysis revealed that endorsement of system-justifying belief systems predicted holding more harmful attitudes toward the environment. In other words, believing more strongly in the Protestant work ethic and opposing equality (or justifying inequality) predicted one's likelihood of prioritizing economic development over environmental protection and agreeing that people should dominate rather than coexist with nature. As before, holding a more conservative political orientation was associated with more negative attitudes toward the environment, and this relationship was partially mediated by endorsement of both types of system-justifying beliefs. Participants with higher levels of educational attainment were less likely to endorse the Protestant work ethic, and they also held more positive attitudes toward the environment. Furthermore, the effect of educational attainment on environmental attitudes was partially mediated by differences in system justification tendencies. With respect to behavioral intentions, endorsing system-justifying beliefs and holding more harmful environmental attitudes were both associated with an unwillingness to provide financial assistance to environmental causes. The holding of harmful environmental attitudes completely mediated the dampening effect of system justification on environmental intentions.

In sum, evidence from three studies involving diverse samples and measures indicates that system justification tendencies exert a powerful negative effect on the acknowledgment of environmental concerns and on the likelihood of developing environmentally beneficial attitudes (*see* Fig. 26-1). A system justification perspective also sheds light on the psychological underpinnings of known group differences in environmental attitudes, including differences associated with political ideology, gender, and educational attainment. Overall, our research suggests that bringing about attitudinal and behavioral change in the environmental domain requires overcoming a set of self-regulatory challenges posed by system justification needs.

Denial Can Facilitate System Justification: An Experimental Demonstration

In the last study we detail in this chapter, we set out to better understand the relationship between system justification, engaging in denial of environmental problems, and the setting of pro-environmental goals. In addition, we wanted to provide further evidence that beliefs about environmental conditions and global warming are related to abstract, ideological beliefs about society and the social system. In particular, we sought to examine whether engaging in denial of environmental realities could influence attitudes about the socioeconomic system and thus serve a system-justifying function.

Using an experimental paradigm, we investigated the hypothesis that providing a situational means of denying environmental problems would allow people to maintain a view of the general social system as legitimate, just, and beneficial, thereby satisfying the system justification goal. Specifically, we exposed New York University students to a fictitious newspaper passage suggesting that the state of the environment was more positive than it really is; that steps were being taken to address environmental problems; and that concerns about environmental problems were decreasing. The passage was designed to create an opportunity for participants to engage in denial of environmental problems; we examined whether it affected the degree to which people viewed the socio-economic system as fair and legitimate. We expected that providing participants with an opportunity to deny environmental problems would allow them to maintain a more positive view of the social system, thereby satisfying the need to justify the system, and that it would weaken their resolve to improve environmental conditions.

In this study, participants were randomly assigned to one of two experimental conditions. In the "denial" condition, participants read the following fictitious excerpt allegedly taken from a newspaper article:

> These days, many people feel satisfied with the steps being taken to address environmental issues, such as increasing efforts to curb greenhouse gas emissions and reduce pollution. Many scientists believe that with dedicated effort we can stabilize global warming and create a cleaner natural environment. People feel safer and more secure about the environment than they used to, and there is a sense of confidence and optimism regarding the climate and the natural environment. It seems that, compared with prior generations, current efforts to help the environment are contributing to a reduction in global warming and pollution. Fewer and fewer people feel a need to be concerned about environmental conditions in the future.

Participants assigned to the control condition were not exposed to this (or any other) excerpt. All participants completed Kay and Jost's (2003) measure of general system justification.

Environmental denial was assessed using three items selected from the New Environmental Paradigm scale, including: "The so-called 'ecological crisis' facing humankind has been greatly exaggerated" and "If things continue on their present course, we will soon experience a major environmental catastrophe" (reverse-coded). Participants also completed a measure of behavioral intentions concerning the environment, which elicited agreement or disagreement with items such as: "I intend to give money to organizations that help the environment" and "I intend to better understand how different policies and practices impact the environment."

As hypothesized, participants who read the article that encouraged denial of environmental problems scored higher on the measure of general system justification than participants assigned to the control condition (*see* descriptive statistics in Table 26–1). In addition, participants who read the excerpt from the newspaper article engaged in greater environmental denial than participants who were assigned to the control condition. They also formed fewer intentions to learn about the impact of individual and societal practices and policies on environmental problems or to engage in helpful behaviors. Moreover, the relationship between exposure to the passage and the degree to which participants intended to help the environment was fully mediated by the denial of environmental problems.

These results indicate that, as hypothesized, the denial of environmental problems appears to serve a system-justifying function and to hinder the formation of pro-environmental intentions. If this is the case, then the need to engage in system justification is likely to disrupt the process of acknowledging environmental problems; admitting that there are human causes of global warming and environmental deterioration; and developing and pursuing intentions to take constructive action. The findings from our experimental study suggest that the media may play a crucial role, insofar as newspaper articles and other journalistic efforts can either encourage

TABLE 26-1. ENVIRONMENTAL GOAL SETTING: EXPERIMENTAL MEANS (AND STANDARD DEVIATIONS)

	System Justification	Denial	Intentions
Denial condition	5.04	2.56	3.08
	(0.26)	(0.19)	(0.20)
Control condition	4.06	1.91	3.73
	(0.27)	(0.19)	(0.21)

Note: We conducted three analyses of variance to examine the effects of the denial manipulation on system justification, denial of environmental problems, and setting of environmentally helpful goals, adjusting for gender and family income. As demonstrated in this table, participants exposed to the denial manipulation justified the system more, $F(1,44) = 8.53$, $p < 0.05$; engaged in greater denial of environmental problems, $F(1,44) = 3.78$, $p < 0.05$, and expressed fewer intentions to help the environment, $F(1,44) = 3.71$, $p < 0.05$. The relationship between exposure to the denial passage and failure to form intentions to help the environment was fully mediated by the denial of environmental problems, Sobel statistic = 1.96, $p < 0.05$.

people to engage in denial or acknowledgement of environmental realities.

Implications of a Self-Regulatory Perspective on Environmental Attitudes

Based on a model of system justification as conscious and nonconscious goal pursuit (Jost et al., 2007), we hypothesized that system justification goals would interfere with the self-regulatory process of setting and fulfilling goals to help the environment. When faced with threatening information about environmental destruction and climate change, people may be tempted to deny environmental realities. Such a response may satisfy the immediate goal to feel more certain and secure about the societal status quo and allow individuals to avoid the psychologically difficult process of facing up to threatening circumstances; questioning extant practices, institutions, and authorities; and resolving to take corrective action. Moreover, we have predicted (and found) that failures to engage in pro-environmental behavior would be especially common when people are either chronically or temporarily high in the motivation to justify the social system.

Explaining Partisan Differences in Environmental Attitudes

Although most Americans support abstract proposals to improve environmental conditions in general, support is significantly greater among those who identify as Democrats than among those who identify as Republicans (Saad, 2007a, 2007b). Public opinion surveys show that Republicans are far less concerned than Democrats about various forms of environmental destruction, including global warming, damage to the ozone layer, water pollution, and species extinction (Carroll, 2006). Republicans are also much less likely than Democrats to believe that environmental destruction is a serious problem, that there is solid scientific evidence of global warming, and that the cause of global warming is human activity (Pew Report, 2006). Moreover, there are prominent differences in response to global warming among liberal and conservative elites, as the August 13, 2007 "Global Warming is a Hoax*" issue of *Newsweek* made clear. The asterisk on the magazine cover leads the reader to the following qualification: "*Or so claim the well-funded nay-sayers who still reject the overwhelming evidence of climate change." The inside article by Begley (2007) documents coordinated efforts among politically conservative business leaders, politicians, think tanks, and others to deny that global warming is a serious concern, and it quotes several liberal activists and politicians who are pushing for collective action to address environmental challenges.

Our analysis of the role of system justification processes in the denial of environmental problems may elucidate public policy debates concerning environmental issues. We found that conservative respondents consistently exhibited stronger system justification tendencies, more denial of environmental problems, and less willingness to engage in behaviors that would help the environment, compared with liberal respondents. Most importantly, we have observed that system justification motivation significantly mediates the effect of political orientation on environmental attitudes and behavior. That is, one of the reasons why conservatives are less likely than liberals to support environmental initiatives is that they are more strongly motivated to defend and justify the societal status quo.

What Can Be Done? Social Psychological Contributions

The daunting problem of resistance to pro-environmental action aimed at slowing down environmental destruction and climate change will require the collaborative effort of researchers and activists. NASA scientist James Hansen, a leading climate researcher, warned that we have only a brief window of time (approximately 10 years) to act (Hansen, 2007). It is crucial that we understand and address perceptions and behaviors related to climate change as soon as possible. Psychology has much to contribute in elucidating the underlying dynamics of denial and complacency. The emerging field of conservation psychology (Clayton & Brook, 2005; Saunders, 2003) has recently begun to address this issue, although national and international forums in which problems of environmental behavior and environmental psychology can be adequately addressed are still lacking (Ehrlich & Kennedy, 2005). Given that the United States produces more of the emissions that threaten the environment than any other country, investigation into the many factors that affect our environmental ethics and behavioral decision-making has special national significance.

What suggestions for environmental goal-setting and pursuit, as well as behavioral change, can we glean from the extensive research on self-regulation? With respect to the setting of environmental goals, our research suggests that the very process of learning about environmental problems may elicit a defensive need to justify the system to relieve anxiety and fear, thereby hindering attainment of pro-environmental goals. The design of media messages, then, should consider psychological dynamics when framing and presenting environmental information. For example, if information about global warming and the need for system-level change can be paired with messages that reaffirm the system in alternative ways, then goal-setting may be facilitated. Or, as Feygina et al. (2010, Study 3) found, encouraging people to regard environmental reform as patriotic and consistent with protecting the status quo (i.e., as a case of "system-sanctioned change") can reduce and even eliminate the negative effect of system justification on environmentalism. Moreover, given that system justification goals can operate implicitly (Liviatan & Jost, 2009), it may be necessary to engage in "unconsciousness raising" as well as consciousness raising when it comes to environmental awareness. This may involve exerting deliberate control over impulsive desires (Baumeister & Heatherton, 1996) and cognitive, rational control over emotional and visceral responses (Loewenstein, 1996; Metcalfe & Mischel, 1999). Self-control can also be fostered by emphasizing the importance of long-term outcomes such as environmental destruction over short-term concerns to preserve the status quo (Trope & Fishbach, 2000). Similarly, utilizing global, integrative, overarching construals of the environmental dilemma should improve self-regulatory efforts (Fujita et al., 2006).

With respect to the pursuit of environmental goals, studies on automatic goal pursuit suggest that goals become linked to the behaviors, people, and situations in which they are realized (Bargh et al., 2001). Increasing the accessibility of means that help attain a goal increases the likelihood that the goal will be activated and pursued. Therefore, when individuals and institutions implement environmentally friendly practices, they contribute to the chronic activation of pro-environmental goals. Moreover, continued activation of a goal increases its chronic accessibility (Bargh et al., 2001). The amount of effort needed to exert control to act in accordance with one's explicit pro-environmental goals, in turn, should greatly diminish with practice. If we exert the effort in the short term to set up situations and practices that take environmental needs into account and pursue environmental goals, then in the long term these more constructive practices will become integrated into the set of automatic behaviors and intentions. Research has also suggested ways to improve self-regulatory efforts in the pursuit of important, albeit difficult, goals. For example, people can make use of implementation intentions that link situational cues to goal-directed behavior, thus helping to increase goal commitment (Gollwitzer & Moskowitz, 1996). Moreover, insofar as goals to justify the system may compete and interfere with environmental goals, people should strive to make use of attention control, emotion control, and motivation control strategies to shield their environmental goals (e.g., Kuhl & Beckmann, 1985).

These self-regulatory strategies should not only be directed at the goals and behaviors of the individual; it is crucial to work toward change at the collective level as well. As citizens become more aware of and more dissatisfied with the status quo when it comes to the environment, they must clearly express their disenchantment to economic and political leaders in order to bring about change in the system. Because politicians, business executives, and other representatives of the current system are especially influential in deciding whether to maintain or transform the status quo, their environmental agendas are of paramount significance. Economic institutions, too, contribute to the norms and structures that people are socialized to follow, thereby shaping the values we hold as well as many of our environmental and other practices. Therefore, the self-regulatory struggle of the individual to recognize problems rather than engaging in system justification and denial has to be carried over to the collective level of organization, so that together we can opt for change rather than perpetuating what has turned out to be a rather destructive state of affairs.

Authors' Note

The writing of this chapter was funded in part by the Center for Catastrophe Preparedness and Risk (CCPR) at New York University and the National Science Foundation (Award # BCS-0617558). Some of the research findings were presented at the 2007 International Society of Political Psychology (ISPP) conference in Portland, OR, the 2008 Society for Personality and Social Psychology (SPSP) conference in Albuquerque, NM, and the 2008 Society for the Psychological Study of Social Issues (SPSSI) conference in Chicago, IL. We thank Ran Hassin and Yaacov Trope for providing extensive feedback on an earlier draft.

References

Bargh, J. A. Auto-motives: Preconscious determinants of thoughts and behavior. In: Higgins, E. T., & Sorrentino, R. M. (Eds.), *Handbook of motivation and cognition*, Vol. 2. New York: Guilford, 1990: pp. 93–130.

Bargh, J. A., Gollwitzer, P. M., Lee-Chai, A., Barndollar, K., & Trötschel, R. The automated will: Non-conscious activation and pursuit of behavioral goals. J Pers Soc Psychol 2001; 81: 1014–1027.

Baumeister, R. F., & Heatherton, T. F. Self-regulation failure: An overview. Psychol Inq 1996; 7: 1–15.

Begley, S. The truth about denial. *Newsweek*. 2007. Available at: http://www.msnbc.msn.com/id/20122975/site/newsweek/-COVER (accessed August 14, 2007).

Carroll, J. Water pollution tops American's environmental concerns: Americans more worried about global warming this year. Gallup Poll News Service 2006.

Carroll, J. Public: Iraq war still top priority for President and Congress. Gallup Poll, 2007. Available at: http://www.galluppoll.com/content/?ci=27103&pg=1 (accessed April 9, 2007).

Clayton, S., & Brook, A. Can psychology help save the world? A model for conservation psychology. Anal Soc Issues Pub Pol 2005; 5: 1–15.

Cottrell, S. P. Influence of sociodemographics and environmental attitudes on general responsible environmental behavior among recreational boaters. Environ Behav 2003; 35: 347–375.

Dawes, R. M., & Messick, D. M. Social dilemmas. Int J Psychol 2000; 35: 111–116.

Dietz, T., Kalof, L., & Stern, P. C. Gender, values, and environmentalism. Soc Sci Quarter 2000; 83: 353–364.

Dunlap, R. E., van Liere, K. D., Mertig, A. G., & Jones, R. E. New trends in measuring environmental attitudes: Measuring endorsement of the New Ecological Paradigm: A revised NEP Scale. J Soc Issues 2000; 56: 425.

Ehrlich, P. R., & Kennedy, D. Millennium assessment of human behavior. Science 2005; 309: 562–563.

Feygina, I., Jost, J. T., & Goldsmith, R. E. System justification, the denial of global warming, and the possibility of "system-sanctioned change." Pers Soc Psychol Bull 2010; forthcoming.

Feygina, I., & Tyler, T. Procedural justice and system-justifying motivations. In: Jost, J. T., Kay, A. C., & Thorisdottir, H. (Eds.), *Social and psychological bases of ideology and system justification*. New York: Oxford University Press, 2009: pp. 351–370.

Fujita, K., Trope, Y., Liberman, N., & Levin-Sagi, M. Construal levels and self-control. J Pers Soc Psychol 2006; 90: 351–367.

Gallup Poll. Environment, 2007. Available at: http://www.galluppoll.com/content/default.aspx?ci=1615&pg=2 (accessed April 9, 2007).

Gollwitzer, P. M., & Moskowitz, G. B. Goal effects on action and cognition. In: Higgins, E. T., & Kruglanski, A. W. (Eds.), *Social psychology: Handbook of basic principles*. New York: Guilford, 1996: pp. 361–399.

Hansen, J. Defusing the global warming time bomb. Sci Am 2004; 290: 68–77.

Hansen, J. Political interference with government climate change science. Presented to the Committee of Oversight and Government Reform on March 19, 2007. Available at: http://www.columbia.edu/~jeh1/20070319105800-43018.pdf (accessed June 22, 2007).

Henry, P. J., & Saul, A. The development of system justification in the developing world. Soc Justice Res 2006; 19: 365–378.

Hardin, G. R. The tragedy of the commons. Science 1968; 162: 1243–1248.

Heath, Y., & Gifford, R. Free-market ideology and environmental degradation: The case of belief in global climate change. Environ Behav 2006; 38: 48–71.

Jost, J. T., Banaji, M. R., & Nosek, B. A. A decade of system justification theory: Accumulated evidence of conscious and unconscious bolstering of the status quo. Pol Psychol 2004; 25: 881–919.

Jost, J. T., Blount, S., Pfeffer, J., & Hunyady, Gy. Fair market ideology: Its cognitive-motivational underpinnings. Res Organ Behav 2003a; 25: 53–91.

Jost, J. T., Burgess, D., & Mosso, C. Conflicts of legitimation between self, group, and system: The integrative potential of system justification theory. In: Jost, J. T., & Major, B. (Eds.), *The psychology of legitimacy: Emerging perspectives on ideology, justice, and intergroup relations*. New York: Cambridge University Press, 2001: pp. 363–388.

Jost, J. T., Glaser, J., Kruglanski, A. W., & Sulloway, F. Political conservatism as motivated social cognition. Psychol Bull 2003b; 129: 339–375.

Jost, J. T., & Hunyady, O. The psychology of system justification and the palliative function of ideology. Eur Rev Soc Psychol 2002; 13: 111–153.

Jost, J. T., & Hunyady, O. Antecedents and consequences of system-justifying ideologies. Curr Dir Psychol Sci 2005; 14: 260–265.

Jost, J. T., Nosek, B. A., & Gosling, S. D. Ideology: Its resurgence in social, personality, and political psychology. Perspect Psychol Sci 2008; 3: 126–136.

Jost, J. T., Pelham, B. W., Sheldon, O., & Sullivan, B. N. Social inequality and the reduction of ideological dissonance on behalf of the system: Evidence of enhanced system justification among the disadvantaged. Eur J Soc Psychol 2003c; 33: 13–36.

Jost, J. T., Pietrzak, J., Liviatan, I., Mandisodza, A. N., & Napier, J. L. System justification as conscious and nonconscious goal pursuit. In: Shah, J. Y., & Gardner, W. L. (Eds.), *Handbook of motivation science*. New York: Guilford, 2007: pp. 591–605.

Jost, J. T., & Thompson, E. P. Group-based dominance and opposition to equality as independent predictors of self-esteem, ethnocentrism, and social policy attitudes among African Americans and European Americans. J Exp Soc Psychol 2000; 36: 209–232.

Kay, A. C., Jimenez, M. C., & Jost, J. T. Sour grapes, sweet lemons, and the anticipatory rationalization of the status quo. Pers Soc Psychol Bull 2002; 28: 1300–1312.

Kay, A. C., & Jost, J. T. Complementary justice: Effects of "poor but happy" and "poor but honest" stereotype exemplars on system justification and implicit activation of the justice motive. J Pers Soc Psychol 2003; 85: 823–837.

Kay, A. C., Jost, J. T., & Young, S. Victim derogation and victim enhancement as alternate routes to system justification. Psychol Sci 2005; 16: 240–246.

Keeling, C. D., & Whorf, T. P. Atmospheric CO_2 records from sites in the SIO air sampling network. In: *Trends: A Compendium of Data on Global Change.* Oak Ridge, TN: Carbon Dioxide Information Analysis Center, Oak Ridge National Laboratory, U.S. Department of Energy, 2005.

Komorita, S. S., & Parks, C. D. *Social dilemmas.* Boulder, CO: Westview, 1997.

Kortenkamp, K. V., & Moore, C. F. Time, uncertainty, and individual differences in decisions to cooperate in resource dilemmas. Pers Soc Psychol Bull 2006; 32; 603–615.

Kramer, R. M. Intergroup relations and organizational dilemmas: The role of categorization processes. In: Cummings, L. L., & Staw, B. M. (Eds.), *Research in organizational behavior*, Vol. 13. Greenwich, CT: JAI Press, 1991: pp. 191–227.

Kuhl, J., & Beckmann, J. *Action control: From cognition to behavior.* New York: Springer-Verlag, 1985.

Lau, G. P., Kay, A. C., & Spencer, S. J. Loving those who justify inequality: The effects of system threat on attraction to women who embody benevolent sexist ideals. Psychol Sci 2008; 19: 20–21.

Liviatan, I., & Jost, J. *A social cognitive analysis of system justification goal striving.* Manuscript submitted for publication, 2009.

Loewenstein, G. Out of control: Visceral influences on behavior. Organ Behav Hum Decis Process 1996; 65: 272–292.

Luhtanen, R., & Crocker, J. A Collective Self-Esteem Scale: Self-evaluation of one's social identity. Pers Soc Psychol Bull 1992; 18: 302–318.

Metcalfe, J., & Mischel, W. A hot/cool-system analysis of delay of gratification: Dynamics and willpower. Psychol Rev 1999; 106: 3–19.

Miller, W. R., & Brown, J. M. Self-regulation as a conceptual basis for the prevention and treatment of addictive behaviours. In: Heather, N., Miller, W. R., & Greeley, J. (Eds.), *Self-control and the addictive behaviours.* New York: Macmillan, 1991: pp. 3–61.

Nordhaus, W. D., & Yang, Z. A regional dynamic general-equilibrium model of alternative climate-change strategies. *Am Econ Rev* 1996; 86: 741–765.

Oreskes, N. The scientific consensus on climate change. Science 2004; 306: 1686.

Ostman, R. E., & Parker, J. L. Impact of education, age, newspapers and television on environmental knowledge, concerns and behaviors. J Environ Educ 1987; 19: 3–9.

Pew Research Center for the People & the Press. Little consensus on global warming: Partisanship drives opinion, 2006. Available at: http://people-press.org/report/280/little-consensus-on-global-warming (accessed December 1, 2009).

Saad, L. To Americans, the risks of global warming are not imminent. Gallup Poll News Service 2007a.

Saad, L. Most Americans back curbs on auto emissions, other environmental proposals. Gallup Poll 2007b. Available at: http://www.gallup-poll.com/content/?ci=27100&pg=1 (accessed April 9, 2007).

Samdahl, D. M., & Robertson, R. Social determinants of environmental concern: Specification and test of the model. Environ Behav 1989; 21: 57–81.

Saunders, C. D. The emerging field of conservation psychology. Hum Ecol Rev 2003; 10: 137–149.

Scheffer, M., Carpenter, S., Foley, J. A., Fokes, C., & Walker, B. Catastrophic shifts in ecosystems. Nature 2001; 413: 591–596.

Scott, D., & Willits, F. K. Environmental attitudes and behavior. Environ Behav 1994; 26: 239–260.

Sidanius, J., & Pratto, F. *Social dominance: An intergroup theory of social hierarchy and oppression.* New York: Cambridge University Press, 1999.

Stangor, C., & Jost, J. T. Individual, group, and system levels of analysis and their relevance for stereotyping and intergroup relations. In: Spears, R., Oakes, P. J., Ellemers, N., & Haslam, A. S. (Eds.), *The social psychology of stereotyping and group life.* Malden, MA: Blackwell Publishing, 1997: pp. 336–358.

Thompson, J. C., & Gasteiger, E. L. Environmental attitude survey of university students: 1971 vs. 1981. J Environ Educ 1985; 17: 13–22.

Tice, D. M., Bratslavsky, E., & Baumeister, R. F. Emotional distress regulation takes precedence over impulse control: If you feel bad, do it! J Pers Soc Psychol 2001; 80: 53–60.

Trope, Y., & Fishbach, A. Counteractive control processes in overcoming temptation. J Pers Soc Psychol 2000; 79: 493–506.

Tyler, T. R. Psychological perspectives on legitimacy and legitimation. Ann Rev Psychol 2006; 57: 375–400.

Tyler, T. R., & Degoey, P. Collective restraint in social dilemmas: Procedural justice and social identification effects on support for authorities. J Pers Soc Psychol 1995; 69: 482–497.

Ullrich, J., & Cohrs, J. C. Terrorism salience increases system justification: Experimental evidence. Soc Justice Res 2007; 20: 117–139.

Van Liere, K. D., & Dunlap, R. E. The social bases of environmental concern: A review of hypotheses, explanations, and empirical evidence. Publ Opin Quarter 1980; 44: 181–197.

Wakslak, C., Jost, J. T., Tyler, T. R., & Chen, E. Moral outrage mediates the dampening effect of system justification on support for redistributive social policies. Psychol Sci 2007; 18: 267–274.

Watson, R., & the Core Writing Team (Eds.). Climate Change 2001: Synthesis Report: Third Assessment Report of the Intergovernmental Panel on Climate Change. Cambridge, UK: Cambridge University Press, 2002.

Weart, S. The discovery of global warming. Cambridge, MA: Harvard University Press, 2004.

Weber, J. M., Kopelman, S., & Messick, D. M. A conceptual review of decision making in social dilemmas. Pers Soc Psychol Rev 2004; 8: 281–307.

Webster, P. J., Holland, G. J., Curry, J. A., & Chang, H-R. Changes in tropical cyclone number, duration, and intensity in a warming environment. Science 2005; 309: 1844–1846.

Zelezny, L. C., Chua, P., & Aldrich, C. Elaborating on gender differences in environmentalism. J Soc Issues 2000; 56: 443–457.

CHAPTER 27

Teleological Behaviorism and the Problem of Self-Control

Howard Rachlin

ABSTRACT

Problems of self-control as well as social cooperation may be seen as conflicts not between internal spiritual or neurological entities, but between highly valued overt behavioral patterns of differing temporal extents or social distances. For example, an alcoholic must choose between having a drink now—valuable in the short run—and being healthy, performing well at work, maintaining satisfying social relationships, etc. —valuable in the long run. The essential question addressed in this chapter is how the latter may come to dominate the former within a person's lifetime. A behavioral evolutionary process is proposed by which valuable temporally or socially extended behavior patterns evolve over an individual lifetime from simpler, shorter patterns. It is argued that complex patterns, such as social cooperation over long periods, arise from simpler patterns in behavioral evolution analogously to the way complex structures, such as the human eye, arise from simpler structures in biological evolution. The idea that complex, long-term behavioral patterns in conflict with short-term patterns, if they are not inherited *in toto*, must be generated by an internal and autonomous spiritual, neurological, or cognitive process is compared to creationism in biological evolution. The role of delay and social discount functions in measuring the extent of coherent behavioral patterns is explicated. Finally, the chapter examines several implications of teleological behaviorism for practical behavioral control.

Keywords: behaviorism, behavioral evolution, behavioral patterns, biological evolution, delay discounting, self-control, social cooperation, social discounting, teleological behaviorism

TELEOLOGICAL BEHAVIORISM AND THE PROBLEM OF SELF-CONTROL

Teleological behaviorism views self-control as an *external* phenomenon—as a conflict between a relatively immediate, particular, high-valued act (such as smoking a cigarette) and an element of a long-range, more abstract, high-valued pattern of acts (such as healthy behavior). When the conflict is resolved in favor of the latter, behavior is said to be self-controlled; when the conflict is resolved in favor of the former, behavior is said to be impulsive. The present chapter contrasts this view, stemming from Aristotle, with the currently dominant view of self-control as an *internal* phenomenon (Rachlin, 2000). The dominant view stems from Descartes' initial conception of self-control as a conflict taking place within the brain between a neural impulse leading to the particular act and willpower—a spiritual force leading to the high-valued pattern.

Like Descartes, many modern thinkers locate the arena of conflict somewhere within the brain, but for them the conflict takes place between neural impulses coming from different places rather than between a neural impulse and a spiritual force. A message from a "lower" brain area dictating a certain action (a neural impulse to smoke the cigarette) conflicts with a message from a "higher" brain area dictating an opposite action (a neural impulse to refuse the cigarette). These neural impulses (internal representations of intentions) fight it out in some area of the brain and the behavior that actually emerges signals the winner.

Teleological behaviorism rejects this view. It says that the fundamental conflict between self-controlled and impulsive actions takes place not among representations of intentions in some specific location in the brain, but in the person's overt behavior over a period of time. From the present viewpoint, self-control is not a battle between internal intentions to do one thing and internal impulses to do another. It is fundamentally a temporal conflict—between behavior that maximizes value over a short time period and behavior that maximizes value over a long time period.[1]

The initial sections of the chapter are theoretical. They contrast the teleological and behavioral view of self-control with that of both modern cognitive psychology and neuropsychology, and they relate teleological behaviorism to evolutionary biology and economics. The internal (Cartesian) and external (Aristotelian) viewpoints are not mutually exclusive. On the one hand, all complex human and animal behavior patterns are controlled by *internal* mechanisms. On the other hand, all complex human and animal behavior patterns exist because of their function in the interaction of the person with the *external* environment. The issue is essentially semantic; should we use the vocabulary of self-control to refer to the interaction between *internal* causes and behavior patterns (as Descartes advised), or should we use that vocabulary to refer to the function of behavior patterns in the *external* environment (as Aristotle advised)? But a semantic issue is not necessarily a pointless one. In this case, the two viewpoints have practical as well as theoretical consequences. Accordingly, the latter sections of the chapter attempt to show how a behavioral viewpoint in general, and a teleological behavioral viewpoint in particular, make a difference in addressing real-world problems of self-control.

THE SELF-CONTROL CONFLICT

The gestalt psychologists claimed that the whole is greater than the sum of its parts (Koffka, 1955). The teleological behavioral extension of that gestalt dictum would say that the value of an activity may be greater than the sum of the values of its parts. As an illustration, suppose you are driving from New York to Chicago. Your car has a CD player and you take along some CDs to play on the trip. You like both classical and popular music, so you take along several symphonies and several pop CDs. Suppose (perhaps contrary to fact) that your tastes are such that the following two inequalities apply:

1. You prefer listening to a 60-minute symphony to spending that 60 minutes listening to twenty 3-minute popular songs.
2. You prefer listening to a popular song for 3 minutes to spending that 3 minutes listening to a 3-minute section of a symphony.

This is a paradigm case of a self-control problem. The problem is that to listen to the whole symphony (which by assumption you prefer to do) you must listen to the first 3 minutes of it (which you prefer not to do). If you just do what you prefer at the moment (assuming your preferences remain constant throughout the trip), you will drive the whole way from New York to Chicago playing only popular songs whereas (again, by assumption) you would have been happier if you had played only symphonies.

Similarly, an alcoholic prefers to be sober, healthy, socially accepted, and to perform well at his job than to be drunk, unhealthy, socially rejected, and perform poorly at his job. However, over the next few minutes, he prefers to have a drink than to not have one. If over successive brief intervals he always does what he prefers, he will always be drinking.

The problem, for both the driver and the alcoholic is how to make choices over the longer time span and avoid making choices on a case-by-case basis. The reason why we find it difficult to avoid making short-term choices is that the value of the immediate alternative (listening to a popular song; having a drink now) is greater than that of a fraction of the longer activity (listening to 3 minutes of a symphony; being sober now). The reason why we *should* make choices over the longer time-span is that the value of the longer activity (listening to the whole symphony; being generally sober) is greater than the sum of all the short-term values.

Each 3 minutes of symphony-listening, each drink refusal, has virtually no value in itself. Moreover, to use behavioristic language, these pieces of the more-valuable behavioral pattern are never reinforced. They are not immediately reinforced, they are not conditionally reinforced, and they are not reinforced after a delay. Clearly a drink refusal is not immediately reinforced. It should also be clear that no single drink refusal is reinforced after a delay. If the alcoholic refuses a single drink, she does not wake up three weeks later suddenly healthier and happier. To realize the value of a drink refusal, she must put together a long string of them—just as to realize the value of a symphony you must listen to the whole thing (or at least a whole movement).

The fact that longer behavioral patterns may have a value greater than the sum of the values of their parts is not unique to these very broad patterns. Each broad behavioral pattern is nested in still broader patterns and contains narrower patterns within it. Listening to a symphony over an hour is nested within the pattern of listening to a mixture of symphonies and popular songs for a day. At the other extreme, listening to a single verse is nested within listening to a whole popular song. Even a seemingly unitary reinforcer, such as a 3-second food delivery to a pigeon, may be seen as a behavioral pattern.

Let us zoom in on a pigeon in an operant laboratory. The pigeon pecks a lit button and a food pellet immediately drops into a feeder below. After the pellet drops, the pigeon has to rebalance itself, orient its body, move its head down to the feeder, move its head into the feeder, open its beak, close its beak around the pellet, and swallow the pellet. Then it has to use the energy from the food get more food, to groom itself, to hunt around its cage, etc. Where, in all this, does reinforcement actually occur? Is there really an instant, as some physiologists tell us, when a section of the pigeon's brain lights up, before which the pigeon's peck was unreinforced and after which it was reinforced? The answer must be no. The teleological behaviorist does not deny that physiological events underlie behavioral changes but does deny that reinforcement itself is an internal event wherein a neural connection or a neural representation of an act is somehow strengthened.

Eating a single food pellet, in this close examination, is a complex sequence of sub-events; the sum of the values of those events is less than the value of the sequence as a whole. So, even an apparently unitary high-valued reinforcer, such as a pigeon eating a food pellet, has the same form as a problem in self-control. And eating *would* be a problem in self-control for pigeons were it not for Mother Nature. Pigeons have evolved to ignore the individual microscopic values of each muscular movement and behave as if the pattern of acts involved in eating were a single act. Humans too are built to ignore the values of the components of eating and choose according to the value of the pattern—or to quickly develop the ability to do so.

To summarize, the concept that the value of a behavioral pattern may be greater than the sum of the values of its parts is not just an empty slogan borrowed from gestalt psychology. It is the very basis of a consistent approach to the problem of self-control—that of teleological behaviorism.

TELEOLOGICAL BEHAVIORISM

Behaviorists disagree with each other about whether complex behavioral patterns of whole organisms may usefully be labeled by terms from our mental vocabulary. Skinner (1990) thought not. I believe, on the contrary, that mental terms are useful in prediction, control, and understanding of behavior (Rachlin, 1994). As an illustration of how teleological behaviorism approaches the study of the mind [derived from the concept of mind of philosophers Ryle (1949) and Wittgenstein (1953)], consider what it means to *imagine* something. Imagination may be conceived not as an image in your head but as a functional mode of behavior—behaving in the absence of some state of affairs as you normally would in its presence.

Suppose two people in a room are both asked to imagine a lion. The first person closes her eyes and says, "Yes, I see it; it has a mane and a tail." The second person runs screaming from the room. The first person is imagining a picture or a movie of a lion, but the second is imagining a lion itself. What is the function of such behavior? Imagination is a necessary part of perception. If perception (as distinct from sensation) is current discrimination of complex, temporally extended sequences of stimuli (as distinct from simpler, immediate stimuli), then the immediate discriminative response, especially if made early in the sequence, involves a sort of gamble—behaving as if the extended sequence had occurred. For example, at any given moment I treat my wife as the person she is in the long run, not as the particular bundle of sensations she presents to me at that moment. It is in connection with such premature but necessary discrimination (the universal arising out of particular instances) that Aristotle (tr., 1941) gives his famous analogy of soldiers turning and making a stand: "It is like a rout in battle stopped by first one man making a stand and then another, until the original formation has been restored" (p. 185). The function of the soldiers' behavior is to create an abstraction (the renewed formation) out of individual actions. The first soldier to turn is behaving as he would if all the others had already turned; he is imagining that they had already turned. His imagination is what he does, not what his nervous system is doing. The functions of our ordinary imaginations allow us to get around in the world on the basis of partial information. We do not have to carefully test the floor of every room we walk into.

From this point of view, good imagination is not only essential for an actor on the stage and for the creation of art, but it is also a necessary part of self-control. One cigarette refusal by a smoker is utterly worthless, like only one soldier in a rout turning and making a stand. Refusal of an individual cigarette is never reinforced—not now, not later, not symbolically, not internally. Only an extended series of cigarette refusals is reinforced. Refusal of the first cigarette is thus an act of imagination—behaving as you would if a state of affairs existed when it does not (yet) exist. To explain self-control, we do not have to rely on creators of our actions, placed inside our heads—as Donner, the thunder god, is placed in Valhalla—because we cannot find them in the outside world. As I will argue in the following section, the complex long-term overt acts involved in self-control may evolve from simpler acts.

The Evolution of Self-Control

Consider a squirrel saving nuts during the fall so as to have food available during the coming winter. Because it does not conflict with any other behavior, the squirrel's nut storing is not self-controlled. The squirrel just wakes up one fall morning, eats a bit, and goes about doing what it feels like doing, collecting nuts and placing them in the hollows of trees or burying them. Mother Nature, in the form of evolution by natural selection, arranged it in the past; squirrels that failed to store nuts in the fall tended to die during the coming winter whereas squirrels that stored nuts (and remembered where they were) tended to live and reproduce. Over generations, squirrels evolved so that they tended to value nut-storing behavior for its own sake rather than for the sake of having food to tide themselves over the winter months. But Mother Nature has not had enough time to work the same sort of magic on humans living in a modern society; it is a rare morning when we wake up and just feel like putting our money in the bank. What she has arranged in our case is a capacity to learn that consistently putting money in the bank is generally a good thing to do even though we might not feel like doing it at the present moment.

It was noted by post-Darwinian psychologists that habits may evolve over the lifetimes of individuals just as the structure of a species evolves over generations. Thorndike's (1898) "law of effect" described an evolutionary process whereby reward strengthens habits within an individual's lifetime just as survival strengthens physical structure (including that of the brain) over generations. If a pigeon is hungry and gets a bit of food each time it pecks a lit button, its pecking rate increases. The food (the reinforcer) allows the act it follows to survive whereas other acts, not reinforced, die out.

This simple act of pecking the button and eating the food may be seen as self-control in miniature. Without the food contingent on the button-peck, the pigeon would peck the button only infrequently; a button-peck *as such* is less valuable to the pigeon than not pecking the button (and maybe preening its feathers instead). If the food is made contingent on the button peck and the pigeon increases its pecking rate, it must be because the sequence: peck the button and eat the food is more valuable than the sequence: do not peck the button and do not eat the food. And this is true even though the initial act in the sequence (peck the button) is *less* valuable than its alternative (do not peck the button). Seen this way, pecking the button and eating the food is a pattern of self-control. By pecking and eating, the pigeon is engaging in a valuable longer-term pattern of behavior and rejecting a (relatively) shorter-term act in conflict with it.

Once the button peck appears, it may be molded or "shaped." Just as the giraffe's long neck evolved by the enhanced survival of long-necked giraffes (as they were enabled to reach higher and higher into trees for food), so the pigeon's button-peck may evolve. If the number of pecks required to obtain the pellet is gradually increased, the pattern of pecking changes. Just as giraffes would have become extinct if, in the distant past, the height of the edible leaves on the trees had suddenly jumped from ten to twenty feet, so a pigeon's button-pecking would die out (would "extinguish") if the number of pecks required (the "ratio" requirement) suddenly increased by 100. To shape pecking in numbers near the limit of the pigeon's ability, the ratio has to be raised in small stages. The natural variation of the pigeon's pecking at each stage must include patterns that will be reinforced at the next stage. Selection by reinforcement, like natural selection, works only on material that is already present as a consequence of variation at a previous stage. That is, selection by reinforcement works on top of natural selection. Natural selection provides the mechanism for selection by reinforcement to work. Taking the Darwinian and the Thorndikian processes of evolution together, if we observe any consistent pattern in the behavior of a human or non-human animal, that pattern must have evolved by natural selection over the history of the species or have evolved by reinforcement over the history of the individual.[2]

Much of the time, these two evolutionary processes work together—as in the normal development of language. But, in many cases, especially in our civilized society, they work at cross-purposes. Such cases constitute a very fundamental problem of self-control. In the past history of the human race, eating as much as you could whenever food became available was a valuable habit; it might be a long time before food became available again. Now, with food continuously available, eating whenever you have the chance is bad in many ways (Logue, 1988).

There is an apparent problem in extending the analogy with natural selection to self-control. Where is the reward when I pass by the bakery even though I am hungry and the smell wafting onto the street is extremely enticing? What in general reinforces self-control? It is clear enough what reinforces lack of self-control. For an alcoholic, drinking is highly rewarding in itself. Drinking, in fact, may reinforce virtually any act the alcoholic is capable of doing. But refusing a drink is far from rewarding in itself and is rarely followed by anything pleasant; in fact it may be followed by a great deal of pain. An alcoholic's single act of drink-refusal is never externally reinforced, neither immediately nor later. To repeat, if an alcoholic refuses a single drink, she does not wake up two weeks later healthier or happier, nor is a single drink refusal intrinsically rewarding—just the opposite. The value of drink refusal—in the form of health, social acceptability, job performance, and so forth—seems to depend on an already established habit of refusing drinks; to attain its high value, drink refusal must occur consistently over a period of time. To attain their high value, bakery avoidance and its equivalents must also occur consistently over a period of time. It seems as if self-control would be impossible because the pattern you need to establish (perhaps a very complex pattern) must exist before its value is attained. Given this problem, how can a complex pattern of behavior evolve by reinforcement?

Let us consider a similar question as it appears in evolution by natural selection: How can very complex structures, such as the human eye, evolve by natural selection? Each part of the eye depends on other parts. It seems as if, until each part develops, the other parts would be useless. The retina cannot function without a lens to focus an image on it; the lens cannot function without a pupil to regulate the light (and what is the function of the pupil if there is no retina?). Finally, an optic nerve has to develop simultaneously with the eye to encode and transmit the image to the brain. And once encoded in the brain, then what? How does a pattern of neural states get converted into a pattern of visual discriminations? On its face, it is hard to conceive how the human eye could have developed in stages from primitive structures. For creationists, such development is impossible to conceive. *God had to have done it*, they think. But evolutionary biologists, studying primitive eyes in other organisms, have been able to trace a convincing story of how the human eye developed. Similarly, regardless of how complex and seemingly interdependent the parts of a behavioral pattern may be, they can be traced in stages to simpler patterns in development. To argue that the immense complexities observed in human self-controlled behavior could not have evolved through reinforcement is a kind of creationist argument. The creator, this time, is not God but something within the person—the soul, the self (as an internal entity), or the nervous system (see Baum, 2005, for a detailed, convincing argument for the evolution of complex behavior over a person's lifetime).[3]

Another apparent problem with the concept of behavioral evolution of self-control lies in the abstract nature of the pattern. The alcoholic who cuts down from a quart of whisky a day to a glass of wine each night with dinner is not doing anything in particular during the time she is now not drinking; her drinking is just occurring at a much slower rate than before. Each day, the time spent not drinking may be filled in with different activities. Every one of these activities cannot individually be more valuable than having a drink. What sustains the low drinking rate? The answer is that the low drinking rate itself is highly valuable in the life of the reformed alcoholic. But the reduced drinking rate is an abstract entity that has no existence at any one moment. You cannot point to it as you would point to an eye. Some alcoholics do learn to abstain or drink moderately. How do the highly abstract patterns in their behavior hold together, and how do they resist being broken up by temptation?

To answer this question, consider an evolutionary process called *group selection*. Imagine a basketball league where the only factor that determined whether a player was retained on a team was points scored by that player (individual selection), and the only factor that determined whether a team was retained in the league (and the franchise not moved to another city) was games won by the team (group selection). It is generally recognized that, all other factors held equal, a team that plays unselfishly will win more games than a team that plays selfishly. Yet, at the same time, the selfish players on each team will score more points than the unselfish players. Thus, individual selection works, in this fictional league, to promote selfishness whereas group selection works to promote unselfishness; the two selection processes are in conflict. Which will win out depends largely on the relative rate of selection. If players are replaced on teams more frequently than teams are replaced in the league (as is the case in almost any real basketball league), then selfishness will come to dominate among players. However, if teams were replaced in the league faster than players were replaced on teams, then unselfishness would come to dominate.[4]

Imagine an environment, say a jungle, with tribes of people all competing against each other for limited natural resources. Just as with basketball teams, tribes within which individuals act altruistically (i.e., possess genes that generate altruistic behavior) will tend to out-compete tribes within which individuals act selfishly. But at the same time, within each tribe, selfish individuals (where selfishness is that of the gene—including sacrifice to preserve the lives of blood relatives) will out-compete unselfish individuals. That is, a conflict exists between group selection for altruism and individual selection for selfishness. Because, in any human environment, individuals are replaced within social groups much faster than social groups are replaced within the environment, by the above reasoning, innate selfishness will tend to predominate over innate altruism in human society.

Now consider selection by reinforcement over the lifetimes of individuals. Does selection by reinforcement act exclusively on individual actions, or can it act on groups of actions? Ainslie (2001), in accord with Skinner (1938), believes that individual extrinsic reinforcers (including reinforcers discounted by their delays from the present moment) may strengthen individual discrete responses but not abstract patterns of responses (the kind constituting self-control and usually thought of as rules of personal conduct). For Ainslie, abstract patterns arise by a process of *internal* bargaining among long-term, medium-term and short-term interests of the individual. I believe, however, that abstract patterns of behavior may be selected as such by a process akin to group selection of individual organisms (Rachlin, 2002). Is there such a thing as group selection of behavior over the lifetime of an individual organism? Again, the answer depends on the relative rate of selection of individual actions and groups of actions. As I argued in the section above, even an individual button-peck by a pigeon may be blown up, as it were, and seen as a temporally extended behavioral pattern. So, even for Ainslie and Skinner, reinforcement acts on behavioral patterns, shaping them into different patterns.

Among nonhumans there is considerable evidence that reinforcement selects not individual responses but patterns of individual responses. For example, if a rat is rewarded for running faster or slower down an alley or for pressing a lever faster or slower, it changes its rate not by adjusting its speed of running or pressing the lever but by periodically pausing more or less frequently between bursts of running or lever-pressing which, within bursts, remain constant in rate. Similarly, a person on a diet, when she eats, may not eat any slower than one not on a diet. The question is, can longer patterns—those constituting human self-control—evolve from simpler patterns over a person's lifetime just as the complex human eye evolved from simpler light-sensitive organs? It is not possible to prove that every complex behavioral pattern we exhibit evolved over our lives from simpler patterns and was not, as in the typical explanation, created by an internal logic mechanism overcoming an equally internal "visceral" force (Loewenstein, 1996). But there is considerable evidence that when behavior is organized into coherent patterns, self-control increases. For example, in choosing between immediate monetary rewards and a non-immediate but larger overall reinforcement rate, with no 1:1 relationship to any particular choice, people forced to make choices in groups of four chose the larger overall rate significantly more frequently than people choosing on a case-by-case basis (Rachlin, 1995). That is, we do know that reinforcement may act directly on patterns of responding (without necessitating that chains of responses be built up from individual responses each followed by its own conditioned reinforcer).

Teleological Behaviorism and Economics

Teleological behaviorism is a psychological version of modern microeconomics. The teleological behaviorist's principle that organisms behave within the reinforcement contingencies so as to maximize reinforcement value (over some time period) is equivalent to the economist's principle that people behave so as to maximize utility within the constraints set by budget and price (Rachlin et al., 1981). The question for self-control is, what is the duration over which reinforcement (or utility) is maximized? When the duration is short, behavior is impulsive; when the duration is long, behavior is self-controlled.

Delay discount functions: One way to measure the duration over which reinforcement is maximized by any person in any situation is to obtain a delay discount function. Delay discount functions may be obtained for individuals by asking them a series of questions about their preferences between smaller amounts of money available now and larger amounts available at various future times (Raineri & Rachlin, 1993). Delay discount functions have been obtained in many studies (see review by Green & Myerson, 2004). The shape of the functions almost universally found is hyperbolic as in Equation 1:

$$v_D = \frac{V_D}{1 + k_{delay} D^s} \quad (1)$$

where V_D is the magnitude of the delayed reward to be obtained after D time units, v_D is the magnitude of an immediate reward equal in value to the delayed reward (the discounted value of the delayed reward) and s and k_{delay} are positive parameters inversely proportional to degree of self-control. When s and k_{delay} are large, a delayed reward is worth only a small present amount; when s and k_{delay} are small, delay has less of a deteriorating effect on reward value. When s and k_{delay} are zero, delay has no effect at all on reward value. Addicts (alcoholics, cigarette smokers, cocaine users, heroin users) have significantly higher individual discount parameters than non-addicts (Bickel & Marash, 2001).

Negative and positive addictions: The economists Stigler and Becker (1977) distinguished between negative and positive addictions. Negative addictions are the ones we usually identify as addictive—smoking cigarettes, drinking alcohol, taking heroin or cocaine, overeating to obesity. The reason we call negative addictions negative is that they are generally bad for one's health. Beyond a certain point, the more one drinks, takes heroin, overeats, etc., the worse their health gets, the worse their social life gets, the worse their job or school performance gets. The crucial theoretical property of negative addictions, however, is that, as their rate increases, tolerance builds up. (In economic terms, as consumption rate increases, marginal cost increases.) An amount of heroin that would kill a non-addict would be barely enough to satisfy the craving of a heroin addict. Certainly tolerance is mediated by physiological events inside the person. But, for the economist, as well as the teleological behaviorist, the defining property of a negative addiction is: *the higher its overall rate, the less the reinforcement-value of the marginal unit of the activity*. At the same time, the demand for the negatively addictive behavior is inelastic (the addict "has to have a certain amount of it") so that to keep up the same level of satisfaction, the rate of the behavior must increase.

Positive addictions, on the other hand, are activities that many people spend a lot of time doing but are not usually identified as addictions—physical exercise, social activity, reading, listening to music, stamp collecting, doing crossword puzzles. The reason we call positive addictions positive is that up to a point they are generally considered to be good for you—or at least not bad. For behavior theory, however, the crucial property of positive addictions is opposite to tolerance. The more you do them, the more satisfying they become. Positive addictions all involve a practice effect. The defining property of a positive addiction is opposite to that of a negative addiction: *the higher its overall rate, the more the reinforcement-value of the marginal unit of the activity*. At the same time, the demand for positive addictions is usually *elastic*, so they get cheaper when you do them more and more (i.e., you "buy more" of them).

As a person engages in more social activities—goes to more parties for instance—the more enjoyment she tends to get from each one. As she collects more stamps, the more fun stamp collecting becomes. Some positive addictions may not be good for us beyond a certain point. In fact, unless they do go beyond that point, we do not usually identify them as addictions at all. Playing computer games, for example, is technically a positive addiction; as the player gains skill, she comes to enjoy the game more; as she plays less frequently, she loses skill and enjoys the game less. Even exercise and social activity may reach a point where we call them addictive. For our purposes, however, negative and positive addictions are defined not by whether they are bad or good for us generally, but whether they are characterized by tolerance or practice—that is, whether a unit of an act decreases or increases in value as its rate increases.

Social cooperation and self-control: Many situations in human life put the benefit of an individual into conflict with the benefit of a group. The individual basketball player choosing whether to pass the ball to a teammate closer to the basket (maximizing the chance of winning the game and benefiting the team) or take a shot himself from further out (maximizing the chance of gaining individual glory) is a case in point. In the laboratory such conflict is often studied by means of a *prisoner's dilemma game*. For example, in a two-player prisoner's dilemma game, each player chooses to cooperate or defect. If both players cooperate, each gains a moderately high reward; if both players defect, each gains a moderately low reward. Obviously, it is better for the group if both cooperate. However, if one player cooperates and the other defects, the defector gets a very high reward whereas the cooperator gets a very low reward. Each player thus benefits maximally by defecting: If the other player defects, you should defect too, otherwise you would be left with the lowest reward; if the other player cooperates, you should also defect; then you will get the very highest reward. In either case it is to your individual benefit to defect. When there are more than two players (as on a basketball team), it is similarly better for each player to defect and better for the group if all cooperate.

Self-control may be viewed in an analogous way to social cooperation. A single person may be envisioned as a point moving along a line stretched out in the future and the past. There is a problem in self-control when the most valuable behavior at this point (smoking a cigarette now) conflicts with the most valuable behavior over a period of time (smoking no cigarettes over a period of several months). A person may therefore be conceived as playing a prisoner's dilemma game with her wider self—analogous to a group of people at various distances from her in the past and future. What we call "self-control" is behavior that benefits the group thus conceived—that benefits the person now together with the same person tomorrow and the day after that and the day after that, and so forth—at the expense of the person now alone. Self-control is the same as cooperation with your own wider self; failure of self-control is the same as defection against your own wider self. According to this picture we go through life repeatedly faced with prisoner's dilemma conflicts of both kinds: ourselves versus other people (social dilemmas) and ourselves at this very moment versus ourselves at wider temporal extents (self-control dilemmas).

Social discount functions: When we say that a person is controlling herself, we mean that she is maximizing utility over a wide temporal extent; when we say that a person is generous or altruistic or socially cooperative, we mean that she is maximizing utility over a wide social space. Degree of self-control and degree of social cooperativeness may be measured in the laboratory by delay and social discount functions. Social discount functions may be obtained for individuals by asking a series of questions about their preferences between smaller amounts of money for themselves and larger amounts to be given to other people at various social distances (Jones and Rachlin, 2006; Rachlin and Jones, 2008).

A method for obtaining social discount functions has recently been developed using a parallel procedure to that used for obtaining delay discount functions. Jones and Rachlin (2006) asked participants to imagine that they had rank-ordered the 100 people closest to them (without actually doing so). Then they were asked to indicate their preferences between smaller amounts of money given to them and larger amounts given to other people at various positions (N-values) on their list. As with delay discounting, social discounting is hyperbolic:

$$v_N = \frac{V_N}{1 + \cdot N} \qquad (2)$$

where V_N is the magnitude of the reward given to the Nth person on the list, v_N is the magnitude of reward to the participant at the point of indifference between reward to himself and a reward of V_N to person-N, and k_{social} is a positive parameter inversely proportional to degree of social discounting. Equation 2 is the same as Equation 1 except that the exponent s for social discounting has been found to equal 1.0. Because k_{delay} and k_{social} are measured in different units, it is not possible to directly compare degree of social and delay discounting. However, the exponent s in the delay discounting equation is often less than 1.0 (typically about 0.8). That is, the relation between v_N and N is (inversely) proportional whereas the relation between v_D and D is (inversely) "subproportional"; to reduce v_D by half, for example, you would need to more than double D. In this sense, *delay discounting is less extreme than social discounting*. To put it another way, as we know, people tend to be more generous to their own future selves than they are to other people.

This difference between cooperation with one's future self and cooperation with others has been borne out in experiments comparing individual and social games (e.g., Brown & Rachlin, 1999). Multi-person prisoner's dilemma game players maximize group returns by cooperating but maximize individual returns by defecting. Each player benefits most by defecting whereas all the other players cooperate (*as when everyone but you contributes to public television resulting in programs of high quality that you may watch without cost*), benefits somewhat less when all players cooperate (*everyone contributes and programs are still of high quality*), benefits much less when all players defect (*no one contributes and the quality of the programs suffers*), and benefits least by cooperating when the other players defect (*programs are still bad and you have lost your contribution*). Rates of cooperation among players in laboratory prisoner's dilemma games are normally low; however, players do learn to cooperate over repeated games if other player(s) play a strategy called "tit-for-tat." Players playing tit-for-tat simply reflect a player's choice on the next trial (*imagine everyone is watching you and contributes to public television in year $N + 1$ whatever you contributed in year N*). Because your choice is automatically reflected by the other players' future choices, you are essentially playing against your future self. On any given year it is to your benefit to defect, but over a course of years it is much more to your benefit to cooperate (*In the case of public television, if others are playing tit-for-tat with you, and you do not contribute during the first year, you will be able to watch high quality programs that year but suffer in future years*).

Participants in laboratory studies of two-person prisoner's dilemma games tend to cooperate when playing against tit-for-tat (that is, against their future selves) but they tend to defect when playing against other participants freely choosing between cooperation and defection. Nevertheless, social cooperation significantly improves when players are required to make four successive choices at a time (that is, to choose in patterns of four) rather than individually on a case-by-case basis (Brown & Rachlin, 1999). That is, just as patterning (making several sequential choices at once) is better than case-by-case choices in generating self-control (Rachlin, 1995) so patterning is better than case-by-case choices in generating social cooperation.

From the viewpoint of teleological behaviorism, in agreement with several modern economists (Akerlof, 1997; Becker, 1981; Simon, 1995), social discounting and delay discounting, although different in degree, are not essentially different in kind. The currency that unites social and delay discounting is common interest. Given the high mutual dependency among people in modern society, there is no *a-priori* reason why people should be more generous towards their future selves than they are to other people with whom they have common interests. We have common interests with others as well as with our future selves. Even though social discount functions are generally more sensitive to social distance than delay discount functions are to time, close social rewards may outweigh distant temporal ones; a person may prefer, when they conflict, the current benefit of his close friends or immediate family to his own benefit in the distant future.

Altruism, therefore, is no more mysterious than self-control. Although it is not possible to sacrifice your life in the present for your future benefit, it is possible to sacrifice your life for the sake of maintaining a consistent pattern of behavior in which you have heavily invested—as Socrates did who preferred to die rather than disobey the laws of Athens. Individual acts of altruism that seem inexplicable when considered on a case-by-case basis, such as the New York City man who recently risked his own life by jumping onto the subway tracks to save a stranger's life, are no more mysterious than a person stopping for a red light when there are no cars in the cross-street and no police in sight. In both cases a valuable pattern is being maintained. Making temporal or social decisions on a case-by-case basis is generally a bad policy because we tend to overestimate the value of a single reward to ourselves now relative to the value of patterns of behavior spread out in time or social space.[5]

IMPLICATIONS OF TELEOLOGICAL BEHAVIORISM FOR TREATMENT OF ADDICTION AND EVERYDAY PROBLEMS OF SELF-CONTROL

Although the primary purpose of this article is to describe and defend a particular conception of self-control, any such approach is ultimately defensible only to the extent that it leads to practical procedures. Here I will outline some implications of teleological behaviorism for self-control problems in general. I will introduce theoretical concepts as needed but not attempt to trace each one's connection to teleological behaviorism. For a detailed account of implications of teleological behaviorism for self-control in general, see Rachlin (2000). For application to control of childhood obesity, see Rachlin (2008); for application to control of teenage violence, see Rachlin (2004).

Habit Change Before Consumption Change: Some bad habits, like gambling, are all behavioral and do not involve consuming a substance; but many do. Where an addictive habit *as* behavior can be separated from the habit *as* consumption, treatment should be directed to the behavior. The methods used to manipulate consumption directly tend to focus on immediate discrete consequences of each act. For example, switching from natural to artificial low-calorie food such as lo-cal sodas or beer allows the dieter to maintain former patterns of behavior reinforced by the undiminished taste and feel of food in the mouth, while still losing weight. Such a diet may help a person lose weight, but the habit of overeating will be maintained (or even made worse as the dieter compensates by eating more of the lo-cal foods) and reappear as soon as the exigencies of everyday life (such as eating out with friends) break the diet. Similarly, methadone replacement therapy and nicotine patches serve only as temporary crutches for fundamental change in behavior patterns. On the other hand, learning how to bring one's behavior under control of temporally wide contingencies is a skill that may be easily relearned after a relapse.

Maintenance Before Reduction: All smokers know at this moment whether they are smoking a cigarette or not. But they often do not know their rate of cigarette consumption over a week or a month—even though cigarettes are easily countable individually. Similarly, dieters are often unaware of (and often underestimate) how much they are eating over a week even though they are aware at every moment whether and what they are eating. Yet, to bring such habits under control, it is overall rate that has to be changed. Learning to monitor and maintain a habit's overall rate is extremely difficult, yet most habit control programs confound such learning with habit *reduction*. Before learning to *reduce* a bad habit, the addict should be taught to *maintain* it at a steady level. The goal, at early stages of treatment, should be to learn to monitor behavior and to reduce its variability—not to change it. The problem with changing a habit first and then trying to maintain the new habit is that the point where the person has just reached his goal is the most vulnerable point of habit reduction. If, at this vulnerable point, the person has already developed maintenance skills, they can more easily be applied now than if they have to be newly learned.

Habits, Both Bad and Good, Are Situation-Bound: We tend to believe that our cravings arise from inside of us and that we carry them around with us wherever we go. A reader may grant that addiction is behavior and not an internal event, but surely, the reader may argue, the craving itself is inside of us. Hunger is strictly a condition of our stomachs conveyed to our nervous systems—or so it appears. But this is only partially true. Consider the prime example of what seems to be an internally generated motive—an addict's craving for heroin. Heroin addiction was very common among American soldiers in Vietnam (Robins, 1974). When the war ended, officials became concerned that the returning soldiers would bring their habits home with them. But that did not happen. Most of the soldiers were able to slough off their addiction along with their uniforms—no violent withdrawal effects, no deep anguish. Why? Because hormonal effects within the bodies of the soldiers as well as overt behavior caused by contingencies of reinforcement are strongly controlled by environmental context. And the contexts provided by army life in Vietnam and family life in the United States could not have been more different: in one case, living without family like an animal in the jungle; in the other, living with family in a civilized society.

Situational control may be even stronger for eating. It is common for people who cannot control their eating under normal conditions to easily control it on religious fast days such as Yom Kippur, Lent, or Ramadan. The special conditions of the holiday isolate it from the rest of life. The selection pressure of reinforcement on behavior may differ in different situations and allow the habits of the isolated situation to develop differently from those of everyday life—just as the flora and fauna of an isolated group of islands may evolve differently from those on the mainland.

The situational specificity of habit development allows a person to get rid of a bad habit in one situation at a time—divide and conquer. The differing stimulus conditions across two situations constitute a wall (permeable though it may be) between habits. Nevertheless, a bad habit—perhaps overeating—may develop in both situations. You cannot just assume that if you succeeded in arranging your life so that you manage to stop desiring chocolate-covered almonds when you enter a movie theater, you will not feel a sudden desire for a hot dog and a beer at the ball park. Some situations are especially conducive to overeating, the movie theater and the ballpark being two of them. The distinct situations in any dieter's life have to be tackled one at a time and bad eating habits corralled to a specific places and times (say, only weddings and bar mitzvahs) where they will do no harm.

Economic Substitutability. The classic examples of economically substitutable commodities are Coke and Pepsi. The more Coke you drink, the less Pepsi you are going to buy and vice-versa. Increasing the price of Coke (while keeping the price of Pepsi constant) will *increase* consumption of Pepsi. The opposite of substitutability is complementarity. Examples of economically complementary commodities are beer and pretzels. Increasing the price of beer (while keeping the price of pretzels constant) will *decrease* consumption of pretzels (because people will buy less beer and will not need as many pretzels to go with the beer).

Consumption of Coke and Pepsi or beer and pretzels may or may not be harmful. But suppose there were, among generally beneficial positive addictions, some activity that substituted for a generally harmful negative addiction. Just as, over a wide range, negative addictions become more immediately valuable the *less* you do them (you build up an appetite for them), so positive addictions become more immediately valuable the *more* you do them (you become better at them with practice). If a positive addiction was an economic substitute for a negative addiction, anything that increased the rate of the positive addiction would also decrease the rate of the negative addiction (the more Pepsi you drink, the less Coke you will drink).

Are there positive addictions that substitute for negative addictions? Surprisingly, perhaps, there are. Considerable evidence exists that in adults, social activity substitutes for cigarette smoking and heroin and cocaine consumption. For example, training in social skills is the single most effective treatment for addiction to cigarettes, cocaine, heroin, and alcohol (Green & Kagel, 1996). The effectiveness of exercise in weight control may not be so much in the calories exercising consumes as in its economic substitutability for overeating.

A Soft-Commitment Program for Self-Control. In general, commitment devices and strategies are ways of ensuring that choices are based on abstract patterns of behavior rather than on particular individual actions. If an obese person could ignore his momentary preference between one bite of food and one bit of exercise and instead chose among ratios of food to exercise over the next week or month, he would eat less and exercise much more frequently. It may be possible to use what I call "soft commitment" (Siegel & Rachlin, 1996) in establishing such changes. Let us follow along with the example of weight control.

In this procedure, a person previously trained in monitoring and maintenance behaves freely on one day each week. She may eat a lot, she may eat a little; she may exercise a lot, she may exercise a little, just as she desires. But on each of the next 6 days, she is obliged to eat the very same number of calories and exercise the same number of minutes—no more *and* no less—as on the first day. Each Twinkie eaten on Day 1 would thus entrain 6 more Twinkies or their equivalent over the next 6 days. Each block walked on Day 1 would similarly entrain 6 more blocks over the next 6 days. Such a restriction keeps the responsibility on the dieter to choose her own behavioral pattern over the course of the first day of each week and also compels her to attend to and record that pattern. This gives heavy weight to choices on the first day. It focuses attention on eating and exercise and provides an exemplar over the rest of the week. Her commitment to keep to the same number of calories and blocks walked each day is a soft commitment in that obedience is not compelled. The advantage of such a commitment procedure is that she sets her own pattern, not by imposition from an authority but by her own behavior. She knows she *can* keep to the chosen ratio because she has already kept to it on the first day. And, if she should be uncomfortable with her self-imposed pattern but sticks to it anyway, she will get an opportunity to set a new pattern next week. Moreover, such a method uses previously learned monitoring skills and would blend in smoothly with initial and subsequent maintenance programs. After a relapse, it could easily be reinstituted. In theory, therefore, this program should be effective in weight reduction and, with modifications, for any kind of self-control.

Conclusions

Taking a teleological and behavioral view of self-control problems goes contrary to our normal way of thinking about self-control. The very term *self-control* seems to imply a force arising from inside—a force opposed to control by the environment. This chapter argues, however, that a change in such thinking—self-control as adaptation to abstract aspects of the environment, impulsiveness as adaptation to particular aspects of the environment—may be worth the effort involved. Teleological behaviorism conveys clarity and transparency to psychological theory; it fits psychology neatly into evolutionary biology by extending Darwinian evolution to behavior change within a person's lifetime.[6]

In psychology, when a theory reaches the limits of what it can explain, further development traditionally proceeds by postulation of normally unobserved events within the organism. To explain the existence of temporally extended acts such as dessert refusal, cognitive and neuropsychological theories of self-control postulate often elegant and intuitively compelling current internal representations. Such representations provide a currently existent blueprint as it were for the construction of the acts over time and avoid causation at a temporal distance. Even Skinner (1974) speculated that *covert* behavior might be *covertly* reinforced. For him, all non-reflexive learning must be explained in terms of immediate reinforcement, either primary or conditioned. Some specific reinforcing event had to follow every action or that action could not increase in rate. According to Skinner, a person who refuses a second dessert might be covertly telling herself what a virtuous person she is and by so doing strengthen her act of dessert refusal.

Teleological behaviorism, in such circumstances, goes in the other direction. When it cannot explain some aspect of behavior in terms of the narrow contingencies of the present environment, teleological behaviorism looks not into the interior of the person but into the contingencies of the person's wider, long-term environment. A person's current dessert refusal is seen as part of a pattern of similar acts maintained over the long run (if it is maintained) by social approval, increased health, and so forth, none of which stand in a 1:1 relation with any specific act. When it cannot explain some aspect of individual behavior—an altruistic act for instance—even in those terms, teleological behaviorism looks to the contingencies of reinforcement of the social group (as discussed in the section on social discounting).

More important than its behavioral purity, however, are the implications of teleological behaviorism for dealing with the practical problems of self-control that currently plague our society. The suggestions in the section "Implications of Teleological Behaviorism for Treatment of Addiction and Everyday Problems of Self-Control" have in common their reliance on methods of external control of external behavior over time. The success or failure of these methods and others like them will determine whether the sacrifice of our common way of thinking about self-control demanded by teleological behaviorism is worth making.

Acknowledgments

This article was prepared with the assistance of a grant from The National Institute of Health.

Notes

1 Of course people have intentions. (An obese person, for example, might resolve to eat less food.) They then either carry out or fail to carry out those intentions. But, according to teleological behaviorism, an intention is itself behavior—a temporally extended pattern of behavior—that may be extrinsically reinforced even when it is not carried out to its end. For example, an obese person may be rewarded, socially, for agreeing to join a weight-reduction program and then for actually joining the program—independent of the other longer-term rewards contingent on losing weight. Such extrinsic rewards may increase the frequency or intensity with which the intention is expressed but they cannot increase the strength of the intention. The strength of an intention is not given by the intensity or determination with which it is expressed or the intensity of a neural discharge or the brightness of a brain-area image in an MRI scan. The strength of an intention is rather the likelihood of its being followed; that likelihood, in turn, may be measured by the frequency with which prior, similar, intentions have been followed by that person and the duration of the prior behavioral patterns that the person intended to follow. Someone who promises to stop drinking now and has promised to stop drinking many times in the past but not stopped, or stopped for a short time, has only a weak intention to stop unless future events prove, retrospectively, that the intention was strong. In this respect an intention is like probability; you can only know for sure that a coin is unbiased if you flip it a sufficient number of times and the overall frequency of heads approaches ½. An apparent intention that cannot be measured in this fashion (such as a purely internal intention) is not really an intention at all. The mechanisms underlying intentions, like the mechanisms underlying all behavior, are surely internal. But the intentions themselves occur in behavior over time, not inside the head. If they are never exhibited in overt behavior you may want to call them potential or aborted intentions but they are not intentions.

2 When my wife berates me for some blunder by saying, "Why in the world did you do that (horrible, or disgusting, or stupid, or all three) thing?" I usually reply, "Heredity and environment." Of course I am evading her question. What she is asking for is the more immediate final cause, nested within heredity or environment. She is certainly not interested in my naming of the act's remote final causes—the much wider patterns that all of my acts fit into.

3 One version of the creationist solution to the problem of self-control has been posited by some behaviorists themselves. Failing to find extrinsic reinforcers for each component of self-controlled behavior, they posit events inside the person, usually inside the brain, representing reinforcement. In an alcoholic, for instance,

such concrete internal representations would immediately reinforce individual drink refusals and counteract the innate immediate value of the drink. Another way to put it is that the reinforcers of self-controlled behavior become "internalized." It is as if, each time he refuses a drink, the alcoholic internally pats himself on the back for a job well done. An internal vision of the alcoholic's future enhanced health, for instance, wells up inside him and somehow reinforces each drink or cigarette or dessert refusal. But the concept of self-reinforcement, internal or external, has numerous empirical and theoretical problems. According to the most well-supported concept of reinforcement (Premack, 1965), a reinforcer is a highly valued activity contingent on the performance of a less-valued activity. An internal or even an external activity that may be initiated or withheld at any time cannot be contingent on any other activity. Patting yourself on the back, internally or externally, has all the reinforcing effect of taking a dollar bill out of your left pocket and putting it into your right pocket. Such "reinforcement" may serve to enhance proprioceptive feedback – like counting the number of cigarettes smoked – that may be part of an effective self-control program, but it does not function as a reinforcer.

4 The reason why players in real leagues, such as the NBA, *sometimes* come to play unselfishly despite the very slow rate of team replacement is probably the frequent rate of coach replacement. The coach will be replaced if the team keeps losing games; he has an interest in the team's performance as a whole. The coach and the individual players are therefore to some extent in conflict.

5 Choosing in the present so as to be consistent with past choices rather than current assessments—that is, refusal to choose on a case-by-case basis—is usually considered by economists to be an error. It is called the "sunk-costs" fallacy. However, as argued here, that so-called fallacy may underlie self-control. For a further discussion of the role of sunk costs in self-control see Rachlin (2000). For a demonstration of the sunk costs fallacy in pigeons see De la Piedad & Rachlin (2006).

6 Teleological behaviorism also relates to behavioral economics by showing how apparent errors and biases are not "externalities" to economic theory but fit within it; apparent errors and biases actually maximize utility—but over a narrow temporal span. In that respect, failure to maximize utility is essentially a failure of self-control. See Rachlin (1989, 2000) for arguments to this effect.

REFERENCES

Ainslie, G. *Breakdown of will.* Cambridge, UK: Cambridge University Press, 2001.

Akerlof, G. A. Social distance and social decisions. *Econometrica* 1997; 65: 1005–1027.

Aristotle. *The basic works of Aristotle.* In: McKeon, R. (Ed.), New York: Random House, 1941.

Baum, W. M. *Understanding behaviorism: Behavior, culture and evolution.* Oxford: Blackwell, 2005.

Becker, G. S. *A treatise on the family.* Cambridge, MA: Harvard University Press, 1981.

Bickel, W. K., & Marsch, L.A. Toward a behavioral economic theory of drug dependence: Delay discounting processes. Addiction 2001; 96: 73–86.

Brown, J., & Rachlin, H. Self-control and social cooperation. Behav Proc 1999; 47: 65–72.

De la Piedad, X., Field, D., & Rachlin, H. The influence of prior choices on current choice. J Exp Anal Behav 2006; 85: 3–21.

Green, L., & Kagel, J. H. *Advances in behavioral economics. Vol. 3; Substance use and abuse.* Norwood, N.J.: Ablex Publishing Co., 1996.

Green, L., & Myerson, J. A discounting framework for choice with delayed and probabilistic rewards. Psychol Bull 2004; 130: 769–792.

Jones, B., & Rachlin, H. Social discounting. Psychol Sci 2006; 17: 283–286.

Koffka, K. *Principles of Gestalt psychology.* Oxford: Routledge & Kegan Paul, 1955.

Loewenstein, G. Out of control: Visceral influences on behavior. Organ Behav Hum Decis Process 1996; 65: 272–292.

Logue, A. W. Research on self-control: An integrating framework. Behav Brain Sci 1988; 11: 665–679.

Premack, D. Reinforcement theory. In: Levine, D. (Ed.), *Nebraska symposium on motivation.* Lincoln: University of Nebraska Press, 1965.

Rachlin, H. *Judgment, decision, and choice: A cognitive/behavioral synthesis.* New York: Freeman, 1989.

Rachlin, H. *Behavior and mind: The roots of modern psychology.* New York: Oxford University Press, 1994.

Rachlin, H. The value of temporal patterns in behavior. Curr Dir 1995; 4: 188–191.

Rachlin, H. *The science of self-control*. Cambridge, MA: Harvard University Press, 2000.

Rachlin, H. Altruism and selfishness. Behav Brain Sci 2002; 25: 239–296.

Rachlin, H. Ten messages for weight control from teleological behaviorism. In: O'Donohue, W., & Moore, B. A. (Eds.), *Pediatric and adolescent obesity treatment: A comprehensive handbook*. New York: Taylor & Francis, 2008.

Rachlin, H., & Jones, B. Social discounting and delay discounting. Behav Decis Mak 2008; 21: 29–43.

Raineri, A., & Rachlin, H. The effect of temporal constraints on the value of money and other commodities. Behav Decis Mak 1993; 6: 77–94.

Robins, L. The Vietnam drug user returns. *Special Action Office Monograph: Series A, Number 2*. Washington DC: Government Printing Office, 1974.

Ryle, G. *The concept of mind*. London: Hutchenson House, 1949.

Siegel, E., & Rachlin, H. Soft commitment: Self-control achieved by response persistence. J Exp Anal Behav 1996; 64: 117–128.

Simon, J. Interpersonal allocation continuous with intertemporal allocation. Rational Soc 1995; 7: 367–392.

Skinner, B. F. *The behavior of organisms: An experimental analysis*. New York: Appleton-Century-Crofts, 1938.

Skinner, B. F. *About behaviorism*. New York: Knopf, 1974.

Skinner, B. F. Can psychology be a science of mind? Am Psychol 1990; 41: 1206–1210.

Stigler, G. J., & Becker, G. S. De gustibus non est disputandum. Am Econ Rev 1977; 67: 76–90.

Wittgenstein, L. *Philosophical investigations*. Translated by G. E. M. Anscombe. New York: Macmillan, 1958.

AUTHOR INDEX

Aarts, H. 274, 280, 281, 287, 300, 358, 359, 360, 394, 410, 422
Abelson 338
Abrams, D. 449, 453, 457, 461, 465, 481
Abramson, L. Y. 266
Ach, N. 202, 203, 204, 209
Achtziger, A. 280, 281, 282, 284
Ackerly, S. 34
Ackerman, P. L. 164
Adarves-Yorno, I. 458
Adolphs, A. 203
Adolphs, R. 81, 435
Ahmed, L. 209, 210, 211, 212
Ahmetzanov, M. V. 107
Ainslie, G. 33, 142, 147, 153, 389, 390, 393, 394, 402, 409, 415, 418, 420, 421, 431, 512
Akerlof, G. A 515
Akrami, N. 225
Alain, C. 252
Albery, I. P. 101
Aldrich, C. 496
Aldwin, C. 323
Alexander, C. N. 94
Alexander, M. P. 27, 28, 35
Alexander, S. C. 464
Alfert, E. 80
Alfieri, T. 234
Algom, D. 99, 100, 106
Alicke, M. D. 40, 42, 269, 271
Allen, A. 298
Allen, C. K. 166
Allen, V. L. 453, 457, 458, 460
Allison, S. T. 478
Alloy, L. B. 266
Allport, A. 179, 180, 205, 206, 207, 208, 209, 210, 211, 214
Allport, D. A. 179, 182, 184
Allport, G. W. 49, 354, 355, 356
Alony, R. 412, 414, 422
Altmann, E. M. 205, 206, 212

Amir, N. 100
Amodio, D. M. 22, 50, 52, 55, 58, 59, 60, 62, 63, 64, 65, 66, 68, 225, 250, 251, 253, 411, 422
Andersen, S. 481
Andersen, S. A. 367
Anderson, C. J. 264
Anderson, C. P. 40
Anderson, J. R. 167, 184, 208, 209, 214
Anderson, M. C. 143, 149, 150, 151, 152, 153, 154, 168
Anderson, P. D. 452
Andrade 339, 343
Anes, M. D. 97
Arbuthnott, K. D. 209, 215
Arias, I. 195
Ariely 343
Ariely, D. 301, 394, 418
Aristotle 506, 507, 509
Arkes, H. R. 226
Armitage, C. J. 281
Aron, A. R. 55, 127, 143, 144, 145, 148, 149, 151, 153, 154
Aronson, E. 455
Aronson, J. 64
Arriaga, X. B. 438
Arrington, C. M. 206, 207
Arrow, H. 450, 451, 454
Asch, S. E. 457, 458, 460
Asher, T. 451
Aspinwall, L. G. 323
Assad, J. A. 29, 34, 123
Aston-Jones, G. 65
Atkin, R. 459
Atkinson, J. W. 117, 396
Atkinson, R. C. 167, 168
Avnet, T. 324
Ayduk 335, 345
Ayduk, O. 85, 291, 298, 384, 398, 399, 414, 421, 432, 435, 438, 439, 440, 441–442
Azrin 343

523

Bacon, P. L. 198
Baddeley, A. D. 50, 211
Badre, D. 28
Bagozzi, R. 267
Baker, N. 313, 411, 414, 421, 434
Bakker, M. 460
Baldwin, G. C. 114
Balota, D. A. 99
Banaji, M. R. 49, 63, 222, 226, 252, 355, 358, 381, 493, 494
Bandura, 346
Bandura, A. 116, 320, 322, 398, 431, 442
Banich, M. T. 15, 96
Banse, R. 225
Barbas, H. 28, 32, 68
Barcelo, F. 82
Barch, D. M. 82, 94
Bargh 339
Bargh, J. A. 185, 191, 244, 246, 288, 298, 299, 305, 306, 308, 316, 317, 318, 355, 359, 360, 367, 394, 395, 399, 410, 414, 422, 493, 501
Barkley, R. A. 203, 289
Barndollar, K. 394, 410, 493, 501
Baron, J. 29, 264
Baron, R. S. 461
Barr, A. 477
Barrett-Howard, E. 477, 478, 480
Barsky, J. 360
Bartholow, B. D. 54, 62, 65, 66, 252, 253
Bass, B. M. 456
Batchelder, W. H. 247
Bates, E. 29
Baudelaire, C. 222
Baum, W. M. 511
Baumeister 337, 343, 348
Baumeister, R. F. 3, 41, 70, 117, 132, 151, 153, 180, 189, 190, 194, 195, 200, 203, 268, 274, 282, 299, 300, 313, 376, 377, 378, 379, 380, 381, 382, 383, 384, 389, 390, 401, 404, 410, 416, 419, 420, 422, 424, 431, 449, 450, 457, 486, 493, 494, 501
Baumgartner, H. 267
Baumhart, R 40
Baxter, M. G. 32
Bayer, U. C. 280, 282, 283, 301, 394
Baylis, G. C. 170
Baylis, L. L. 34
Bazerman, M. H. 475
Beaton, A. 169
Beauregard, M. 87
Beaver, J. 130
Bechara, A. 30, 31, 32, 44, 125, 145
Becker, G. S. 513, 515
Becker, H. S. 390
Beckers, T. 186
Beckmann, J. 177, 390, 398, 502
Beer, J. S. 39, 40, 41, 42, 43, 44, 45, 249
Begley, S. 500
Beier, M. E. 164

Bell, T. 206, 207
Bem, D. J. 297
Ben-Shakhar, G. 107
Berckmoes, C. 98
Berdahl, J. L. 450, 451
Berg, E. A. 290
Berkman, E. T. 152
Berlyne, D. 433
Bernal, S. 436
Berns, G. S. 123
Bernstein, E. 204
Berntson, G. G. 78
Berridge, K. C. 120, 121, 122, 125
Berscheid, E. 476
Besner, D. 94, 181
Bettman, J. R. 273, 415, 421
Beukeboom, C. J. 420
Bhanji, J. F. P. 42, 43
Bianco, A. T. 315, 324
Bickel, W. K. 33, 153, 513
Biederman, I. 204, 212, 213
Biernat, M. 453
Bird, B. 285
Bishop, S. 17, 18, 43
Bjork, R. A. 150
Blackstone, T. 461, 462
Blader, S. 477, 481, 482, 486
Blair, I. V. 233, 355, 358, 367
Blakemore, S. J. 77, 89
Blanton, H. 226, 453
Blascovich, J. 64
Block, J. J. 39, 298
Bloom, P. 269, 270
Blount, S. 494, 495
Bobo, L. 49
Bodenhausen, G. V. 50, 53, 239, 367, 410
Boeckmann, R. 476, 477, 479
Boies, S. 214
Boland, J. E. 191
Bonanno, G. A. 437
Bond, C. F. Jr. 452
Bongiolatti, S. 128, 132
Bonino, S. 203
Bonson 339, 344
Boone, A. L. 200, 376, 419
Born, D. G. 166
Boroughs, J. M. 188
Bos, M. 360, 367
Botivinick, M. M. 43
Bottom, W. P. 286
Botvinick, M. M. 6, 7, 8, 13, 15, 59, 60, 63, 82, 94, 96, 102, 108, 143, 154, 202, 214, 249
Bourgouin, P. 87
Boutilier, C. 181
Bowlby, J. 320
Bown, N. 453
Boyle, M. O. 164
Bradley, M. M. 83

AUTHOR INDEX

Bramel, D. 460
Brandstaetter, V. 400
Brandstätter, V. 185, 280, 281, 289, 423
Brandstatter, V. 302
Branscombe, N. R. 453
Brass, M. 183, 207, 210
Bratslavsky, E. 117, 132, 299, 420, 494
Braver, T. S. 13, 14, 94, 127, 128, 129, 130, 132, 204, 211
Bray, R. M. 452
Brehm, J. W. 274, 394, 396, 461
Brehm, S. S. 461
Brekke, N. C. 50, 52, 230, 235, 357
Brendl, C. 395
Brennan, M. 380
Breslin, F. C. 195
Brewer, M. B. 50, 58, 246, 356, 358, 460, 461, 479
Brewin, C. R. 168, 169, 170
Brindle, N. A. 122
Briñol, P. 222, 381
Broadbent, K. 168
Brockner, J. 282, 485
Brodbeck, F. C. 461, 462
Brook, A. 501
Brown, J. 515
Brown, J. D. 39, 40, 41, 396
Brown, J. M. 494
Brown, J. W. 13, 14, 210, 214
Brown, R. M. 53, 441
Brown, S. M. 214, 436
Brownell 349
Brozoski, T. J. 441
Bruner, J. S. 354, 357
Brunsman, B. 461
Bruun, S. 459
Bruyneel, S. 385
Bryck, R. L. 211, 212
Buchel, C. 82
Buckner, R. L. 4
Buehler, R. 396
Bundesen, C. 206, 208
Bunge, S. A. 78, 143, 152, 154, 202
Bunney, B. 124
Bunting, M. F. 166, 167
Burgess, D. 493
Burgess, P. W. 288, 289, 290
Burke, C. 420
Burkell, J. 202, 212
Burkley, M. A. 238
Burklund, L. 152
Burns, K. C. 57
Burnstein, E. 459
Burrows, L. 185
Burt, J. S. 106
Busceme, S. 461, 462
Bush, G. 17, 18, 43, 82, 94, 99, 103, 104, 105
Butera, F. 463
Butler 340
Butler, R. 417

Butter, C. M. 31, 144
Butz, D. A. 379
Buunk, B. P. 477

Cabeza, R. 82
Cacioppo, J. T. 65, 77, 78, 88, 246
Caggiula, A. R. 453
Cahill, L. 81
Cairns, E. 106
Calder, A. J. 125
Calkins, S. D. 465
Camerer, C. 270
Cameron, J. 116
Campbell, D. T. 64, 487
Campbell, W. K. 377, 382
Canham, L.181 188
Cantor, J. 163, 164, 168
Cantor, J., 167
Cantor, N. 313, 314, 315, 323, 326, 431
Caputo, D. 39
Cardinal, R. N. 147
Carlsmith, J. M. 21
Carlson, R. A. 205, 212, 214
Carpenter, P. A. 164
Carpenter, S. 491, 492
Carretié, L. 288
Carroll, J. 492, 500
Carter, C. S. 5, 6, 7, 13, 43, 64, 82, 94, 96, 180
Carullo, J. 163, 164, 168
Carver 335, 336, 337, 338, 339, 341, 342, 343, 350
Carver, C. S. 41, 80, 115, 117, 118, 131, 298, 314, 316, 318, 319, 323, 378, 450
Casbon 349
Casey, B. J. 143
Caspi, A. 134
Cassiday, K. L. 100
Cattell, J. M. 95
Cattleino, E. 203
Caul, W. F. 122
Chai, A. L. 410
Chaiken, S. 80, 245, 246, 412, 413, 420
Chajut, E. 99
Chamberlain, S. R. 35, 291
Chambers, C. D. 144
Chambers, J. R. 39
Chance, Z. 267
Chang, H-R. 491, 492
Channon, S. 106
Chapleau, K. M. 233
Chartrand, T. L. 274, 288, 317, 359, 360, 367, 394, 410
Chattarji, S. 436, 439, 441–442
Chatterjee, A. A 29
Chee, M. W. 249
Chemers, M. M. 450
Chen, E. 494
Chen, M. 185, 395
Chen, S. 305

Cheng, C. 324, 329
Chevrier, A. D. 144
Chistrom, J. T., Jr. 452
Choi, H.-S. 454, 462
Chorev, Z. 214
Christensen, P. N. 463
Christie, C. 453
Chua, H. F. 191
Chua, P. 496
Chun, W. 306, 307
Church, M. A. 314, 315, 319, 325, 329
Cialdini, R. B. 63, 451
Ciarocco, N. J. 337, 381, 449
Clark, L. F. 144, 323
Clark, M. S. 481
Clark, S. 188
Clarke, C. 290
Claus, E. D. 170
Clayton, S. 501
Clore, G. L. 420
Coan, J. A. 125
Cohen, A. 107, 181
Cohen, A-L. 214, 287
Cohen, G. L. 234
Cohen, J. A. 280
Cohen, J. D. 3, 4, 11, 43, 58, 65, 82, 94, 96, 127, 153, 154, 202
Cohen, J. R. 146
Cohen, M. X. 145, 148
Cohrs, J. C. 494
Coifman, K. 437
Cole, A. B. 438
Cole, B. 453
Coleman, J. 453
Coles, M. G. H. 12, 13, 102, 179
Collette, F. 202
Collins 347, 348
Collins, L. R. 379
Collins, P. 130, 133
Collins, R. L. 385
Colvin, C. R. 39
Compton, R. J. 104, 105
Connell, J. P. 63, 64
Connelly, S. L. 253
Conners, C. K. 290
Conrey, F. R. 59, 60, 244, 249, 250, 253
Converse, B. A. 389, 404
Conway, A. R. A. 164, 165, 166, 170
Conway, M. A. 168
Cook 339
Cook, C. 101
Cook, E. H., Jr. 441
Cools, R. 35, 145, 148
Cooper, J. 268
Cooper, P. J. 100
Copper, C. 459
Corbetta, M. 129, 131, 142, 155
Corneille, O. 53

Correa, M. 122
Correll, J. 62, 355, 365
Costa, P. T. 298
Costell, R. M. 457
Cottrell, S. P. 496, 497
Courtney, S. M. 142, 155
Covassin, T. 284
Cowan, N. 166
Cowan, W. B. 50, 55, 252, 290
Cox, V. C. 140
Craft, J. L. 196
Craik, F. I. M. 82
Crandall, C. S. 451, 452
Critchley, H. D. 11, 78
Crocker, J. 497
Crosby, F. 63, 463
Cross, S. E. 189, 190, 191
Crowe, E. 321
Cullen, F. T. 376, 377
Cunningham, W. A. 34, 55, 65, 68, 131, 226
Curry, J. A. 491, 492
Custers, R. 359, 360
Cuthbert, B. N. 83

Dagher, A. 124
Dahl, M. 97
Daichman, A. 209, 214
Dalgleish, T. 107, 163, 169, 170
Dalley, J. W. 441
Daly, G. 441
Damasio, A. R. 29, 34, 43, 51
Daneman, M. 164
Darley, J. 451, 452, 453
Darley, J. M. 50, 234, 270
Davidson, M. 86
Davidson, R. 118, 130
Davidson, R. J. 55, 77, 119, 151
Davies, J. C. 463
Davis, A. 481
Davis, J. H. 458, 459
Davis, K. D. 105
Davis, K. E. 50, 271
Davis, M. 435
Davis-Stitt, C. 370
Daw, N. D. 125
Dawes, R. 460
Dawes, R. M. 236, 492
De Dreu, C. K. W. 462
De Houwer, J. 186
De Jong, R. 196, 205, 214, 215
De la Piedad, X. 520
de Moura, G. R. 461
de Vader, C. L. 456
de Vries 347
De Vries, N. K. 462
Deacon, T. W. 154
Deakin, J. 125

AUTHOR INDEX

deCharms, R. 452
Deci, E. 116, 117
Deci, E. L. 63, 316, 319
DeCoster, J. 50
DeCremer, D. 460, 484
DeCremer, S. 484
Degoey, P. 476, 477, 479, 492
Degoey, S. 479, 482, 483
Dehaene, S. 5
DeJong, G. M. 477
Delgado, M. 123
DeMartino, B. 44
Denenberg, V. H. 120
Denno, D. W. 261, 272
Depue, R. 133
Derrfuss, J. 213
Desimone, R. 129
DeSmet, A. 437
D'Esposito, M. 44
Deutsch, M. 451, 457, 458, 476, 477
Deutsch, R. 50, 80, 255, 410
Devine, D. J. 459
Devine, P. D. 250
Devine, P. G. 50, 51, 52, 54, 55, 56, 58, 60, 63, 64, 65, 69, 222, 225, 244, 246, 250, 253, 355, 356, 364, 410, 411, 422
Devinsky, O. 17
DeWall, C. N. 382
Dhar 343
Dhar, R. 305, 401
Dhar, R. W. 401
di Pellegrino, G. 11
Dias, R. 31, 81, 144
Dickson, W. J. 465
Dickter, C. L. 66, 252
Dietz, T. 496
DiGirolamo, G. J. 78
Dijksterhuis, A. 52, 185, 273, 274, 306, 360, 367, 370, 394
Dijksterhuis, A. P. 280, 360
Ditto 349
Dixon, P. 212
Dolan, R. 125
Dolan, R. J. 44, 63, 81, 435
Dolski, I. 435
Donchin, E. 102
Donders, F. C. 176
Donzella, B. 440
Doris, J. M. 270
Dosenbach, N. U. F. 213, 214
Dougherty, D. D. 17
Dougherty, M. R. P. 41
Dougill, M. 453
Dovidio, J. 358
Dovidio, J. F. 53, 64, 252, 253, 356, 357
Downey, G. 437, 438, 441
Dreezens 347
Dreisbach, G. 208, 212

Dreuy, M. 214
Drevets, W. C. 4, 16
Duckworth, A. L. 177, 376
Dugoni, B. L. 462
Dumontheil, I. 288
Dunbar, K. 58, 96
Duncan, J. 129
Dungelhoff, F. J. 97, 108
Dunlap, R. E. 496, 497
Dunn, B. D. 30, 146
Dunn, M. A. 367
Dunnette, M. D. 476
Dunning, D. 39, 40
Dunton, B. C. 63, 224, 225, 367
Durgin, F. H. 96, 108
Durston, S. 152
Duval, S. 452
Dweck, C. 117, 133, 319
Dweck, C. S. 417, 431

Eagleman, D. 45
Easdon, C. M. 252
Easterbrook, J. A. 64
Ebbesen 344
Ebbesen, E. B. 313, 432, 433
Eberhardt, J. L. 65, 233
Eelen, P. 186
Egeth, H. E. 96
Egner, T. 9, 10, 15, 16, 18, 97, 101, 102, 104, 108, 438
Ehlers, A. 100
Ehrlich, P. R. 501
Eibach, R. 439
Eibach, R. P. 414, 421
Eich, E. 439
Eidelman, S. 453
Eigsti, I. 291, 432
Einstein, G. 132
Eisenberg, J. 465
Eisenberger, N. 436
Eisenberger, N. I. 21
Eitam, B. 360
Ekehammar, B. 225
Ekman, P. 189
Elliot, A. 117
Elliot, A. J. 60, 314, 315, 316, 318, 319, 320, 325, 329
Elliott, R. 44, 63, 143, 145, 146, 154
Ellis, A. L. 459
Ellis, J. 185
Ellis, L. K. 436
Elsner, B. 186, 188
Elster, J. 390
Emerson, M. J. 211
Engle, R. W. 163, 164, 165, 166, 167, 168, 169, 170
Epley, N. 270, 272
Epstein, J. N. 290
Epstein, S. 80, 246, 435
Erber, R. 357, 410

Eriksen, B. A. 179
Eriksen, C. W. 179
Erkanli, A. 290
Ernst, M. 146
Eshel, N. 146
Eslinger, P. J. 29, 34
Estes, W. K. 433
Etkin, A. 10, 18, 97, 101, 102, 104, 105, 108
Evans, A. C. 124
Evans, G. W. 384
Evans, N. 252, 357
Evenden, J. L. 27, 142
Everitt, B. J. 125, 441
Exner, S. 182, 188
Eyal, T. 412, 413, 420

Faber, R. J. 268, 376, 385
Fagot, C. 204, 205, 211
Falkenstein, M. 6
Fan, J. 94
Farah, M. J. 10, 27, 31, 33, 34, 81, 144, 146, 170
Faucheux, C. 477, 478
Faust, D. 236
Fazio, R. H. 52, 53, 63, 222, 223, 224, 225, 244, 246, 253, 255, 367, 399
Feather, N. T. 396
Fedorikhin, A. 272, 410
Fehr, E. 460, 477, 478
Fein, S. 367, 370
Feindler 342, 343, 345, 346, 350
Feldman Barrett, L. 82
Feldman, S. 313, 431, 437
Fellows, L. K. 10, 29, 30, 31, 32, 33, 34, 35, 81, 152, 170
Ferguson, M. 420
Ferguson, M. J. 299, 394, 414, 422
Fernald, R. D. 77
Festinger, L. 21, 221, 274, 460
Feyerabend, C. 100
Feygina, I. 494, 496, 501
Fichtenholtz, H. M. 105
Field, D. 520
Fielding, K. S. 451, 452, 453
Fiez, J. A. 128
Fillmore, M. T. 152
Finkel, E. J. 377, 383
Finkenauer, C. 382
Finn, P. R. 170
Fischhoff, B. 270
Fishbach 337, 339
Fishbach, A. 29, 46, 54, 225, 298, 301, 302, 303, 304, 307, 308, 313, 358, 360, 361, 362, 389, 390, 391, 393, 394, 395, 396, 397, 398, 399, 400, 401, 402, 403, 404, 410, 411, 414, 418, 419, 422, 493, 494, 501
Fishman, D. J. 169
Fiske, S. 263, 268, 272
Fiske, S. T. 50, 58, 63, 68, 246, 356, 411

Fitzgerald, M. 441
Fitzsimons, G. M. 317, 318, 359
Flew, A. 297
Foa, E. B. 440
Fokes, C. 491, 492
Foley, J. A. 491, 492
Folkman, S. 323
Fong, C. 367
Ford, T. E. 300
Foreyt 343
Forgas, J. P. 457
Forman 343, 350
Forster 339
Förster, J. 56, 274, 315, 321, 359, 367, 413, 414
Forstmann, B. U. 206, 207
Forsyth, D. R. 450, 464
Foster 349
Foti, R. J. 456
Fox, M. D. 102
Fox, N. 118
Franco, F. M. 274
Franco-Watkins, A. M. 170
Frank, J. 209, 215
Frank, L. R. 34
Frank, M. J. 32, 35, 170
Frazier, T. W. 289
Frederick, S. 45, 269, 271, 272, 409, 415, 421
Freitas, A. 118, 119, 438
Freitas, A. L. 301, 325, 413, 414, 415, 417, 418, 419, 421
Frensch, P. A. 183
Freud, S. 51, 297, 313, 318, 432
Frey, D. 461, 462
Fricke, K. 53
Friedman 339
Friedman, L. N. 384
Friedman, N. 202, 213
Friedman, R. 54, 306, 308, 314, 316, 318
Friedman, R. S. 54, 274, 301, 302, 321, 324, 325, 358, 367, 391, 400, 410, 413, 414, 422
Friend, R. 460
Frijda 338, 340
Fristoe, N. 409, 421
Frith, C. D. 4, 44, 63, 68, 185, 194
Frith, U. 63, 77, 89
Fujita 343, 348
Fujita, K. 298, 304, 305, 314, 315, 320, 330, 384, 398, 399, 409, 412, 413, 418, 419, 420, 422, 423, 493, 494, 501
Fuller, V. A. 268
Funder D. 39, 40
Fussell, P. 460

Gable, S. 117
Gable, S. L. 314
Gabriel, S. 190, 191
Gabrieli, J. D. 78
Gächter, S. 460

AUTHOR INDEX

Gade, M. 206, 214
Gaertner, S. L. 49, 64, 253, 357
Gaffan, D. 34
Gailliot, M. T. 132, 300, 378, 379, 380
Gaillot, M. T. 331
Galanter, E. 314
Galinsky, A. D. 357, 369, 370
Gallagher, H. L. 82
Gallucci, M. 460
Galton, F. 176
Gangestad, S. W. 451
Garavan, H. 144, 145, 148, 149
Garcia-Larrea, L. 81
Gardner, W. 190, 191, 193
Gardner, W. L. 190, 194, 195
Gaschler, R. 183
Gasper, K. 420
Gasteiger, E. L. 497
Gatenby, J. C. 435
Gaunt, R. 435
Gavinsky, I. 270, 271
Gawrilow, C. 279, 290, 291, 292, 293
Gawronski, B. 50, 222, 225, 229, 384
Gehring, W. J. 5, 6
Genshaft 343
George, M. S. 104
Gerard, H. B. 451, 455, 457, 458
Geschke, D. 225
Ghashghaei, H. T. 32, 68
Gibson, J. J. 207
Gifford 343
Gifford, R. 491
Gifford, R. K. 355
Gigerenzer, G. 41
Gilbert 349
Gilbert, A. M. 128
Gilbert, D. 116
Gilbert, D. T. 42, 50, 52, 53, 65, 68, 269, 271, 272, 358, 435
Gilbert, S. J. 62, 209, 210, 288, 289
Gilboa-Schechtman, E. 107, 108
Giles, M. 106
Gill 349
Gill, M. 441
Gillath, O. 151
Gillihan, S. J. 27
Gilovich, T. 266, 269
Gilovich, T. D. 222
Giner-Sorolla, R. 411, 414, 422
Gintis, H. 477, 478
Glaser, J. 365, 494, 503
Glaser, M. O. 97, 98, 108
Glaser, W. R. 97, 98, 108
Glass, D. C. 384
Glimcher, P. W. 33, 147
Goethals, G. R. 450
Gold, M. 479
Goldman, P. S. 441

Goldsmith, R. E. 496, 501
Golinkoff, R. M. 94, 97
Gollwitzer 343
Gollwitzer, P. 115, 116, 132, 291, 292, 293
Gollwitzer, P. M. 54, 62, 185, 273, 274, 279, 280, 281, 282, 283, 284, 285, 286, 288, 289, 290, 291, 292, 293, 298, 299, 301, 302, 308, 316, 355, 359, 389, 394, 400, 410, 413, 414, 415, 417, 418, 423, 442, 493, 501, 502
Gondola, J. C. 284
Gonsalkorale, K. 243, 252, 253
Goodrick 343
Gopher, D. 205, 206
Gordon, R. D. 202, 207
Gore, J. C. 435
Goschke, T. 195, 210, 214
Gosling, S. D. 494, 497
Gotler, A. 206, 211, 212, 214
Gotlib, I. H. 16
Goto, S. G. 191
Gottfredson, M. R. 376, 377
Gottlieb, B. H. 464
Grace, A. 124
Graf, P. 42
Grafman, J. 202
Gramzow, R. H. 39
Granon, S. 441
Grant, H. 299, 326, 327, 328
Grant-Pillow, H. 325
Grasby, P. M. 4
Gratton, G. 6, 102, 108, 187
Gray, J. A. 55, 117, 130, 318, 435
Gray, J. R. 130
Gray, K. 260, 269
Gray, W. D. 206, 212
Green, C. 149, 152
Green, L. 393, 409, 421, 513, 517
Greene, J. D. 20
Greenwald, A. G. 49, 222, 223, 244, 245, 246, 247, 251, 252, 255, 272, 381
Gregg, A. P. 41, 42
Grice, G. R. 188
Griffin, D. 396
Griffin, D. W. 408, 411
Grigorenko, E. L. 440
Grillon, C. 435
Grimshaw, G. M. 98
Groenenboom, A. 460
Gronau, N. 107
Gross 335, 336, 337, 338, 339, 340, 341, 342, 343, 346, 348
Gross, J. J. 19, 57, 58, 67, 78, 80, 103, 143, 148, 150, 154, 202, 323, 420, 435, 439, 441
Gross, P. H. 50
Gruenfeld, D. H. 461
Gschneidinger, E. 439
Gschwendner, T. 225
Guastello 343

Guglielmi, R. S. 64
Gully, S. M. 459
Gunnar, M. R. 440
Gurerk, O. 477
Gurr, T. R. 463
Gusnard, D. A. 102
Gutsell, J. N. 180
Gwaltney 346

Haas, B. W. 97, 98, 104
Haber, S. N. 35
Hadland, K. A. 16
Haggard, P. 188
Hahn, S. 208
Haidt, J. 268, 269
Hall, J. 170
Hallett, P. E. 165
Halperin, J. M. 290, 292
Hamann, S. 148, 435
Hamann, S. B. 81
Hambrick, D. Z. 165
Hamilton, D. L. 355, 371
Hamilton, R. W. 307
Hampton, A. N. 32
Han, H. A. 412, 413
Hanegby, R. 284
Hannover, B. 53, 191, 192
Hansen, J. 492, 501
Happe, F. 82
Harackiewicz, J. 117
Harackiewicz, J. M. 325
Harasty, S. A. 58
Harber, K. 53
Hardin, C. D. 355
Hardin, G. R. 490, 491, 492
Harenski, C. L. 148
Hariri, A. R. 134, 148, 149, 153
Harkins, S. G. 459
Harleβ, E. 186
Harmon-Jones, C. 22, 23
Harmon-Jones, E. 22, 23, 52, 55, 65, 250,
    268, 411, 422
Harmon-Jones, E. A. 55, 58
Harris, H. J. 452
Harris, L. T. 63, 68
Harris, M. B. 195
Harris, V. A. 269, 271
Hart, C. M. 460, 461, 464
Hart, H. L. A. 260, 266, 271
Hartley, A. A. 205, 206
Hasbroucq, T. 179
Hasher, L. 95, 253
Haslam, S. A. 459
Hassin, R. R. 305, 316, 317, 318, 329, 359, 360
Hastie, R. 371, 458
Hausmann, L. 454
Hawi, Z. 441

Hayes, A. E. 205
Hayward, A 106
Heath, C. 396
Heath, R. G. 120
Heath, Y. 491
Heatherton, T. A. 41
Heatherton, T. F. 190, 202, 300, 376, 377, 378, 382, 384,
    389, 404, 410, 414, 420, 421, 424, 493, 501
Hebl, M. 437
Heckhausen, H. 177, 316
Heggestad, E. 117
Heider, F. 50, 354
Heitz, R.P. 164
Helgeson, V. S. 464
Hembree, R. 319
Henderson, M. D. 282, 412, 420
Henning, S. 436
Henrich, J. 477
Henriques, J. 118
Henry, P. J. 493
Hensley, V. 452
Herlocker, C. E. 478
Herman, C. P. 384
Hermann, A. D. 381
Hermsen, S. 358
Heron, S. 441
Herrington, J. D. 104
Herrnstein, R. J. 409
Hertel, G. 460
Herwig, A. 186, 188
Heuer, H. 206, 209, 212
Hewstone, M. 461, 462
Hick, W. E. 176
Higgins, E. T. 69, 78, 118, 189, 301, 314, 315, 316, 318,
    319, 320, 321, 322, 324, 325, 326, 329, 376, 429
Higgins, T. 118
Hinson, J. M. 170
Hinton, D. 285
Hirsch, J. 9, 10, 102, 438
Hirschi, T. 376, 377
Hirst, W. 177
Hixon, J. G. 42, 53, 358
Hoch, S. J. 377, 385
Hodgkins, S. 281, 302
Hoeksma, J. B. 152
Hoff rage, U. 41
Hoffman, J. E. 179
Hoffman, K. L. 382
Hoffmann, J. 186, 187
Hofmann, W. 222, 223, 225, 384
Hogg, M. 481
Hogg, M. A. 451, 452, 453, 456, 457, 458, 465
Hohnsbein, J. 5
Holland, G. J. 491, 492
Holland, R. 191
Holland, R. W. 281, 287, 360
Hollander, E. P. 452, 456, 458
Hollander-Blumoff, R. 479

AUTHOR INDEX

Hollingshead, A. B. 464
Holmes, A. 134
Holroyd, C. B. 12, 13
Holt, R. 459
Holtzman, L. 272
Holzberg, A. D. 40
Homack, S. 290
Hommel, B. 175, 178, 179, 180, 181, 182, 183, 184, 186, 188
Hong, J. 327
Hönig, G. 281
Honig, G. 302
Honoré, T. 260, 266, 271
Hornak, J. 17
Hornsey, M. J. 452, 453, 458, 461
Horton, R. S. 41, 42
Hoshkikawa, Y. 11
Howell, A. J. 319
Hoza, B. 289
Hsee 338
Hsieh, S. 184
Hübner, M. 213, 214
Hübner, R. 207, 208, 210, 214
Hughes, B. L. 42, 43
Hulbert, L.G. 460
Hull, C. L. 119, 432
Hull, J. G. 378
Hunt, A. R. 205, 214
Hunt, M. K. 146
Hunter, S. 64
Hunyady, Gy. 494, 495
Hunyady, O. 493, 494, 497
Huo, Y. J. 476, 477, 479
Hutchison, P. 453, 461

Ichheiser, G. 271
Ickes, W. 64
Ignarri, C. 367
Igou, E. R. 420
Ikemoto, S. 121, 124
Illingworth, K. S. S. 327
Insel, T. R. 77
Inzlicht, M. 180
Irlenbusch, B. 477
Isbell, L. M. 57
Isenberg, D. J. 459
Isenberg, N. 104, 105
Ishii, K. 98, 108, 191
Israel, J. 453
Ito, T. A. 65
Iversen, S. D. 144
Iwata 342, 343, 345, 346, 350
Izquierdo, A. 29, 34

Jaccard, J. 226
Jackson, D. 452

Jackson, D. C. 77, 435
Jackson, J. R. 367
Jacobs, A. 455
Jacobs, J. W. 435
Jacobson, B. L. 96
Jacobson, L. 230
Jacoby, L. L. 59, 231, 232, 233, 236, 246, 250
Jadhav, S. 439, 441–442
Jahanshahi, M. 185
James, W. 164, 268, 297, 420, 428, 439
Jameson, T. 170
Jameson, T. L. 170
Janis, I. L. 316, 453
Jensen, J. 125
Jersild, A.T. 204, 205, 211
Jetten, J. 239, 410, 452, 453, 458, 461
Jimenez, M. C. 493
Johansson, P. 266, 267
John 340
John, O. P. 41
Johns, M. 64
Johnson, E. 452
Johnson, E. J. 273
Johnson, M. K. 131
Johnstone, T. 85
Jonas 339
Jones 344
Jones, B. 31, 414, 421, 514
Jones, C. R. 78
Jones, D. M. 202
Jones, E. E. 50, 269, 271, 452, 458
Jones, J. T. 381
Jones, R. E. 496
Jones-Chesters, M. H. 100
Jones-Gotman, M. 124
Jonides, J. 98
Jonides, J. S. 82, 84
Jorgenson, D. O. 476
Jost, J. T. 49, 493, 494, 495, 496, 497, 499, 501, 503
Judd, C. M. 53, 54, 233, 355
Just, M. A. 212

Kőpetz, C. 297, 307
Kőpetz, C. E. 299, 307
Kaarbo, J. 462, 463
Kable, J. W. 33, 147
Kadish, S. H. 272
Kagel, J. H. 517
Kahan, D. 384
Kahneman, D. 45, 50, 269, 271, 272, 316
Kahnemann, D. 418, 420
Kakolewski, J. W. 140
Kalenscher, T. 147
Kalin, N. H. 77
Kalisch, R. 87, 143, 148, 151
Kalof, L. 496
Kalogeras, J. 188

Kalyanaraman, S. 418, 420
Kameda, T. 458
Kan, K. J. 100, 106, 108
Kane, M. J. 163, 164, 165, 167, 168, 169
Kanfer, R. 117
Kaplan, M. F. 458
Karau, S. J. 459
Karnath, H. O. 29
Kaspi, S. P. 100
Kasri, F. 190
Kasser, T. 319
Kastenbaum, R. J. 33
Katz, L. B. 371
Kaufman-Gilliland, C. M. 460
Kavanagh 339, 343
Kawagoe, R. 122, 123
Kawakami, K. 53, 251, 253, 358
Kay, A. C. 493, 494, 496, 497, 499
Kazén, M. 380
Kedem, P. 420
Keele, S. W. 205, 209, 212, 215
Keeling, C. D. 491, 492
Keller, P. 186, 187
Kelley, H. H. 50, 269
Kelley, W. M. 82, 84
Kelly, G. A. 80
Keltner, D. 118
Kennard, M. A. 16
Kennedy, D. 501
Kennedy, J. L. 441
Kennedy, K. A. 414, 421
Kerns, J. G. 8, 9, 10, 16, 20, 96, 102, 108
Kerr, N. L. 451, 453, 457, 458, 459, 460
Kerschreiter, R. 461, 462
Kessler, Y. 202, 207, 213
Keysar, B. 270
Khan 343
Khan, S. 53
Khan, U. 305
Khouri, H. 438
Kieras, D. E. 184, 202
Kihlstrom, J. F. 314, 315
Kim, M. 438, 441
Kim, P. H. 461
Kim, S. H. 148
Kinsella, G. 290
Kirby 343
Kirby, K. N. 409
Kitayama, S. 98, 108, 189, 191
Kivetz, R. 411, 414, 420
Klandermans, B. 463
Klauer, K. C. 252, 253
Klayman, J. 39, 41, 42
Kleiman, T. 360
Klein, R. M. 205, 214
Kleinbolting, H. 41
Kleinsorge, T. 209
Klem, A. 324

Kliegl, R. 205, 206, 214
Klimchuck, D. 261
Knight, R. T. 44, 82
Knowles, E. D. 365
Knowles, E. S. 380
Knowlton, B. J. 27
Kobayashi, S. 125
Kober, H. 142, 155
Koch, I. 186, 204, 206, 207, 208, 209, 210, 212, 214, 215
Kochanska, G. 202
Koechlin, E. 128
Koffka, K. 507
Komorita, S. S. 460, 492
Konishi, S. 144, 152
Koo, M. 401
Koob, G. F. 121
Kooken, K. 190
Koole, S. L. 360, 365, 367
Kopelman, S. 491, 492
Kornblum, S. 179
Kortenkamp, K. V. 491
Kosslyn, S. M. 78
Kostopoulos, P. 142, 155
Kotz, S. A. 98
Kozak, M. J. 440
Krain, A. L. 146, 148, 154
Kramer, R. M. 460, 479, 492
Krasowski, M. D. 441
Krawczyk, D. 129
Kray, J. 204
Krigolson, O. E. 13
Kringelbach, M. L. 123, 124, 145, 148, 154
Kross, E. 85, 86, 88, 384, 398, 399, 414, 421, 437, 438, 439, 440, 441–442
Kruer, J. L. 97
Kruger, J. 266
Kruglanski 339
Kruglanski, A. W. 54, 297, 299, 300, 301, 302, 304, 306, 307, 308, 314, 315, 316, 317, 318, 329, 358, 359, 391, 394, 398, 400, 410, 414, 422, 453, 494, 503
Krull, D. S. 39, 41, 50, 271
Krysan, M. 49
Kudadjie-Gyambi, E. 418, 420
Kuhl, B. A. 16
Kuhl, J. 380, 390, 398, 502
Kühnen, U. 191, 192
Kuijer, R. G. 477
Kukish, K. S. 94, 97
Kumar, P. A. 371
Kunda 339
Kunda, Z. 309, 355, 367, 411
Kunde, W. 101, 108, 186, 187
Kurzban, R. 451

LaBar, K. 125
LaBar, K. S. 63, 81
LaHoste, G. J. 441

Lalande, K. 437
Lambert, A. J. 53, 64, 65, 231, 232, 233, 236, 250
Lane, K. A. 226
Lane, R. 20
Lane, R. D. 19, 20, 82, 239
Lang, P. J. 83, 318
Langan, C. J. 458
Lange, J. E. 355
Langendam, D. 281, 287
Langer, E. J. 264, 266
Larrick, R. P. 396
Larsen, R. J. 99, 106
Larwood, L. 40
Latané, B. 459
Latham, G. P. 396, 402
Lau, G. P. 494
Laughlin, J. E. 170
Laughlin, P. R. 458, 459
Laurent, B. 81
Lavezzary, E. 360
Lawrence 338
Lawrie, S. M. 17
Lazarus 338, 340
Lazarus, R. S. 80, 323, 439
Leary, M. 439
Leary, M. R. 190, 449, 451, 457
Lebiere, C. 184
Leck, K. 461
LeDoux, J. 435
LeDoux, J. E. 81
Lee 339, 343, 344
Lee, A. 191
Lee, A. Y. 327
Lee-Chai, A. 394, 493, 501
Lee-Chai, A. Y. 305
Leggett, L. 417
Leh, S. E. 154
Leiderman. P. H. 457
Leiman, B. 185
Leirer, V. O. 371
Leith, K. P. 153
Lejuez, C. W. 145, 146
Lengfelder, A. 185, 280, 289, 302
Leon, M. I. 124, 125, 126
Lepore, L. 53
Lepper 339
Lerner, J. 118
Lesch, K. 134
Lessing, E. E. 33
Lev, S. 99
Levenson, R. W. 57, 58, 67
Leventhal 344
Levesque, J. 87, 143, 148
Levin-Sagi 343, 348
Levin-Sagi, M. 298, 304, 314, 409, 493, 494, 501
Levine 339
Levine, J. M. 452, 454, 458, 460, 461,
    462, 463, 465

Levy, B. J. 149, 152
Levy, S. R., 413
Lewin, K. 313, 318
Lewis, M. 456
Lewis, M. D. 152
Lewis, S. A. 458
Leyens, J.-P. 453
Li, P. 365, 366
Libby, L. 439
Libby, L. K. 414, 421
Liberman 343, 348
Liberman, N. 56, 274, 298, 301, 304, 314, 359, 367, 408,
    409, 411, 412, 413, 414, 417, 419, 420, 421, 422,
    434, 493, 494, 501
Libet, B. 188
Lichtman, R. R. 39
Lickel, B. 53, 64
Liddle, P. F. 144
Lieberman, M. D. 21, 55, 65, 66, 68, 77, 78, 80, 127, 129,
    143, 148, 149, 153, 154, 435, 441
Lieberrman, M. D. 152
Lien, M.-C. 208, 213, 214
Lijffijt, M. 152
Lin, E. L. 419
Lin, M. H. 58
Lind, E. A. 479, 480, 484
Lind, S. 480
Lindenberger, U. 204
Lindsay, D. S. 246
Linn, J. A. 380
Linville, P. W. 323
Liu, X. 96
Liviatan, I. 494, 497, 500, 501
Livingston, R. W. 356, 358
Locke 346
Locke, E. A. 396, 402
Locke, H. S. 127, 129
Lodewijkx, H. F. M. 455
Loewenstein 337, 349
Loewenstein, G. F. 29, 270, 377, 385, 389, 402,
    409, 411, 415, 418, 420, 421, 431, 475, 493,
    494, 501, 512
Logan, G. D. 50, 55, 96, 143, 152, 164, 165, 176, 177,
    180, 202, 203, 205, 206, 207, 208, 209, 211,
    212, 252, 290
Logue, A. W. 510
Lombardo, M. V. 42, 43
London, B. E. 438
Lord 339
Lord, R. G. 456
Los, S. A. 204, 214
Lotze, R. H. 186
Lovallo, D. 418, 420
Lowe 339
Lu, C. H. 178
Luce, M. F. 273
Luck, S. J. 166
Lucki, I. 125

Luhtanen, R. 497
Lui, L. 479
Lundgren, S. 461, 462
Luo, Q. 20
Luria, R. 205, 207, 208, 209, 214, 215
Lust, S. A. 54
Luty 343
Luu, P. 43, 82, 94, 130
Lynch, M. E. 253

Maass, A. 274, 306
McAuley, E. 190
McAuliffe, B. J. 452, 453
McCarthy, G. 63
McClelland, J. L. 30, 50, 58, 96
McClure, S. M. 33, 143, 147, 154
McCollough, A. W. 166
McConahay, J. B. 239
McConnell, A. R. 255
McCrae, R. R. 298
McCulloch, K. C. 301, 394
McDaniel, M. 132
MacDonald 349
MacDonald, A. W., 3rd, 7, 8
MacDonald, G. 458, 465
MacDonald III, A.W. 209
MacDonald, P. A. 94, 95
McElreath, R. 477
McEwen, B. S. 65, 439, 441–442
McGhee, D. E. 223, 244
McGrath, J. E. 450, 451, 453, 454
McGregor, H. 117, 319
Machizawa, M. G. 166
McIsaac, H. K. 439
MacKay, D. G. 101, 107
McKenna, F. P. 99, 100, 101, 107, 108, 109
Mackie, D. M. 450
McLaughlin, J. P. 49
MacLeod, C. M. 94, 95, 97, 149, 178
McNally, R. J. 100, 106
McNaughton, N. 55
Macrae, C. N. 53, 54, 56, 57, 63, 168, 239, 357, 367, 410
Maddox, W. T. 114
Maddux, W. W. 189
Madson, L. 190, 191
Magee, L. E. 97
Magen 335, 343, 346, 348
Magen, E. 420
Maia, T. V. 30
Majka, K. 453
Malkoc, S. A. 415, 421
Malone, P. S. 269, 271
Mandisodza, A. N. 497, 500
Manes, F. 31, 145
Mann 343, 349
Mann, L. 316
Mann, T. 300, 382

Marciano, H. 205, 207, 212
Markman, A. B. 114, 395
Markus, H. 189
Marques, J. M. 453, 457, 461
Marriott 342, 343, 345, 346, 350
Marsch, L. A. 33, 153, 513
Martijn 347
Martin, J. 463
Martin, L. L. 52, 246, 367
Martin, R. 461, 462
Martinez-Pons, 346
Marwell, G. 477
Masicampo, E. J. 268
Mason, M. F. 102
Mason, R. 78
Masson, M. E. J. 210, 212, 213
Masuda, T. 189, 191
Matheson, K. 453
Mathews, A. 94
Matsumoto, D. 189, 190, 194, 195
Matthews, S. C. 11, 102, 143, 146
Mauer, N. 101, 108
Mauss 339
May 339, 343
Mayberg, H. S. 16
Mayer, D. 396
Mayr, U. 9, 10, 205, 206, 207, 209, 211, 212, 213, 214, 215
Meacham, J. A. 185
Mead, N. L. 377
Meehl, P. E. 236
Meek, D. 459
Mehta, A. D. 441
Meiran, N. 202, 204, 205, 206, 207, 208, 209, 210, 211, 212, 213, 214, 215
Mendes, W. B. 64
Mendoza, S. A. 52, 62
Mendoza-Denton, R. 438
Menon, V. 144
Mercer, K. A. 99
Merckelbach 347
Mertig, A. G. 496
Messick, D. M. 479, 491, 492
Mesulam, M. M. 28
Metcalfe 343
Metcalfe, J. 39, 80, 147, 301, 313, 389, 398, 399, 400, 411, 414, 422, 435, 439, 493, 494, 501
Metzner, R. 431
Meuter, R. F. 209
Meyer, D. E. 109, 184, 202
Meyerowitz, J. A. 40
Michealis, B. 438
Midden, C. 280, 360
Mikulincer, M. 420
Milàn, E. G. 212
Milgram, S. 239, 267
Milham, M. P. 15
Miller, C. E. 452, 458
Miller, D. T. 266, 271

AUTHOR INDEX

Miller, E. K. 3, 4, 82, 96, 127, 153, 154, 202
Miller, G. A. 206, 314
Miller, J. 179
Miller, M. 436
Miller, N. E. 313, 318
Miller, R. S. 383
Miller, W. R. 494
Milliken, B. 54
Mills, J. 268, 455
Milne 343
Milne, A. B. 53, 239, 367, 410
Milne, S. 281, 302
Miltner, W. H. 12, 13
Minsk, E. 264
Mirenberg, M. C. 381
Mirenowicz, J. 124
Mische, 335, 345l Mischel 342, 343, 345
Mischel 343, 344
Mischel, W. 80, 81, 85, 147, 150, 291, 292, 297, 298, 301, 313, 376, 384, 389, 390, 398, 399, 400, 411, 414, 419, 421, 422, 428, 429, 430, 431, 432, 433, 434, 435, 436, 437, 438, 439, 440, 441–442, 493, 494, 501
Mishkin, M. 31, 144
Mitchell, D. G. 152, 153
Mitchell, J. P. 63, 143, 150, 151, 154
Mitchell, R. L. 98
Mitra, R. 439, 441–442
Miyake, A. 34, 202, 211, 213
Mizon, G. A. 206
Mobini, S. 147
Mohanty, A. 104
Mojzisch, A. 461, 462
Moll, J. 358
Monsell, S. 100, 106, 179, 192, 194, 203, 205, 206, 208, 209, 211, 212, 214, 215
Montada, L. 477
Montague, P. R. 123
Monteith, M. J. 55, 56, 58, 62, 251, 368, 369
Monteith, M. M. 52, 60
Monterosso 147
Monterosso, J. R. 142, 148, 152, 153
Moore 343
Moore, B. 313, 414, 421, 433, 434
Moore, C. F. 491
Moran, J. M. 42, 43
Moray, N. 166
Moreland, R. L. 450, 451, 452, 454, 456, 461, 462, 463, 465
Moreland, R. M. 454
Morewedge, C. K. 260, 264, 265, 267, 269, 270, 272, 273
Morf, C. C. No Cross-reference Morgenstern, O. 262
Morilak, D. A. 65
Morris, M. L. 189
Morris, N. 202
Morsella, E. 299, 308
Morwitz, V.G. 273

Moscovici, S. 460, 461
Moskowitz, G. B. 54, 225, 253, 280, 299, 317, 354, 355, 357, 359, 360, 363, 364, 365, 366, 367, 368, 369, 370, 390, 394, 410, 502
Mosso, C. 493
Mougios, V. 438
Mowrer, O. H. 318, 320
Mufson, E. J. 16
Mullen, B. 459
Muller, C. J. 435
Muraven 343, 347, 348
Muraven, M. 3, 117, 132, 141, 151, 180, 190, 282, 299, 300, 304, 379, 385, 404, 410, 416, 422
Muraven, M. R. 376, 378, 383, 384, 385
Murgraff, V. 285
Murias, M. 441
Murphy, G. L. 419
Murphy, K. 106
Murray, A. 463
Murray, E. A. 31, 32, 182
Myerson, J. 409, 421, 513
Myrseth, K. 398
Myrseth, K. O. R. 401

Nagai, Y. 11
Nail, P. R. 458, 465
Napier, J. L. 497, 500
Nathan, M. 285
Nattkemper, D. 183
Nederkoorn, C. 152, 153
Neely, J. H. 399
Neill, W. T. 54, 180, 193
Neinhaus 347, 348
Neisser, U. 177, 438, 439
Nemeth, C. J. 461
Neter, E. 39, 41, 46, 417, 420
Neuberg, S. L. 50, 58, 246, 356, 451
Nicholls, L. 44
Nickerson 339
Nicoletti, R. 178
Nielsen-Bohlman, L. 82
Nienhaus, K. 385
Nier, J. A. 230
Nieuwenhuis, S. 13, 14, 64, 180, 214
Nigro, G. 438, 439
Niki, H. 144
Nisbett, R. E. 50, 189, 191, 230, 246, 268, 269, 306, 307
Noel, J. G. 453
Nolen-Hoeksema, S. 438, 439
Nordgen, L. F. 273
Nordgren 349
Nordgren, L. 360, 367
Nordhaus, W. D. 492
Norem, J. K. 326, 327
Norman, D. 50
Norman, D. A. 208

Norton, M. I. 234, 267
Nosek, B. 227
Nosek, B. A. 225, 252, 253, 367, 493, 494, 497
Nussbaum, S. 412, 420
Nyberg, L. 82

O'Boyle 345
Ochsner Gross 87
Ochsner, K. 58, 86
Ochsner, K. N. 11, 18, 19, 77, 78, 79, 81, 82, 83, 84, 86, 103, 143, 148, 149, 154, 202, 435–436, 438, 441
O'Connor, K. J. 435
O'Doherty, J. 29, 84, 145, 148, 435
O'Doherty, J. P. 123, 126
O'Donoghue, T. 390, 409, 415, 421
Oettingen, G. 279, 280, 281, 282, 291, 292, 293, 302, 317, 396, 423
Olds, J. 120
O'Leary, K. D. 195
Olekalns, M. 286
Olivola, C. Y. 414, 421
Olson, C. 126, 134
Olson, C. R. 34, 147
Olson, M. A. 53, 222, 223
Olsson, A. 77
Ongur, D. 154
Oosterlaan, J. 290, 441
Orbell 343
Orbell, J. 460
Orbell, S. 281, 302
Oreskes, N. 491, 492
Osman, A. 179
Ostman, R. E. 497
Ouellette, J. A. 461, 462
Ouwerkerk, J. W. 460
Oyserman, D. 191, 192

Packer, D. J. 461
Padoa-Schioppa, C. 29, 34, 123
Paez, D. 453
Pailing, P. E. 130
Painter 339, 343, 344
Pak, H. 293
Palmiter, R. D. 120
Pandya, D. N. 16
Panksepp, J. 77, 121, 124
Papa, A. 437
Papies, E. 410, 422
Papies, E. K. 300, 358
Paradiso, S. 82
Pardo, J. V. 4, 127
Parent 343, 349
Park, B. 54, 355
Park, I. H. 97, 98, 102, 104, 109
Parker, J. L. 497
Parks, C. D. 460, 492

Parks-Stamm, E. J. 280
Pashler, H. 202
Passetti, F. 441
Passingham, R. E. 187
Paton, J. J. 125
Patterson 342, 343, 345
Patterson, C. J. 291, 442
Paty, J. A. 379
Paulhus, D. L. 39, 40, 41, 42
Paulus, M. P. 34
Payne, B. K. 59, 60, 221, 224, 225, 231, 232, 233, 236, 237, 238, 246, 249, 250, 253
Payne, J. W. 273
Paz, D. 420
Peake, P. 376, 432, 437, 438
Peciña, S. 122
Pelham, B. W. 39, 41, 50, 381, 493, 497
Penley, J. A. 324
Perner, J. 270
Pero, S. 284
Peruche, B. M. 365
Pervin, L. A. 314
Petersen S. 127
Petersen, S. E. 4
Peterson, B. 82
Peterson, J. L. 464
Petrides, M. 142, 154, 155
Petry, N. M. 33
Petty, R. E. 52, 53, 222, 246, 459
Peyron, R. 81
Pfeffer, J. 494, 495
Phaf, R. H. 100, 106, 108
Phan, K. L. 18, 87, 148, 149, 249
Phelps, E. A. 65, 249, 435
Philipp, A. M. 207, 210, 214
Piaget, J. 474
Pickett, C. L. 65
Pierce, W. D. 116
Piéron, H. 176
Pieters, J. M. 104
Pieters, R. 267
Pietrzak, J. 497, 500
Pizarro, D. A. 269, 270
Plant, E. A. 63, 64, 69, 225, 250, 364, 365, 379
Pleydell-Pearce{,} C. W. 168
Pochon, J. B. 128, 129
Poehlman, T. A. 49
Poldrack, R. A. 55, 79, 127, 143, 144, 145, 148, 149, 151, 155
Polivy, J. 384
Pool, G. J. 461
Posner M. 127
Posner, M. I. 4, 43, 50, 78, 82, 94, 95, 214, 436
Pösse, B. 180
Postmes, T. 449, 461
Poston 2nd 343
Pratt, T. C. 376, 377
Pratto, F. 496

AUTHOR INDEX

Preacher, K. J. 226
Premack, D. 520
Preston, J. 268, 273
Preston, S. D. 97, 98
Pribram, K. H. 314
Price, J. L. 35, 154
Principe, A. 421
Prinz, W. 175, 178, 186, 188, 207
Prior, M. 290
Prislin, R. 463
Pritchard, D. 476
Proctor, R. W. 178
Pronin, E. 222, 266, 269, 414, 421, 439
Purvis, D. 461

Rabin, M. 390, 420
Rachlin, H. 163, 170, 389, 390, 393, 402, 405, 414, 418, 420, 421, 431, 506, 509, 512, 513, 514, 515, 516, 518, 520
Rachman, S. 440
Radin, D. I. 265
Radley, J. J. 436
Rafferty, Y. 460
Raghunathan, R. 46, 420
Raichle, M. E. 4, 102
Raineri, A. 513
Randsley de Moura, G. 461
Ranelli, C. J. 452
Ranganath, K. 227
Rankin, R. E. 64
Raphelson, A. 117
Rassin, E. 150
Rauch, W. 384
Raven, J. C. 284
Rawls, J. 475
Ray, R. D. 435–436
Raye, C. L. 86, 131
Raz, A. 94
Read, D. 418, 420, 431
Reason, J. T. 177
Reicher, S. 464
Reis, H. 476
Remijnse, P. L. 145, 151
Reyes, J. A. 98, 191
Reynolds, S. M. 125
Rhee, E. 192
Rholes, W.S. 78
Riccio, C. A. 290
Richard, F. D. 452
Richards 340, 348
Richards, A. 100
Richeson, J. A. 55, 64, 65, 68, 237, 253
Ridderinkhof, K. R. 154
Riefer, D. M. 247
Riemann, B. C. 106
Riggs, D. S. 195
Riordan, M. A. 54

Risko, E. F. 181
Rizzuto, D. S. 142, 155
Robbins, T. 35
Robbins, T. W. 55, 81, 127, 441
Roberts, A. C. 81
Roberts, J. C. 418, 419, 420
Robertson, E. R. 435–436
Robertson, R. 497
Robins, L. 517
Robins, R. W. 39, 40
Robinson 340, 349
Robinson, J. A. 439
Robinson, S. 120
Robinson, T. E. 120, 121
Rocher, A. B. 436
Rockenbach, B. 477
Rodriguez, M. L. 80, 292, 390, 411, 414, 419, 422, 428, 429, 430, 432, 436, 437, 438, 439
Roeder, U. 191
Roesch, M. 126, 134
Roesch, M. R. 34, 147
Roethlisberger, F. J. 465
Rogers, R. D. 43, 179, 192, 194, 205, 206, 208, 211, 212, 214, 215
Rollock, D. 64
Rolls, E. T. 31, 34, 44, 78, 81, 123, 144, 145, 148, 154
Rorden, C. 29
Rosen, V. M. 167, 169
Rosenbaum, D. A. 175
Rosenthal, R.130 127, 129, 230
Roseveare, T. A. 168
Rosinski, R. R. 94, 97
Ross 339
Ross, L. 50, 222, 269, 408, 411, 439
Ross, L. D. 269
Ross, M. 266, 396, 421
Rosvold, H. E. 290, 441
Rota, G. 98, 104
Roth, J. 266
Rothbart 345
Rothbart, M. 370
Rothbart, M. K. 436
Rothman, A. J. 355
Rotter, G. S. 457
Rotton, J. 384
Ruback, R. B. 452
Rubia, K. 143, 144, 145, 148, 149, 290
Rubin, O. 204, 211
Rubinstein, J. S. 205, 208, 211
Rudell, A. P. 178
Rumelhart, D. E. 50
Runcimann, W. G. 463
Rusbult, C. E. 438, 481
Rush, C. R. 152
Russin, A. 358
Russo, A. 306
Rust, R. T. 307
Ruthruff, E. 208, 212

Ryan, R. 116, 117
Ryan, R. M. 63, 64, 316, 319
Rydell, R. J. 255
Ryle, G. 509

Saad, L. 500
Sadovsky, A. 436, 465
Sagristano, M. 412, 420
Sahakian, B. J. 291
Sakagami, M. 144
Sakurai, M. M. 458
Salamone, J. D. 122, 125
Salomon, A. R. 54, 253, 359
Salovey, P. 413, 417, 419, 421
Samdahl, D. M. 497
Samper, A. 273
Sanderman, R. 477
Sanders, A. F. 176
Sandstrom, S. M. 120
Sanfey, A. G. 125
Sani, F. 464
Sapolsky, R. M. 65
Sarbin, T. R. 453
Sarter, M. 78, 120, 128
Sasota, J. A. 422
Sassenberg, K. 359, 369
Saul, A. 493
Saults, J. S. 54
Saunders, C. D. 501
Saver, J. L. 43
Saxe, L. 452
Sayette, M. A. 100
Schaal 343
Schaal, B. 54, 281, 285, 288, 302, 355, 394, 410
Schachter, S. 453
Schacter, D. L. 77
Schaefer, S. M. 87
Scheff, T. 451
Scheffer, M. 491, 492
Scheier 335, 336, 337, 338, 339, 341, 342, 343
Scheier, M. 41, 115, 117
Scheier, M. E. 378
Scheier, M. F. 298, 314, 316, 318, 319, 323, 450
Schein, E. H. 455
Schelling, T. C. 390, 393, 394
Schirmer, A. 98
Schkade, D. 269
Schlesinger, A. M. Jr. 458
Schmader, T. 64
Schmeichel, B. J. 313, 377, 378, 380, 383, 385, 404, 422, 486
Schmitt, D. R. 477
Schmitt, M. 225
Schmitt, M. T. 453
Schmittlein, D. 273
Schneider, A. 477
Schneider, D. W. 205, 206, 207, 208, 209

Schneider, R. 485
Schneider, W. 50, 95, 177
Schnetter, K. 293
Schnyer, D. M. 44
Scholer, A. A. 314, 315
Schooler, J. W. 53, 169, 274
Schrock, J. C. 164, 165, 170
Schroeder 343
Schroeder, C. E. 441
Schubert, B. 191
Schuch, S. 209, 210, 214, 215
Schulhofer, S. J. 272
Schultz, W. 34, 123, 124
Schulz, K. P. 290, 292
Schulz-Hardt, S. 461, 462
Schuman, H. 49
Schvaneveldt, R. W. 109
Schwartz, B. 263
Schwartz, J. K. L. 223
Schwartz, J. L. K. 244
Schwarz, N. 52, 272, 439
Schweiger Gallo, I. 296
Schweighofer, N. 35
Scott, D. 491, 492
Sedikides, C. 39, 41, 382
Seeley, E. A. 190, 194, 195
Segalowitz, S. J. 130
Sīlānanda, Ven. U. 169
Sekaquaptewa, D. 355
Self, E. A. 396
Seligman, M. E. P. 177, 376, 431
Sembi, S. 292
Semin, G. R. 420
Sentis, K. 459
Sergeant, J. A. 290
Serodio, R. G. 453
Sestir, M. A. 66, 252
Seymour, B. 125
Shadlen, M. N. 124, 125, 126
Shaeffer, E. M. 414, 421
Shaffer, L.H. 204, 205, 206, 212
Shah, J. Y. 54, 302, 308, 314, 315, 316, 317, 318, 322, 323, 324, 325, 329, 359, 360, 363, 394, 398, 410, 411, 414
Shallice, T. 50, 208, 209, 210
Shamosh, N. A. 163, 170
Shapiro, T. 94
Sharma, D. 99, 100, 101, 107, 108, 109
Sharot, T. 42, 43
Shaver, K. G. 269
Sheeran 343
Sheeran, P. 280, 281, 282, 284, 302
Shefrin, H. M. 389, 390, 393, 394
Sheldon, K. M. 315, 316, 319, 320, 325, 329
Sheldon, O. 493, 497
Shelton, J. N. 64, 253, 254
Sherman, J. W. 50, 56, 57, 59, 243, 244, 245, 250, 251, 252, 253, 255

AUTHOR INDEX

Sherman, S. J. 256
Sherrington, C. S. 187
Sherrod, D. R. 384
Shiffman 348
Shiffman, S. 379
Shiffrin, R. 50
Shiffrin, R. M. 95, 167, 168, 177
Shimamura, A. P. 44
Shin, L. M. 16
Shiv, B. 272, 410
Shoben, E. J. 419
Shoda, Y. 80, 292, 376, 390, 411, 414, 419, 422, 428, 429, 430, 432, 436, 437, 438, 439, 440, 442
Shoup, R. 181
Showers, C. 314, 323, 326, 327
Showers, C. J. 70
Shulman, G. L. 142, 155
Sidanius, J. 496
Siegel, E. 516, 518
Siegrist, M. 107
Silver, L. A. 253
Silvia, P. J. 453
Simon, B. 463
Simon, H. A. 263
Simon, J. 515
Simon, J. R. 178, 196, 287
Simons, J. S. 62, 288
Simonson, I. 305
Singelis, T. M. 190, 194
Singer 344
Singer, J. E. 384
Singer, T. 63, 125
Sisson 343
Skinner, B. F. 319, 508, 512, 518
Skurnik, I. 357
Sleeth-Keppler, D. 306
Slessareva, E. 304
Sloman, S. A. 50, 246
Small, D. M. 124, 126, 129, 131
Smallwood, J. 169
Smart, L. 169, 170
Smeets, K. 360
Smith, C. 227
Smith, C. L. 436, 465
Smith, C. M. 462
Smith, D. M. 451
Smith, E. E. 82, 84, 98, 103
Smith, E. R. 50
Smith, H. J. 476, 477, 479, 484
Smith, J. 292
Smith, M. C. 97
Smith, P. K. 412, 414, 422
Smith, R. 429
Snyder, C. R. R. 50, 95
Snyder, H. N. 452
Snyder, M. 190
Sobotka, S. S. 119
Sohn, M. H. 205, 208, 209, 212, 214

Solarz, A. K. 302, 395
Sommer, K. L. 190, 195, 381
Sondak, H. 479
Sonnentag, S. 380
Sonuga-Barke, E. J. S. 292
Spangler, W. J. 17
Sparrow, B. 268
Spears, R. 449
Speilberger, C. 319
Spelke, E. 177
Spencer, S. J. 355, 359, 367, 370, 494
Spengler, S. 62
Spiegel, S. 325
Spinrad, T. L. 436, 465
Spranca, M. 264
Sprecher, S. 477
Springer, A. 192, 193, 194
Sriram, N. 370
Sroka, K. R. 452
Srull, T. K. 57, 359
Stacy, A. W. 410
Stadler, G. 293
Stalnaker, T. A. 32
Stangor, C. 355, 451, 452, 492
Stansfield, R. B. 97, 98
Stapel, D. A. 52
Stasser, G. 458, 459
Staub, B. D. 271
Staub, E. 431
Steeh, C. 49
Steele, C. M. 64
Steele, J. D. 17
Stein, M. A. 441
Stein-Seroussi, A. 42
Steinhauser, M. 207, 210
Stenberg, G. 97, 98
Stenger, V. A. 96
Stephan, C. W. 64
Stephan, E. 408, 411, 412, 413, 421
Stephan, W. G. 64
Stern, P. C. 496
Sternberg, S. 176
Stewart, B. D. 237, 238
Stigler, G. J. 513
Stokes, M. B. 238
Stokes-Zoota, J. J. 452
Stoltz, J. A. 181
Stolz, J. A. 94, 181
Stout, J. C. 170
Stover, E. R. S. 146
Strack, F. 50, 53, 80, 253, 410, 439
Strahan, E. J. 359
Strange, B. A. 435
Strauman, T. J. 322, 328
Strayhorn 335, 348, 350
Stroebe, W. 300, 358, 410, 422
Stroessner, S. J. 57, 314
Stroobant, N. 98

Stroop, J. R. 4, 93, 165, 178, 244, 290, 366
Strotz, R. H. 393
Strube, M. J. 39
Stucke, T. S. 382
Studt, A. 286
Sturmer, S. 463
Stuss, D. T. 27, 28, 35
Styles, E. A. 184
Subic-Wrana, C. 239
Suchner, R. W. 452
Sudevan, P. 179, 206, 213
Sullivan, B. N. 493, 497
Sulloway, F. 494, 503
Sumner, P. 209, 210, 211, 212
Sunohara, G. A. 441
Sunshine, J. 482
Sunstein, C. R. 269
Sutton, S. 115, 117, 130
Swann, W. B. 39, 41, 42
Swanson, J. M. 441
Swanson, K. K. 439
Swets, J. A. 315
Swick, D. 10, 11
Syroit, J. E. M. M. 455

Tajfel, H. 463
Tamir 340
Tanabe, J. L. 96
Tanaka, S. C. 35
Tangney, J. P. 200, 376, 419
Taylor, C. M. 54, 253, 359
Taylor, D. A. 179, 206, 213
Taylor, E. 292
Taylor, S. E. 39, 40, 41, 78, 323, 396, 411
Taylor, S. F. 126, 128, 131, 249
Taylor, T. J. 106
Tenbèult 347
Tenenbaum, G. 284
Teong, S. K. 238
Tesser, A. 367
Testa, A. 438
Tetlock, P. E. 226
Teves, D. 132
Thagard, P. 309, 367
Thaler, R. H. 389, 390, 393, 394, 409, 415, 421
Thibaut, J. 477, 478, 479, 486
Thomas, K. L. 441
Thomas-Hunt, M. C. 461
Thompson 336, 338, 340, 341, 342
Thompson, D. V. 307
Thompson, E. P. 494, 497
Thompson, J. C. 497
Thompson, L. 475
Thompson, M. M. 300
Thompson-Schill, S. 129
Thornhill, R. 451
Thorpe, S. J. 144

Thrash, T.M. 318
Tice 343, 348
Tice, D. M. 117, 132, 180, 190, 300, 376, 379, 382, 389, 416, 420, 494
Tindale, R. S. 462
Tipper, S. P. 193, 363
Tobler, P. N. 146
Todman, J. 464
Tomaka, J. 324
Tomarken, A. J. 151
Tonry, M. 233
Tornay, F. J. 212
Trafimow, D. 191
Tranel, D. 81, 435
Trawalter, S. 64, 237
Tremblay, L. 34, 123
Triandis, H. C. 191
Troetschel, R. 394, 410
Trolier, T. K. 355
Trope 337, 343, 348
Trope, Y. 46, 50, 80, 225, 234, 245, 298, 301, 303, 304, 308, 313, 314, 360, 361, 390, 391, 393, 394, 398, 399, 408, 409, 411, 412, 413, 414, 415, 417, 418, 419, 420, 421, 422, 434, 435, 493, 494, 501
Trost, M. R. 63, 451
Trötschel, R. 286, 493, 501
Tsuchida, A. 35
Tucker, D. 130
Tuholski, S. W. 170
Tulving, E. 4
Turken, A. U. 10, 11
Turner, J. 463
Turner, M. L. 164
Tversky, A. 50, 272, 316
Tyler, J. M. 57
Tyler, R. B. 252, 357
Tyler, T. R. 476, 477, 478, 479, 480, 481, 482, 483, 484, 485, 492, 494

Uhlmann, E. 49, 269, 270
Uhlmann, E. L. 234
Ulbert, I. 441
Uleman, J. S. 50
Ullrich, J. 494
Ullsperger, M. 10
Umiltà, C. 178
Unsworth, N. 163, 164, 165, 167, 168, 170

Vagg, P. 319
Valenstein, E. S. 140
Vallacher 338
Vallacher, R. R. 314, 413, 418, 419, 423
van Baaren, R. B. 191, 360, 367
van den Bos, K. 484
Van Den Heuvel, C. E. 214
van der Linden, M. 202

AUTHOR INDEX

van der Pligt 349
van Dijk, E. 460
van Harreveld 349
van Hooff, J. C. 101
van Kamp, S. 53
van Knippenberg, A. 52, 360, 185, 367, 370
van Knippenberg, D. 457
Van Lange, P. 481
van Lange, P. A. M. 460
van Liere, K. D. 496, 497
Van Maanen, J. 455
Van Selst, M. 42
van Veen, V. 6, 13, 15, 21, 22, 64, 96, 180
Van Vugt, M. 460, 461, 464
Vandello, J. A. 234, 461
Vargas, P. 355
Verbruggen, F. 214
Verfaellie, M. 44
Vermunt, R. 460
Vescio, T. K. 53
Vieth, A. Z. 322, 328
Viki, G. T. 461
Vingerhoets, G. 98
Virzi, R. A. 96
Visser, P. S. 65, 271
Vogel, E. K. 166
Vogel-Sprott, M. 252
Vogt, B. A. 17
Vohs 337
Vohs, K. 414, 421, 422
Vohs, K. D. 202, 268, 274, 300, 313, 376, 377, 378, 379, 380, 381, 382, 383, 384, 385, 389, 390, 404, 449, 450, 486
Voils, C. I. 368
von Hippel, W. 253, 355, 457
von Neumann, J. 262
Vorauer, J. D. 64
Vrana, S. R. 64
Vuilleumier, P. 17, 18
Vyas, A. 436, 439, 441–442

Wadden 349
Wager, T. D. 79, 81, 102, 103, 152
Wakeman, E. A. 78
Wakslak, C. 494
Wakslak, C. J. 411, 412, 413, 414, 422
Walker, L. 479, 486
Wallace, R. J. 178
Walster, E. 476
Walster, G. W. 476
Wann, D. L. 453
Wansink 339, 343, 344
Ward 343, 349
Ward, A. 300
Ward, L. M. 207
Wasel, W. 54, 355, 394, 410
Waszak, F. 179, 180, 181, 183, 186, 187, 188, 204, 210

Watanabe, M. 120
Waters, A. J. 100, 101, 108, 109
Watson, D. C. 319
Watson, R. 491
Wayment, H. A. 39, 41
Weart, S. 491, 492
Webb, T. L. 280, 282
Weber, J. 86
Weber, J. M. 491, 492
Weber, M. 270, 297
Webster, D. M. 307, 318, 453
Webster, P. J. 491, 492
Wegener, D. T. 52, 53, 246
Wegner 338
Wegner, D. M. 56, 57, 67, 150, 169, 190, 221, 246, 260, 262, 264, 265, 266, 268, 269, 272, 274, 314, 357, 358, 410, 413, 418, 419, 423
Weiner 350
Weiner, B. 269, 272
Weintraub, J. K. 323
Weiss 344
Welford, A. T. 176
Wellman, C. L. 436
Wells, G. L. 270, 271
Wenk-Sormaz, H. 94
Wenke, D. 183
Wenzlaff, R. M. 150
Wertenbroch 343
Wertenbroch, K. 301, 394, 409, 418
Wertz, J. M. 100
West, R. 252
Westling 343, 349
Westphal, M. 437
Whalen, P. J. 17, 18, 94, 99, 103, 104, 107
Wheatley, T. 268
Wheatley, T. P. 262, 265
Wheeler, E. Z. 32
Wheeler, L. 453
Wheeler, S. C. 381
White, P. 53
White, T. L. 118
Whitney, D. J. 459
Whitney, P. 170
Whorf, T. P. 491, 492
Wickens, D. D. 166
Wiebe, J. S. 324
Wiers, R. W. 410
Wigal, S. B. 441
Wiking, S. 97
Wilbur, C. J. 222
Wilke, H. 479
Wilke, H. A. M. 460
Willard, G. 39
Williams, C. J. 367
Williams, J. M. G. 94, 99, 168, 169, 170
Williams, K. D. 453, 457, 459
Willis, R. H. 461
Willits, F. K. 491, 492

Wilson 349
Wilson, A. E. 421
Wilson, T. 116
Wilson, T. D. 50, 52, 53, 230, 235, 246, 255, 268, 269, 272, 306, 307, 357
Wimmer, H. 270
Windes, J. D. 94
Windschitl, P. D. 39
Winkielman, P. 272
Winston, J. 89
Winston, J. S. 435
Wise, S. P. 182
Wit, A. P. 460
Witkin, H. A. 191
Wittenbrink, B. 54, 355
Wittgenstein, L. 509
Woike, B. A. 360
Wolfe, C. T. 367
Wood, J. N. 202
Wood, J. V. 39
Wood, W. 458, 461, 462
Woodworth, R. S. 182
Woolfolk, R. L. 270
Wright, R. A. 396
Wright, S. C. 463
Wu, G. 146, 396
Wughalter, E. H. 284
Wyer, N. A. 57
Wyer, R. S. 57, 359
Wyland, C. L. 143, 150, 151, 155
Wylie, G. 179, 206, 208, 210
Wylie, G. R. 213

Yamamoto, Y. 11
Yamasaki, H. 63

Yang, Z. 492
Ybema, J. E. 477
Yehene, E. 205, 208
Yen, Y. 438
Yeung, N. 11, 13, 208, 209, 213
Yin, H. H. 27
Young, S. 494
Yuki, M. 189
Yves von Cramon, D. 98
Yzerbyt, V. Y. 453

Zacks, R. T. 95, 253
Zajonc, R. B. 459
Zanna, M. P. 268, 359
Zatorre, R. J. 124
Zauberman, G. 415, 421
Zayas, V. 432, 438
Zbrodoff, N. J. 165
Zdaniuk, B. 460, 461
Zeigarnik, B. 367
Zeiss, A. 313, 414, 421
Zeiss, A. R. 433, 434
Zeitlin, S. B. 100
Zelazo, P. 421
Zelezny, L. C. 496
Zellmann, P. 186
Zemke, P. 479
Zhang, Y. 391, 395, 396, 397, 398, 399, 401, 402, 403
Zheng, Y. 411, 414, 420
Ziaie 345
Ziller, R. C. 454
Zimerman, S. 317
Zimmerman 346
Zysset, S. 98

SUBJECT INDEX

Academic test performance, implementation intentions effects in, 283. *See also* Action control mechanisms
   intelligence test, 283–84
   math test, 283
ACC. *See* Anterior cingulate cortex (ACC) activity
Accuracy, 41, 43
Action control mechanisms, 174, 175, 176, 177
   ideomotor theory, 185–94
   impact of self on, 192
Action-effect integration, 188
Action identification theory
   versus construal level analysis, of self-control, 423
Actors' perceptions, premeditated decisions, 262
   consistency, 266–67
   effort, 268–69
   exclusivity, 267–68
   multiple potential courses of action, 263–65
   temporal priority, 265–66
Actual premeditation, 262
Actus reus. *See* Guilty act
Addictive behavior, and self-control, 384
   overeating, 384
   drugs and alcohol abuse, 385
   consumer behavior, 385
"Affective shifting," 31
Affect misattribution procedure (AMP), 224–25
Affect versus cognition models, 411, 422
Aggression, and self-control, 382
Altruism, 516
Amygdala activity, 18
Antecedent-focused emotion-regulation, 340
Anterior cingulate cortex (ACC) activity, and cognitive control, 3–9, 142
   activation, 6
   in automatic regulation, 11–12
   conceptualization, 5
   conflict monitoring versus control implementation, 5–7, 12
   conflict theory. *See* Conflict theory
   and DLPFC, 8–9
   emotion processing, 16–20
   error detection, reinforcement learning theory, 12–13
   error response theory, 13–15
   finger movements, 4
   free recall, 4
   lesion studies, 10–12
   response conflict, 5
   selection-for-action, 6
   and social cognition, 20–23
   social exclusion trials, 21
   specification, 5
   stem completion, 4
   verb generation, 4
   word generation, 4
Arbitrary visuomotor mappings, 182
Associational-Propositional Evaluation (APE) model, 50
Asymmetrical interference, 96
Asymmetric counteractive process, 392
Asymmetric self-control strategies, 391–92
Attention-Deficit/Hyperactivity-Disorder (ADHD), children with, 289–90
   BOLD signal, 152
   gratification performance delay, 291–92
   response inhibition, 290–91
   set-shifting and multitasking, 292–93
Attention deployment, 345
Attention processing, 55
Attribution, 50
Autobiographical knowledge, 168–69
Automatic associations, 243
   self-control dilemmas, 244–45
   in social behavior, 245–47
   Quad model, 247–49
Automaticity accounts, 95, 96
Automatic versus effortful control models, 410

Balloon Analogue Risk Task (BART), 146
Barratt Impulsiveness Scale, 146

Behavioral Activation System (BAS), 117
Behavioral inhibition, and egalitarian goals, 368–69
Behavioral Inhibition System (BIS), 55, 117
Behavioral performance, 8, 104, 127
Behavioral responses, of implementation intentions, 287
Berridge model, 122
BESA 2000 source localization, 6
Biology, of self-control, 380
Bipolar depression, 16
"Black sheep effect," 453
Blind auditions, 235
Bottom-up control, 28
Botvinick model, 7
Breach of expectation, 32
Broadmann areas, 17

Caudodorsal ACC (cdACC), 96
Change signal task, 13, 14
Chinese pictographs, 228
Choice bracketing, 418
Choice paradigm, multifinality in, 306–7
Chronic egalitarianism, 54
 and stereotyping, 363–65
Cingulate lesions, effect on cognitive task, 11
Cingulotomy, 16–17
Cognitive–Affective Processing System (CAPS), 429
 hot/cool systems within, 434
  balance influencing factors, 436
  interactions between systems, 435–36
  interactions in everyday life, 437
 self-regulatory dynamics, 430f
Cognitive affective units (CAUs), 429
Cognitive and neural substrates of cooling strategies, identification of, 441
Cognitive change
 meaning of, 345
 of self-control, 345–46
Cognitive control, 3–4
 role of anterior cingulate cortex. See Anterior cingulate cortex (ACC) activity
Cognitive dissonance, 21–23
Cognitive Evaluation Theory (CET), 116
Cognitive negative variation (CNV), 11
Cognitive neuroscience perspective, 126–31
 brain
  processing in, 127–28
  task-relevant activity, 128–29
 continuous performance task, 127
Cognitive reappraisal, 57–58
 neurodynamics of, 82–83
  decrease trials, 83
  increase trials, 83
  look trials, 83
Cognitive responses, of implementation intentions, 288
Cognitive task, 208

Cognitive wanting, 121
Collective action, in groups, 463–64
Collective intervention, in global warming, 492
Collective tactics versus individual tactics, 455
"Color conflict," 107
Color-naming response, 4
Color-word Stroop task, 4, 9, 177–78
 cognitive control study, 4
 response conflict, 4
Common sense, 185
Compatible trial (cI), 6, 10
Completed goals condition, 307
Components of self-control, ego depletion, 378
 monitoring, 378
 motivation, 379
 standards, 378
 strength, 378–79
Conflict adaptation effects, 10, 15
Conflict and control, 312
 goals and control, 316
  multiple goals, managing conflicts between, 317–18
  regulatory focus and control, 320–23
  regulatory reference and control, 318–19
  self-concordance and control, 319–20
 hierarchies, of self-regulation, 314–16
 strategies and control, 323
  defensive pessimism, 326–27
  flexible control of strategies, 327
  regulatory fit theory, 324–26
 targeting control, in self-regulatory hierarchy, 327–30
Conflict monitoring, ACC activity. See Anterior cingulate cortex (ACC) activity
Conflict theory, 4–9
 criticism on, 9–15
 implications of, 16–23
 types and control, 15–16
Conformity, in groups, 457–58
Congruent trials, 4
Conjunction analyses, 152
Connectivity analyses, 18
Conservative minorities, 463
Conservative tactics, 315, 316
Construal level analysis, of self-control, 408, 411–14
 versus action identification theory, 423
 construals effect
  on self-control, 414–18
  on prospective self-control, 418–19
 and extant models, 421–23
  automatic versus effortful control, 410
  hot versus cool, 411
  temporal discounting, 409–0
 versus implementation intentions, 423
 implications, for self-control, 419–21
Consumer behavior, and self-control, 385
Context-specific control mechanism, 15–16
Continuous Performance Test (CPT), 5, 127, 290

SUBJECT INDEX

Control, definition, 51
Control agents, groups as, 450
   external dynamics, 450
   group structure, 450–51
   internal dynamics, 450
   norms, 451
      injunctive norms, content of, 451–52
      norm violations, reaction to, 452–53
   roles, 453
      control mechanisms, 454
      group membership, phases of, 454
      group socialization theory, 454
      leaders, 456–57
      new members, 455–56
      prospective members, 454–55
Controlled processing, definition, 51
Control processes, 211
Cool/hot systems, within CAPS, 301, 411, 434
   factors influencing balance of, 436
   interactions between systems, 435–36
   interactions in everyday life, 437
Correction, 52–53
Correct-related negativity (CRN), 61
Correlational analysis, 149
Counteractive control, 390–91
   in stereotyping, 360–62
Counteractive optimism, 395–96, 397
"Covert modification," 346
Creativity goals, in stereotyping, 369
Cross-talk, 179
Curse of knowledge, 270
Cybernetic control model, 336, 337, 338–39, 341
   cybernetic process model, 341
      attention deployment, 345
      cognitive change, 345–46
      response modulation, 346–47
      situation modification, 344
      situation selection, 342–44

Decision making, definition of, 275
Defensive pessimism, in strategic control, 326–27
   negative possibilities condition, 327
   positive possibilities condition, 327
Delay discounting, 513
Delayed intention, 185
Delayed knowledge condition, participants in, 265
Delay of gratification, 293
   ability to, 432
      attentional processes, 432–33
      reconstrual processes, 433–34
   in children with ADHD, 291–92
   decision to, 431–32
Deliberation, definition of, 275n
Descriptive norms, 451
Detection and overcoming bias
   implicit attitude, 249
      aging, 252–53

      alcohol intoxication, 252
      biased associations, 251–52
      intergroup interactions quality, 253
      motivation-based individual differences, 250–51
      public versus private contexts, 250
Diet-saliency condition, 300
Diffusor Tension Imaging (DTI), 153
Direct self-regulation, 465
Disjunctive tactics versus serial tactics, 456
"Disordered self-control," 31
Distributive justice
   effectiveness of, 476
      consensus, 476
      willingness to enforce justice rules, 477, 478
      willingness to follow justice rules, 476–77
   meaning of, 475
   principles of, 476
   versus procedural justice, 479–80
   and self-control, 475–78
   social interaction, 478
Divestiture tactics versus investiture tactics, 456
Dopamine (DA), 119
   Berridge model, 122
   haloperidol, 122
   hedonic response, 121
   liking, 121
   Panksepp model, 122
   in reward and motivation, 120–21
   Salamone model, 122
   wanting, 121
Dorsal ACC activity, 60, 180
   cognitive dissonance, 21–23
   emotional and nonemotional interference, 104–5
   in reappraisal, 18–20
Dorsolateral prefrontal cortex (DLPFC) activity, 4, 7–8, 18, 128, 142
   and ACC, 8–9
   color-naming response, 8
   word-reading response, 7
Dorsomedial prefrontal medial cortex activity, 18
Drive theory, 119
Drugs and alcohol abuse, 385
Dual-process model, 246
Dual route logic, 176–81

Economics and teleological behaviorism, 512–16
   altruism, 516
   delay discount functions, 513
   negative addictions, 513–14
   positive addictions, 513–14
   prisoner's dilemma game, 514
   social cooperation and self-control, 514
   social discount functions, 514–15
Economic substitutability, and self-control, 517
Effective goal hierarchy, 299, 300
Effortful versus automatic control models, 410
Egalitarianism, stereotyping conflicts, 357

Egalitarianism, stereotyping conflicts, (*Cont.*)
   and behavioral inhibition, 368–69
   chronic egalitarian goals, 363–65
   counteractive control, 360–62
   goal priming, 359–60
   goal shielding, 362–63
   goal systems and goal shielding, 360
   implicit goals, 359
   temporary creativity goals, 369
   temporary egalitarian goals, 365–68
   temporary perspective taking goals, 369–70
   temporary self-esteem goals, 370
Ego depletion, 375, 377, 380, 410, 422, 423
   components, of self-control, 378
      monitoring, 378
      motivation, 379
      standards, 378
      strength, 378–79
   executive function of self, 380–81
   importance of self-control, 376–77
   inner motivational conflicts, resolving, 376
   interpersonal interactions, 381
      addictive behavior, 384–85
      aggression, 382
      emotion regulation, 383–84
      esteem threat, 382
      rejection and ostracism, 381–82
      relationships, 382–83
      self-presentation, 381
      stress, 384
   and limited resource model of self-control, 375, 379
      biology, of self-control, 380
   meaning of, 404
Egoistic relative versus fraternal relative deprivation, 463
Embedded Figures Test, 191
Emotional conflict, 107
Emotional counting Stroop, 17
Emotional distinct hypothesis, 103
Emotional facial Stroop, 18
Emotion processing, ACC in, 16–20
Emotion-regulation, antecedent-focused, 340
Emotional responses
   of implementation intentions, 287–88
   and regulatory focus, 322–23
Emotional Stroop task
   conflict
      automaticity account, 97
      Glaser and Glaser model, 97
      prosodic and lexico-semantic dimension, 98
      translational account, 97
      valence categorization task, 98
      word emotion, 97
   distraction
      arbitrariness, 99
      carry-over effects, 100
      emotional word trials, 99
      gender-identification task, 102
      neutral word trials, 99
      slow effects, 100
      subsequent trial control, 101
      subsequent trial interference, 100–101
Emotion regulation, 148–49
   and self-control, 383–84
   process model of, 336, 337–38, 340–41
   juxtaposition of, 341f
Emotion suppression, 57
Empathy gap, 349
Energy, at response level, 7
Environmental attitudes
   ideological differences in
      and system justification, 496–97
   partisan differences in, 500–501
   self-regulatory perspective on, 500
Environmental problems, denial of
   and system justification, 496, 498–500
Equality, opposition to
   and system justification, 498
Eriksen flanker task, 6, 181
   compatible trial, 6
   incompatible trial, 6
Error commission, 5
Error likelihood, ACC response to, 13–15
Error negativity, 5
Error positivity, 64
Error-related negativity (ERN), 6, 11, 180
Error trials, 5
   response conflict, 5
Essential goal hierarchy, 299, 300–301
Esteem threat, and self-control, 382
Eureka errors, 268
Event-related potential (ERP) studies, 5–6, 10–11, 14, 60–62
Executive functions, 28, 380–81
Expectancy, 116
Expectation, breach of, 32
Expectation–outcome discrepancy, 463
Explicit, definition of, 52
Explicit regulation, of stereotype control, 357
Explicit test, 227
Explicit versus implicit strategies, of self-control, 302–3
Extant approaches, to self-control
   automatic versus effortful control, 410
   hot versus cool systems model, 411
   temporal discounting, 409–10
External factors, of self-control interventions, 348–49
External motivation to respond without prejudice (EMS), 364, 365
Extrinsic motivation, 116
Eysenck Impulsiveness Scale, 146

Facial Stroop task, 9–10
   conflict adaptation effects, 10
   congruent trials, 9
   incongruent trials, 9

SUBJECT INDEX

Facilitated communication, 268
Factional conflict, in groups, 462–63
Failures to exercise self-control, 401–4
Failures to recognize self-control, 401–4
Fast-learning system, 50
Fast reaction time, 9
Feedback ERN, 12,13
Feeling thermometer, 226
Fixed tactics versus variable tactics, 455
Flanker task, 6, 181, 152
Flexible control, of strategies, 327
Font color identification, 95
Forced response trials, 233
Formal tactics versus informal tactics, 455
Fraternal relative deprivation versus egoistic relative deprivation, 463
Free response trials, 233
Frontal lobe damage
  decision making, 29
  self-control, 29–30
Frontopolar cortex, 128
Functional MRI, 7
  incongruent trial, 7
Future thinking, 33

Galvanic skin response (GSR), 11
Gender differences, in system justification, 496
General model, of self-control, 335
  cybernetic process model, 341
    attention deployment, 345
    cognitive change, 345–46
    response modulation, 346–47
    situation modification, 344
    situation selection, 342–44
  existing models, 337
    cybernetic control model, 337, 338–39
    emotion-regulation, process model of, 337–38, 340–41
    extrapersonal factors, 337
    intrapersonal factors, 337
  self-control interventions, selecting and designing, 347
    external and internal contextual factors, 348–49
    high-leverage versus low-leverage methods, 347–48
  temptations, and self-control, 336–37
Global warming
  collective intervention, need for, 492
  environmental concerns, nature and severity of, 491
  and environmental destruction, 491
"Global Warming is a Hoax," 500
Go/NoGo task, 290
"Go" system, 301, 435
Goal, 164, 316
  meaning of, 315
  choice, actual process of, 309
  intention of, 116
  multiple goals, managing conflicts between, 317–18
  regulatory reference and control, 318–19
  regulatory focus and control, 320–23
  self-concordance and control, 319–20
  by temptations, 302, 303–5, 389
    asymmetric self-control, 391–92
    counteractive control theory, 390
    instrumental self-control, 391
    self-control dilemma, 389
    self-control failure, 401–4
    self-control problems, 390
    strategies of self-control, 392
Goal-directed process, 108
Goal priming, in stereotyping, 359–60
Goal setting, 211
Goal shielding, 360, 362–63
Goal-striving, self-control by, 279–93
Goal value, 115
Grand unifying theory, 28
Gratification task, 80
Group decision making and problem solving, 458–59
Group engagement model, 482–83, 484, 485
Group membership
  investigation phase, 454
  maintenance phase, 454, 456
  remembrance phase, 454
  resocialization phase, 454
  socialization phase, 454, 455–56
Groups, self-control in, 449
  as agents of control, 450
    norms, 451–53
    roles, 453–57
  individuals as targets of control, 457
    members' capitulation to group pressure, 457–60
    members' resistance to group pressure, 460–64
Group socialization theory, 454
Guilty act, 261
Guilty mind, 261
Gut reactions, 227

Habits
  change of, 516
  situation-bound development, 516–17
Habitual responses, overcoming, 286
  behavioral responses, 287
  brain, implementation intentions in, 288–89
  cognitive responses, 288
  emotional responses, 287–88
Heart rate variability (HRV), 11
Hick's law, 176
High-level versus low-level construals, 412, 413
High-leverage versus low-leverage methods, 347–48
High versus low priority goals, 298
Horizontal self-control conflicts, 312, 314
Hot/cool processes, 301
  in negative emotion regulation, 437–40
    interpersonal conflict, coping with, 437–38

Hot/cool processes, (Cont.)
    intrapersonal conflict, coping with, 438–40
    within CAPS, 434
        factors influencing balance of, 436
        interactions between systems, 435–36
        interactions in everyday life, 437
Hot system versus cool systems, 411

Ideal premeditation, 262
Ideomotor action, 185–87
    common-sense concept, 185
    goal-based acquisition, 188
    and sensorimotor action, comparison of, 187–89
    stimulus-based acquisition, 188
    test phase, 186
"If, then" guidelines, 237
Imagining unrelated distant future events, 413
Imagining unrelated near future events, 413
Immediacy factor, 299–301
Implementation intentions, 62–63, 116, 185
    versus construal level analysis, of self-control, 423
    in hardships, 282
        academic test performance, 283–84
        children with ADHD, 289–93
        habitual responses, overcoming, 286–89
        opponents, dealing with, 284–86
    formation of, 280–81, 301–2
    links created by, 280–81
    and overcoming problems of goal-striving, 281
        getting started, 281
        halt, calling, 282
        not overextending oneself, 282
        staying on track, 281–82
    process-dissociation analyses, 62
    Shooter Task, 62
Implicit, definition of, 52
Implicit Association Test (IAT), 223, 245
Implicit goal conflict, 359
Implicit motivation to control prejudice (IMCP), 365
Implicit test, 227
Importance, of self-control, 376–77
    at individual level, 376
    at interpersonal level, 376–77
    at societal level, 377
Impulsivity, 142
Incentive salience, 121
Incompatible trial (iI), 6, 10
    high adjustment, 9
    low adjustment, 9
Incongruent trial, 4, 7
Indirect self-regulation, 464–65
Individual differences, 129–31, 189
    error-related negativity, 130
    error variance, 129
Individual mobility, 463
Individuals, as targets of group control, 457
    members' capitulation to group pressure, 457
        conformity, 457–58

group decision making and problem solving, 458–59
mixed-motive settings, behavior in, 459–60
members' resistance to group pressure, 460
    collective action, 463–64
    factional conflict, 462–63
    minority influence, 461–62
    nonconformity, 460–61
Individual tactics versus collective tactics, 455
Individual variables, 115
Individuating information, 58
Inferior frontal cortex (IFC), 144
Inferior frontal gyrus (IFG), 144
Informal tactics versus formal tactics, 455
Informational influence, in groups, 458
Information processing conflicts, 96
Inhibition, 53
    behavioral inhibition, 55–56
    hierarchical inhibition, 54–55
    lateral inhibition, 53–54
    mental suppression, 56
    suppressing race-biased affect, 57
    suppressing stereotypic thoughts, 56–57
Injunctive norms
    content of, 451–52
    meaning of, 451
Inner motivational conflicts, and self-control, 376
Institutions and values, evolution of, 487
Instrumental self-control, 391
Insula activity, 125
Intact cingulate, 10
Intelligence test, 283–84
Intergroup approach, 49
    dualistic philosophy, 51
    theoretical roots, 50–51
    theories of control, 59–66
Internal and External Motivation to Respond Without Prejudice Scale, 225
Internal contextual factors, of self-control interventions, 348–49
Internally versus externally driven control, 63–64
Internal motivation to respond without prejudice (IMS), 364, 365
International Affective Picture System, 83
Interpersonal conflict, coping with, 437–38
Interpersonal interactions, and self-control, 381
    addictive behavior, 384–85
        consumer behavior, 385
        drugs and alcohol abuse, 385
        overeating, 384
    aggression, 382
    emotion regulation, 383–84
    esteem threat, 382
    relationships, 382–83
    self-presentation, 381
    social rejection and ostracism, 381–82
    stress, 384
Intrapersonal conflict, coping with, 438–40

SUBJECT INDEX

Intrinsic motivation, 116
   free-choice situation, 117
Investigation phase, of group membership, 454
Investiture tactics versus divestiture tactics, 456
Iowa Gambling Task (IGT), 30, 145

Justice, as social self control, 473
   distributive justice and self-control, 475–78
   functions of, 484–86
   institutions and values, evolution of, 487
   meaning of, 475
   motives, transformation of, 480–84
   procedural justice and self-control, 478–80
Justice rules, quality of, 475
Justified consumption condition, 298, 299

"Know" cognitive system, 301
"Know" system, 411

Lateral prefrontal cortex (LPFC), 18, 43, 124
"Law of effect," 510
Leaders, in groups
   acceptance ceremonies, 456
   image management perspective, 457
   maintenance phase, of group membership, 456
Lesion and ACC activity studies, 10–12, 29
Levels of Emotional Awareness Scale (LEAS), 19
Limited resource model, of self-control, 375, 376, 378, 379
   biology, of self-control, 380
Logan-Burkel paradigm, 212
Long term versus short term goals, 298, 299
Loss framed negotiations, prevailing in, 285–86
Loss-of-function techniques, 29
Low-level versus high-level construals, 412, 413
Low versus high priority goals, 298

Maintenance phase, of group membership, 454, 456
Mastery goals, 117
Mathematical approach, 59–60
   process-dissociation procedure, 59
   quadruple-process model, 59
Math test, implementation intention effects, 283
Means generation, multifinality constraints on, 307–8
Medial frontal cortex, 63
Medial prefrontal cortex (mPFC), 96, 142
Mediating emotion-related behavior, 145
Members' capitulation, to group pressure, 457
   conformity, 457–58
   group decision making and problem solving, 458–59
   informational influence, 458
   mixed-motive settings, behavior in, 459
      group performance, 459–60
      social dilemmas, 460
   normative influence, 457–58

Members' resistance, to group pressure, 460
   collective action, 463–64
   factional conflict, 462–63
   minority influence, 461–62
   nonconformity, 460–61
   reciprocal influence, 462
Memory inhibition, 149–50
   forget, 149
   no-think, 149
   remember, 149
   think, 149
Mens rea. See Guilty mind
Mental arithmetic test, 11
Mental contamination, 230, 357
Mental correction, 53, 67, 357
Mental load, 42
Mental set, 207
   backward inhibition, 209
   inertia, 209
   lingering inhibition, 209
   reversed asymmetry, 209
   switch asymmetry, 209
Mental suppression, 56
   monitoring process, 56
   operating process, 56
   rebound, 56
Mesencephalic dopamine system, 12
Meta-cognitive thinking, 238
Methamphetamine, 148
Mindfulness-based cognitive therapy, 170
Mind wandering, 169–70
Minority influence, in groups, 461–62
"Miswanting," 116
Mixed-motive settings, behavior in, 459
   group performance, 459–60
   social dilemmas, 460
Modernist minorities, meaning of, 463
Modern Racism Scale, 224
Mood regulation, 41
Motivated behavior, 32
   amygdala, 32
   hypothalamus, 32
Motivational properties, of system justification, 494
Motivation influence, 114
   cognitive neuroscience perspective, 126–31
   neuroscience perspective, 119–26
   social and psychological perspective, 115–19
Motivation to control prejudiced reactions, 224, 225
Motivation to respond without prejudice, 225, 364, 365
Motives, transformation of, 480–84
Motor response inhibition, 143
   go/no-go task, 143
   reversal learning, 144
   stop signal task, 143
Multifinality quest, 309
   choice paradigm, 306–7
   and means generation, 307–8
   as response to self-control dilemma, 305–8

Multiple goals, managing conflicts between, 317–18
Multitasking, in children with ADHD, 293
"Myopia for the future," 30

N2, 6, 11
Nationally representative sample, generalization to, 497–98
 and system justification, 497–98
Negative addictions, 513–14
Negative affect, 115
Negative feedback, response to, 13, 32, 431, 485
Negative multifinality, 304
Negative slow wave (NSW), 252
Neural alarm system, 21
Neural circuitry
 punishment motivation, 124–26
 reward motivation, 122–24
Neuroscience approach, 60, 119–26
 drive-increasing stimuli, 119
 drive-reducing stimuli, 119
 integration of, 65
  labeling task, 66
  passive face viewing task, 65
Neutral trial types, 15
Newcomers, in groups, 455–56
 entry ceremonies, 455
 negative treatment, 455
 positive treatment, 455
 socialization phase, of group membership, 455–56
 tactics, 455–56
New Environmental Paradigm scale, 496, 499
Newsweek, 500
Nonconformity, in groups, 460–61
Nondeliberative mechanisms, 60–61
 Black-tool trials, 60
 Black-gun trials, 60
Non-emotional counting Stroop, 17
Nontemptations, and temptations, 337t
Normative influence, in groups, 457–58
Norm violations, group reaction to, 452–53
Nucleus accumbens (NAcc), 120
"Number conflict," 107

Observers' perceptions, premeditated decisions, 269
 consistency, 270–71
 effort, 272
 exclusivity, 271–72
 multiple potential courses of action, 270
 priority, 270
Opponents, dealing with, 284
 loss framed negotiations, prevailing in, 285–86
 winning tennis competitions, 284–85
Orbitofrontal cortex (OFC) activity, 28, 43, 44
Ostracism, and self-control, 381–82
Overconfidence, 41
Overeating, and self-control, 384

Override, 58
 competing tendency, 58
 response tendency, 58

Panksepp model, 122
Parallel-constraint-satisfaction theory, 367
Parallel search process, 50
Partisan differences, in environmental attitudes, 500–501
Perceptions. *See* Actors' perceptions, premeditated decisions; Observers' perceptions, premeditated decisions
Perceptuomotor integration, 188
Performance goals, 117
 performance-approach goals, 117
 performance-avoidance goals, 117
Personal self-control, problems of, 474
Person perception, 50
Perspective taking goals, in stereotyping, 369–70
Pharmacological manipulations, 29
Pictographs, Chinese, 228
Picture-naming task, 179
Pieron's law, 176
Planchette, 265
Planning, power of
 goal-striving, self-control by, 279–82
  implementation intention. *See* Implementation intention effects
"Pleasure chemical," 119
Positive addictions, 513–14
 substituting negative addictions, 517–18
Positive affect, 115, 118
Positive feedback, response to, 32, 431
Positive illusions, 39
 accuracy, 41
 adaptive value, 41
 short-term benefits, 41
Positive multifinality, 304
Positive self-views, 39
Posterior cingulate cortex (PCC) activity, 126
Posttraumatic stress disorder (PTSD), 16
Potential bloomers, 230
Predictions, 116
Preference judgments, 34
Prefrontal cortex (PFC) activity, 28
Prejudice and stereotyping, 50
 context of control, 50–52
 control models, 52–59
Premeditation, 260
 actors' perceptions, 262
  consistency, 266–67
  effort, 268–69
  exclusivity, 267–68
  multiple potential courses of action, 263–65
  temporal priority, 265–66
 actual premeditation, 262
 definition of, 275n

SUBJECT INDEX

effects of, 272–74
ideal premeditation, 262
implications, for self-control, 274
observers' perceptions, 269
    consistency, 270–71
    effort, 272
    exclusivity, 271–72
    multiple potential courses of action, 270
    priority, 270
Premotor cortex, 126
Prepared reflex logic, 181–82
Prevention-focused individuals, 118–19
Prior knowledge condition, participants in, 265
Prisoner's dilemma game, 514
Proactive control, 132
Procedural justice, 486
  authorities and institutions, link to, 480
  versus distributive justice, 479–80
  meaning of, 475
  and self-control, 478–80
Process Dissociation Model of memory, 246
Process Dissociation Procedure (PDP), 288
Processes, in self-control, 430–37
  ability to delay gratification, 432
    attentional processes, 432–33
    reconstrual processes, 433–34
  decision to delay gratification, 431–32
  hot/cool systems, within CAPS, 434
    factors influencing balance of, 436
    interactions between systems, 435–36
    interactions in everyday life, 437
Processing reward-related information, 145
Progressive minorities, 463
Promotion-focused individuals, 119
Prospective members, in groups, 454–55
Prospective memory (PM) paradigm, 289
Prospective self-control, construals effect on, 418–19
Protestant work ethic, endorsement of, 497–98
Proximity factor. *See* Immediacy factor
Punishment motivation, 124–26
Putative component processes, 29

Quadruple Process Model, 244
  application, 249
  Associations (AC), 247
  behavioral and neurological evidence, 249
  Detection (D), 247
  Guessing, 247
  Overcoming bias, 247
Quasi-experimental design
  emotional distraction Stroop task, 105

Racial bias, 54
  backfire, 237
  control of
    intergroup anxiety, 65
    interracial interactions, 64–65
    self-reported anxiety, 65
    subjective anxiety, 64
    weapons identification task, 65
  patterns, 230
Random tactics versus sequential tactics, 455
Rational action, 261
Raven's Advanced Progressive Matrices intelligence test, 283, 284
Reactionary minorities, meaning of, 463
Reactive control, 132
Real attitude, 226
Real-life social interactions, 69
Reappraisal, 18
  dorsal ACC activity in, 18–19
Reciprocal influence, in groups, 462
Reflective and Impulsive Determinants model, 50
Regulatory fit theory, 324–26
Regulatory focus, 320–23
  and emotions, 322–23
  and task demands, 321–22
Regulatory reference, and self-control, 318–19
Reinforcement, of self-control, 510–11, 512, 513
Reinforcement learning theory, of ACC error detection, 12–13, 32
Rejection sensitivity (RS), 85, 437–38
Relationships, and self-control, 382–83
Relative saliency of goals and temptations, 301–2
  goals by temptations, activation of, 302
  implementation intentions, formation of, 301–2
Repetition priming effects, 9–10
Residual switch cost, 214
Response conflict, 5–6, 15
  co-activation response, 5
Response ERN, 12, 13
Response-focused emotion-regulation, 340
Response incongruent (RI), 15
Response inhibition, in children with ADHD, 290–91
Response-locked ERP, 187
Response-locked incorrect trials, 5
Response modulation
  meaning of, 346
  of self-control, 346–47
Retributive justice, 475
Retroactive adjustment, 210
Reversal learning, 31, 144
Reward representation, 123
Reward motivation, 116, 122–24
Reward signaling, 123
Reward value, 123
Right inferior frontal cortex activity, 127
Right inferior prefrontal cortex activity, 55
Right lateral prefrontal cortex activity, 127
Risk-taking behavior, 145–46
  Balloon Analogue Risk Task, 146
  Iowa Gambling Task, 145
  risky decision making, 146
  safe decision making, 146

Risky tactics, 315, 316
Roles, in groups
   control mechanisms, 454
   group membership, phases of, 454
   group socialization theory, 454
   leaders, 456–57
   meaning of, 453
   new members, 455–56
   prospective members, 454–55
Rostral ACC, 17
   in emotional conflict detection, 17–18
   in emotional awareness, 19–20
Rostrodorsal ACC (rdACC), 96
   emotional interference, 104
Rules of distributive justice, 476
Rumination, 85

Salamone model, 122
   activational wanting, 122
   directional wanting, 122
Scrambled Sentence Test, 185
Selection-for-action, 6
Self-concordance, and self-control, 319–20
Self-Construal Scale, 190, 194
Self-control, 38–39, 76, 163. *See also individual entries*
   behavioral evolution of, 511
   biological and environmental influences, on self-control ability, 440–41
   and conflict theory, 16
   cognitive and neural substrates of cooling strategies, identification of, 441
   construals effect on, 414–18
   definition of, 3, 141, 298, 337, 375
   essential and effective goal hierarchies, 299
   evolution of, 509–12
   as goal, 302–3
   immediacy factor, 299–301
   potential component processes, 28
   problems of, 390
   and teleological behaviorism, 506–7
   self-regulatory training interventions, 441–42
   and social cooperation, 514
   strategies for, 235
      cognitive connoisseurs, 238
      potential for bias, 235–36
      proactive control, 236–38
   teleological behaviorism, implications of, 516
      economic substitutability, 517
      habit change, before consumption change, 516
      habit development, situational specificity of, 517
      habits, situation-bound, 516–17
      maintenance before reduction, 516
      positive addictions substituting negative addictions, 517–18
      situational control, 517
      soft-commitment program, for self-control, 518
Self-control conflict, 507–8
   definition of, 312
   outcome of, 377
Self-control dilemma, 298, 308, 389
   and resolution modes, 297
      self-control, definition of, 298
      strategies, of self-control
Self-control failure, 401
   failures to recognize versus failures to exercise, 401–4
   temptation indulgence, effects of, 404
Self-control interventions, selecting and designing, 347
   external and internal contextual factors, role of, 348–49
   high-leverage versus low-leverage methods, 347–48
Self-control strength, 180
Self-discrepancy theory
   actual self, 118
   ideal self, 118
   ought self, 118
Self-distanced perspective versus self-immersed perspective, 439
Self Efficacy Theory, 116
Self-enhancement, 40
Self-esteem goals, in stereotyping, 370
Self-immersed perspective versus self-distanced perspective, 439
Self-imposed punishment adoption, 419
Self-memory system, 168–69
Self-Monitoring Scale, 190
Self-present, 223, 381
   implicit and explicit racial attitudes, 223
   racial bias. *See* Racial bias
Self-regulation, 164
   hierarchies of, 314–16
      conservative bias, 315, 316
      goals, 315
      risky bias, 315, 316
      strategies, 315
      tactics, 315
Self-regulatory failure, in commons dilemma, 492–93
Self regulatory processing system, 429–30
Self-regulatory training interventions, 441–42
Self-reported distress, 21
Self-serving monitoring, 41
Self-serving perceptions, 42
Self-system theory (SST), 322
Semantic conflict, 15
Sensation Seeking Scale, 146
Sensorimotor action
   automaticity, 177
   automatic response activation, 182–84
   dual route logic, 176–81
   goal-based, 175
   intelligence, 185
   prepared reflex logic, 181–82
   simple reaction-time task, 176
   stimulus-based, 175
   stupidity, 185

SUBJECT INDEX

Sequential tactics versus random tactics, 455
Serial search process, 50
Serial tactics versus disjunctive tactics, 456
Serotonin (5-HT), 125
Serotonin transporter (5-HTT), 134
Set-shifting, in children with ADHD, 292–93
Shake-saliency condition, 300
Short term versus long term goals, 298, 299
Simon task, 8–9, 178
   conflict adaptation, 8
Single-cell recording, 144
Situation modification
   meaning of, 344
   of self-control, 344
Situation selection
   meaning of, 342
   of self-control, 342–44
Six Elements Test (SET), 290, 293
Slower reaction time, 9
Slow-learning system, 50
Social and psychological perspective, on cognitive control, 115–19
   active self, 117
   anger, 118
   expectancy, 116
   high positive affect, 118
   low positive affect, 118
   predictions, 116
   rapid depletion, 117
Social Cognitive and Affective Neuroscience (SCAN), 77–79
   challenges, 79
   functional inferences, 78
   implementation, 79-
   integration, 78
   psychological inferences, 78
Social competition, 463
Social cooperation, and self-control, 514
Social creativity, 463
Social discounting, 514–15
Social exclusion, and ACC activity, 21
Social identity theory, 453, 463
Social inclusion, and ACC activity, 21
Social intelligence, 474
Social interaction, key issues of, 473–74
Socialization phase, of group membership, 454, 455–56
Social movements, 463
Social norms, meaning of, 451
Social psychological contributions
   and system justification, 501–2
Social rejection, and self-control, 381–82
Soft-commitment program, for self-control, 518
Spider phobia, 287–88
Source-modeling data, 13
Statistical prediction rules, 235
Status quo, motivation in service of, 493
   system justification
      individual and situational differences in, 494–95

      motivational properties of, 494
      potential to disrupt environmental self-regulation, 495
Stereotype, definition of, 355
Stereotype activation, implicit control of, 354
   egalitarian goals, stereotyping conflicts with, 357
   and behavioral inhibition, 368–69
   chronic egalitarian goals, 363–65
   counteractive control, 360–62
   goal priming, 359–60
   goal shielding and (counteractive) control over stereotype activation, 362–63
   goal systems and goal shielding, 360
   implicit goals, 359
   temporary creativity goals, 369
   temporary egalitarian goals, 365–68
   temporary perspective taking goals, 369–70
   temporary self-esteem goals, 370
   goal conflict, 355
   implications of, 356
   explicit control, problems with, 357
   implicit nature of, 355–56
Stereotype rebound, 56
Stimulus-based condition, 187
Stimulus control, to self-control, 428
   willpower, understanding, 428
      hot/cool processes in negative emotion regulation, 437–40
      integrated understanding of self-control, 440–42
      processes, in self-control, 430–37
      self regulatory processing system, 429–30
Stimulus evaluation, 5
Stimulus incongruent (SI), 15
Stimulus-outcome expectation, 32
Stimulus repetitions, 9–10
Stimulus-response (S-R) translation, 177
Stimulus-response task, 152
Stimulus reversal, 10
Stimulus-set binding, 210
Stop-change task, 14
Stop-signal paradigm, 212
Stop-signal reaction time (SSRT), 143
Stop Signal Test (SST), 290
Strategies, of self-control, 323, 392
   attainment expectations, changing, 395
      goal engagement, 396–97
      optimistic predictions, 397–98
      temptation engagement, 396–97
   choice situation, changing, 392
   goals. *See* Goals
   defensive pessimism, 326–27
   explicit versus implicit strategies, 302–3
   flexible control of, 327
   goals versus temptations, 303–5
   multifinality quest as response to, 305–8
   psychological meaning of choice, 398
      activating goals and inhibiting temptations, 400–401

Strategies, of self-control, (Cont.)
    bolstering goals and devaluing temptations, 398–99
    heating up goals and cooling down temptations, 399–400
    regulatory fit theory, 324–26
    relative saliency of goals and temptations, manipulating, 301–2
Stress, and self-control, 384
Stroop/Simon combination task, 15
Stroop task, 290
    automatic process, 95
    conflict, 96–99
    distinct, 103
    distraction, 99–100
    for emotion regulation, 93
    experimental model, 95
    font color identification, 95
    reliability, 94
    translational accounts, 96
    versatility, 94
    word identification, 95
Subjective construals, 411
Subjective Group Dynamics (SGD) model, 453
Subjective value, 29, 33–34
Subthalamic nucleus (STN), 144
"Sunk-costs" fallacy, 520n
System justification, and environmental goal-setting disruption, 490
    collective level of analysis, 494
    denial, facilitating system justification, 498–500
    environmental attitudes
        partisan differences in, 500–501
        self-regulatory perspective on, 500
        ideological differences in, 496–97
    and environmental problems denial, 491–92, 496
    gender differences in, 496
    global warming, 491
    individual and situational differences in, 494–95
    motivational properties of, 494
    nationally representative sample, generalization to, 497–98
    potential to disrupt environmental self-regulation, 495
    self-regulatory failure in commons dilemma, 492–93
    short-term palliative effects, 493–94
    social psychological contributions, 501–2
    status quo, motivation in service of, 493–95
    theoretical approach, 496
        empirical support for, 496–502

Tactics, 315
Targeting control, in self-regulatory hierarchy, 327–30
Task-cuing effect, 211
Task decision, 211
    and goal maintenance, 211
Task demands, and regulatory focus, 321–22

Task instruction, methods of, 205
    alternating runs paradigm, 205
    instruction from memory, 205–6
    instructions in each trial, 206
    position-in-run effect, 206
    restart cost, 206
    self-selected tasks, 206–7
    task-chunk, 206
    task preparation, 206
    task repetition, 206
Task-relevant activity, 128–29
    fusiform gyrus activity, 129
    parahippocampus activity, 129
    spatial attention task, 129
    visual attention, 129
Task set, 207
Task Set Inertia (TSI), 182
Task switching, 203
    associative equivalent, 204
    bivalent stimuli, 204
    cognitive task, 208–9
    episodic memory tasks, 205
    historic foundations, 203–4
    information filtering, 213–14
    inhibition, 212–13
    mental sets and task sets, 207–8
    mixing cost, 204
    monitoring, 214
    odd-even judgment, 205
    preparation, 214
    processes, 209–15
    same-different judgments, 205
    semantic memory tasks, 205
    single step versus multiple step tasks, 205
    switch cost, 204
    task differences, 205
    task instruction, methods of, 205–7
    task-repetition trials, 204
    univalent stimuli, 204
    vowel-consonant judgment, 205
Task variables, 115
Teleological behaviorism, 506, 508–9
    and behavioral view of self-control, 507
    and economics, 512–16
    economic substitutability, 517
    evolution, of self-control, 509–12
    external phenomenon, 506
    habit change, before consumption change, 516
    habit development, situational specificity of, 517
    habits, situation-bound, 516–17
    internal phenomenon, 506–7
    maintenance before reduction, 516
    positive addictions substituting negative addictions, 517–18
    and problem of self-control, 506–7
    self-control conflict, 507–8
    situational control, 517
    soft-commitment program, for self-control, 518

SUBJECT INDEX

for treatment of addiction and everyday problems of self-control, 516–18
"Temporal discounting," 33, 146–48, 409–10
  "cool" cognitive system, 147
  "hot" emotional system, 147
Temporary egalitarian goals, in stereotyping, 365–68
Temptation indulgence, effects of, 404
Temptations
  definition of, 336
  versus goals, 303–5
  and nontemptations, 337t
  and self-control, 336–37
Theoretical approach, of system justification, 496
Thought suppression, 57, 150
  white bear, 150
Time estimation task, 13
"Tit-for-tat," 515
Top-down control, 28, 51, 68
Traditional automatic versus effortful control models, 422
"Tragedy of the Commons," 490–91
  self-regulatory failure in, 492–93
Transcranial magnetic stimulation (TMS) studies, 29, 144
Translational accounts, 96

Uncompleted goals condition, 307
Unconscious beliefs, 222
  explicit attitudes, 223
  implicit attitudes, 223
  indirect measures, reliability of, 226
  measurement error, 226
Unipolar depression, 16
Unjustified consumption condition, 298, 299
Unrealistically positive self-evaluations, 39
  behavioral evidence, 42
  neural evidence, 42
  overconfidence, 41
Utilitarian action, 20

Valence-dependent side, 135
Valence-independent side, 135
Variable tactics versus fixed tactics, 455–56

Ventral ACC activity, 12, 43
Ventrolateral prefrontal cortex (VLPFC) activity, 21, 142
Verbal instructions, 182
Verbal praise, 116
Vertical self-control conflict, 312, 314
Visual search task, 14
Voluntary action, 175, 182, 185–86

Weapon identification task, 231, 250
Willpower dynamics, 428
  hot/cool processes in negative emotion regulation, 437–40
  integrated understanding of self-control, 440–42
  processes, in self-control, 430–37
  self regulatory processing system, 429–30
Winning tennis competitions, 284–85
Wisconsin Card Sorting Test (WCST), 152, 290
Within-run slowing, 212
Eord identification, 95
Word-reading response, 4
Word-reading task, 179
Working memory, 82, 163
Working memory capacity (WMC), 164
  antisaccade, 165
  complex span task, 164
  controlled retrieval, 166
    fan interference, 167
    proactive interference, 166–67
  dichotic listening, 166
  and impulsive decision making, 170
  and mind wandering, 169–70
  operation span task, 164
  as primary and secondary memories, 167–68
  reading span task, 164
  and self-memory system, 168–69
  short-term memory, 164
  simple word span task, 164
  Stroop, 165–66
  successive visual arrays, 166
  and suppressing unwanted thoughts, 169

Alliant International University
Los Angeles Campus Library
1000 South Fremont Ave., Unit 5
Alhambra, CA 91803